Petroleum Geoscience: From Sedimentary Environments to Rock Physics

Knut Bjørlykke

Petroleum Geoscience: From Sedimentary Environments to Rock Physics

With contributions from Per Avseth, Jan Inge Faleide,
Roy H. Gabrielsen, Nils-Martin Hanken, Kaare Høeg, Jens Jahren,
Martin Landrø, Nazmul Haque Mondol, Jenø Nagy
and Jesper Kresten Nielsen.

 Springer

Knut Bjørlykke
Department of Geosciences
University of Oslo
0316 Oslo
Blindern
Norway
knut.bjorlykke@geo.uio.no

ISBN 978-3-642-02331-6   e-ISBN 978-3-642-02332-3
DOI 10.1007/978-3-642-02332-3
Springer Heidelberg Dordrecht London New York

Library of Congress Control Number: 2010921863

Cover design: deblik, Berlin

Printed on acid-free paper

Springer is part of Springer Science+Business Media (www.springer.com)

Preface

Petroleum geology is not a well-defined academic subject. It includes many different aspects of the Earth sciences which are used in petroleum exploration and production. Nearly all types of insight can in some cases be useful in petroleum exploration, but there are some disciplines that are most relevant. Since petroleum is formed and hosted in sedimentary rocks, sedimentology is critical. Palaeontology is important for dating rocks and carbonate reservoirs may consist mostly of fossils. Structural geology and basin analysis are also vital for reconstructing the migration and trapping of petroleum. Geochemistry and petroleum chemistry are also important.

Geophysical methods are essential for logging and seismic exploration, and recently electromagnetic methods have also been used in exploration and production.

Most universities do not offer specific courses in petroleum geology/geophysics and only a few have a Masters or PhD programmes in this field. Oil companies therefore recruit many geologists with little training in these subjects.

In this book we have tried to give a basic introduction to disciplines relevant to petroleum exploration and we have also included some aspects of petroleum production.

Since so many different disciplines are included in this book it is clear that it has not been possible to make in-depth treatments of each of these. This book provides a relatively condensed and precise presentation of the basic facts in each subject and it was therefore necessary to limit the number of field examples and cases.

We have attempted to write a book which requires only a limited background in geology and geophysics. Some of the chapters are therefore relatively basic, but others are more advanced and we have then included more discussions and references to original research papers.

The reference to the original literature had to be limited because of the wide range of disciplines. In the past textbooks often included very extensive lists of references which were very useful when searching for relevant literature. With the electronic data bases available now it is easy to search for relevant references and new textbooks.

In this textbook we want to bridge the gap that often seems to exist between geophysical and geological disciplines and there is an emphasis on sediment compaction, fluid flow and rock physics.

The skills required from a petroleum geologist have changed greatly over the years. Traditionally the main task was to identify reservoir rocks, structures with closure and the proximity of a mature source rock. We are running out of "the easy to find" and "easy to produce" oil and gas, and exploration and production technology is becoming more advanced.

Production of unconventional oil (tar sand, oil shale) and also tight gas reservoirs and gas shale, requires a stronger background in mineralogy, chemistry and physics.

The geophysical methods have become increasingly sophisticated and it is now often possible to detect the presence of gas and oil prior to drilling based on seismic data. Electromagnetic methods that were primarily used in mineral prospecting are also used to find oil. As conventional oil is becoming more scarce more geologists are becoming involved with exploration and production of heavy oil, oil shales and shale gas.

This requires a stronger background in the chemistry and physics of petroleum and also in mineralogy and rock mechanics (rock physics). Physical and chemical modelling is also very important.

Even if alternative sources of energy are being developed the world will require fossil fuels for several decades. It is a great challenge to limit the environmental consequences of such production.

Although I have the main responsibility for this book, it includes separate contributions from:

Per Avseth (Rock Physics), Odin Petroleum, Bergen, Norway.

Jan Inge Faleide (Regional Petroleum Geology – mainly from offshore Norway), University of Oslo, Oslo, Norway

Nils-Martin Hanken (Carbonate Rocks), University of Tromsø, Tromsø, Norway

Roy Gabrielsen (Structural Geology), University of Oslo, Oslo, Norway

Martin Landrø (4 D Seismics), NTNU Trondheim, Trondheim, Norway.

Nazmul Haque Mondol (Exploration Geophysics), University of Oslo, Oslo, Norway

Jenø Nagy (Stratigraphy), University of Oslo, Oslo, Norway

Kaare Høeg (Rock Mechanics), Department of Geosciences, University of Oslo, Oslo, Norway

Jens Jahren (Sandstone Reservoirs), Department of Geosciences, University of Oslo, Oslo, Norway

Jesper Kresten Nielsen (Carbonate Sedimentology), SINTEF, Trondheim, Norway.

Adrian Read has been of great assistance as Text Editor in the preparation of this book.

Several of our PhD students have helped out with illustrations etc:

Tom Erik Maast

Øyvind Marcussen

Olav Blaich

Brit Thyberg

Delphine Croizé, Masaoki Adachi and Jon Reierstad have made many of the figures in the book and Jan Petterhold and Tove Midthun have produced illustrations for Chapter 5.

Olav Walderhaug has kindly made useful comments to Chapter 4 (sandstone reservoirs and clastic diagenesis).

Statoil has provided funding which has helped the preparation of this book, and we are grateful for this support.

Fugro Geoscience Division has kindly provided good seismic data from offshore Norway and also from other parts of the world.

Oslo, Norway   Knut Bjørlykke
March 2010

Contents

Contributors

Per Avseth Odin Petroleum, Bergen; NTNU, Trondheim, Norway,
pavseth@yahoo.com

Knut Bjørlykke Department of Geosciences, University of Oslo, Oslo, Norway,
knut.bjorlykke@geo.uio.no

Jan Inge Faleide Department of Geosciences, University of Oslo, Oslo, Norway,
j.i.faleide@geo.uio.no

Roy H. Gabrielsen Department of Geosciences, University of Oslo, Oslo, Norway,
roy.gabrielsen@geo.uio.no

Nils-Martin Hanken University of Tromsø, Tromsø, Norway,
Nils-Martin.Hanken@uit.no

Kaare Høeg Department of Geosciences, University of Oslo, Oslo, Norway,
Kaare.Hoeg@geo.uio.no

Jens Jahren Department of Geosciences, University of Oslo, Oslo, Norway,
jens.jahren@geo.uio.no

Martin Landrø NTNU Trondheim, Trondheim, Norway, martin.landro@ntnu.no

Nazmul Haque Mondol Department of Geosciences, University of Oslo;
Norwegian Geotechnical Institute (NGI), Oslo, Norway, nazmul.haque@geo.uio.no

Jenø Nagy Department of Geosciences, University of Oslo, Oslo, Norway,
jeno.nagy@geo.uio.no

Jesper Kresten Nielsen SINTEF, Trondheim, Norway, jesper.nielsen@sintef.no

Chapter 1

Introduction to Petroleum Geology

Knut Bjørlykke

Petroleum geology comprises those geological disciplines which are of greatest significance for the finding and recovery of oil and gas. Since most of the obvious and "easy to find" petroleum already has been discovered it is necessary to use sophisticated methods in the exploration of sedimentary basins. These include advanced geophysical techniques and basin modelling. There is also much more emphasis now on enhanced recovery from the producing fields. Petroleum technology has made great progress and many new tools and modelling programs have been developed, both in exploration and production.

It is however important to understand the geological processes which determine the distribution of different sedimentary rocks and their physical properties. This knowledge is fundamental to being able to successfully apply the methods now available.

It is difficult to know where to start when teaching petroleum geology because nearly all the different disciplines build on each other.

This introductory chapter will provide a short and rather simple overview of some aspects of petroleum geology to introduce the subject and the problems. Most of the other chapters will then expand on what is presented here to provide a better background in relevant subjects.

Since practically all petroleum occurs in sedimentary rocks, *sedimentary geology* forms one of the main foundations of petroleum geology. *Sedimentological models* are used to predict the location of different facies in the sedimentary basins, and from that the likely presence of source rocks with a high content of organic matter, reservoir rocks and cap rocks. The distribution and geometry of potential sandstones or carbonate reservoirs requires detailed sedimentological models, and sequence stratigraphy has been a useful tool in such reconstructions.

The *biostratigraphic correlation* of strata encountered in exploration wells is achieved by *micropalaeontology* (including palynology), a field developed very largely by the oil industry. Due to the small size of the samples obtained during drilling operations one cannot rely on macrofossils; even in core samples the chance of finding good macrofossils is poor. By contrast a few grams of rock from the drill cuttings may contain several hundred microfossils or palynomorphs. These also usually provide better stratigraphic resolution than macrofossils.

Reservoir rocks are mostly sandstones and carbonates which are sufficiently porous to hold significant amounts of petroleum. The composition and properties of other rock types such as shales and salt are also important.

The sedimentary environments (sedimentary facies) determine the distribution of reservoir rocks and their primary composition. Sediments do, however, alter their properties with increasing overburden due to diagenesis during burial.

Diagenetic processes determine the porosity, permeability and other physical properties such as velocity, in both sandstone and limestone reservoirs. Chemical processes controlling mineral reactions are important.

K. Bjørlykke (✉)
Department of Geosciences, University of Oslo, Oslo, Norway
e-mail: knut.bjorlykke@geo.uio.no

K. Bjørlykke (ed.), *Petroleum Geoscience: From Sedimentary Environments to Rock Physics*, DOI 10.1007/978-3-642-02332-3_1, © Springer-Verlag Berlin Heidelberg 2010

Organic geochemistry, which includes the study of organic matter in sediments and its transformation into hydrocarbons, has become another vital part of petroleum geology.

Tectonics and structural geology provide an understanding of the subsidence, folding and uplift responsible for the creation and dynamic history of a basin. The timing of the folding and faulting that forms structural traps is very important in relation to the migration of hydrocarbons.

Seismic methods have become the main tool for mapping sedimentary facies, stratigraphy, sequence stratigraphy and tectonic development. Marine seismics recorded from ships have become very efficient and seismic lines are shot at only a few 100 m spacing or less. Because of the rapid improvement in the quality of seismic data processing techniques, geological interpretation of seismic data has become an entirely new and expanding field. Seismic and other geophysical data are often the only information we have, particularly for offshore exploration where drilling is very costly. Shooting seismic lines with a close spacing allows high resolution 3D seismic imagery to be produced for critical parts of sedimentary basins. By repeating a 3D reservoir seismic survey during production, one can observe how the gas/oil and oil/water contacts move as the reservoir is depleted. This is called 4D seismic because time provides the fourth dimension.

Geophysical measurements may include gravimetry and magnetometry; electromagnetic methods that were used mostly in ore exploration have also been applied to oil exploration. Electromagnetic methods have been used to detect sediments with low resistivity due to the presence of oil instead of saline water. This method requires a few 100 m of water and relatively shallow accumulation.

Seismic surveys are more expensive on land than by ship at sea because geophones have to be placed in a grid, often on uneven and difficult land surfaces. Drilling on land, however, costs much less than from offshore rigs and a much denser well spacing can be used during both exploration and production.

Indirect methods of *mapping rock types* employing geophysical aids are becoming increasingly important in petroleum geology, but it is still necessary to take samples and examine the rocks themselves. A petroleum geologist should have a broad geological training, preferably also from field work.

Geophysical *well-logging* methods have developed equally rapidly, from simple electric and radioactive logs to highly advanced logging tools which provide detailed information about the sequence penetrated by the well. Logs provide a continuity of information about the rock properties which one can seldom obtain from exposures or core samples. This information makes it possible to interpret not only the lithological composition of the rocks and the variation of porosity and permeability, but also the depositional environment. Image logs make it possible also to detect bedding and fractures inside the wells.

Practical petroleum geology is not only based on many different geological and geophysical disciplines. A good background in basic chemistry, physics, mathematics and computing is also required, particularly for different types of basin modelling.

1.1 A Brief Petroleum History

There are many places where oil seeps out of the ground. Bitumen produced from such naturally occurring crude oil has been collected and used since ancient times, both for lighting and medicine, and by the Greeks even for warfare. In some places, for example Germany in the 1800s, small mines were dug to get at the oil. Before 1859 oil was also recovered from coal for use in kerosene lamps. It was not until Edwin Drake's exploits in 1859 at Oil Creek near Titusville in West Pennsylvania that oil was recovered in any quantity from boreholes. He drilled a well about 25 m (70 ft) deep which produced 8–10 bbl/day, a huge production rate compared with anything earlier. A few years later there were 74 wells round Oil Creek, and the USA's annual production had risen to half a million barrels. Outside the USA the calculated total production at that time was maximum 5,000 bbl. In 1870 production had increased tenfold, with 5 million bbl from the USA, and 538,000 bbl from other countries. In southern California oil production started early in 1864 (in Santa Paula), but for many years oil was mined by driving shafts into the oil-bearing strata because it was so heavy and biodegraded that it would not flow in a well.

In its infancy, oil exploration consisted largely of looking for oil seepage at the surface and drilling in the vicinity, which did not require much geological knowledge. It was then realised that oil and gas occur where layers of sedimentary rocks form domes or anti-clinal structures since petroleum is less dense than water and a low permeability (seal) layer is needed to prevent the oil and gas from rising and escaping. This led to extensive geological mapping of anticlines and domes visible at the surface, particularly in the USA. It was also found that oil fields had a tendency to lie along structural trends defined by anticlines or faults and this "rule" was used in prospecting. This is also often the case with salt domes, which became important prospecting targets.

Oil production developed rapidly up to the end of the nineteenth century, and more systematic geological principles for prospecting were gradually developed. The geological information which one obtains at the surface is often not representative of the structures deeper down. Structures which are not visible at the surface could be mapped by correlation between wells using logs and cuttings from the drilling. One method was to measure the depth of particularly characteristic strata through analysis of cuttings in different wells. Improved electrical measurements (logs) from wells, developed during the 1920s and 1930s, made the whole effort much simpler because they provided continuous vertical sections through the rocks. The first logs were simple recordings of how well rocks conduct electrical currents (resistivity), and later also gamma logs recorded the gamma-radiation emitted by the different sedimentary rocks.

The USA maintained its position as the major world producer of oil and gas well into the twentieth century. Americans thus became leaders in the development of oil technology, which today is strongly reflected in the industry's terminology. USA also rapidly became the world's greatest consumer of oil and gas, and now has to import at least 60% of its oil consumption despite still having a large home production (8 million bbl/day).

The US consumption (21 million bbl/day) is a very large fraction of the total world consumption (90 million bbl/day)

It was first in the 1930–1940 period that the industry became aware of the vast oil resources of the Middle East, which now account for about 60% of world reserves. Since then this region has dominated oil production.

In the 1950s and 1960s prospecting for oil and gas was extended onto the continental shelves, opening up new reserves. Until the development of modern seismic it was not possible to effectively explore deep below the seafloor in sedimentary basins offshore. As long as there was an abundance of oil to be found in onshore basins there was little incentive to develop costly drilling rigs for offshore exploration and platforms for production.

Since the 1970s an increasingly large share of international prospecting has taken place offshore, helped by improved seismic methods. Advanced well log technology in particular made it possible to gain optimal information from each well. Before the development of powerful computers, seismic recordings were based on analogue methods which produced results very far removed from the standard of modern seismic data. Recording of seismic data from ships is much less expensive than onshore where geophones have to be laid out manually. Onshore, extensive drilling may be cheaper than expensive seismic mapping.

Rising oil prices and new technology have made exploration financially attractive in areas which previously were of little interest, including in very deep waters. High oil prices can also pay for more enhanced hydrocarbon recovery from reservoirs. There are now relatively few sedimentary basins in the world that have not been explored and it is getting increasingly difficult to find new giant fields (>500 million bbl).

There is now increasing interest in heavy oil, tar sand and oil shale.

Oil shale is a source rock exposed near the surface. If the source rock (shale) is mature it will have a characteristic smell of hydrocarbons, but it may not be mature so that hydrocarbons have not been generated. If the oil shale is mature much of the oil has escaped by primary migration. Since the hydrocarbons are thoroughly disseminated in the fine-grained sediment, oil cannot be produced in the same way as from sandstone or carbonate reservoirs. The hydrocarbons can only be obtained by breaking and crushing the shale and heating to distill off the interspersed hydrocarbons. Shales can however contain gas which can be produced when there is a network of small fractures. Gas shale is expected to be an important source of petroleum in the years to come, particularly in the US. Very large amounts of fossil fuels are stored in organic-rich mudstones or shales that have not been buried deeply enough for the organic matter to be converted to petroleum. In this case very little hydrocarbon has escaped but these deposits must be mined and heated to 400–500°C in ovens to generate petroleum (pyrolysis).

The Tertiary Green River Shale in Colorado, Utah and Wyoming represents one of the largest petroleum reservoirs in the world. This is a lake deposit, and the organic matter consisted mainly of algae.

Although very large quantities of petroleum can be produced from oil shale, production costs are at present too high compared to conventional oil. There are also serious environmental problems involved in production from oil shale, and the process requires very large quantities of water, a resource which is not always plentiful.

The oil reserves in such deposits exceed conventional oil reserves, but the expense and environmental issues involved with production from these types of reservoirs clearly limit their exploitation. This is particularly true of production from oil shale.

1.2 Accumulations of Organic Matter

It is well documented that oil accumulations are of organic origin and formed from organic matter in sediments. Methane can be formed inorganically and is found in the atmosphere of several other planets, but inorganic methane from the interior of the earth is likely to be well dispersed and thus not form major gas accumulations in the earth's crust.

The organic matter from which petroleum is derived originated through photosynthesis, i.e. storage of solar energy (Fig. 1.1).

Sunlight is continuously transformed into such energy on Earth but only a very small proportion of the solar energy is preserved as organic matter and petroleum. The oil and gas which forms in sedimentary basins each year is thus minute in comparison with the rate of exploitation (production) and consumption. In practice petroleum must therefore be regarded as a non-renewable resource even though some petroleum is being formed all the time.

Most of the organic materials which occur in source rocks for petroleum are algae, formed by photosynthesis. The zooplankton and higher organisms that are also represented grazed the algae and were thus indirectly dependent on photosynthesis too. The energy which we release when burning petroleum is therefore stored solar energy. Since petroleum is derived from organic matter, it is important to understand how and where sediments with a high content of organic matter are deposited.

The total production of organic material in the world's oceans is now 5×10^{10} tonnes/year. Nutrients for this organic production are supplied by erosion of rocks on land and transported into the ocean. The supply of nutrients is therefore greatest in coastal areas, particularly where sediment-laden rivers discharge into the sea. Plant debris is also supplied directly from the land in coastal areas.

Biological production is greatest in the uppermost 20–30 m of the ocean and most of the phytoplankton growth takes place in this zone. In clear water, sunlight penetrates much deeper than in turbid water, but in clear water there is usually little nutrient supply. At about 100–150 m depth, sunlight is too weak for photosynthesis even in very clear water.

Phytoplankton provides nutrition for all other marine life in the oceans. Zooplankton feed on phytoplankton and therefore proliferate only where there is vigorous phytoplankton production. Organisms sink after they have died, and may decay so that nutrients are released and recycled at greater depths.

Basins with restricted water circulation will preserve more organic matter and produce good source rocks which may mature to generate oil and gas (Fig. 1.2a, b).

In polar regions, cold dense water sinks to great depths and flows along the bottom of the deep oceans towards lower latitudes. This is the thermal conveyor belt transporting heat to higher latitudes and it keeps the deep ocean water oxidizing. In areas near the equator where the prevailing winds are from the east the surface water is driven away from the western coast of the continents. This generates a strong upwelling of nutrient-rich water from the bottom of the sea which sustains especially high levels of primary organic production (Fig. 1.3). The best examples of this are the coast of Chile and off West Africa.

Through photosynthesis, low energy carbon dioxide and water are transformed into high energy carbohydrates (e.g. glucose):

$$CO_2 + H_2O \rightarrow CH_2O \text{ (organic matter)} + O_2$$

The production of organic matter is not limited by carbon dioxide or water, but by nutrient availability.

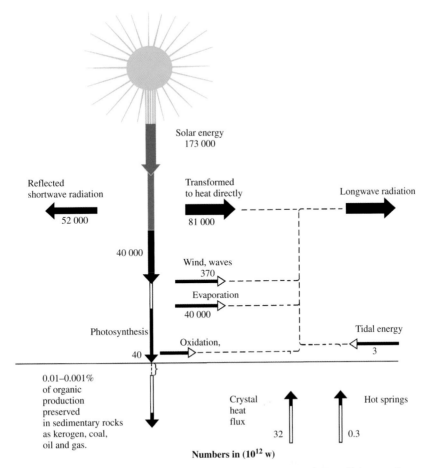

Solar energy
173 000

Reflected
shortwave radiation

52 000

40 000

Transformed
to heat directly

81 000

Longwave radiation

Wind, waves
370

Evaporation
40 000

Photosynthesis

40

Oxidation,

Tidal energy

3

0.01–0.001%
of organic
production
preserved
in sedimentary rocks
as kerogen, coal,
oil and gas.

Crystal
heat
flux

32

Hot springs

0.3

Numbers in (10^{12} w)

Fig. 1.1 Transformation of solar energy to fossil fuels by photosynthesis. Only a small fraction of the solar energy is used for photosynthesis and most of the produced organic matter is oxidised. As a result very little organic matter is buried and stored in sedimentary rocks and very little of this is concentrated enough to become a potential source rock

Phosphorus (P) and nitrogen (N) are the most important nutrients, though the supply of iron can also be limiting for alga production. It is this process of photosynthesis, which started 4 billion years ago, that has built up an atmosphere rich in oxygen while accumulating reduced carbon in sedimentary rocks as oil, gas and coal. Most of the carbon is nevertheless finely divided within sedimentary rocks, for example shales and limestones, in concentrations too low to generate significant oil and gas.

Energy stored by photosynthesis can be used directly by organisms for respiration. This is the opposite process, breaking carbohydrates down into carbon dioxide and water again, so that the organisms gain energy.

This occurs in organisms at night when there is no light to drive photosynthesis. Also when we burn hydrocarbons, e.g. while driving a car, energy is obtained by oxidation, again essentially reversing the photosynthesis equation quoted above. Oxidation of 100 g glucose releases 375 kcal of energy. Carbohydrates that are produced but not consumed by respiration can be stored as glucose, cellulose or starch in the cell walls. Photosynthesis is also the biochemical source for the synthesis of lipids and proteins. Proteins are large, complex molecules built up of condensed amino acids (e.g. glycine (H_2NCH_2-COOH)).

Dried phytoplankton contains 45–55% carbon, 4.5–9% nitrogen, 0.6–3.3% phosphorus and up to 25% of both silica and carbonate.

Planktonic algae are the main contributors to the organic matter which gives rise to petroleum. Among the most important are diatoms, which have amorphous silica (opal A) shells.

Deposition of source rocks

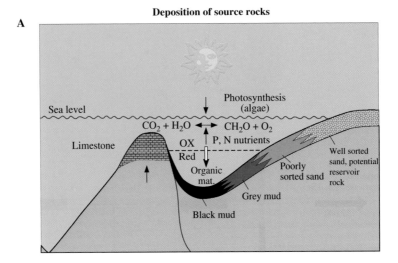

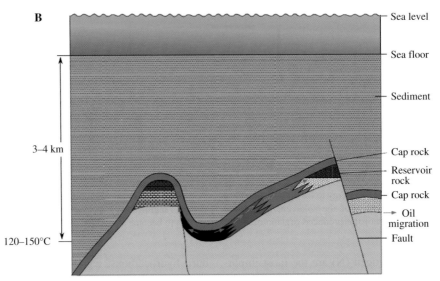

Fig. 1.2 (a) Depositional environments for potential source and reservoir rocks. Depressions on the sea floor with little water circulation provide the best setting for organic matter to be accumulated before it is oxidised. (b) Migration of petroleum from source rocks into reservoir rocks after burial and maturation. The carbonate trap (e.g. a reef) is a stratigraphic trap, while the sandstone forms a structural trap bounded by a fault

Diatoms are most abundant in the higher latitudes and are also found in brackish and fresh water. Blue–green algae (cyanobacteria) which live on the bottom in shallow areas, also contribute to the organic material in sediments.

In coastal swamps, and particularly on deltas, we have extensive production of organic matter in the form of plants and trees which may avoid being oxidised by sinking into mud or bog. The residues of these higher land plants may form peat, which with deeper burial may be converted into lignite and bituminous coal. But such deposits are also a potential source of gas and oil. Plant matter, including wood, also floats down rivers and is deposited when it sinks to the bottom, usually in a nearshore deltaic environment. When the trees rot they release CO_2 and consume as much oxygen as the plant produced during the whole period when it was growing. There is thus no net contribution of oxygen to the atmosphere. This also applies to the bulk of the tropical rainforests. Where trees and plants sink into

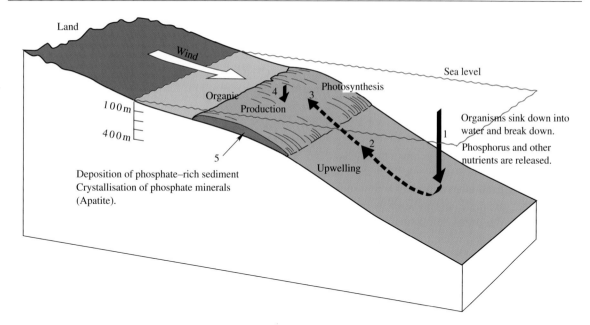

Fig. 1.3 Upwelling of water rich in nutrients on a continental margin with deposition of organic-rich mud

Trophic level 1	Trophic level 2	Trophic level 3	Trophic level 4
1,000 kg	100 kg	10 kg	1 kg
Phytoplankton	Zooplankton	Crustaceans	Fish

black mud, preventing them from being oxidised, there is a net contribution of oxygen to the atmosphere and a corresponding reduction of CO_2 in the atmosphere.

All animal plankton (zooplankton) live on plant plankton, and in turn are eaten by higher organisms in the food chain. At each step in the food chain, which we call a trophic level, the amount of organic matter (the biomass) is reduced to 10%.

Ninety percent of the production of organic matter is therefore from algae. This is why algae and to some extent zooplankton account for the bulk of the organic material which can be transformed into oil. Larger animals such as dinosaurs are totally irrelevant as sources of oil.

The most important of the zooplankton which provide organic matter for petroleum are:

1. Radiolaria – silica shells, wide distribution, particularly in tropical waters.
2. Foraminifera – shells of calcium carbonate.
3. Pteropods – pelagic gastropods (snails) with a foot which has been converted into wing-shaped lobes; carbonate shells.

This is the second lowest level within the marine food chain. These zooplanktonic organisms are eaten by crustaceans which themselves are eaten by fish.

The total amount of organic matter that can be produced in the ocean is dependent on the nutrient supply from rivers, but river water does not only carry inorganic nutrients. It also contains significant amounts of organic matter, in particular humic acid compounds, lignin and similar substances formed by the breakdown of plant material which are weakly soluble in cold water. When the river water enters the sea, there is precipitation due to the increased pH and lower surface temperature in the ocean.

Other plant materials, like waxes and resins, are more chemically resistant to breakdown and are insoluble in water. Such organic particles tend to attach themselves to mineral grains and accompany sediment out into the ocean.

Most of the oil reservoirs which have been formed since the Palaeozoic have been uplifted and eroded, and over time vast quantities of oil have flowed (seeped) out onto the land or into the sea. In this sense, oil pollution is a natural process. Only a small proportion of the petroleum that has been formed in source rocks has actually become trapped in a reservoir. One might expect this seepage to have provided a source of recycled petroleum in younger sediments, but petroleum breaks down extremely rapidly when subjected to weathering, oxidising to CO_2, and the nutrients (P, N) that were required to form the organic matter are released and may act like a fertilizer.

On land, evaporation will remove the lighter components while bacteria will degrade the heavier components. Fossil asphalt lakes consist of heavy substances which neither evaporate nor can be easily broken down by bacteria. In the ocean, the lighter components will dissolve quite rapidly, while the heavier asphalt fraction will sink to the bottom and be degraded and recycled.

In uplifted sedimentary basins like the Ventura Basin and the Los Angeles Basin in Southern California there are abundant natural oils seeps both onshore and offshore.

On the beaches from Santa Barbara towards Los Angeles there are many natural oil seeps.

1.3 Breakdown of Organic Matter

Almost all (>99%) of the organic matter which is produced on land and in the oceans is broken down through direct oxidation or by means of microbiological processes. If oxygen is present, organic matter will be broken down in the following manner:

$$CH_2O + O_2 \rightarrow CO_2 + H_2O$$

Where oxygen is available, organic matter is oxidised relatively rapidly both on land and in the sea. As organisms die, organic material suspended in seawater sinks through the water column consuming oxygen. If water circulation is restricted due to density stratification of the water column, the oxygen supply will be exhausted. Instead, the bound oxygen in sulphates or nitrates is used by sulphate-reducing and denitrifying bacteria which decompose organic material in an anoxic environment. The first few centimetres below the seabed are usually oxidised, while reducing conditions prevail 5–30 cm below the sea floor. Below this redox boundary where there is no free oxygen,

sulphate-reducing bacteria react with organic matter as indicated below:

$$2CH_2O + 2H^+ + SO_4^- \rightarrow H_2S + 2CO_2 + 2H_2O$$
$$NH_3 + H^+ + SO_2 \rightarrow NO_3 + H_2S + H_2O$$

H_2S is liberated, giving stagnant water and mud a strong smell. Through denitrification we get

$$5CH_2O + 4H^+ + 4NO_3^- \rightarrow 2N_2 + 5CO_2 + 7H_2O$$

When the rate of accumulation of organic matter exceeds the rate of oxygen supply the redox boundary will be in the water column, separating the oxidising surface water from the reducing bottom water.

This is typical of basins separated from the deep ocean by a shallow sill, like the Black Sea and some of the deep Norwegian fjords. Fresh or brackish surface water floating on more saline water also helps to maintain a stable water stratification with little vertical mixing. Lakes may have good water stratification because warm surface water is less dense than the colder bottom water. Black mud deposited at the bottom of lakes may produce good source rocks. In cold climates, however, the water in the lakes overturns in the winter because the maximum water density is at $4°C$, preventing the stable stratification required to form source rocks.

1.4 Formation of Source Rocks

All marine organic material is formed near the surface of the ocean, in the photic zone, through photosynthesis. For the most part this is algae. Some phytoplankton are broken down chemically and oxidised and some are eaten by zooplankton. Both types of plankton are eaten by higher organisms which concentrate the indigestible part of the organic matter into fecal pellets which may be incorporated into sediments. Plankton is made up of very small organisms which sink so slowly that they are in most cases almost entirely degraded (oxidised) before they reach the bottom. Pellets, on the other hand, are the size of sand grains and sink more rapidly, and this organic matter is more likely to be preserved in the sediments.

On the bottom, organic matter will be subjected to breakdown by micro-organisms (bacteria). It will also be eaten by burrowing organisms which live in the top portion of the sediments. The activity of these organisms contributes to reducing the organic content of the sediments because most of the organic matter is digested when the sediment is eaten. Bioturbation also stirs up the sediments, exposing them more to the oxygen-bearing bottom water. However, if the bottom water is stagnant, the lack of oxygen and the toxicity of H_2S will exclude most life forms. The resultant lack of bioturbation will thus preserve more organic matter in the sediment together with perfect, undisturbed, lamination. Stagnant, or anoxic, conditions are defined by an oxygen content of <0.5 ml/l water. Sulphate-reducing bacteria, however, can use a good deal of organic matter and precipitate sulphides (e.g. FeS_2). If the sediments contain insufficient soluble iron or other metals which could precipitate sulphides, more sulphur will be incorporated in the organic matter and will eventually be enriched in the oil derived from such source beds.

Except where the water is completely stagnant, slow sedimentation rates will result in each sediment layer spending longer in the bioturbation and microbiological breakdown zones, and consequently less organic matter will be preserved in the sediment. Rapid sedimentation leads to more of the deposited organic matter being preserved but from the outset it will be highly diluted with mineral grains. Consequently an intermediate sedimentation rate in relation to organic production (10–100 mm/1,000 years) results in the best source rocks.

As we have seen, the net accumulation of organic matter in sediments is not so much a function of the total productivity, but rather of the relationship between productivity and biogenic breakdown and oxidation. In areas with powerful traction currents, most organic matter will be oxidised. An important source of oxygen-rich water in the deep ocean is the cold surface water which sinks to the bottom of the ocean in polar regions and flows along the ocean floor towards equatorial regions. This flow balances the surface flow to higher latitudes like the Gulf Stream in the Atlantic.

These bottom flows are of considerable magnitude during glacial periods, when large amounts of cold water are sinking near ice sheet peripheries. In warm periods, for example during the Cretaceous, the poles were probably ice-free and there was much less cold surface water available to sink down and drive the ocean conveyor system. The deeper parts of the Atlantic experienced stagnant bottom conditions during such periods.

Limited water circulation in semi-enclosed marine basins due to restricted outflow over a shallow threshold is a common cause of stagnant water bodies (Fig. 1.2a). The Black Sea is a good example. In response to an abundant freshwater supply from rivers and a relatively low evaporation rate, a low salinity surface layer leads to density stratification in the water column and a consequent reduction in circulation. In basins with little precipitation and where there is net evaporation, the surface water will have higher salinity and density than the water below it, and will sink down. This circulation brings with it oxygen from the surface and can give oxidising bottom conditions with little chance for organic matter to survive to form source rocks.

Lakes or semi-enclosed marine basins often have a temperature- or salinity-induced density stratification so that oxygenated surface water does not mix with water in the deeper part of the basin. This leads to anoxic conditions and a high degree of preservation of the organic matter produced in the surface waters. This aspect is therefore of considerable interest in exploration for petroleum in freshwater basins, particularly in Africa and China. The open oceans have normally had oxygenated water, but during the Cretaceous most of the Atlantic Ocean is believed to have been stagnant during so-called "anoxic events", and substantial amounts of black shale were deposited in the deeper parts of the ocean during these periods.

1.5 Early Diagenesis of Organic Matter

Microbiological breakdown of organic matter in sediments is due to the activity of bacteria, fungi, protozoa, etc. and under oxidising conditions these are extremely effective. However, the porewater quickly becomes reducing if the oxygen is not replenished. In relatively coarse-grained sediments (sand), oxygen may diffuse to depths of 5–20 cm below the seabed, while in clay and fine-grained carbonate mud the boundary between oxidising and reducing water (redox boundary) may be a few millimetres below the seafloor. The pores in the sediments here are so small that water circulation and diffusion are insufficient to replace the original oxygen in the porewater as it gets used up by oxidation of organic matter. Clay-rich sediments soon become a relatively closed system, and the downward diffusion of oxygen from the seabed is very slow in fine-grained sediments.

Aerobic breakdown is therefore much more effective in coarse-grained sediments than in fine-grained ones. In anaerobic transformation bacteria use organic matter, e.g. short carbohydrate chains. Cellulose is broken down by fungi, and finally by bacteria. The end products are methane (CH_4) and carbon dioxide (CO_2). Methane, however, is the only hydrocarbon produced in any quantity at low temperatures by bacteria close to the surface of the sediment. Gas occurring at shallow depths (shallow gas) therefore consists largely of methane (dry gas) unless there has been addition from much deeper strata. Biogenic gas may form commercial accumulations, as in Western Siberia and also in the shallow part of the North Sea basin. The presence of abundant shallow gas may represent a hazard in the form of blowouts and fire during drilling. Gas occurring at shallow depth may also have a deeper source generated from a gas-prone source rock (coaly sediments) or by cracking of oil, but such gas has a very different isotopic signature than biogenic gas.

1.6 Kerogen

As organic material becomes buried by the accumulation of overlying sediments, water is gradually expelled during compaction.

Complex organic compounds like proteins are broken down into amino acids, and carbohydrates into simpler sugar compounds. These are able to recombine to make larger compounds, for example by amino acids reacting with carbohydrates (melanoid reaction). As this type of polymerisation proceeds, the proportion of simpler soluble organic compounds diminishes at depths of a few tens of metres down in the sediment. It is these newly-formed complex organic structures which are called kerogen.

Kerogen is a collective name for organic material that is insoluble in organic solvents, water or oxidising acids. The portion of the organic material soluble in organic solvents is called bitumen, which is essentially oil in a solid state.

Kerogen consists of very large molecules and is a kind of polymer. When it has been exposed to sufficient time and temperature these large molecules will crack into smaller molecules, mostly petroleum. When the temperature is about 100°C a long period of geological time is required. In rapidly subsiding basins the exposure time is shorter and oil generation may only start at about 140–150°C. In the North Sea basin the "oil window" may typically be between 130 and 140°C.

1.7 Migration of Petroleum

Petroleum migrates from low permeability source rocks into high permeability reservoir rocks from which the petroleum can be produced (Fig. 1.2b).

The main driving force for petroleum migration is buoyancy because it is less dense than water. The forces acting against migration are the capillary forces and the resistance to flow though rocks with low permeabilities

Migration of oil and gas will therefore nearly always have an upwards component.

We distinguish between *primary migration*, which is the flow of petroleum out of the source rock and *secondary migration*, which is the continued flow from the source rock to the reservoir rock or up to the surface (Fig. 1.4).

Oil and gas may also migrate (leak) from the reservoir to a higher trap or to the surface. Hydrocarbons are relatively insoluble in water and will therefore migrate as a separate phase. Solubility varies from as little as 24 ppm for methane to 1,800 ppm for benzene. Other compounds, such as pentane, are even less soluble (2–3 ppm). However, solubility increases markedly with pressure. Many hydrocarbons have solubilities of less than 1 ppm in water

It is difficult to envisage oil being dissolved in water and transported in an aqueous solution, both because of the solubility and the low flow rates. It would also be difficult to explain how the oil would come out of solution in the reservoirs (traps).

Gas, in particularly methane, has a fairly high solubility in water, especially under high pressure. If methane-saturated water rises to lower pressures, large quantities of methane can bubble out of a solution.

It is therefore necessary to assume that oil is mostly transported as a separate phase. Oil is lighter than

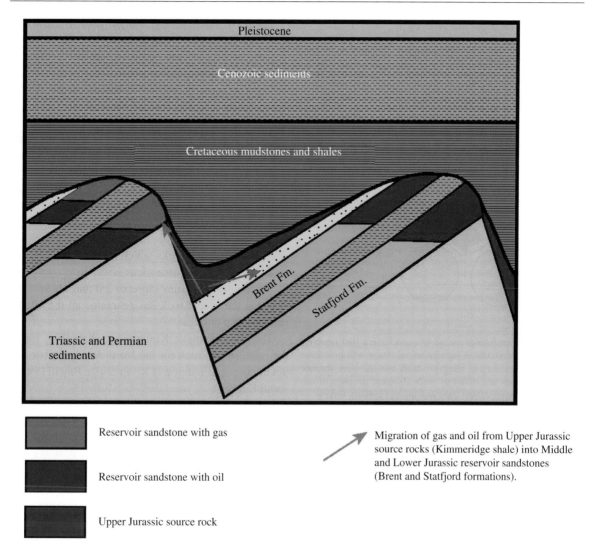

Reservoir sandstone with gas

Reservoir sandstone with oil

Upper Jurassic source rock

Migration of gas and oil from Upper Jurassic source rocks (Kimmeridge shale) into Middle and Lower Jurassic reservoir sandstones (Brent and Statfjord formations).

Fig. 1.4 Schematic illustration of primary migration (expulsion) of petroleum from a source rock and secondary migration into a reservoir (trap). This example is from the northern North Sea where rifting in Upper Jurassic time produced good conditions for the formation of a source rock and also traps on the uplifted fault blocks

water, and oil droplets would be able to move through the pores in the rocks but the caplliary restance is high for separate oil drops in a water-wet rock (Fig. 1.5). In order to pass through the narrow passage between pores (pore throat), the oil droplets must overcome the capillary forces. When the pores are sufficiently small in a fine-grained sediment, these forces will act as a barrier to further migration of oil. The small gas molecules, however, can diffuse through extremely small pores and thus escape from shales which form tight seals for oil.

Oil can therefore not migrate as small discrete droplets, but moves as a continuous string of oil where most of the pores are filled with oil rather than water (highly oil-saturated). The pressure in the oil phase at the top is then a function of the height of the oil-saturated column (string) and the density difference between oil and water.

The rate of migration is a function of the rate of petroleum generation in the source rocks. This is a function of the temperature integrated over time (Fig. 1.6).

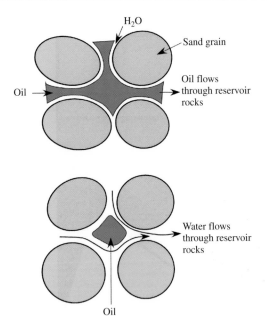

Fig. 1.5 Most sandstones are oil-wet and have a thin layer of water around the grains. A continuous oil phase will flow easily if the permeability is relatively high and the pore throats between the pores are relatively wide. Isolated droplets of oil will, however, be prevented from moving by capillary forces

The temperature history is a function of the burial depth and the geothermal gradients.

Deep burial over long time will cause all oil to be decomposed (cracked) into gas.

The degree of alteration of organic matter can be measured in different ways. Plant material is altered from a dull material to a material which becomes more shiny with increasing temperature. This can be quantified by measuring the amount of light reflected from a piece of plant material (vitrinite) under the microscope. A vitrinite reflectivity of 1.2 indicates that the source has generated much of the oil that can be generated. We will say that the source rock is in the middle of the "oil window" (Fig. 1.7). Values below 0.7–0.8. are found in source rocks which have not been heated enough (immature source rocks).

Vitrinite reflectivities close to 2.0 and above indicate that the source rock has generated all the oil and can generate only gas.

The Upper Jurassic Kimmeridge shale (Fig. 1.8) is the main source rock for the North Sea basin but it is not mature at its outcrop at Kimmeridge Bay in Dorset, south England.

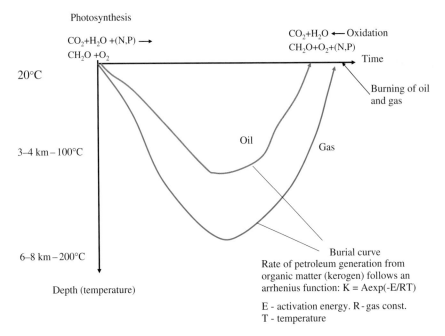

Fig. 1.6 Burial curve for source rocks determining the transformation of kerogen to oil and gas depending on time and temperature (burial depth)

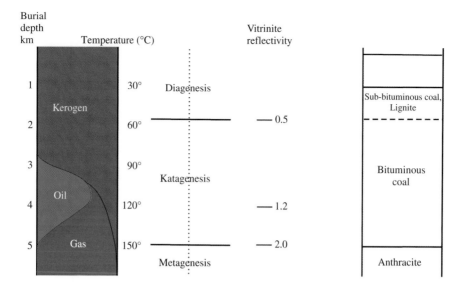

Fig. 1.7 Alteration (maturation) of organic matter and generation of oil and gas as a function of temperature. The maturation is also a function of time and this can be determined by measuring the vitrinite reflectivity. Coals become more shiny and have higher reflectivity with increasing temperature. The depth (temperature) range where oil is generated is called the "oil window". At higher temperatures oil will be altered into gas by cracking

Fig. 1.8 Kimmeridge Clay (Upper Jurassic) unconformably overlain by the lower Chalk (Upper Cretaceous). This is a very good source rock and equivalent shales are the main source rocks for oil and also much of the gas in the North Sea basin and also further North in mid Norway (Haltenbanken) and the Barents Sea. A layer of red Chalk marks the transition from black shale facies to carbonate facies. From South Ferriby, Yorkshire, England

1.8 Hydrocarbon Traps

Traps consist of porous reservoir rocks overlain by tight (low permeability) rocks which do not allow oil or gas to pass. These must form structures closed at the top such that they collect oil and gas, which is lighter than water. We can think of an oil trap as a barrel or bucket upside down (Fig. 1.9) which can then be filled with petroleum which rises through the water until it is full. The point where the petroleum can leak from this structure is called the *spill point*. *The closure* is the maximum oil column that the structure can hold before leaking through the spill point (Fig. 1.10).

The cap rock may not be 100% effective in preventing the upward flow of hydrocarbons, but these will still accumulate if the rate of leakage is less than the rate of supply up to the trap. *Cap rocks* are usually not totally impermeable with respect to water, but may be impermeable to oil and gas due to capillary resistance in the small pores.

Traps can be classified according to the type of structure that produces them. We distinguish between:

(1) *Structural traps* that are formed by structural deformation (folding, doming or faulting) of rocks.

(2) *Stratigraphic traps* which are related to primary features in the sedimentary sequences and do not require structural deformation like faulting or folding. This may be sandstones pinching out in shales due to primary changes in facies (Fig. 1.11).

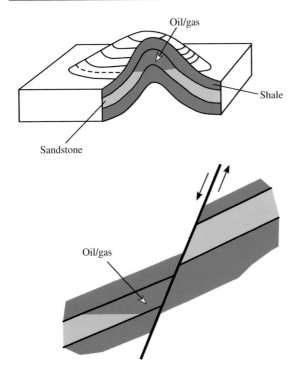

Fig. 1.9 Examples of structural traps. Simple anticlinal trap and a fault-controlled trap

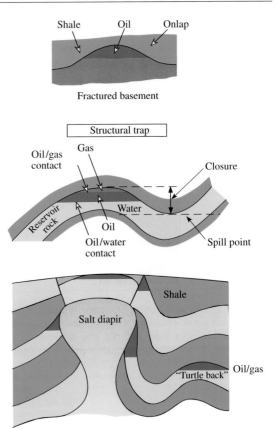

Fig. 1.10 Structural traps related to salt domes and anticlinal folds. A basement high can also be a trap when it is covered by a black shale (source rock). The basement may have some porosity due to fractures or a thin sediment cover

Carbonate reefs tend to form primary structures which function as stratigraphic traps.

There are also combinations between stratigraphic and structural traps (Fig. 1.12).

It is important to establish when structural traps were formed in relation to the migration of the petroleum. Structures formed after the main phase of source rock maturation and associated migration will not be effective traps. In some cases traps formed late can collect gas which normally is generated and migrates later than oil.

Stratigraphic traps, by contrast, have been there all the time, and the timing of the migration is not so important. They may however depend on slight tilting of the strata involved.

A simple anticline is not sufficient to trap oil. Anticlines with an axial culmination are needed to provide four-way closure. This means that the fold axis must be dipping in both directions (Fig. 1.9).

Anticlinal traps can form in association with faulting. This is especially true in connection with growth faults (roll-overs) (see below), but also with thrust zones.

1.8.1 Structural Traps

(a) Anticlinal Domes

Domes formed by diapirism or other processes may form closures in all directions (four-way closure).

(b) Salt Domes

Salt domes are formed because salt (specific gravity c.1.8–2.0) is lighter than the overlying rock, and the salt therefore "floats" up due to buoyancy. The quantitatively most important salt minerals are halite (NaCl – density 2.16 g/cm^3), gypsum (CaSO$_4$. 2H$_2$O – density

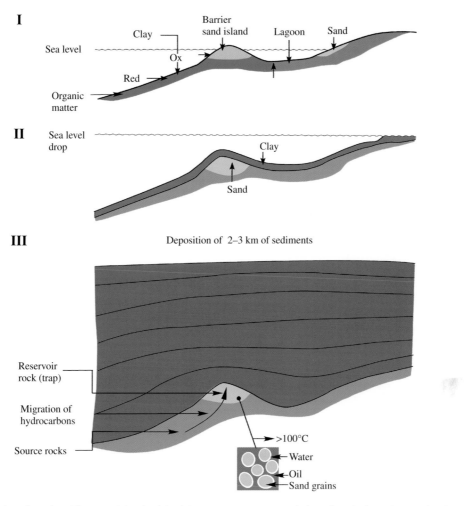

I — Sea level, Clay, Ox, Red, Organic matter, Barrier sand island, Lagoon, Sand

II — Sea level drop, Clay, Sand

III — Deposition of 2–3 km of sediments, Reservoir rock (trap), Migration of hydrocarbons, Source rocks, >100°C, Water, Oil, Sand grains

Fig. 1.11 Examples of stratigraphic traps. A barrier island forms a separate accumulation of sand where the associated mudstones (shales) may represent both source rocks and cap rocks

2.32 g/cm^3). Anhydrite (CaSO$_4$ – density 2.96 g/cm^3) is too dense to contribute to the formation of diapers.

In order for the salt to move upwards and form a salt dome, a certain thickness of overburden is required and the salt beds themselves must be at least 100–200 m thick. The upward movement of salt through the overlying sequence, and the resultant deformation of the latter, is called halokinetics or salt tectonics.

The rate of salt movement is extremely slow and a dome may take several million years to form. Movements of the earth's surface may, however, also be recorded in recent history as is the case onshore Denmark. Salt may break right through the overlying rocks and rise to the surface, or form intrusions in younger sediments. If gypsum has been deposited,

this will be altered into anhydrite at about 1 km burial depth, with a consequent 40% compaction and the increase in density will remove the buoyancy relative to the surrounding sediments. A comparable expansion occurs when rising anhydrite comes into contact with groundwater and reverts to gypsum.

Traps may be created (1) in the layers above the salt dome, (2) in the top of the salt dome (cap rock), (3) in the beds which are faulted and turned up against the salt structure and (4) through stratigraphic pinching out of beds round the salt dome. Reservoirs may form by solution and brecciation at the top of salt domes.

Salt tectonics is of great importance in many oil-bearing regions where there are thick salt deposits in

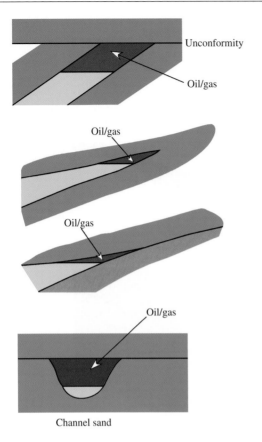

Unconformity

Oil/gas

Oil/gas

Oil/gas

Oil/gas

Channel sand

Fig. 1.12 Combination of stratigraphic and structural traps. A reef forms a trap due to the primary relief and also due to later compaction of the mud around the reef

the passive margin sequences of the South Atlantic and the Gulf of Mexico. In the eastern USA we find extensive tracts with Silurian salt, and in Texas and New Mexico we have Permian salt.

Salt layers are the ideal cap rock because of salt's low permeability and ductile properties, which prevent fracturing and leakage.

Salt deposits are particularly common in the Permo-Triassic around the Atlantic. This is because prior to the opening of the Atlantic there were vast areas with fault-controlled basins (rifts) in the middle of a supercontinent (America + Europe, Asia and Africa) with little precipitation. We find similar conditions today around the Red Sea and the Dead Sea. The Permian Zechstein salt in Germany and Denmark continues below the North Sea, and halotectonic movements have formed dome structures in the Chalk, for example in the Ekofisk area.

(c) Growth Anticlines

These are dome-like structures formed when part of a basin subsides more slowly than its surroundings, resulting in least sedimentation on the highest part. The sediment thickness decreases towards the dome centre, which also compacts less than the adjacent thicker sediments and thus contributes to the formation of an anticlinal structure. Growth anticlines form contemporaneously with sediment accumulation, not through later folding.

Growth anticlines can be formed above salt domes, reefs or buried basement highs, through differential compaction.

(d) Fault Traps

In fault traps, the fault plane forms part of the structure trapping the oil and hindering its further upward migration. The fault plane must therefore be sealing for vertical flow in order to function as a barrier and a cap rock for the reservoir rocks. If the reservoir rock is juxtaposed against a sandstone or other permeable rocks the fault must also be impermeable for flow across the fault plane. Most frequently, however, the reservoir rock is faulted against a tight shale or mudrock and the fault is then in most cases sealing (Fig. 1.9). When there is sandstone on both sides of the fault plane, the permeability across it will amongst other things depend on how much clay has been smeared along the junction. At greater depths (>3–4 km) there may be diagenetic changes such as quartz cementation, which can make the fault plane tighter.

There are many different types of faults:

Normal faults – often in connection with graben (rift) structures.
Strike-slip faults.
Reverse faults formed by tectonic stress.
Growth faults.

The displacement along faults can be both vertical (normal faults) and horizontal (strike slip faults). Reverse faults are faults where the hangingwall is moved upwards relative to the footwall below the fault plane. These are typical of areas with high horizontal stresses i.e. due to converging plate movements. Growth faults are driven by gravity-sliding along curved (listric) fault planes and are typical of sedimentary sequences such as deltas deposited with relatively rapid sedimentation. The fault plane is often (though not always) sealing and can stop oil and gas from migrating further upward. However, oil traps are equally often formed in anticlines on the upper side of the fault plane. These are rollover anticlines. Because the faulting is active during sedimentation, the layers on the downthrown side will be thickest. The name growth fault goes back to the early days of oil exploration without seismic data. It was noticed that the layers had "grown" in thickness in the wells on the downthrown side of the fault. The displacement of the beds decreases upwards along the fault plane. Smaller, antithetic faults often develop in the opposite direction in the beds which are turned inwards towards the main fault plane. Growth faults tend to have low permeability and may contribute greatly to reduced porewater circulation in sedimentary basins, and we often find undercompacted clay, which can turn into clay diapirs in association with growth faults.

1.8.2 Stratigraphic Traps

These are traps which are partially or wholly due to facies variation or unconformities, and not primarily the result of tectonic deformation. Porous and permeable sands which pinch out up-dip in less permeable rocks, e.g. shale (Figs. 1.11 and 1.12) are good examples. Barrier islands often form stratigraphic traps because they may be separated from the coast by fine-grained lagoonal facies. The main types are:

(a) Fluvial channel sandstones may be isolated and surrounded by impermeable clay-rich sediments, or they may be folded so that we obtain a combination of stratigraphic and structural traps (Figs. 1.9 and 1.10).

(b) Submarine channels and sandstone turbidites in strata rich in shale. Here we will often find pinchout of permeable layers up-dip from the foot of the continental slope. This will result in stratigraphic traps without any further folding being necessary.

(c) Reefs often form stratigraphic traps. A reef structure projects up from the sea bed and often has shale sediments surrounding it, so that oil could migrate from the shale into the reef structure.

Fig. 1.13 Natural oil seep at Carpenteria State Beach, California. Oil is flowing on land, on the beach and also offshore on the sea floor

(d) Traps related to unconformities. Sandstones or other porous rocks may be overlain with an angular unconformity by shales or other tight sediments, forming a trap underneath the unconformity (Fig. 1.12). Topographic highs in the basement overlain with shales can also provide good traps in fractured basement rocks. Remember that oil can migrate upwards into stratigraphically lower rocks. In China there are numerous examples of this type of trap.

Much of the oil generated in sedimentary basins has not been trapped in reservoirs but reached the surface on the seafloor or on land. There it is then broken down by bacteria and becomes heavy oil, which is not very toxic. In California there are many examples of natural oil seeps which can be observed along roads, on the beach (Fig. 1.13) and also offshore.

1.9 Other Types of Trap

More unusual kinds of trap can be encountered. If the porewater in a sedimentary basin has sufficiently strong flow of meteoric water into the basin, the oil/water contact may diverge markedly from a horizontal plane due to the hydrodynamic stresses. This has implications for calculations of oil volumes within a structure, and in some instances oil can accumulate without being sealed in, within a so-called hydrodynamic trap. The circulation of fresh (meteoric) water down into oil-bearing rocks will, however, lead to biodegradation and the formation of asphalt. Asphalt can then become a tight cap rock for the oil.

At greater depths, beyond the reach of meteoric water, water movement is limited and any deflection of the oil/water contact is more likely to reflect pressure differences within the reservoir. Water will also flow then because of the pressure gradient, but unless there are low permeability barriers the pressure will soon equalise. Tectonic tilting will also tilt the oil/water contact.

Reservoir Geology is not a well-defined discipline. It includes many aspects of geology that are of special relevance to the production of petroleum. It is also linked to engineering aspects of petroleum production. *Reservoir geophysics* has in recent years become very important and is now well integrated with reservoir geology.

1.10 Porosity and Permeability

Any rock with sufficiently high porosity and permeability may serve as a reservoir rock provided that there is a source of petroleum, a structure, and a tight cap rock.

Sediments consist of solid grains and of fluids which for the most part are water but may be oil and gas.

Porosity (φ) is an expression of the percentage (or fraction) of fluids by volume (V_f) compared to the total rock volume with fluids (V_t), so that $\varphi = V_f / V_t$. Porosity is often expressed as a percentage, but in many calculations it is easier to express it as a fraction, for example 0.3 instead of 30% porosity.

The *void ratio* (VR) is the ratio between pore volume (φ) and the volume of the grains ($1-\varphi$).

$$\text{VR} = \varphi / (1 - \varphi)$$

Void ratio is often used in engineering and it has certain advantages in some mathematical expressions.

If we assume that we know the density of the mineral grains, the porosity can be found by measuring the density of a known volume of the sediment. The density of the sediments (ρ_s) is the sum of the density of the grains, which are mostly minerals ρ_m, and the density of the fluids (ρ_f).

$$\rho_s = \varphi \rho_f + \rho_m (1 - \varphi)$$

Well sorted, rounded sand grains are almost spherical in shape. If we have grains of the same size, which are all quite well rounded and with a high degree of sphericity, we will be able to pack the grains so as to get minimum porosity. Rhombic is the densest packing, resulting in 26% porosity, but this can not be obtained naturally. Cubic packing, where the grains are packed directly one above another, results in about 48% porosity and this does not occur in nature either. Most well sorted sandstones have a porosity which lies between these two values, typically around 40–42%. Poorly sorted sand may have lower primary porosity and will also compact more at moderate burial depths. Clay-rich sediments have a much greater porosity immediately after deposition, typically 60–80%. This means that immediately following deposition a sand bed is denser than a bed of clay or silt. However, clay and silt lose their porosity more rapidly with burial. Porosity may be classified into different types depending on its origin.

Pore space between the primary sediment grains is often referred to as *primary porosity. Intergranular porosity* simply means porosity between the grains whereas *intragranular porosity* means porosity inside the sediment grains. The latter may be cavities in fossils, e.g. foraminifera, gastropods, molluscs, but also partly dissolved feldspar and rock fragments. Pore space formed by dissolution or fracturing of grains is called *secondary porosity*.

Cavities formed by selective solution of sediment grains or fossils are classified as *mouldic porosity*. A typical example is when dissolution of aragonite fossils like gastropods leaves open pore spaces (moulds).

Particularly in carbonates we may also have porosity on a large scale i.e. as caverns (karst) and in reefs.

Pore space produced by fracturing is called *fracture porosity*.

Permeability is an expression of the ease with which fluids flow through a rock. It will depend on the size of the pore spaces in the rocks, and in particular the connections between the pore spaces. Even thin cracks will contribute greatly to increasing the permeability.

Permeability can be measured by letting a liquid or gas flow through a cylindrical rock sample under pressure. The pressure difference $P_1 - P_2$ between the two ends of a horizontal cylinder is ΔP, the cylinder length L, and the flow rate of water (or another fluid) through the cylinder, is Q (cm³/s). A is the cross-section and μ the viscosity of the fluid

$$Q = \frac{k \cdot A \cdot \Delta P}{L \cdot \mu}$$

where k is the permeability, which is expressed in Darcy.

The volume of water which flows through each surface unit in the cross-section A is thus equal to the flux $F = Q/A$. F can be measured in cm³/cm²/s or in m³/m²/s. This is equal to the Darcy velocity which is m/s.

Well-sorted sandstones may have a permeabilities exceeding 1 Darcy and values between 100 and 1,000 mD are considered to be extremely good. Permeabilities of 10–100 mD are also considered to be good values for reservoir rocks. Permeabilities of 1–10 mD are typical of relatively dense sandstones and limestones, so-called tight reservoirs. There are also examples of rocks with even lower permeabilities being exploited commercially for oil production, for example in the Ekofisk Field where the generally low permeability of a chalk matrix is enhanced by fractures which increase the overall permeability.

In the great majority of rocks, the permeability differs according to flow direction. In sedimentary rocks the permeability is much higher parallel to the bedding compared with normal to the bedding. Channel sandstones can also have a marked directional impact on the permeability.

In well-cemented sandstones and limestones, and also in certain shales, the matrix permeability is extremely low and the effective permeability may be mostly controlled by fractures if they are present.

Claystones and shales have very low permeability and can be almost completely tight. In the laboratory shale permeabilities as low as 0.01 nanodarcy have been measured. Samples from cores or outcrops can contain minute fissures formed in response to unloading during retrieval to the surface and these must be closed to replicate the *in situ* permeability prior to unloading.

Most rocks are far from homogeneous. We may measure the porosity and permeability of a hand specimen or core plug, but it is not certain that these are representative of a larger volume. Fractures occur at varying intervals, and range in size from large, open joints down to microscopic cracks which can barely be seen in a microscope.

Rocks with low porosity and permeability may fracture and sufficiently increase their porosity, and particularly permeability, to form large oil reservoirs. This means that reservoirs may be good producers despite relatively low porosity.

Occasionally we find petroleum in fractured metamorphic and igneous rocks but reservoirs normally consist of sedimentary rocks. Sandstones make up about 50–60% of the reservoirs in the world and carbonate reservoirs may account for almost 40%. Many of the reservoirs in the Middle East are carbonate rocks but in the rest of the world the percentage of carbonate reservoirs is lower.

The most important aspects of reservoir rocks include:

(1) The external geometry such as the thickness and extent of the reservoir rock in all directions.
(2) The average porosity, pore size and pore geometry.
(3) The distribution of permeability in the reservoirs, particularly high permeability conduits and low permeability barriers to fluid flow.
(4) Mineralogy and wettability of the pore network.

The properties of sandstones and carbonate reservoirs are primarily linked to the depositional environment, the textural and mineralogical composition and the burial history. A good background in general sedimentology, facies analysis and sequence stratigraphy is therefore important.

Nevertheless, many of the important properties of reservoir rocks linked to changes in facies and smaller faults are below the vertical resolution of exploration seismic (15–30 m) and it is important to establish relationships between facies models, diagenetic processes and reservoir properties. The properties of faults are also very important factors determining oil flow during production.

1.11 External Geometry of Reservoir Rocks

The external geometry of reservoir rocks is largely determined by the depositional environments, but faulting and diagenesis may define the lateral or vertical extent of a reservoir.

Fluvial sandstones typically represent point bar sequences in a meandering river system. The lateral accretion of the point bar will deposit a sandstone layer extending to the width of the meander belt in the valley. The thicknesses of channel sandstones are limited by the depth of the river. The primary thickness at the time of deposition is, however, reduced by 10–30% or more by compaction.

The overbank muds will become tight shales which will reduce the vertical permeability. Fluvial channels are characterised by fining-upwards sequences with the highest permeability near the base. This makes it more likely that water will break through along the basal part and the oil will be by-passed in the finer-grained upper part during production.

Braided stream facies will tend to have higher sand/shale ratios and will normally have better lateral and vertical permeabilitites on a larger scale.

The ratio between the intervals with high enough porosity and permeability to be produced (net or pay), and the total sequence (gross), will be mostly determined by the primary facies relationships. The net/gross ratio is often taken to be approximately equal to the sand/shale ratio but even at moderate burial depths many sandstones are not reservoir rocks, due to poor sorting or carbonate cement.

Aeolian dunes also have specific external geometries and there are many different types. Here the net/gross will be very high. Aeolian sand is often reworked by transgressions, accumulating as marine sediments in drowned topographic depressions (valleys).

Marine sandstones deposited as delta mouth bars, shoreface accretion and barrier islands have thicknesses controlled by the wave energy (wave base depth). In protected environments, particularly inter-distributary bays, the shoreface sandstones may be very thin. Each shallow marine unit has a limited thickness controlled by fair-weather wave base. Local subsidence or transgressions can increase the thickness of these sand deposits.

The tidal range is very important in determining the thickness and the length of tidal channel sandstones. Tidal channels and also fluvial channels in deltas tend to be oriented perpendicular to coastlines.

Drilling into shallow marine sandstones, it would be very important to determine whether it was a barrier island which would represent an elongated reservoir parallel to the coastline, or a tidal sandstone which tends to be oriented perpendicular to the coastline. In some cases dipmeter logs could help to determine the orientation of cross-beddding and progradation direction of sand bars.

Turbidites may be laterally very extensive, but may also be confined to narrow submarine channels. In either case they may form very thick sequences because there is ample accommodation space. We may have very thick sequences of stacked sandstone reservoir rocks in slope and deepwater facies and this may compensate for the lower porosity and permeability compared to beach deposits.

Turbidites and fluvial sandstones form fining-upwards units while marine shoreface and mouth bar sandstones are coarsening-upwards. This becomes very significant during production because oil and gas will be concentrated in the upper part. Coarsening-upwards sandstones therefore have the best properties for flow of oil and gas during production.

1.12 Changes in Rock Properties During Burial and Uplift (Diagenesis)

The changes in properties are due to increased burial and also to uplift. Both sandstone and carbonate reservoirs undergo diagenesis, which will cause a reduction in porosity and permeability as a function of increasing burial.

The reduction in porosity (compaction) may be mechanical in response to increased effective stress from the overburden, or chemical as a result of the dissolution and precipitation of minerals. The porosity of reservoir sandstones or carbonates may increase with depth in certain intervals, but this is because of the changes in the primary sediment composition. Each lithology has a different porosity depth curve. In a uniform primary lithology the porosity and the density will be reduced as a function of burial depth (temperature and stress). Overpressure causes reduced effective stress resulting in mechanical compaction. Near the surface, meteoric flow may cause dissolution and a net increase in the porosity in carbonates (karst) and even, to a certain extent, in sandstones.

In continuously subsiding basins, open faults and fractures will be rare because of the progressive compaction processes.

During uplift and erosion (exhumation, unloading) the rocks will be subjected to extension, and extensional fractures will be produced. The porosity will not increase significantly but the extensional fractures will increase the permeability and thus improve the reservoir properties. Unfortunately the cap rock may also fracture during unloading, causing leakage from the reservoir.

To understand the properties of reservoir rock we need to integrate what we know about the sedimentology (depositional environments) of reservoir rocks, sediment composition (provenance), diagenesis and the structural geology.

Prior to drilling exploration wells, nearly all our knowledge about a reservoir is based on geophysical data. Even after data has been acquired from exploration wells, and also from production wells, prediction of the reservoir properties continues to be mostly based on geophysical methods and extrapolation between wells.

Geophysical methods including 3D and 4D seismic now provide a much more detailed picture of the reservoir than only 10–20 years ago. The methods for detecting fluid contacts, not only gas/water contacts but also oil/water contacts, from seismic data have improved greatly.

Reefs can form long continuous barriers as in Australia (Great Barrier Reef). In the US much oil was found by following Jurassic and Cretaceous reef trends around the Gulf of Mexico.

High energy beach deposits on carbonate banks may consist of well-sorted carbonate sand (grainstones). Ooid sands (ooliths) are formed as beach and shoreface deposits and may have limited vertical thickness, reflecting the wave base. They may however stack up and form thicker sequences of such rocks. Ooids may also be transported from the shelf into deeper water as turbidites and other slope deposits, but the sorting and reservoir quality is then reduced.

Ooids are rather stable mechanically during burial, but at 2–3 km porosity may be strongly reduced by cementation in the intergranular pore space. The calcite cement may be derived by dissolution along stylolites. Stylolites have a thin layer of clay and other minerals that are not soluble and may present a barrier during oil migration and production.

Carbonate muds are very fine grained and do not have high enough primary porosity and permeability to form reservoir rocks, but may gain porosity and permeability by fracturing and become fractured reservoirs. They can also have mouldic porosity from dissolved aragonite fossils.

1.13 Carbonate Reservoirs

Reefs stand up as positive structures and may be draped by mud and form a stratigraphic trap determined by the size of the reef structure.

Carbonate reefs, and other carbonate deposits which can be reservoirs, form in a wide range of environments but all require clear water without much clay sedimentation. Reefs are deposited in high energy environments along coastlines exposed to high wave energy. The Bahamas carbonate platform has well-developed reefs on the exposed eastern side but not on the more protected western side. Coral reefs also require warm water (>20°C) and do not form where cold water is upwelling, e.g. along the coast of West Africa.

Reefs build up on the seafloor and may be buried beneath mud during transgressions. The reef then becomes a perfect stratigraphic trap, often with good permeability both vertically and horizontally.

1.14 Drilling for Oil and Gas

Drilling for oil is a costly process, especially offshore. The object of a well is to prove the presence of, or produce, oil or gas. Sometimes wells are also drilled to inject water, chemicals or steam into the reservoir during production. Even a well which fails to find hydrocarbons (a dry well) is still of great value, because of the information it provides about the rocks in the area. This information forms part of the basis for the geological maps and profiles which are used in further exploration for oil and can be sold or exchanged for data from other companies. This is the reason why oil companies wish to keep the geological results of oil drilling confidential for some years after a well has been completed.

The first well in a new area is called a *wildcat* well, while *appraisal* wells are drilled to estimate the extension of an oil field. They may also become *production* wells. Stratigraphic wells are drilled mainly to obtain stratigraphical information from the basin.

Oil drilling used to be carried out largely on land, but now offshore drilling takes place not only on our continental shelves, but also in deep water (1–3 km). This type of drilling is many times more costly than drilling in shallow water or on land. This has led to increased efforts to gain maximum information from wells. The cost of analysing samples and logs is small in relation to the cost of drilling the well. We shall not go very deeply into the technical aspects of drilling for oil here, but merely look briefly at some of the most important principles.

When drilling commences at the surface, the diameter of the well may be 20"–30" (50–75 cm), but decreases downwards to 3"–6" (7–15 cm) at great depths. Normally a roller bit is used, which crushes the rock into small pieces (about 2–5 mm) called cuttings. Core samples are only taken when drilling through especially important rock strata (usually reservoir rocks) where large intact samples are needed for detailed examination. A circular diamond core drill bit must then be used. This takes time and costs a lot more per running metre, as the entire drill string has to be recovered to get each core section to the surface. Only the most critical sections are therefore cored.

Drilling mud is pumped down through the drill string into the well during drilling. This mud has several functions. When one drills several hundred or a 1,000 m down into rock, one encounters water, gas or oil which may be under high pressure. The drilling mud acts as a counterweight to prevent the uncontrolled gush of water or petroleum into the well and up to the surface in a blow-out. The pressure exerted by the drilling mud must exceed the pressure of oil and water in the surrounding formation. Heavy minerals such as barytes are frequently added to increase density; the main components of drilling mud are montmorillonite (smectite) containing clays, with a large number of different additives. The drilling mud also serves to cool the drill bit, and cuttings are brought back to the surface suspended in the circulating mud. The cuttings are then washed out from the drilling mud onto a sieve (shale shaker) and the mud can be used again.

The cuttings are continuously analysed on the drilling platform by a geologist who logs the composition of the cuttings, making a preliminary description of the rocks which are being drilled, and their mineralogy. Samples of cuttings are usually taken every 10–30 ft drilled. More detailed analyses are carried out in the laboratory. The cuttings are often poorly washed and need extra cleaning to get rid of the drilling mud; in weakly indurated mudstones it may be difficult to separate the drilling mud from the soft cuttings. Organic additions to drilling mud may cause problems when analysing cuttings, and sometimes oil-based rather than water-based drilling mud is used, which confuses analyses for oil. The fossil content of the rock fragments, largely microfossils, is used to determine the age of the strata (see Biostratigraphy).

Not all the cuttings which come up with the drilling mud have necessarily come from precisely the strata being drilled through at that time. There may also be contamination due to the *caving in* of overlying strata into the rising drilling mud. This means we can find material from younger rocks with a different composition mixed in with the formation being drilled, together with younger fossils. This demands considerable care when making stratigraphic interpretations based on microfossils identified in cuttings. The safest way is to register the first occurrences of a species when proceeding downwards from the top in the well. The last occurrence of a fossil may be the result of cave-ins from younger strata.

Since the pressure of the drilling mud is being monitored and adjusted to prevent oil and gas from penetrating into the well, significant oil and gas occurrences may be drilled through without being registered. This should be detected on the logs but it may be advisable to carry out special tests in the most promising strata to find out if there is petroleum present, and in what quantities.

As drilling proceeds, the well is lined with steel casing to prevent rock and loose sediment falling into it, but prior to casing, each section of the well has to be logged with different logging tools which require physical contact with the wall of the well. Radioactive logs, however, can also be run after the casing has been installed. It is useful to note when the different casings are installed, because that limits the strata which could have "caved in" and contaminated the cuttings. If the well is going to produce oil, a production pipe is used and installed running through the petroleum-bearing strata. It is then perforated by shooting holes in the steel casing (in the oil column) so that petroleum can flow into the well.

For an oil field to be capable of full production, several wells are normally required.

Up to 1990–1995 most of the exploration wells and production wells were nearly vertical. Horizontal barriers due to changes in facies or faults could then be critical barriers during production. It was difficult to produce oil from thin sandstones or carbonate beds because of vertical flow of water from below or gas from above. Onshore this could be compensated for by having a dense well spacing but offshore that would be too expensive. Horizontal drilling has revolutionised oil production.

It is now possible to follow thin oil columns laterally and to make complex wells to drain different compartments in the reservoir. Large oil fields with thin oil columns like the Troll field in the North Sea would have been difficult to produce without horizontal drilling. Horizontal and deviated wells may extend 8–9 km from the drilling platform, enabling production from relatively small reservoir compartments away from the platform.

Earlier, geophysics was used mainly to define structures but the quality of the seismic data was not good enough to provide much detail. 3D seismic based on 50 m line spacing allows the construction of a three-dimensional cube of geophysical data, which provides much more detailed information.

By repeating seismic surveys during production at 2–5 year intervals, the effect of changes in the fluid composition on the seismic data can be seen, thus adding the time dimension to 3D seismics. 4D seismics has made it possible to follow the depletion of an oil or gas field by monitoring the GWC and often also the OWC during production. By this means, parts of a reservoir that have not been drained by the production wells can be detected. If the isolated reservoir compartments are large it may be economic to drill additional wells to drain them.

1.15 Oil Reserves – How Long Will Conventional Oil Last?

In the last 40–50 years we have had a discussion about how long the oil reserves will last and the famous geologist M. King Hubbert predicted in 1956 that oil production would peak in the United States between 1965 and 1970, and later Colin Campbell predicted

that world production would peak in 2007 (see the Peak Oil movement). There is, however, a great deal of uncertainty in the estimation of reserves since this will depend on advancements in exploration technology as well as production efficiency.

The price of oil and taxation policies will also determine which type of oil accumulations can be exploited economically and thus reckoned as reserves.

At present the world reserves are estimated to be 1.2×10^{12} bbl $\left(2 \times 10^{11} \mathrm{m}^3\right)$ and nearly 60% of this is located in the Middle East. Europe has only got 1% and the whole of Asia about 3%.

Canada had up to recently very small reserves but after the heavy oil and tar sand in Alberta was included it has now nearly 15% and is second only to Saudi Arabia.

If oil shales were to be included, however, the USA would be the country with the largest reserves.

The world consumes about 85 million bbl/day $\left(85 \times 10^6 \mathrm{bbl/day}\right)$; the US consumption $(20.7 \times 10^6 \mathrm{bbl/day})$ makes up nearly 25% of that.

The most important producers are Saudi Arabia $(10.2 \times 10^6 \mathrm{bbl/day})$ and Russia $(9.9 \times 10^6 \mathrm{bbl/day})$. The US is also a major producer $(7.5 \times 10^6 \mathrm{bbl/day})$ but this is still only about 37% of the country's consumption.

China has become a major importer of oil with a consumption of 7.6×10^6 bbl/day, while their production is 3.9×10^6 bbl/day.

It is clear that consumption of oil in Asia will rise and it will be very difficult to meet this demand.

Norway's oil production was 122 million Sm^3 in 2008 and about the same amount of gas. This corresponds to 0.8×10^6 bbl or 2.1 million bbl of oil/day. However since the domestic consumption is only about 10% of this (0.22 million bbl/day), Norway is a major exporter.

It will probably be difficult to meet the demand for conventional oil in the next decades. There are, however, very large reserves of fossil fuels in terms of gas, heavy oil, tar sand and also coal. All these types of fossil fuel can be used for heating and transport. Particularly in North America there is much oil shale and also gas shale. Gas in fine-grained siltstones and shales is expected to be a major source or energy in the years to come.

In recent years coal methane and shale gas have become important sources of such energy. Gas in solid form (gas hydrates) may also represent a future source of hydrocarbons. There are, however, many environmental problems connected to the production and utilisation of these resources and this represents a great challenge, including for geoscientists.

It will probably take a long time before fossil fuels can be replaced by other sources of energy. As the demand increases, oil exploration and production will become more and more sophisticated technologically and also geologically.

A broader background in geological and engineering disciplines will also be required to reduce the environmental problems with the exploitation of fossil fuels.

Storage of carbon dioxide requires expertise from petroleum geologists.

The exploitation and burning of fossil fuels releases large amounts of CO_2 into the atmosphere which is an addition to what is part of the natural carbon cycle (Fig. 1.14). The CO_2 in the atmosphere is dissolved in seawater to H_2CO_3 and then precipitated as carbonate. Another part is taken up by plants, including algae, and may be stored as reduced carbon.

The total amounts of calcite precipitated is equal to the amounts of Ca^{++} released by weathering of silicate rocks (e.g. plagioclase) and transported into the ocean by rivers.

1.16 The Future of Petroleum Geoscience

Petroleum geoscience is geology and geophysics applied to petroleum exploration and production. In this book we will try to show the wide range of disciplines that are relevant and useful for this purpose.

Many of the disciplines in the geosciences are highly specialised and there is often too little communication between the different fields. Most researchers naturally focus on a very small area because of the requirements with respect to methods and analytical techniques, and the demands of following the literature. Petroleum geoscience requires a broad overview of substantial parts of geology and geophysics and

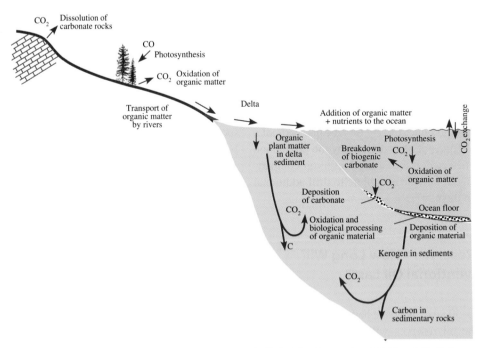

Fig. 1.14 Illustration of the carbon cycle. Carbon from organic matter and carbonate rocks are the major sinks for carbon (CO_2). The rate of precipitation of carbonate in the ocean by organisms is limited by the supply of Ca^{++} and Mg^{++} from weathering of silicate rocks brought in by rivers

provides good training in the integration of very different types of data and models. These skills are also applicable in many types of environmental research and when solving practical environment problems.

The petroleum industry employs a large percentage of the world's geologists and geophysicists and funds much of the research in this field.

Most of the obvious petroleum-bearing structures have already been found in the explored sedimentary basins and there are now rather few areas that have yet to be explored seismically and by drilling. The large, easy-to-find structures did not usually require very advanced methods and geological skills.

More and more sophisticated methods are therefore used in modern exploration. We are on a global basis not finding enough new oil fields oil to replace the produced oil. Global reserves have however not changed very much because of higher estimates of recovery from existing fields and because unconventional oil like tar sand is now included in the reserves.

There is a major challenge for geoscientists to develop ever better exploration methods and to optimise production.

Even if the global production of conventional oil may be reduced there will be significant production for many decades. This will mostly come from the tail production of giant fields and from small reservoirs, but it is rather labour intensive. This is also the case with unconventional oil (tar sand and oil shale) and also tight gas reserves and shale gas.

Until enough alternative sources of energy are developed it is necessary to extract fossils fuels from these resources. This should be done with as little environmental damage as possible and this requires a new generation of highly skilled geoscientists.

1.17 Summary

In the exploration for oil and gas we need to predict the occurrence and distribution of source rocks which is a function of the sedimentary environment and climate at the time of deposition. The burial history has to be reconstructed to predict the timing of maturation and migration. This involves complex modelling based on the kinetics and temperature history.

Before drilling, the position of a trap and a reservoir rock has to be determined.

The porosity and permeability of the reservoir rock is critical. If the porosity is 30% it may at the best contain about 250 l of oil and 50 l of water for each m^3 of rock, and assuming 50% recovery, 125 l can be produced. If the porosity is only 15% it may contain about 100 l of oil and the recovery may then be lower so that less than 50 l will be produced.

The quality of reservoir rocks depends on the depositional environments and the primary mineralogical and textural composition, and on the diagentic processes that change the reservoir properties during burial.

For sandstone reservoirs the distribution and geometry of sand is critical because much of this is below the resolution of seismic methods, and even after drilling the information from each well has to be extrapolated in 3 dimensions.

For this reason I have included a chapter on sedimentary structures and sedimentary facies and also on carbonates. There are, however, many textbooks which will give a more detailed presentation of clastic sedimentology. To be able to predict the reservoir properties we must understand the principles of sandstone diagenesis and this is discussed in Chapter 4. This also involves chemical reactions driven by thermodynamics and kinetics. Reservoir quality can to a certain extent be modelled as a part of basin modelling, if the primary mineralogical and textural composition is known.

Exploration and production of conventional oil has become well established and technologies have been refined over many years. This book will include an introduction to geophysical methods and also 4D seismic methods. Interpretation of geophysical data requires that the physical properties of sedimentary rocks are known or can be predicted. We have therefore put some emphasis on rock mechanics and rock physics.

Unconventional oil such as heavy oil, tar sand, oil shale and shale gas represents new challenges and requires also different training and background from that of the conventional petroleum geologist. The organic chemistry of petroleum and the detailed mineralogy of the rocks become increasingly important.

Estimates of global petroleum reserves vary greatly with time and are dependent on the price of oil and gas and other energy sources and also on exploration and production technology.

It is however clear that it is becoming more difficult to find new reserves to compensate for production and that there are relatively few remaining unexplored sedimentary basins.

Much updated information can be found on the internet:

US geological survey (http://www.usgs.gov/)

Peak oil (http://peakoil.com/)

Norwegian Petroleum Directorate (http://www.npd.no/)

Geological Society of London (http://www.geolsoc.org.uk)

IFP (http://www.ifp.fr/)

Units for petroleum reserves:

1 Sm^3 (One standard cubic metre) = 6.293 bbl

1 Sm^3 = 35.3 standard cubic feet.

1,000 Sm^3 of gas = 1 Sm^3 o.e

1 Sm^3 of oil = 1 Sm^3 o.e

1 tonne of NGL = 1.3 Sm^3 o.e

o.e – oil equivalents are used to sum up the energy of oil, gas and NGL (Natural Gas Liquids).

Further Reading

Allen, P.A. and Allen, J.R. 2005. Basin analysis. Principles and Applications. Blackwell, Oxford, 549 pp.

Beaumont, E.A. and Foster, N.H. 1999. Exploring for oil and gas traps. Treatise of petroleum geology. AAPG Special Publication 40, 347 pp.

Biju-Duval, B. 1999. Sedimentary Geology. Sedimentary Basins, Depositional Environments, Petroleum Formation. Institut Francais du Petrole Publication, Rueil-Malmaison Editions Technip, 642 pp.

Glennie, K.W. 1998. Geology of the North Sea: Basic Concepts and Recent Advances. Blackwell, Oxford, 636 pp.

Groshong, R.H. 1999. 3D Structural Geology. A Practical Guide to Surface and Subsurface Interpretation. Springer, New York, 324 pp.

Hunt, J.M. 1996. Petroleum Geochemistry and Geology. Freeman and Co., New York, 741 pp.

Miall, A.D. 1996. The Geology of Fluvial Processes, Basin Analyses, and Petroleum Geology. Springer, New York, 582 pp.

Salvador, A. 2005. Energy; a historical perspective and 21st century forecast. AAPG Studies in Geology 54, 207 pp.

Chapter 2

Introduction to Sedimentology

Sediment Transport and Sedimentary Environments

Knut Bjørlykke

Sedimentology is the study of sedimentary rocks and their formation. The subject covers processes which produce sediments, such as weathering and erosion, transport and deposition by water or air, and also the changes which take place in sediments after their deposition (diagenesis). Changes in sedimentary rocks at temperatures of over 200–250°C are called *metamorphic processes* and are not dealt with here. In this chapter we shall discuss primarily transport and deposition of clastic sediments and sedimentary environments. These processes determine the distribution and geometry of reservoir rocks in a sedimentary basin and also the changes in rock properties during burial. Accumulation of organic-rich sediments which may become source rocks is also an integral part of sedimentological models.

Like all natural sciences, sedimentology has an important descriptive component. In order to be able to describe sedimentary rocks, or to understand such descriptions, it is necessary to familiarise oneself with quite an extensive nomenclature. There are specialised names for types of sedimentary structures, grain-size distributions and mineralogical composition of sediments. We also have a genetic nomenclature, which names rock types according to the particular way in which we think they have formed. Examples of these are fluvial sediments (which are deposited by rivers) and aeolian (air-borne) sediments. The descriptive nomenclature is used as a basis for an interpretation of how the rock was formed. When we are reasonably confident about their origin we may use the genetic nomenclature.

Sedimentology covers studies of both *recent* (modern) sediments and older sedimentary rocks. By studying how sediments form today we can understand the conditions under which various sedimentological processes take place. From such observations we may be able to recognise older sediments which have been formed in the same way. This is called using the principle of uniformitarianism, which has been of great importance in all geological disciplines since its proposal by James Hutton (1726–1797).

Conditions on the surface of the Earth have fluctuated widely throughout geological history and the principle of uniformitarianism cannot be applied without reservation. One important aspect of sedimentological research is attempting to reconstruct changes in environments on the Earth's surface throughout geological time. This applies particularly to climate, vegetation and the composition of the atmosphere and the oceans.

Palaeontology is important to sedimentology, not only for dating beds, but also because organisms are an important component of many sedimentary rocks (particularly limestones), and organic processes contribute to weathering processes and the precipitation of dissolved ions in seawater. Many organisms make very specific demands of their environment, and fossils are consequently a great help in reconstructing the environment in which the sediments were deposited. Palaeoecology is the study of ecological conditions as we are able to reconstruct them on the basis of remains or traces of plants and animals in rocks. Traces of animals in sediments have proved to be very useful environmental indicators.

K. Bjørlykke (✉)
Department of Geosciences, University of Oslo, Oslo, Norway
e-mail: knut.bjorlykke@geo.uio.no

K. Bjørlykke (ed.), *Petroleum Geoscience: From Sedimentary Environments to Rock Physics*,
DOI 10.1007/978-3-642-02332-3_2, © Springer-Verlag Berlin Heidelberg 2010

Studies of recent and older sedimentary rocks provide a fruitful two-way exchange of information in sedimentology. From studies of recent environments we can learn about the conditions that particular processes require. In older rocks, however, we can study sedimentary sections which encompass many millions of years of sedimentation, offering us a completely different record of the way sedimentological processes can vary as a function of geologic time. As a result, studies of older rocks also contribute to our understanding of the recent environment and offer non-uniformitarian explanations.

When we study rocks, we should attempt to give *objective descriptions* of the composition, structure etc. of the rocks, and on the basis of these try to *interpret* how they were formed. However, it is impossible to give a completely exhaustive, objective description of a rock. Nevertheless it is often most fruitful to have a theory or hypothesis against which to test our observations. Data collection can then be focused on observations and measurements which can support or disprove the hypothesis.

We know from experience that we have a tendency to observe what we are looking for, or what we anticipate finding.

Early descriptions of sedimentary sequences contain few observations about sedimentary structures which we would consider fairly conspicuous and important today. We observe sedimentary structures because we have learned to recognise them and understand their genetic significance.

Many sedimentologists use a standard checklist for what they should observe in the field, so that their descriptions are as comparable as possible. Nevertheless, it is important that field observations do not become too much of a routine. Facies analysis should be a creative process and the various depositional models should be kept in mind when making the observations. It is also desirable to quantify field observations as far as possible, for example by surveys in the field and a range of laboratory analyses. These might be texture analyses (e.g. grain-size distribution), microscope analyses (perhaps using a scanning electron microscope) or chemical analyses. Pure descriptions of sedimentary rocks are useful because they increase the data base on which we can build our interpretations.

Systematic studies of recent environments of sedimentation are important to find connections between the environment and the sediments which accumulate. The environment governs the sedimentological processes which determine what sort of sediments are formed and deposited. The connection we are trying to understand in modern environments is thus: environment → process → sediment.

Today a large number of modern environments of sedimentation have been studied in great detail. These include aeolian and fluvial environments, deltas, beach zones, tidal flats and carbonate banks. Deep sea environments have naturally not been so easy to study, but modern sampling and remote sensing equipment and underwater TV cameras have made it easier to gather observations from this environment, too. In recent years systematic drilling through sedimentary layers on the ocean floor (Deep Sea Drilling Program – DSDP, and Ocean Drilling Program – ODP) has provided an entirely new picture of the geology of the ocean depths. Specially constructed diving ships (e.g. ALVIN) make it possible for geologists to observe the ocean floor directly at depths of up to about 3,500 m and take samples of surface sediments. In addition geophysical, particularly seismic, surveys provide one of the most important bases for understanding the stratigraphy and geometry of sedimentary basins.

In studying older rocks we base our approach on certain features which we can observe or measure, and attempt to interpret the processes that produced them. Particular variations in grain size and sedimentary structures in profiles can be interpreted as having been formed through particular processes, e.g. aeolian, tidal or deltaic processes. The recognition of such sedimentary processes helps us to reconstruct the environments. The sequence of interpretation in studies of older sedimentary rocks is thus: description of sedimentary rock → processes → environment.

Applied geology has always been important, even for purely scientific research. The interests of economic recovery of raw materials from sedimentary rocks create a demand for sedimentologists and sedimentological research. Exploration for and recovery of raw materials also provide important scientific information. Sedimentary rocks contain raw materials of considerably greater value than those we find in metamorphic and eruptive rocks. The most important are oil, gas and coal deposits, but a very large amount of the world's ore deposits is also found in sedimentary rocks and many types of ore have been formed through sedimentary processes. Limestone, clay, sand

and gravel are also important raw materials which require sedimentological expertise.

The petroleum industry employs a very high percentage of the world's professional geologists. This industry has a particular need for research, and also has the financial capacity to invest in it. Because oil and gas are found largely in sedimentary rocks, exploration for and recovery of hydrocarbons is based to a large extent on sedimentology. Much of what we now know about the world's sedimentary basins and their regional geology is derived from seismic profiles which have been shot in connection with oil exploration and drilling for oil and gas. The oil industry has also helped to stimulate pure sedimentological research, and significant contributions to research in this area are published by the research laboratories of the oil companies. Research based on economic interests is also useful from a purely scientific point of view, because it often focuses on particular questions which may be quite fundamental.

Petroleum geology requires close teamwork between reservoir engineers and geologists, to establish in great detail the geometry and distribution of porosity and permeability in reservoir rocks. We also need very much to know more about the physical and chemical properties of reservoir rocks, for reasons which are discussed at the end of the book. Geophysical methods provide most of the information used in petroleum exploration and production and many petroleum geologists rarely examine real rocks in cores and cuttings. It is however important to know something about the textural and mineralogical composition of the sedimentary sequences. The geophysical data rarely provide unique solutions when inverting seismic and log data to rock properties.

2.1 Description of Sedimentary Rocks

2.1.1 Textures

The *textures* of clastic sediments include external characteristics of sediment grains, such as size, shape and orientation. These properties can be described relatively objectively and say a great deal about the origin and conditions of sediment transport and deposition.

By grain size we normally mean grain diameter, but the two are only strictly synonymous in the case of completely spherical particles. Most grains are not spherical, however, and it is difficult to identify a representative diameter, particularly in the case of elongated or flat grains. For this reason we have adopted the concept "nominal" diameter (d_n), defined as the diameter of a spherical body which has the same volume as the grain. In practice we are seldom in a position to measure the volume of individual grains, and we therefore use indirect methods to measure the distribution of grain size within a sample.

Sand and gravel can most simply be analysed by means of *mechanical sieving*. A bank of sieves consists of sieves with mesh sizes which decrease downwards. A sample is put in the uppermost sieve and the bank of sieves is shaken (Fig. 2.1). Grains which are larger than the mesh size will remain, while smaller grains will fall through and perhaps remain lying on the next sieve. By weighing the fraction of the sample which remains on each sieve, we can construct a grain-size distribution curve. The lower practical limit for sieve analyses is 0.04–0.03 mm; finer particles exhibit much more cohesion, which makes it difficult for them to become separated and pass through the finer sieves.

Fine silt and clay fractions can be analysed in a number of ways. Most classic methods are based on measurements of settling velocity in liquids, and are based on Stokes' Law:

$$v = cgR^2 \Delta\rho/\mu$$

Here c is a constant (2/9) and μ is the viscosity of the water. R is the radius (cm) of the grain and $\Delta\rho$ is the density difference between the grain and the fluid (water).

When the settling velocity of grains (falling through water, for example) is constant, the resistance to the movement (friction), which acts upwards, must be equal to the force of gravity, which acts downwards (Fig. 2.2).

$$6\pi R v \mu \text{ (friction)} = 4/3\pi g R^3 \Delta\rho \text{ (gravity)}$$

$$v = c\, gR^2 \Delta\rho/\mu$$

$$\text{Log } v = 2\log R + c \text{ (a constant)}$$

The settling velocity is sensitive to temperature variations, which affect the viscosity of the water (μ).

A

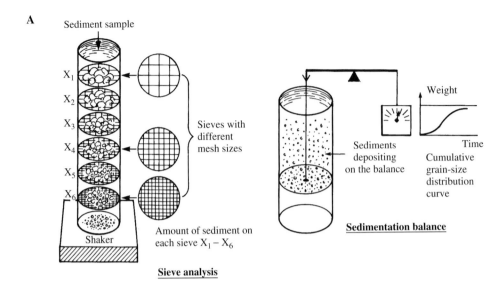

Sieve analysis

B   Wentworth Scale

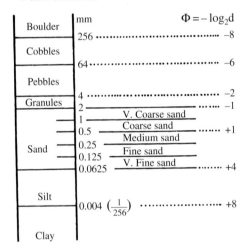

Fig. 2.1 (**a**) Sketch showing the principles involved in sieve analysis and use of a sedimentation balance. Sieve analysis is usually used for grain sizes down to 0.03–0.02 mm, but with wet-sieving even finer sediment grains can be sieved. The sedimentation balance gives us a direct expression of settling velocity, i.e. weight increase as a function of time. This is therefore a cumulative grain-size distribution. (**b**) Grain-size classification of clastic sediments. The grain size (*d*) is often described in terms of φ values ($\varphi = -\log_2 d$)

We can measure the settling velocities of sediment grains indirectly by measuring the density of the water with suspended sediment sample with a hydrometer, which registers the fluid density. We disperse the sample in a cylinder with a mixer so that at a start time T_0 we have an even distribution of all grain sizes, and therefore of density, throughout the cylinder. The individual sediment grains then sink to the bottom at a rate which is a function of their size. The change in the fluid density as progressively fewer grains remain in suspension is therefore a function of the grain-size distribution. This applies for small particles, where the flow of the liquid around the grain is laminar and the concentration of grains is low.

When the grains are larger than about 0.1–0.5 mm, the settling velocity increases, and turbulence develops

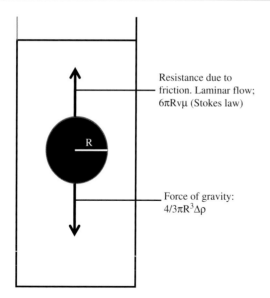

Fig. 2.2 The velocity of a falling grain in water is controlled by the gravity forces directed *downwards* and the resistance to the flow around the grain which is directed *upwards* ($6\pi R v \mu$). The force of gravity is a function of the volume of the grains ($4/3\pi g R^3$) and the density difference ($\triangle\rho$) between the grain and the fluid ($\rho_g - \rho_f$)

around the grains. The frictional resistance therefore increases, and in the case of larger grains (>1 mm) the settling velocity increases approximately in proportion to the square root of the radius. It is not practicable to measure each grain, but the settling velocity can be measured indirectly. A hydrometer floating in a suspension of sediments and water measures the density of the suspension through time. The rate of density reduction in the suspension is a function of the grains' size. By taking successive readings of the density we may plot a curve which expresses density reduction as a function of time. Since density reduction is a function of settling velocity, this curve can be calibrated to give a grain-size distribution curve.

When we analyse fine-grained sediments with a large clay fraction, or separate out clay fractions, it may be useful to use a centrifuge. We then increase the acceleration term **g** in Stokes' Law. There are also "sedimentation balances". Sediments suspended in a cylinder fall through the water column and accumulate on a balance pan at the bottom of the cylinder. This balance records and writes out the increase in weight, which is the precipitation from suspension, as a function of time. This gives a direct cumulative curve.

Other methods are based on the refraction or dispersion of a laser beam passed through suspensions producing a characteristic "scatter" which is calibrated against samples of known grain size. These machines use very small samples and have a high degree of repeatability. Equipment has also been developed which uses X-rays instead of light to produce the characteristic scatter patterns.

It is important to note that *no* method measures the nominal diameter. In methods which measure settling velocity, grain shape is a significant factor. A large, thin mica flake has a settling velocity which corresponds to that of a considerably smaller spherical grain. The diameter of a spherical grain with the same volume and settling velocity is called the *effective diameter* (d_e). With the scatter method, flaky grains are assigned a different, probably greater, diameter than that indicated by the settling velocity method.

2.2 Grain-Size Distribution in Solid Rocks

Lightly-cemented sandstones can be disintegrated by means of ultrasound in the laboratory and then analysed as loose sediment in the normal manner. Carbonate-cemented rocks may be disintegrated using acids. However, we must bear in mind that new clay minerals may have formed through post-depositional alteration (diagenetic processes), and that some of the original minerals may have been broken down mechanically or dissolved chemically. Consequently it is not certain that we are dealing with the original grain distribution. Diagenesis must be taken into account.

Well-cemented rocks must be analysed in thin section by means of a petrographic microscope. It is difficult to analyse the finer fractions (fine silt and clay) in this manner and we must always remember the "section effect", i.e. that in most cases we will not be seeing the greatest diameter of the grains. With spherical grains, the relation between the real diameter, d_r, and the observed diameter, d_o, can be expressed statistically: $d_r = 4d_o/\pi$.

2.3 Presentation of Grain-Size Distribution Data

Grain-size distribution is one of the many types of natural data which must be presented on a logarithmic scale for convenience. Wentworth's scale is based on logarithms to the base 2, and this is now the one most widely found in geological literature.

For the sake of convenience, these data are commonly plotted against a linear scale. The phi (φ) scale, where $\varphi = -\log_2 d$, allows convenient interpolation of graphic data. The reason this negative logarithm is used is that normally most of the sediment grain diameters (d) are less than 1 mm, so these will have a positive phi value (Fig. 2.3). It is convenient to plot grain-size distribution data as a function of phi values, especially on cumulative curves. In normal descriptions of grain size, however, it is more helpful to state grain size in mm, so the reader does not have to calculate back from phi values.

The simplest, and visually most informative, way of presenting grain-size distribution data is by means of histograms (Fig. 2.3). These show the percentage, by weight, of the grains falling within each chosen subdivision of the size range. It is then easy to see how well sorted the sediments are, and whether the distribution of grain sizes is symmetrical, or perhaps bi- or polymodal, i.e. with two, or more, maxima.

A *cumulative* distribution curve shows what percentage of a sample is larger or smaller than a particular grain size. The steeper the curve, the better the sorting. Note that engineers use the inverse term "grading", whereby well graded = poorly sorted.

If we use probability paper, distributions which are lognormal (following a logarithmic distribution) will plot as straight lines and the slopes of these will reveal the degree of sorting. Even if the whole distribution is not lognormal, it often appears that the curve can be regarded as a composite of 2–3 lognormal grain-size populations. These populations generally overlap, so that some sections of the curve represent a combination of parts of two populations, each of which may be lognormal. Each population may represent a different mode of grain transport, for example saltation, rolling (bedload) or suspension (Fig. 2.4).

It is important that we collect representative samples for grain-size distribution analysis, i.e. each sample only has material from one bed. This ensures that it represents deposition by a single sedimentary process. If we take a sample at the boundary between two beds, we will often get false bimodal distributions which can easily be mistaken for naturally produced bimodal sediments, leading to interpretation errors.

2.4 Grain-Size Distribution Parameters

Phi (φ) $= -\log_2 d$ (after Folk and Ward 1957) where d is the grain diameter in millimetres (as previously defined). The percentage of grains larger than a certain grain size (φ) is called the *percentile*. $\varphi 30$ means that 30% of the grain population by weight is larger than the grain size. For $\varphi = 4$ the grain size is 0.0625 mm so that 30% of the sample is sand or larger grains.

2.5 Significance of Grain-Size Parameters

The *mean* diameter is an arithmetically calculated average grain size. The *median* diameter is defined by the grain size where 50% by weight of the sample grains are smaller, and 50% are larger. Only in the case of completely symmetrical distribution curves will the mean diameter (M) and the median diameter (Md) coincide. The mean will otherwise shift further than the median in the direction of the "tail" of the distribution. If the sample has a wide spread (tail) towards the fine grain sizes (larger phi values) and a relatively sharp delimitation at the large grain-size end, we say that the sample has positive skewness. This will be typical of fluvial sediments. There will be a fairly definite upper limit to the grain sizes that rivers can transport as bedload, while there will be no sorting of the fine fractions which are transported in suspension. Major variations in flow velocity, for instance during floods, will give poorer sorting.

Aeolian (wind) deposits are very well sorted (Fig. 2.5). They also have positive skewness because there is an upper size limit to the grains which can be transported. Although the finest particles may be removed selectively, there will still be a "tail" of fine

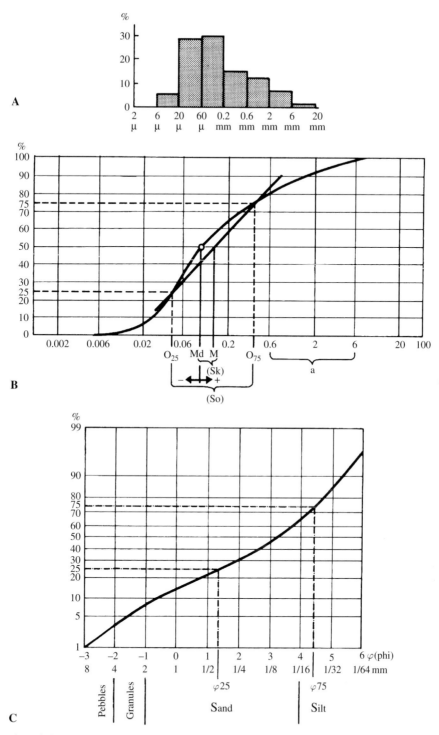

Fig. 2.3 Presentation of the same grain-size distribution data as (**a**) histogram, (**b**) cumulative curve, and (**c**) cumulative curve on probability paper. When plotted on probability paper, a logarithmic normal distribution, like a Gaussian curve, will plot as a *straight line*

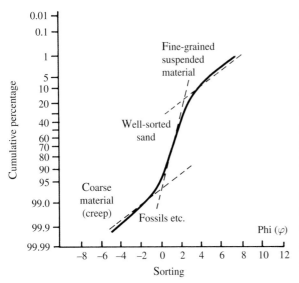

Fig. 2.4 Grain-size distribution curve presented as a function of grain size in φ values against a logarithmic cumulative percentage

material. The fine material in dunes may also be protected by a cover of larger particles *(lag)* against further erosion and transport. Beach sand deposits, on the other hand, are clearly negatively skewed, i.e. the distribution curve shows a definite lower limit, while there is often a "tail" of larger particles, i.e. granules and pebbles. The hydrodynamic conditions on a beach are such that each wave brings some sediment in suspension. Whereas sand grains, particularly medium to coarse sand, will rapidly settle from suspension and be deposited on the beach again, fine sand, silt and clay will remain in suspension longer. This finer material will be transported further out and at a depth of some metres (1–50 m), depending on how strong the waves

are (and consequently the depth of the wave base), we will have poorer sorted deposits because the fine fractions will be deposited and mixed with coarser-grained sediments which are carried out during storms.

Sediments deposited from suspension have poor sorting and positive skewness. This is very typical of turbidites. Clay suspensions which are deposited on land as high-density suspensions (mud flows), have negative skewness because they often contain large clasts.

Kurtosis is an expression for the spread of the extreme ends of a grain-size distribution curve in relation to the central part. This distribution parameter is used somewhat less than the others, but relatively small amounts of silt and clay can have a significant effect on the properties of coarse-grained sand. Pebbles and stones in otherwise fine-grained sediment may also be important.

One of the most important points to bear in mind when interpreting grain-size distributions is the availability of grain sizes supplied to the area where the process is taking place. Strong currents or wave energy will only be able to deposit coarse-grained sediments if there is coarse-grained sediment present in the area. A source area where sediments are generated by erosion and weathering will often supply specific grain sizes. Chemical weathering of acid rocks (granites), for example, will lead to the formation of sediments consisting of quartz grains corresponding in size to the quartz crystals in the granite, and clay (kaolinite, smectite, illite) formed through weathering of the feldspars. Weathering of basic rocks (basalts and gabbros) will produce almost exclusively clay minerals, and practically no sand grains.

Fig. 2.5 Sorting and skewness are grain-size distribution parameters which are suitable for distinguishing between sediments deposited through various processes and in different environments of deposition. Turbidites and fluvial sediments have a tail of small sediment particles while beach sand may have coarser grains such as pebbles

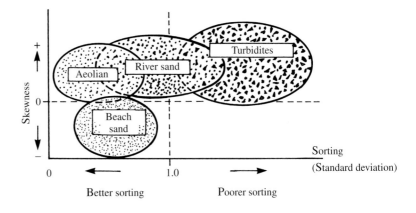

The grain-size distribution in a depositional area also depends on the transport mechanisms carrying sediment to the area. The grain-size distribution we observe in a sediment therefore reflects both the hydrodynamic conditions and the grain-size population of sediments available from the source area. Furthermore, although a particular grain-size distribution may be characteristic for a type of deposit, it does not point unambiguously to a particular environment of deposition because similar hydrodynamic conditions can exist in different environments. Statistical comparison of high resolution (minimum $\frac{1}{2}$ Phi sampling interval) grain-size distributions, applying pairs of Folk and Ward or Krumbein parameters, can provide some diagnostic criteria for hydrodynamic interpretations.

Sediments may change their grain-size distribution by diagenetic processes at quite shallow burial as well as at greater depth. The formation of the clay minerals kaolinite and smectite from feldspar and rock fragments (volcanics) in a sediment certainly results in very different grain-size distributions from those at the time of deposition.

2.6 Grain Shape

We distinguish between three parameters:

1. *Roundness* is a property of surface shape – whether it is smooth or angular. A visual scale is most commonly used.
2. *Sphericity* is an expression for how much a particle deviates from a spherical form, and is defined as the ratio between the diameter of a circumscribed circle round the grain and the diameter of a sphere of the same volume (the *nominal diameter*).
3. We also use various expressions for grain shape such as (a) *discoid* or *bladed* for grains which are flat, (b) *Prolate* or *roller* for grains with one dimension considerably greater than the two others, (c) *equant* for grains with three relatively equal dimensions and (d) *oblate* for grains with one large, one medium and one small dimension.
4. *Surface textures* are concerned with the nature of the surface itself, whether it is rough, smooth, pitted, scratched etc. Some textures are diagnostic

of specific modes of transport, and superimposed texture features may reveal the transport history of a grain. The surface texture of grains can best be studied under the scanning electron microscope. Aeolian sand grains may develop fine pitting on their surfaces due to the collisions of grains during transport, clearly visible under a binocular microscope.

Large grains become rounded far more rapidly than smaller ones because the impact energy released in collisions with other grains declines in proportion to the cube root of the radius. Blocks may be rounded after only a few hundred metres or several kilometres of transport. Grains less than 0.1 mm in diameter undergo little rounding even when carried very long distances in water, for example by tidal currents.

2.7 Sediment Transport

Sedimentary grains can be transported by water or by air. In order to understand the transportation processes we must know a little about the hydrodynamic (or aerodynamic) principles involved. When a liquid or gas flows in a channel or pipe it exerts a force (shear stress) against the walls or bottom. This force is counteracted by friction from the walls.

Pure water without suspended sediment is a Newtonian fluid which obeys Newton's law:

$$\tau = \mu \, dv/dh$$

A Newtonian fluid has no shear strength, so it will be deformed even by an infinitely small shear stress (dv/dh).

$\tau = shear\ stress$, which is an expression of force per unit area (N/m^2). μ is the dynamic viscosity expressed in poise ($g/cm/s$ or 0.1 N s/m^2), dv/dh is the change in velocity (dv) or velocity gradient (deformation velocity) as a function of distance from the boundary (dh). The viscosity of pure water decreases with increasing temperature. Suspended material may also affect viscosity, but the concentration of suspended material must be quite high (15–25%) before the viscosity increases significantly. If the water contains a large percentage of swelling clay minerals (smectite), however, the viscosity will increase at lower concentrations. The kinematic viscosity v is the dynamic viscosity (μ) divided by density ρ, i.e. $v = \mu/\rho$ and units are cm^2/s.

We distinguish between laminar flow, where each point in the liquid moves along a straight line parallel to the bed, and turbulent flow, where each point follows an irregular path so that eddies form (Fig. 2.6) Reynold's number (Re) is a dimensionless number which describes flow in channels and pipes. It is defined as:

$$Re = vh\rho/\mu$$

Here v is the mean velocity, h is the depth of a channel or the diameter of a pipe in which fluid is flowing, ρ is the fluid density and μ its viscosity. If Reynold's number exceeds a certain value, about 2,000, the flow changes from laminar to turbulent. The density of water is 1 g/cm^3 and the viscosity is one centipoise (0.01 Ps $= 0.01$ $g/cm/s$). We see that the boundary between laminar and turbulent flow corresponds to 20 cm/s.

This means that for the flow of water to be laminar the product of velocity (cm/s) and depth (cm) must not exceed 20. If the velocity is 1 cm/s, there will be turbulence if the depth (h) is greater than 20 cm. In practice, then, flow in rivers and channels is always turbulent. For turbulent flow the expression for shear stresses which applies to laminar flow, is no longer adequate. The shear stress in turbulent flow will then increase as a function of velocity because of the eddies which produce an *eddy viscosity* (η) (Fig. 2.6). The total shear stresses will then be: $\tau = (\eta + \mu)/dv/dh$

2.8 Flow in Rivers and Channels

For all types of water flow the forces acting on the water must be in equilibrium. In most cases it is the force of gravity which balances bed frictional forces. In order to understand geological processes in connection with the erosion, transport and deposition of sediments, it is important for us to be aware of the relationships which govern the flow of water in channels.

If the channel has a cross-section A and we look at a stretch L of the channel, the force of gravity will be:

$$F_1 = \rho \cdot g \cdot L \cdot A \cdot \sin\alpha,$$

where ρ is the density of water, g is the force of gravity (constant) and α is the angle of slope of the channel. The resistance to flow consists of frictional forces against the bed and against the air. If we disregard friction against the air, the frictional forces are:

$$F_2 = \tau \cdot L \cdot P$$

where $\tau =$ shear stress (force per unit area) and $L \cdot P$ is the area of the bed on which the forces are acting. P is

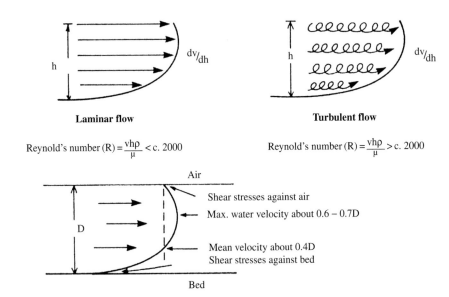

Fig. 2.6 Diagram showing principles of turbulent and laminar flow and the shear stress against the underlying bed

Flow in a channel

Force of gravity $= F_1 = \rho g \cdot L \cdot A \cdot \sin \alpha$

Frictional forces $= \tau \cdot P \cdot L$

Flow in a pipe

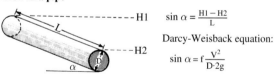

$$\sin \alpha = \frac{H1 - H2}{L}$$

Darcy-Weisback equation:

$$\sin \alpha = f \frac{V^2}{D \cdot 2g}$$

Fig. 2.7 Flow of water in channels is controlled by the ratio between the gravitational forces and the shear stress against the bottom of the channel

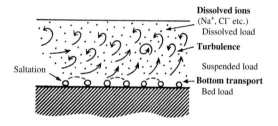

Fig. 2.8 Different forms of transport in water. Sediment grains may be carried in suspension if the vertical component of the turbulence is equal to the falling velocity of the grains. Larger grains are carried along the bottom due to the shear stress

the wet perimeter and L is the length of a line along the bed in a section along the channel. If the water flow has a steady velocity, the force of gravity F_1 will just equal the frictional force F_2 (Fig. 2.7). Consequently:

$$\tau \cdot L \cdot P = \rho g \cdot L \cdot A \cdot \sin \alpha$$

or

$$\tau = \rho \cdot g \cdot \frac{A}{P} \cdot \sin \alpha$$

A/P is the cross-section of the channel divided by the wet perimeter, and we call this the hydraulic radius, R. For flow in a pipe:

$$R = D/4$$

The shear stresses (τ) $R = D/4$ increase in proportion to the square of the velocity ($\tau = c \cdot v^2$).

This relation between shear stress and flow velocity can also be used for flow in channels where we have bedload transport (Fig. 2.8). Solving the two equations above with respect to the velocity (v) we obtain:

$$v = C \left(R \sin \alpha\right)^{1/2}$$

This is the Cherzy equation and C is the Cherzy number.

The value of C depends on the roughness of the bed and on the *shape* of the channel, particularly its *sinuosity*.

Often used in engineering for calculating the velocity of water in channels, is Manning's formula:

$$v = R^{2/3} \cdot (\sin \alpha)^{1/2} / \eta$$

where n is the coefficient of roughness of the bed: $n = 0.01$ corresponds to a smooth metal plate and $n = 0.06$ to a shifting bed of gravel. It is of great practical importance to be able to calculate water velocity and thereby the erosion potential of artificial channels.

The Froude number is a parameter which is often used to describe water flow:

$$F = v/(g \cdot h)^{1/2}$$

where v is the average velocity, h is depth of water and g the force of gravity. The Froude number is the ratio between the kinetic energy of the water masses (which is proportional to the square of the velocity) and the force of gravity, which is proportional to the depth, h. For low Froude numbers the water flows out of phase with the bedforms, and current ripples or cross-bedding develop. This is called the lower flow regime. When the velocity, v, becomes high in relation to the depth of water, h, rapid or shooting flow develops where the waves come into phase with the boundary irregularities (Fig. 2.9), which represents the upper flow regime.

The transition between lower and upper flow regimes corresponds to a Froude number of 0.6–0.8.

2.9 Sediment Transport Along the Bed Due to Water Flow

What actually gives flowing water the capacity to carry sediment, and how are sediment particles transported?

We have seen that flowing water exerts shear forces against the stream bed. Frictional forces are converted into turbulence in the overlying water, and have the effect of transporting sediment particles along stream

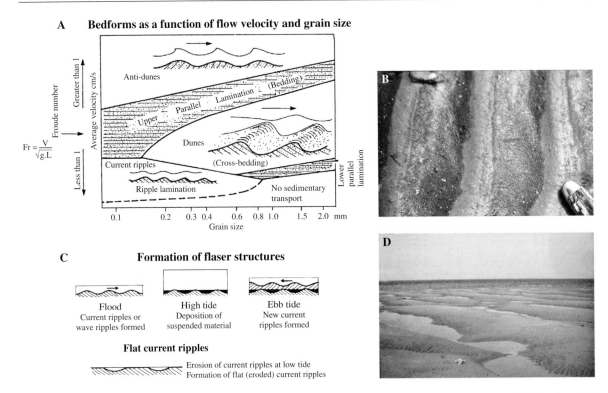

Fig. 2.9 (**a**) Sedimentary structures as a function of flow velocity, grain size and water depth. The Froude Number (F) is an expression of the velocity as a function of depth. (**b**) Ripples with clay pellets accumulating between the ripple crests. (**c**) Formation of current ripples and truncated ripples in a tidal environment. (**d**) Dunes formed on a coastline with high wave energy

bed. Under moderate flow conditions the largest particles will be transported along, or just above, the bed as bedload (Fig. 2.8). This takes place partly through rolling or slow creep, partly through saltation, i.e. the grains jump along the bed.

Saltation can be partly explained through Bernoulli's equation:

$$P + g \cdot h + \frac{v^2}{2} = C(\text{constant})$$

Here P = pressure, h = height above the stream bed, and v = velocity. We see that water which flows over a sediment grain on the bed will have a greater velocity than water which flows under the grain. Bernoulli's equation predicts that the pressure above the grain must be less than the pressure adjacent to the grain (P), and when this difference becomes sufficiently great it will be possible to lift the grain from the stream bed. This "airplane wing effect" does not work once the grain is in the water above the stream bed, and the grain will then drop to the bed again.

The condition for sediment grains being transported in suspension is that their settling velocity must be less than the upward vertical turbulence component. This means that the grain must be transported upwards through the water at least as fast as it falls downwards. The magnitude of the vertical turbulence upwards will be a function of the horizontal velocity (about 1:8). Under normal flow conditions (< about 1 m/s) only clay and silt will be transported in suspension. Under high flow energy conditions, e.g. during floods, sand and gravel may also be transported – at least partly – in suspension.

Erosion and transport are a function of shear stress against the stream bed. This in turn is a function not only of water velocity but also of depth. There is a connection between the flow velocity and the size of the sediment grains which can be transported but the water depth is also important. Hjulstrøm's diagram (Fig. 2.10) applies to channels about 1 m deep. Other factors which complicate these relationships are the viscosity (and hence temperature) of the water and the density and shape of the sediment grains.

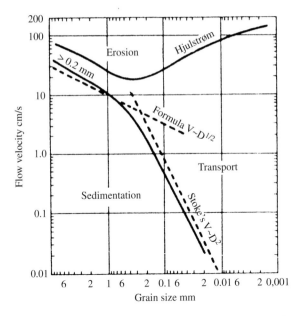

Fig. 2.10 Hjulstrøm's curve showing relations between grain size, flow velocity, erosion and sedimentation. Fine-grained sediments stay in suspension and are transported at low velocities, but require higher velocities to erode than silt due to cohesion

With small grain sizes (silt and clay) the flow velocity required to sustain transport is far less than the velocity needed to erode a particular grain. This is because cohesion between sediment particles, particularly clay, is such that once they are deposited, it is difficult to erode them again.

Note that fine sand (about 0.1 mm) is the easiest sediment to erode. On the other hand, finer-grained particles remain in suspension for a long time at low velocities. Flocculation of small clay particles to form larger ones in seawater increases the settling velocity of clays. Also "pelletisation", through clay being eaten by organisms, is important for the formation of many fine-grained clay sediments.

2.10 Different Types of Sediment Transport

We have shown that water or air flowing over a surface exerts shear forces on the substratum so that sediment can be transported by what we refer to as *traction currents*. When we have a relatively low concentration of sediment in water (or air) there is little increase in the density and viscosity of the fluid phase. The flow will then still have approximately the same characteristics as it had without the sediment.

Another type of sediment transport is due primarily to the density difference between a water mass carrying suspended sediments and the clear water outside the suspension. We call this phenomenom *gravity flow* (Fig. 2.11) and it includes turbidity currents and debris or mass flows. The force of gravity causes movement of sediment/water mixtures because they have a higher density than their surroundings, i.e. they are not in equilibrium with the ambient clear water mass. Gravity flow is thus distinguished from traction currents by the fact that it can take place in otherwise still water. However, there are transitions between these two fundamentally different processes and we often have combined effects.

2.11 Turbidity Currents

Sediment in suspension will be carried down submarine slopes because the suspension is heavier than the surrounding clear water, forming a *turbidity current*.

It may be started by river water containing suspended material entering a sedimentary basin. In marine basins the difference in density between river (fresh) water and salt seawater is so great that even if river water carries sediment in suspension, it will in most cases not be denser than seawater. In consequence there will not be a positive density contrast, which is the prerequisite for the formation of turbidity currents; instead the flow may become an overflow plume. In lakes, on the other hand, river water is often heavier, due both to suspended material and to its being colder than the lake surface water. It will then be able to follow the bed downslope, and become a turbidite.

Submarine slides may also quickly evolve into turbidity currents. River sediment entering a marine basin will mix with the seawater so that clays flocculate and are deposited on the delta slopes. If the slope angle becomes too steep, we get slides, which may result in turbidity currents. When fine-grained sediments are deposited on slopes from suspension they have a very high water content. Compaction, sometimes caused by earthquakes, will cause upward flow of porewater and may result in liquefaction. This causes the sediments

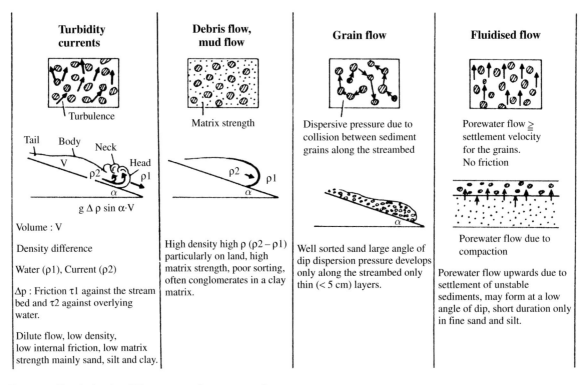

Fig. 2.11 Sketch showing different types of transport on slopes

to begin to flow even on gentle slopes because friction is reduced, and they may then turn into turbidity currents.

The forces driving a turbidity current are:

$$F_1 = g \cdot \Delta\rho \cdot V \sin\alpha$$

where g is the gravity constant, $\Delta\rho$ is the difference between the density of the current and that of the surrounding water, V is the volume of water along a certain length of the channel, with the cross-section (A) of a turbidity current with length L, and α the angle of the slope. Acting against the movement are frictional forces (F_2) which, as long as the current is not accelerating, must be equal to the gravitational forces. These are shear forces against the bed, τ_1, and against the overlying water, τ_2, plus internal friction and turbulence within the current which keep the sediments in suspension. In order for the sediment grains to remain in suspension, the turbulence must be sufficiently strong to have an upward component which corresponds at least to the settling velocity of the coarsest grains. Turbulence is greatest near the

bottom of the current, where change in velocity as a function of height above the bottom (velocity gradient) is greatest. The largest grains in suspension will thus be concentrated near the bottom of the current. Near the bed, in addition to turbulence, we also have shear stresses which will transport the grains in virtually "pseudo-laminar" flow in a thin layer over the bottom. If the concentration of large sand grains along the bed becomes large, we also get *dispersive energy* because of collisions between the grains (see Sect. 2.13).

We therefore find that both the concentration of sediment in suspension, and maximum grain size in suspension, decrease upwards from the bottom. If we disregard internal friction, we obtain

$$g \cdot \Delta\rho \cdot V \sin\alpha = (\tau_1 + \tau_2) \cdot A$$

where A is the area of contact with the bottom and the overlying water. The ratio between the volume (V) and the contact area A is approximately the thickness of the flow H. The shear stresses are proportional to the square of the velocity $\left(\tau = cv^2\right)$.

The velocity (v) of a turbidity current is then:

$$v = c \cdot (g \cdot \Delta\rho \cdot H . \sin\alpha)^{1/2}$$

Here the coefficient c includes the coefficient of resistance for friction against the seafloor and against the overlying water. This corresponds to Chezy's number for fluvial flow, so in many ways we can regard a turbidity current as an underwater river.

We see from the above equation that thick turbidity flows will have a higher velocity than thin ones and that thick flows can flow on gentler slopes than thinner ones. This is because the shear stress against the bottom and the overlying water is nearly independent of the thickness of the flow. The flow velocity also increases with increasing density of the sediment-water mixture in the flow, but high density flows will have higher internal friction and require higher velocities to keep the material in suspension.

A turbidity current can be divided into head, neck, body and tail. The sediment particles in the head area move somewhat faster than the front of the current itself. This leads to sediment being swept upwards and then backwards towards the neck, where it mixes with water from the overlying water mass. From there it is carried backwards to the body and tail, where we find a finer-grained, thinner suspension. When the turbidity current loses velocity, the largest particles in the head will settle out of suspension first because of reduced turbulence. Gradually smaller and smaller grains will settle and we get deposition of a bed which is fairly massive, without internal structure, but which becomes finer upwards. In most cases, apart from in proximal turbidites, we also find deposition of some fine material in this layer, so there is poor sorting. An example of this is Unit A of the Bouma Sequence (Fig. 2.12).

As settlement from suspension slows down, the water will have time to sort the grains further, and we find lamination and bedding structures. The B Unit exhibits parallel lamination which may be due to flow just above the *upper flow regime* boundary. The C Unit exhibits current ripples and convoluted laminae and represents further velocity reduction, with deposition in the lower flow regime. The D Unit has parallel lamination and was probably deposited from the tail, which consists of very fine-grained sediment. The E Unit consists mainly of pelagic material, fine-grained clay and fossils that accumulated on the seabed during the long periods (often thousands of years) between turbidity flows (Fig. 2.12). The E Unit is therefore not necessarily a part of the turbidite sequence.

The Bouma sequence, first described by Arnold Bouma in 1962, is an ideal sequence in the sense that in most cases we do not find all the units developed. In some sequences, particularly those thought to have been deposited close to the base of submarine slopes where the gradient is still fairly steep, we will find only the coarsest parts of a turbidity current deposited, Unit A or a sequence of A + B. We call these *proximal turbidites*. The finest-grained fractions of a turbidity current tend to be deposited beyond the foot of the slope or out on the ocean abyssal plain. In these areas we often only find alternations between C-D-E, or just D-E. These are called *distal turbidites*. In many cases it

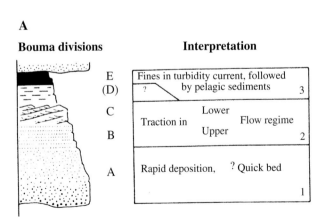

Fig. 2.12 (**a**) Bouma sequence in turbidites. (**b**) Turbidites in a Late Precambrian sequence at Lillehammer, Norway. The sequence is younging to the *right*

may be difficult to distinguish between distal turbidites and alternations between silt and clay formed by traction currents at great depths. Proximal turbidites which are well sorted may also resemble coarser sediments deposited by powerful traction currents in submarine channels.

At the base of turbidity sequences, particularly at the base of the A Unit, we often find well-developed erosion structures, particularly *flute casts* and *groove casts*. Flute casts are formed by the turbulence of turbidity currents when they pass over a substratum which consists of finer-grained sediments. The structures, which point upcurrent, are produced by vortices in the turbulent flow eroding into the sediment surface. Groove casts are formed by larger grains being dragged along the bottom. Even though flute casts and groove casts are typical of turbidites, they cannot be used as proof that we are dealing with turbidites because similar structures can also be formed by various types of traction currents where there is turbulence and transport along the bottom, e.g. in fluvial environments. Sequences resembling Bouma sequences may also be produced by processes other than turbidity flows, for example rapidly accelerating fluvial flows. To assist our interpretation we should therefore look at the entire sequence and also try to obtain palaeo environmental information from fossils or oriented grains. On submarine slopes there may also be very swift traction currents, particularly in submarine valleys due to the focusing of the tidal forces, and relatively strong currents may go up the canyon as well. These traction currents are capable of transporting coarse sand and at times even coarser material.

2.12 High Density Mass Flows – "Debris Flows" and "Mud Flows"

Debris flows occur both on land and under water, and represent a type of mass transport where the sediment/water ratio is very much greater than in turbidity currents, resulting in high viscosity and high internal friction during flow. This also means that the density of the mass is from 1.5 to 2.0 g/cm^3, while most turbidity currents have a density of 1.1–1.2 g/cm^3 or less. The high density of a debris flow means that all clasts have increased buoyancy because of the dense matrix. The high viscosity of the matrix also means

that large stones do not sink rapidly towards the bottom of the flow. In flows with high density and viscosity, blocks may remain near the surface of the flow until it solidifies, through loss of water or reduced gradients, and becomes quite rigid as a result of increased density and cohesion. The shear strength of the matrix is often referred to as matrix strength. Debris flows may be rich in stones and other coarse material because of the matrix strength. Mud flows typically have a more clay-rich matrix. But there is no clear distinction between the two.

Because of the high matrix strength, little sorting takes place in debris flows (Fig. 2.11). Large blocks are often concentrated at the front or on the sides of the flows, and there may be less coarse material near the base of the flow because of the shear movements.

On land, the density difference ($\Delta\rho$) between the flow and its surrounding fluid (air) is much higher than under water. Flows with a particularly low water content and high shear strength will only flow slowly down a slope, and we get transitions into what we call *solifluction* (creep). Sediment flows with a higher water content can move faster, however. In debris flows with their high internal friction most of the shear forces will be released along the bottom of the flow, so that the overlying mass moves more or less as a coherent mass with little internal deformation. This helps to reduce the total frictional resistance to movement. Shear strength and viscosity tend to decrease with increasing rate of shear, and this means that when a flow first gets going it will tend to accelerate.

Debris flows and mud flows normally have *thixotropic* properties: the shear forces must reach a critical threshold before deformation (shear) takes place, and the material loses much of its shear strength following deformation. In clay containing smectite (montmorillonite) this property is particularly well developed. Under shear stress, water will be released. The house-of-cards packing of clay minerals is destroyed, the clay particles tending to develop parallel alignment, with the release of water which will reduce the shear strength causing shear weakening. Some of the water in the bottom layer in which deformation is taking place will be lost to the other sediments, however, and friction will mount again. If large clasts enter the basal shear zone, friction will also increase and the flow may stop.

The stability of mud on slopes and the flow properties of mud flows depend on the clay mineral composition of the mud and on the geochemistry of the porewater and the ions adsorbed on the clay minerals. The presence of potassium and sodium tend to stabilise the mud.

Debris flows are often described in continental deposits, but are also found on submarine slopes. In water the density difference between sediment and surroundings is far less than on land, so that the angle of the slope must be greater for flows with the same internal friction. Submarine mud flows, on the other hand, will not dry up and they can easily take up more water as they move.

Debris flows are particularly common in desert deposits. This is because of the often powerful rainstorms which mobilise sediments that in a wetter climate would have been transported by fluvial processes. In addition, there is little vegetation in deserts to bind the sediments, so they are more easily set in motion. Also of great significance is the fact that the clay mineral smectite is formed particularly through weathering in desert environments. Clay containing smectite will expand when it begins to rain, preventing the water from filtering rapidly through the soil profile. Instead, water will be bound to the sediments and the viscosity may be reduced enough for mud flows with thixotropic properties to form. In continental environments with freshwater the content of stabilising salt (K^+, Na^+) is low.

The term "quick clays" is used for extremely thixotropic clays. Undisturbed clays have a relatively high shear strength, but after shaking or some other type of deformation they can flow like liquids, with a very low internal friction. In Scandinavia, Holocene marine clays which have been uplifted by glacio-isostatic rebound have been slowly weathered by percolation of rainwater so that sodium has been leached out, making them more prone to landslides and to form mud flows.

Systematic surveys using modern coring and remote sensing techniques, such as underwater cameras, side scan sonar and 3D seismic time slices, have shown that large-scale debris flows are rather common on continental slopes.

On the eastern slope of the Norwegian Sea a huge slide (the Storegga slide) occurred about 8000 years ago, involving about 3500 km^3 of sediment. The slope scar stretches for nearly 300 km and parts of the flow extended up to 800 km across the deep ocean floor. The transport mechanism was chiefly debris flow, where the sediments were riding as a plug on a wedge of water that reduced the friction against the bottom.

2.13 Grain Flow

Grain flow is flow of relatively well-sorted sediment grains which remain in a sort of suspension above the substratum due to collisions between the grains. We see this if we make a little landslide in a dry sandpit or pour sugar out of a bag. Grain flow can develop only when the initial flow is near the angle of repose (about $34°$). Bagnold (1956) described how collisions between sediment grains led to *a dispersive stress*. However, this stress is only significant near the base of a flow, where we have rapid variation in flow velocity as a function of height above the base (dv/dh). Here grains with very different velocities will strike one another, and transfer velocity components to one another. Higher up in the flow the dispersive stress due to collisions between grains will be considerably less as the grains have far more similar velocities, despite turbulence. The dispersive pressure developed near the base cannot support a thick layer of overlying sediment and therefore grain flows have an upper thickness limit of about 5 cm.

Sand grains which avalanche down the lee side of sand dunes form small grain flows and are probably one of the few significant examples of natural pure grain flow. Grain flow may also occur on beaches and in shallow marine environments.

2.14 Liquefied Flow

Liquefaction is the name given to a process whereby sediments lose most of their internal friction, and consequently act almost like fluids. This is the case when the pore pressure is equal to the weight of the overburden. When sediments are deposited, they have a high water content and the sediment grains are packed in an unstable manner. As the overburden increases, the stress on the grain contacts increases, and the framework of the sediment grains may collapse suddenly. Earthquakes produce tremors which may cause this structure to collapse, but it can also take place purely as a result of stress (loading). When the packed framework of grains which was formed during deposition is destroyed, the grains can pack more closely together. For this to be able to happen, however, water must flow out of the bed as the porosity decreases. This leads to an upward flow of pore-water and fine sediment particles, which may be as great as or greater than the settling velocity of the grains. This process is called liquefaction. The force of gravity, acting on the contact between the grains, is therefore neutralised, and friction between the grains tends towards zero, resulting in liquefaction. If we measure the pressure in the porewater, we find that it increases during settlement (compaction) when the unstable grain framework is destroyed. At one stage the pore pressure will be approximately as great as the weight of the overlying sediments. We can then use Coulomb's Law:

$$\tau = C + (\sigma_v - P) \tan \varphi,$$

where τ = shear strength, C = cohesion, σ_v is the weight of the overlying sediment, P = pore pressure and φ is the angle of friction (about $34°$). $(\sigma_v - P)$ is the effective stress. When the pore pressure, P, approaches the weight of the overlying sediments (σ_v), the friction component $(\sigma_v - P) \tan \varphi$, approaches zero. Fine-grained sediments like clay have considerable cohesion (C), and this will often prevent clay sediments from sliding even if there is little friction. However, once a deformation plane forms, there will often be movement mainly along it due to cohesion in the rest of the clay. If we have coarse-grained sediments, i.e. coarse sand and gravel, compaction will lead to excess water flowing out so rapidly that the overpressure will drop very quickly, assuming the high permeability has allowed it to build up properly in the first place (Fig. 2.13). It is therefore silt and fine sand, the fractions most susceptible to liquefaction, that are likely to generate high-velocity subsea flows. Liquefaction can, as already mentioned, be triggered by tremors, e.g. earthquakes, and stress. Stresses on sediments (soils) due to buildings, fills, etc. can lead to collapse of the grain frameworks and cause liquefaction. Lowering of the groundwater table on a slope, for example down towards the coast, has a similar effect because of reduced buoyancy in part of the sediment column. Extremely low tides or a combination of a strong ebb and a land wind can trigger a slide in otherwise stable coastal sediments. This is because the effective stress in the sediments increases when the sea level is low.

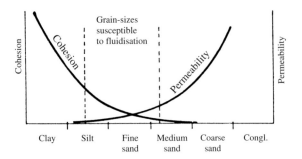

Fig. 2.13 Relation between permeability and cohesion in sediments. In coarse-grained sediments water escapes quickly, preventing build-up of overpressure, and in fine-grained clayey sediments the cohesion prevents mobilisation

The stability of slopes can be estimated by calculating the gravitational forces acting on a particular volume of sediment in relation to the frictional forces. During construction work, slides may sometimes be prevented by drilling wells which release the excess pore pressure so that the effective stress and the friction increases.

2.15 Sedimentary Structures, Facies and Sedimentary Environments

It is difficult to observe or take measurements of rocks entirely objectively and consistently. Most types of measurements and observations have a considerable degree of inherent uncertainty, and the validity and usefulness of results often depend on the experience and skill of those carrying out the field work. It has turned out to be very difficult to observe structures which one does not recognise and understand the significance of.

A good description of a stratigraphic profile depends on good theoretical knowledge of sedimentary processes, and of experience from studies of similar rocks.

It is easy to forget to record or measure some of the properties of a rock. In order to obtain a more comprehensive description and avoid forgetting anything, it may be a good idea to have a well-established routine or even a checklist. Photographs from outcrops or cores may help when writing final reports. If our investigation has a definite and limited objective, we measure only the properties we think will be relevant.

A list of features which can be observed (measured or registered) in sedimentary rocks:

1. Textures – grain size, sorting, grain shape etc.
2. Grain orientation – fabric.
3. Sedimentary structures and their orientation.
4. Fossils.

   A Preservation or impressions, casts, or the fossils themselves, and their mode of occurrence.
   B Trace fossils.

5. Colour.
6. Resistance to weathering and erosion.
7. Composition (a) Mineral (b) Chemical.
8. Thickness and geometry of beds.
9. Variations in texture and composition within a bed, e.g. increase in grain size upwards or downwards in the bed (grading or inverse grading).
10. Type of contact between beds (e.g. erosional contact, conformable contact, gradational contact).
11. Association or any tendency to statistical periodicity in the features of the strata in a profile – bed types, structures.

These observations form the basis for defining facies, which are a synthesis of all the data listed above which can be used to group certain types of rocks. They may be genetic facies, i.e. strata which one assumes have formed in the same manner. All strata which contain criteria which indicate that they were deposited in shallow water can be described (in reality interpreted) as shallow water facies. In the same way we have fluvial facies, deep water facies, evaporite facies, and so on. The facies concept can also be used to distinguish between different rock compositions (lithologies), e.g. carbonate facies, sandstone facies.

2.16 Sedimentary Structures

By sedimentary structures we mean structures in sedimentary rocks which have formed during or just after deposition. We distinguish between *primary* structures which are formed at the time of deposition of the sediments, and *secondary* structures which are formed after deposition.

2.17 Layering and Lamination

Most sedimentary rocks exhibit some lamination or bedding, but we also have massive (unlaminated) rocks. Lamination records variations in the sediment composition as successive depositional layers are draped over the contours of the sedimentary surface. The variations may reflect different grain sizes, sorting, mineral composition or organic matter. *Laminae* are less than 1 cm thick. Units greater than 1 cm are called *beds*. A bed will contain sediments which have been deposited by the same sedimentary processes. Some sedimentary processes, for example deposition of a turbidite bed, may be fairly rapid. Migration of a sand dune to give cross-bedding takes somewhat longer, while deposition of a clay bed from suspended material may take a very long time.

Graded beds have a grain size which tends to decrease upwards within the bed. The opposite is called *inverse grading. Normal grading* may be due to deposition from suspension, when the largest particles tend to fall to the bottom first, as with turbidites, or to flow velocities dropping off during deposition in a river.

Inverse grading may be due to increasing flow velocity but if the increase in velocity is too high, the result will be erosion. The supply of coarse material during transport and deposition may also produce inverse grading. In high density sediment currents (debris flows) we may get inverse grading; smaller particles sink more easily to the bottom between the large particles. Massive beds, or at any rate reasonably massive beds without visible lamination or bedding, may be formed during very rapid deposition of sediments from suspension. X-ray photography of apparently massive sediments nevertheless usually reveals the presence of weak lamination even in sand. Massive sand beds occur due to rapid fall out from suspension. In mudstones the primary lamination may be destroyed by intense bioturbation.

On the surface of laminated sediment we may find erosion structures.

Water running over a plane surface, e.g. beach sand, will produce small-scale, branched (dendritic) erosion marks called *rill marks*. The flow of water may be due to runoff from big waves or from groundwater seepage at low tide. These structures are good indicators of inter- or supratidal environments, but they are seldom preserved because they are usually destroyed when the water level rises again. Raindrop imprints are often preserved on bedding surfaces, and are good indicators of subaerial exposure.

2.18 Bedforms

Bedforms are morphological features resulting from the interaction between particular types of flow and the sediment grains on the bottom. A specific type of bedform will only form within a limited flow velocity range and is also dependent on the availability of grain sizes which can be moved by that flow.

Current ripples form in fine-grained sand when the velocity exceeds the lower limit for sediment movement.

Ripples and dunes have a stoss side upstream where erosion takes place and a lee side on which deposition takes place (Fig. 2.14). They therefore migrate as a result of the combined effects of erosion and deposition. Current ripples form in fine-grained sand when the velocity exceeds the lower limit for sediment movement. Sections through ripples show inclined foreset laminae, a structure called "small-scale cross-lamination". Ripples may also form in coarser sand but there is an upper limit of 0.6–0.7 mm for the grain diameter. Ripples are less than 3–5 cm high, and may

Fig. 2.15 Aeolian dunes several metres high in the Navajo Sandstone (Jurassic)

have a wavelength of up to 40 cm. The ripple index is an expression of the ratio of the wavelength divided by the wave height, and varies between 10 and 40. Waves may generate oscillatory flow that produces symmetrical ripples, which have foreset laminae pointing in both directions.

Dunes are similar to ripples in shape and structure, and form in coarse-, medium- and fine-grained sand, but require significantly higher flow velocities for their formation. Dunes range in height from 5 cm up to several metres, and wavelengths may exceed 10 m (Fig. 2.15).

Cross-bedding (large-scale cross-stratification) is seen in cross-sections through dunes (Fig. 2.16). Each lamination is called a *foreset bed* and represents the lee-side surface of a migrating dune. If the dunes have straight crests the foreset beds on the lee side will form a straight transverse plane (*tabular cross-bedding*). Curved dunes have rounded foreset beds, and in *trough cross-bedding* the laminae have a rounded surface which is concave in the downstream direction. The foreset lamination may form a relatively sharp angle with the underlying bed, or may have a more tangential contact. The latter is typical of trough-shaped sets.

Aeolian dunes may be many metres high, and their cross-bedding will then be correspondingly large (Fig. 2.15).

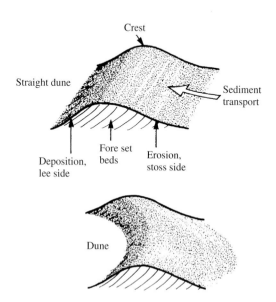

Fig. 2.14 Principle illustration of dune and cross-bedding

Fig. 2.16 Cross-bedded sandstone which represents the filling of a fluvial channel above a coal bed. The grain size fines upwards somewhat. It is largely trough cross-bedding represented here. From the Tertiary sequence of Spitzbergen. (Photo A. Dalland)

Whatever the size of the bedform, ripple movement provides the basic mode of migration. Seabed survey profiling records clearly show smaller ripples climbing the stoss sides of larger ones (so-called "megaripples", i.e. small dunes), which in turn are climbing the stoss sides of sandwaves.

Cross-sections through dunes show "large-scale cross-stratification", which is often referred to as "cross-bedding". This can be observed in real time on seabed video recordings made during periods of strong tidal current flow, where gradual forward movement of "megaripples" is seen, caused by sand cascading down the lee side after reaching the crest.

"Plane beds" (upper stage) may form when the shear stress against the bed exceeds the values which produce dunes. In cross-section we only see planar lamination, which is an internal structure, but on the bedding surface we may see very small ridges, which define a lineation called primary current lineation, parallel to flow.

At even higher velocities in relatively shallow water, standing waves may produce antidunes when the Froude number exceeds 0.8. The antidunes which are produced when standing waves are in phase with the bedforms develop resulting in low-angle cross-lamination which dips up-current.

Both ripples and dunes are formed through sand being transported along the bottom and deposited in sloping strata on the lee side of the structure. In consequence they always have dipping laminations *(foreset beds)* which may lie at an angle (angle of repose) of up to 35° to the surface of the bed, though such high angles are rather rare. Current ripples in plan view may be straight, or form curved patterns *(sinuous crests)*. Ripples with a symmetrical cross-section (symmetrical ripples) are formed by waves as a rule. Asymmetrical ripples are formed by a predominantly unidirectional current and their steeper side faces downstream. Ripples with a high sinuosity are also asymmetrical in most cases.

Tongue-shaped *(linguoid)* ripples have a very high sinuosity and asymmetry and are usually formed in shallower water or under higher velocities than straight-crested types. Wave ripples in particular may split laterally into two ripples. This is called bifurcation. In intertidal zones ripples formed at high tide may be eroded at low tide, and the crests become flattened. When the tidal flat is submerged at high tide it may also be below the normal wave base, resulting at slack water in deposition of clay which tends to collect in the ripple troughs. Current ripples with thin lenses of clay between them constitute *flaser bedding*. Wind may generate waves moving in different directions, particularly at very low water, so that we find two or more sets of ripples at an angle to each other *(interference pattern)*. In most cases each bed with current ripples represents a sort of equilibrium with deposits reflecting current patterns. Isolated sand lenses in clay are called *lenticular bedding*.

Normally current ripples form completely horizontal beds. In some cases, however, we find examples of current ripples appearing to climb downstream in relation to the horizontal plane. They form several sets of cross-laminated beds delimited by erosion boundaries, but with small internal erosion planes. These are called *climbing ripples* and are due to sedimentation taking place so rapidly that, in contrast to normal ripples, there is no equilibrium between erosion and

Fig. 2.17 Hummocky
stratification (After Harms
et al. 1975, Walker 1982)

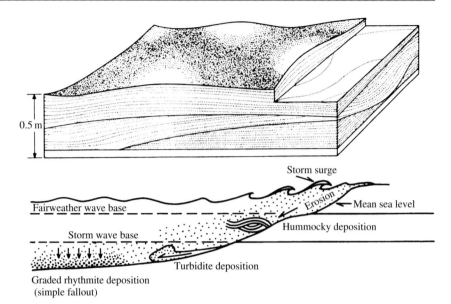

sedimentation. Climbing ripples are therefore typical of environments with rapidly declining flow velocity and consequently a high rate of sedimentation.

Sandwaves are large-scale transverse bedforms, generally 2–15 m high, with a wavelength of 150–500 m. They may be formed at flow velocities of 65–125 cm/s. Cross-stratification may be symmetrical or asymmetrical on both sides of the sandwave, depending on the relative strength of the opposing currents.

At velocities of less than 1 m/s *sand ribbons* may be deposited – longitudinal bedforms developed parallel to the currents. Sand ribbons are typical of subtidal environments (20–200 m), and they may be up to 20 km long, 200 m wide and less than a metre thick.

Relatively low-energy environments (< 50 cm/s) are characterised by *sand patches* and mud. The sand forming the patches probably only moves during storms. Lateral structures are typically current ripples and small dunes.

Sandwaves often have current ripples ("smaller dunes") on their surface. This may also be the case with dunes. Bedforms are a function of both flow velocity and grain size. Current ripples are formed only in silt and medium-grained sand, while dunes require medium to coarse sand. Upper flow regime plane beds which have an internal structure of planar lamination are formed when the Froude number is about 0.6–0.8. Plane beds typically develop in beach sand.

A well-developed lineation on the bedding surface parallel to the direction of sediment transport is typical of plane beds (Fig. 2.9). Antidunes are formed by higher velocities in the upper flow regime; the absolute flow velocities required are lower when the water is shallow.

Below the ordinary, fair-weather wave base we find different types of sedimentary structures from those formed through constant wave action. During storms in nearshore areas traction currents may develop and carry fine material (fine sand) from the beach out to greater depths. This material is deposited as *hummocks* consisting of parallel laminae which form dome-shaped structures (Fig. 2.17). Because deposition of beds of this type takes place during short periods during storms, they are often reworked by bioturbation near the top. Hummocky stratification is common in sections through some shallow marine deposits and wave-dominated deltas.

2.19 Erosion Structures on the Underside of Sand Beds *(Sole Structures)*

When sand is deposited rapidly over finer sediments (silt and clay) there will also in very many cases be an initial erosional phase, so that erosion hollows are formed in the substrate. These hollows fill with sand, and we get a sort of *cast*. We see these erosion structures on the underside (sole) of sandstone and conglomeritic strata. We seldom observe these surfaces unless the bed is steeply inclined or inverted. Wherever there are overhanging rocks, it is important to study the underside of the bed.

Flute casts are formed by static vortices in the water eroding into the underlying sediments. Their sharp end points upstream and they broaden in the downstream direction. Flute casts (Fig. 2.18) are good indicators of current direction, which is measured along an axis of symmetry through the structure. They are typical on the underside of turbidite beds, but also occur in fluvial sandstones as a result of fluid turbulence.

Similar structures are formed around objects which project up from the bottom, such as stones and large fossils.

Gutter marks are longitudinal grooves up to 20 cm deep and rather narrow (less than 20–30 cm) with a spacing of 1 m or more. They are the result of channelised flow, and flute casts or groove casts may be found along their margin.

Ridges and furrows may also be formed through erosion of the substratum. Some are U-shaped or V-shaped channels. Erosion structures formed by objects which are transported along the bottom are called *tool marks*. The objects may be fossil fragments or larger sediment particles.

Groove casts are long, narrow erosion furrows due to something being dragged along the bottom.

Chevron marks are erosion furrows with V-shaped structures in the clay sediments on either side. The V-structure closes downstream, and is due to a cast forming in cohesive clay.

Prod marks show where an object has dug down into the clay and then been plucked out again by the current. As a result the steep side of prod marks is the downstream side.

Bounce marks are rows of more symmetrical marks due to objects being swept or bounced along the bottom.

2.20 Deformation Structures

Sediments are often unstable immediately after deposition, and later movements will deform the primary structures.

Deformation may be caused by four main factors:

1. Shear stress due to water or sediment movement, e.g. convolute lamination.
2. Expulsion of porewater (liquefaction, dewatering), e.g. dish structures, clastic dykes.
3. Heavier beds above lighter beds (inverse density), e.g. load casts and ball-and-pillow structures.
4. Gravitational deformation. Gravitation-induced sliding, folding and faulting on a slope, e.g. slumping.
5. Shrinkage, e.g. due to dessication, or permafrost (ice wedges).

It is important to distinguish between these types of deformation structures because they have completely different implications for interpreting depositional environments. They are not mutually exclusive, however, and may be found in close association in the same sequence.

Convolute lamination forms in fine sand or silt, and is due to laminae, e.g. current ripples, being deformed almost as they develop. Folded and inverted (overturned) lamination is typical, and the structure is deformed in the downstream direction due to the stress induced by the water movement and the instability (almost a state of liquefaction) of the sediments (Fig. 2.19).

Dish structures (Fig. 2.19) are thin, clay- and silt-enriched dish-shaped laminae in sandy sediments. Their structure is due to the porewater which flows

Fig. 2.18 Flute casts on the lower surface (sole) of a coarse-grained sandstone bed in the Ring Formation, Rena, South Norway. Note the small flute casts on the large flute cast structure. Flute casts are casts of the erosion structures formed in the finer-grained underlying bed by vortices

Convolute lamination

Convolute lamination
structures within a bed
formed during deposition.
Erosion of these structures
before the next bed is deposited.

Slumping

Folding of a number of beds
simultaneously due to gravity-
induced sliding following
further deposition.

Dish structures

Clay- and silt-enriched, rounded
dish-shaped laminae formed by
deposition from upward-flowing
water immediately following
deposition.

Fig. 2.19 Sedimentary structures due to liquefaction and soft
sediment deformation during the deposition process. See text

Fig. 2.20 Load cast structures at the base of a sandstone bed
in the Late Precambrian Ring Formation at Rena, Southern
Norway. The picture covers an area of 3 × 4 m. (Bjørlykke et al.
1976)

upwards immediately after deposition, transporting
clay and silt which become trapped in these thin
laminae. *Sand dykes (or clastic dykes)* are intrusions
of sand upwards into cracks in a finer-grained sediment, due to porewater overpressure. The overpressure
reduces the friction between the grains and injects
water with sand into vertical fractures produced by
high overpressures or follows bedding planes as sills.
Overpressurised porewater with sand may rise right to
the surface and form *sand volcanoes*.

   Well-sorted sand has at the time of deposition an
initial porosity of 40–45%, whereas recently deposited
mud contains 50–80% water. When sedimentation
is rapid, the mud has little time to lose its excess
water, and a sand bed deposited on top may then
sink down into the underlying silt and clay and form
load structures. On the lower surface of a sand layer
we often see pillow-shaped depressions surrounded by
clay which has oozed up around them. If this process
continues, it will form isolated sand pockets in the
underlying clay: *ball-and-pillow structures*. Primary
structures such as flute casts often sink in the underlying mud and are deformed by *loading* (Fig. 2.20).

   This mechanism also operates on a larger scale, for
example channel sand will sink down into surrounding clay (see "deltas"). Poorly compacted clay and silt
will be lighter than surrounding sediments, and flow
upwards to form clay diapirs.

   Gravitational deformation occurs in sediments
which are deposited on slopes. These forces can be
resolved into a vector normal to the bedding, and a

shear stress parallel to it. The vector which acts along
the bedding is proportional to the sine of the angle of
dip, and acts as a compaction force which can slide
and fold the beds. In the upper part of the slide, tensional deformation is prominent, producing faulting,
while compression occurs near the base of the slide
and produces folding. The result is called *slumping*
(Fig. 2.21). Slumping occurs most readily where we
have rapid sedimentation and therefore relatively thick
beds with a high water content and low shear strength.
Deformation takes place when the shear stress exceeds
the shear strength. The shear stress increases with the

Fig. 2.21 Sediment beds which have been folded immediately
after deposition (through slumping) due to sliding on submarine
slopes. Note that the overlying beds are undeformed showing
that this is not tectonic folding. From the Ridge Basin (Miocene-
Pliocene), California. Scale, John Crowell

thickness of the unconsolidated sediments, but there is not usually a corresponding increase in shear strength with increasing thickness. Slumping may resemble convolute lamination, but normally affects more than one bed. Gravitational deformation also leads to faults on a greater or lesser scale. Sliding of large volumes of sediment down slopes produces slope scars. *Growth faults* and other types of "listric" faults are a result of large-scale gravitational deformation in the upper part of an area which is under tension (p. 84).

Dessication cracks or mud cracks are examples of contraction or shrinkage of sedimentary beds due to dehydration. Cracks frequently form regular polygons, often hexagons or orthogonal sets. Dessication cracks form only in clay and silt, and the cracks often become filled with sand, resulting in good contrast. The formation of dessication cracks requires that the beds be exposed to the air so that the sediments can dry out. In certain cases shrinkage structures may also form underwater through dehydration of clay minerals (smectite) as a result of variation in the salinity of the porewater *(syneresis)*. Shrinkage structures are less regular than desiccation cracks and are not usually interconnected. In the smectite-rich Eocene sediments from the North Sea basin and the Norwegian Sea, seismic data show large (several hundred metres wide) polygons which have been interpreted as shrinkage cracks.

When the porewater in sediments freezes to ice and remelts, we also find expansion and contraction which results in polygonal surface marks and associated vertical ice wedges.

2.21 Concretions

Concretions are round, flat or elongated structures which consist of cement which has been chemically precipitated in the pores of the sediment. The most common types of concretion are carbonate (calcite and siderite) and silica (chert). Sulphides, particularly pyrite, also form concretions.

A characteristic feature of concretions is that any laminations in the sediment pass through the concretions. This shows that the concretion has been formed through passive filling of its pores. As the overburden increases, the sediments around the concretion will be subject to compaction, while the concretion cannot

be compressed because the pores are full of cement. A concretion therefore has a cement content which corresponds to its porosity at the time of formation. Carbonate concretions in clay may have a carbonate content reflecting 50–70% porosity if the matrix does not contain carbonate. Concretions in calcareous rocks, i.e. marls, contain clastic or biogenic carbonate in addition to carbonate cement, and therefore have more carbonate than the matrix. In these cases the carbonate content cannot be taken as an indication of the porosity at the time of formation. Concretions often contain fossils, showing no sign of compaction while the same fossils are severely deformed and sometimes also dissolved outside the concretion. In carbonate sediments, particularly chalk, there are silica concretions (chert). These are formed through precipitation of finely divided amorphous silica to form a type of chert called *flint*. The source of the silica is usually amorphous biogenic silica, frequently sponge spicules.

2.22 Trace Fossils

Trace fossils are structures in sedimentary rocks which have been left by organisms that lived on and/or burrowed in the sediment. Such organisms are extremely sensitive to changes in the composition of the nutrient content, the sedimentation rate and bottom currents, and are therefore useful indications of the environment (Fig. 2.22a,b). Trace fossils are therefore good indicators of the depositional environments. They can be classified taxonomically, i.e. according to the animal which left the traces, but it may not be possible to determine which animal was responsible. The same species may form several different types of trace depending on the sediment composition and on its mode of life. Furthermore, different animals may leave traces which are so similar that they are classified as one trace fossil. For these reasons a descriptive, morphological classification of trace fossils is used.

Trace fossils can also be classified according to where they occur in relation to the bed:

1. On top of beds (e.g. a thin sand or carbonate layer Epichnia).
2. Within the bed (Endichnia).
3. On the lower surface of the bed (Hypichnia).
4. Outside the bed (Exichnia).

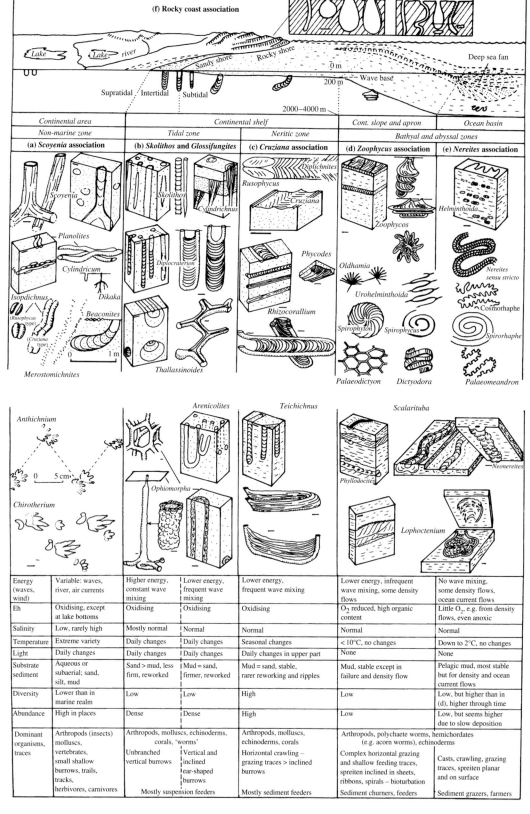

Fig. 2.22 Common trace fossils in different sedimentary facies (from Collinson and Thompson 1982)

The impression made by an animal may create a *mould* or a *cast* in the overlying bed. Organisms which burrow into sediments often secrete a cement which ensures that the walls of their burrows do not collapse. These secretions also contribute to preservation.

When worms eat sediment, for example, it passes through their digestive organs and their burrow refills with a sediment which has a somewhat different composition from the surrounding sediments. This is particularly noticeable in burrows at the interface between two strata with different compositions. The amount of bioturbation reflects nutritive conditions and the sedimentation rate. With very rapid sedimentation there will be less time for organisms to burrow through the sediments. Where we have very slow sedimentation or an hiatus, the sediments will often be thoroughly churned up by bioturbation, and thereby homogenised. Bioturbation is most widespread in marine environments, but can also be found to a lesser extent in freshwater sediments. In poorly oxygenated environments there is, however, little bioturbation.

In high energy environments, e.g. above the fairweather wave base, we only have vertical trace fossils, e.g. *Skolithos* or *Diplocraterion*. The high current velocity prevents these organisms from crawling around on the bottom; they have to burrow down into the sediment and live by filtering nutrients out of the water. To remain at the same level beneath the sediment surface they must move upwards or downwards in their holes, depending on whether erosion or sedimentation is proceeding in the area. Structures or fillings which reflect such adjustments are called "spreiten".

In modern marine environments we can study a number of organisms which create bioturbation structures. The most common are worms like *Arenicola* which make U-shaped traces in fine-grained sand and silt. Such traces (arenicolites) are also to be found in older rocks. Burrowing bivalves create various types of trace as well. Arthropods like crabs and prawns make burrowing structures in beach sand (*ophiomorpha*). These vertical trace fossils are grouped together in an ichnofacies called the *Skolithos* facies. *Thallassinoids* are more horizontally aligned networks of arthropod burrows.

Below the wave base and in other protected environments, for example the intertidal zone, we find traces in the horizontal plane from organisms which live off blue-green algae and other organic material

on the surface of the sediment. In this *neritic* zone (Fig. 2.22a,b) we find a number of different types of traces from organisms which eat their way through sediments, and which form different patterns. This is called the *Cruziana* facies. *Rusophycus* and *Cruziana* are typical of the neritic zone and represent horizontal traces of arthropods which feed on the sediment surface. *Rhizocorallium* and *Teichichnus* are other trace fossils that occur below the *Cruziana* facies. In deeper water where wave and current energy is even lower, we find *Zoophycus* and *Nereites* facies. It is important to remember that these environments are primarily a function of wave and current energy, and cannot simply be correlated with absolute depth. In shallow enclosed seas with a shallow wave base, e.g. the modern Baltic, we find an effective wave base of only 5–10 m in many areas, while elsewhere we may have stronger currents along parts of the deeper trenches. In the epicontinental Cambro-Silurian marine sedimentary sequence of the Oslo area we find *Nereites* facies in the shales, but the water depth was probably not more than 100–200 m, perhaps even less. In the deep oceans the *Nereites* facies may correspond to a depth of several thousand metres.

Trace fossils are very useful facies indicators and should be noted whenever sedimentary sections are examined for facies interpretations. Certain trace fossils can be linked with animals that have fairly specific environmental requirements.

2.23 Facies and Sedimentary Environments

The word "facies" is used in a number of geological disciplines. A term such as "metamorphic facies" is thoroughly entrenched. Sedimentary facies have also long been identified in sedimentology to distinguish between sedimentary rocks which differ in appearance and have formed in different ways. The term facies can be used both descriptively and genetically. We use terms like "sandy facies", "shaly facies", "carbonate facies" when we are describing properties of the rock that can be observed or analysed objectively. We use terms such as "shallow water facies", "deep water facies", "turbidite facies", "deltaic facies", "intertidal facies", "aeolian facies", "reef facies" etc., depending on which environment we believe the rocks represent.

In these examples, the word "facies" represents an interpretation, and is therefore not very suitable for describing sedimentary rocks objectively. For this reason it is important that we define the objective criteria (observations) on which we are basing our interpretations.

What we can do, then, is first describe and take measurements on a series of beds in the field, and on the basis of certain criteria divide the series into facies. The criteria will generally be texture, sedimentary structures, mineral composition, and, if present, also fossils. Interpreting a facies in terms of depositional environment is often very difficult, especially since few criteria are unambiguously diagnostic of one particular environment. In some cases it may be useful to use statistical methods for distinguishing between different facies and for describing facies sequences.

What we observe and measure is a selection of the properties of the rock. When we describe a sedimentary rock, we observe the results of processes which have acted in the environment in which the rock was deposited. The sedimentary structures we observe tell us something of the hydrodynamic conditions during deposition. Organic structures and fossil content help us reconstruct the ecological conditions at the time of deposition in the basin. Evidence of chemical processes such as weathering, diagenesis and precipitation of authigenic minerals also supply important information about the depositional environment, as well as its early post-depositional history. Figure 2.23 is a diagrammatic representation of the major depositional environments on a continent and the various transition stages to deep water.

Following an interpretation of the facies of each bed or sequence of beds, we may try to group these into what we call *facies associations*. These associations of facies can be related to large-scale processes in particular environments. An association of facies coarsening upward from marine shales into delta front sand or into coal beds of the delta plain facies may represent a facies association typical of delta progradation.

Other facies associations may represent transgressions or regressions in a shallow marine environment.

If we have a continuous sequence without breaks (unconformities) the vertical succession represents the lateral succession of environments. This, in essence, is Walther's Law, named after the German geologist Johannes Walther.

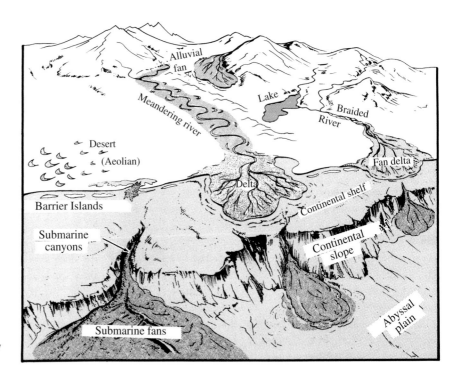

Fig. 2.23 Schematic representation of sedimentary facies on a passive margin

2.24 Alluvial Fans

Alluvial fans are accumulations of sediment which has been transported by fluvial processes or different forms of mass transport (e.g. mud flows and debris flows) and deposited in valleys or on slopes fairly near the erosion area. The fans form in areas that have considerable relief and are usually associated with faults which keep uplifting the erosion area in relation to the area of deposition. This tectonic subsidence of the alluvial fans is a necessary condition for their being preserved in the geological series. Alluvial fan deposits in older rocks may therefore supply important information about tectonic movements during deposition. Alluvial fans which have formed in areas with a high relief but little tectonic activity, will merely represent a stage in the transport of the sediments. For example, in glaciated areas we may find alluvial fans on slopes steepened by glacial erosion. They will be eroded again if they are not rapidly covered by a transgression.

Erosion on the uplifted block will form V-shaped valleys (or canyons), and these will drain into the valley. The apex of the alluvial fan is usually near the main fault plane and sediment transport from here will tend to follow the steepest slope downwards so that the sediments will be spread out in a fan (Figs. 2.24 and 2.25). If there is only a short distance between adjacent valleys, and consequently between fan apices, the fans will coalesce. If major drainage systems develop, larger fans will form further apart from one another. In areas with a relatively humid climate, fluvial processes will account for sediment transportation even high up

Fig. 2.24 Alluvial fan. Death Valley, California

on the fan. In arid climates the water table under the fan will be deep down, and when it rains the water will rapidly filter down into the upper part of the fan. The slope of the fan may then increase due to deposition on the upper part. As a result these sediments may be transported with a high sediment/water ratio as debris flows or mud flows. The upper part of the fan may consist of large blocks or cobbles which form an open network system through which finer-grained sediments can pass. In this way the sediment is sieved, and *sieve deposits* are formed.

Downslope on the fan, channels usually split into a number of smaller channels. This reduces the hydraulic radius of the channels and in consequence their velocity, and hence capacity for carrying sediment, is lowered. Sediment will therefore become finer-grained downslope, even if there is no reduction in the gradient.

The water table will be deepest at the top of the fan, and shallowest at the foot. Alluvial fans are good groundwater reservoirs, easy to tap because of their porosity and permeability. The circulation of groundwater through an alluvial fan leads to strong oxidation of at least the upper part of the sediments, giving them a red colour due to iron oxides. In most cases any organic material will be completely oxidised.

In arid climates there will be a great deal of evaporation from the groundwater which emerges from the fan, and the ions in solution will be precipitated as carbonate *(caliche)* and iron oxide. Water flows beyond the arid alluvial fans in *ephemeral rivers*, which only exist after heavy rain, and may then collect in *playa* lakes which dry up each year, forming evaporite deposits (Fig. 2.25).

Humid fans will be dominated on the lower part by fluvial channels. These may drain a major portion of the fan, and thus have a large hydraulic radius and a greater capacity for transporting sediment, even if the slope of the fan is not very great. In the lower part of a humid fan cross-bedding will be fairly pervasive, while the upper part will tend more to consist of massive conglomeratic beds. The foot (distal end) of the fan will merge into the other sediments which cover the valley floor. These may be lacustrine or fluvial deposits.

A characteristic feature of transport and sedimentation across and around arid zone alluvial fans is that sedimentation takes place during short periods in connection with the rains. However, the water rapidly disappears into the ground, increasing the

Fig. 2.25 An alluvial fan
developed along a basement
fault. The degree of
progradation of the fan into
the basin depends on the relief
along the fault, but is also
dependent on the rainfall and
the catchment area on the
upthrown block

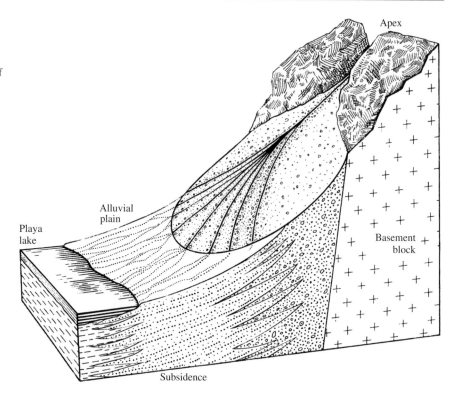

surface sediment/water ratio, and eventually dries up to leave poorly sorted conglomeratic beds. It is important to remember that the sedimentary structures and sorting we observe are only representative of the final deposition phase.

The grain size of the sediments on the fan is a function of weathering and erosion in the source area and of transport capacity outwards across the fan. With fluvial transport, the velocity **v** and thereby transport capacity is both a function of the slope and the depth (hydraulic radius) of the channels (see the Cherzy equation).

The velocity is therefore much higher during flooding.

2.24.1 The Water Budget

A part of the total rainfall will sink into the ground and supply the groundwater. Another part will evaporate on the surface or flow on the surface into rivers (Fig. 2.26). In hot, desert areas where the air is very dry, evaporation may correspond to 2–3,000 mm/year, i.e. more than the normal amount of precipitation in most places. Plants also contribute to water loss through transpiration, and in areas with vegetation, evaporation and transpiration may reach up to 2,000 mm/ year. Without vegetation we find a significant amount of evaporation only when there is free water or damp ground right up to the surface. Even if the water table is shallow (less than 1 m), there is little direct evaporation, but some water will be drawn up by capillary forces and evaporate.

Plants are important in connection with evaporation of groundwater as tree roots may penetrate more than 10 m below the surface. Trees can be a major drain on groundwater reserves because they use water which could otherwise have seeped into wells, rivers or lakes. Dense forests may use water corresponding to 200–500 mm of precipitation. In areas with limited water reserves it may consequently be necessary to limit vegetation, particularly of varieties of trees and bushes with a relatively high transpiration rate and no useful function. However, vegetation is important for stabilising the topsoil and thereby preventing erosion. Much of the water evporated from vegetation may return as rain locally. Vegetation may also have an effect on the *albedo*, i.e. the amount of sunlight which is reflected from the earth's surface, and this in turn may affect the climate.

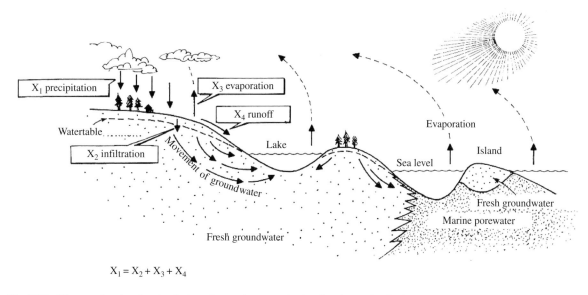

$$X_1 = X_2 + X_3 + X_4$$

Fig. 2.26 Diagram showing the circulation of groundwater on land and under marine basins

Runoff is the fraction of precipitation which reaches the rivers. The percentage of precipitation which filters down to and recharges the groundwater is of particular importance. Rivers also have groundwater added to them if their surface is below the water table. If their surface is above the water table they will lose water. In consequence, rivers cannot be considered in isolation from groundwater reserves.

Infiltration of surface water into the ground depends on the permeability of the soil cover and on the vegetation. Clayey sediments allow far slower infiltration of surface water than sand or gravel. Where there are soil types with swelling clay (smectite group), which are very common in desert areas, the first rains will cause these minerals to swell, further reducing their permeability. Consequently, infiltration is reduced, and the runoff which creates floods increases. If the earth is dry above groundwater level, water has to overcome the capillary forces which act against water percolating down through dry, fine-grained sediments. We may find a layer with air in the pores between the groundwater table and the water filtering down from above (trapped air).

Alluvial fans are porous and have a relatively good potential as water reservoirs, and we often find springs at the foot of fans. In dry areas the depth to the water table can be most simply charted through seismic recordings.

Because freshwater is lighter than saltwater, it will flow over saline porewater and form pockets of freshwater under islands. Freshwater may also flow from the continents, following permeable beds beneath the continental shelf. Freshwater of high quality (little pollution) is scarce in many regions, and in recent years in particular a large number of geologists have been involved in mapping groundwater reserves.

2.25 Desert

Deserts are areas with little or no vegetation and we can distinguish between different types of desert:

1. Deserts due to low precipitation and high evaporation in hot climates.
2. Deserts due to low precipitation in cold climates.
3. Deserts due to soil erosion.

The first two types of desert are largely governed by meteorological factors. Areas which lie on the lee-side of major mountains will be dry. At about 30°N and 30°S high pressure areas predominate, resulting in low precipitation. Since the prevailing wind at these latitudes will come from the east, we find most of the deserts on the western side of mountain chains. This is true of Asia Minor, the Sahara,

California and Nevada in the northern hemisphere, and Australia, South Africa and Chile in the southern hemisphere. Inland areas surrounded by mountains tend to have a desert climate. During the glacial periods there was less evaporation and therefore less rainfall. North Africa and the Sahara were then very much drier than the present day.

Iceland has good examples of desert regions which are due to a cold climate, and has problems with soil erosion due to lack of vegetation in many areas.

Shortly after the withdrawal of the ice across Scandinavia there was a desert with considerable aeolian deposits, which formed before the vegetation cover developed. Loess deposits are fine-grained sediments which form largely through aeolian erosion and transport from deserts and also from glacial sediments exposed after glacial retreat. Large areas of China, for example west of Beijing, are covered by up to several hundred metres of Quaternary loess.

The lack of vegetation naturally means that aeolian transport and deposition are important in deserts, but only a relatively small part of the desert areas are covered by wind-blown sand.

Large areas consist of bare rocks and mountains with little sediment. In other regions there is only wind erosion (deflation), which leaves the ground covered (*armoured*) by a layer of stones which protect it from further erosion. Even in desert regions like the Sahara many areas are dominated by a fluvial drainage pattern. Although several years may pass between rains in this area, a heavy rainstorm may transport so much sediment that the fluvial drainage pattern survives for many years.

Large expanses of wind-blown sand are called *ergs*. They may be formed by the coalescence of different aeolian bedforms. *Barchans* are crescent-shaped isolated aeolian dunes with a convex erosion side and a concave lee side (Fig. 2.27). Barchans are found

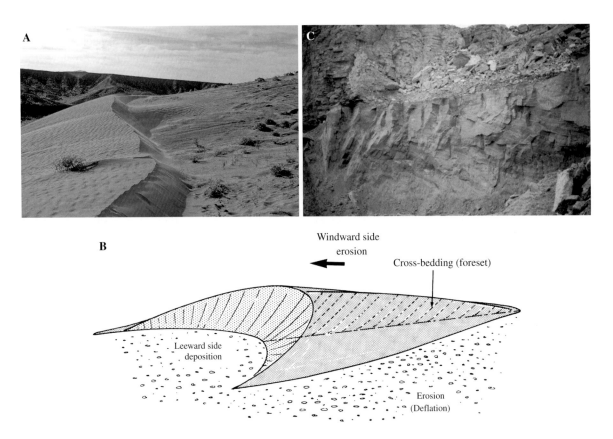

Fig. 2.27 (**a**) Aeolian dune from the desert in Death Valley, California. (**b**) Schematic representation of a an aeolian dune (barchan). (**c**) Aeolian cross-bedding 4–5 m high in the Permian "*Yellow Sand*" of northeast England (Old Quarrington Quarry, Durham). This sand is equivalent to the "Rothliegendes" sandstone of the North Sea

mainly on the edges of the erg area, where there is not enough sand to form a continuous thick cover. Transverse dunes are common within ergs. *Seif dunes* are long, straight sand dunes which may occur within an erg but are also found in areas with incomplete sand cover. In the central zones very large dunes form which may be over 100 m high and 1 km long.

Ancient aeolian deposits are recognised by:

1. Their large-scale cross-bedding, up to 20–30 m and often with wind ripple marks on top of slanting cross-bedded surfaces (foreset beds).
2. High degree of oxidation which gives a red colour.
3. Good sorting – largely medium- to fine-grained sand.
4. Lack of fossils or organic material.

The lack of thin silt or clay beds between the cross-bedded layers, and the general association with overlying and underlying sediments, are also important criteria. One of the best-known examples of an aeolian sandstone is the Navajo Sandstone (Jurassic) in the Rocky Mountains. In Northern Europe there are aeolian sandstones in the Lower Permian (Rotliegendes) and the equivalent Yellow Sand in Eastern England (Fig. 2.27).

In pre-Devonian times all continental environments were deserts in the sense that they lacked vegetation. Also in these rocks, however, we can distinguish between dry and wet climates, largely from the nature of the fluvial or aeolian deposits.

Sand dunes are always moving and may end up in the sea, still partly preserving an aeolian sorting and grain-size distribution.

2.26 Lacustrine Deposits

How Are Lakes Formed?

In principle we have three types of lake:

1. Lakes of glacial origin:

   (a) lakes formed through glacial erosion,
   (b) lakes formed through damming by moraines or by the glaciers themselves.

2. Lakes of tectonic origin:

   (a) lakes formed in areas of rapid tectonic subsidence (rifting) or more uniform subsidence,
   (b) lakes formed as a result of damming by horsts which have been elevated through faulting, or damming by lava etc.

3. Lakes formed by sedimentary processes, e.g. *oxbow lakes* in fluvial environments, and delta top lakes.

Lakes of glacial origin will have a relatively short lifetime by geological standards. In 10,000–100,000 years most of the lakes in Scandinavia and North America will have filled up with sediment, if we do not have another glaciation.

Lakes of tectonic origin, however, will continue to subside. If the rate of subsidence keeps pace with the rate of sedimentation, a lake will continue to exist.

Extensive carbonate beds (freshwater carbonates) and diatom deposits are also common in lacustrine basins. Some of the largest lakes formed by rifting are found in East Africa, where a number of lakes occur in the actual rift valley system. Lake Victoria, which lies between two rift valley systems, was formed by tectonic movements about 100,000 years ago.

In humid climates all lakes will have outlets in the form of rivers or via the groundwater, but in arid climates lakes develop into evaporite basins without outlets. Normal, non-saline lakes differ from ocean basins in several ways:

1. Low salinity leads to slower flocculation of clay particles than in marine basins, producing more distinct lamination.
2. Low wave and tidal energy and weaker currents mean less erosion and resedimentation. Seasonal variation in the influx of sediments may produce annual cycles in lacustrine sediments.
3. Limited water circulation makes it easier for the water to develop layering based on temperature (thermocline) and therefore also density (pycnocline) between dense layers at the bottom and less dense water near the surface. This may restrict oxidation of organic material and promote the development of organic-rich sediments (source rocks) in the deeper parts of the lakes. In warm climates the temperatures of the lake waters are always above 4°C which makes the water stratification rather stable. In colder climates the water column will be inverted when the temperatures falls to 4°C which is the highest density. This results in an oxygen supply to the lake bottom.

4. River water will generally have approximately the same density as lake water, and will therefore mix well and deposit sediments rapidly. Cold (glacial) river water or river water with a lot of suspended material will be heavier than lake water, however, and will form turbidity currents along the bottom.

5. Lake sediments have a distinctly different fauna from marine basins. The geochemistry of lake sediments and the composition of carbonates and evaporites are also different from those of marine sediments.

By comparison with marine deltas, lacustrine deltas are among the most constructive of all (e.g. the Gilbert Delta) because erosion in the basin is so limited. Nevertheless, wind conditions, the size of the lake and the composition of the sediments will be crucial factors. In large lakes we might have relatively high wave energy which could erode sediments from the river mouth and deposit them laterally on beaches.

Wind will also cause water to well up from the bottom, increasing the circulation of oxygen along the bottom.

In dry regions we may get evaporitic lakes and lakes which dry up after each flood *(ephemeral lakes)*.

Lake deposits are characterised by:

1. Lack of marine fossils, but diatoms and algae may be important components.
2. Very fine laminations in clay and silt sediments.
3. Bioturbation structures may be found in lacustrine sediments, but they are less common than in marine sediments.
4. Chemical analyses of freshwater sediments will reveal low concentrations of the trace elements which are enriched in seawater (e.g. Cl, Br and B). Minerals formed in freshwater lakes may have characteristic isotopic compositions. Evaporites in lakes consist mainly of carbonate minerals; sulphates and chlorides are normally less common.
5. The content of sulphur-bearing authigenic minerals must be relatively low since there is little sulphate in freshwater. However, organic material in lakes contains a fair amount of sulphur which can be precipitated as sulphides by sulphate-reducing bacteria. Sulphur may also be derived from volcanic sources.
6. Lack of tidal structures and similar marine structures is also an important indicator.

Amongst the most important ancient lacustrine deposits are the Karoo deposits of South and East Africa, from the Carboniferous to the Jurassic periods. These sediments contain important coal deposits. In eastern China there are large areas of Cretaceous and Tertiary lacustrine sediments, which are also important petroleum-bearing sediments.

The Green River Formation (Eocene) of Colorado, Wyoming and Utah, is a lacustrine sediment of great extent which represents one of the most important petroleum source rocks in the world.

Rifting associated with the opening of the Atlantic Ocean in Mesozoic times resulted in a large number of lacustrine basins, and sediments deposited in such basins now underlie the continental shelves. In dry regions these turned into evaporite basins.

2.27 River Deposits

Transport of sediments in rivers depends on the gradient and cross-section of the river, which determine the flow rate, and on the composition and concentration of sediment. In rivers which flow only during and directly after the rains, but are otherwise dry *(ephemeral streams)*, equilibrium between flow and sediment load is not achieved, so that even fine-grained sediment is deposited in the channel when it dries up, leaving a mud-cracked surface. Intensive oxidation of sediments, including silt and clay beds, is typical of such fluvial deposits. There is little organic production in areas with a dry climate, and clay and silt particles will therefore contain little organic material which could act as a reducing agent. A low water table during dry periods also contributes to oxidation of organic matter.

Most rivers have marked seasonal and annual flow variations. The flow velocity increases appreciably with flow volume because the friction per unit volume of water is inversely proportional to the water depth. When water flow is greater than the capacity of the channel, the water flows out over the banks. The water outside the channel will normally be very shallow and have a low flow velocity and fine-grained sediments, which have been transported in suspension, are then deposited as what we call *overbank sediments*. Overbank sediment may build up into elevated banks called *levees*. The resulting soil along modern rivers is often very rich and ideal for cultivation. In post-Devonian sedimentary rocks we often find traces of plants, commonly roots, in levees. Levee deposits have parallel lamination and in some cases also current ripples, particularly *climbing ripples*, which are typical of rapid sedimentation. The primary sedimentary structures, however, will often be partly or wholly destroyed by traces left by plant roots.

Major floods may cover the areas beyond the levees as well, and clay and silt will be deposited on these *flood plains*. Many rivers which carry a large amount of suspended sediment gradually build up their beds through deposition in their channels and on the levees, so that the surface of the water in the channels may be considerably higher than the surrounding plain. During floods the water may then break through the levee and flow down from the channel to the plain, forming temporary lakes several metres deep. The lower Mississippi River is considerably higher than the surrounding area, including New Orleans. The major Chinese rivers have also built themselves up above the surrounding countryside, so floods can cause very great damage.

However, the stability of the levées will determine how much a channel can build itself up, and clay-rich sediments form stronger levees than sand and silt because of their cohesiveness. When the levees give way, sand will flow out of the channel and deposit sandy sediments in fans or *crevasse splays* on the flood plain. They are characterised by thin sand beds, often fining upwards, with an erosional base in the proximal part (nearest the channel). However, we may also find small sequences which coarsen upwards as a result of progradation of this fan, because grain size decreases towards the edge of the fan.

2.27.1 Channel Shapes

We distinguish between relatively straight fluvial channels and curved channels. Curved channels are shaped rather like a sine curve. To describe the degree of curvature we speak of high and low sinuosity. Sinuosity is defined as the ratio between the length along the channel and the length of a straight line through the meander belt, i.e. the length of the river valley. *Meandering streams* have a single channel and by definition a sinuosity greater than 1.5.

Braided streams have a branched course, but in most cases the river channels are fairly straight. These rivers are branched because the river channel is not very stable, and because sediment is deposited in the middle of the channel, forming small islands or bars. Braided streams typically develop in coarse sediments containing little clay or silt which can form stable levées. A higher stream gradient gives greater energy and greater erosion of the sides of the channel. An abundant supply of sediment leads to sediment being deposited more rapidly than it can be eroded and transported onwards, resulting in the formation of deposits in the channel. Braided streams are therefore also typical of areas where the velocity declines somewhat, e.g. when a river widens onto a plain after passing through a narrow valley. The velocity and flow will in most cases vary considerably with time. When the river is low it will flow round bars of sand and gravel, while at high water and with strong currents sand bars will migrate as they are eroded on the upstream side and accumulate deposits on the downstream side. Gravel bars, on the other hand, have rather complex patterns of erosion and deposition.

Meandering rivers have a wavelength which is a function of the breadth of the river. The wavelength is approximately 11 times the breadth or 5 times the radius of the river. There is also a relatively regular ratio (about 7:1) between the breadth and depth of the river.

Meandering rivers move in loops, with the greatest velocity at the outer bank so that erosion is concentrated there. The velocity along the inner bank is much lower, so sediment is deposited there. The flow velocity and shear forces against the bottom also decrease upwards towards the top of the bank on the inner side of the bend, and sediments are deposited in *a point bar* which reflects the hydrodynamic conditions. At the lowest point there is a lag conglomerate followed by large-scale cross-bedding due to the flow being in the upper part of the lower flow regime. Then come current ripples, and finally fine-grained sand, silt and clay sediments corresponding to the lower part of the lower flow regime at the top of the profile (*overbank sediments*). This produces a fining-up sequence from sand to silt and clay (Fig. 2.28).

The secondary flow, in a vertical section at right angles to the downstream flow direction, moves from the outer bank where erosion takes place along the

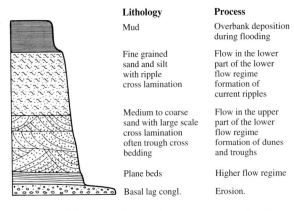

Lithology	Process
Mud	Overbank deposition during flooding
Fine grained sand and silt with ripple cross lamination	Flow in the lower part of the lower flow regime formation of current ripples
Medium to coarse sand with large scale cross lamination often trough cross bedding	Flow in the upper part of the lower flow regime formation of dunes and troughs
Plane beds	Higher flow regime
Basal lag congl.	Erosion.

Fig. 2.28 Fining-upwards point bar sequence deposited by a meandering river

bottom, and up the inner (point bar) bank where deposition takes place. This is a result of a difference in hydrostatic pressure because the surface of the water slopes inwards towards the inner bank due to centrifugal forces. If we combine this movement with the main flow of water down-river, we find a corkscrew or *helical* movement (Fig. 2.29). At each bend in the river the helical flow reverses direction. The point bar becomes asymmetrical, with coarser material on the upstream side, so that a perfect fining-upward profile is not developed. The upper fine-grained part and the overbank deposit will also be lacking. Variations in the depth of water in the channel will also lead to departures from the ideal fining-upward sequence.

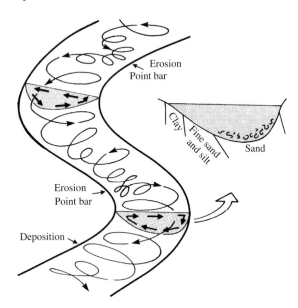

Fig. 2.29 Diagram showing how the flow in a vertical section reverses direction in each bend (helical flow)

During floods the water may flow over the point bar and form a little channel, or *chute*. This may be widened by further erosion to become the main channel. This process is called *chute cut-off*. The sinuosity may also become so high that erosion cuts a channel through a narrow neck (neck cut-off), making a straight course to a lower bend in the river. The whole meander will then be abandoned by the river, and a small, curved *oxbow lake* is left, which will fill with clay, silt and organic matter.

Rapid subsidence and low sand/mud ratios will increase the stability of the channels and we may find an anastomising channel distribution where there is little or no lateral acretion of the channel (Fig. 2.30).

When the whole river channel shifts course (avulsion) the abandoned channel will fill up with mud and form a clay plug.

2.28 Summary of Fluvial Sedimentation

Fluvial processes are fairly simple in principle. We know the physical laws which govern the flow of water in channels and transport of sediment in water. The great variation in composition, structure and geometry demonstrated by fluvial deposits is due to all the variables which influence transport and sedimentation. As we have seen, the most important factors are:

1. Climate, particularly precipitation and seasonal distribution of precipitation
   (a) in catchment areas
   (b) along river valleys – vegetation stabilises river banks.

2. The drainage area
   (a) size
   (b) type of rock being supplied
   (c) topography – tectonic uplift.

3. Subsidence of the alluvial plain.

In order for thick fluvial series to be deposited and preserved, the fluvial plain must be located in a tectonically subsiding area.

Catchments with a lot of shale and other fine-grained sedimentary rocks will produce sediments with a high clay and silt content. The products from weathering of eruptive rocks will contain a considerable

Fig. 2.30 Different types of fluvial channels. The type of fluvial channel determines the distribution and geometry of sand bodies and depends on the sediment composition and slope variations in the water flow

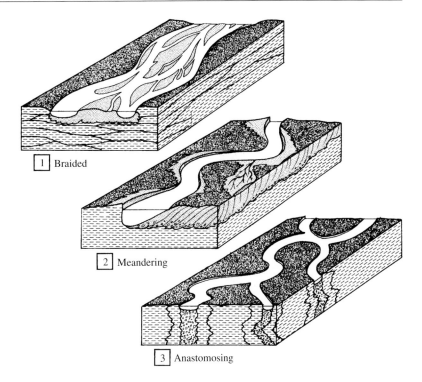

1 Braided

2 Meandering

3 Anastomosing

amount of clay, mainly kaolinite, illite and smectite, which increases the cohesiveness of the sediments and stabilises the fluvial channels. The climate along the river plain may be very different from that in the drainage area but some vegetation can be supported by the groundwater close to the river also in dry climates. In humid climates the vegetation will help to stabilise the river channel and reduce the velocity of the floodwater outside the channel.

The flow of rivers on the river plain will also depend on the groundwater table. The river will contribute water to the groundwater if it is lower than the surface of the river, and groundwater will flow into the river if the reverse is the case.

2.29 Delta Sedimentation

Large deltas require drainage areas with sufficient precipitation to produce high runoff. They are formed along passive plate margins where extensive drainage systems can unite into large rivers. The greater part of the drainage from North and South America and Africa flows into the Atlantic Ocean. Only minor rivers enter the Pacific Ocean from the American continent, and

relatively little of the drainage from Africa enters the Indian Ocean. The major rivers follow old drainage systems, the main features of which have existed since the Mesozoic, and which often seem to have been governed by Mesozoic rifting along the Atlantic continental margin.

As mentioned in connection with fluvial sediments, the erosion area will determine the composition and grain size of the sediments which are transported by rivers and deposited on the delta. The Mississippi, for example, drains extensive tracts of Palaeozoic and younger sediments which contain a large percentage of shale, and the river therefore transports a high percentage of clay and silt in suspension. Areas of metamorphic and acid eruptive rocks will give sandier sediments because of their quartz and feldspar contents. Deltas prograde outwards into a sedimentary basin and form a surface called a *delta top* near the lake or ocean level. The waves break against the *delta front*, beyond which is the *delta slope*.

The formation of a delta can be depicted as a battle between the fluvial development of the delta and its erosion by marine forces. We therefore distinguish first between river-dominated deltas, which prograde far out into the basin and consist largely of fluvial sediments, and deltas which are eroded more rapidly by

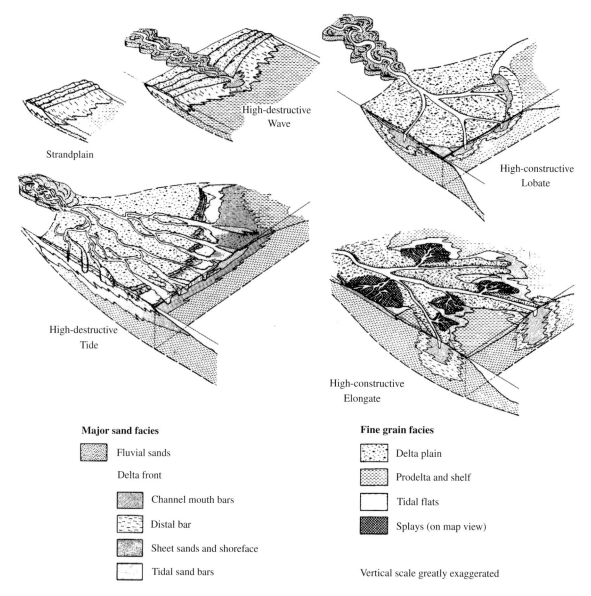

Major sand facies

| | Fluvial sands |

Delta front

	Channel mouth bars
	Distal bar
	Sheet sands and shoreface
	Tidal sand bars

Fine grain facies

	Delta plain
	Prodelta and shelf
	Tidal flats
	Splays (on map view)

Vertical scale greatly exaggerated

Fig. 2.31 Classification of delta types based on the relative strength of waves and tidal energy in relation to the rate of sediment input. After Fisher et al. 1974

marine forces (tide and waves) and consist largely of marine sediments (Fig. 2.31).

River-dominated deltas are often referred to as constructive deltas, since they tend to prograde more rapidly into the basin without much marine reworking. Tide- and wave-dominated deltas are often called destructive deltas.

We may therefore distinguish between three main types of deltas: (1) river-dominated, (2) tide-dominated and (3) wave-dominated. Most deltas fall somewhere between these three extremes. The Mississippi, however, comes near the river-dominated end in a marine environment because the sediments are fine-grained and rather cohesive. The delta front is also protected from big ocean waves by a shallow shelf.

High rates of sediment supply also favour fluvially-dominated deltas. The Rhine is fairly wave-dominated as well, because the sediments are rather coarse-grained and the sedimentation rate is lower. The Ganges and Mekong deltas are typical tidal deltas.

The difference in density between the river water and the water in the marine basin plays a major role. In most cases river water will be lighter than saltwater, even if it contains a good deal of suspended material. In deltas, river water will therefore flow far out over the salt water (*hypopycnal flow*) before salt- and fresh-water mingle. If the river water and the water in the basin have the same density (*homopycnal flow*) there will be more rapid mixing of the water masses in *axial flow*, and the sediments will settle out of suspension more rapidly. *Hyperpycnal flow*, where the river water is denser than the water in the basin, takes place only in lakes as a rule and leads to the flow of river water along the bottom of the delta slope, with erosion of the delta front and formation of turbidites on the basin floor.

In studies of modern deltas we must also take into account the fact that most deltas are rather out of balance as regards progradation in relation to sea level because the Holocene transgression after the last glaciation raised sea level by more than 100 m. In the Niger delta we find beyond the present delta front deposits which are 12–25,000 years old, which corresponds to the last advance of the last glaciation.

2.30 River-Dominated Deltas (Mississippi Type)

The Mississippi drains a huge area of the North American continent (about 3.2×10^6 km^2) with an average precipitation of 685 mm/year. It carries vast quantities of sediment (about 5×10^8 tonnes/year) with a high clay and silt content, and the gradient of the lower part of the river is extremely low (about 5 cm/km).

In the 200–300 years during which bathymetric measurements have been taken in the area, it has been possible to record considerable progradation. The high clay and silt content gives the sediments great cohesion, making the fluvial channels relatively stable. Clay and silt sediments which are deposited on the delta plain have a high porosity and water content (60–70%), but they loose much of their porosity by compaction at shallow depth. Sand will be deposited chiefly in the channels, and in mouth bars were the channels enter the sea. Well-sorted sandy sediments, which have only about 40–45% porosity immediately after deposition,

will then sink into the underlying clay because of their higher density. The fluvial channels therefore sink into the mud and this contributes to their stabilisation, with the consequence that channels change course (*avulsion*) less frequently. Because of this subsidence while sedimentation is continuing, the channel sand may be thicker than the depth of the channel. The long strings of sand which may then be preserved in the mud-rich environment are called *bar-finger sands*.

During floods large quantities of silt and clay are deposited in overbank areas between channels, and these sediments are stabilised by vegetation. In the Mississippi delta, sedimentation is very rapid so that the channel and the levees build up above their surroundings. This means that the average gradient of the channel decreases. Sooner or later the channel will have to find a new, shorter route (through avulsion) to the ocean, and which therefore has a somewhat greater slope. At any given time most of the sediments are deposited in one delta lobe prograding into the ocean until the slope becomes to low. A major avulsion will then start the formation of a new delta lobe.

We can see that the Mississippi has constantly shifted course, and has consequently been a focus of sedimentary activity in historic as well as modern times (Fig. 2.32). The present course has extended the modern delta far out and should have been abandoned for a shorter course towards the southwest, to the Atchafalaya basin. However the flow in this direction has been artificially limited because of its importance for transport to and from towns like New Orleans which lie on the present channel.

When a delta lobe is abandoned, it slowly subsides because of compaction and tectonic subsidence, while sedimentation takes place elsewhere. The abandoned lobe may sink below sea level before the fluvial supply returns to this part of the delta and another delta lobe is deposited in the same area, allowing intervening deposition of thin marine beds. Thin layers of carbonate or shale represent periods of local transgression (*abandonment facies*) which are time equivalent with regressions in the prograding delta lobes. Between the delta lobes in the interdistributary bay facies wave energy is very low and there may be little or no beach (sand) deposits between the marine mud and mud deposited above sea level.

In vertical profile we observe alternations of fluvial channel sediments: levee deposits, crevasse splays and possibly marine sediments (Figs. 2.33 and 2.34).

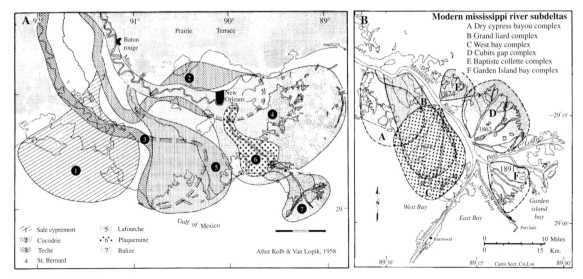

Fig. 2.32 Sedimentation in the modern Mississippi delta. When the river water breaks through the levees, crevasse channels and splays are formed, which help to fill the areas between the channels (Coleman and Prior 1980). (**a**) Delta lobes which show how the sedimentation has changed during the past 7,000 years. Each lobe of the delta appears to be active for 1,000–1,500 years (Coleman and Prior 1980). (**b**) Sedimentation in the last few 100 years

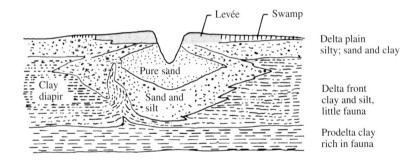

Fig. 2.33 Fluvial channels in a clay-rich delta environment. The channel sand is denser than the clay and will sink into the underlying mud. Clay diapirs may form as a result

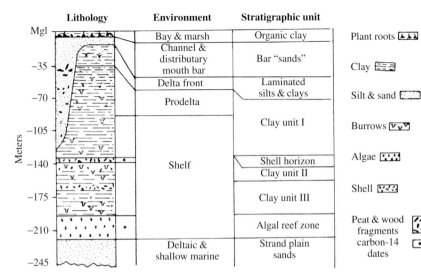

Fig. 2.34 Vertical section through modern Mississippi delta sediments. From Coleman and Prior (1980)

Even small variations in sea level have a strong influence on delta sedimentation. In a section where a fluvial facies gives way to a marine bed, it is often difficult to know whether this represents a transgression due to a rise in sea level or whether this part of the delta was abandoned by the fluvial system and is subsiding. Only if we can correlate a transgressive bed over the whole area can we assume that it is due to changes in sea level. Transgressive carbonate or thin sandstone beds are the most useful for regional correlation.

2.31 Delta Front Sedimentation

At the channel mouth the fluvial water flow rapidly loses its energy, and all the sand which has been transported along the bottom (the *bed load*) is deposited in the form of a *channel mouth bar* (Fig. 2.34). Most of the suspended material is also deposited fairly rapidly, at a relatively short distance from the mouth. While the river water, which is lightest, flows out of the channel, salt and brackish water flow back in along the bottom, and may encroach a long way up the channel when the rate of fluvial flow is low. Marine fossils can therefore be transported quite a way into the fluvial environment. During floods the saltwater wedge is forced back over the bank at the mouth. The wedge reduces the cross-section of the river, and the velocity increases somewhat in consequence (Fig. 2.35). Spring tides will force the river water up the channel and may cause it to overflow its banks.

The channel mouth bar itself is a very characteristic deposit which grades upwards from delta mud to well-sorted sand on top. Because of brackish water and the high rate of sedimentation, there are few organisms which live right at the mouth of the channel, but bioturbation may occur in the surrounding sediments.

The delta front is the area where fluvial and marine forces meet, and we can virtually quantify the marine influence on the delta. Tidal forces are a function of the tidal range, and from wave measurements we can also estimate the *wave power*. While wave power near the beach is estimated to be 0.034×10^7 erg/s on average for the Mississippi delta, it is 10×10^7 erg/s for the Nile, and 20.6×10^7 erg/s for the Magdalene River (Wright 1978). In rivers like the Mississippi, which has a shallow profile beyond the delta front, most of the wave power dissipates before it reaches the delta front. Deltas which develop into deep water right by the continental slope are exposed to the greatest wave power, because waves are not damped before they reach the delta front. The River Magdalene in

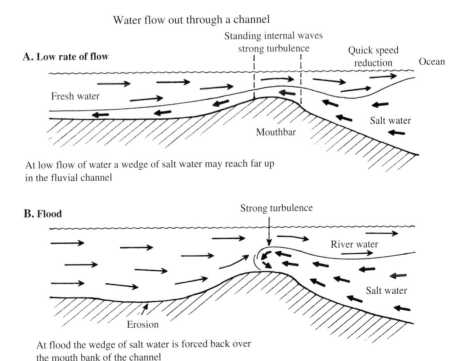

Fig. 2.35 Flow of fresh and salt water at the mouth of a river channel (after Wright and Coleman 1974)

Venezuela is a very typical example. The thickness and the lateral extent of the delta-front sand is a measure of wave power. Powerful storms may, however, erode vast quantities of sediment, far more than evenly distributed wave power would have achieved.

Wave-dominated delta front facies have a typical coarsening-upward sequence: from pro-delta clay to increasingly sandy sediments and finally well-sorted cross-bedded sediments with a low angle of dip, i.e. a *beach facies* profile. We often find traces of plant roots at the top of such sequences, and this shows that the sand bank has had vegetation right down to the shoreline. It was probably protected by shoreline barriers which absorb most of the wave energy. Vegetation can offer protection against both fluvial and tidal erosion, and mangrove swamps such as those in the Niger delta are particularly effective. There will always be some erosion on a delta front. Whether the delta progrades or is broken down by marine forces depends on the supply of sediment. The delta front will be fed with sediment – particularly sand which migrates from the channel mouth bar – along the beach in the wave zone. We often call this "strike feeding" because sediment transport is parallel with the strike of the shoreline, i.e. along a horizontal line. This is in contrast to the transport in channels, which is parallel with the dip of the deposit.

If there is little erosion of the channel mouth bar, the channels will extend far into the sea, and little sediment, least of all sand, will be supplied to the rest of the delta front.

2.32 Stability in a Delta

Sediments which are deposited in a delta possess very high porosity, and clay- and silt-sized grains form a very unstable structure after deposition. Clay and silt have a low permeability, however, and will expel water only very slowly. If sedimentation is rapid the sediment load will increase faster than the water can flow out, and overpressure will develop in the porewater. This means that more of the overburden is carried by the porewater, reducing the effective stresses between the sediment grains and as a result also the compaction. The friction between grains, which is a function of the effective stresses, is greatly diminished, and in consequence so is the shear strength of the sediments.

If the pore pressure attains the pressure exerted by the overlying sediments, the effective intergranular stresses will be equal to zero. There is then no friction between the grains, and the sediments can flow like a liquid *(liquefaction)*. The resulting instability may cause diapirs of mud to be squeezed up into the sand bed. The Mississippi delta is characterised by rapid sedimentation, and we find diapirs of overpressurised clays. Clay diapirs rise like salt diapirs because they are less dense than the more compact clay, silt or sand, which have lower water contents.

The stability of sediments also depends on the chemical composition of the porewater. In freshwater, clay mineral particles with negatively charged surfaces will repel one another. In saltwater these surface charges will be neutralised by cations (Na^+, K^+ etc.) so that clay minerals flocculate.

Deltas which prograde out into deep water will develop a slope which may vary greatly. The force of gravity acts on the sediments in the delta front and on the slope so that shear stresses develop in the sediments. When these stresses exceed the shear strength of the sediments, the sediments will be deformed by some sort of gravity-governed process. This may take place through sliding and slumping which in turn may generate turbidity currents, or steep fault planes may develop, i.e. *growth faults* (Fig. 2.36). The name was introduced during early oil exploration and refers to the fact that beds would thicken (grow) on the downfaulted side. The growth fault plane gradually deflects and flattens out with depth.

The growth fault may form near the delta front and lead to this part of the delta subsiding rapidly, resulting in very thick deposits with delta front facies stacked on top of one another (Fig. 2.37).

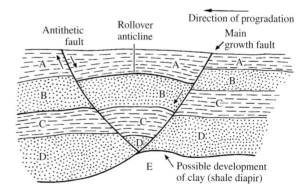

Fig. 2.36 Main features of growth faults. The rollover anticlines may be traps for oil and gas

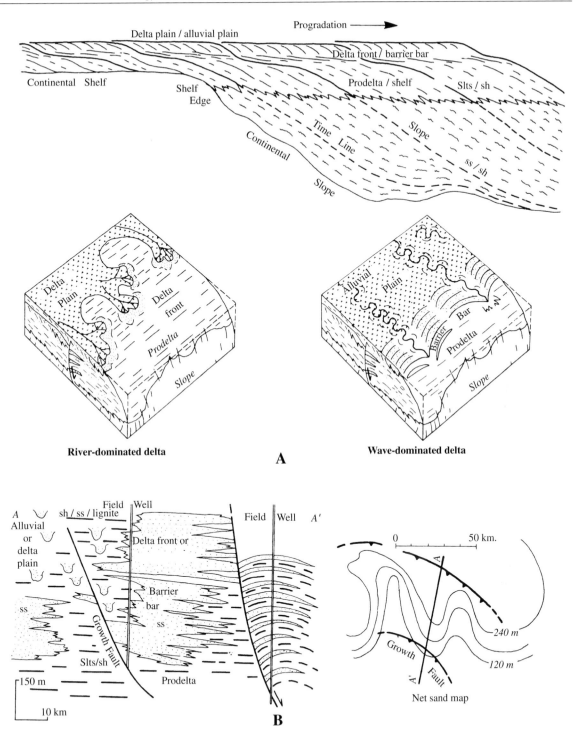

Fig. 2.37 Section through a prograding delta system showing potential traps for oil and gas (Brown and Fisher 1977)

2.33 Tide-Dominated Deltas

A large tidal range affects delta sedimentation very strongly and in a number of ways. The lower part of a major river has a very low gradient, so when the sea level rises and falls by 6–8 m it causes great fluctuations in the gradient, with the tidal range often affecting river flow as much as 50–100 km upstream. During flood tide the river surface gradient is reversed and a great deal of sand (bedload) and biological material will be transported up-river. The difference between ebb and flood decreases gradually up the river, and the periods of rising water (when the tide comes in) will be shorter, right down to 2–3 h. This must be compensated by higher flow velocities. Consequently it is often only the flow up the river channel during high tide which gives rise to shear forces against the bottom strong enough to transport a bedload. Sand deposited in such rivers has structures *(bedforms)* which indicate transport up the river. It is very important to remember this when studying older rocks.

Rivers in areas with high tidal ranges broaden markedly as they approach the sea because of the ever greater volumes of water to be transported out and in. At the outermost point a broad estuary forms, with clay and sand banks cut by channels which transport tidal water. The Thames and the Rhine are examples of modern rivers with well-developed tidal estuaries.

The mixing of fresh and marine water also causes flocculation of mud and deposition in the more protected parts of the estuary between sand bars. Estuarine deposits are therefore characterised by a mixture of sandstone and mudstone, often with marine fossils which may be transported several kilometres upstream by the salt wedges along the bottom. Tidal flats are often found in the inner parts of the estuary. Tidal channels are oriented at a high angle to the coastline, but in the outer part waves may produce elongated bars which are oriented parallel to the coastline. The geometry and orientation of the sandbars are very different from those associated with wave-dominated deltas and coastlines, where the sand bars are parallel to the coast.

Tide-dominated deltas are characterised by the fact that the erosion due to tidal currents is very strong compared to erosion by waves. The fluvial channels have cross-sections which are too small to allow transport of water out and in, and the result is the development of a broad belt of tidal channels, which farthest out are separated only by narrow ridges of sand and clay. Very often separate ebb and flow channels

develop, and in each channel one transport direction will dominate. Signs of transportation in two opposite directions have often been used as a criterion for tidal deposits. This will apply to the whole area, but we cannot always expect to find cross-bedding with opposite current directions in the same channel deposits.

2.34 Wave-Dominated Deltas

Wave-dominated deltas have a vertical sequence very like that of a beach. Little of the fluvial part of the delta is preserved, and instead of distinct deposits at each channel mouth (distributary mouth bars), the waves distribute the sediments in a continuous *beach ridge* along the delta. The result is a coarsening-upward profile which is not always easy to distinguish from a beach deposit outside a delta. In a wave-dominated delta, however, beach ridges form right in front of the delta top deposits, which often contain coal beds from vegetation, whereas beach profiles formed by barrier islands have a lagoon behind them. The first unambiguous proof that one is dealing with a delta, however, is if the beach ridges are found to be cut by fluvial channels. In addition to the Rhone and the Niger deltas, the Nile delta is also a good example of a wave-dominated delta Fig. 2.38. Here sediment supply to the delta has been greatly reduced through the building of the Aswan Dam, which traps sediments, and the balance between sediment supply and wave erosion has been disturbed so that the delta is now retreating. This is a modern example of how progradation or destruction (transgression) of a delta depends on a very sensitive balance between sediment supply and erosion.

2.35 Coastal Sedimentation Outside Deltas

Most sediment transport to the sea takes place via rivers which drain into deltas. In fluvially-dominated deltas, the greater part of the sediment is deposited there.

On destructive deltas, a great deal of sediment is eroded and transported out into the sea or along the coast. Coastal areas between deltas are very largely supplied with sediment which has been eroded from deltas and transported along the coast by waves and coastal currents. Since the surface of the coast slopes

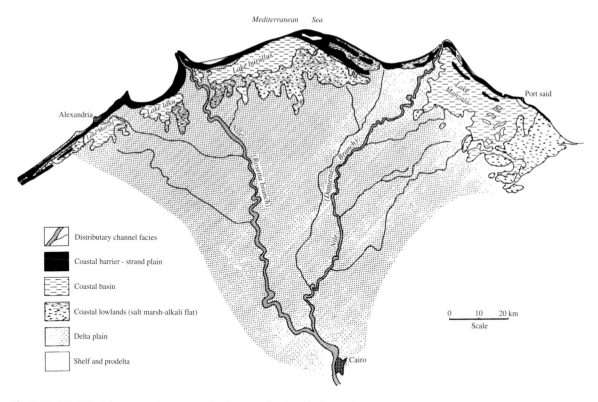

Fig. 2.38 The Nile delta system is an example of a wave-dominated, destructive delta system. (Fisher et al. 1974)

outwards, the direction parallel with the shoreline can be defined as the *strike* of the deposit, and the direction normal to the coast is called the *dip*.

Sediment supply parallel with the coast is often called "strike feeding". If the supply of sediment along the coast is greater than the erosion rate, the coast will build out into the sea and we will get regression through coastal progradation. If sediment supply is less than the rate of erosion, the coast will retreat, and we will get transgression.

Because there is a finite amount of sediment being transported along any stretch of the coast, engineering constructions which are sometimes built to accumulate sediment and prevent erosion at a particular spot (groynes) will as a rule lead to increased erosion along other parts of the coast. This has great practical significance in areas with high coastal erosion. Beach sedimentation must therefore be seen, not in isolation, but in the context of the overall sediment budget along the coast.

The Mississippi delta is the greatest source of sediment to the Gulf of Mexico, although there are a number of minor rivers in Texas. Here sediment transport takes place along the coast from east to west. Because the active part of the Mississippi delta lies as far east as it does today, less sediment arrives at the Texas coast. The fact that this coastline is subsiding at the same time leads to transgression of the coastline. This example shows that if the supply of sediment due to strike feeding cannot match subsidence, the result will be a transgression.

2.36 The Shore Zone

The shore zone is where land and sea meet, and we can distinguish between different types:

1. *Rocky Beach.* The beach is covered with pebbles and blocks or solid rock representing ancient, resistant bedrock. Much of the coast of Norway is of this type, but only 2% of the coastline of North America. Rocky beaches form in areas which have been uplifted tectonically, and where the coast erodes metamorphic or eruptive rocks.

2. *Sand or Gravel Beach.* This is the most common form of beach zone where we have active sedimentation or erosion of older sediments (makes up 33% of the coastline of North America).

3. *Barrier Island.* This is a beach which is separated from the main coastline by a lagoon. Common in North America (22%), but less common in other areas.

4. *Muddy Coastlines.* Beach zones which consist basically of clay are formed where we have sediments with a very low sand content, and where there is little wave activity to wash out or enrich any sand the sediments might contain. We find this in parts of clay-rich deltas and estuaries and where there is abundant vegetation along the beach zone which protects the clay sediments against erosion (e.g. mangrove forest).

5. *Cheniers* are isolated sandy beach ridges on coastal mud flats. They are abundant along the coastline around major deltas like the Mississippi and the Amazon. They require low tidal ranges, moderate wave energy and abundant mud, and a limited amount of sand.

The composition of beach sediments varies greatly, depending on the materials available and the grade of mechanical and chemical breakdown of the minerals. Quartz sand is the most widespread because quartz is the most stable of the minerals we find in most beach sediments. But on volcanic islands, which consist largely of basalt, there is no quartz, and we get sand derived from basalt or volcanic glass. Carbonate sand is formed locally from the broken skeletons of carbonate-secreting organisms. This does not only happen in tropical regions, e.g. the Bahamas, but also along coasts with cold climates. In many parts of Norway, for example, the beach sand consists very largely of carbonate sand from molluscs, barnacles, bryozoans and calcareous red algae.

Beach profiles are formed by wave power acting on the coast and depend on a number of different factors:

1. The composition of the available sediments.
2. Wave power

    a. Average wave height
    b. Size and frequency of storms
    c. Angle between the commonest orientation of the waves and the beach.

3. Tidal range.
4. Vertical profile off the beach.
5. Supply of sediment from land or along the coast.

The morphology of the beach zone and nearshore areas is the result of interaction of various features related to wave activity. As waves approach land, they will be affected by friction against the bottom. The depth at which this starts depends on wave height and length. When the depth becomes less than about half the wave length, friction against the bottom will be great, the circular wave motion (orbit) will be distorted and oscillatory sediment transport will affect the seabed, producing ripple marks. The depth at which this occurs is called the *wave base*. Dunes are formed of sediments which are deposited when the waves break, and are at the same time the reason for the waves breaking precisely there. There is thus an interaction between the wave regime and the bottom geometry in beach sediments. In addition to breaking on the *foreshore* itself, we often find that waves break at two or three places offshore, and at each of these places we find a sand bar.

In the French Riviera (e.g. Nice) the beaches are often full of rounded pebbles with less sand. This is because they are near mountains supplying coarse clasts and also close to a shelf edge. Sand is then easily eroded from the beaches during storms while the coarse clasts remain.

2.37 Prograding Beach and Barrier Sequences

The progradation of a beach will produce a characteristic coarsening-up sequence (Fig. 2.39) which will obey Walther's Law of facies succession. The vertical sequence will represent the environments from the shelf to the shoreline. The transition between shelf mud and fine-grained sandstones with ripples may represent the wave base. Isolated sand layers in mud may have been deposited near the *storm wave base*, whereas the transition to continuous fine-grained sand may represent the *fair-weather wave base*. The thickness of the sequence from the fair-weather wave base to the foreshore is an expression of the wave energy at the coastline when the sequence was deposited.

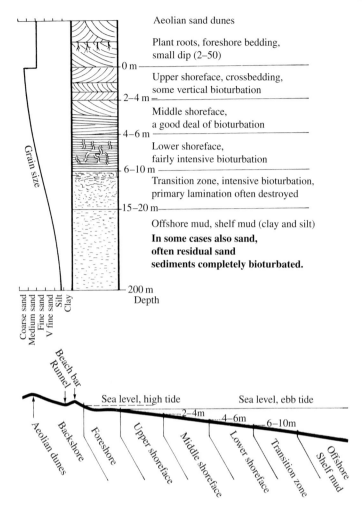

Fig. 2.39 Diagrammatic section through a sand deposit with some typical sedimentary structures. Just under normal fair-weather wave base, 5–20 m, "hummocky" stratification occurs

Bioturbation occurs in the lower part of this sequence. Below the wave base it usually takes the form of horizontal feeding traces, and in the lower and middle shorefaces, where there is relatively high wave energy, as vertical traces (*Skolithos* facies). As erosion and reworking intensify, the preservation potential of bioturbation is reduced and it becomes less frequent. The formation of sand bars and erosion surfaces on the upper shoreface results in cross-bedding, usually trough cross-bedding, representing flow in the upper part of the lower flow regime.

In the breaker zone there is an upper flow regime, producing a planar facies which in vertical section will appear as very low-angle cross-bedding. On the beach we often have a beach bar which is flooded only during storms, and a depression behind it called a runnel, which helps to drain the backshore area.

Because the exposed beach is a rich source of sand, the wind will tend to blow sand from the beach and redeposit it as aeolian dunes, usually where it is trapped by vegetation. Aeolian sediments therefore often cap ancient beach profiles, but the aeolian dunes may also be eroded and not be preserved in the geological record.

2.38 Barrier Islands

Barrier islands are beach deposits which are separated from the mainland by a lagoon. They form long, thin sand ridges which are often only a few hundred metres to a couple of kilometres broad, and which rise up to 5–10 m above sea level.

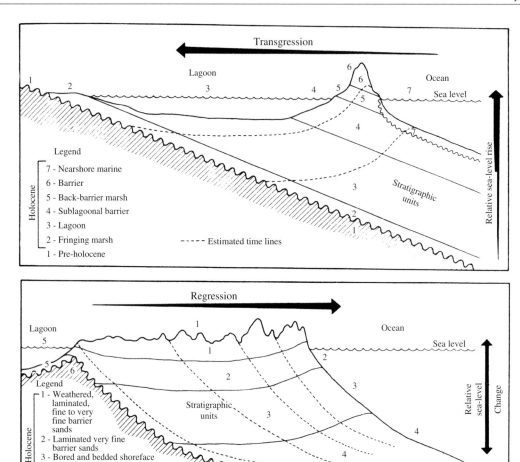

Fig. 2.40 Transgressive and regressive barrier systems. (from Kraf and John 1979)

A vertical section through the part of a barrier island facing the sea resembles an ordinary beach deposit (Fig. 2.40). We find a coarsening-upwards sequence from marine clay to beach sand, often with vegetation on top. On the lee side, facing the lagoon, there is little wave power and clay, mud and often oyster reefs are deposited in the lagoon. Barrier islands are very well developed along long stretches of the coast of North America, particularly off Texas, Georgia and North Carolina. There are several indications that barrier islands are due to transgressive conditions such as those of Holocene times. This is certainly the case along the eastern coast of the southern North Sea. When the ocean rises in relation to the land, beach deposits can continue to grow through gradual deposition of sand in the beach zone, but the areas behind them sink below sea level and form a protected lagoon with clay sedimentation. More localised transgressions may be caused by compaction and subsidence of sediments along a coastline where the sediment supply is insufficient to keep pace with the subsidence.

One prerequisite for forming barrier islands is an adequate supply of sand, so that the island can grow and keep pace with the transgression. This sand cannot be transported across the lagoon, and must be added along the length of the islands parallel to the coast (strike-feeding from deltas or eroding coastlines).

During storms or hurricanes the sea level may rise due to wind stress, and waves may break over and through the barrier island. An erosion channel is then formed through the island, and a *washover fan delta* develops at the rear, out into the lagoon. A delta of this type can form in a matter of hours during a hurricane. Barrier islands may extend for tens of kilometres, but they will not form a continuous belt along the coast. Water has to circulate between the lagoons and the ocean through gaps between barrier islands. The gaps are called *tidal inlets*, and the distance between them will be a function of the tidal range. Both seaward and landward of the inlets small sandy deltas may develop in response to ebb and flood currents respectively (Fig. 2.41). Flow at ebb tide will normally be stronger than that at flood tide. This is due to the profile of the lagoons. At high water in the lagoon the volume of water which must flow out to compensate for a specific lowering of sea level is greater than the volume needed to raise the water level in the lagoon correspondingly at low tide when the area of the lagoon is smaller. The tidal inlets are therefore capable of transporting more sediment out during the ebb, and structures indicating this flow direction may predominate (Fig. 2.41). Inlets are characterised by an erosional base, and lateral migration of inlets produces a characteristic fining-upwards sequence. The strong currents in tidal inlets often generate sand waves which tend to migrate in the ebb direction, but they may also be modified by flood currents.

Ebb-tidal deltas consist of a channel dominated by ebb currents with smaller flood-tide channels on the sides. At the ocean end of the channel sediment is deposited in a sand ridge which is similar to a channel mouth bar in an ordinary delta. This sand ridge, which is called *a terminal lobe, is* subject to wave erosion, and smaller *swash bars* may form, which reach above sea level. In areas with strong wave power, ebb-tidal deltas will be less obvious because of erosion and further transport along the barrier ridges. Ebb-tidal deltas will be characterised by greater water depths than flood tidal deltas.

Flood tidal deltas form inside the lagoon and are well protected from wave erosion. Here the water flows into flow channels which branch inwards in a flood-tidal delta, where the sediments are deposited on a tidal flat. Ebb currents move back along the edges of the outer side of this delta and may form small *spillover lobes* when ebb-currents penetrate over the edge of the flood-tidal delta. Flood-tidal deltas are associated with shallower channels than ebb-tidal deltas and are not eroded very much by waves.

Tidal channels fill with sand which forms an upward-fining sequence overlain by tidal flat sediments (Fig. 2.42). In areas with carbonate sediments or cohesive clays, erosion due to lateral migration of tidal channels results in intraformational breccias.

Barrier island deposits thus consist of a long, thin body of sand. The thickness of the sand layer will correspond to the depth of the wave base plus a few

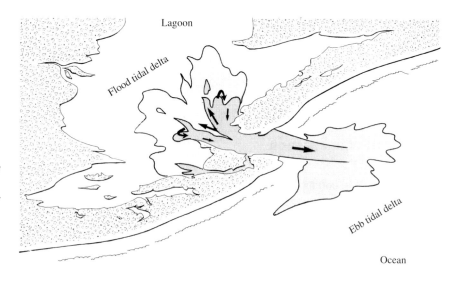

Fig. 2.41 Tidal channels with tidal deltas forming between barrier islands. The barrier islands and the channels will migrate laterally and deposit channel facies sediments by lateral accretion. Note that the ebb-tidal delta outside the barrier is much more exposed to waves than the flood-tidal delta in the lagoon

Lagoon

Flood tidal delta

Ebb tidal delta

Ocean

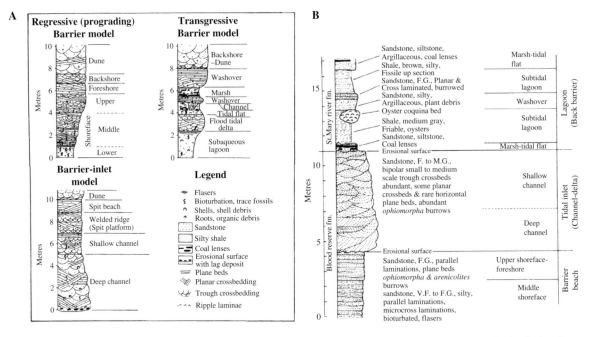

Fig. 2.42 (**a**) Interpretation of a vertical section in a tidal environment (From Walker 1979). (**b**) Section through a barrier beach cut by a tidal inlet channel into the lagoonal facies

metres which correspond to the height it builds up to above sea level.

In areas with a larger tidal range, this lateral migration will be rather pronounced, and fining-upwards sequences will also be common.

If the barrier islands are drowned by a transgression, a carpet of clay and silt will be deposited over these sandstone deposits. This represents the ideal stratigraphic trap for oil and gas. Compaction or tectonic tilting will cause the sandstone deposits to interfinger with mud from the lagoon deposits, which are a good source rock. Oil will be able to collect in the top of the barrier ridge sand or in flood-tidal delta deposits (or *washover fans)* which represent *pinch-outs* in the muddy lagoon sediments.

2.39 Tidal Sedimentation

Tidal range is an important factor in coastal sedimentation. We distinguish between:

1. Microtidal environment (tidal range less than 2 m).
2. Mesotidal environment (tidal range 2–4 m).
3. Macrotidal environment (tidal range greater than 4 m).

The average tidal range in the open sea is only about 50 cm. Along the coasts, however, we often get increased interference by tidal currents. This is particularly true around large islands where tidal waves converge on the lee side and can build up, and also in bays along the coasts. In long, narrow bays we may get a high degree of *resonance*. This occurs if the bay has a length and depth which cause tidal waves which are on the rebound to reinforce the next incoming tidal wave. The highest tidal range which has been measured is 16.3 m in the Bay of Fundy in Canada. In some areas around the British Isles the tidal range may be up to about 12 m, and there are also large ranges in the German Bight and adjacent parts of the North Sea. Tidal waves also move anticlockwise in N hemisphere around centres with zero tidal range (amphidromic points), with the tidal range increasing radially with the distance from the centre. This is typical of the tidal pattern in the North Sea.

The width of the continental shelf plays a major role in determining tidal ranges. When tidal waves enter shallow water, their velocity is reduced due to friction against the bottom. When the water depth and the velocity decline, the height of the tidal wave will increase, so that the total energy flux is maintained. The tidal range, which is thus the height of the tidal wave, therefore increases inwards across the shelf. Where there are embayments along the coast, tidal waves become focused, so that the tidal range increases towards the middle of the bay. This is true of the east coast of the USA, off Georgia, and of the German Bight in Europe.

If the shelf becomes even wider, as in an epicontinental sea, tidal power will gradually be exhausted due to the high frictional resistance. This is the case on the present Siberian continental shelf where the tide is very low (<10–20 cm). Wind stress may set up rather large storm tides. When the water is very shallow, as on the Bahamas Bank, friction damping takes place over much shorter distances.

The great epicontinental seas, in Ordovician and Cretaceous times for example, were also characterised by low tidal ranges.

About 1/3 of the world's coasts have tidal ranges greater than 4 m (macrotides), 1/3 have mesotides (2–4 m) and 1/3 microtides (less than 2 m). Small inland seas, like the Mediterranean, the Black Sea and the Baltic Sea, are too small to keep pace with the attraction of the moon and the sun, and have small tidal ranges. This is also true of lakes.

1. *Tidal Channels*. The infills may resemble fluvial channels or submarine channels in that they form fining-upwards sequences. Channels formed in estuaries in fact are often connected to fluvial channel systems. Tidal channels which are not part of a river delta, however, tend to be filled with sandy sediments from the surrounding tidal flat, because they have no supply from land. Channels will often erode their banks and cause them to collapse, and this may result in the formation of intraformational conglomerates if the sediments are slightly lithified.

Lateral migration of tidal channels may produce typical epsilon cross-bedding which is the result of lateral accretion of point bars in the tidal channel.

Tidal channels often contain a bed of marine fossils at the base, and marine trace fossils. Channels on tidal flats which are not associated with deltas (i.e. those in estuaries) will not receive much clastic material from land. Conglomerates and breccias in these tidal channels will therefore typically be of the intraformational type, derived by local reworking of tidal flat sediment. Because of their

early lithification, carbonate beds in particular can be reworked to form intra-formational conglomerates and breccias.

Tidal cycle duration is about 12 h and 25 min, with currents switching direction every 6 h, and we sometimes find good examples of cross-bedding with opposite current directions. This is not always the case in tidal environments, however. Some tidal channels are dominated by ebb flow and others by flood currents. This is because the ebb and flood often find different dominant pathways. Bipolar cross-bedding is therefore not an essential feature of tidal channels. Storms with opposing wind directions may also produce some form of bipolar cross-bedding.

In a regressive sequence tidal channels will be overlain by lagoonal sediments (Fig. 2.42).

2. *Flaser Bedding*. Consists of clay laminae in a matrix of sandstone with ripple cross-lamination. The clay occurs mainly as infill in ripple troughs, but may also drape over the ripple crests as well. Flaser bedding forms as a result of alternating periods of currents or wave activity, and slack water. The clay settles out in the slack water periods in a tidal environment, though this type of bedding may also form in other environments where there is rhythmic sedimentation, such as in certain fluvial environments. On tidal flats, clay and silt will settle out at high tide to be deposited between ripples formed by the ebb and flood currents. Here the fine sediment may consist partly of clay pellets (faecal pellets from marine organisms) which settle out faster than clay-sized particles. Flaser bedding belongs to a type of structure which we get with mixtures of sand and clay. *Lenticular bedding* represents isolated laminae or lenses of sand in mud.

The inner parts of a tidal shelf often have embayments consisting of very muddy sediments, usually with abundant bioturbation. Mollusc shells in the mud (Fig. 2.43) are often eroded and deposited as shell lag.

3. *Tidal Bundles*. The best identifying feature for tidal environments is regular lamination consisting of fine sand and mud, making up tidal couplets that each represent a tidal cycle. Both modern and ancient tidal sediments show a regular variation in the thickness of such couplets, reflecting the energy levels of spring and neap cycles. Regular laminations reflecting tidal cycles are often called tidal bundles (Fig. 2.44).

Fig. 2.43 (**a**) Tidal channel on a tidal flat in the muddy facies of the inner part of a tidal flat (near Wilhelmshaven, Germany). Erosion by tidal channels in this mud produces a lag of mollusc shells. (**b**) Mollusc in living position

Fig. 2.44 Tidal bundles. These are bundles of laminae which reflect the tidal cycles between spring tides. From the Late Precambrian Wonoka Formation, Patsy Spring, Flinders Ranges, Australia

2.40 Shallow Marine Shelves

The shelf extends from the nearshore environment to the shelf edge, where there is a rather abrupt increase in slope, usually at a depth of 200–500 m. The width of the shelf varies considerably, and may exceed 1,000 km. Continental shelves are generally very flat areas which may be cut by deeper channels transporting sediments across the shelf from nearshore or deltaic environments.

The study of sedimentation on modern continental shelves is complicated by the fact that sea level was more than 100 m lower only 10,000 years ago. This means that most shelf areas have not yet reached an equilibrium with respect to the modern environment. Sandy shelf deposits are much more difficult to core than muddy sediments and the commonly used gravity corer has to be replaced by vibro-core equipment.

Most shelf areas are below the wave base for normal waves (fair-weather wave base) and sedimentation is governed largely by tidal currents and storms.

When the wind is landward, waves will usually approach the beach obliquely, resulting in wave refraction effects, particularly on relatively steep beaches with high wave energy. This will produce *rip currents*, where the water piling up against the beach has its return flow seawards. Rip currents are often rich in suspended material and transport material from the beach towards the shelf. Another component of wave energy is transmitted parallel with the beach as longshore currents, contributing to the longshore drift of sediment transport.

During onshore storms the sea level near the coast may be raised by several metres due to the combined effect of wind stress, tides and the low barometric pressure associated with storms. The increased potential (elevation) of the coastal water due to these *storm surges* will result in strong bottom currents which are capable of transporting sediment further out onto the shelf. Storm surges may transport fine sand and mud in suspension, but not as true turbidity currents. In this case the increased potential of the coastal water is the driving force and not the density difference between the current and the surrounding water, as with turbidites. Hummocky cross-bedding is a characteristic sedimentary structure produced by the deposition of coarse particles from suspension during storms, and rounded, undulating sand surfaces tend to form. Finer particles will be transported further, into areas with lower energy.

In many modern shelf areas tidal currents are important transport mechanisms, in combination with rare strong storm currents. Sediments deposited in this environment are characterised by rather abrupt transitions from well-sorted sand to mud and from bioturbated to non-bioturbated strata.

The main characteristic of tidal shelves is the mobility of the clastic sediments, on scales ranging from the diurnal tidal cycle to annual (storm augmented) cycles and gradual long-term movement of the largest bedforms. In shelf areas with relatively strong tidal currents (> 150 cm/s) we may get *furrows* and *gravel waves*. At velocities of less than 1 m/s *sand ribbons* may be deposited – longitudinal bedforms developed parallel to the currents. Sandwaves are large-scale transverse bedforms, generally 2–15 m high, with a wavelength of 150–500 m. Sandwaves require current velocities exceeding 60–70 cm/s. Cross-stratification may be symmetrical or asymmetrical on both sides of the sandwave, depending on the relative strength of the opposing currents.

Relatively low-energy environments (< 50 cm/s) are characterised by *sand patches* and mud. The sand forming the patches probably only moves during storms.

Shelf sediments characteristically accumulate at relatively low overall sedimentation rates (1–10 mm/1,000 years).

2.41 Continental Slopes

Continental slopes are the areas between the edge of the mostly very flat continental shelf which commonly lies at a depth of 200–500 m and the *continental rise*, where the ocean deep begins at a depth of 2–4,000 m. The continental slope gradient is typically 2–6°, and

it is 20–100 km broad. The gradient is a function of a number of different factors, but the stability of the shelf edge constitutes a major control factor. The steepest slopes are therefore to be found off carbonate banks with well-cemented coral reefs and carbonate beds which have high shear strength. In areas with rapid sedimentation, loose sediments have little shear strength and submarine slides, slumping and formation of turbidites occur, maintaining relatively gentle slopes (1–2°). Where sedimentation is slower the sediments have more time to consolidate, and will be more stable. The steepest submarine slopes in clastic sediments (greater than 10°) are therefore found in submarine canyons, where erosion cuts into older, well-consolidated sedimentary strata.

Along passive continental margins the continental slope is associated with the transition from continental crust to oceanic crust. In areas with a large supply of sediment, the shelf may have prograded beyond this boundary.

2.42 Organic Sedimentation on the Slope of the Continental Shelf

The continetal slopes are enriched in organic matter compared to the shelf and the deep ocean.

This is because the slope is where we have the greatest upwelling of nutrients from the deep. We also find low oxygen content in the water column on continental slopes, allowing much of the organic matter produced to be retained in the sediments. There will also be high productivity on shallower slopes in front of deltas because of the large nutrient supply from river water, but the organic matter may be greatly diluted by rapid clastic sedimentation. On the continental shelf the supply of nutrients is small and the prevalence of stronger currents and turbulence means that most of the organic production there will be oxidised.

Out in the open ocean basin organic production is relatively low, due to a limited supply of nutrients. Much of the planktonic organic matter is oxidised near the seafloor by deep currents of cold, oxygenated water from the polar areas.

Sediments deposited on the continental slope are therefore more promising as source rocks for oil than shelf and deep-water facies.

2.43 Sediment Transport on Submarine Slopes

Gravitational processes are naturally important on submarine slopes.

Gravity forces can be represented by forces (vectors) normal to, and parallel to, the slope. The component parallel to the slope consists of shear forces which may overcome the shear strength of the sediment, causing slumping. Sliding of large volumes of sediments downslope produces extensional faulting in the upper part of the slope and compression in the lower part. Gravitational instability on the slopes may also develop into debris flows and turbidity currents. Collapse of the sediment grain framework may cause sudden compaction and liquefaction of the slope sediments.

Traction currents may, however, also play a part on the submarine slopes, particularly in canyons but also near the toe of the slope, where we may have *contourites* – deposited by currents flowing parallel to the slope contours.

2.44 Submarine Canyons

These are valley-shaped depressions which extend from the top to the bottom of the slopes, down to 2,000–4,000 m. In some cases they may start in shallow water near the beach, in others close to the edge of the shelf. The height from the bottom of the canyon to the top of the slope on each side may be up to 2,000 m. We are dealing with enormous topographical features, which would have been very impressive indeed if they had been on land, towering structures on the scale of the Grand Canyon.

Shepard et al. (1979) systematically gathered data on currents and sediment transport in submarine canyons and found that tidal currents are of great importance *also* at great depths in submarine canyons. Current meters have shown that currents flow *both up and down* the submarine canyons, and that they switch every 6 h like tidal currents in shallow water. Current velocity is often only 10–20 cm/s, but in many canyons velocities of up to 40 cm/s occur sometimes, powerful enough to transport fine to medium-grained sand. The flow velocity tends to be greatest in the upper part of the canyon and diminish downvalley. Most sediment transport takes place during these episodic

and unusually high flow velocities which may be linked with storms which create wind-induced shear forces which sweep the water up against the coast *(storm tides)* and may cause currents to develop along the bottom and down the submarine canyons. However, high flow velocities have also been measured without it being possible to associate them with storms or wind stress. The most powerful flow velocity is most commonly directed downvalley, but upvalley-directed streams have been observed with velocities of up to 90 cm/s.

Detailed measurements in the submarine canyons off the coast of California reveal that the highest flow rates are oriented up the canyon in a way which seems to indicate that they are generated by internal waves from the ocean basin, and not by gravitational forces. Currents may then develop when the waves "break" against the coast or the continental shelf.

Powerful currents in submarine canyons are capable of transporting sand, sometimes in rare instances even coarser material. These are frequently not turbidity currents, but *traction currents*, which transport and deposit better-sorted material. Turbidity currents have been observed in submarine canyons as well, but definite (observed) examples were only low-velocity, low-density turbidity currents with a maximum velocity of 70–100 cm/s. In submarine canyons we thus have both traction currents, which are controlled partly by tidal forces, and turbidity currents. Downward-moving currents driven by tidal forces may, if they contain much suspended material, turn into turbidity currents. The downward-moving currents have both a component of traction and gravitation.

The relief of submarine canyons is due partly to erosion down into the underlying sediments, and partly to lack of deposition in the canyon while the adjacent beds were being deposited. During low sea level stands rivers may prograde closer to the shelf edge and hence supply more sediment to the submarine canyons, thus feeding submarine fans. At sea level highstands, currents in the submarine canyons and the shelf may erode shelf and slope sediments and deposit pure sand onlapping an erosional unconformity at the toe of the canyon.

Most of the canyon itself is an area of sediment transport and erosion. Deposition takes place where there is a change of slope near the basin floor. Here the channel defined by the canyons splits up into several channels which build depositional lobes called *suprafan lobes* (Fig. 2.45).

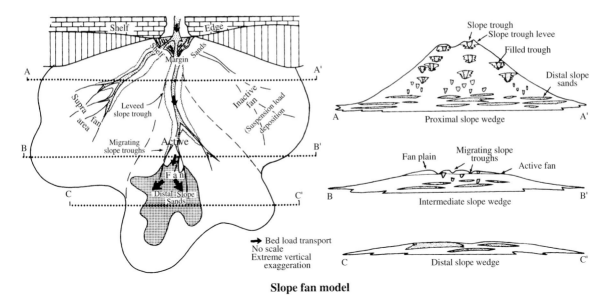

Slope fan model

Fig. 2.45 Model of submarine fans deposited at the foot of submarine slopes. Submarine fans have channels which sometimes meander. On the sides of the channels we find fine-grained levée deposits resembling those along fluvial channels. The pattern of shifting depositional lobes resembles that of deltas. (Brown and Fisher 1977)

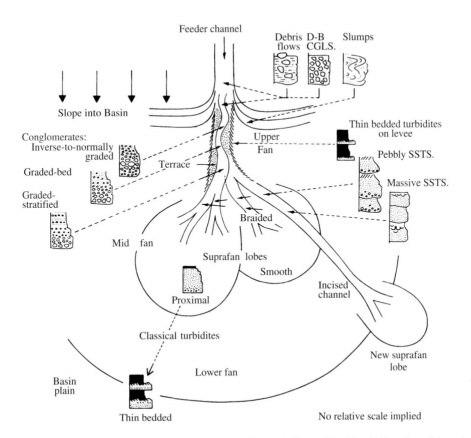

Fig. 2.46 Submarine fan model showing progradation and shifting of lobes similar to delta lobe shifting. One of the main differences between submarine fan and delta facies is the absence of wave reworking. (Walker 1984)

As the lobes build up, the gradient of the slope is reduced and a new channel will form in a part of the fan where there is a steeper slope (Fig. 2.46). This produces lobe-shifting similar to that observed in fluvially dominated deltas. Each lobe will tend to build a fining-upwards sequence, with conglomerate and coarse sand near the base. On the sides of the channels fine-grained material in suspension is deposited as thin-bedded turbidites. The levee builds up on both sides of the channel, and resembles a river levee. On the submarine Amazon delta slope there are well-developed meandering channels. The distal fan is also dominated by fine-grained sediments deposited as thin, graded fine sand, silt and clay. Progradation of submarine fans may also produce an upwards-coarsening sequence (Fig. 2.47).

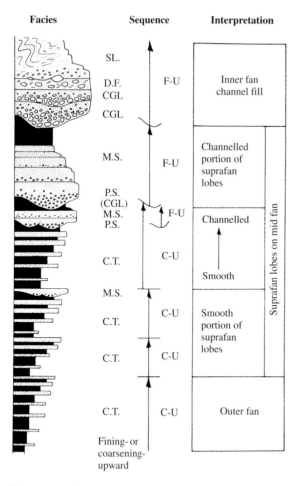

Fig. 2.47 Vertical sequence through a submarine fan (Walker 1984)

2.45 Sedimentation Along Continental Margins and in Epeiric Seas

As discussed above most of the sediment derived from land is deposited in river deltas and distributed relatively close to the shore line by longshore transport. There is, however, also significant sedimentation further away from the deltas and the coastlines. Turbidites and debris flows can in some cases transport sediments hundreds of kilometres offshore. Debris flows represent a very efficient type of sediment transport which can carry sediment far beyond the continental slopes because there is little internal shear deformation, only at the base and the top of the flow.

In fine-grained shales we may find evidence of ripples indicating traction currents. Storms and spring tides may produce rather high velocities (20–40 cm/s) on the seafloor far away from land and these currents are capable of transporting silt and fine sand. Clay should not produce ripples, but the clay is often composed of pellets of clay aggregates, often also with some organic matter. These clay particles behave like silt but they have lower densities and may be transported by traction currents forming small-scale ripple cross-lamination.

There may also be significant contributions of aeolian dust from deserts and of volcanic ash. In Palaeozoic times, before there were many land plants, aeolian dust was more important, but even today much of the sedimentation in the South Atlantic and Pacific Ocean is aeolian.

2.46 Sedimentation Along Island Arcs and Submarine Trenches

Submarine trenches form along converging plate boundaries where oceanic lithospheres are disappearing into a subduction zone. Along these converging plate boundaries sediment basins with very special deposition environments are formed. There are three main sources of sediment:

1. From the continent.
2. From island arcs, which may consist of continental crust, oceanic crust and/or volcanic rocks.
3. Pelagic sediment, including biogenic sediment and wind-blown volcanic ash.

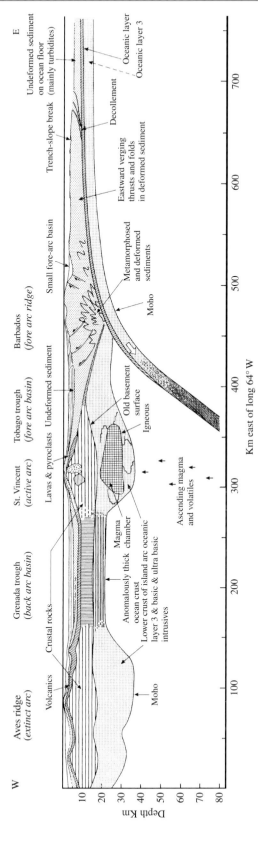

Fig. 2.48 Section through an area with converging plate boundaries in the Caribbean Sea. Note that sediments deposited on oceanic crust are scraped off the descending plate and imbricated in the accretionary prisms (see Fig. 2.49). From Leggett (1982)

These have quite different compositions. Sediment which is added from the continent is deposited in deltas and in turn fills the basin behind the arc *(back-arc basin)*. Sediments which are formed on the island arcs are rich in volcanic material, and this will characterise deposits in small basins on island arcs and *fore-arc basins*. Back-arc basins may also receive volcanic sediments from the island arcs. During the initial subduction phase fore-arc basins will tend to be characterised by turbidites deposited in relatively deep water, but sediment filling may convert them into a shallow water environment which may include carbonate sediments. A structural high separates the fore-arc basin from the actual slope down to the submarine trench.

Beyond the deep sea trenches a significant amount of pelagic sedimentation takes place on a relatively flat ocean crust. Some of the sediments get scraped off the subducting volcanic crust and stacked up in what are called "accretionary prisms" (Figure 2.48). Some sediment may also be carried down with the subducting plates. The supply of sediment to the deep-sea trenches themselves is often very limited, which is why they do not fill up with sediment. They represent the greatest depths in the ocean (up to 10 km) and this can be explained isostatically by the fact that the oceanic plate which is undergoing *subduction* is cold, and therefore heavy. The downward movement acts against the direction of heat flow, resulting in low geothermal gradients and therefore dense crust.

The sediments may be pelagic oozes or distal turbidites. Since the oceanic plate is moving towards the island arc, the sediments on the oceanic crust have been deposited further away from the sediment source, and in consequence we do not normally have very thick sedimentary sequences in the subducting plate. The accretionary prism consists of a series of sliding faults which are steepest near the surface and have a lower gradient downwards. They are often draped with a blanket of pelagic sediments (Fig. 2.49).

Listric faults of this type are similar to those we find in plate boundaries with tension (rifting) but the relative movements are in the opposite direction (reverse faults). Sediments which are still not very consolidated tend to deform along the imbricated faults and develop various kinds of *drag folds*. Continued movements of the imbricated fault planes cause this slope to become very steep locally, and conglomerates and fan deposits may become unstable and slide. Lithified carbonates and sandstones will break up and form large blocks in a more clay-rich matrix. Volcanic rocks may also be included in this package of broken-up sediments and be incorporated into coarse conglomerates with large blocks called *olistostromes*. The blocks, which lie in a matrix of clay sediments, may be from a few metres up to several hundred metres across. The result is called a *tectonic melange*.

The sediments in an accretionary prism are subjected to strong tectonic deformation prior to deeper

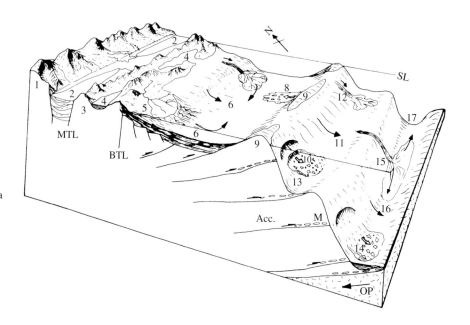

Fig. 2.49 Sedimentation in a submarine trench (accretionary prism) near Japan. Sedimentation is very much influenced by the relative movements (thrusting) of the rock units piled up in the prism (from Taira et al. 1982)

burial, and soft sediment deformation is a very characteristic feature of such deposits. If we look at the total package of imbricated wedges, there is a younging in the opposite direction, towards the subduction zone. This is a feature that can be used to recognise this depositional environment.

2.47 Summary

The study of sedimentary processes and facies relationships is important for the prediction of the distribution of different facies and rock properties.

We are interested in the geometry and distribution of sedimentary facies and also the internal properties of the sediments as they change during burial.

We therefore need to reconstruct sedimentary basins with respect to water depth, wave and tidal energy, and climate at different gelogical times.

The grain-size distribution and the mineralogy are to a large extent controlled by the source area being eroded and by transport processes.

The primary sediment compostion is very important with respect to the diagenetic process during burial. This will deterime the rock properties recorded seismically, and also reservoir quality.

Further Reading

Brown, L.F. and Fisher, W.L. 1977. Seismic – Stratigraphic interpretation of depositional systems. Examples from the Brazilian rift and pull-apart basins. In: Payton, C.E. (ed.), Seismic Stratigraphy – Application to Hydrocarbon Exploration. AAPG Memoir 26. Tulsa, OK, pp. 213–248.

Coleman, J.M. and Prior, D.B. 1980. Deltaic Sand Bodies. Continuing Education Notes Series 15. AAPG, Tulsa, OK, 171 pp.

Collinson, J.D. and Levin, J. 1983. Modern and Ancient Fluvial Systems. International Association of Sedimentologists. Blakwell, Oxford, Special Publication 6, 575 pp.

Collinson, J.D. and Thompson, D.B. 1982. Sedimentary Structures. Allen and Urwin, New York, NY, 194 pp.

De Blasio, F.V., Engvik, L., Harbitz, C.B. and Elverhøi, A. 2004. Hydroplaning and submarine debris flows. Journal of Geophysical Research 109, C01002, doi:10.1029/2002JC001714.

Leggett, J.K. 1982. Trench-Forearc Geology: Sedimentation and tectonics on modern and ancient plate margins. Geological Society Special Publication 10, 576 pp.

Kraft, J.C. and John, C.J. 1979. Sedimentary patterns and geologic history of Holocene marine transgression. Geological Society of America Bulletin 63, 2145–2163.

Miall, A.D. 1984. Principles of Sedimentary Basin Analysis. Springer, New York, NY, 490 pp.

Nichols, G. 2009. Sedimentology and Stratigraphy. Wiley-Blackwell, Chichester, 411 pp.

Nichols, G., Williams, E. and Paolola, C. 2008. Sedimentary Processes, Environments and Basins. Wiley-Blackwell, Chichester, 648 pp.

Mutti, E., Bernoulli, D. and Ricci Lucchi, F. 2009. Turbidites and turbidity currents from alpine 'flysch' to the exploration of continental margins. Sedimentology 56, 267–318.

Reading, H.G. 1996. Sedimentary Environments. Processes, Facies and Stratigraphy. Blackwell, Oxford, 688 pp.

Taira, A., Okao, A.H., Whitaker, J.H.McD. and Smith, A.J. 1982. The shimants belts at Japan. Cretaceous lower Miocene active margin sedimentation. In: Legget, J.K. (ed.), Trench-Forearc Geology: Sedimentation and tectonic on modern and ancient plate margin. Geological Society Special Publication 10, pp. 5–26.

Talling, P.J., Lawrence, A., Amy, A. and Wynn, R.B. 2007. New insight into the evolution of large-volume turbidity currents: Comparison of turbidite shape and previous modelling results. Sedimentology 54, 737–769.

Tinterri, R., Drago, M., Consonni, A., Davoli, G. and Mutti, E. 2003. Modelling subaqueous bipartite sediment gravity flows on the basis of outcrop constraints: First results. Marine and Petroleum Geology 20, 911–933.

Tucker, M.E. 2001. Sedimentary Petrology: An introduction to the origin of sedimentary rocks. Blackwell, Oxford, 262 pp.

Walker, R.G. 1979. Facies Models. Geoscience Canada. Reprint Series 1. 211 pp.

Walker, R.G. 1984. Facies models: Geoscience Canada. Reprint Series. Toronto 317 pp.

Wright, L.D. and Coleman, J.M. 1974. Mississippi River mouth processes: Effluent dynamics and morphology developments. Journal of Geology 82, 751–778.

Chapter 3

Sedimentary Geochemistry

How Sediments are Produced

Knut Bjørlykke

The composition and physical properties of sedimentary rocks are to a large extent controlled by chemical processes during weathering, transport and also during burial (diagenesis). We can not avoid studying chemical processes if we want to understand the physical properties of sedimentary rocks. Sediment transport and distribution of sedimentary facies is strongly influenced by the sediment composition such as the content of sand/clay ratio and the clay mineralogy. The primary composition is the starting point for the diagenetic processes during burial.

We will now consider some simple chemical and mineralogical concepts that are relevant to sedimentological processes.

Clastic sediments are derived from source rocks that have been disintegrated by erosion and weathering. The source rock may be igneous, metamorphic or sedimentary. The compositions of clastic sediments are therefore the product of the rock types within the drainage basin (provenance), of climate and relief. The dissolved portion flows out into the sea or lakes, where it is precipitated as biological or chemical sediments. Weathering and abrasion of the grains continues during transport and sediments may be deposited and eroded several times before they are finally stored in a sedimentary basin.

After deposition sediments are also being subjected to mineral dissolution and precipitation of new minerals as a part of the diagenetic processes. For the most part we are concerned with reactions between minerals and water at relatively low temperatures. At temperatures above 200–250°C these processes are referred to as metamorphism which is principally similar in that unstable minerals dissolve and minerals which are thermodynamically more stable at certain temperatures and pressures precipitate.

At low temperatures, however, unstable minerals and also amorphous phases may be preserved for a long time and there may be many metastable phases.

Many of the reactions associated with the dissolution and precipitation of minerals proceed so slowly that only after an extremely long period can they achieve a degree of equilibrium.

Reactions will always be controlled by thermodynamics and will be driven towards more stable phases. The kinetic reaction rate is controlled by temperature.

Silicate reactions are very slow at low temperature and this makes it very difficult to study them in the laboratory.

Biological processes often accompany the purely chemical processes, adding to the complexity. Bacteria have been found to play an important role in both the weathering and precipitation of minerals. Their chief contribution is to increase reaction rates, particularly during weathering.

In this chapter we shall examine the processes between water and sediments from a simple physical-chemical viewpoint. A detailed treatment of sediment geochemistry is however beyond the scope of this book.

K. Bjørlykke (✉)
Department of Geosciences, University of Oslo, Oslo, Norway
e-mail: knut.bjorlykke@geo.uio.no

K. Bjørlykke (ed.), *Petroleum Geoscience: From Sedimentary Environments to Rock Physics*,
DOI 10.1007/978-3-642-02332-3_3, © Springer-Verlag Berlin Heidelberg 2010

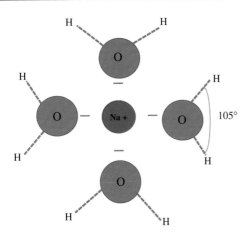

Fig. 3.1 The strong dipole of water molecules causes them to be attracted to cations which thereby become hydrated. Small cations will be most strongly hydrated and less likely to be adsorbed on a clay mineral with a negative charge

Water (H_2O) consists of one oxygen atom linked to two hydrogen atoms, with the H-O-H bonds forming an angle of 105° (Fig. 3.1). The distance between the O and the H atoms is 0.96 Å, and between the hydrogen atoms 1.51 Å. Water molecules therefore have a strong dipole with a negative charge on the opposite side from the hydrogen atoms (Fig. 3.1). This is why water has a relatively high boiling point and high viscosity, and why it is a good solvent for polar substances. Another consequence of this molecular structure is that water has a high surface tension, important for enabling particles and organisms to be transported on its surface. The capillary forces which cause water to be drawn up through fine-grained soils are also a result of this high surface tension.

A number of concepts are particularly useful for describing and explaining geochemical processes:

1. Ionic potential
2. Redox potential Eh
3. pH
4. Hydration of ions in water
5. Distribution coefficients
6. Isotopes

3.1 Ionic Potential

Ionic potential is a term introduced by V.M. Goldschmidt to explain the distribution of elements in sediments and aqueous systems. It must not be confused with ionisation potential. Recent authors have proposed the term "hydropotential" for the concept, to avoid confusion.

Ionic potential (I.P.) may be defined as the ratio between the charge (valency) Z and the ionic radius R:

$$IP = \frac{Z}{R}$$

The ionic potential is an expression of the charge on the surface of an ion, i.e. its capacity for adsorbing ions. Small ions carrying a large charge have a high ionic potential while large ions with a small charge have a low ionic potential (see Fig. 3.2).

Ions with low ionic potential are unable to break the bonds in the water molecule and therefore remain in solution as hydrated cations (e.g. Na^+, K^+). This means that the ion is surrounded by water molecules with their negative dipole towards the cation (Fig. 3.1).

This is because the O–H bond is stronger than the bond which the cation forms with oxygen (M–O bonding, M = metal); this is particularly true of alkali metal ions (Group I) and most alkaline earth elements (Group II, I.P. <3). Metals with an ionic potential only slightly lower than that required to form M–O bonds, namely Mg^{2+}, Fe^{2+}, Mn^{2+}, Li^+ and Na^+, will be the most strongly hydrated. The hydration strongly affects the chemical properties of the ion and its capacity to be adsorbed or enter into the crystal structure of a mineral. Since the ions are surrounded by water molecules, we can use the expression "hydrated radius" to describe the space occupied by the ion and its water molecules within a crystal structure (Fig. 3.3).

If the M–O bond is approximately equal in strength to the O–H bond (I.P. 3–12), the metal ion replaces one of the hydrogen atoms to form very low solubility compounds of the type $M(OH)_n$ (see Fig. 3.2). Examples of these so-called hydroxides that we commonly encounter in sedimentary rocks as a result of weathering are $Fe(OH)_3$, $Al(OH)_3$ and $Mn(OH)_4$. These hydroxides have very low solubility.

Ions with high ionic potential (>12) form an M–O bond that is stronger than the H–O bond, giving soluble anion complexes such as SO_4^{--}, CO_3^{--}, PO_4^{3-} and releasing both of the H^+ ions into solution.

This approach can be used to explain the behaviour for elements on both sides of the Periodic Table (electropositive and electronegative) which form ionic bonds. The elements in the middle, however, have a greater tendency to form covalent bonds in which the

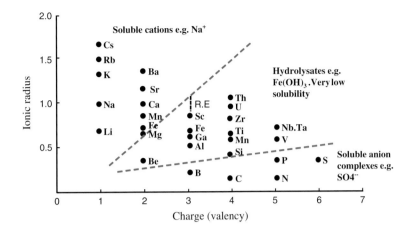

Fig. 3.2 Ionic radius and charge (valence) for some geochemically important elements. Ions with low ionic potential are soluble as cations (e.g Na+, K+) while ions with intermediate ionic potentials will bond with OH− groups and have very low solubility, forming hydrolysates (e.g Al(OH)₃), Fe (OH)₃). High ionic potentials make soluble cation complexes like CO_3^{--} and SO_4^{--}. The ratio between these parameters - the ionic potential - can be used to explain their behaviour in nature

Fig. 3.3 Ionic radius (in Ångstrom units) of hydrated and non-hydrated ("naked") ions of alkali metals and alkaline-earth metals. The smaller ions have higher ionic potentials and form stronger bonds with water molecules so that they become hydrated. This hydration effect is reduced with increasing temperature

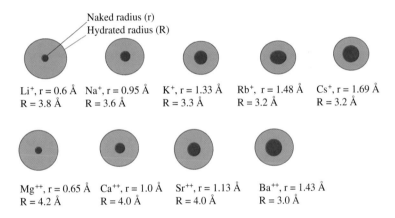

strength of the M-O bond is not merely a function of the valency and radius, and the picture becomes far more complex. The concept of ionic potential is nevertheless still useful; we see that during weathering, elements with low ionic potential remain in solution along with the anionic complexes of metals and non-metals with high ionic potential. This is reflected in the composition of seawater. The hydrolysates, on the other hand, become enriched on land as insoluble residues or through weathering (Al^{3+}, Fe^{3+}, Mn^{4+}, Ti^{4+}, etc.). Note also that Fe^{++} and Mn^{++} which occur in reducing environments have lower ionic potential and are much more soluble that Fe^{3+} and Mn^{4+}.

The most soluble ions remain in the seawater until they are precipitated as salt when seawater is concentrated during evaporation. In addition to the chlorides (i.e. NaCl, KCl), these are mainly salts of cations with low ionic potential, and of anions with high ionic potential, e.g. $CaSO_4 \cdot 2H_2O$, Na_2CO_3 and carbonates such as $CaCO_3$ (calcite), $CaMg(CO_3)_2$ (dolomite) and $MgCO_3$ (magnesite).

The principle of ionic hydration and the size of the ionic radius are capable of explaining a whole range of geochemical phenomena. Among the Group I elements of the Periodic Table, we know that Li^+ and Na^+ are enriched in seawater. This because the strong hydration prevents absorption on clay minerals which usually have a negative surface charge. K^+, Rb^+ and Cs^+, on the other hand, have larger ionic radii and consequently are less strongly hydrated. This leaves them with a more effective positive surface charge which facilitates their adsorption onto clay minerals, etc.

This is demonstrated in nature during weathering and transport. While similar amounts of potassium and sodium are dissolved during weathering of basement rocks, the potassium concentration in the sea is much lower (K/Na ratio of only 1:30). This is because K^+ is more effectively removed by adsorption because it is less protected by hydration. The same is true to an even greater extent for Rb^+ and Cs^+, which are adsorbed even more readily. These ions therefore have a relatively short residence time in seawater, between being delivered by rivers and then removed by accumulating sediment.

With regard to Group 2 elements, Mg^{++} for example will be more strongly hydrated than Ca^{++} because it is a smaller ion. As a result, Mg^{++} has a greater tendency to remain in solution in seawater. However, despite the fact that the Mg/Ca ratio in seawater is 5, it is calcium carbonate which is the first to form through chemical and biological precipitation. Dolomite or magnesite do not precipitate directly from seawater and this is in part due to the strong hydration of Mg^{++}. Normally, if we had naked (unhydrated) ions, $MgCO_3$ and $FeCO_3$ would be more stable than $CaCO_3$ because Mg^{++} and Fe^{++} have greater ionic potentials and stronger bonding to the CO_3^{2-} ion. However with increasing temperature the hydration declines because the bonds with the dipole of the water molecules become weaker. Mg^{++} is then more likely to be incorporated into the carbonate mineral structures. Therefore during diagenetic processes at 80–100°C, magnesium carbonates precipitate more readily even if the Mg^{++}/Ca^{++} and Fe^{++}/Ca^{++} ratios are low. Even if Mg is preferred in the carbonate structure and also in the clay minerals, very little magnesium is usually available in the deeper parts of sedimentary basins except in the presence of evaporites with Mg salts.

3.2 Redox Potentials (Eh)

Oxidation potential (E) is an expression of the tendency of an element to be oxidised, i.e. to give up electrons so it is left with a more positive charge. This potential can be measured by recording the potential difference (positive or negative) which arises when an element functions as one electrode in a galvanic element. The other electrode is a standard one, normally hydrogen. The oxidation potential of

the reaction $H_2 = 2H^+ + 2e$ (electrons) is defined as $E^0 = 0.0$ V at 1 atm and H^+ concentration of 1 mol/l at 20°C. Different conventions have been used to assign plus and minus values. In geochemical literature, metals with a higher reducing potential than hydrogen are assigned negative values, e.g. $Na = Na^+ + e^- = -2.71$V, while strongly oxidising elements are given a positive sign, e.g. $2F^- = F_2 + 2e = 2.87$V. A list of redox potentials shows which elements will act as oxidising agents, and which will be reducing agents. Reactions which result in a negative oxidation potential (E) will proceed spontaneously, while those which have positive voltage will be dependent on the addition of energy from an outside source. We can predict whether a redox reaction will occur by using Nernst's Law (see chemistry textbooks).

3.3 pH

The ionisation product for water is $[H^+] \cdot [OH^-] = 10^{-14}$. The concentration of H^+ in neutral water will be 10^{-7}. pH is defined as the negative logarithm of the hydrogen ion concentration, and is therefore 7 for neutral water (at 25°C). However, the ionisation constant (product) varies with temperature, e.g. at 125°C the ionisation constant for water is $[H^+] \cdot [OH^-] = 10^{-12}$. In other words, neutral water then has a pH of 6. It is important to remember this when considering the pH of hot springs or in deep wells, for example oil wells.

In nature the pH of surface water mostly lies between 4 and 9. Rainwater is frequently slightly acid due to dissolved CO_2, which gives an acid reaction:

$$H_2O + CO_2 = H_2CO_3 \text{ (carbonic acid)}$$
$$H_2CO_3 = H^+ + HCO_3^- = 2H^+ + CO_3^{2-}$$

Humic acids may give the water in lakes and rivers a low pH. Sulphur pollution from burning oil and coal gives SO_2, which is oxidised in water to sulphuric acid:

$$2SO_2 + O_2 + 2H_2O = 2H_2SO_4$$

In areas with calcareous rocks or soils this sulphuric acid is immediately neutralised and the water becomes basic, as is the case across much of Europe. By contrast, in areas with acidic granitic rocks as in the south of Norway and large areas of Sweden, the rock does not have sufficient buffer capacity to counteract acid rain or acidic water produced by vegetation (due to humic acids). Organic material also contains a certain amount of sulphur, and drainage of bogs, or drought, can produce an acidic reaction. This is because H_2S from organic material is oxidised to sulphate when the water table is lowered, allowing oxygen to penetrate deeper in these organic deposits.

The water near the surface of large lakes and the sea can have a high pH because CO_2 is consumed due to high organic production (photosynthesis). If the organic material decomposes (oxidises) on its way to the bottom, CO_2 is released again, causing the pH to decrease with depth since the solubility of the CO_2 increases with the increasing pressure.

CO_2 is also less soluble in the warm surface water than in the colder water at greater depth.

Seawater is a buffered solution, with a typical pH close to 8, though this varies somewhat with temperature, pressure and the degree of biological activity.

Eh and pH are important parameters for describing natural geochemical environments, and the diagram obtained by combining these two parameters is particularly useful.

The lower limit for Eh in natural environments is defined by the line $Eh = -0.059\,pH$, because otherwise we would have free oxygen, and the upper limit corresponds to $Eh = 1.22 - 0.059\,pH$, beyond which free oxygen would be released from the water. If we also set pH limits at 4 and 9 in natural environments, we can divide the latter into four main categories:

1. Oxidising and acidic
2. Oxidising and basic
3. Reducing and acidic
4. Reducing and basic

Variations of pH and Eh are the major factors involved in chemical precipitation mechanisms in sedimentary environments where there is not strong evaporation (evaporite environments).

The solubility of many elements is highest in the reduced state and they are precipitated by oxidation. This is particularly characteristic of iron and manganese, whereas others such as uranium and vanadium are least soluble in the reduced state.

3.3.1 Distribution Coefficients

When a mineral crystallises out of solution, the composition of the mineral will be a function of the composition of the solution and the temperature and pressure. Trace elements which are incorporated in the mineral structure are particularly sensitive to variations of these factors. With constant temperature and pressure, the concentration of an element within a mineral which is being precipitated, is proportional to its concentration in the solution. The ratio between the concentration of an element in the mineral and its concentration in the solution (water) is called the distribution coefficient.

A number of elements substitute for Ca^{++} in the calcite lattice: Mn^{++}, Fe^{++} and Zn^{++} have distribution coefficients $(k) < 1$. This means that they will be captured, so that the mineral becomes enriched in these elements relative to the solution.

$$Mn^{++}/Ca^{++}(\text{mineral}) = k \cdot Mn^{++}/Ca^{++}(\text{solution})$$

k here is about 17, that is to say the manganese concentration in the calcite is 17 times greater than in the solution.

At low temperatures (25°C) Mg^{++}, Sr^{++}, Ba^{++} and Na^+ have distribution coefficients <1. This means that the mineral phase will contain proportionately less of these elements than the aqueous phase. For Sr^{++}, k is about 0.1 (0.05–0.14) in calcite, such that the Sr content in calcite is relatively low. The Sr content in aragonite is considerably higher because the Sr^{++} ion, which is larger than the Ca^{++} ion, is more easily accommodated within the lattice. By analysing trace elements in minerals like calcite we can infer something about the environment when the minerals precipitated. Limestones with a high content of strontium may have had much primary aragonite which was replaced by calcite. Calcite containing significant amounts of iron must have precipitated under reducing conditions because only Fe^{++} would be admitted into the calcite structure.

3.4 Isotopes

A number of elements occur in nature as different isotopes: the atomic number (protons) is constant but there are different numbers of neutrons. They therefore have the same chemical properties although their masses are slightly different. Isotopes which are radioactive (unstable) break down at a specific rate characteristic for the isotope species (the disintegration constant). By analysing the reaction products formed in the minerals they can be dated. The $^{87}Rb - ^{87}Sr$ and the $^{40}K - ^{40}Ar$ methods are the ones most commonly used in determining the age of rocks. The ratios between lead isotopes can also be employed because of the $^{235}U - ^{207}Pb$, $^{238}U - ^{206}Pb$ and $^{238}Th - ^{208}Pb$ reactions.

Dating sedimentary rocks is a complicated procedure and the results are often difficult to interpret. The main problem is that clastic sediments are comprised of fragments and minerals which have been eroded from older rocks and the measured radiometric age may be strongly influenced by the age of these source rocks. Separating the newly formed (authigenic) mineral to be dated, can be particularly challenging.

The fact that isotopes have different masses causes fractionation to take place through both chemical and biological processes. The simplest example is water, H_2O, which contains two oxygen isotopes and two hydrogen isotopes. The oxygen isotopes are fractionated through evaporation, with more $H_2^{16}O$ evaporating than $H_2^{18}O$. This is because the ^{18}O isotope has greater mass and a phase change from fluid to vapour therefore requires more energy. $H_2^{16}O$ has higher vapour pressure than $H_2^{18}O$. This is the reason why rainwater and ice contain less ^{18}O than seawater.

Isotope fractionation is a function of temperature, however, and is much more effective with evaporation at low temperatures than at high ones. The explanation for this is that at high temperatures the energy of the molecules are so great that the difference in mass between ^{18}O and ^{16}O is of less consequence. At low temperatures the isotopic separation evaporation is much more selective so that the water evaporated is more enriched in ^{16}O. When water vapour condenses to rainwater, molecules with ^{18}O are most stable. Rain and snow becomes enriched in the heavier isotope (^{18}O), so that the water vapour remaining in the air becomes more enriched in ^{16}O. Most of the evaporation takes place at low latitudes and the water vapour in the air has a progressively lower ^{18}O-content towards higher latitudes as the air cools and it rains. The concentration of oxygen isotopes is expressed in relation to a standard:

$$\delta^{18}O = \left(^{18}O/^{16}O_{sample}/^{18}O/^{16}O_{std} - 1\right) \cdot 1000$$

This standard may be the average composition of seawater, called SMOW (Standard Mean Ocean Water). Another commonly used standard is PDB (Pee Dee Belemnite), which is the composition of calcite in a Cretaceous belemnite. The calcite ($CaCO_3$) was precipitated in the sea and its composition was in equilibrium with the seawater at normal temperatures (15–20°C). There is more ^{18}O in calcite than in the water (positive fractionation), but with higher temperatures the less effective fractionation of oxygen lowers the $\delta^{18}O$ values. The relationship between the two standards is:

$$\delta^{18}O_{SMOW} = 1.031 \cdot \delta^{18}O_{PDB} + 30.8$$

PDB values are preferred for carbonate minerals while the SMOW scale is mainly used for water samples and silicate minerals.

Hydrogen has two stable isotopes, ^1H and ^2H (deuterium), and an unstable one, ^3H (tritium), which has a half-life of 12 years. The hydrogen isotopes are even more strongly fractionated than oxygen isotopes during evaporation. Water molecules with deuterium (heavy water) have lower vapour pressure that water moles with hydrogen.

In meteoric water there is a linear relation between the deuterium/hydrogen ratio (D/H) and the $\delta^{18}O$.

The isotopic composition of seawater has varied through geological time, though not so much during the last 200–300 million years. During glacial periods, seawater acquires more positive $\delta^{18}O$ values because the water bound as ice has more negative $\delta^{18}O$ values. Rainwater (meteoric water) has normal $\delta^{18}O$ values from –2 to –15. The values become more negative towards higher latitudes, and near the poles one can measure $\delta^{18}O$ values of about –50 and δD (^2H) values close to –350 (see Fig. 3.4). Minerals that form in seawater show decreased $^{18}O/^{16}O$ ratios with increased ambient temperature during formation. The $\delta^{18}O/^{16}O$ ratio in carbonate secreting marine organisms, for example, is thus a function of both temperature and salinity. The seawater changes its $\delta^{18}O$ values by around 1–1.5‰. Isotopes can thus provide important proxy evidence for palaeoclimate studies.

Cold freshwater gives strongly negative $\delta^{18}O$ values, whereas evaporites are enriched in ^{18}O isotopes

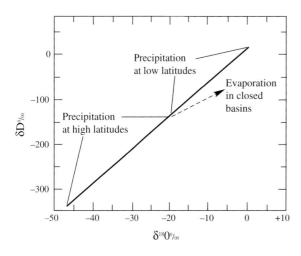

Fig. 3.4 Ratio between the isotopic composition of seawater and freshwater. Evaporites will deviate from the mixing line between these endmembers

(positive $\delta^{18}O$). Shallow marine carbonates that are diagenically modified by freshwater, give lower $\delta^{18}O$ values than marine carbonates deposited in deeper water.

Stable oxygen isotope analyses were first used by Urey, in 1951, to demonstrate past temperature changes in seawater. By taking samples through a cross-section of a belemnite it was possible to register annual variations in seawater temperature from 150 million years ago (Fig. 3.5).

The precipitation of newly formed (authigenic) minerals gives an oxygen isotope composition which is a function of the composition of the porewater in which the mineral is precipitated, and the temperature. If the porewater isotope composition is known, the temperature (T) can be calculated, and vice versa.

The calcite precipitation formula is:

$$T = 16.9 - 4.38(^{18}O_{carb} - ^{18}O_{water}) +$$

$$0.1(^{18}O_{carb} - ^{18}O_{water})^2$$

Here the values for calcite are given in PDB and for water in SMOW. We see that if the $\delta^{18}O$ value for calcite is 0 (PDB) and seawater has 0 (SMOW), the temperature is 16.9°C, which may have been a typical sea temperature when the standards were precipitated.).

The above formula can be expressed graphically, enabling the temperature to be read off a curve as a function of the isotopic composition of the calcite,

which is the assumed composition of the porewater during precipitation (Fig. 3.6). Similar calculations can be done for other precipitated minerals, for example for quartz using the $\delta^{18}O$ fractionation as a function of the temperature for quartz.

Carbon has two stable isotopes (^{12}C – 98.9% and ^{13}C – 1.1%). During photosynthesis a greater proportion of $^{12}CO_2$ than $^{13}CO_2$ forms organic compounds, because $^{12}CO_2$ has a smaller mass. Organic material is therefore enriched in ^{12}C relative to atmospheric CO_2 and HCO_3^- in seawater. The isotopic composition of carbon is expressed as $\delta^{13}C$ values:

$$\delta^{13}C = [^{13}C/^{12}C(sample)/^{13}C/^{12}C(std) - 1] \cdot 1000$$

All samples are compared against a standard of marine calcite, the PDB belemnite, which by definition has $\delta^{13}C = 0\%o$ PDB. The isotopic composition of dissolved carbon (CO_2) has been relatively constant during the last 300–400 million years, but limestones can nevertheless be dated and correlated using differences due to variation in the composition of seawater. Towards the end of the Precambrian the composition of seawater seems to have been more variable, and there this type of correlation is particularly valuable since there are no fossils. In large massive limestones the isotope composition does not change significantly during diagenesis, because the volume is so great. Atmospheric CO_2 has $\delta^{13}C = -7\%o$. Land plants have an average $\delta^{13}C$ value of –24 (–15 to –30‰), and marine organisms have a similar range of values. Freshwater containing CO_2 released by the breakdown of organic matter, and groundwater filtered through a soil profile, will take up CO_2 with negative $\delta^{13}C$ values from roots and organic material.

Bacterial fermentation of organic material ($2CH_2O = CH_4 + CO_2$) forms gas (methane) which is very strongly enriched in ^{12}C ($\delta^{13}C = -55$ to –90‰) and CO_2 which is positive ($\delta^{13}C = +15$).

Thermal breakdown (thermal decarboxylation) of organic matter produces $\delta^{13}C$ values of –10 to –25.

The strontium isotope ratio ($^{87}Sr/^{86}Sr$) in seawater has varied considerably through geological time (Fig. 3.7). This is because there are two radically different sources of strontium in seawater. Continental weathering supplies much ^{87}Sr to the sea since granitic rocks contains relatively high concentrations of rubidium which can decay to ^{87}Sr. Dissolution of basalt at

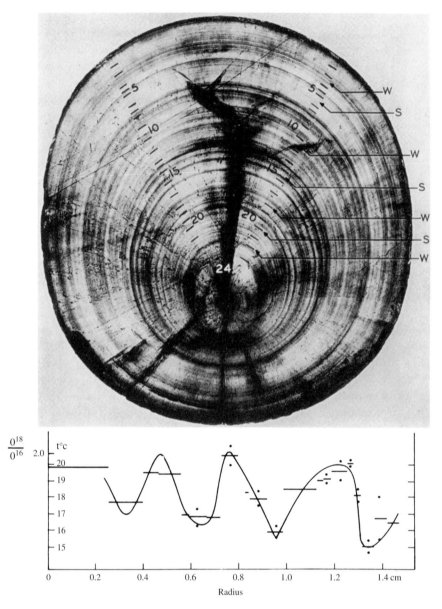

Fig. 3.5 Analyses of oxygen isotopes in a Jurassic belemnite from the centre to the outermost layer. Colder water during the winter is recorded by lower $^{18}O/^{16}O$ ratios. We can see that the belemnite lived for 4.5 years and died in the spring. (From Urey et al. 1951)

the mid-oceanic ridges will supply strontium with a relatively low $^{87}Sr/^{86}Sr$ ratio because basalt contains a little potassium and also rubidium.

When there is rapid seafloor spreading a great deal of water passes through the mid-oceanic ridges, so that the seawater receives much Sr with a low $^{87}Sr/^{86}S$ ratio. During such periods, for example in the

Jurassic – Cretaceous, the creation of new warm seafloor will lead to a transgression onto the continents. This reduces the gradients and hence transporting capacity of rivers, limiting the supply of clastic material to the ocean.

Since the Jurassic, the $^{87}Sr/^{86}Sr$ ratio has risen almost continuously, and by analysing marine calcitic

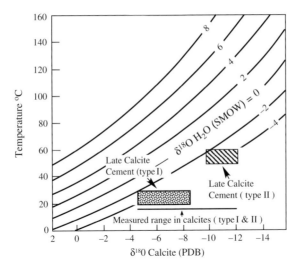

Fig. 3.6 Relation between the isotopic composition of porewater and carbonate cement, as a function of temperature (from Saigal and Bjørlykke 1987)

Fig. 3.7 $^{87}Sr/^{86}Sr$ ratio in seawater from the Cambrian to present. Based on MacArthur et al. (2001). This ratio reflects the relative contribution form weathering of continental rocks with high contents of ^{87}Sr and exchange with basaltic rocks with low contents of ^{87}Sr on the oceanic spreading ridges

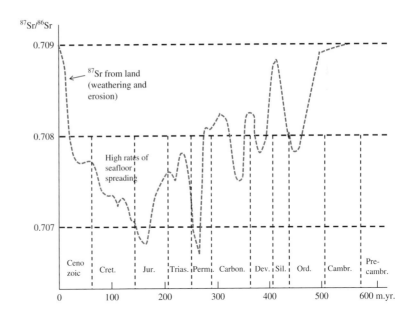

fossils such as foraminifera, one can obtain rather accurate age determinations. This applies particularly to the Tertiary period, when the rise in the $^{87}Sr/^{86}Sr$ ratio was particularly rapid (Fig. 3.7).

The isotopic composition of clastic sediments can also be used for stratigraphical correlation. Then it can be more useful to employ isotopes which do not go into solution and react with water, but retain the original age of the rocks from which they were eroded. In the North Sea and on Haltenbanken the ratio between the rare earth elements samarium and neodymium

$(^{147}Sm/^{143}Nd)$ was used to correlate reservoir rocks both in-field and regionally.

3.5 Clay Minerals

A number of minerals are referred to as clay minerals because they predominantly occur in the finest grain-size fraction (clay fraction) of sediments and sedimentary rocks. However, this is not an accurate definition, because the clay fraction contains many other minerals than those we call clay minerals, and because the clay minerals themselves are often larger than 4 μm (0.004 mm). By "clay minerals" we usually mean sheet silicates which consist chiefly of oxygen, silicon, aluminium, magnesium, iron and water (H_2O, OH^-). Clay minerals in sedimentary basins are partly derived from sheet silicate minerals occurring in metamorphic and eruptive rocks (e.g. biotite, muscovite and chlorite), but during weathering and transport these clastic minerals are typically altered from their initial composition in the parent rock.

Mica (muscovite and biotite) lose some potassium which is replaced by water (H_2O, H_3O^+) to form illite (hydro mica). Clay minerals are also formed through weathering reactions, for example by the breakdown of feldspar and mica. Clay minerals which are formed by the breakdown of other minerals within the sediment, are called authigenic.

Sheet silicates have a structure consisting of sheets of alternating layers of SiO_4 tetrahedra and octahedra. In the tetrahedral layers, silicon or aluminium atoms are surrounded by four oxygen atoms. In the octahedral layers the cation is surrounded by six oxygen or hydroxyl ions. Both bi and trivalent ions can act as cations in the octahedral layer. In sheet silicates with trivalent ions (e.g. Al^{3+}) only two of the three positions in the octahedral layer are occupied, and such minerals are therefore called dioctahedral. With bivalent ions (Mg^{++}, Fe^{++}) all three positions must be filled to achieve a balance between the positive and negative charges, so these minerals are called trioctohedral.

The main method of identifying clay minerals is X-ray diffraction (XRD), by which the thickness of the sheet silicates is determined using X-rays which are diffracted according to Bragg's Law: $n\lambda = 2d \sin \varphi$. Here λ is the wavelength of the X-ray, φ the angle of incidence and d the thickness of the reflecting silicate layers; d is thus a function of angle φ.

Sheet silicates may also be identified by means of differential thermal analysis (DTA), which records characteristic exothermal or endothermal reactions.

Figure 3.8 shows the structure of some of the main clay minerals. Illite consists of sheets with two layers of tetrahedra and one of octohedra, bonded together by potassium. This ionic bonding is relatively weak so the mineral cleaves easily along this plane. The bonds within the tetrahedral and octahedral layers are

more covalent and stronger. The potassium content in mica corresponding to the formula of mica is about 9% K_2O, while illite has a greater or lesser deficit of potassium. Smectite (montmorillonite) has the same structure except that most of the potassium is replaced by water (H_3O^+), other cations or organic compounds (e.g. glycol). There are strong indications that smectite consists of small particles of 10 Å, plus water.

Illite is most likely comprised of several layers of these small 10 Å particles stacked on top of one another.

In an atmosphere of glycol vapour, smectite will swell from 14 to 17 Å, while illite is unable to expand because there are numerous layers bonded together with K^+ or other cations, for example NH^+. Smectite has a very high ion-exchange capacity and to some extent can exchange ions in the octahedral layer. The stability of smectite declines in aqueous solutions with high K^+/H^+ (Na^+/H^+) ratio and with increasing temperature, and it converts to illite. Vermiculite has a structure reminiscent of the smectites, and also undergoes ion exchange and thus charge deficit in the tetrahedral layer, so that the bonding between each layer is too strong for much swelling to occur. Vermiculites are mostly trioctahedral, containing mostly Mg or Fe in the octahedral layer.

Glauconite is a green mineral which forms on the seabed. It is a potassium and iron bearing silicate somewhat similar to illite and contains both di - and trivalent iron. It is therefore formed right on the redox boundary, and during periods with little or no clastic sedimentation this can result in relatively pure beds of glauconite.

Kaolinite consists just of a tetrahedral layer and an octahedral layer and is very stable at low temperatures. There are no positions in the structure where exchange can precede easily, which gives kaolinite a much lower ion exchange capacity than smectite. At higher temperatures kaolinite becomes unstable and will convert to illite if K-feldspar or other sources of potassium are available (at 130°C) or pyrophyllite ($Al_3Si_4O_{10}(OH)_2$) at higher temperatures.

Kaolinite is part of the kaolin mineral group, which includes dickite which tends to form at slightly higher temperatures (100°C).

Chlorite is a mineral which consists of two tetrahedral layers and two octahedral layers, totalling 14 Å. The octahedral layer is filled with Mg^{++} and Fe^{++}. Magnesium-rich chlorites are typical of high

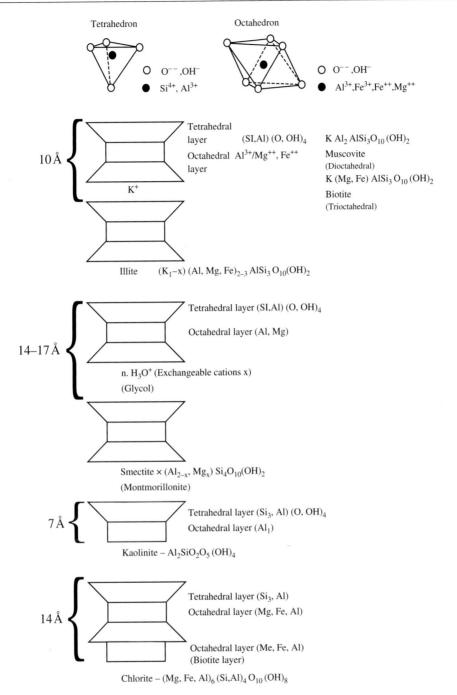

Fig. 3.8 Simplified illustration of the main groups of clay minerals. Their physical properties can be explained by their crystal structure. The chemical bonds between SiO_4^{4-} and O^{2-} in the tetrahedral structure are very strong. In the octahedral layers the bonds are weaker because Mg^{++} is surrounded (co-ordinated) with 6 oxygen. In the illite (mica) structure potassium is co-ordinated with 12 oxygen molecules, resulting in weak bonds that produce a strong cleavage

temperature metamorphic rocks, while iron-rich ones may form authigenically in sediments near the seafloor or at shallow depth.

Chamosite is an iron-rich chlorite mineral which often forms close to the sediment surface in reducing conditions. Together with siderite, chamosite is an important mineral in sedimentary ironstones which, especially in England, earlier were used as iron ore.

Clay minerals have a number of properties which distinguish them from most other minerals. Because of their very large specific surface area they have a great capacity for adsorbing ions, which is increased by the fact that clay minerals have negatively charged edges due to broken bonds. In water with a low electrolyte content, clay minerals will therefore repel each other. If cations are added, clay minerals will then accumulate a layer of positive ions (a double layer), and repulsion between negatively charged clay minerals declines as the strength of the electrolyte increases. Van der Waal's forces will therefore cause flocculation more easily in saltwater, where repulsion due to the negative charge is reduced. This is why clays transported by rivers flocculate into larger particles which sink more rapidly to the seafloor.

A colloid solution of clay in water is called a *sol*, which can be regarded as a Newtonian fluid. Flocculated clay is called a *gel* and has thixotropic properties. This means that the shear strength decreases with increasing deformation so that it changes from a gel to a sol which consists of dispersed colloidal particles in water (hydrosol). After a time without deformation, it regains its strength. This is typical for smectitic clays.

Sediments which increase their volume when they are deformed, are called dilatant. When the original packing of the grains is destroyed, the new packing may be less effective and the volume then increases. Walking on a beach the deformation of the sand causes increased porosity so that water is sucked into the sand at the surface.

Norwegian clays deposited in the sea when the ice sheet retreated about 10,000 years ago, consist of crushed rock fragments and have approximately the same composition as the parent rock. During deposition these clays acquired a markedly porous structure, with the minerals stacked like a card house. Saltwater helped to hold this structure together because it neutralised the negative charges. When the clays are elevated above the sea by isostatic recovery of the land, they are exposed to meteoric water. Even if the clay has a low permeability, freshwater will slowly seep through it and remove the saline water. This reduces the strength of the clay's structure and hence its stability and the clay becomes "quick". This is caused by overpressure and weak effective stress between

the grains and hence little friction. The card house structure may then collapse, releasing the interstitial water to produce a low viscosity clay slurry that will flow even down very gentle gradients. The addition of salt (NaCl or KCl) binds the clay particles so that the shear strength is increased, and this is employed when the ground has to be stabilised for buildings and construction.

3.6 Weathering

The composition of clastic rocks in sedimentary sequences depends to a large extent on the supply of sediments from source areas undergoing weathering and erosion. The physical properties of sedimentary rocks are controlled by the primary sediment composition and changes during burial (diagenesis). We shall here briefly look at the processes producing sediments.

Mechanical weathering is the physical breakdown of rocks into smaller pieces which can then be transported as clastic sediment.

Chemical weathering involves the dissolution of minerals and rocks and precipitation of new minerals which are more stable at low temperatures and high water contents. Parts of the parent rock will then be carried away in aqueous solution by the groundwater and rivers into the ocean.

Erosion is the combined result of the disintegration of rock and the removal of the products.

As we shall see, biological processes are not only important in connection with chemical weathering but also with respect to mechanical weathering. Both mechanical and chemical weathering, which are essentially land surface processes, are due to the chemical instability of rocks that were formed or modified under other conditions (at greater depths, higher pressures or temperatures, or in different chemical environments). They are no longer stable when exposed to the atmosphere, water and biological activity.

3.6.1 Mechanical Weathering

Igneous and metamorphic rocks which have been formed at many km depth at higher temperatures and pressures are not stable when exposed at the surface.

When uplifted and unloaded the rocks expand, mostly in the vertical direction, producing horizontal fractures (*sheeting*) parallel to the land surface. This is because as the rocks near the surface are unloaded by the reduction in overlying rock, they can expand vertically but not horizontally. In this way the vertical stresses become less than the horizontal ones, and joints develop normal to the lowest stresses. We see this most clearly in granites, which are homogeneous, while expansion in metamorphic and sedimentary rocks occurs along bedding surfaces, along tectonically weak zones with crushing, or along fractures that were formed at great depth. Joints opened by stress release in turn provide pathways for groundwater to circulate, increasing the surface area of rock exposed to chemical weathering.

In areas that experience freeze-thaw cycles, frost weathering becomes very important. When water freezes in cracks in rock, it expands by 9% and can generate very high stresses, further widening cracks near the surface. The surfaces of exposed rocks are also subjected to daily temperature fluctuations which cause greater expansion of the outer layers relative to the rest of the rock. Desert regions in particular experience very wide daily temperature ranges, though the importance of this process for mechanical weathering has been questioned. The roots of plants and moss can also contribute to mechanical weathering as they grow into fractures, take up water and expand.

3.6.2 Biological Weathering

Rocks are a source of nutrients for plants, and plants are capable of dissolving and breaking down the major rock-forming minerals. Moss, which consists of algae and fungi living in symbiosis, produces organic compounds that can slowly dissolve silicate minerals. Even in the earliest stages of weathering, we see that fungus hyphae penetrate into microscopic cracks.

Plant roots produce CO_2 which helps to lower the pH and dissolve minerals such as feldspar and mica, thus freeing an important plant nutrient, potassium. Plants also produce humic acids, which likewise strongly influence the solubility of silicate minerals, and also affect the stability of clay minerals. The production of humic acids is perhaps the major factor influencing the rate of weathering. In heavily vegetated areas, such as rain forest in the tropics, the weathering rate is exceptionally high because so much humic acid is produced.

Bacteria and fungi, which are found in almost all soil types, are active in breaking down minerals. Animals also contribute to weathering, and certain marine organisms such as mussels are able to bore into solid rock (see Chap. 8). Microbiology has become a key area of research in the quest to understand how minerals are dissolved and precipitated.

3.6.3 Chemical Weathering

There is no sharp demarcation between biological and chemical weathering, because we find biological activity in almost all soils and rocks near the surface. The chemical environment in water at the surface of the earth is very much affected by local biological activity, and in most cases it is biological processes that cause weathering to continue after rainwater has been neutralised through reaction with minerals. We will therefore use the term "weathering" here for both chemical and biological processes.

3.6.4 Weathering Profiles (Soil Profiles)

Both chemical and biological weathering are to a large extent controlled by climate. The crucial factor is the ratio between precipitation and evaporation in an area. In areas where precipitation far exceeds evaporation, *podsol profiles* develop in which there is a net transport of ions down through the soil profile as minerals are dissolved. In other words, we get weathering due to the fact that rainwater is slightly acidic (on account of its CO_2 and H_2SO_4 content) and contains oxygen. Rainwater is initially undersaturated with respect to all minerals. Some minerals are only very slightly soluble, others more soluble in this slightly acidic, oxidising water. Dissolved ions are transported down to the water table, but ferrous iron liberated from iron-bearing minerals will be oxidised and precipitated as ferric iron ($Fe(OH)_3$). Vegetation at the top of the soil profile produces CO_2 from roots and organic compounds, particularly humic acid, which will increase the solubility of silicate minerals. Similarly, aluminium derived from

Fig. 3.9 Simplified representation of soil profiles as a function of rainfall (precipitation) and evaporation

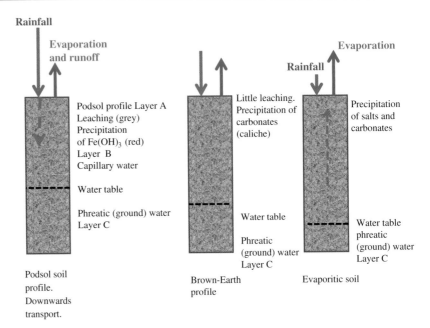

a solution of feldspar and mica, for example, precipitates as $Al(OH)_3$ but is less noticeable because aluminium hydroxide is white. The uppermost part of the soil profile, where dissolution due to undersaturated rainwater and organic acids dominates, is called the A-horizon. Some of the dissolved salts and particularly iron hydroxide is precipitated in the B-horizon below (Fig. 3.9). These may develop into a layer of solid rock (hard-pan) cemented with iron and aluminium oxides and hydroxides.

Where precipitation is approximately equal to evaporation, there is less leaching within the soil profile. At a certain depth (about 0.5–1 m) carbonate will be precipitated and form an indurated layer (calcrete) which may be eroded and from conglomerates.

The organic content is greater in the B-horizon which is brown due to less oxidation of organic matter, hence the term *brown-earth profiles*. If evaporation is greater than precipitation there will be a net upward transport of porewater, causing dissolved salts from the groundwater to be precipitated high in the soil profile.

3.6.5 *What Factors Control Weathering Rate and Products?*

Because weathering is the most important sediment-producing process, we are interested in understanding how the rate of weathering is related to rock type, precipitation, temperature, vegetation, relief etc. We also try to establish correlations between weathering products, particularly clay minerals, and these factors. By studying sediments from older geological periods, we can learn something about weathering conditions at those times. Weathering products will also bear the stamp of the rocks undergoing weathering. The stability of a mineral during weathering is largely a function of the strength of the bonds holding the cations in the crystal lattice. Potassium (K^+) in mica is held by weak bonds (low ionic potential) which are responsible for the pronounced cleavage. In biotite, the Mg^{++} and Fe^{++} in the octahedral layer will also be weakly bonded. During weathering cations like K^+, Na^+, Ca^{++}, Mg^{++} and Fe^{++} can be attacked by protons (H^+) which will replace them and send them into solution. Chain silicates like hornblendes and pyroxenes will also be relatively unstable and rapidly weather. In feldspars the alkali ions are dissolved so that the whole mineral disintegrates. Stability is lowest in calcium-rich plagioclase, while pure albite (sodium feldspar), orthoclase and microcline (potassium feldspar) are more stable. The breakdown of these silicate minerals will primarily liberate alkali cations. Silicon and aluminium have very low solubility and form new silicate minerals, largely clay minerals, though some silicic acid (H_4SiO_4) goes into solution.

1. $2K(Na)Al_2 AlSi_3O_{10}(OH)_2$ (muscovite) $+ 2H^+$
   $+ 3H_2O = 3Al_2Si_2O_5(OH)_4$ (kaolinite)
   $+ 2K^+(Na^+)$
2. $2K(Mg, Fe)_3AlSi_3O_{10}(OH)_2$ (biotite)
   $+ 12H^+ + 2e + O_2 =$
   $Al_2Si_2O_5(OH)_4$ (kaolinite) $+ 4SiO_2$ (in solution) $+$
   $Fe_2O_3 + 4Mg^{++} + 6H_2O + 2K^+$
3. $2K(Na) AlSi_3O_8$ (feldspar) $+ 2H^+ + 9H_2O =$
   $Al_2Si_2O_5(OH)_4$ (kaolinite) $+ 4H_4SiO_4 + K^+(Na^+)$

We see that potassium has been replaced by hydrogen ions in the new silicate minerals. The same applies to sodium in albite. The equations show that the reactions are driven to the right by low K^+/H^+ and Na^+/H^+ ratios.

The degree of weathering depends on how undersaturated the water is with respect to the minerals comprising the rock, and on the volume of water flowing through the rock. If the reaction products in the solution, K^+, Na^+ and silica (H_4SiO_4), are not removed by water flow, the reactions will cease. This is why weathering is always found to begin along cracks where water can penetrate (Fig. 3.10). The vertical and horizontal joints that often develop in response to pressure release when previously deeply buried rock is exposed at the land surface provide the initial pathways. As the weathering process spreads outward from joints, blocks of unweathered rock are gradually isolated. They have rounded corners and may become entirely round (spheroidal weathering) (Fig. 3.11a,b,c). In desert areas, where there is little rainfall, weathering proceeds much more slowly. Illite and montmorillonite may be formed under higher K^+/H^+ and Na^+/H^+ ratios than kaolinite, and they are frequently formed where there is less water percolation and the removal of potassium or sodium is slower.

There is often also a high silica content in the water in desert areas due to frequent silica algae (diatom) blooms and because silica is concentrated by water evaporation. This helps enhance the stability of smectite.

Granites subjected to weathering over a very long period often develop a special topography. Fractures and fault zones weather fastest and form valleys where the groundwater collects, which further accelerates the weathering. The more massive granitic areas will stand out in the terrain, and because precipitation runs swiftly down into the depressions between the elevated portions, the topographic difference will become more and more pronounced. Granites surrounded by sedimentary rocks will, because of their high content of feldspar, normally weather faster than the sediments which contain more quartz and other stable minerals. Particularly if the sediments consist of quartzites and shales, the granite will form a depression in the terrain. The weathering products from granites will normally be quartz grains which form sand grains the same size as the quartz crystals in the granite, and clay consisting of kaolinite, and possibly also some illite and smectite formed from feldspars and micas. We have at the outset a bimodal grain-size distribution with sand and clay, but very little silt.

Basic rocks (e.g. gabbro) will weather far more rapidly than granite because basic plagioclase (Ca-feldspar), pyroxenes and hornblende are very unstable and dissolve faster than silica-rich (acid) minerals. During progressive weathering sodium, potassium, magnesisum and calcium will be removed by the groundwater but some potassium may be adsorbed on clay minerals. The weathering residue will be enriched in elements with low solubility such as Ti, Al, Si and Mn (Fig. 3.12). In a normal, oxidising weathering environment, all the iron will precipitate out again as iron oxide ($Fe(OH)_3$) while the magnesium will tend to remain in solution. In areas with high rainfall the concentration of ions like K^+, Na^+ and silica will be

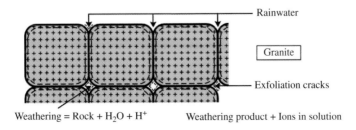

Weathering = Rock + H_2O + H^+ Weathering product + Ions in solution

Fig. 3.10 Weathering of granites along extensional fractures

Fig. 3.11 (**a**, **b**) Weathered granite (North of Kampala, Uganda) showing different stages of weathering. (**c**) Concentric – spheroidal weathering in a basic intrusive rock, Weathering is much faster in the absence of quartz

diluted and kaolinite will precipitate. Where porewater circulation is slower we may get a higher build-up of Mg^{++}, Ca^{++} and silica concentrations in the water, so that smectite (montmorillonite) or chlorite precipitates. Smectite requires porewater with a relatively high silica concentration (Fig. 3.13) and is therefore often found in sediments derived from volcanic rocks that contain glass or soluble silicate minerals. Biogenic sources of silica (diatoms, radiolaria) will also increase

the silica concentration in porewater because amorphous silica is much more soluble than quartz. In desert environments evaporation of water after rainfalls will concentrate silica in the porewater and make smectite stable. Figure 3.12 shows analyses of rocks at various stages of transformation due to weathering. Weathering proceeds particularly rapidly in amphibolites: Na^+, Mg^{++} and Ca^{++} are quickly leached out, while Al^{3+}, Fe^{3+} and Ti^{4+} become enriched. K^+ is

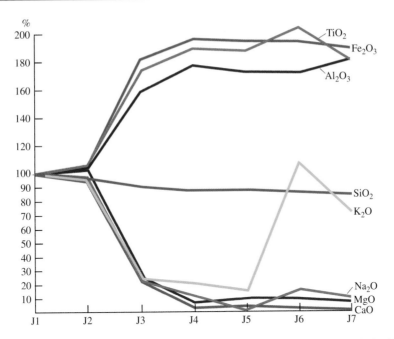

Fig. 3.12 Chemical analyses of changes in the chemical composition of samples representing progressive weathering of an amphibolite compared to an unweathered sample. The stable elements (Ti, Fe and Al) are enriched while there is a strong depletion of Na, Mg and Ca. Potassium is depleted but is adsorbed on clay minerals in the soil

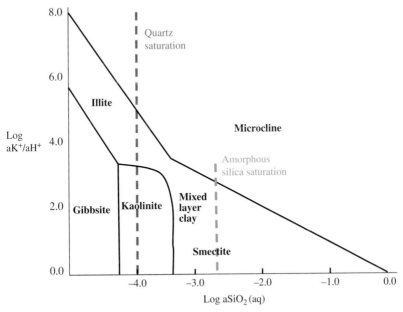

Fig. 3.13 Activity diagram showing the stability of some minerals as a function of silica and the K^+/H^+ ratio (after Aagaard and Helgeson 1982). Weathering reactions are characterised by reduction in both these two parameters towards the *lower left* of the diagram

however to large extent adsorbed on clay minerals and much more Na^+ than K^+ is therefore supplied to the oceans with the rivers.

After the alkali cations have been dissolved out of the silicates and kaolinite has been formed, extremely slow leaching of quartz commences. When the concentration of silica in the porewater is sufficiently low (see Fig. 3.13), kaolinite will be unstable and be replaced by gibbsite $Al(OH)_3$ or $(Al_2O_3 \bullet 3H_2O)$. Since gibbsite cannot form as long as the porewater is in equilibrium with quartz, all the quartz must have dissolved first or become encapsulated (e.g. in a layer of iron oxides). Clearly, gibbsite will form far more rapidly during the weathering of basic rocks than of granites, since the initial silica content is considerably lower. It takes a very long time to dissolve all the quartz in a granite. The solubility of quartz at surface temperatures and pH 7–8 is about 5 ppm, increasing at higher pH values. Alkaline (basic) water can therefore increase the solution rate of quartz. The end product of the weathering process is *laterite*, which consists of gibbsite and iron oxides or hydroxides. Under atmospheric (oxidising) conditions with a neutral pH, aluminum hydroxide and iron oxides may for practical purposes be regarded as insoluble.

At low pH values, e.g. under the influence of humic acids in humid tropical climates, aluminum is more soluble than iron and selective leaching of Al^{3+} will produce iron-rich laterites. Aluminum hydroxide can be dissolved at low (acidic) or high (basic) pH values and may reprecipitate as aluminum oxide, which has a lower iron content and is thus a higher grade aluminium source (bauxite).

3.7 Distribution of Clay Minerals and other Authigenic Minerals as a Function of Erosion and Weathering

3.7.1 What Determines the Type of Clay Minerals We Find in Sediments and Sedimentary Rocks?

When rocks are subjected to erosion and weathering, clastic minerals are broken down and perhaps somewhat altered, with respect to the minerals in the parent rocks. We can also get new minerals formed in the source rock itself, and precipitation of new minerals through weathering.

The more rapidly erosion and transport take place compared to the rate of weathering, the closer the composition of sediments is to the source rock. Glacial sediments represent one end of the scale in terms of sediment composition. Because glaciations are characterised by very high rates of erosion and low temperatures, chemical weathering will be very weak, and Quaternary sediments – including clays – will have a composition which essentially represents the average of the rocks which have been eroded. Sediments deposited in fault-controlled basins (e.g. rift basins) have a short transport distance between the site of erosion and deposition and the sediments have little time to weather on the way. Clastic chlorite and biotite break down relatively rapidly during weathering and are therefore likely to be preserved in addition to feldspar in rift basins. These unstable minerals are good indicators of rapid erosion and/or cold climates, since otherwise they are unlikely to survive.

An analysis of the distribution of clay minerals in modern sediments shows that we find clastic chlorite almost exclusively in high latitudes, except around islands of basic volcanic rocks (e.g. basalts). Clastic chlorites from metamorphic rocks or altered basic rocks are more magnesium-rich than authigenic chlorites, which are iron-rich (chamositic). In temperate areas with moderate to high precipitation, weathering proceeds relatively rapidly. In desert areas weathering proceeds very slowly because all weathering reactions require water. The type of weathering and the distribution of clay minerals are clearly related to latitude (Fig. 3.14).

Smectite and illite are typical of deserts because the weathering is much slower when water is nearly absent. When feldspar and other unstable minerals are altered (weathered) gradually in a dry climate, alkali ions and alkaline earth ions such as K^+, Na^+, Ca^{2+} and Mg^{2+} will not be removed rapidly enough due to little percolation of fresh rainwater.

Rainwater will cause some dissolution of minerals and, during the periods of drying, the silica concentration may be higher so that smectite becomes stable.

As a result we will usually get illite or smectite formed as authigenic minerals because they are stable in the presence of high K^+/H^+ ratios in the porewater.

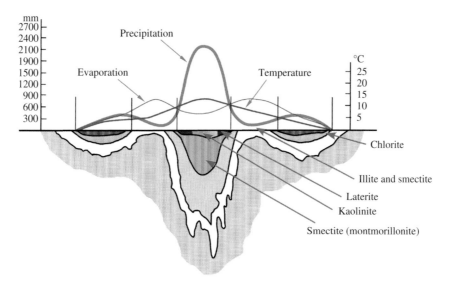

Fig. 3.14 Simplified diagram showing the distribution of weathering and common clay minerals as a function of latitude and rainfall. Cold areas and deserts are characterised by little weathering and more mechanical erosion

Smectite (montmorillonite) is thus a common clay mineral in desert areas, and its ability to swell when wet renders sediments very plastic during floods. This expansion of smectite also lowers its permeability and may be the reason why water can flow over the surface for a long time before sinking into the ground. In addition, capillary forces will prevent rapid percolation of water through dry soil. On the ocean floor near desert regions we find that illite and smectite are typical minerals, brought there by aeolian transport.

In tropical areas where precipitation is relatively high, the rate of weathering will be very rapid. This is not only because weathering processes accelerate with temperature, but also because vegetation produces large amounts of organic acids (humic acids) which are very effective in breaking down silicate minerals. Microbiological organisms such as fungi and bacteria also help in the breakdown process by producing CO_2 which forms carbonic acid, H_2CO_3.

Gibbsite ($Al_2O_3 \cdot 3H_2O$) and iron oxides (haematite, goethite, $Fe_2O_3 \cdot 3H_2O$) are constituents of the laterite which we find only in tropical areas with rapid weathering and slow erosion. Whereas iron oxides are also found at higher latitudes, gibbsite occurs almost exclusively in humid tropical areas.

Laterisation is a very slow process and takes millions of years, even in tropical regions with rapid weathering. It is therefore primarily in tropical

areas that we find bauxite for the aluminum industry. Iron-rich laterites may have an iron content of over 50%, and in some areas (e.g. India) have been exploited as iron ore. Laterite forms a very hard cement-like crust over the weathering profile and is also virtually devoid of nutrients, so crop cultivation is impossible. In East Africa (especially Uganda), however, erosion has incised through a layer of Tertiary laterites. While the laterite cover remains on flat elevated surfaces, fresher, more fertile, rocks and weathering material is exposed in the valley sides. The vegetation in some tropical areas is more abundant, even if the soils are very poor in nutrients, because the vegetation recycles those nutrients which are available. If the vegetation is removed and organic material is no longer produced, oxidation and the absence of humic acids (increased pH) will lead to precipitation of oxides and hydroxides which make the soil hard and uncultivable.

Volcanic ash consisting of glass and unstable volcanic mineral assemblages may alter to smectite on land or on the seafloor. In deep sea sediments zeolites like phillipsite are common.

Areas with volcanic rocks, particularly amorphous material (volcanic glass), will often form zeolites. These require a high concentration of both silica and alkaline ions in the water, which is the situation when glass dissolves. Zeolites, particularly phillipsite, are formed authigenically on the Pacific Ocean bed and are also found in lakes (e.g. in East Africa).

In summary we can say that the factors which determine which types of clay minerals are "produced" in the various areas are:

1. The rocks which are eroded/weathered (source rocks)
2. Rate of erosion
3. Temperature
4. Precipitation
5. Vegetation
6. Permeability of source rocks and sediments (percolation of water).

Typical distribution of various minerals:

1. Chlorite and biotite – high latitudes (cold climate) – rapid erosion.
2. Kaolinite – humid temperate and humid tropical regions – good drainage.
3. Smectite (montmorillonite) – low precipitation or poor drainage. Typical of desert environments, but also formed in impervious, e.g. basaltic, rocks in more humid environments. Typically formed from volcanic rocks.
4. Gibbsite – tropical humid climate – long weathering period.
5. Zeolites – formed in areas with volcanic material and restricted porewater circulation. Require a high concentration of silica and alkali ions.

3.8 Geochemical Processes in the Ocean

The ocean can be regarded as a reservoir of chemicals dissolved in water. It looks as though the composition of seawater has not altered radically throughout the geological ages from the early Palaeozoic until the present day, although there have certainly been some variations.

The supply of elements to ocean water from rivers and by water circulation at the spreading ridges must be balanced by a removal of the same elements from the ocean water (Fig. 3.15). The annual addition of salts dissolved in river water is about 2×10^9 tonnes/year. The figure was probably less in the geological past because vegetation was sparse or absent. The development of land plants that produce humic

acids, which in turn produced more rapid weathering, has probably increased the supply of salts (since Devonian times). This trend was sometimes slowed by periods with higher sea level that caused widespread transgressions and converted huge tracts of coastal land areas into continental shelf (e.g. during the Upper Cretaceous) reducing the weathering and the supply of salts and nutrients to the ocean.

Some ions like potassium (K^+) are adsorbed to clay minerals supplied by rivers (B in Fig. 3.15). Sodium (Na^+) is so strongly hydrated that it has a tendency to remain in solution, while potassium will be far less hydrated and can be more easily adsorbed onto clay minerals and rapidly removed from seawater.

Most of the elements which the rivers bring to the ocean are precipitated by organic processes. Organisms can build their own internal chemical environment, and use their energy to precipitate minerals which are not normally stable in seawater. Carbonate-secreting organisms, e.g. foraminifera, molluscs etc., will precipitate aragonite or calcite even when the water is cold and undersaturated with respect to these minerals. Diatoms are so effective in precipitating silica (amorphous silicon dioxide, SiO_2), that in most places the seawater near the surface in the photic zone (photosynthetic zone) is much more depleted with respect to silica than to quartz. Organisms, when they die, will in most cases start to break down by oxidation of organic matter and by dissolution of the mineral skeletons. In reducing environments much of the organic matter will however be preserved. In shallow tropical waters, like on a carbonate bank, the seawater may be saturated with respect to carbonate (calcite), but carbonate also accumulates in cold water like the Barents Sea because the rate of dissolution is slower than the rate of precipitation.

The more efficient organisms are at building skeletons despite undersaturation, the more rapidly they will dissolve. Diatoms, for example, dissolve to the extent of 99–99.9% before they have sunk to the seabed. Only a very small proportion is therefore preserved in sediments.

Photosynthesising organisms in the surface water use up CO_2 and produce oxygen and organic matter:

$$CO_2 + H_2O + nutrients = CH_2O + O_2$$

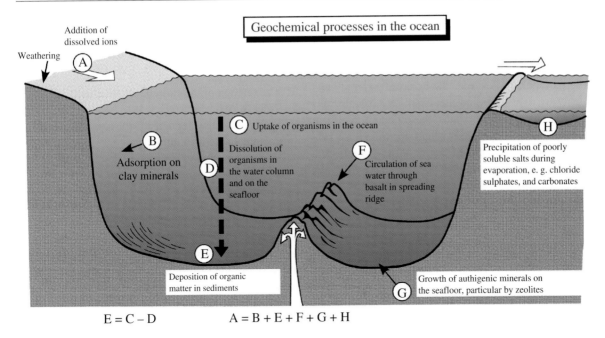

Fig. 3.15 The chemical composition of the seawater remains nearly constant over geologic time. The supply of ions in solution from rivers and from spreading ridges must therefore be equal to the removal of dissolved components by precipitation of minerals and adsorption on clay minerals. The most soluble components (Na, Cl, KCl) are only removed by evaporation

This helps to keep the pH high so that carbonates are stable or dissolve slowly. At depth in the absence of sunlight, respiration and oxidation prevail, releasing CO_2 and thus lowering the pH and increasing the solubility of carbonate.

The depth at which the solubility of carbonate increases relatively rapidly is called the *lysocline*. The depth where the rate of solution is greater than the rate of carbonate sedimentation is called the *carbonate compensation depth* (CCD). Near the equator the CCD may be at 4–5 km, becoming shallower at higher latitudes. Carbonate can therefore still accumulate on the seafloor even when the water column is undersaturated, if the rate of supply exceeds the rate of dissolution. This is analogous to snow accumulating when the temperature is above freezing so long as it falls quicker than it melts This solution of organic material liberates nutrients which can then be returned to the surface through upward flow. In this manner they are repeatedly recycled. The annual biological production in the ocean is therefore many times greater than the supply of nutrients from the land. That fraction of the organic production which has been removed from the ocean by preservation in seabed sediments must be

replaced with nutrients, mostly from land. They cannot be returned to the ocean before the sediments are elevated and subjected to erosion and weathering. The amount of organic matter which is deposited in sediments is a function of the rate of production minus the rate of solution.

The growth of authigenic (newly formed) minerals on the seabed (C in Fig. 3.15) is also an important process in removing elements from seawater. The most significant are the zeolites, which can develop where sediments on the seabed have a high silicate or aluminium content, particularly from volcanic material (glass). They may remove Na^+, K^+ and Ca^{2+} from seawater, but growth can also proceed very much at the expense of elements already present in the sediment (e.g. in the Pacific Ocean). This applies particularly to phillipsite, heulandite, clinoptilite and analcite.

Apart from this there is little direct chemical precipitation from seawater with normal salinity. This is because biological precipitation is more efficient in many cases and prevents the build-up of sufficient concentrations of the elements needed for chemical precipitation. Sulphate-reducing bacteria are active in the uppermost few centimetres of the sediment, however, removing sulphur from seawater in the form of sulphate and reducing it to sulphides which are then precipitated (e.g. iron sulphides, FeS and FeS_2).

In areas that are almost cut off from the open ocean, where evaporation exceeds the freshwater rainfall and supply by rivers, we find that even very soluble salts are being precipitated. Under these conditions, there is enrichment of elements which are otherwise precipitated only to a limited degree through biological or chemical processes, such as Na, Cl, S, Mg and trace elements such as B and Br. The amount of salt thus precipitated in evaporites has probably varied very markedly throughout the geological ages.

Biological and chemical precipitation, as described above, are not sufficient alone to account for the geochemical balance of the ocean. Important geochemical reactions are taking place in the spreading ridges in the oceans (F in Fig. 3.15). Heat from the basalt that is flowing up along the spreading ridge, drives convection cells which cause ocean water to flow through the basalts and up along the ridge. The seawater reacts with hot basalt (basic rock melt) and disolves minerals containing iron and other metals. Since seawater contains sulphur (as SO_4^-), this leads to precipitation of sulphides, for example iron sulphides and copper sulphides. When hot water flows up pipe-like *chimneys* in the ocean floor near spreading ridges, it mixes with the seawater and again sulphides are precipitated. When water is oxidised, iron oxides and manganese oxides are precipitated around the spreading ridges.

3.8.1 Residence Periods for Different Elements in the Ocean

How long does an element spend in the sea after being delivered by rivers, before it is chemically or biologically precipitated? This *residence* period is an expression of how rapidly an element is removed compared to its concentration in the ocean. For example, sodium has the longest period of residence (about 200 million years) because little sodium is removed through biological or chemical precipitation except in evaporite basins. The residence time for potassium is about 1 million years because it is more rapidly adsorbed onto clay minerals and thereby removed. Rare earths have periods of only a few 100 years.

3.9 Circulation of Water in the Oceans

Ocean currents are driven by:

1. The rotation of the earth (Coriolis effect)
2. Tidal forces
3. Differences in water density due to variations in salt content and temperature
4. Wind forces due to atmospheric circulation.

Ocean currents are extremely important for redistributing heat from low latitudes to higher latitudes. This circulation is strongly dependent on the topography of the ocean floor and the distribution of ocean and continents. Bottom currents in the ocean basins are very different from those at the surface, and often flow in opposite directions. While warm surface water flows from the equator to the poles, cold surface water sinks at the poles and flows along the ocean floor to the equator. Both currents are deflected by the Coriolis effect, towards the right in the northern hemisphere and towards the left in the southern hemisphere. While surface currents (like the Gulf Stream) will be deflected eastwards, deeper currents will be deflected towards the west of the oceans (e.g. the Atlantic Ocean). These deep-sea currents which follow the depth contours, may be strong enough to transport silt and fine sand, and the resultant deposits are called *contourites*.

The vertical circulation of seawater is highly sensitive to variations in temperature and salt concentration. In periods with glaciation at the poles, the temperature gradient in the surface water flowing from the equator to the poles is far greater than in non-glacial periods (such as the Mesozoic). Oxygen-rich cold water now flowing down from the polar regions into the ocean basins is important for maintaining oxidising conditions in the deep ocean basins.

Animals and bacteria use oxygen from seawater continuously (for respiration), and oxidation of dead organic material also requires oxygen. If we did not have this downward flow of cold surface water, the water in the ocean basins would be reducing. Some ocean basins are isolated from this circulation, and we may then have a more permanent layering of water based on temperature and salt content. Warm surface water with low density can flow over heavier, colder basal water without the water masses mixing to any major extent. The boundary between warm and cold water masses is called a *thermocline*. If the density difference is largely due to salt content, we call the boundary a *halocline*. A *pycnocline* is the boundary between two water masses with different densities, without a specified cause. Lakes in temperate and cold regions have good circulation. This is due to water attaining maximum density at 4°C, below which temperature density inversion causes turnover.

The addition of freshwater to a basin (e.g. Baltic Sea, Black Sea) also leads to stratification of the water due to salt concentration because brackish water flows on top of marine water with higher salinity. Evaporite basins produce water which is heavy due to its high salt content and therefore forms a layer which flows along the bottom. If a salinity stratification becomes established, it will weaken or destroy the circulation and lead to reducing conditions in the bottom layer. If the density contrast due to salt concentration is greater than that due to temperature, this will impede or prevent the downward flow of cold, oxygen-rich surface water (Fig. 3.16).

There are indications that in previous geological periods (e.g. the Cretaceous) the bottom water in the ocean basins was warm, salty water (about 15°C) compared with the present situation with cold basal water (2–3°) and a normal salinity. A higher average temperature in the oceans leads to reduced CO_2 solubility and a deeper carbonate compensation depth (CCD). The volume of water welling up from deeper water layers to the surface corresponds to the amount of downflow. If we have basal water with high salinity, the reduced density contrast results in less downward flow and consequently less upwelling, so less nutrients are added to the surface water.

We have seen that a number of different processes control the geochemical equilibrium of the ocean. There must also be an equilibrium between the addition and removal of chemical components for the ocean water composition to remain relatively constant. The conditions in which this equilibrium was maintained, however, have varied through geological time. During the first part of the Earth's history, up to about 2.5 billion years ago, the atmosphere was reducing. Most geochemical processes acted very differently then from the way they do now. Weathering was less

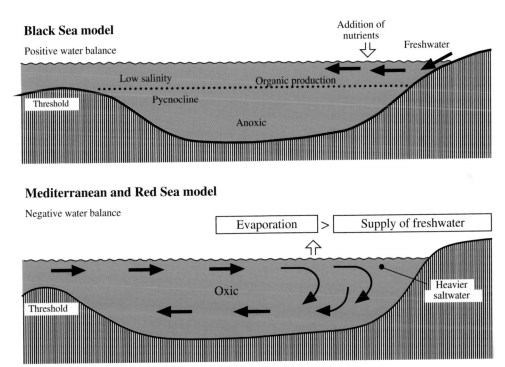

Fig. 3.16 Circulation of water in basins with a surplus of freshwater input compared to evaporation (Black Sea model). This results in poor vertical circulation and anoxic (reducing) bottom waters. The Mediterranean and Red Sea model represents excess evaporation compared to freshwater input. The surface water will then have the highest salinity and density and sink to the bottom. This increases the vertical circulation and helps to maintain oxic conditions

efficient because of the low oxygen concentration in the atmosphere and limited biological weathering. The supply of ions to the ocean via rivers was consequently less. On the other hand, seafloor spreading was most probably faster with more seawater circulated through the spreading ridges. Isotope studies ($^{87}Sr/^{86}Sr$) of seawater in early Precambrian rocks indicate that at that time the composition of seawater was more strongly controlled by circulation through the basalt on the spreading ridges. We can say that the chemical composition of the ocean was buffered by material from the spreading ridge, i.e. mantle material (Veizer 1982).

3.10 Clastic Sedimentation in the Oceans

Clastic sediments are produced chiefly on the continents and are brought to ocean areas through fluvial or aeolian transport. Island arcs associated with volcanism may produce large amounts of sediment compared to their area because they are tectonically active, which leads to elevation and accelerated erosion. Volcanic rocks are for the most part basic and weather quickly, forming large quantities of sediments around volcanic island groups, while fine-grained volcanic ash becomes spread over wide areas. Submarine volcanism may also produce some sediment, for example along the Mid-Atlantic Ridge, but this is very limited.

The main supply of clastic sediment is fed into the ocean through deltas, then transported along the coast and down the continental slope to the abyssal plains. Around Antarctica there is a significant amount of deposition of clastic, glacial sediments. In areas in the middle of the Atlantic Ocean, far from land, the rate of sedimentation is as low as 1–10 mm/1,000 years.

The Atlantic receives a relatively large supply of clastic sediment, in particular from seven major rivers: the St. Lawrence, Mississippi, Orinoco, Amazon, Congo, Niger and Rhine. Exceptionally high sedimentation rates characterise the Gulf of Mexico, where rapid deposition of thick sequences from the Mississippi delta has prevailed since Mesozoic times.

The South American and African continents drain mainly into the Atlantic. The water divide between the Atlantic and the Pacific Oceans lies far to the west in South America, and that with the Indian Ocean in Africa is far to the east (Fig. 7.16).

The Pacific Ocean is surrounded by a belt of volcanic regions and island arcs. There are relatively few rivers that carry large amounts of clastic sediment directly into the Pacific Ocean, in contrast to the Atlantic. Sediment which is eroded, for example on the Asian continent, is deposited in shallow marine areas (marginal seas) such as the Yellow Sea and China Sea. The sediments are cut off from further transport by the island arc running from Japan and southwards. The Pacific Ocean is therefore dominated by volcanic sediments.

Volcanic sedimentation takes the form of volcanic dust and glass, which may be transported aerially over long distances. After sedimentation, volcanic glass will turn into *palagonite*, an amorphous compound formed by hydration of basaltic tuff. Palagonite may then be further converted into montmorillonite or zeolite minerals. The zeolite phillipsite is very widely found in the Pacific, but is scarce in the other oceans. Pumice is also a volcanic product, and may drift floating over great distances. The eruption of volcanoes in the Pacific Ocean area in historic times has shown that large eruptions produce 10^9–10^{10} tonnes of ash, and much the same amount of pumice and agglomerates.

Submarine volcanism, by contrast, produces very little ash to form sediment. The lava which flows out onto the seabed will solidify as an insulating crust on

Table 3.1 Review of the ratio between mechanical and chemical denudation of the different continents (After Garrels and Machenzie 1971)

Continent	Annual chemical denudation (tonnes/km)	Annual mechanical denudation (tonnes/km)	Ratio mechanical/chemical denudation
North America	33	86	2.6
South America	28	56	2.0
Asia	32	310	9.7
Africa	24	17	0.7
Europe	42	27	0.65
Australia	2	27	10.0

contact with the water (often forming pillow lava), so that little volcanic matter goes into suspension.

Weathering and erosion processes are responsible for the entire volume of sediment which can be deposited in sedimentary basins. Material added by rivers takes the form of clastic and dissolved matter. The ratio between the quantities of these two forms of sediment addition is a function of precipitation, temperature and relief. Dry areas, like Australia, produce mainly clastic material, while the African continent produces mainly dissolved material because of the intensive weathering in some parts of the continent (Table 3.1).

Further Reading

Garrels, R.M. and Machenzie, F.T. 1971. Evolution of Sedimentary Rocks. W.W. Norton & Co Inc., New York, NY, 397 pp.

Kenneth 1982. Marine Geology. Prentice Hall. Englewood Cliffs. 813 pp.

Chamley, H. 1989. Clay Sedimentology. Springer, New York, 623 pp.

Chester, R. 1990. Marine Geochemistry. Unwin Hyman, London, 698 pp.

Eslinger, E. and Pevear, D. 1988. Clay Minerals for Petroleum Geologists and Engineers. SEPM Short Course 22.

Garrels, R.M. and Christ, C.L. 1965. Solutions, Minerals and Equilibria. Harper and Row, New York, 450 pp.

Manahan, S.E. 1993. Fundamentals of Environmental Chemistry. Lewis Publ., Chelsea, MI, 844 pp.

Saigal, G.C. and Bjørlykke, K. 1987. Carbonate cements in clastic reservoir rocks from offshore Norway – Relationships between isotopic composition, textural development and burial depth. In: Marshall, J.D. (ed.), Diagenesis of Sedimentary Sequences. Geological Society Special Publication 36, 313–324.

Veizer, J. 1982. Mantle buffering and the early Oceans. Naturvissenshaffen 69, 173–188.

Velde, B. 1995. Origin and Mineralogy of Clays. Springer, Berlin, 334 pp.

Weaver, C.E. 1989. Clays, muds and shales. Developments in Sedimentology 44, 819 pp.

Chapter 4

Sandstones and Sandstone Reservoirs

Knut Bjørlykke and Jens Jahren

4.1 Introduction

About 60% of all petroleum reservoirs are sandstones; outside the Middle East, carbonate reservoirs are less common and the percentage is even higher. The most important reservoir properties are porosity and permeability, but pore geometry and wetting properties of the mineral surfaces may also influence petroleum production. Sandstones provide reservoirs for oil and gas, but also for groundwater which is a fluid that is becoming increasingly valuable.

The outer geometry and distribution of sand bodies is determined by the depositional environment and the reservoir properties. The internal properties (porosity, permeability) are, however, critical for petroleum recovery.

The properties of sandstone reservoirs are functions of the primary composition, which is controlled by the textural and mineralogical composition (provenance), of the depositional environment and of the diagenetic processes near the surface and during burial.

Sand and sandstones are rocks which consist largely of sand grains, i.e. sedimentary particles between 1/16 and 2 mm in diameter. However, sandstones also contain greater or lesser amounts of other grain sizes and we find transitions to more silt- and clay-rich rocks. Most sandstones have a well-defined upper grain-size limit with variable contents of silt and clay. If they have a significant content of coarser grains (pebbles) we call them conglomeratic sandstones. Most classification systems are based on the relationship between the relative quantity of sand-sized grains, the composition of the sand grains, and the clay and silt content (matrix).

If we use a four-component diagram we can distinguish between clay, and sand grains which consist of quartz, feldspar and rock fragments (or unstable rock fragments, U.R.F.) (Fig. 4.1). Rock fragments consisting of microcrystalline (or cryptocrystalline) quartz (including chert) are usually classified together with the quartz mineral grains. Sandstones with more than 25% feldspar and a low content of rock fragments are called *arkoses*. If the percentage of rock fragments is high, we speak of *lithic sandstones* which are normally derived from very fine-grained sedimentary rocks or basalts and intrusive igneous rocks where one sand grain often consists of several minerals.

Quartz arenite or *orthoquartzite* are the terms for pure quartz sandstones which contain less than 5% feldspar or rock fragments. Sandstones with moderate feldspar contents (5–25%) are called *subarkoses*. When granitic rocks and other coarse- to medium-grained rocks are broken down by weathering or erosion, they form sand grains consisting for the most part of single minerals, mostly quartz and feldspar.

The prerequisite for forming arkose is that not too much of the feldspar in the source rock has been weathered to clay minerals (e.g. kaolinite). Arkoses are therefore formed if granites and gneisses are eroded rapidly in relation to weathering, and the sediments are buried in a basin after a relatively short sediment transport. In the great majority of cases arkose is therefore associated with sedimentary rift basins formed by faulting in gneissic and granitic rocks, i.e. continental crust. In cold climates the rate of weathering is reduced, thus preserving feldspar and unstable rock fragments.

K. Bjørlykke (✉)
Department of Geosciences, University of Oslo, Oslo, Norway
e-mail: knut.bjorlykke@geo.uio.no

K. Bjørlykke (ed.), *Petroleum Geoscience: From Sedimentary Environments to Rock Physics*,
DOI 10.1007/978-3-642-02332-3_4, © Springer-Verlag Berlin Heidelberg 2010

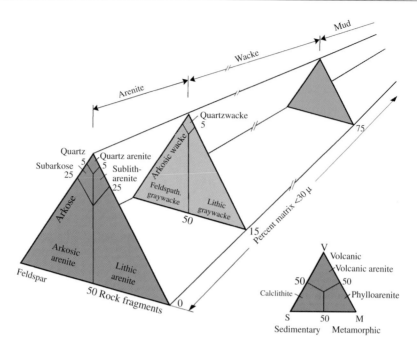

Fig. 4.1 Classification of sandstones (from Dott 1964)

In tectonically stable areas with mature relief there is far more time for weathering of both *in situ* bedrock and sediment in transit. In such environments sand particles (grains) are deposited and eroded many times before reaching their final deposition site. A greater proportion of the feldspar grains will then break down during transport and subarkoses or orthoquartzites will be deposited. Orthoquartzites (or quartz arenites) are the purest quartz sandstones, formed when weathering has eliminated virtually all the unstable minerals to leave a concentration of quartz and some heavy minerals. This is particularly true of beach sand where nearly all the clay particles have been removed. Sand deposited by rivers (fluvial sand) is less well sorted and may contain more clay. Sandstones with more than 15% matrix are called *greywackes*.

Sand which is transported in suspension or through mass flow (e.g. turbidites) can have poorer sorting, a high matrix content, and grade into sandy mudstones and form greywackes. Clay minerals in sandstones may form after deposition during diagenesis from alteration of feldspar, mica and rock fragments. This makes the sediments less well sorted than at the time of deposition.

Basic rocks like gabbro and basalt and minerals like amphibole and pyroxene are inherently unstable both mechanically and chemically. Primary sand grains of volcanic or basic rocks may break down to become

part of the matrix and it is then difficult to distinguish this material from the primary matrix. Greywacke is therefore typical of areas where the sand grains are derived from volcanic or basic rocks along converging plate boundaries (fore-arc, inter-arc, back-arc basins). Weathering of basic rocks will produce nearly exclusively clay since there are no quartz grains.

We have seen that the various types of sandstone reflect different source rocks and areas with varying tectonic stability. Studies of different types of sandstone and their mineralogical maturity can therefore be used as palaeo-indicators of relief and climate, and also of tectonic deformation in the geological past.

4.2 Prediction of Reservoir Quality

The properties of all reservoir rocks are continuously changing, from the time the sediments are deposited through to their burial at great depth and during any subsequent uplift. This is a combined function of mechanical compaction and of chemical processes involving dissolution and precipitation of minerals.

At any given burial depth the properties depend on the composition of the sandstones when at shallow depth, and on their temperature and stress history during burial. Practical prediction of the porosity and permeability during exploration and production is only possible if the processes that change these parameters are understood.

It should be realised that the starting point for the diagenetic processes is the initial sandstone composition. This is a function of the rocks eroded (provenance), transport, and depositional environments. Diagenetic models must therefore be linked to weathering and climate, sediment transport, facies models and sequence stratigraphy, and should be integrated in an interdisciplinary *basin analysis*.

Diagenesis is often considered a rather specialised field of sedimentology and petroleum geology, but it embraces all the processes that change the composition of sediments after deposition and prior to metamorphism. The most important factor in predicting reservoir quality at depth is the primary clastic composition and the depositional environment (Fig. 4.2). The diagenetic changes also determine the physical properties of sandstones, such as seismic velocities (V_p and V_s) and the compressibility (bulk modulus, see Chap. 11). This is also critical when predicting physical rock changes during production (see 4D seismic, Chap. 19).

The main diagenetic processes are:

(1) Near-surface diagenesis. Reactions with fresh groundwater (subsurface weathering). In dry environments, with saline water concentrated by evaporation. Sand may also be cemented with carbonate cement near the seafloor.

(2) Mechanical compaction, which reduces the porosity by packing the grains closer together and by grain deformation and fracturing, increasing their mechanical stability. Mechanical compaction is a response to increased effective stresses during burial and follows the laws of soil mechanics.

(3) Chemical diagenesis (compaction), which includes dissolution of minerals or amorphous material and precipitation of mineral cement. The clastic minerals in the primary mineral assemblage are not in equilibrium, and there is always a drive towards thermodynamically more stable mineral assemblages. Kinetics determine the reaction rates, which for silicate reactions are extremely slow so temperature plays an important ole.

(4) Precipitation of cement (i.e. quartz cement) will increase the strength of the grain framework and prevent further mechanical compaction. The sandstone is then overconsolidated – not due to previously higher stress, but due to cementation. Further compaction will then mostly be controlled by the rate of dissolution and precipitation.

In the following the diagenetic processes typical of different burial depths will be discussed.

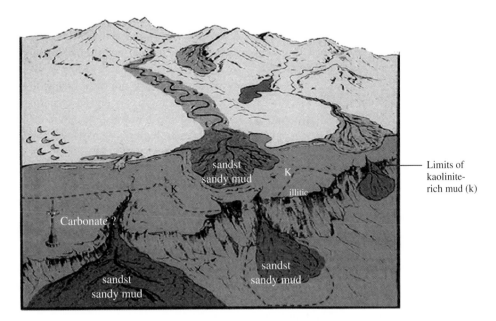

Fig. 4.2 Schematic illustration of a sedimentary basin on a continental margin. The primary composition of the sediments is a function of the provenance, transport and depositional environment. Fluvial, deltaic and shallow marine sediments will be flushed by meteoric water after deposition, particularly in humid climates

4.3 Early Diagenesis

As soon as sediments are deposited, early diagenetic reactions start to modify the primary sediment composition. At very shallow burial depth (<1–10 m), sediments have the maximum potential to react with the atmosphere or water, both by fluid flow and diffusion. Transport of dissolved solids by diffusion and fluid flow (advection) is most efficient near the surface; in the case of diffusion within about 1 m of the seabed. The potential for sediments to change their bulk composition after deposition is therefore much higher at shallow depth than at greater burial. Near the surface on land, and also in the uppermost few centimetres of the seabed, the conditions may be oxidising, while at greater depth in the basin they are always reducing.

Precipitation of minerals due to porewater concentration by evaporation can only occur on land or at shallow depth within enclosed basins (see Chap. 6).

On land, sediments are exposed to air and fresh (meteoric) water. Weathering is partly due to reactions with oxygen in the atmosphere and partly due to dissolution of minerals in freshwater, which is initially undersaturated with respect to all the minerals present. These are soil-forming processes which can be considered to be examples of early diagenesis.

In desert environments groundwater and occasional rainwater may become concentrated through evaporation, causing precipitation of carbonates and also silicates. Coatings of red or yellow iron oxides and clays frequently form on desert sand and this may subsequently retard or prevent quartz cementation at greater depth.

In the sea, the water above the seabed is normally oxidising. Only where there is poor water circulation (poor ventilation) is the lower part of the water column likely to be reducing, though the phenomenom is more widespread in lakes and inland seas like the Black Sea. However, even below well-oxygenated water, oxidising conditions extend in most cases for only a few centimetres into the sediments, since oxygen is quickly consumed by the oxidation (decay) of organic matter in the sediment. This is for the most part facilitated biologically by bacteria. Accumulating sediments normally contain sufficient organic matter to serve as reducing agents in the porewater. This organic matter is comprised of both the remains of bottom fauna and of pelagic organisms, including algae, accumulating on the seafloor, and also often includes terrestrial plant debris transported into the basin.

4.4 Redox-Driven Processes on the Seafloor

Across the *redox boundary* there is a high gradient in the concentration of oxygen and sulphate, and of ions that have very different solubilities in oxygenated and reduced water. The redox boundary is usually just 1–20 cm below the seafloor and represents equilibrium between the supply of oxygen by diffusion, and its consumption by the (mostly biological) oxidation of organic matter. The oxygen content in the porewater thus decreases rapidly below the water/sediment interface, providing a concentration gradient for the downward diffusion of oxygen into the uppermost sediments.

The rate of downward diffusion of oxygen is a function of the concentration gradient of oxygen in the porewater and the diffusion coefficient in the sediments. The diffusion coefficient in coarse-grained sand is higher than in mud and therefore sand tends to have a deeper redox boundary than mud.

Oxygen can also be consumed in the sediments by the oxidation of elements like iron and manganese in minerals, but this is rare in marine environments and more common in continental deposits. In most marine environments there is enough organic matter to serve as reducing agents and therefore little oxidation of iron in minerals takes place, which explains why marine sediments do not normally acquire a red colour. A notable exception is red oxidised mud which may form in marine environments characterised by slow sedimentation rates and low organic productivity. These muds are not very common but occur in some deep-water facies and also in shallower water environments with low sedimentation rates.

Uranium is highly soluble in seawater as uranyl (UO^{2+}) and precipitates as reduced uranium oxide (UO_2) on organic matter in the water column and below the redox boundary. There is thus a strong concentration gradient transporting uranium from the seawater down into the sediments. The adsorption of uranium onto organic matter settling on the seafloor coupled with restricted ventilation of the water above the sediment at the time of deposition, makes source rocks like the Kimmeridge shale strongly enriched in uranium, causing peaks on the gamma ray well log curve.

Iron and manganese may be transported upwards through the sediments in the reduced state by diffusion and then precipitate on the seafloor as oxides because of their reduced solubility in the oxidised state. Iron may also be precipitated below the redox boundary as iron sulphides or iron carbonate (siderite), though iron carbonates will not form during sulphate reduction. This is because all the iron will be precipitated as sulphide, which has a much lower solubility than siderite. Manganese is not precipitated as sulphides because of the high solubility of Mn-sulphides, but may be precipitated as Mn-carbonate in the reduced zone.

At a depth where there is practically no more free dissolved oxygen in the porewater, sulphates can be used by sulphate-reducing bacteria. The reduction of sulphates produces sulphides such as pyrite.

The composition of clastic sediments is modified by the addition of new components produced locally within the basin:

(1) Biogenic carbonates and silica.
(2) Authigenic minerals precipitating near the seabed such as carbonates, phosphates, glauconite, chamosite, sulphides and iron and manganese minerals.
(3) Meteoric water-flushing causing leaching of feldspar and mica and precipitation of kaolinite beneath the seafloor.

4.5 Importance of Biogenic Activity

Bioturbation plays an important role in changing the textural composition of the sediments after deposition. The burrowing organisms eat mud and thereby oxidise organic matter and physically destroy the primary lamination. Sediments overturned by bioturbation become more exposed to oxidation at the sea or lake bottom. Bioturbation may reduce the porosity and permeability of sandy laminae by mixing clay with clean sand.

Bioturbation will also destroy thin clay laminae, which may significantly increase the vertical permeability and this may be very significant for reservoir quality. Undisturbed primary lamination may be evidence of rather rapid sedimentation giving little time for a burrowing bottom fauna to become established, or alternatively indicate strongly reducing conditions restricting the fauna. Black shales usually have well preserved lamination due to lack of burrowing organisms. The presence or absence of burrowing also influences the physical properties, particularly the difference in velocity and resistivity parallel and vertical

to bedding (anisotropy) and this may be important for geophysical modelling.

Burrowing worms produce faecal material which may develop into smectite-rich clays, which in turn may develop into chlorite coatings, thus improving reservoir quality. Early diagenetic formation of coatings on quartz grains is extremely important due to its role in preserving porosity at greater depth.

Most clastic depositional environments have some organisms producing organic matter which, at least in part, is incorporated within the sediments. Both sandstones and mudstones nearly always contain significant amounts of biogenic material from calcareous, and sometimes also siliceous, organisms and this may later be an important source of carbonate and silica cement at greater burial depth.

Marine organisms composed of aragonite dissolve during relatively shallow burial and calcite precipitates either as replacements within the fossils (neomorphism) or as cement in pore space between the grains.

Carbonate cement in sandstones may form layers or concretions and in most cases is derived from biogenic carbonate, particularly from organisms composed of aragonite. Siliceous organisms composed of opal (e.g. diatoms or siliceous sponges) may be an important source of micro-quartz coatings on quartz grains at greater depth.

Carbonate cements in both mudstones and sandstones are mostly due to dissolution and reprecipitation of biogenic carbonate or early aragonite cement. There are usually no other major sources of carbonate cement. In the sulphate-reducing zone, carbonate concretions form, often with a negative $\delta^{13}C$ due to the CO_2 produced during sulphate reduction. Carbonate concretions in cores may be mistaken for continuous carbonate layers but it is possible to recognise that they are concretions (Walderhaug and Bjørkum 1998). Even if CO_2 is generated from organic matter, there are few Ca^{2+} sources available in sandstones or mudstones for making calcite. Leaching of plagioclase can supply some Ca^{2+}, which can be precipitated as calcite, but this can only account for very small amounts of the calcite observed in such sediments. The distribution of carbonate cement is related to facies and sequence stratigraphy.

The evolution of pelagic planktonic calcareous organisms in the Mesozoic drastically increased the supply of carbonate on the seafloor, including in deeper waters. Before then most of the carbonate was produced by benthic organisms restricted to shallow water facies. Upper Jurassic and younger sandstones often contain abundant calcite cement due to the "rain" of calcareous algae, foraminifera and other planktonic organisms settling on the seafloor. Silica-producing organisms may also be important for diagenesis and reservoir quality at greater burial. Organisms like siliceous sponges are composed of amorphous silica which at higher temperatures will be dissolved and replaced by opal CT and quartz. Diatoms and radiolarians may also be a major source of silica which will be precipitated as quartz. Diatoms appeared during the Cretaceous and have been a major source of amorphous silica during the Cainozoic. Diatoms can produce pure siliceous rocks like the Tertiary Monterey Fm of California, which is both a source rock and a fractured reservoir rock.

Biogenic carbonate is in most cases the main source of calcite cement. The distribution of such cement must therefore be linked to sedimentary facies, more specifically to biological productivity relative to the clastic sedimentation rate. Environments with low clastic sedimentation rates, particularly submarine highs, often have high organic carbonate production.

4.6 Meteoric Water Flow and Mineral Dissolution

Meteoric water is rainwater which infiltrates the ground. Initially this water is distilled water and therefore undersaturated with respect to all minerals. The

reactions between meteoric water and the land surface are an important part of the weathering process. Rainwater contains carbon dioxide (CO_2) and sulphur dioxide (SO_2) from the air and is therefore slightly acidic, producing carbonic acid (H_2CO_3) and sulphuric acid (H_2SO_4).

Some of the rainwater seeps down to the groundwater, and as long as the groundwater table is above sea level, meteoric water will flow along the most permeable beds into the basin. Meteoric water will first dissolve carbonates and then slowly dissolve unstable minerals like feldspar and mica (Fig. 4.3).

Decaying organic matter in the ground produces CO_2 which is added to the groundwater, making it more acid. Humic acids generated by decaying plants also hasten the weathering reactions. At the same time this acidity is neutralised by weathering reactions with silicate minerals like feldspar and the dissolution of carbonates which consume protons (H^+). As the groundwater reacts with minerals and in some cases with amorphous phases, it will approach equilibrium with many of the minerals present and this will happen first with carbonates. In the case of silicate minerals these reactions are very slow so the porewater may remain under- or supersaturated for a long time with respect to silicate minerals like quartz and feldspar.

Depending on the elevation of the groundwater table and the distribution of permeable layers (sandstones), the flow of meteoric water can extend beneath the seafloor far out into sedimentary basins. Reactions between meteoric water and minerals occur in the

Fig. 4.3 Diagenetic processes in shallow marine environments. Sandstones deposited in these environments will be flushed by meteoric water flow and/or from the delta top, causing dissolution of feldspar and mica. Calcareous fossils and early carbonate cement may be a very important addition to the composition of the sandstones. The occurrence of siliceous organisms such a sponges can strongly influence reservoir quality at depth

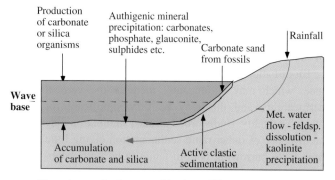

The primary clastic composition is modified by:
1) Meteoric water leaching and precipitation of kaolinite.
2) By addition of biogenic carbonate and silica.
3) By precipitation of authigenic minerals on the the seafloor.

ground and are a kind of subsurface weathering along the groundwater flow paths. Leaching by meteoric water is generally strong in fluvial and alluvial sediments. Even within dry river beds there is a focused flow of groundwater.

The groundwater level represents the head (potentiometric surface) for groundwater flow and groundwater therefore has a potential to flow through sediments or other aquifers far below sea level.

The rates of leaching of minerals like feldspar and mica and the precipitation of kaolinite are functions of the flux of groundwater flowing through each rock volume per unit of time. These are in principle weathering reactions similar to those which take place during normal weathering in a humid climate. Cations like Na^+ and K^+ are stripped from silicate minerals like feldspar and mica and brought into solution.

These reactions can be written as below:

$$2K(Na)AlSi_3O_8 + 2H^+ + 9H_2O = Al_2Si_2O_5(OH)_4 + 4H_4SiO_4 + 2K^+(2Na^+)$$

Feldspar   Kaolinite   dissolved silica   dissolved cations

$$2KAl_3Si_3O_{10}(OH)_2 + 2H^+ + 3H_2O = 3Al_2Si_2O_5(OH)_4 + 2K^+$$

Muscovite   Kaolinite

We see from these reactions that low K^+/H^+ ratios will drive the reactions to the right. Dissolution of feldspar and mica and precipitation of kaolinite require that the reaction products, Na^+, K^+ and silica, are constantly removed and that there is a supply of new freshwater which is undersaturated with respect to feldspar and mica. Without a through flow of water these reactions stop because the reaction products on the right hand side of the equations are not removed. Groundwater must flow into the ground and up to the surface again or on to the seafloor. A clay coating on feldspar often remains and the dissolved aluminium and some of the silica is precipitated as kaolinite, so there is a rather small increase in porosity and reduced permeability (Fig. 4.4a). The pores between the kaolinite crystals may be too small (Fig. 4.4b, c) to be filled with oil so that the oil saturation is reduced in kaolinite-rich sandstones (see Chap. 20).

The silica released from feldspar dissolution can normally not be precipitated as quartz because of the low temperature near the surface, but remains in solution even if the porewater is highly supersaturated with respect to quartz. Silica must, however, also be removed along with potassium by the flowing water. If the silica concentration in the porewater increases too much, kaolinite is no longer stable and smectite will precipitate instead. This happens in sediments rich in volcanic material or biogenic silica and where the flux of meteoric water is low. In a desert environment evaporation of groundwater may increase the silica concentration and make smectite more stable.

The porewater does not have to be acidic for kaolinite to form, but the K^+/H^+ ratio must be low. If the pH is high the K^+ concentration has to be correspondingly lower. Authigenic kaolinite may also form in impure limestones as a result of meteoric water flushing, and the porewater is then certainly not acidic. Even if there is only a small amount of carbonate it will buffer the composition of the porewater.

The average groundwater flux is determined by the rainfall and the percentage of water infiltration into the ground. In moderately humid climates the rainfall may be 1 metre/year and the infiltration in the order of 0.1–0.3 metres/year. High-permeability subsurface pathways (aquifers) focus the flow. The aquifers may be sand or gravel beds in muddy sediments. Meteoric water may penetrate deeply into sedimentary basins because of the potential created by the head of the groundwater table but the flux of meteoric water decreases strongly in the deeper parts of basins.

The meteoric water will gradually become less undersaturated with respect to minerals like feldspar and mica and its leaching capacity will gradually diminish. The most intense mineral leaching will therefore occur near the surface or at relatively shallow depth beneath the seafloor. In areas with low sedimentation rates the total flow of water through the sediment will be higher because the sediments remain longer at shallow depth. If the sediment stays in the zone intensively flushed by meteoric water, the amount of feldspar leaching will be high. In basins with high sedimentation rates, synsedimentary faulting

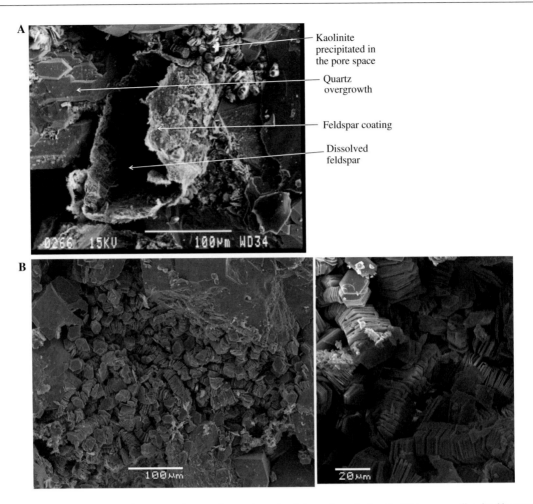

A

— Kaolinite
precipitated in
the pore space

— Quartz
overgrowth

— Feldspar coating

— Dissolved
feldspar

B

Fig. 4.4 (**a**) Scanning electron microscope picture of a sand-stone (Brent Group) from the North Sea. The scale is 0.1 mm (100 μm). In the centre of the picture we see a cavity left by a dissolved feldspar grain. A clay rim around the feldspar remains undissolved, outlining the primary grain morphology. In the *upper part*, authigenic kaolinite crystals are forming small (10–20 μm) booklets. They have formed from the silica and

aluminium released when the feldspar was dissolved by meteoric water. To the *left*, authigenic quartz is growing on clastic quartz. Note the relatively large pores between quartz and feldspar grains and the small pores between kaolinite crystals. (**b**) Pore-filling authigenic kaolinite. We see that the pores between the kaolinite crystals are very small – only 1–2 μm (from T.E. Maast unpublished)

(i.e. growth faults) may disconnect sand bodies from the main freshwater aquifers. The degree of feldspar leaching could then be used as an indication of the conductivity in the reservoir.

River water and groundwater is usually supersaturated with respect to quartz but undersaturated with respect to amorphous silica. About 10–30 ppm dissolved silica is common in groundwater and shallow porewater while the solubility of quartz at 20°C is only 4–5 ppm, showing that quartz does not form at low temperatures. In very alkaline water (i.e. E. African lakes), quartz may precipitate at low temperatures.

The early burial history of sandstones is not well studied for the simple reason that cores are not normally taken at depths shallower than 1–1.5 km in offshore basins, while onshore, erosion may have removed most of the youngest sequence. Looking at sandstone thin sections one can often get the impression that kaolinite is precipitated at a relatively late stage because it is a pore-filling mineral that may subsequently be surrounded by quartz cement. More detailed textural studies and isotopic evidence indicate that the kaolinite is formed early and may be enclosed in quart cement. Pore-filling kaolinite must, however, have been hanging to the pore wall and may have been pushed aside by growing quartz cement.

The isotopic composition ($\delta^{18}O$) of kaolinite suggests that it precipitated at relatively low temperatures, in the range 30–60°C, depending on the assumptions made about the isotopic composition of the porewater (Glasmann 1989). These temperatures are little higher than should be expected during meteoric water flushing and it is possible that some of the kaolinite may be recrystallised at a higher temperature, resetting the isotopic composition. Much of what has been described or analysed as kaolinite has turned out to be dickite, which has the same composition but often with thicker, more blocky crystals. Studies have shown that dickite often replaces some of the kaolinite when temperatures exceed 100°C.

Another possibility is that kaolinite may form diagenetically from other precursor minerals such as gibbsite ($Al(OH)_3$) or amorphous aluminium compounds. Kaolinite could then form without meteoric water flushing since such reactions do not produce any other cations like K which would have to be removed. In the North Sea basin abundant authigenic kaolinite is found in the shallowest reservoirs (1.5–2 km) where there is very little or no quartz cement in sandstones and this is the best evidence that most of the kaolinite formed early at shallow depth. The fact that kaolinite is much more abundant in shallow marine and deltaic sandstones than those deposited on submarine slopes is also evidence that kaolinite forms at shallow depth.

4.7 Consequences for Reservoir Quality

Meteoric water flushing dissolves feldspar and mica and precipitates authigenic clay minerals, most commonly kaolinite. This dissolution produces holes which are secondary pore spaces (secondary porosity) but the precipitation of clay minerals like kaolinite reduces the porosity, so that there is little net gain in pore space. Authigenic kaolinite tends to occur as pore-filling minerals and this reduces the permeability. Clean well-sorted sand may increase its specific surface and pore size distribution due to the authigenic kaolinite. The smaller pores (<0.005 mm) in between the authigenic kaolinite crystals may be too small to be filled with oil because of the high capillary entry pressure necessary to infiltrate these pores. The total water saturation will consequently then be higher in the reservoir rock.

Authigenic kaolinite usually occurs as clusters and is rarely pervasive through the sandstones, allowing oil to flow between and around the most densely kaolinite-cemented pores. However, if the kaolinite is altered to illite at greater depth, the damage to the reservoir may be much more severe, due to permeability reduction.

4.8 Mechanical Compaction of Loose Sand

During the first part of its burial history (0–2 km) well-sorted sand is generally still loose if it is not carbonate-cemented. Mechanical compaction may nevertheless be very significant. Experimental compaction of loose sand with an initial porosity 40–42% shows that, depending on grain strength and grain size, the porosity may be reduced to 35–25% at stresses of 20–30 MPa corresponding to 2–3 km of burial for normally pressured rocks (Fig. 4.5). The experimental data show that well sorted coarse-grained sand is more compressible than fine-grained sand (Chuhan et al. 2002, 2003). Overpressure reduces the effective stress and will then preserve porosity due to reduced mechanical compaction.

In sedimentary basins with normal geothermal gradients, quartz cementation will stabilise the grain framework and prevent further mechanical compaction at about 2 km burial depth (80–100°C) (Fig. 4.6). At greater depth, compaction is not primarily a function of effective stress but temperature. In cold sedimentary basins (low geothermal gradients) quartz cementation may not start before 4–6 km burial depth and porosity loss will then occur by mechanical compaction and severe grain crushing up to about 50 MPa effective stress.

The degree of porosity loss by mechanical compaction determines the intergranular volume (IGV) at the onset of chemical compaction (quartz cementation).

The IGV measured in some North Sea sandstones varies from about 38 to 28% (Walderhaug 1996) and the net porosity after precipitation of 10% quartz cement will then be very different.

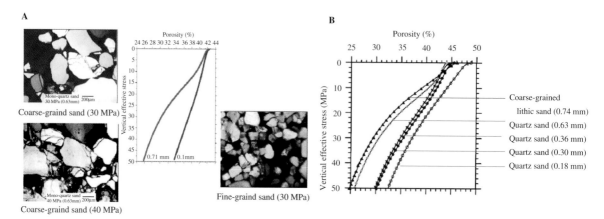

Fig. 4.5 (**a**) Experimental compaction of fine-grained and coarse-grained sand showing that well sorted fine-grained sand is less compressible than coarse-grained sand. (**b**) The porosity loss as a function of grain size due to more grain crushing (from Chuhan et al. 2007)

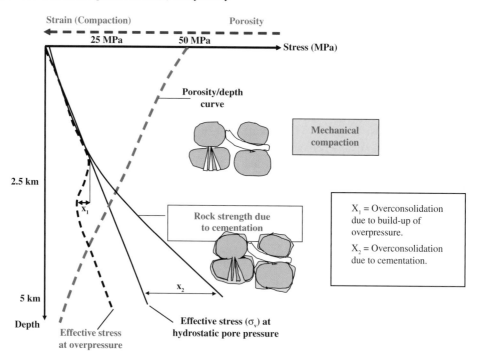

Fig. 4.6 Before sandstones become cemented (at 80–100°C) they compact mechanically as a function of effective stress (depth) by grain reorientation and grain breakage. Relatively small amounts of quartz cement (2–4%?) make the sandstone stiffer and "overconsolidated" so that there is little mechanical compaction (strain) at greater depth (higher stresses)

4.9 Sandstone Reservoirs Buried to Intermediate Depth (2.0–3.5 km, 50–120°C)

In basins where there has been mostly continuous subsidence, reservoirs buried to depths shallower than about 2.0–2.5 km are still loose or only poorly cemented, except where there is carbonate cement or high geothermal gradients. This is well documented in most of the North Sea basin and parts of the Gulf Coast Basin (Sharp and McBride 1989, Bjørlykke et al. 1992). In the Statfjord Field where the Middle Jurassic Brent sandstone is buried to 2.5–3 km, there are intervals that are so poorly cemented that it is difficult to obtain good cores because the sandstone disintegrates in the core barrel. Loose sand grains may then be produced with the oil (sand production).

Prior to quartz cementation or other types of cementation the sand grains compact mechanically by sliding and reorientation. Sand grains may also fracture under the overburden stress; coarse-grained sand compacts more due to grain crushing than well sorted fine-grained sand (Fig. 4.5). In basins like the North Sea the quartz cementation increases the rock strength at 2–3 km burial depth (80–100°C) but coarse-grained sand may additionally show significant compaction due to grain fracturing.

Poorly sorted sandstones and sand with rock (lithic) fragments lose much of their porosity at rather shallow (1–2 km) depth. Quartz cementation strengthens the rocks at a faster rate than the increase in vertical stress from the overburden. Only 2–4% quartz cement will in most cases effectively shut down further mechanical compaction in sandstones, so that further compaction is mainly chemically controlled by the rate of mineral dissolution and precipitation (Fig. 4.6). The mechanical compaction is important because it determines the intergranular volume (IGV) which is the porosity prior to quartz cementation. This is typically 25–30% or even more for well sorted quartz-rich sandstones. Sand with even relatively small amounts of detrital clay will compact more than clean sand. Lithic (rock) fragments also compact more readily (Pittman and Larese 1989) and this is reflected in their lower IGV.

Generally, in relatively well-sorted quartz arenites and feldspathic sandstones the porosity is to a large extent destroyed by quartz cementation (Fig. 4.7a).

The amount of quartz cement is mainly a function of the grain surfaces available for quartz precipitation and the time-temperature integral (Walderhaug 1994). High geothermal gradients and slow subsidence rates will therefore tend to increase the amount of quartz cement at a specific depth.

In several of the Upper Jurassic reservoir rocks from the North Sea, amorphous silica from *Rhaxella* sponges dissolved to produce high supersaturation of silica relative to quartz. This caused precipitation of a coating of minute quartz crystals on the surface of clastic quartz grains (Fig. 4.7b, c). This coating of micro-quartz has prevented or retarded the precipitation of later quartz cement and is the main reason for high porosity and good reservoir quality at great depth (up to 5 km) in these reservoir rocks. The micro-quartz precipitated at low temperature (60–80°C), when the

porewater was highly supersaturated with respect to quartz through the dissolution of Opal A or Opal CT and while the quartz growth rate was low. At higher temperatures when unstable silicates like Opal A, Opal CT and smectite have dissolved, the porewater will only be slightly supersaturated with respect to quartz, insufficient to precipitate quartz on the micro-quartz surfaces which requires higher supersaturation than normal quartz (Aase et al. 1996). *Rhaxella* had not evolved before the Upper Jurassic and so older sandstones like the Middle Jurassic Brent sandstones do not have this type of micro-quartz.

At temperatures above about 100–120°C some of what we have called kaolinite has recystallised to dickite, which has the same chemical composition. Dickite often occurs as slightly thicker crystals and can also be distinguished from kaolinite on XRD scans. Kaolin or kandite may be used as a common name for these clay minerals.

Carbonate-cemented intervals may be effective barriers to fluid flow. This can be detrimental to the reservoir quality, though in some cases may be useful if they are laterally extensive. Such low permeability layers may then prevent the flow of gas from below the oil/water contact and from above the gas/oil contact into an oil-producing well. This is called coning.

The replacement of K-feldspar or plagioclase by albite is often observed in sandstone buried to about 3 km or more and is referred to as *albitisation*. Albite becomes more stable than K-feldspar because Na^+ is normally the dominant cation in the porewater while the potassium concentration is reduced due to removal by the clay mineral reactions. Albitisation is normally observed as a partial replacement of K-feldspar or plagioclase grains, which does not change the reservoir properties very much. Albitisation of plagioclase will, however, release some Ca^{2+} that may then precipitate as calcite (Boles 1982), though the amount is rather limited.

Smectite may be present in some muddy, and particularly volcanic, sandstones which have been flushed with limited amounts of meteoric water. At temperatures from about 70 to 80°C smectite dissolves and is replaced by mixed-layer minerals and illite. Sandstones containing smectite normally have poor reservoir quality.

Dissolution of smectite and precipitation of illite and quartz will cause a sharp increase in the seismic velocity and rock density and this mineral transition may therefore show up as a horizontal reflector on

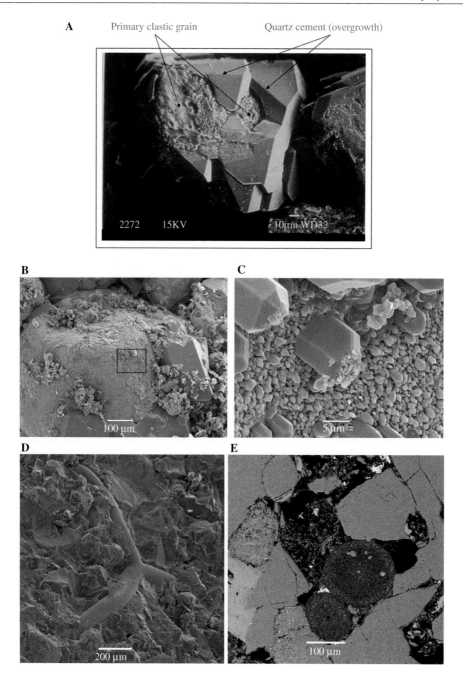

Fig. 4.7 (a) SEM image of quartz cement which has grown on detrital quartz grains into the pore space. Note that the authigenic quartz crystals have smooth crystal surfaces while the primary quartz grain has an irregular abraded surface from weathering and transport of the sand grain. Scale bar 10 μm (0.01 mm). Oil inclusions may be found in between the primary grains and the cement, and also in the cement. Sandstone from the Brent Group, North Sea. (b) A thin layer of small quartz crystals covering clastic quartz grains has been precipitated at high silica supersaturatation caused by dissolution of organic silica (sponge spicules). (c) is an enlargement of Fig. 4.4b showing micro-quartz crystals that are not overgrown at greater depth. From T.E. Maast (unpublished). (d) Pictures (SEM) of the siliceous sponge (*Rhaxella*) which is the source of much of the early quartz cement. To the *left* a broken surface. The sponges have a *circular* outline in cross-section (*thin section*), see Fig. (e). Upper Jurassic sandstone from the North Sea (from T.E. Maast unpublished)

seismic and could be mistaken for a fluid contact i.e. gas/oil or oil/water contact (Thyberg et al. 2009).

4.10 Deeply Buried Sandstones (>3.5–4 km, >120°C)

Once quartz cementation has started and quartz overgrowth has formed, quartz cementation does not stop until nearly all the porosity is lost, unless the temperature drops below 70–80°C.

In most sedimentary basins we find there is a rather strong reduction in porosity and permeability in sandstone reservoirs from about 3–3.5 km to 4–4.5 km burial depth, corresponding to a temperature range from about 120 to 160°C. This is due in most cases to precipitation of quartz cement and diagenetic illite. The rate of quartz cementation increases as an exponential function of temperature and we estimate that the rate may increase by a factor of 1.7 for every 10°C temperature increase (Walderhaug 1996). The precipitation of quartz is also a function of the surface area available for quartz cementation and as quartz cement is filling the pores the surface area available for further quartz growth decreases (Fig. 4.8). The crystal surfaces have different solubilities and potential

for crystal growth and thereby for porosity reduction (Lander et al. 2008).

The temperature history of the sandstones at this stage becomes rather critical in terms of modelling and predicting the amount of quartz cement and remaining porosity. Between 100 and 140°C the rate of quartz precipitation may double four times, i.e. increase 16-fold. By contrast the effective stress increases linearly with depth (under hydrostatic conditions) and thus increases only by 30–40% through the interval from 3 to 4 km of burial. Temperature is therefore by far the main factor controlling the rate of quartz precipitation. This allows the amount of quartz cement and the porosity to be modelled as an exponential function (Arrhenius equation) of the temperature integrated over time and proportional with the surface area available for quartz precipitation. Commercially available programs (Exemplar, Touchstone) have been developed for this purpose based on Walderhaug (1996). Dissolution at grain contacts requires stress, so the process is often called "pressure solution", but the degree of stress needed is only relatively moderate and in the case of silicate minerals temperature is the most important factor. Contacts between mica or illitic clay and quartz are preferred areas of dissolution. The rate-limiting process in quartz cementation seems to be the rate of nucleation and precipitation in the pore space and the reaction is then surface-controlled (Bjørkum et al. 1998). There is no stress on the grain surfaces facing the pores where quartz overgrowth forms. Quartz cementation is therefore relatively insensitive to variations in effective stress in the grain framework. If the dissolution process had been rate-limiting, quartz cementation would have been more sensitive to the effective stress in addition to temperature and surface properties of the minerals. The silica dissolved at grain contacts or along stylolites is transported by diffusion to the grain surfaces where the quartz overgrowth forms (Fig. 4.9).

If the transport of silica was the rate-limiting factor for quartz cementation (transport-controlled) we would expect to observe a concentration of quartz cement close to stylolites of thin clay laminae, where dissolution occurs and where the concentration in the porewater would be highest. When dissolution is concentrated along stylolites with relatively large spacing (>50–100 cm), studies suggest that the amount of quartz cement may decrease away from the stylolites, indicating some degree of transport control. The

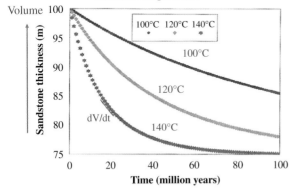

Compaction due to quartz cementation at constant temperature

Fig. 4.8 Modelling of quartz cementation and chemical compaction due to quartz dissolution and cementation as a function of time and temperature (from Walderhaug et al. 2001). We see that the rate of porosity loss (compaction, dv/dt) is highest at high temperatures and also when the porosity is still relatively high. When the porosity is reduced the surface area available for quartz cementation becomes smaller so that the rate of cementation slows down

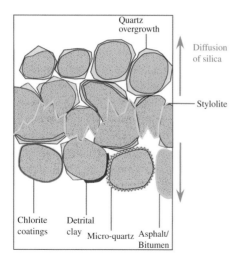

Fig. 4.9 Schematic illustration of a stylolite. The dissolved silica is transported away from the clay-rich stylolite by diffusion. This makes more long distance and advective transport of silica difficult. The rate of precipitation of quartz cement is a function of the surface area available. Grain coatings such as chlorite, illite, detrital clay, iron oxide (haematite), micro-quartz and bitumen prevent or retard quartz cementation

stylolites represent a barrier for fluid flow both during compaction and during production, but may not be completely continuous laterally.

We have seen that the rate of quartz cementation is a function of temperature, the degree of supersaturation and the surface area available for quartz cementation. A consequence of a surface-controlled quartz precipitation model is that quartz cementation will

continue as long as the temperature is above the threshold temperature for quartz growth (70–80°C) and there is remaining porosity in the sandstone. It is important to note that quartz cementation, and hence sandstone compaction, will continue also during basin inversion and uplift, but at a slower rate (Fig. 4.10). When uplifted to shallower depths (temperature <70–80°C) the sandstone is unloaded and without any chemical compaction there will be net extension due to elastic expansion due to reduced stress.

If temperature is higher than 70–80°C the cementation process will proceed but at a slower rate, modifying the normal porosity/depth relation found elsewhere in the basin. During progressive burial quartz cementation must continue until all available porosity is filled and the sandstone becomes a well-cemented hard quartzite after exposure to 200–300°C for several million years.

If the surfaces of sand grains are coated with other minerals, or with substances like petroleum or bitumen, quartz overgrowth is hindered or at least stopped for some time (Fig. 4.9). A thin layer of authigenic chlorite has proven to be very effective in preventing quartz overgrowth. This has been described from many places around the world like the Tuscaloosa sandstones of Southern Louisiana (Pittman and Larese 1992).

Sandstones with grain coatings may remain uncemented down to 4–5 km burial depth and be subjected to 40–50 MPa effective stress, causing pervasive grain crushing (Chuhan et al. 2002).

Fig. 4.10 Diagenetic processes, mainly quartz cementation, as a function of temperature and time. Note that quartz cementation will continue also during uplift as long as the temperature exceeds 70–80°C

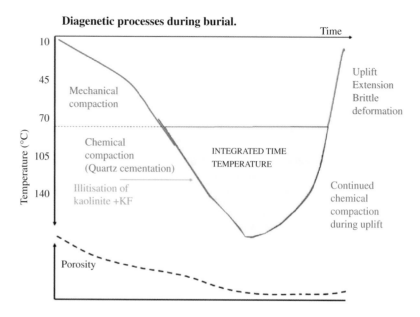

A

Quartz cementation of fractured grains from
Smørbukk Field, Haltenbanken, offshore mid-Norway

B

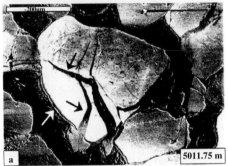

Natural fractures in reservoir sandstone (Tilje Fm,
Smørbukk Field). The quartz grains are chlorite-coated
but quartz cement has grown from fractured quartz. From
Chuhan et al. 2002

Fig. 4.11 (**a, b**) Cathodoluminescence pictures showing that
quartz grains have been subjected to fracturing prior to
quartz cementation. Quartz cementation was delayed by chlorite coatings but grain fracturing exposed fresh quartz surfaces from which quartz could grow into the pore space.
Authigenic quartz is darker than the clastic high temperature
quartz

Under the microscope, and particularly when using
cathodoluminescence, we can see that some of the sand
grains have been fractured and later healed by quartz
cement (Fig. 4.11a, b). The fractures in the clastic
grains do not usually continue through the quartz overgrowth, demonstrating that they predate the quartz
cementation.

At Haltenbanken, offshore mid-Norway, the
Jurassic Tije and Garn formations have abundant
chlorite coatings and also illite coatings (Ehrenberg
1993, Storvoll et al. 2002) (Fig. 4.12).

This is the main reason for the good reservoir
properties in petroleum discoveries on Haltenbanken,
where the porosity sometimes exceeds 25% at more
than 5 km depth.

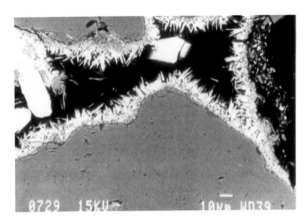

Fig. 4.12 Quartz grains coated with chlorite preventing
quartz overgrowth. Note pore-filling authigenic kaolinite. Tilje
Formation, Haltenbanken

To predict the distribution of such high porosities at
great depth we need to understand what controls the
development of authigenic chlorite. The chlorite found
in marine sandstones is most probably an alteration
product of an earlier iron silicate phase (precursor)
formed on the seafloor and which may be linked to
facies. This may be iron- and magnesium-rich smectites coating the primary quartz grains. Illite may also
form an effective coating preventing quartz overgrowth
(Heald and Larese 1974, Storvoll 2003) and this may
have formed from smectite.

It is typical that clean well-sorted sandstone has the
best porosity down to about 3.5–4 km but then loses its
porosity rapidly due to quartz cementation. Sandstones
with a moderate clay content lose more porosity during
mechanical compaction compared to the clean sandstone, but may be the winners at greater depth due to
retarded quartz overgrowth.

4.11 Fluid Inclusions in Quartz Cement

Inclusions of small drops of oil may be trapped in
quartz cement and show up well in fluorescent light
(Fig. 4.12a). Fluid inclusion data from quartz helps
to constrain the temperature for quartz cementation
and in sandstone reservoirs it has been demonstrated
quite clearly that quartz cementation continues after
oil emplacements in a reservoir (Karlsen et al. 1993,
Walderhaug 1990).

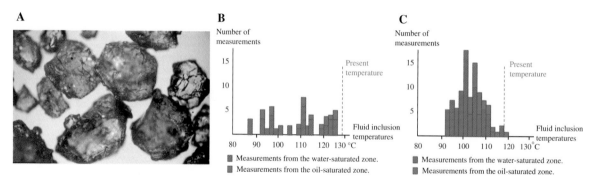

Fig. 4.13 (**a**) Inclusions of oil in quartz cement on sand grains. (**b**) Temperatures from oil inclusions in the Fulmar Field, North Sea (Saigal et al. 1992). (**c**) Fluid inclusion temperatures from Jurassic sandstones, Haltenbanken (offshore mid-Norway) (Walderhaug et al. 1990b)

The lowest fluid inclusion temperatures in quartz cement indicate an onset of quartz cementation close to 70–80°C (Burley et al. 1989, Walderhaug 1994a). In the Ula Field (North Sea basin), fluid inclusion temperatures in quartz cement range from 86°C to 126°C (Fig. 4.13b), the highest temperature being close to the present day reservoir temperature which is also the maximum burial depth and temperature (Saigal et al. 1992). Fluid inclusions from 24 samples from 11 different reservoir units from the North Sea and Haltenbanken also show temperatures from about 80°C to values close to the present reservoir temperatures (Walderhaug 1994a). This shows that quartz cementation occurs as a continuous process at a rate controlled by the temperature (Walderhaug 1994b) (Fig. 4.13c).

There is no evidence that quartz cementation is episodic, controlled by the supply of silica, or that quartz cementation stops after the sandstones have become oil-saturated (Walderhaug 1990, Saigal et al. 1992) (Fig. 4.13). In a water-wet reservoir precipitation can still continue in the remaining water around the grains. At high oil saturation, the transport of silica by advection as well as by diffusion becomes much less efficient. The continued growth of quartz cements after oil emplacement results from the closed system nature of quartz cementation in sandstones.

In an oil-wet system, however, quartz can not precipitate on the grain surfaces and oil or bitumen may become very effective coatings. Asphaltic oil and bitumen formed by biodegradation or other processes may preserve good reservoir quality, particularly if the heavy oil only occurs as a grain coating.

In summary, quartz cementation is controlled by the slow kinetics (high activation energy) for quartz

cementation and normally a minimum temperature of 70–80°C is required. This is however also dependent on the pH. In sedimentary basins marine porewater starts out with a pH close to 7 but quickly becomes more acid due to the build of CO_2 and other reactions with the minerals present. At 3–4 km depth the pH may typically be 4.5–5 but at 120°C the pH is close to neutral (see Chap. 3). The rate of quartz cementation is then lowered by the pH but increased by higher temperatures. At very high pH quartz cementation may occur at the surface and silcrete is fine-grained quartz formed in soils due to concentration of porewater by evaporation, and quartz is also forming in some alkaline African lakes.

Authigenic illite consists of thin hair- or plate-like minerals and it is fairly obvious that they would have a detrimental effect on reservoir quality by reducing the permeability (Fig. 4.14a, b). SEM images are routinely taken from dried-out cores where the illite appearance is no longer representative of its morphology in the reservoir. When cores are dried without destroying the delicate illite morphology the pore space often looks like it has been filled with rockwool. Illite can often be seen to grow at the expense of kaolinite and may also form by alteration of smectite. Although authigenic illite may also be observed on fractures and other places where there are no obvious precursor minerals, it is most commonly found as a replacement of an earlier Al-rich mineral phase.

Because of the low Al-solubility in porewater, illite will in most cases precipitate where the source of Al is available locally from a dissolving mineral. Calculations suggest that the solubility of aluminium is only about 1 ppm at 150°C and that organic acids

Fig. 4.14 (a) Pore-filling illite replacing pore-filling kaolinite preserving the kaolinite textures in Jurassic sandstone, Haltenbanken, offshore mid-Norway. (b) Pore-filling illite probably altered from smectite. Triassic sandstone from the North Sea

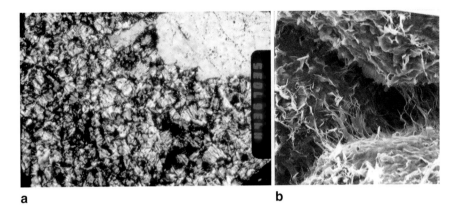

a   b

have little effect in terms of increasing its solubility (Bjørlykke and Aagaard 1992).

The formation of illite from smectite via mixed-layered minerals is well known and occurs in sandstones in the temperature range of 70–100°C. Sandstones with abundant smectite are poor reservoir rocks at the outset, and illitisation of such rocks may itself slightly improve reservoir quality as illite has a lower specific surface area than smectite. In better-sorted and potentially good reservoir rocks kaolin minerals (kaolinite or dickite) are the most important precursors for illite. However, the formation of illite requires potassium, and K-feldspar is usually the only significant source present in the sediment.

$$KAlSi_3O_8 + Al_2Si_2O_5(OH)_4 = KAl_3Si_3O_{10}(OH)_2 + 2SiO_2 + H_2O$$

$$\text{K-Feldspar} + \text{Kaolinite} = \text{Illite} + \text{Quartz}$$

This is a simplified presentation of this reaction. High temperature detrital K-feldspar contains some sodium and the illite formed contains less potassium than indicated here, where the formula for muscovite is used.

The reaction between K-feldspar and kaolinite occurs at about 130°C and above this temperature these two minerals are no longer thermodynamically stable together. In the North Sea basin and at Haltenbanken this corresponds to a burial depth of about 3.7–4 km. A sharp increase in the illite content in sandstone reservoirs is observed at this present day depth (Bjørlykke et al. 1986, Ehrenberg 1992). From thin sections and SEM images the illite can be seen to replace kaolinite, and potassium then has to diffuse from the K-feldspar to the kaolin. If the matrix is well-cemented the rate of diffusion is reduced, and the minerals are then able to co-exist at higher temperatures if they do not occur close together.

In sandstones with little or no K-feldspar, however, kaolin remains stable at greater depth as it is not dissolved and replaced by illite. The formation of illite can therefore be predicted from the sandstone provenance with respect to K-feldspar supply, and from the early diagenesis and freshwater flushing with respect to the distribution of kaolinite. If a sandstone is derived from an albite-rich gneiss the K-feldspar content is likely to be too low and much of the kaolinite would then not be illitised.

Similarly, not much illite will be formed in sandstones with little kaolinite or smectite. Both in Haltenbanken and the North Sea there are Jurassic reservoirs where plagioclase is the dominant feldspar and where the low K-feldspar content is unable to supply the necessary potassium for illitisation of kaolinite. The low illite content in such reservoirs preserves better permeability. This is a direct function of the provenance and could be due to erosion of albite gneisses rather than granitic gneisses.

The distribution of authigenic illite in sedimentary basins like the North Sea and Haltenbanken shows that illite formation is strongly controlled by the present day burial depth and temperature. The increase in illite content at about 3.7–4.0 km is usually very sharp, indicating a temperature-controlled reaction rather than a high kinetic reaction rate when the association of kaolinite and K-feldspar becomes thermodynamically unstable. Basin loading from thick Pleistocene sequences in these areas suggests that the illite formed recently.

K-Ar dating of illite gives variable ages for the formation of illite. This is probably because even very small amounts of detrital (older) mica or feldspar will produce too-old ages (Hamilton et al. 1989).

4.12 Porewater

In a sedimentary basin the amount of solids dissolved in porewater is very small compared to the volume of solids. During burial diagenesis, significant precipitation of authigenic minerals must be accompanied by dissolution of other minerals or the same mineral, as in the case of pressure solution. Even if the porewater is supersaturated with respect to a certain mineral, only very small amounts can precipitate before the porewater attains equilibrium. Precipitation of new minerals requires that other minerals dissolve because the porewater has very little capacity to store ions. This has been quantified using chemical modelling (Giles 1997) from which it can be concluded that extremely large fluid fluxes are required through the pores in order to dissolve or precipitate significant amounts of cements like calcite or quartz.

The porewater reacts with the minerals it is in contact with, and with increasing temperature the composition becomes more and more in equilibrium with these minerals. This is because the kinetics of the mineral reactions become faster. The porewater in sedimentary basins consists of solutions buffered by the minerals present. The pH is controlled partly by the carbonate reactions (pCO_2), and the pH will decrease from 7.5–8 in the seawater to 4–5 at 3–4 km depth. As temperature rises above 100°C, silicate reactions become increasingly important, e.g. between K-feldspar, kaolinite and illite which will determine the K^+/H^+ ratio.

Organic acids are weak acids which can not significantly change the pH in the strongly buffered porewater. The buffering capacity of organic acids has been shown to be orders of magnitude lower than for the carbonate and silicate systems (Hutcheon and Abercrombie 1990). Organic acids generated in source rocks like the Kimmeridge Clay Fm in the North Sea are likely to be neutralised by reactions with calcite which is commonly present in these

source rocks. The limited effect of organic acids and CO_2 on mineral dissolution and diagenesis has also been shown experimentally (Barth and Bjørlykke 1983).

When porewater moves it will nearly always cause some dissolution and precipitation but this is rarely significant due to low velocities. The volume of minerals dissolved or precipitated can be calculated using the following equation:

$$V_c = F\,t\sin\alpha\,(dT/dZ)\,\Delta S/\rho$$

The volume of precipitated mineral V_c is a product of the fluid flux integrated over time (t), the angle of fluid flow relative to the isotherms (α), the geothermal gradients (dT/dZ), the solubility as a function of the temperature (ΔS) and the density of the mineral (ρ).

The solubility gradient (ΔS) is 1–3 ppm/°C, depending on the temperature (Wood 1986). This means that at 100–150°C about 2 ppm of quartz precipitates for each degree the porewater is cooled. This gives a solubility gradent of 2.10^{-6}/°C. If the geothermal gradient is 30°C/km (3×10^{-2}°C/m), porewater must move upwards more than 30 m to reduce the temperature by 1°C and the quartz cement is distributed through these 30 m of sandstone. Assuming vertical flow (sin $\alpha = 1$) and geothermal gradients close to 30°C/km (3.10^{-2}°C/m), a solubility gradient of 2.10^{-6}/°C and a quartz density of 2.65 g/cm³ we obtain:

$$Vc = F \cdot t \cdot 2.3 \times 10^{-8}$$

From the above equation we see that each 1 m³/m² will precipitate about 2.3×10^{-8}m³ of quartz. To precipitate 10% quartz cement ($Vc = 0.1$) requires a total flow (integrated flux over time)) of ($F \cdot t$) of about $4 \cdot 10^6$ m³/m². This assumes that the porewater is in equilibrium with the mineral phases which is true at depth with temperatures exceeding 80–100°C. At 30% porosity a water column of 1,200 km must pass through a sandstone layer to introduce 10% quartz cement.

This is clearly impossible in sedimentary basins. In addition compaction-driven porewater is not flowing upwards in relation to the surface and is therefore normally not subjected to cooling which would cause precipiation of quartz (see Chap. 10).

At shallow depth the temperature is low and the fluid flow rate high, so in the zone of meteoric water flushing this may not be true. The porewater may then be undersaturated or supersaturated, particularly with respect to silicate minerals. When temperatures exceed 100°C the porewater will approach equilibrium with the minerals, both because of higher reaction rates and low flow rates.

Small amounts of calcite are nearly always present, at least in marine sediments. Calcite has a retrograde solubility meaning that the solubility normally decreases with increasing temperature. The solubility also depends on the pressure, but in most cases it is the temperature effect which is strongest. Upwards (cooling) porewater flow, which should precipitate quartz, will dissolve calcite at a rate which is 30–100 times faster (Bjørlykke and Egeberg 1993). We may therefore conclude that if calcite was present in a sandstone very little quartz could have precipitated until all the calcite had been dissolved.

Thermal convection is probably not very significant in sedimentary basins except where there are hydrothermal heat sources (Bjørlykke et al. 1988). If thermal convection did occur at a significant rate, however, quartz could precipitate because the same water could be used over again, precipitating quartz and dissolving calcite on the way up, and dissolving quartz and precipitating calcite on the way down when the porewater is heated. All the calcite would then be dissolved and quartz would be precipitated by this process.

4.13 Effect of Oil Emplacement

When oil migrates into a reservoir rock the water content is reduced to a percentage of the porosity corresponding to "irreducible water saturation" if the rock is water-wet. This may vary from 10% water content in clean sand to 50% or more in clay-rich sandstone, the value depending on the amount and type of clay present. The traditional assumption has been that the emplacement of oil stops, or at least slows down, the rate of diagenetic processes and hence the rate of porosity reduction.

If the transport of silica by either diffusion or advection was rate-limiting for the quartz cementation one would indeed expect the rate of quartz cementation to be very much reduced. Fluid inclusions in quartz cement, however, clearly demonstrate that in fact quartz continues to grow after oil emplacement in sandstone reservoirs (Walderhaug 1990). The explanation for this is that silica is transported along the thin film of water between the mineral grains and the oil phase. It is possible that the rate of quartz cementation could be slower after oil emplacement but this would imply that the quartz cementation would no longer be surface controlled, but transport controlled. It is very difficult to prove that a higher porosity in the oil-saturated part of a reservoir is due to the introduction of oil. There are usually so many other variables such as facies that influence the final reservoir porosity. Since the oil is emplaced gradually and the oil/water contact moves downward over time, a sharp difference in porosity should therefore not be expected right at the present OWC if quartz cementation was a function of oil emplacement.

In gas reservoirs the saturation may be rather high in clean sand, perhaps reducing the water film around the grains and the quartz overgrowth, but the degree to which this might apply is uncertain.

Biodegraded and asphaltic oil or bitumen will stick to the grain surfaces and effectively prevent quartz overgrowth but then some of the porosity may be lost to the bitumen and heavy oil.

4.14 Prediction of Reservoir Quality

It is important to distinguish between sandstones of different primary composition:

Volcanoclastic sandstones may vary greatly in composition depending on the volcanic source and the depositional environments. Basic volcanic rocks in particular have a very low content of stable grains like quartz, but a high content of basic feldspar and pyroxenes which break down rapidly, both mechanically and chemically. Matrix-rich sandstones like greywackes may have had a higher sand content at the time of deposition because many of the grains were unstable during diagenesis and became effectively part of the matrix. What were deposited as grains of volcanic rock fragments may be squeezed so that they become a chlorite-rich matrix.

Volcanoclastic sandstones lose most of their porosity at rather shallow depth (<1–2 km) and therefore make poor reservoir rocks. However, the geothermal gradients in volcanic regions may be high, causing source rocks to mature at shallow burial depth and thus increase the potential for migration into shallow structures.

Lithic sandstones have a high content (>10%) of rock fragments. Normal and coarse-grained granites and gneisses produce grains that mostly consist of a single mineral while sandstones derived from finer-grained igneous and metamorphic rocks are mostly comprised of rock fragments. Rock fragments are generally weaker than quartz and feldspar grains, as has been demonstrated experimentally (Pittman and Larese 1991).

Arkoses contain more than 25% feldspar and such sandstones are typical of tectonically active basins like rift basins where the erosion, transport and deposition of basement derived rocks is fast, leaving little time for feldspar to weather. Temperature and rainfall also play a role here. Arkoses compact more mechanically than quartzitic sandstones, leaving a smaller intergranular volume to be cemented with quartz at greater depth. The area available for quartz cementation is also reduced since quartz does not grow on feldspar.

Feldspathic sandstones and quartzites are the most common sandstone reservoir rocks. The feldspar content is usually a function of the source climate and the relief in the drainage area. On tectonically stable cratons sediments are repeatedly eroded and deposited and some feldspar and mica is dissolved during each cycle.

Palaeozoic quartzites typically occur as transgressive sheet sands on cratons. On the North American craton there are good examples of this in the Lower Palaeozoic sequence. Such clean shallow marine sandstones have extremely good reservoir properties at shallow to moderate burial depth. This is likewise the case with aeolian sandstones. Fluvial sandstones are also normally well-sorted in such environments because they are often reworked aeolian sands.

Carbonate cement in shallow marine sandstones is mostly derived by recrystallisation of calcareous organisms. Meteoric water will dissolve aragonite and precipitate calcite in sandstones, producing early cement.

In modern environments, particularly in beach and shoreface settings, fragments of crushed calcareous organisms are quite common. We find less carbonate cement in fluvial sandstones because of the lower biogenic carbonate production in freshwater. Carbonate cement has a local source in most cases, but may be redistributed and concentrated by diffusion. The range of effective diffusion is generally small (<1 m) because the porewater is in equilibrium with calcite and there are small concentration gradients. Advective flow will transport dissolved carbonate but can not precipitate tight carbonate cement because the permeability decreases as precipitation proceeds. The advective flow will then tend to bypass the volume where carbonate cementation has started. In the case of compaction-driven upwards-directed (cooling) porewater, the solubility of calcite will increase, causing dissolution rather than precipitation.

Aeolian sandstones and other desert sandstones generally show less evidence of meteoric water flushing than fluvial and shallow marine sandstones. Sandstones like the Permian Rotliegend from the southern North Sea have relatively low amounts of kaolinite and more smectite or illite as pore-filling cement. However, even deserts have groundwater so some leaching occurs. Fluvial sediments will normally be flushed by groundwater after deposition and in most cases show ample evidence of feldspar leaching. Reworking of such sediments will bring authigenic kaolinite into the clastic clay fraction. Continental sandstones often have haematite or manganese oxide coatings on quartz grains and this may inhibit quartz overgrowth.

4.15 Turbidites

Turbidites form important reservoirs in many basins and although the reservoir quality may be less favourable than within shallow marine sandstones, they often form extensive vertically-stacked reservoir sequences and this may compensate for the lower porosity.

Turbiditic sandstones generally have a higher clay content than shallow marine sandstones. There is however a wide range of clay contents in turbidites from rather clean, usually proximal and channel facies, to more clay-rich distal and overbank facies.

Sandstones with clay contents higher than 10–15% lose their porosity rapidly with mechanical compaction, because the detrital clay acts as a lubricant in the compaction of the quartz grains. Turbiditic sandstones of Palaeocene and Eocene age form very important reservoirs in the North Sea. The Frigg sandstone in the Frigg Field in the Norwegian Sector is an example, where the reservoir quality is rather good despite representing a distal facies relative to the Shetland Platform where the sand originated.

Because turbidites are deposited further away from land, they are less exposed to flushing by meteoric water. Turbidites normally contain less evidence of feldspar dissolution and authigenic kaolinite precipitation than shallow marine and fluvial sandstones. In proximal turbiditic facies, however, the sandstones may have good contact with the meteoric water lens. In the North Sea, Tertiary turbidites generally have a relatively low content of authigenic kaolinite, but it may be higher in proximal Cretaceous turbidities which formed around islands produced by local uplift.

The total flow of meteoric water through sandstones is a function of meteoric water flux and the sedimentation rate. At low sedimentation rates a given volume of sand spends more time in the shallow zone of meteoric water flushing.

Cretaceous and Tertiary turbidite sandstones are often very tight due to pervasive carbonate cement. This may be due to pelagic carbonate organisms being mixed in with the turbidite sand and recrystallising into carbonate cement. This makes many Tertiary sandstones hard and indurated even if they have not been buried very deeply.

Many of the planktonic carbonate organisms developed during the Jurassic, so before then the sources of carbonate cement in deep sea sandstones were more restricted.

4.16 Practical Predictions of Reservoir Quality and Porosity Depth Curves

The oil industry has a practical need to be able to predict the properties of reservoir rocks ahead of drilling. When planning petroleum production the rock properties between the wells must also be estimated. Particularly in the deeper reservoirs, porosity is the most important factor determining the economic viability of a prospect. The main diagenetic processes with quartz cementation are summarised in Fig. 4.15.

In a relatively mature basin the porosity/depth functions of the different reservoir rocks can be treated statistically so that the uncertainty of the estimates can be expressed. The estimates based on the statistical averages can also be adjusted up or down as a function of temperature, stress etc., depending on what the interpreter considers most significant. The Middle Jurassic Brent sandstone in the North Sea has been intensively studied and a relatively linear trend found between burial depth and porosity (Giles et al. 1992, Bjørlykke et al. 1992, Ramm et al. 1992, Wilson 1994). Porosity predictions will depend on the primary sediment composition and the subsequent compaction processes.

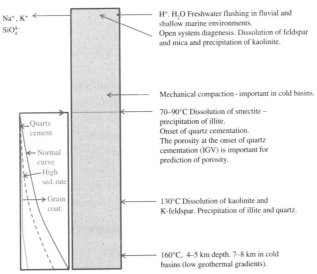

Fig. 4.15 Summary of the most important processes in clastic diagenesis. Dissolution of feldspar and mica requires a through flow of meteoric water removing K$^+$ (Na$^+$) and silica before kaolinite can be precipitated. This process is most dominant in fluvial and shallow marine sediments at shallow burial. At greater depth diagenetic reactions are nearly isochemical at a scale of 1–10 m distance. The composition of the dissolved material is equal to what is precipitated. Carbonate cements are usually derived from biogenic carbonate or clastic carbonate grains. Quartz cementation is controlled by temperature (geothermal gradients), subsidence rates and the presence of grain coatings

Above we have discussed some of the processes that cause reductions in porosity and permeability. All these processes are driven towards denser packing of grains and thermodynamically more stable mineral assemblages as the stress and temperature increases during burial. The kinetics of mineral reactions determines the rate of thermodynamic equilibration, which increases as an exponential function of temperature.

The rate of compaction as a function of stress can be measured experimentally using rock mechanics testing procedures. However, the reactions involved in chemical compaction are so slow, particularly in silicate rocks, that it is difficult to reproduce them in the laboratory although in some cases this is now becoming possible.

In the field of clastic diagenesis, petrographic observations about mineralogy and textural relationships are used to interpret the sequence of dissolution and mineral precipitation and its relationship to changes in porosity and permeability. It is however important to consider the geochemical constraints on diagenetic reactions. During burial the reaction must add up so the dissolution is balanced by precipitation because there are strong limitations with respect to supply and removal of solids dissolved in the porewater.

Changes in mineralogy or porosity with depth may provide useful depth trends within an area, but in a sedimentary basin the initial mineral compositions may vary laterally. This is also the case for early diagenetic processes like meteoric water flushing and marine cementation. The observed changes with depth may therefore also reflect some of these factors and not only the burial depth.

Based on the theory that the rate of quartz cementation is controlled only by temperature, time and the grain surface available for quartz precipitation, the amount of quartz cement and consequently the porosity can be modelled (Walderhaug 1996). The presence or absence of clay or other coatings is the most critical input for this modelling because it determines the area available for quartz cementation. Prediction of reservoir properties must start from sedimentological facies models. The depositional environment and the provenance of the clastic sediments determine the starting composition for the diagenetic processes.

Early diagenetic processes like marine carbonate cementation and meteoric water flushing are also linked to facies and they strongly influence the burial diagenesis and the porosity reduction at depth. The precursor minerals controlling the growth of chlorite coating are also probably to a large extent controlled by facies. The distribution of silica organisms (like *Rhaxella*) which can produce micro-quartz coatings is linked to the environment and ecology.

A broad geological background is therefore required to synthesise all the factors that have to be considered before modelling or making semi-quantitative predictions of reservoir quality. The capacity of porewater to keep solids in solution is always rather limited and mineral dissolution and precipitation must therefore balance.

At greater burial depth (3–4 km) the solubility of silicate minerals increases but the volume of porewater is low and the potential for supersaturation and undersaturation strongly reduced. Precipitation of new mineral phases must therefore be linked to the dissolution of other minerals or of the same mineral.

Assuming that the burial diagenetic processes are relatively isochemical, the reservoir properties can be predicted from depositional facies and provenance studies which define the starting composition for burial diagenesis. Both observations and theoretical arguments suggest that advective transport can not significantly change the rock compostion below the reach of meteoric water flow. Short distance transport by diffusion may nevertheless be important.

Quartz cementation was often interpreted to occur as events of relatively short duration (approximately <10 million years) that could start and stop late in a sandstone's burial history (Robinson and Gluyas 1992). This would imply that the quartz cementation was controlled by advective transport of silica in solution and the source of silica from other reactions. Fluid inclusion data, however, shows that quartz cementation occurs throughout the temperature range corresponding to the burial history.

Modelling of quartz precipitation (Walderhaug 1996, Olkers et al. 2000, Walderhaug et al. 2000) is based on the assumption that this is a continuous process controlled by the kinetics and therefore by temperature. This is the basis for the practical models widely adopted by the petroleum industry (e.g. Exemplar and Touchstone).

The assumption that the quartz precipitation is the rate limiting factor may not always be strictly true in clean quartz arenites where silica sources (stylolites) are widely spaced, resulting in decreasing quartz cementation away from the stylolites (Walderhaug 2003). The modelling does, however, require that the burial curve and the temperature as a function of geological time are known. It is also very sensitive to changes in the primary sediment composition, which strongly influence both the mechanical compaction and the chemical reactions. The porosity loss and increased sediment density resulting from chemical compaction as a function of temperature cause basin subsidence (Bjørkum et al. 1998, 2001; Bjørkum and Nadeau 1998). Temperature-driven chemical compaction results in a volume reduction (shrinkage) and the strain is then independent of effective stress. As a result, differential stresses in siliceous rock will be relaxed by the compaction processes during basin subsidence as long at the temperature exceeds about 80°C (Bjørlykke 2006).

4.17 Porosity/Depth Trends in Sedimentary Basins

Data from wells in sedimentary basins which have undergone almost continuous subsidence can be regarded as records of a natural compaction experiment. We can use the log porosity and in the cored intervals we have data from core plugs.

At depths shallower than about 2 km (80–100°C) we can compare the log porosities with experimental compaction of similar sands in the laboratory. There will then be a marked effect of overpressure reducing the effective stress. Poorly sorted sands will lose most of their porosity at relatively shallow depth but well sorted sand may have 30–35% porosity at 2–3 km depth, which corresponds to experimental compaction at about 20–30 MPa effective stress. This suggests that there is little creep over long geologic time.

Data from deep wells will always show a trend towards lower porosities with depth but there may also be intervals where this trend is reversed. This does not mean that net porosity has been created by diagenetic processes. Because of the very low solubility of silica and even more so of aluminium in porewater it is very difficult to explain how minerals in several metre thick sandstones can be dissolved without precipitation of other minerals in the same sandstones. Each lithology will have a characteristic porosity depth trend (Fig. 4.16) and increases in porosity reflect variations in the primary composition.

As discussed above the rate of quartz cementation can be modelled if the surface area available for quartz

cementation and the time-temperature history during burial are known.

At about 4.0 km burial depth (120–140°C) the amount of quartz cement may be 10–15% so the remaining porosity may be only 10–15%. We do however find good reservoir quality (>20% porosity) at greater depths and higher temperatures but this is due to grain coatings. Prediction of porosity at great depth therefore requires that the occurrence of coating of chlorite, illite, haematite or micro-quartz can be predicted. Such prediction of the primary sediment composition must again be linked to facies and provenance.

Prospects at great depth called HTHP (High Temperature, High Pressure) are expensive to drill and involve high risks, but in many mature basins nearly all the shallower prospects have been drilled already.

An extensive study of reservoir sandstones in the Gulf of Mexico showed that temperature and time are the main factors controlling reservoir quality (Nadeau et al. 2008).

4.18 Practical Prediction of Reservoir Quality

The most important factor controlling reservoir quality at depth is the primary composition of the sandstones. Sedimentological and sequence stratigraphic analyses are normally used primarily to predict reservoir geometry, but for diagenetic processes grain size, sorting and mineralogical composition are more critical. It is imperative to establish changes in provenance since the primary mineralogy places important constraints on diagenetic reactions at depth.

Reconstructions of facies and climate will provide a basis for predicting the degree of feldspar dissolution and precipitation of pore-filling kaolinite. Biogenic components like calcareous and siliceous organisms will control the distribution of carbonate cements and opal A, which will be altered to opal CT and quartz. Primary aragonite may cause extensive calcite cementation that occludes much of the primary porosity. Organic silica, e.g. from siliceous sponges, may serve as a precursor to grain-coating micro-quartz preserving porosity at depth.

Porosity loss due to mechanical compaction can vary greatly as a function of textural and mineralogical composition. Experimental compaction of loose sands

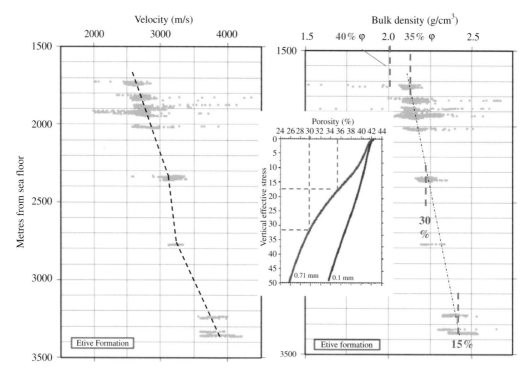

Fig. 4.16 Density/depth and velocity/depth trends for the Etive Fm (Brent Group) showing that a single lithology has a nearly linear trend with depth, based on Marcussen et al. (2009). The calculated porosities show that the compaction down to about 2 km depth are mechanical and similar to experimental data inserted from Chuhan et al. (2002). At greater depth, compaction is mostly chemical and higher compared with mechanical compaction

with different grain size and sorting provides a good basis for prediction of porosity and inter-granular volume before quartz cementation. In cold sedimentary basins (low geothermal gradient) sand may be buried to 4–5 km before there is significant quartz cementation and in the absence of overpressure it can be subjected to 40–50 MPa effective stress.

Well log data from distinct lithologies buried to different depths may also provide a useful database for predicting the porosity loss due to mechanical compaction.

4.19 Summary

Porosity loss due to quartz cementation can be modelled as a function of temperature, time and surface area available for quartz cementation. This is sensitive to grain size and grain coatings, which must be predicted from primary facies evaluation. The presence of pore-filling illite depends on precursors which may be smectite or kaolinite. Dissolution of kaolinite

and precipitation of illite requires temperatures above 130°C and the local presence of K-feldspar. Sandstone containing mostly plagioclase does not develop pore-filling illite as there is insufficient supply of potassium. Prediction of reservoir quality can thus be based on provenance.

The examples of reservoir quality predictions listed above are based on the assumption that burial diagenetic reactions are essentially isochemical. Open system diagenesis allowing large scale import and export of solids in solution violates constraints from mineral solubilities and fluid flow rates and therefore provides a poor basis for prediction.

4.20 How Much Oil Can Be Produced from Sandstone Reservoirs?

Petroleum exploration requires predictions about the reservoir properties ahead of drilling to justify the investment that a well represents. The reservoir properties determine the percentage of recoverable petroleum in each volume of rock. Even after drilling several exploration wells and also production wells in the development of an oil or gas field, the porosities and permeabilities are only known from the cores. Data from cores, cuttings and well logs must be extrapolated to produce a 3D model of the large volumes of rock between the wells. This must be based on predictions from facies distribution, distribution of faults and fault properties. Changes in reservoir properties as a function of depth require diagenetic models which can predict changes in porosity as a function of effective stress and temperature/time.

To calculate the producible oil in a prospect the total rock volume (Gross – G) in the structure must be estimated as well as the percentage of sand which can be produced (Net Sand – N).

The oil in place (V_p) is:

$$V_r \cdot \text{N/G} \cdot \varphi \, O_{\text{sat}}$$

Here V_r is the volume of rock between the oil/water contact (OWC) and the reservoir cap rock or the gas/oil contact. N/G (net/gross) is the ratio between the fraction of the reservoir rock that can be produced and the total volume of the reservoir rock. φ is the average porosity of the producible part of the reservoir (net volume). O_{sat} is the average saturation of oil; typically about 80–85% of the pores in sandstone are filled with oil. The remaining portion is water, which in siliciclastic rocks occupies the mineral surfaces and the smallest pores where the capillary entry pressure is too high for oil.

If the porosity is low the permeability will in most cases also be very low so that the flow of oil from the rock formation to the well becomes too slow to be economical. About 10% porosity may be the minimum porosity for defining the producible (net) part of the reservoir. In fractured reservoirs the permeability may be high even when the porosity is below 10%.

In the planning of production from an oil field, data from cores and logs from wells must be extrapolated into a 3D model of the flow properties of the reservoir.

The producible percentage of the oil in place is called the *recovery factor*, which may range from 20–30% up to 40–60%. In sandstones with good reservoir quality, improved production technology has in some cases (e.g. Statfjord and Gullfaks fields, offshore Norway) boosted the recovery factor to close on 70%. Recovery is limited both by the amount of oil remaining in the pores of the drained sandstones and the presence of undrained compartments within the reservoir where oil is bypassed. Reservoir quality is a very important factor in the financial risk assessment calculations for a prospect.

Sediment input and burial diagenesis

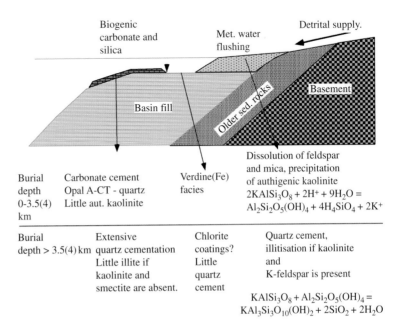

Fig. 4.17 Summary diagram for clastic diagenesis. The primary composition of the sand depends on the provenance, weathering, transport and depositional environment. Early diagenesis including meteoric flushing or marine cementation, is controlled by the depositional environment. Burial diagenesis is controlled by mineral stability (thermodynamics) and reaction rates (kinetics)

| Burial depth 0-3.5(4) km | Carbonate cement Opal A-CT - quartz Little aut. kaolinite | Verdine(Fe) facies | Dissolution of feldspar and mica, precipitation of authigenic kaolinite $2KAlSi_3O_8 + 2H^+ + 9H_2O = Al_2Si_2O_5(OH)_4 + 4H_4SiO_4 + 2K^+$ |
| Burial depth > 3.5(4) km | Extensive quartz cementation Little illite if kaolinite and smectite are absent. | Chlorite coatings? Little quartz cement | Quartz cement, illitisation if kaolinite and K-feldspar is present $KAlSi_3O_8 + Al_2Si_2O_5(OH)_4 = KAl_3Si_3O_{10}(OH)_2 + 2SiO_2 + 2H_2O$ |

4.21 Conclusions

Diagenetic reactions are driven towards higher mechanical and chemical stability. Reactions in sandstones are driven by the effective stress from the overburden causing reduced porosity (volume) at temperatures below 70–80°C. At greater depth (higher temperatures) compaction is mostly chemical and mineral reactions are controlled by thermodynamics and kinetics. Because of low kinetic reaction rates (high activation energies) silicate reactions are very sensitive to temperature. The precipitation of quartz has high activation energy, and temperature is the main control on quartz cementation causing much of the porosity loss in well-sorted sandstones. The dissolution of K-feldspar and kaolinite at about 130°C occur because the mineral assemblage illite and quartz is more stable.

Because of the low solubility of silicate minerals and the limited flow of porewater in the deeper parts of sedimentary basins, burial diagenetic reactions must be nearly isochemical. Significant amounts of solids can not be exported from a sandstone and the porosity of a single reservoir rock will only decrease and can not increase during progressive burial.

Prediction of reservoir quality at great burial depth depends on the initial sediment composition (provenance), sedimentary facies (Fig. 4.17), early diagenetic processes and the subsequent burial history. This should therefore be linked to facies models and provenance in addition to basin subsidence and heat flow. Modelling of quartz cementation has proved to be very useful for the prediction of porosity as a function of the temperature history of the reservoir sandstones, but the results are sensitive to the primary textural and mineralogical composition of the sandstones and the presence of grain coatings.

Further Reading

Aagaard, P., Egeberg, P.K., Saigal, G.C., Morad, S. and Bjorlykke, K. 1990. Diagenetic albitization of detrital K-feldspars in Jurassic, Lower Cretaceous and Tertiary Clastic Reservoir Rocks from Offshore Norway, II. Formation Water Chemistry and Kinetic Considerations. Journal of Sedimentary Petrology 60(4), 575–581.

Aase, N.E., Bjørkum, P.A. and Nadeau, P.H. 1996. The effect of grain coating microquartz on preservation of porosity. AAPG Bulletin 80, 1654–1673.

Barth, T. and Bjørlykke, K. 1993. Organic acids from source rock maturation; generation potentials, transport mechanisms and relevance for mineral diagenesis. Applied Geochemistry 8, 325–337.

Boles, J.R. 1982. Active albitization of plagioclase, Gulf Coast Tertiary. American Journal of Science 282, 165–180.

Boles, J.R. 1998. Carbonate cementation in Tertiary sandstones of the San Joaquin Basin. In Morad, S. (ed.), Carbonate Cementation in Sandstones. International Association of Sedimentology, Heidelberg, Special Publication 26, 261–284.

Bjørkum, P.A., Oelskers, E.H., Nadeau, P.H., Walderhaug, O. and Murphy, W.M. 1998. Porosity prediction in quartzose sandstones as a function of time, temperature, depth, stylolite frequency, and hydrocarbon saturation. AAPG Bulletin 82, 637–648.

Bjørkum, P.A., Walderhaug, O. and Nadeau, P.H. 2001. Thermally driven porosity reduction: Impact on basin subsidence. In: Shannon, P.M., Houghton, P.D.W. and Corcoran, D.V. (eds.), The Petroleum Exploration of Ireland's Offshore Basins. Geological Society Special Publication 188, 385–392.

Bjørlykke, K., Ramm, M. and Saigal, G.C. 1989. Sandstone diagenesis and porosity modification during basin evolution. Geologische Rundschau 78, 243–268.

Bjørlykke, K. and Egeberg, P.K. 1993. Quartz cementation in sedimentary basins. AAPG Bulletin 77, 1538–1548.

Bjørlykke, K., Mo, A. and Palm, E. 1988. Modelling of thermal convection in sedimentary basins and its relevance to diagenetic reactions. Marine and Petroleum Geology 5, 338–351.

Bjørlykke, K. and Aagaard, P. 1992. Clay minerals in North Sea sandstones. In: Houseknecht, D.W. and Pittman, E.D. (eds.), Origin, Diagenesis, and Petrophysics of Clay Minerals in Sandstones. SEPM, Tulsa, OK, Special Publication 47, 65–80.

Bjørlykke, K., Nedkvitne, T., Ramm, M. and Saigal, G. 1992. Diagenetic processes in the Brent Group (Middle Jurassic) reservoirs of the North Sea – An overview. In: Morton, A.C., Haszeldine, R.S., Giles, M.R. and Brown, S. (eds.), Geological Society Special Publication 61, Geology of the Brent Group, 263–287.

Bjørlykke, K., Aagaard, P., Egeberg, P.K. and Simmons, S.P. 1995. Geochemical constraints from formation water analyses from the North Sea and Gulf Coast Basin on quartz, feldspar and illite precipitation in reservoir rocks. In: Cubitt, J.M. and England, W.A. (eds.), Geological Society Special Publication 86, The Geochemistry of Reservoirs, 33–50.

Bjørlykke, K., Chuhan, F., Kjeldstad, A., Gundersen, E., Lauvrak, O. and Høeg, K. 2004. Modelling of sediment compaction during burial in sedimentary basins. In: Stephansson, O., Hudson, J. and King, L. (eds.), Coupled Thermo-Hydro-Mechanical-Chemical Processes in Geo-Systems. Elsevier, Amsterdam, pp. 699–708.

Burely, S.D. 1986. The development and destruction of porosity in the Upper Jurassic reservoir sandstones of the Piper and the Tartan oilfields. Outer Moray Firth, North Sea. Clay Minerals 19, 403–440.

Chuhan, F.A., Kjeldstad, A., Bjørlykke, K. and Høeg, K. 2002. Porosity loss in sand by grain crushing. Experimental evidence and relevance to reservoir quality. Marine and Petroleum Geology 19, 39–53.

Chuhan, F.A., Kjeldstad, A., Bjørlykke, K. and Høeg, K. 2003. Experimental compression of loose sands simulating porosity reduction in petroleum reservoirs during burial. Canadian Geotechnical Journal 40, 995–1011.

Ehrenberg, S.N., Nadeau, P.H. and Steen, Ø. 2008. A megascale view of reservoir quality in producing sandstones from the offshore Gulf of Mexico. AAPG Bulletin 92, 145–164.

Ehrenberg, S.N. 1990. Relationship between diagenesis and reservoir quality in sandstones of the Garn Formation, Haltenbanken, mid-Norwegian continental shelf. AAPG Bulletin 74, 1538–1558.

Ehrenberg, S.N. 1997. Influence of depositional sand quality and diagenesis on porosity and permeability: Examples from Brent Group reservoirs, northern North Sea. Journal of Sedimentary Research 67, 197–211.

Ehrenberg, S.N. and Nadeau, P.H. 2005. Sandstone versus carbonate petroleum reservoirs: A global perspective on porosity-depth and porosity-permeability relationships. AAPG Bulletin 89, 435–445.

Giles, M.R. 1997. Diagenesis; A Quantitative Perspective. Kluwer Academic Publ., Dordrecht, 526 pp.

Giles, M.R., Stevenson, S., Martin, S.V., Cannon, S.J.C., Hamilton, J.D., Marshall, J.D. and Samways, G.M. 1992. The reservoir properties of the Brent Group. A regional perspective. In: Morton, A.C., Haszeldine, R.S., Giles, M.R. and Brown, S. (eds.), Geological Society Special Publication 61, 289–327.

Hamilton, P.J., Kelley, S. and Fallick, A.E. 1989. K-Ar dating of illite in hydrocarbon reservoirs. Clay Minerals 14, 215–231.

Heald, M.T. and Larese, R.E. 1974. Influence of coatings on quartz cementation. Journal of Sedimentary Petrology 44, 1269–1274.

Hesthammer, J., Bjørkum, P.A. and Watts, L. 2002. The effect of temperature on sealing capacity of faults in sandstone reservoirs – Examples from the Gullfaks and Gullfaks Sør Fields, North Sea. AAPG Bulletin 86(10), 1733–1751.

Hoveland, M., Bjørkum, P.A., Gudemestad, O.T. and Orange, D. 2001. Gas hydrate and seeps – Effects on slope stability: The "hydraulic model". ISOPE Conf. Proceedings, Stavanger, 471–476, ISOPE (International Society for Offshore and Polar Engineering), New York.

Karlsen, D.A., Nedkvitne, T., Larter, S.R. and Bjørlykke, K. 1993. Hydrocarbon composition of authigenic inclusions – application to elucidation of petroleum reservoir filling history. Geochemica et Cosmochemica Acta 57, 3641–3659.

Lander, R.H., Larese, R.E. and Bonell, L.M. 2008. Toward more accurate quartz cement models; The importance of euhedral versus noneuhedral growth rates. AAPG Bulletin 92, 1537–1563.

Lichner, P.C., Steefel, C.I. and Oelkers, E.H. 1996. Reactive transport in porous media. Reviews in Mineralogy 34, 438. Mineralogical Society of America.

Marcussen, Ø., Maast, T.E., Mondol, N.H., Jahren, J. and Bjørlykke, K. 2009. Changes in physical properties of a reservoir sandstone as a function of burial depth - The Etive Formation, Northern North Sea. Marine and Petroleum Geology. In press.

Nadeau, P.H., Bjørkum, P.A. and Walderhaug, O. 2005. Petroleum system analysis: Impact of shale diagenesis on reservoir fluid pressure, hydrocarbon migration, and biodegradation risks. In: Dore, A.G. and Vining, B.A. (eds.), Petroleum Geology: Northwest Europe and Global Perspectives: Proceedings of the 6th Petroleum Geology Conference, Geological Society, 1267–1274.

Oelkers, E.H., Bjorkum, P.A. and Walderhaug, O. et al. 2000. Making diagenesis obey thermodynamics and kinetics: The case of quartz cementation in sandstones from offshore mid-Norway. Applied Geochemistry 15, 295–309.

Ortoleva, P., Al Shaieb, Z. and Puckette, J. 1995. Genesis and dynamics of basin compartments and seals. American Journal of Science 295, 345–427.

Pittmann, E.D. and Larese, R.E. 1992. Compaction of lithic sands: Experimental results and applications. AAPG Bulletin 75, 1279–1299.

Ramm, M. 1992. Porosity-depth trends in reservoir sandstones: Theoretical models related to Jurassic sandstones offshore Norway. Marine and Petroleum Geology 33, 396–409.

Ramm, M., Forsberg, A.W. and Jahren, J. 1998. Porosity-depth trends in deeply buried Upper Jurassic reservoirs in the Norwegian Central Graben: An example of porosity preservation beneath normal economic basement by grain-coating micro-quartz. In: Kupecz, J.A., Gluyas, J. and Bloch, S. (eds.), Reservoir Quality Prediction in Sandstones and Carbonates: AAPG Memoir 69, 177–199.

Saigal, G.C., Bjørlykke, K. and Larter, S.R. 1992. The effects of oil emplacements on diagenetic processes – Examples from the Fulmar reservoir sandstones, central North Sea. AAPG Bulletin 76, 1024–1033.

Storvoll, V., Bjørlykke, K., Karlsen, D. and Saigal, G. 2002. Porosity preservation in reservoir sandstones due to grain-coating illite: A study of the Jurassic Garn Formation from the Kristin and Lavrans Fields offshore Mid-Norway. Marine and Petroleum Geology 19, 767–781.

Thyberg, B., Jahren, J., Winje, T., Bjørlykke, K., Faleide, J.I. and Marcussen, Ø. 2009. Quartz cementation in Late Cretaceous mudstones, northern North Sea: Changes in rock properties due to dissolution of smectite and precipitation of micro-quartz crystals. Marine and Petroleum Geology 1–13.

Thyne, G., Bjørlykke, K. and Harrison, W. 1996. Chemical reaction and solute transport rates as constraints on diagenetic models. Annual Meeting Abstract. AAPG and SEPM 5, 140.

Wilson, M.D. 1994. Reservoir Quality Assessments and Prediction in Clastic Rocks. SEPM Short Course 30, 432 pp.

Walderhaug, O. 1990. A fluid inclusion study of quartz-cemented sandstones from offshore mid-Norway: Possible evidence for continued quartz cementation during oil emplacements. Journal of Sedimentary Petrology 60, 203–210.

Walderhaug, O. 1994. Temperatures of quartz cementation in Jurassic sandstones from the Norwegian Continental shelf – Evidence from fluid inclusions. Journal of Sedimentary Petrology 64, 311–323.

Walderhaug, O. 1994b. Precipitation rates for quartz cement in sandstones determined by fluid inclusion microthermometry and temperature history modelling. Journal of Sedimentary Research A64, 324–333.

Walderhaug, O. 1996. Kinetic modeling of quartz cementation and porosity loss in deeply buried sandstone reservoirs. AAPG Bulletin 80, 731–745.

Walderhaug, O. and Bjørkum, P.A. 1998. Calcite cement in shallow marine sandstones: Growth mechanisms and geometry. International Association of Sedimentologists, Special Publication 26, 179–192.

Walderhaug, O., Bjørkum, P.A., Nadeau, P. and Langnes, O. 2001. Quantitative modelling of basin subsidence caused by temperature-driven silica dissolution and reprecipitation. Petroleum Geoscience 7, 107–113.

Walderhaug, O. and Bjørkum, P.A. 2003. The effect of stylolite spacing on quartz cementation in the Lower Jurassic Stø Formation in well 7120/6–1, southern Barents Sea. Journal of Sedimentary Research 73, 146–156.

Walderhaug, O., Oelkers, E.H. and Bjørkum, P.A. 2004. An analysis of the roles of stress, temperature, and pH in chemical compaction of sandstones – Discussion. Journal of Sedimentary Research 74, 447–450.

Chapter 5

Carbonate Sediments

Nils-Martin Hanken, Knut Bjørlykke, and Jesper Kresten Nielsen

Carbonate sediments are a part of the carbon cycle (Fig. 1.14). CO_2 in the atmosphere dissolves in water and makes carbonic acid (H_2CO_3) which reacts with Ca^{2+} or Mg^{2+} to precipitate $CaCO_3$ or $MgCO_3$. This process is an important sink for CO_2. The rate of carbonate sedimentation globally is controlled by the supply of cations (mostly Ca^{2+} and Mg^{2+}) into the ocean from rivers. This again is a function of the rate of weathering of Ca-bearing silicate minerals like plagioclase.

Weathering of limestones will release CO_2 and therefore does not contribute as a sink for CO_2. During contact metamorphism when limestones are heated (to 550–600°C), $CaCO_3$ reacts with SiO_2 to form $CaSiO_3$ (wollastonite) and CO_2. Large amounts of CO_2 formed this way are released through volcanoes. When minerals like wollastonite are weathered, Ca^{2+} is released into rivers and the ocean again. Another part of the CO_2 is reduced by plants and stored as carbon or organic compounds which make up petroleum. These may also be oxidised to CO_2.

Carbonate sediments are for the most part formed (born) within sedimentary basins even if there is a clastic supply of carbonate sediments. The sedimentology of carbonates therefore differs in many respects from siliceous sand and mud. Carbonates may precipitate chemically from the seawater but most of the limestones are composed of calcareous organisms. The properties of limestones are therefore closely linked to their biological origin and the mineralogy of the carbonate skeletons. The primary composition is very important for their alteration during burial (diagenesis) and consequent reservoir properties (porosity and permeability), which to a large extent is controlled by chemical processes. Since both biologically and chemically precipitated carbonate sediments are composed of minerals we will first examine the mineralogy and geochemistry of carbonates. This is important for understanding diagenetic reactions and prediction of reservoir quality.

5.1 Geochemistry of Carbonate Minerals

Carbonate minerals consist of CO_3^{2-} and one or more cations. The most common cations in carbonate minerals together with their mineral names are listed in Table 5.1. The common rock-forming carbonate minerals are either rhombohedral (calcite) or orthorhombic (aragonite) in crystal habit. Where cations with small ionic radii are incorporated into carbonates a trigonal

Table 5.1 Mineralogy of the most common carbonate minerals

Carbonate sediments formed in normal marine environments consist of three main minerals:

Low-Mg calcite $CaCO_3$ (<4% $MgCO_3$) (hexagonal)
High-Mg calcite (Ca,Mg) CO_3 (>4% $MgCO_3$) (hexagonal)
Aragonite ($CaCO_3$) (orthorhombic)
Other common carbonate minerals are:
Siderite $FeCO_3$
Magnesite $MgCO_3$
Strontianite $SrCO_3$
Rhodochrosite $MnCO_3$
Smithsonite $ZnCO_3$
Ankerite $Ca(Mg,Fe)(CO_3)_2$
Dolomite $CaMg(CO_3)_2$

N.-M. Hanken (✉)
University of Tromsø, Tromsø, Norway
e-mail: Nils-Martin.Hanken@uit.no

K. Bjørlykke (ed.), *Petroleum Geoscience: From Sedimentary Environments to Rock Physics*,
DOI 10.1007/978-3-642-02332-3_5, © Springer-Verlag Berlin Heidelberg 2010

(rhombohedral) crystal lattice is formed, while larger cations result in orthorhombic unit cells. Ca^{2+} has an ionic radius close to 1 Å which is intermediate between small and large cations and near the limit of sixfold co-ordination. Thus $CaCO_3$ is dimorphous forming either rhombohedral or orthorhombic structures.

Cations smaller than 1 Å such as Fe^{2+}, Mn^{2+}, Zn^{2+} and Mg^{2+} can be incorporated in the calcite lattice. These metals all have an ionic radius of about 0.6–0.7 Å, and therefore calcite can contain considerable concentrations of these cations. The calcite group of minerals, such as siderite ($FeCO_3$), rhodochrosite ($MnCO_3$), smithsonite ($ZnCO_3$) and magnesite ($MgCO_3$), all have the same crystal structure as calcite.

Two types of calcite are recognised, depending on the magnesium content: low-Mg calcite (<4 mol% $MgCO_3$) and high-Mg calcite (>4 mol% $MgCO_3$). Biologically secreted calcite is high-Mg calcite and typically ranges between 11 and 19 mol% $MgCO_3$. Low-Mg calcite (in most cases simply called calcite) is more stable than high-Mg calcite, and fossil fragments originally composed of high-Mg calcite are converted to low-Mg calcite during diagenesis. The alteration of high-Mg calcite to low-Mg calcite takes place by a process of leaching of Mg^{2+} ions, which leaves the microarchitecture of the grain unaffected. The exsolved Mg^{2+} may form microdolomite rhombs that are sometimes seen as inclusions in calcitised high-Mg calcites (quite common in fragments of echinoderms and calcareous red algae).

The orthorhombic lattice has an arrangement of CO_3^{2-} anions where cations larger than 1 Å (such as Sr^{2+}, Ba^{2+} and Pb^{2+}) are preferred. Analogous with this aragonite crystal structure are strontianite ($SrCO_3$), witherite ($BaCO_3$) and cerrusite ($PbCO_3$). Sr in particular is an important trace element in aragonite. Aragonite crystals forming in marine environments today contain 5,000–10,000 ppm Sr. Aragonite is unstable and after some time will be replaced by calcite which still retains relatively high concentrations of strontium. Aragonite may occasionally be preserved, particularly in dense shales, even in Mesozoic rocks.

Iron is only very weakly soluble in the oxidised state, forming hydroxides $Fe(OH)_3$ and oxides (Fe_2O_3), but in the reduced state it occurs as soluble Fe^{2+}. Reduction of iron normally takes place within the microbial sulphate reduction zone where high concentrations of sulphur will cause available Fe^{2+} to be precipitated as sulphides (pyrite, FeS_2), so that very little is available to enter the calcite structure. The principal environment in which Fe^{2+} can enter the calcite lattice to form ferroan calcite is thus in the reducing porewater below the sulphate reduction zone. The ferroan calcite may contain a few thousand ppm of iron.

Dolomite ($CaMg(CO_3)_2$) is a carbonate mineral in which layers of $CaCO_3$ alternate with layers of $MgCO_3$. Fe^{2+} is commonly found substituting for Mg^{2+} in dolomite, and a complete series extends to ankerite ($Ca(Fe,Mg)(CO_3)_2$). Dolomite formed early in diagenesis is fine-grained and can often have a magnesium deficit in relation to calcium [e.g. $Ca_{55}Mg_{45}(CO_3)_{100}$]. This is called protodolomite which during burial may be transformed into a regular dolomite.

5.2 Carbonate – CO_2 Systems in the Sea

Even if most carbonate precipitation occurs biologically it is important to understand the chemical constraints on carbonate reactions. Carbon dioxide concentration is the factor which has the greatest influence on pH and the solubility of carbonates in water. CO_2 dissolves in water to form carbonic acid (H_2CO_3), which dissociates into bicarbonate (HCO_3^-) and carbonate ions (CO_3^{2-}).

$$CO_2 + H_2O = H_2CO_3 = H^+ + HCO_3^- = CO_3^{2-} + 2H^+$$

Calcite solubility depends on the solubility product $\left[Ca^{2+}\right] \cdot \left[CO_3^{2-}\right]$. We see that when the water is relatively acid (high H^+ concentrations) the reaction is driven towards the left and the carbonate ions concentration $\left[CO_3^{2-}\right]$ will be low. With high pH (i.e. low H^+ activity), the bicarbonate $\left(HCO_3^-\right)$ concentration will be lower.

Mineral solubility is not strictly a function of the ionic concentration, but of the activity (a) which is influenced by the temperature and other ions present.

The dissociation constants for H_2CO_3 and HCO_3^- are:

$$K_1 = \left(aH^+ \cdot aHCO_3^-\right) / \left(aH_2CO_2^-\right)$$

Fig. 5.1 Solubility of carbonate ions in seawater as a function of pH

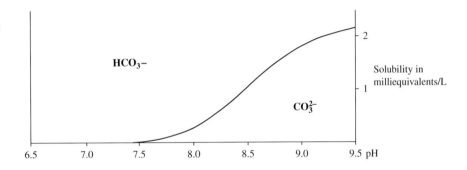

and

$$K_2 = aH^+ \cdot aCO_3^{2-}/aHCO_3^-$$

This in turn requires that aCO_3^{2-} also must be higher in order to satisfy the dissociation equation. At low pH values the equilibrium will shift to the left, giving more free CO_2 and H_2CO_3 (Fig. 5.1). CO_2 is found in both water and the atmosphere, and is exchanged between them. Statistically, the residence time in the atmosphere is c. 8 years, while it is c. 600–1,000 years in the ocean. The CO_2 in the ocean is partly removed when precipitated as organic matter in the sediments, and partly by precipitation as carbonates.

The solubility of CO_2 in water is greatest at low temperatures and high pressures, decreasing as the temperature rises and pressure decreases. Since it is largely the CO_2 concentration which determines the pH of water, the pH is highest (8.0–8.5) in the warm surface layer at low latitudes, and lowest in polar areas (7.5–8.0).

Photosynthesis also contributes to the consumption of CO_2, increasing the pH of the surface water. pH decreases with water depth, not only because CO_2 is no longer removed by photosynthesis, but also because of the lower temperature and higher pressure.

Whereas photosynthesis involves the removal of CO_2 from the water, respiration adds CO_2.

$$H_2O + CO_2 = CH_2O + O_2$$

CH_2O is a general formula for sugar. During photosynthesis this reaction will go to the right. The reverse reaction is respiration. We see that while photosynthesis raises the pH, respiration will lower it. The water below the photic zone will gain CO_2 from the respiration of zooplankton, and the breakdown

(i.e. oxidation) of organic matter which sinks down through the water column will also produce CO_2 and lower the pH. In shallow water, a daily variation in pH has been registered as a result of the fact that photosynthesis takes place only during the day, increasing the pH, while respiration continues at night reducing the pH. Respiration by organisms in the water below the photic zone contributes further to the pH declining downwards through the water column. The recent increase in CO_2 content in the atmosphere (from 280 to 380 ppm) will also influence the ocean water, making it slightly more acidic. Ocean water is, however, strongly buffered and the amount of carbonate that can be precipitated in the oceans is primarily dependent on the supply of cations, especially Ca^{2+} which is mostly liberated by land weathering of carbonate rocks and calcium silicates such as plagioclase.

5.3 Skeletal Components

5.3.1 Introduction

The evolution of organisms precipitating carbonate skeletons has played an important role in the accumulation of carbonate sediments and their properties. The skeletal material is widely different in size, with diameters ranging from a few micrometres in coccolithophores to more than a metre in some bivalves and sponges. The range and taxonomic diversity of major groups of skeletal organisms are shown in Fig. 5.2. However, in this text book we have limited the descriptions to those groups which are most commonly encountered in carbonate hydrocarbon reservoirs.

The first condition for obtaining relatively pure carbonate deposits is that there must be very little or no

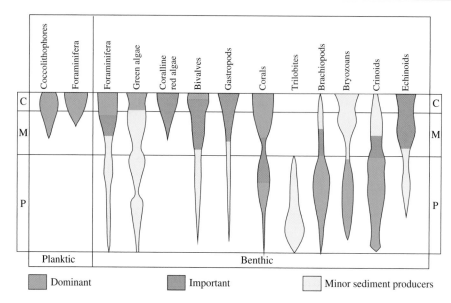

Fig. 5.2 Diversity, abundance and relative importance of various calcareous marine organisms as sediment producers. P = Palaeozoic; M = Mesozoic; C = Cenozoic (modified after Wilkinson 1979)

supply of terrigenous sediments such as sand or clay which otherwise would dilute the carbonate content. Many of the carbonate-secreting organisms require clear water because they filter out nutrients from the seawater. The presence of mud will kill organisms like corals, severely reducing carbonate production. This puts an important limitation on the occurrence of carbonate sediments in sedimentary basins.

5.4 Plants

5.4.1 Stromatolites

"Stromatolite" is the term for lamination in carbonate rocks due to the accumulation or precipitation of carbonate as a result of algal growth. Stromatolites have an intermediate status between fossils and sedimentary structures. They consist of cyanobacteria (earlier called blue–green algae) which have a growth form as unsegmented, micron-sized filaments or unicells (coccooids). The filaments occur in rows or strands within a sticky mucilaginous sheath. Only a few forms produce a biochemically precipitated skeleton, usually tubiform. Other cyanobacteria may generate organic films on the sediment surface, which trap and bind lime

mud to form irregular laminae. Rhythmic variations in algal filament growth produce a laminated structure of alternating light, sediment-rich laminae and dark, organic-rich laminae (Fig. 5.3). The end result is either parallel lamination following algal mats, more complex algal growth structures (stromatolites), or a concentric type of structure (oncoids). However, there are not always algal remains to be found, and the only evidence then is the lamination in the rock.

If the laminae are flat-lying they are referred to as *algal laminated sediments*, but if they form structures with vertical relief they are called *stromatolites*. The different overall shapes of stromatolites range from:

1. Laterally linked hemispheroids
2. Discrete, vertically stacked hemispheroids
3. Discrete hemispheroids

These main types are shown in Fig. 5.4, but combinations can occur.

The lamination commonly follows the outline of the structure, or is terminated at the edge of individual heads or stacks. Individual laminae are often thickest at the centre of the structure and thin laterally towards the periphery. Laminae draping over the edge have often accumulated more steeply than the angle of repose because of the sticky surface of the cyanobacteria.

Fig. 5.3 Diagrammatic representation of the day-night accretion in stromatolites. (**a**) and (**b**) During daytime the cyanobacteria trap and bind sediment and proceed to grow up and around the sediment grains. (**c**) A sticky surface that traps and binds the next sediment layer is produced during the night (modified from Gebelein 1969)

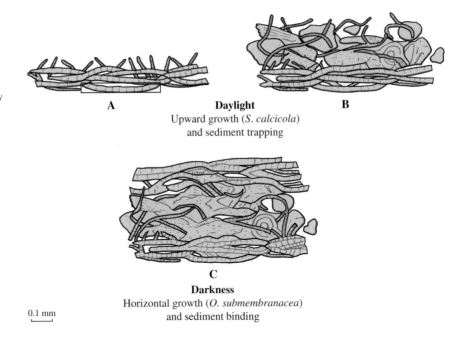

A   **Daylight**   B
Upward growth (*S. calcicola*)
and sediment trapping

C
Darkness
Horizontal growth (*O. submembranacea*)
and sediment binding

0.1 mm

Algal laminae may also form coatings on carbonate grains (e.g. skeletal material), producing structures known as oncoids (Fig. 5.5). These grains become spherical to oval as irregular, sometimes discontinuous, concentric algal laminations consisting of cyanobacteria form a round the nucleus. Once again, the layer of sticky algae on the surface of the grain will trap microscopic sediment grains, or precipitate aragonite because CO_2 is consumed by photosynthesis. In this way the grains will "grow". These "growth layers" will not be continuous around the primary grain, as in oolites, because that requires constant movement. Sediment grains coated in this manner may reach up to 10 cm in diameter. Large oncoids are flattish and often called "algal biscuits". A deposit of oncoid sand is called an oncolite.

5.4.1.1 Ecology

Cyanobacteria have inhabited mostly open marine, shallow water and intertidal environments throughout geological time. However, hypersaline and rare freshwater forms have been recorded too.

Stromatolites are most common in upper intertidal to supratidal environments (Fig. 5.6), but have also occasionally been described from subtidal environments. Intertidal to supratidal stromatolites are often associated with fenestral (birdseye) structures, mud cracks and evaporites. The specific growth form may be used for environmental interpretation. Laterally linked hemispheroids often dominate on protected mud flats where wave action is slight. Exposed, intertidal mud flats, where the scouring action of waves and other interacting factors prevent the growth of algal mats between stromatolites, are characterised by vertically stacked hemispheroids. Concentrically arranged spheroids (oncoids) may occur in low, intertidal areas that are exposed to waves, and in agitated water below the low-tide mark. Oncoids are generally formed in shallow subtidal environments of low to moderate turbulence.

5.4.1.2 Geological Range

Cyanobacteria range from the early Precambrian to Recent. Stromatolites were very important in the Precambrian, but gradually decreased during the Phanerozoic due to the evolution of grazing organisms, in particular gastropods. Today, stromatolites are mostly present in stressed environments where high evaporation and therefore enhanced salinity limit grazing species.

Fig. 5.4 Main types of stromatolites (modified from Logan, Rezzak and Ginsburg 1964)

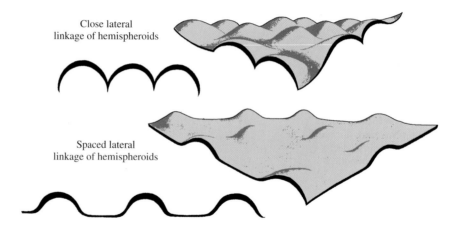

Close lateral linkage of hemispheroids

Spaced lateral linkage of hemispheroids

Discrete, vertically stacked hemispheroids

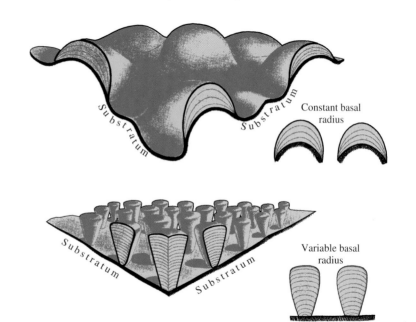

Constant basal radius

Variable basal radius

5.4.1.3 Significance for Petroleum Geology

Some unicellular cyanobacteria synthesise oil and their remains may accumulate to form sapropelic deposits. These deposits may be important source rocks for oil. Oncolites may possess primary interparticle pores, but high porosity is not as common as in oolites because of poor sorting in less turbulent water. Oncolites also form thinner and less persistent beds than oolites and have relatively low reservoir and source rock potential.

A   B
                    2 cm

Fig. 5.5 Recent oncoids from the tidal zone. (**a**) Top surface of a hemisphaerical oncoid. (**b**) Vertical section showing the rough lamination. The flat base reflects stationary growth. Ras Muhamad Peninsula, Sinai

Fig. 5.6 (**a**) Supratidal crust composed mainly of fine-grained dolomite. Bahamas (Photo G.M. Friedman). (**b**) Supratidal environment (sabkha), Abu Dhabi. Photo D. Kinsmann. (**c**) Algal mat with gypsum from supratidal environment (sabkha), Abu Dhabi (Photo D. Kinsmann)

5.4.2 Calcareous Algae

Calcareous algae form a diverse group of plants that inhabit most shallow marine, clear water environments. They prefer normal marine salinity although some forms may be euryhaline. Most calcareous algae are sessile, although a few, such as the coccolithophores, are planktonic. Most important are red algae and green algae, but one species of brown algae (*Palina*) also secretes a calcareous skeleton, which consists of aragonite.

A wide range of growth forms exists, and the most important are:

- **Encrusting calcareous algae**
  These are usually red algae and form smooth or uneven crusts, nodular shapes and rigid branching forms (Fig. 5.7). They vary in size from a few millimetres to many centimetres.

- **Calcareous algae with erect growth morphology**
  Usually segmented, branching algae which may become several centimetres high (Fig. 5.8). Each segment is normally less than a few millimetres, often cylindrical with circular or oval cross-section. Since the segments generally come apart when the alga dies, it is very difficult to reconstruct the original plant body (thallus) of fossil species.

5.4.2.1 Ecology

Red algae are almost exclusively marine and prefer shallow (<25 m) subtidal environments where the light is strong. They may, however, range down to a depth of approximately 250 m. Some of the growth forms can withstand fairly strong agitation and are common on shallow, rocky substrates associated with reef and near-reef environments on platform margins. Red algae are generally less tolerant of salinity variations than green algae and cyanobacteria.

Fig. 5.7 Encrusting calcareous algae comprise many different growth forms including smooth and uneven crusts, nodular shapes and rigid, branching forms. (**a**) Irregular crust. (**b**) Nodule. (**c**) Rigid branching form (modified from Wray 1977)

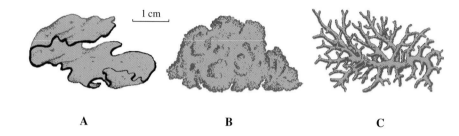

A   B   C

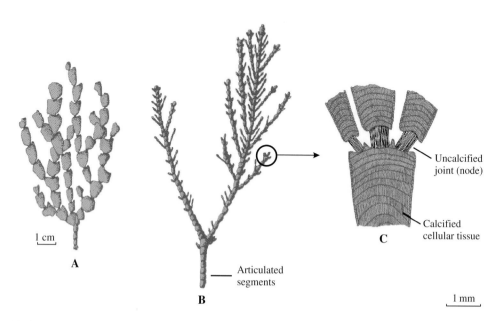

Fig. 5.8 (**a**) Segmentary growth form of the green alga *Halimeda* sp. Each segment is wide and flat with a roughly triangular cross section. The segments are separated by joints. Since there is no calcification near the joints, the algae have some flexibility. (**b**) Typical growth form of articulated coralline algae (*Corallina* sp.). (**c**) Internal structure of calcareous segments as seen in longitudinal section. Between each joint there is a poorly calcified region, giving the plant some flexibility (modified from Wray 1977)

During photosynthesis, algae consume CO_2 and produce oxygen. The absorption of sunlight by water is not, however, equally rapid for the various constituent wavelengths, the red portion of the spectrum (longwave light) being more rapidly absorbed than the blue (shortwave). Red algae can also use blue shortwave light, enabling them to live at greater depths than, for example, green algae, which are dependent on long-wave red light (Fig. 5.9). Areas with carbonate sedimentation normally have little clastic material in suspension, and the water is therefore very clear. The absorption of light drops off exponentially with depth and below about 50 m there is very little light left.

Most recent green algae grow in shallow marine, tropical waters of normal salinity, but some tolerate

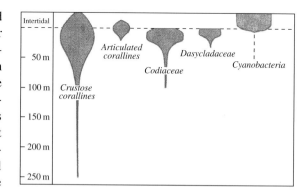

Fig. 5.9 Generalised depth distribution of the major groups of Recent marine calcareous algae. The red algae have been drawn in *red*, the green algae in *green* and the cyanobacteria in *blue* (modified from Wray 1977)

hypersaline or brackish water. Recent green algae are abundant in depths of 5–10 m, but may range down to 100 m and also extend into lagoons and mangrove swamps. Some species can grow on firm substrates, but most grow in sand or mud. They generally prefer environments with only moderate wave energy. The calcareous green algae are major producers of small aragonite crystals which form lime mud. The presence of green algae therefore ensures that lime mud is deposited in the lagoons.

5.4.2.2 Mineralogy

The skeletal material in living algae is either calcite or aragonite, never mixtures of these two minerals in the same species. The extent of calcification is highly variable and may be only partial in some groups while other groups have a pervasive calcification of the whole plant body. In recent calcareous algae, the precipitated calcite (usually high-Mg calcite) or aragonite consists of small crystals (<4 μm). The same is assumed to apply to pre-Quaternary species too. The preservation of the plant bodies depends on the extent of calcification, with the most calcified parts selectively preserved in the fossil record.

Modern red algae deposit either aragonite or high-Mg calcite with a Mg content up to about 30 mol% $MgCO_3$. The skeletal elements in green algae consist only of aragonite. This primary difference in their skeletal composition is an important preservation factor. The solubility of skeletal aragonite in pure water is about twice that of low-Mg calcite, whereas the solubility of skeletal high-Mg calcite is up to 10 times that of low-Mg calcite. Hence, the fine-grained aragonite skeleton of the green algae is replaced by sparry calcite, whereas coralline red algae with a primary calcitic composition of varying magnesium content show different degrees of alteration (often partial dissolution).

5.4.2.3 Geological Range

Calcareous green algae first appeared in the Cambrian, which led to the production of large amounts of lime mud. These algae are still an important factor in modern shallow-water carbonate sedimentation. Red calcareous algae also evolved in the Cambrian. In Recent marine environments they are commonly encountered both in cold and warm water deposits within the photic zone, although most of them live in fairly shallow water (<25 m).

5.4.2.4 Significance for Petroleum Geology

Red algae, particularly those with erect growth forms, may form primary framework porosity in reefs. The algae are susceptible to neomorphic replacement and develop secondary porosity when exposed to extensive freshwater dissolution.

Red algae are common associates of many oil and gas producing reefs, notably stromatoporoid and coral reefs. This includes the Middle Silurian pinnacle reefs (stromatoporoid-coralgal reefs) in the Michigan Basin of southeastern Michigan, USA, the Upper Palaeocene Intisar "D" Field (coralgal reef) in Libya, the Miocene coralgal reefs in the Salwati Basin of Irian Jaya, Indonesia and the Miocene Central Luconia Fields of offshore Sarawak, Malaysia.

Calcareous green algae may form important source rocks for oil.

5.4.3 Coccolithophores

Coccolithophores are an example of free-floating algae with carbonate shells. They are single-celled, microscopic (5–20 μm) plants belonging to the golden-brown algae. A coccolithophore is generally spheroidal or ovoidal and is covered with an external calcareous skeleton (coccosphere) (Fig. 5.10). The coccosphere is composed of a variable number of small (a few μm), commonly round to oval, calcareous plates (coccoliths). The coccoliths are made of tiny, platy low-Mg calcite crystals (0.25–1 μm) with a flattened rhombic form, and are stacked in an imbricate fashion that produces a spiral pattern. The coccolithophores usually disaggregate after death and accumulate as individual coccoliths, sometimes forming extensive chalk deposits.

5.4.3.1 Ecology

Coccolithophores are most common in the photic zone of open seas, although they may also range into

Fig. 5.10 (**a**) Skeletal structure of coccospheres and coccoliths. In this example the spheroidal coccosphere is composed of many round calcareous plates (coccoliths) which consist of tiny, platy calcite crystals stacked in an imbricate pattern. Modified from Hjuler (2001). (**b**) SEM picture of Liège chalk consisting of a mixture of coccoliths and loose platy calcite crystals. (**c**) SEM picture of Upper Cretaceous coccoliths in the Ekofisk field, North Sea. Petroleum occurs between the small plate-like coccoliths. The limestone has 32% porosity and 1 mD permeability (from Bjørlykke 1989)

nearshore lagoons. They swim freely in the surface waters (commonly the upper 100 m) of the ocean, occurring in greatest abundance in temperate zones. Coccolithophores are abundant (commonly as many as half a million individuals per litre of water) in the photic zone of modern oceans. Coccoliths are important constituents of some fairly deep basinal and distal shelf sediments.

Deep-sea drilling has shown that coccolithophores were very much more widespread during the Cretaceous and Lower Tertiary than they are today. The colder climate which began in the Miocene shifted the northern limits for coccolithophore sedimentation southwards. Many pelagic species have very specific temperature requirements and their occurrence provides an important contribution to the palaeoecology of marine sediments of various ages.

5.4.3.2 Mineralogy

The plates of the coccolithophores consist of low-Mg calcite which makes them fairly stable during diagenesis.

5.4.3.3 Geological Range

Coccolithophores range from the Triassic to Recent, but were not common before the Jurassic. They were particularly abundant in the Cretaceous and Lower Tertiary, forming thick, extensive chalks. They comprise approximately 25% of present-day calcareous oozes and up to 90% of some Cretaceous and Tertiary chalks.

5.4.3.4 Significance for the Petroleum Industry

Chalks made up of coccoliths may be extremely porous. The porosity is in the form of primary interparticle micropores and may be as high as 40–50%, even in deeply buried chalks. Although their porosity may be high, permeability is generally low (less than a few mD) because of small pore-throat diameters (generally <1 μm). Hydrocarbon-filled interparticle pores in chalks are therefore productive only in combination with some other pore type, preferably fracture pores, which increases the permeability. Fractured chalks of Maastrichtian (late Cretaceous) to Danian (early Tertiary) age form major hydrocarbon reservoirs in the Central Graben of the North Sea (Ekofisk and associated fields). These are in fact the world's only major oilfield in such rocks, but the low permeability of this fine-grained lithology creates some production problems.

5.4.4 Calcispheres

Calcispheres are of uncertain biological affinity, but many consider them to be algae because there is a great similarity between non-ornamented fossil calcispheres and reproductive cysts of some living green algae (dasycladacean algae). They are generally 40–200 μm in diameter and are composed of a thin (commonly 3–30 μm), well-defined calcite wall, enclosing a single spherical chamber (Fig. 5.11). The wall may be singly layered or have several concentric layers distinguishable by the alternation of uniform micritic texture with a fabric showing radial elements. The wall has radial pores, but there is no aperture. The outer surface may bear spines. The chamber is normally filled with cement or sediment.

5.4.4.1 Ecology

Many fossil calcispheres seem to owe their origin to dasycladacean algae, and as such, they provide a useful palaeo-environmental indicator. Devonian and Permian calcispheres are most common in shallow subtidal environments, especially in restricted

or back-reef environments. In the Cretaceous, calcispheres are especially important in pelagic deposits such as Chalk sediments in the North Sea.

5.4.4.2 Geological Range

Calcispheres are widely known from Upper Palaeozoic limestones, especially Devonian and Carboniferous, but also occur in older and younger strata.

Calcispheres have no particular significance for petroleum geology. However, they may be an important component improving the overall potential of a source rock.

5.4.5 Distribution of Algae in Modern and Older Carbonate Sediments

If one places a profile extending from a basin centre to a coast with reef development (Fig. 5.12), we see the following distribution of algae:

A. The sedimentation basinward consists largely of planktonic algae (coccolithophores), planktonic foraminifera and some other calcareous organisms.
B. In the reef facies we find mainly red algae which build strong, solid structures of carbonate, which in conjunction with the corals can resist waves which break against the reef. In the lagoons landward of the reef we find green algae. These have bush-shaped skeletons and can only grow where the wave energy is moderate. They are major producers of small aragonite crystals which form lime mud. The green algae therefore ensure that lime mud is deposited in the lagoon.
C. In protected parts of lagoons and in the tidal zone, the sea bed is often covered with cyanobacteria which lie like a gelatinous carpet over the sediments. These algae consist of a network of threads which hold the sediment in place and protect it against moderate currents and wave erosion. However, they may be eaten by grazing animals, e.g. snails. In high-energy nearshore environments we find red algae.

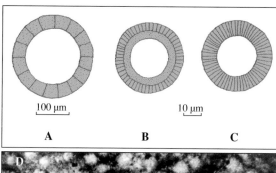
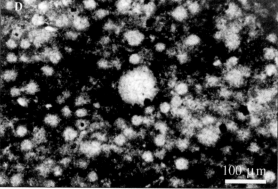

Fig. 5.11 Sections of calcispheres. In most cases, the calcitic wall consists of only one layer (**a, c**), but multilayered walls are also known (**b**). (**d**) Micrograph of the abundant calcispheres, partially replaced by euhedral pyrite as viewed in transmitted light, Ravnefjeld Formation (Upper Permian), East Greenland. See Nielsen and Hanken (2002). **a, b** and **c** are modified from Scoffin (1987)

The age range and taxonomic diversity of the major calcareous algal groups are shown in Fig. 5.13.

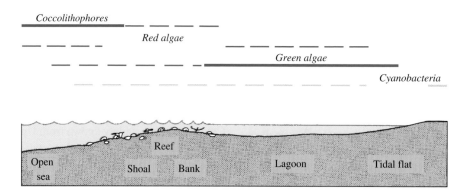

Fig. 5.12 Generalised distribution of recent skeletal and non-skeletal algae across a sediment basin in a low-latitude region (modified from Wray 1977)

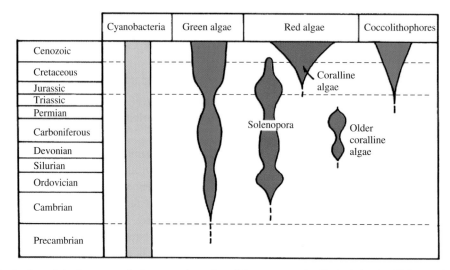

Fig. 5.13 Distribution of the important algal groups through geological time. Cyanobacteria have existed since early Precambrian times, and the oldest stromatolites known are up to about 3 billion (3×10^9) years old. Both green and red calcareous algae first appeared in the Cambrian. The evolution of calcareous green algae led to the production of large amounts of lime mud without chemical precipitation. The Solenoporaceae family (red algae), which was widely distributed in the Palaeozoic, died out during the Cretaceous. Coralline algae appeared in the Jurassic, and still exist. The evolution of the planktonic algae, mainly coccolithophores from the Triassic to present, has been a vital factor in global carbonate sedimentation (modified from Ginsburg et al. 1971)

5.5 Invertebrate Skeletal Fossils

5.5.1 *Foraminifera*

Foraminifera (or forams for short) are a very important group of single-celled animals that secrete chambered, calcareous or sometimes agglutinated tests. Agglutinated tests are made up of foreign particles like mineral grains and/or shell fragments such as coccoliths, small foraminifera and sponge spicules (Fig. 5.14a–c). These particles are attached by calcitic or ferruginous cement to a layer of tectin (an organic compound composed of protein and polysaccharides).

Most foraminifera have an outer carbonate shell with one or more openings (apertures). A few species have only one chamber, but most foraminifera add new chambers as they grow. The individual chambers are arranged in a specific pattern (Fig. 5.14d–g), either in a single row (uniserial) or in two or three alternating

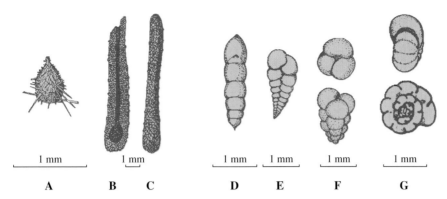

Fig. 5.14 Examples of variations in shape, size, arrangement of chambers and mineral composition in foraminifera. (**a–c**) Agglutinated forms. (**d–g**) Foraminifera with carbonate shells. (**a**) *Reophax* sp., agglutinated with sponge spicules. (**b**) (transverse section) and (**c**) (external view) *Hyperamina* sp., agglutinated with fine-grained detrital material. (**d**) *Nodosaria* sp., uniserial shell. (**e**) *Textularia* sp., biserial shell. (**f**) *Verneuiliana* sp., triserial shell. (**g**) *Lenticulina* sp., planospiral shell. Lowermost is a median horizontal section showing the individual chambers (modified from Shrock and Twenhofel 1953)

rows (biserial and triserial). The shell may also be spirally coiled, either flat (planospiral) or as in a cone (helicoidal). The partitions (septa) between the individual chambers are the former front wall, and the former aperture is then often retained as an opening (foramen) between the chambers (Fig. 5.15).

5.5.1.1 Ecology

Virtually all foraminifera are marine, but a few live in brackish or fresh water. Although they occur at all depths down to the abyssal zone and at all latitudes, they have their greatest abundance and diversity in shallow, tropical, open marine waters. Some foraminifera live in symbiosis with algae living in their protoplasm, and these species are of course limited to the photic zone. Most foraminifera are benthic (living on the sea bed; both vagrant and sessile forms are known), but some are planktonic (freely drifting).

Many species require very specific ecological conditions. Some have left- or right-curved tests, depending on the temperature. Benthic forms are sensitive to current conditions, temperature, salinity, supply of nutrients etc. Foraminifera thus are useful as indicators of palaeoecological conditions. The amount of planktonic foraminifera relative to benthic ones declines in shallower water, and so can be used in reconstructions of palaeobathymetry.

However, only a small percentage of the foraminiferal tests are preserved in the geological record because they are generally fragile and easily broken by the action of scavengers and burrowing organisms. The calcareous tests also dissolve as they

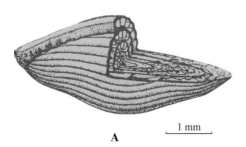

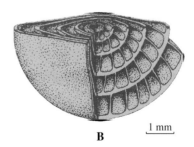

Fig. 5.15 Partially sectioned fusulinid (**a**) and nummulitid (**b**), showing the internal arrangement of chambers. The fusulinids are important index fossils for the Upper Carboniferous/Lower Permian while nummulitids have been widely used for correlation of Miocene marine deposits (modified from Rasmussen 1969)

sink through the deeper and colder waters and will not be preserved if they are deposited below the carbonate compensation depth (CCD).

5.5.1.2 Mineralogy

Benthic foraminifera may have tests of precipitated calcium carbonate, usually calcite, or of small sand grains cemented together (agglutinated shell). Some foraminifera have aragonite tests. The calcite tests vary from high-Mg to low-Mg calcite, so that their solubility in seawater varies greatly. Benthic foraminifera with high-Mg calcite are found especially in warm-water environments, where they are less soluble.

5.5.1.3 Geological Range

Foraminifera range from the Cambrian to Recent. Most of the Lower and Middle Palaeozoic forms had an agglutinated test whereas calcareous forms were most common in late Palaeozoic and later times. Planktonic foraminifera did not appear until the Mesozoic but are important constituents of Cretaceous and younger deep-sea deposits. The large-sized, lens-shaped nummulitid foraminifera (up to about 15 cm) are widely distributed in Early Tertiary carbonate ramp and platform deposits. They are particularly useful as an index fossil in Cenozoic marine deposits of the Mediterranean Sea region. Foraminifera provide the basis for much of the stratigraphic subdivision of the Upper Palaeozoic, Mesozoic and Cenozoic.

Foraminifera commonly have a low reservoir potential due to their small particle size and stable primary mineralogy or neomorphic replacement. Some primary intragranular porosity may be of local significance, but generally connectivity is poor, giving low permeability.

5.5.2 Sponges

Sponges may vary considerably in size, commonly from a few centimetres to more than a metre in diameter. Their external morphology is also highly variable, but plate-shaped, globular, vase- or bowl-shaped forms are common. In its simplest form, the body is sac-shaped with a large internal cavity which opens upwards, the closed end being attached to the substrate. The wall is perforated by numerous small pores leading to the central cavity.

The skeleton is internal and may be spongin (an iodine-bearing protein which is rarely preserved in the geological record), siliceous or calcareous. Some sponges may incorporate foreign particles like sand grains in their skeleton. However, the skeleton typically consists of soft tissue supported by a complex network of spicules (Fig. 5.16). The soft tissue decays on death and spicule-bearing forms are therefore seldom preserved intact. The spicules may occur in two different sizes within a single individual. The large spicules (megascleres) vary from 0.1 to 1 mm and constitute the skeleton proper. The small spicules (microscleres) vary from 0.01 to 0.1 mm and are found isolated in the sponges, especially around the pores. The megascleres may occur as isolated spicules, be in close contact with each other, or be firmly intergrown to form a skeleton.

5.5.2.1 Ecology

Nearly all sponges are marine although a few freshwater forms are known. Sponges live mostly in relatively clear water, from the littoral zone down to abyssal depths. Siliceous sponges are generally found in deep water, whereas calcareous forms are more common in shallow-water environments. Most sponges lived attached to the bottom.

Many sponges are important boring organisms which help to break down calcareous rocks and carbonate skeletons produced by larger organisms into finer-grained material. The holes are 0.2–1 mm in diameter, forming thin holes up to 0.5 cm long.

5.5.2.2 Mineralogy

The main division of the sponges is based on the chemical composition of the spicules and the symmetry of the skeletal elements. Calcispongiea (calcareous sponges) have spicules of calcite or aragonite. These spicules are characterised by having a monocrystalline microstructure where each individual

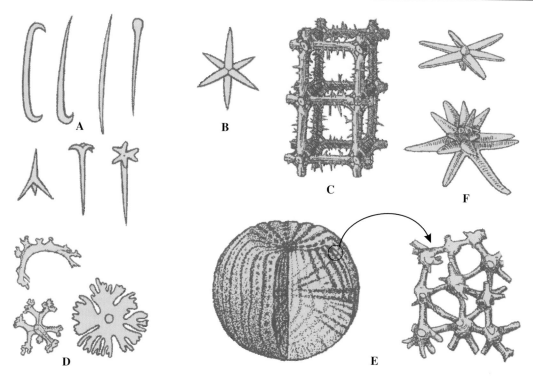

Fig. 5.16 Examples of different forms of sponge spicules. (a) Monaxon spicules. (b) Triaxon spicule. (c) Part of a sponge with triaxon spicules. (d) Pharetone spicule. (e) Various kinds of tetraxon spicules. (f) Part of a sponge skeleton built up of tetraxon spicules (modified from Rasmussen 1969)

spicule extinguishes as a single entity in polarised light. Hyalospongiea (siliceous sponges) have spicules of amorphous silica (opal A, $SiO_2 \cdot nH_2O$), Fig. 5.17. However, opaline silica is relatively unstable and siliceous sponges are commonly replaced by calcite. Demospongiea have spicules which are made exclusively of either spongin or amorphous silica, or of both silica and spongin.

Sponge spicules are locally very abundant, especially in cherts and silicified limestones. Siliceous sponge spicules have often been invoked as the source of silica in such deposits.

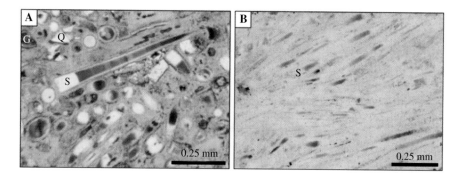

Fig. 5.17 (a) Photomicrograph of porous spiculite consisting of spicules from hyalospongiea (siliceous sponges). Note spicule in longitudinal cross-section (S), quartz (Q) and glauconite grains (G). Pores have been impregnated with blue epoxy. (b) Well cemented spiculite with very low porosity. Kapp Starostin Formation (Upper Permian), Svalbard. Plane light (photos courtesy of Sten-Andreas Grundvåg)

5.5.2.3 Geological Range

Sponges range from the Cambrian (Precambrian?) to Recent. They are minor sediment contributors in modern settings but have been prolific in the past. Siliceous forms were common in the Lower Palaeozoic; calcareous forms first appeared in the late Devonian. Sponges are major constituents of some Upper Carboniferous, Permian, Triassic and Jurassic build-ups.

Sponges have extremely high primary porosity due to their network of internal canals. However, this porosity is rarely preserved during burial. Occasionally, high primary porosity is preserved in hollow spicules, but permeability is generally extremely low and most of the porosity is ineffective.

Secondary porosity may form by dissolution of siliceous and aragonitic spicules, but rarely constitutes a significant part of a reservoir.

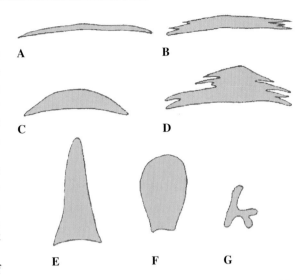

Fig. 5.18 Examples of growth forms in stromatoporoids. (**a**, **b**) Laminar. (**c**, **d**) Low dome-shaped. (**e**) High dome-shaped. (**f**) Hemispherical. (**g**) Branched (modified from Kershaw and Riding 1978)

5.5.3 Stromatoporoids

The systematic position of stromatoporoids has been much debated, but they are now regarded as belonging to the demosponges. They form colonies of variable size and shape. They are commonly from a few centimetres to about a metre in diameter and their shape may be tabular, hemispherical, cylindrical or branched (Fig. 5.18). The skeleton tends to split into concentric layers each a few centimetres thick, and therefore stromatoporoids often occur as larger or smaller fragments if they have been exposed to mechanical abrasion prior to their final deposition.

5.5.3.1 Ecology

Stromatoporoids are sessile, open-marine, colonial organisms. Since the growth form is to a certain degree dependent on the environment, some forms can be used in palaeoecological analysis relating the growth rate vs. sedimentation rate. Stromatoporoids were important constituents of Ordovician to Devonian reefs along with tabulate corals, colonial rugose corals and calcareous algae. These species mostly lived in agitated, shallow, warm seas, while Jurassic and Cretaceous forms were more adapted to life in muddy environments.

5.5.3.2 Mineralogy

The stromatoporoids had a very porous skeleton (coenosta) forming a laminar structure transected by fine vertical pillars. However, the primary structures readily underwent diagenetic modifications and therefore were commonly lost during diagenesis, leaving only a reticulate pattern of anastomosing elements. This often renders primary microstructures difficult to differentiate from secondary modifications. The neomorphic fabric indicates that the original composition of the stromatoporoid coenostae was aragonitic.

5.5.3.3 Geological Range

Stromatoporoids range from the Cambrian to Oligocene, but were most important during the Silurian and Devonian.

5.5.3.4 Significance for Reservoir Quality

Stromatoporoid reefs and associated reef detritus may form important hydrocarbon reservoirs in Silurian and Devonian rocks, e.g. Middle Silurian pinnacle reefs in the Michigan Basin of southeastern Michigan,

USA and Middle to Upper Devonian reefs in Alberta, Canada. Although stromatoporoids may form high primary framework porosity, this porosity is commonly obliterated by synsedimentary infilling and cementation of the pores. Secondary porosity is generally more important and may be related to dolomitisation or freshwater dissolution. Stromatoporoid reefs were topographically higher than adjacent sediments and were therefore more prone to be subaerially exposed and dissolved.

5.5.4 Corals

Most corals are colonial, although solitary forms are also common. Colonial corals may be from a few centimetres to several metres in diameter, but are commonly less than 0.5 m. The external form of the colonies may be extremely variable, but can generally be described as crustose, hemispherical, irregular or shrub-like branched (Fig. 5.19). The colonies grow in size by repeated budding from the side or top of the individual corallites. The newly formed corallites may abut against one another like the cells in a beehive, thus forming compact colonies with polygonal, generally hexagonal corallites (Fig. 5.19). However, some species form more open colonies so that the individual corallites are free-standing and rounded in cross-section.

Solitary corals may be curved or erect, and their fundamental shape appears to be a reversed cone. Depending on the apical angle of the cone and other characteristics, such as the growth form of the mature region and the occurrence of sharp angulations or flattened areas, several main types of growth forms can be recognised (Fig. 5.20).

Individual corallites are usually divided by radially arranged vertical elements (septa), horizontal partitions (tabulae) and small partitions in the lower periphery of the corallite (dissepiments). The presence or absence of these structures is a basis for classifying corals into four major groups: rugosa (horn corals), tabulata and scleractinia (hexacorals), and alcyonaria (octacorals).

Rugose corals: Rugose corals consist of both solitary and colonial growth forms. All rugose corals have both tabulae and septa. The septa are arranged in a radial pattern and generally have a marked bilateral symmetry. Both the form, size and number of septa vary up through the corallite (Fig. 5.21). During growth, the deepest part of the horn (calice), may be cut off by transverse partitions (tabulae) or contracted by small, curved partitions (dissepiments).

Many rugose corals are colonial. There are two main types: fasciculate where the corallites are cylindrical and isolated from each other, and massive where the corallites are closely packed and polygonal in cross-section.

Tabulate corals: All tabulate corals are colonial and individual corallites may be connected by small pores through a common wall, porous tissue or tube-like canals (Fig. 5.19b, c). Most tabulate corals have horizontal tabulae. Septa, as seen in rugose and scleractinian corals, are absent, but some species have small septal spines. The microstructure usually differs little from that of the rugose corals.

Scleractinia (hexacorals): Scleractinian corals consist of both solitary and colonial growth forms. They have an aragonitic skeleton in which septa form a sixfold radial symmetry, and tabulae may form horizontal partitions.

Alcyonaria (octacorals): A fourth group of corals is the alcyonaria. All alcyonarian corals are colonial. Most have an endoskeleton consisting of spicules of sclerites. In its simplest form, the sclerite is a rod with more or less acute ends. More complex forms have irregularly sculptured rods with spines and protuberances. Sclerites may be locally relatively common in Recent sediments, but since they are often difficult to recognise in pre-Quaternary deposits, the fossil record of alcyonarians is sparse.

5.5.4.1 Ecology

Corals are bottom-dwelling, sessile, marine organisms which are most common in shallow, well-oxygenated, tropical to subtropical waters of normal marine salinity. Some forms live in deep water in the photic zone and/or in cold waters. Their growth forms are useful environmental indicators.

Scleractinian corals have been major reef builders since the Triassic and tabulate corals were important constituents of Silurian and Permian build-ups. Some recent scleractinian corals (hermatypic corals) corals live in symbiosis with single-celled algae

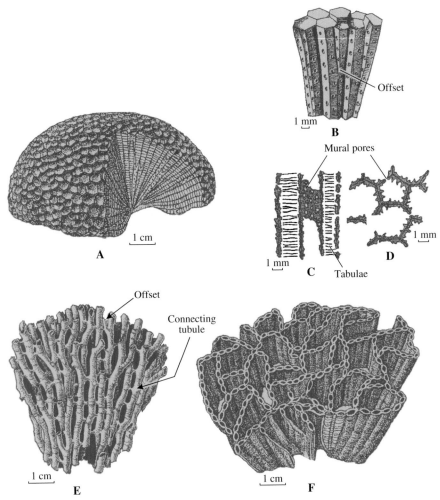

Fig. 5.19 Examples of different growth forms in colonial tabulate corals. (**a**) *Favosites* sp. Massive growth form with polygonal (generally hexagonal) corallites. Part of the *right side* has been removed to expose the corallites in longitudinal section. (**b**) Enlarged section of corallites viewed in longitudinal section. Note the vegetative reproduction in that one of the corallites is dividing into two, giving rise to an offset. (**c**) Enlarged longitudinal section through three corallites where tabulae are clearly visible. Mural pores are seen where the plane of the section passes slightly obliquely through the corallite wall. (**d**) Enlarged transverse section showing mural pores and septal spines. (**e**) *Syringopora* sp. Open growth form with round corallites linked by short, horizontal, connecting tubules. (**f**) *Halysites* sp. Cylindrical corallites which form palisade-like, dendritic rows in an irregular, reticulate pattern with large open spaces (modified from Moore 1956, Rasmussen 1969)

(dinoflagellates or zooxanthella). They do best in very shallow areas (1–20 m deep, but some species might be found down to 90 m) and in warm water (18–29°C, but they flourish best between 25 and 29°C). Most Mesozoic and Cenozoic reef corals are hermatypic, and are therefore restricted to shallow, well-lit warm water. Most of these corals are euryhaline and do not tolerate turbid water.

The other main type of scleractinian corals (ahermatypic corals) do not live in symbiosis with algae, and often occur below the photic zone. They are most abundant down to 500 m, but have been found at water depths as great as 6,000 m. They can live in water with temperatures as low as 0°C. Ahermatypic corals can form reef-like structures at depths of several hundred metres. *Lophelia* reefs or biostromes are found, for example, in deep water settings off the Norwegian coast.

Rugose and tabulate corals were important components of Ordovician and Silurian reef structures.

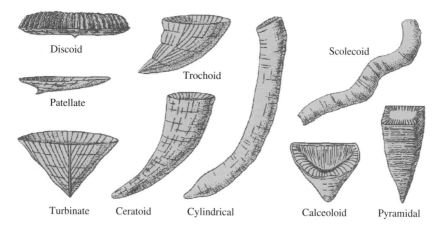

Fig. 5.20 Examples of different growth forms in solitary rugose corals (modified from Moore 1965)

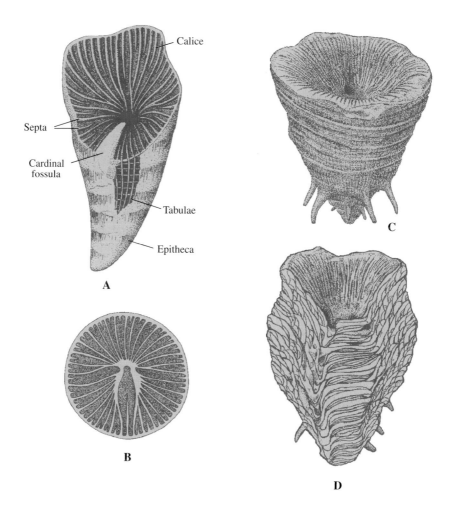

Fig. 5.21 (**a**) The rugose coral *Zaphrentites* with part of the epitheca removed, showing septa and tabulae. (**b**) Transverse section showing the tabulae. (**c**) *Ketophyllum*. Solitary rugose coral with tentacle-like appendages producing tabular holdfasts. (**d**) Longitudinal section through *Ketophyllum* showing tabulae in the middle part and dissepiments along the periphery (modified from Clarkson 1979)

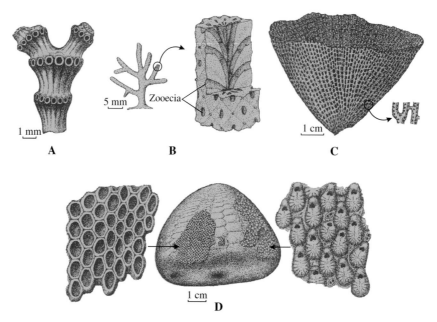

Fig. 5.22 Examples of different growth forms in bryozoa. (**a**) Cyclostomata. *Spiropora* sp. with a dendroid growth form with mouths on separate polypides. (**b**) Trepostomata. Colony and enlarged section showing part of the interior. (**c**) Cryptostomata. *Fenestrellina* sp. with a cornet-shaped colony. All the zooecia are placed on the radial, longitudinal branches, with their mouths inside the cornet. (**d**) Cheilostomata. Encrusting bryozoans on a dead sea urchin (*Membranipora* sp. to the *left* and *Membraniporella* sp. to the *right*) (modified from Rasmussen 1969)

5.5.4.2 Mineralogy

All scleractinaian corals are aragonitic and pre-Holocene specimens are therefore normally preserved as moulds or have been more or less cemented with calcite spar. Rugose and tabulate corals have usually retained their primary microstructure, which indicates that the skeleton consisted of calcite. Recent octacorals have sclerites consisting of calcite or aragonite, and it is believed that pre-Quaternary species also had the same mineralogical variations.

5.5.4.3 Geological Range

Rugose corals range from the Middle Ordovician to the Upper Permian. They were not common until the Silurian, after which their numbers gradually increased to a peak in the Lower Carboniferous. Their numbers and diversity then declined until they finally became extinct by the end of the Permian.

Tabulate corals occur in rocks of Cambrian (Middle Ordovician?) to Permian (and possibly Triassic to Eocene?) age. Scleractinian corals range from the Middle Triassic to Recent, and alcyonarian corals range from the Precambrian to Recent. Scleractinian corals are the dominant coral group today.

5.5.4.4 Significance for Reservoir Quality

Corals may form high primary framework porosity in reefs. Scleractinian corals have an unstable primary mineralogy (aragonite) so may develop high secondary mouldic porosity during dissolution. Corals, in particular scleractinian corals, therefore have relatively high reservoir potential.

5.5.5 Bryozoa

Bryozoans form colonies which are usually 0.5–2 cm in diameter, but colonies up to about 60 cm are known. They live permanently fixed to a substrate, which may be a stone, shell (living or dead), algae or other objects. The colonies have a wide range of external morphologies, including encrusted (flat, hemispherical or irregular), branched (hollow or massive, flattened or circular), and lacy network (Fig. 5.22). Individual colonies may consist of tens to hundreds of circular or polygonal tubes or boxes (zooecia) which contain the individual organisms.

5.5.5.1 Ecology

Bryozoans are widespread, colonial, sessile, suspension feeding organisms which have a limited tolerance for strong wave action, soft sediment surfaces and desiccation. They prefer normal marine salinity but can tolerate short-term salinity variations. Most bryozoans are marine, occurring in shallow to abyssal depths (most common down to 200 m water depth), although a few Recent forms live in freshwater. In spite of their broad environmental distribution, bryozoans are generally most abundant in shallow to moderately deep, marine settings.

5.5.5.2 Mineralogy

There are five orders of bryozoans with a fossil record. In the Cyclostomata, Trepostomata and Cryptostomata, the skeleton is composed of calcite with less than 8 mol% $MgCO_3$. Bryozoans belonging to the Ctenostomata are poorly mineralised with more than 50% organic matter. The most primitive members of the Ctenostomata and Cheilostomata possess a skeleton of calcite only, whereas the more specialised cheilostomes have a secondary thickening of aragonite. However, in general, most Palaeozoic bryozoans had a calcitic skeleton, while many Recent forms are composed of aragonite or mixed aragonite and calcite.

5.5.5.3 Geological Range

Bryozoans range from the Ordovician to Recent. They are relatively rare in modern environments, but were especially important as framebuilders or sediment binders in many Ordovician to Permian build-ups.

5.5.5.4 Significance for Reservoir Quality

Bryozoans may have relatively high primary porosity in the form of interparticle, intragranular and framework porosity, although this porosity is not always preserved during burial. Intragranular pores are usually poorly interconnected giving low permeabilities, and are therefore generally of minor importance in petroleum reservoirs. Framework and interparticle pores are usually better interconnected and may be important in some bryozoan build-ups, e.g. in the Middle Carboniferous of Central Spitsbergen, occasionally forming good petroleum reservoirs.

Bryozoans generally have a low potential for secondary porosity formation due to a stable primary mineralogy. However, bryozoan-rich sediments may still be highly porous when they form build-ups. Bryozoan build-ups formed topographical highs during deposition and were therefore prone to be subaerially exposed during sea level fluctuations. In a humid climate, these build-ups may be pervasively dissolved with the formation of secondary, non-fabric selective porosity. The Lower Permian bryozoan-*Tubiphytes* build-ups in the Midland Basin of West Texas, USA, are examples of this.

5.5.6 Echinoderms

The phylum Echinodermata comprises several classes, of which crinoidea and echinoidea are most common in the fossil record (Fig. 5.23). Modern representatives include the sea urchins (echinoids), starfish and brittle stars (stelleroids), and sea lilies and feather stars (crinoids). However, only crinoidea and echinoidea are dealt with in this book because these groups are most commonly encountered in the field. One of the characteristic features of the echinoderms is that, almost without exception, various degrees of endoskeleton calcification are found in the body wall. A single individual may be composed of a very large number of plates (often several hundred, or even thousands). On death, the individual plates, ossicles and spines disarticulate, unless the organism is quickly buried and thus avoids physical destruction of the skeleton. These plates are fairly sturdy and are often found whole in limey deposits.

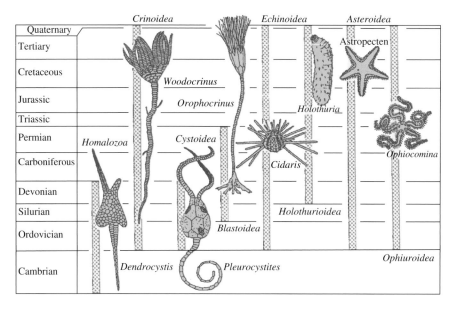

Fig. 5.23 Time range and classification of echinoids with representative genera illustrated (modified from Ziegler 1983)

Because of their microporous texture, echinoderm plates have a low density and may be readily transported into virtually all kinds of environments after death, though a high concentration of plates or the presence of articulated specimens normally indicates deposition within the area they inhabited.

5.5.6.1 Crinoids

The length of crinoids varies greatly, from a decimetre up to several metres (even 25 m). Their general structure is shown in Fig. 5.24. In the upper part is a cup (calyx), which consists of 10–15 plates in two or three rows. Five (or a multiple of five) arms (brachia) extend from the calyx and may have small branches (pinnulae). The mouth is centrally located in the top of the calyx, and the anus is often situated on an outgrowth from the calyx (anal conus). The calyx is situated on a flexible stem consisting of various types of small segments (columnal plates or stem plates). These columnals consist of variously shaped and sized plates, generally arranged in a single series (Fig. 5.25b). Each plate has a central round, or less commonly pentapetoloid, perforation, which together form an axial canal (Fig. 5.25a). The surface separating the stem segments often has a regular radial pattern of

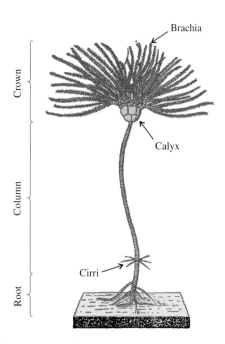

Fig. 5.24 Features of the general crinoid morphology showing a complete crinoid skeleton in living position (modified from Shrock and Twenhofel 1953)

fine chambers and furrows. The stem may have thin lateral branches (cirri) which serve to fasten the crinoid to nearby objects (Fig. 5.25c). Lowermost, the stem may

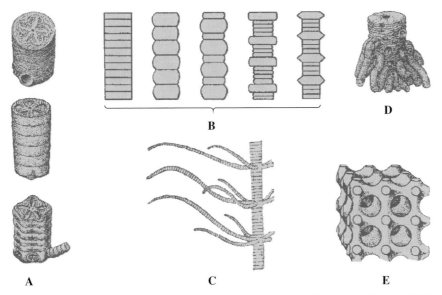

Fig. 5.25 (**a**) Parts of crinoid columnals showing their general morphology with radial grooves for connective tissues on the top plate. Note the round lumen in the centre of the plates. (**b**) Schematic drawing of columnals viewed from the side showing examples of the great variation in stem morphology. (**c**) Part of the cirrate column of *Comastrochinus* sp. (**d**) Part of the root system of *Cyathocrinites* sp. (**e**) Enlarged section of an echinoderm showing the network of large openings which transects the structure (modified from Shrock and Twenhofel 1953, Rasmussen 1969, Scoffin 1987)

form a root-like, branched system that serves as a hold-fast to objects (Fig. 5.25d), or forms an anchor in loose sediments. When the crinoid dies, the column is quite likely to fall apart to a greater or lesser extent. Some fossil crinoids have the entire stem and root system preserved, but fragmental stems and single columnals are much more typical. The skeleton in Recent echinoderms is exceedingly porous (ca. 50%) with a reticulate pore structure (Fig. 5.25e). The pore diameter is about 25 μm.

5.5.6.2 Echinoids (Sea Urchins)

Echinoids have a hemispherical, disc- or heart-shaped endoskeleton consisting of interlocking plates. The outer surface is covered by spines situated on the interambulacral plates which alternate with the spine-free, mostly smaller, ambulacral plates (Fig. 5.26).

5.5.6.3 Ecology

All echinoderms are marine, preferring normal marine salinity and ranging from shallow water to abyssal depths. Crinoids inhabited shallow marine environments during the Palaeozoic, sometimes forming small bioherms or beds dominated by crinoid debris, but since then they have been most common in relatively deep waters (>100 m). In contrast, other echinoderms have commonly inhabited shallow marine environments throughout their geological time range. Cystoids and blastoids are rare fossils but may occasionally be numerous within, or associated with, reefs.

Most echinoderms are benthonic and a few are pelagic. Echinoids are usually vagrant, whereas crinoids are generally sessile.

5.5.6.4 Mineralogy

The skeleton is composed of high-Mg calcite (5–15 mol% $MgCO_3$). Each plate, spine or sclerite has a single-crystal microstructure behaving optically as a single calcite crystal when viewed under the polarising microscope. However, the optically uniform crystals are really a mosaic of submicroscopic crystals whose *c*-axes are aligned almost perfectly parallel.

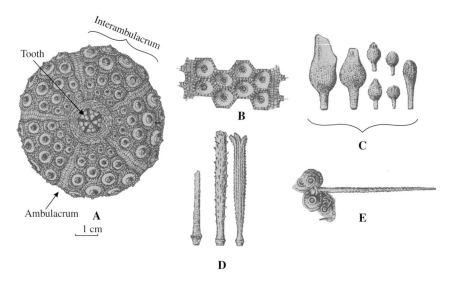

Fig. 5.26 (**a**) Adoral surface of the regular *Archaeocidaris* sp. with terminology. (**b**) Part of the surface showing ambulacra and interambulacra with tubercles. (**c**) Interambulacra with spines preserved. (**d**) Examples of different growth forms of spines from regular echinoids. (**e**) Part of *Archaeocidaris* sp. showing a few ambulacral plates with spines preserved. (modified from Rasmussen 1969, Clarkson 1979, Scoffin 1987)

5.5.6.5 Geological Range

Echinoderms range from the Lower Cambrian to Recent, but were not common before the Ordovician. Their greatest abundance was during the Carboniferous. Crinoids appeared for the first time in Lower Ordovician sediments, but did not become abundant before the Silurian. They remained numerous in the Devonian and Carboniferous, sometimes forming small bioherms or beds dominated by crinoid debris. Their abundance declined in the Permian, but revived in the Mesozoic, although they never equalled their Palaeozoic peak. Crinoids remain prolific today.

Echinoids appeared for the first time in the Upper Ordovician, but are most common in Mesozoic and Cenozoic sediments. They remain a vital part of the invertebrate marine realm and are probably as abundant now as at any time in the past.

5.5.6.6 Significance for Reservoir Quality

Echinoderms are commonly overgrown by syntaxial calcite cement which completely obliterates both intragranular and intergranular porosity in echinoderm-rich rocks. This, along with the common neomorphic replacement of the echinoderm skeleton, gives these rocks low reservoir potential. Dolomitised crinoidal sediments may occasionally be oil producing, as for example in the Silurian crinoid-rich skeletal build-ups in the northeastern Anadarko Basin of Oklahoma, USA (Morgan 1985).

5.5.7 Post-mortem Destruction

As shown in the preceding review, carbonate grains differ widely in size with diameters ranging from a few micrometres in coccolithophores to more than a metre in some bivalves and sponges. Apart from grain type, grain size is also dependent on mechanical abrasion and bio-erosion prior to final burial. These physical, biological and chemical processes can be described as:

- **Mechanical destruction**. The rate of abrasion is highly dependent on the transport mechanism and energy available in the environment for particle movement. Shell fragments and other skeletal particles are rapidly crushed and ground down to become structurally unrecognisable in a high-energy environment such as a surf zone, while in low-energy environments the integrity of the original material may remain more or less intact.

Fig. 5.27 Experimental abrasion of skeletal material. Starting sample (*upper left*) contained fresh specimens of the bivalve *Mytilus* sp., the gastropods *Aletes* sp., *Haliotis* sp. and *Tegula* sp., various species of limpets and enchinoids, the starfish *Pisaster* sp. and the calcareous algae *Corallina* sp. The series of diagrams shows selective destruction of the assemblage by tumbling (modified after Chave 1964)

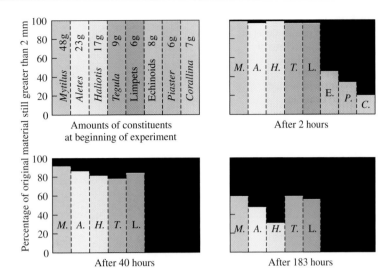

Mechanical destruction of skeletal material has characteristic rates of destruction, depending on shell thickness, microstructure and possible prior bio-erosion. A primary factor is the surface area to weight relationship (Fig. 5.27). These differences in abrasion rate may sometimes give rise to a fossil assemblage that deviates markedly from the original composition (Fig. 5.28). In general, in any given abrasive environment, skeletal material with a small surface area per unit weight is generally more durable than those with a large surface area per unit weight.

- **Biological destruction**. As soon as any organism dies, its skeletal material begins a process of deterioration due to the activity of a great variety of organisms. In particular many species of sponges, worms, bivalves, fungi and algae have the ability to sculpt or penetrate hard calcareous substrates e.g. skeletal material lying exposed on the sea bottom. Skeletal material has a low hardness (e.g. low-Mg calcite has a hardness of 3 and aragonite 3.5–4 on Moh's scale of hardness) and is easily dissolved in weak acids which render these substrates attractive for bioeroding organisms. The density and diversity of bioeroding organisms may therefore be high on carbonate material which has been exposed on the sea bottom for a year or more. Bio-erosion is a self-destructive process, and the sculpture that is produced can only be preserved if covered by sediment, which suffocates the endolithic community. Bioerosion increases the vulnerability of skeletal

material to mechanical destruction, with the result that a significant part of carbonate deposits may consist of a combination of bio-eroded and mechanically fragmented material (Fig. 5.29).

5.5.8 Micritisation

This process, which was first described by Bathurst (1975), is due to the fact that exposed skeletal fragments within the photic zone may serve as suitable substrates for various green algae, cyanobacteria and fungi. These organisms destroy the surface by boring a fraction of a millimetre into the fragment by secreting acids which dissolve tubular boreholes in which the filaments reside. The diameter of each hole varies somewhat, but usually it is around 5–7 μm (maximally 15 μm).

When the bioeroding organisms die, each hole gets filled with tiny crystals (micrite) of aragonite or high-Mg calcite. Which crystal type is precipitated depends on the primary mineralogical composition of the shell. Usually the precipitate is aragonite if the shell is aragonite, and likewise for fragments of high-Mg calcite. However, some bacteria can secrete aragonite, and therefore aragonite-fills in empty boreholes may have been produced by bacteria which lived on the algae. In pre-Quaternary sediments low-Mg calcite has usually replaced both high-Mg calcite and aragonite micrites. Repeated borings followed by

Fig. 5.28 Examples of
breakdown of various skeletal
components by mechanical
abrasion. Echinoderms,
coccoliths and many
bryozoans are particularly
fragile. (**a**) Echinoid;
(**b**) ostracod; (**c**) brachiopod;
(**d**) calcareous red algae;
(**e**) bryozoan; (**f**) bivalve.
Not to scale (modified from
Bromley 1980)

precipitation and infilling of the empty cavities cause the fragments to be replaced by a dark micritic coating (Fig. 5.30). The micritised coating is highly stable and is commonly preserved in both Quaternary and pre-Quaternary sediments. Where aragonite fossils have been micritised, the coating (now consisting of low-Mg calcite) often clearly indicates the original shape of the fossil (Fig. 5.31).

5.5.9 Microcrystalline Lime Mud (Micrite)

Lime mud deposited in areas with carbonate sedimentation has a grain size of about 1–4 μm. This precludes

effective study under an ordinary microscope, only with an electron microscope can each individual grain really be seen.

Carbonate mud (micrite) was previously assumed to be chemically precipitated carbonate, as opposed to fossils or fossil fragments of clearly organic origin. However, a very large amount of modern lime mud, for example from the Bahamas, has been found to consist of aragonite needles from the breakdown of calcareous green algae, particularly *Halimeda*, *Rhipocephalus* and *Penicillus*. This was confirmed partly by studies of the shape of aragonite needles (by electron microscope) and partly through isotope studies of the $^{18}O/^{16}O$ and $^{13}C/^{12}C$ ratios in the aragonite mud. The isotope values for the aragonite mud correspond well with the

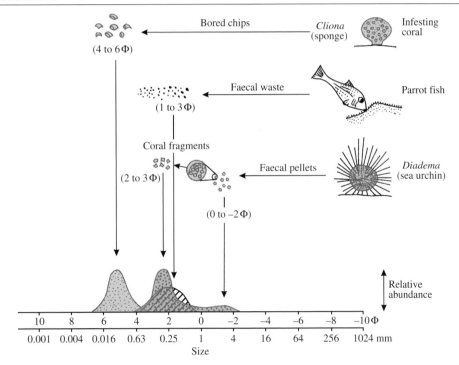

Fig. 5.29 Diagram of the dominant size fractions of grains produced by biological destruction of massive corals by endolitic sponges, parrot fish and sea urchins. The endolitic sponges infest calcareous substrates to create a sheltered site for habitation. Both the parrot fish and sea urchins are important grazers that break down the surface of calcareous substrates in search of epilithic and endolithic plants for food (modified from Scoffin 1987)

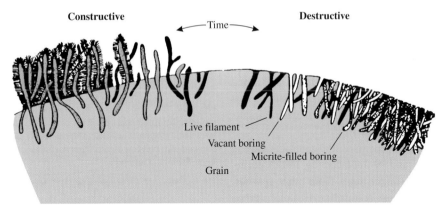

Fig. 5.30 Micritisation of a calcareous grain. The process is related to the activities of microboring algae, cyanobacteria and fungi. A micritic envelope is developed if the micritisation is confined to the outer surface. If the grain is completely micritised, a peloid or cryptocrystalline grain may be the end result. Not to scale (modified from Kobluk 1977)

values one finds for muds formed by various types of algae.

In shallow marine areas near the equator, like the Bahamas, the seawater is often saturated with respect to aragonite. Nevertheless it has not been possible to prove conclusively that any purely chemical precipitation of aragonite takes place in these areas. Sudden *whitings* of the seawater that are often observed in these areas, has been interpreted as spontaneous crystallisation of aragonite, but the phenomenon can also be interpreted as a disturbance of the lime mud from the bottom, for example by shoals of fish which are present in large numbers where there is whiting of the seawater.

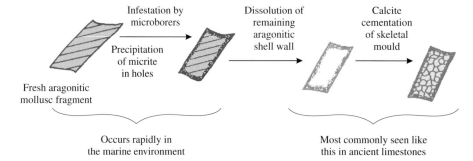

Fig. 5.31 Diagram showing the formation and preservation of a micritic envelope. Skeletal fragments of aragonite may be dissolved and replaced by calcite by one of two processes. *Left*: much of the original structure, such as organic inclusions, may be preserved (albeit not perfectly) through gradual solution of the skeletal material and immediate precipitation of calcite along a thin solution film. *Right*: complete solution of the skeletal fragment. During later diagenetic stage(s), the mould may be partly or completely cemented by calcite (modified from Bjørlykke 1989)

In areas where strong evaporation and high salinity approach evaporite conditions, such as in the Dead Sea and the Persian Gulf, this kind of chemical precipitation does occur, but only when the salinity is very high, about every fifth year. A sudden proliferation of diatoms could also consume so much CO_2 that the pH rises and causes aragonite to be precipitated.

Microcrystalline lime mud may also be formed by mechanical abrasion. Fossils which lie exposed to wave and current activity will be abraded mechanically, and a fine-grained lime mud formed. However, this process will not produce the well-crystallised aragonite needles of which modern lime muds are found largely to consist. We may therefore conclude that mechanical abrasion of fossils or carbonate fragments is not the main source of lime mud.

Through diagenetic processes, the microcrystalline mud will dissolve and be replaced by microsparite, but this is clearly distinguishable from other sparites, not least by its brownish colour due to incorporated organic matter (Fig. 5.32).

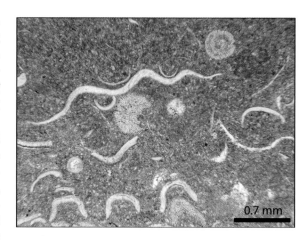

Fig. 5.32 The original microcrystalline mud (micrite) has been replaced by microsparite which consists of a mosaic of small brownish crystals. The skeletal remains are dominated by trilobite and ostracod fragments. Thin section seen in plane polarised light. Silurian, Gotland

5.6 Non-skeletal Grains

Carbonate grains can be classified as skeletal or non-skeletal. Non-skeletal components are defined as grains which do not appear to have been precipitated as skeletal material. However, this neither proves that they could not once have been skeletal, nor that they have an inorganic origin, it only signifies that in their present state no skeletal origin can be ascertained.

5.6.1 Intraclasts

Intraclasts are erosional fragments of essentially penecontemporaneous carbonate sediment. They are of intrabasinal origin, i.e. were eroded from the sea bottom or adjacent tidal flats and deposited within the area of original deposition. Early cementing, e.g. in the beach zone, or freshwater cementation due to regressions, may result in cemented sediments which are later broken up. We may also find early cementation in a marine environment, forming seafloor hard grounds. If these are then eroded by tidal channels, for example, the resulting intraclasts may accumulate as characteristic intraformational conglomerates. Intraclasts can also consist of semi-consolidated pieces of stromatolite.

Intraclasts may be of any size or shape. They are quite common in shallow water limestone facies and are good indicators of early cementation and erosion by currents.

5.6.2 Lithoclasts

Lithoclasts, which encompass both intraclasts and extraclasts (or extrabasinal clasts), are produced by the erosion of exposed older or synsedimentary lithified or partially lithified sediment. On carbonate platforms, lithoclasts are normally derived from erosion of the landward margin, where exposed, cemented limestones are broken up by physical, chemical and/or biological processes. Lithoclasts may be of any size or shape, but have generally become well rounded during transport. Because carbonates are vulnerable to weathering and are mechanically weak, the content of extrabasinal carbonate clasts declines with increasing transport distance, particularly in humid climates.

5.6.3 Faecal Pellets

Faecal pellets are organic excrements generally produced by mud-eating grazing animals such as crustaceans, worms and molluscs. They are generally round, oval or rod-like in shape and commonly consist of carbonate mud. Most faecal pellets are 0.1–0.6 mm in length and 0.1–0.4 mm in diameter, but

can be up to 2 mm long. Animals which eat mud, like snails, bivalves and crustaceans, deposit vast quantities of carbonate mud pellets in carbonate environments, and clay and silt pellets where fine-grained siliciclastic material forms the seafloor. Faecal pellets have a characteristic dark brownish colour imparted by their organic content.

Faecal pellets are produced in all environments, often in well-sorted large quantities forming pelsparite/pelloidal grainstone (Fig. 5.33). Their preservation within a sediment generally requires intragranular lithification, a process occurring in shallow waters which are supersaturated with respect to calcium carbonate. The presence of a low-energy environment is also important to prevent physical destruction. This implies that faecal pellets are most commonly encountered in low-energy, muddy depositional environments in the platform interior.

5.6.4 Peloids

The term peloid is used for a polygenetic group of spherical, ellipsoidal or angular grains with diffuse margins but no internal structure. The grains consist of fine-grained carbonate irrespective of size or origin. Because of the diverse origin of the particles, the term peloid is purely descriptive and does not denote origin. Various origins are possible: faecal pellets, micritised grains, intraclasts, a type of abiogenic precipitate, and a microbially mediated precipitate. Most peloids, however, are formed by extensive micritisation of pre-existing carbonate grains which is related to the activities of microboring algae and fungi within the photic zone. Thus, peloids generally indicate shallow, non-agitated waters.

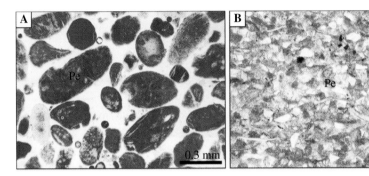

Fig. 5.33 (a) Rounded, rod-shaped pellets. Upper Cretaceous, France. (b) Dark pellets (Pe) in a sandy matrix. Silurian, Oslo Region, Norway. Thin sections as seen in plane polarised light

5.6.5 Ooids

Ooids are round grains of carbonate formed through chemical or biological precipitation by the rolling action of waves on beaches. Sediments consisting of ooids are called oolites. Ooids are by definition less than 2 mm, but most are 0.3–1.0 mm in diameter (Fig. 5.34). They have a concentric structure with layers of carbonate around a core which may consist of small clastic grains or a carbonate fragment. Similar concentric structures larger than 2 mm are called pisolites, and these are usually precipitated by algae.

Modern ooids consist of aragonite with a concentric tangential structure composed of small aragonite crystals (<3 μm) with their *c*-axis parallel to the lamination.

We find ooids only in very warm marine environments and in some saline lakes. The water must be close to saturation with carbonate. Agitation from waves or tidal currents is necessary to roll the ooids around to so that precipitation is even and forms concentric layers. Some studies indicate a thin organic membrane of bacteria on ooids which helps to precipitate aragonite, and which also may trap small aragonite particles in the sediment or suspended in the bottom water. If direct chemical precipitation was involved we would expect the needles to orientate themselves radially on the surface of the ooid, whereas snowball-type growth through the accumulation of small aragonite needles would give the observed concentric layers.

Fig. 5.35 Sand bar of ooids prograding over carbonate mud. The mud is covered with blue–green algae (cyanobacteria) which is cohesive and protects the sediments from erosion to certain extent (photographed from the air)

Because ooids require warm water and frequent wave agitation, they are only formed in very shallow water, normally <2–3 m deep (Fig. 5.35). During storms, however, they may be carried out into greater depths and deposited there. Ooids are typical of the Bahamas, of many areas around the Indian Ocean and on islands in the Pacific Ocean. On the east side of the Atlantic Ocean – along the coast of Africa – the water is too cold to permit sufficient carbonate saturation. Ooids are therefore an important indicator of depositional environment and climate.

5.6.5.1 Mineralogy

Ooids from earlier geological periods with warm climates (greenhouse) have a radial structure and may have been composed of primary Mg-calcite. During cold periods in the Earth's history (icehouse conditions), as now in the Quaternary and during the Permo-Carboniferous glaciation, ooids are initially composed of aragonite (Fig. 5.36).

5.6.5.2 Significance for Reservoir Quality

Ooids may form extensive bank-margin or shoreline sand bodies which can be cemented together to form a sedimentary rock called an oolite. These deposits may have excellent reservoir qualities because of a combination of secondary mouldic porosity related to

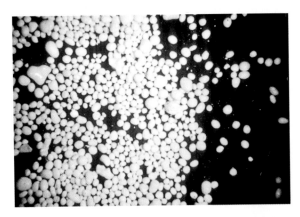

Fig. 5.34 Recent ooids from Bahamas. They are well sorted with grain size mostly between 0.5 and 1 mm

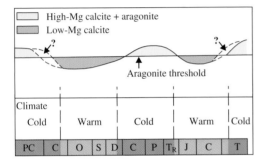

Fig. 5.36 A comparison between the mineralogy of ooids and climate. During warm climate conditions ooids consist of low-Mg calcite while during cold periods they consist of high-Mg calcite or aragonite

the dissolution of ooids and/or primary intergranular porosity (Fig. 5.37). Permeability may vary depending on the degree of cementation or dissolution, but is generally high.

Examples of petroleum production from secondary mouldic pores include the Upper Silurian oolite shoals of the northeastern Anadarko Basin, Oklahoma, USA, the Lower Permian oolite shoals of the Midland Basin, West Texas, USA, the Upper Jurassic oolites of the Smackover Formation of the Gulf of Mexico coastal plain, USA and the Lower Cretaceous oolite shoals of offshore Angola. Petroleum production from primary intergranular pores is exemplified by Lower Carboniferous oolite shoals in the St. Genevieve Formation, Illinois Basin, USA.

5.7 Carbonate Environments

5.7.1 Introduction

Most carbonate deposits are due to the production of skeletal material rather than being precipitated chemically. The composition of the sediments is therefore largely a function of the type of organisms which produced them. Biological precipitation removes $CaCO_3$ from seawater at a sufficient rate that the seawater does not become supersaturated with respect to calcite, except locally. It is mostly in evaporite basins that chemical precipitation is dominant but even there it is often assisted by algae. However, a very large fraction of the carbonate precipitated by organisms is dissolved in the water column or soon after deposition

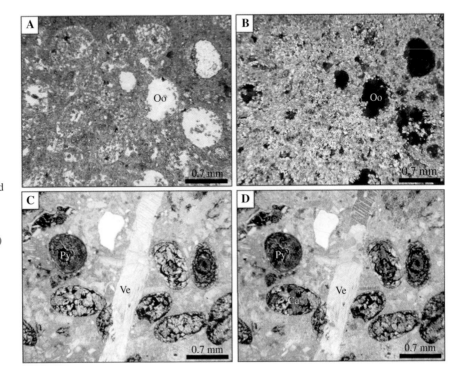

Fig. 5.37 (**a**) and (**b**) Oomoulds (Oo) formed by dissolution of ooids as viewed in transmitted and plane polarised light respectively. The matrix consists of dolomite. Roker, England. (**c**) and (**d**) Calcite vein (Ve) cross-cutting partially cemented oomoulds infilled with calcite cement and pyrobitumen (Py), viewed in transmitted and plane polarised light respectively. Lower Ordovician, Oslo Region, Norway

at the sea bottom and will therefore not be incorporated in the geological record. The total biological production of carbonate is approximately equal to the total amount of carbonate which becomes dissolved in the oceans plus the supply of Ca^{2+} released by weathering on land and transported to the ocean by rivers.

Accumulations of carbonate deposits require:

- A supply of nutrients to feed the carbonate-precipitating organisms.
- Clear water with a low content of suspended clastic material.

The effect of suspended fine-grained clastic material is to dilute the carbonate content, but it also prevents the growth of filter feeding organisms like corals. These sessile organisms require a through-flow of water to obtain enough planktonic organisms for nutrition, and too much suspended clastic material tends to clog up their filtering system.

Carbonates accumulate along coastlines with little clastic supply from rivers. This is often the case when the adjacent land has a dry climate with little runoff. Carbonates also accumulate on submarine highs and carbonate platforms which are surrounded by deeper water that traps the clastic sediments. The Bahamas platform is a good example of such a setting. Reefs commonly grow at the edge of deep waters where they are exposed to light, high wave energies and a steady supply of nutrients through upwelling.

During periods of global transgressions (high-stand) more carbonate is deposited on the continental shelves leaving less carbonate to be deposited in deepwater environments. In the stratigraphic record an abundance of carbonate rocks is seen from the Ordovician to Devonian periods and from the Cretaceous. This is not because more carbonate sediments were precipitated during these periods, but because more of what was precipitated was preserved on the cratons.

During transgressions there is little supply of clastic material that could dilute the carbonate precipitated. These carbonate deposits therefore generally contain little clastic material. However, during sea level low-stand more of the carbonate becomes mixed with clastic sediments to produce carbonate deposits with a fairly high terrigenous content.

5.7.2 Major Controls on Carbonate Sedimentation

Temperature and salinity are the main factors affecting shallow marine carbonate-secreting organisms. On this basis it is possible to distinguish between three principal skeletal grain associations; chlorozoan, foramol and chloralgal associations (Lees and Buller 1972) (Fig. 5.38).

5.7.2.1 Chlorozoan Association

The chlorozoan association is found in warm shallow seas at low latitudes as in the Bahamas and the Persian Gulf. Such warm-water carbonates are mostly derived from algae, or from benthic organisms living in symbiosis with algae. Warm-water carbonates are formed in areas with clear water and little clastic supply. The reef-building corals in warm areas (hermatype corals) live in symbiosis with algae and can only grow in shallow water with abundant sunlight. These corals are very vulnerable to cold (and very warm) waters and tolerate only a quite narrow salinity range. Therefore the chlorozoan association does not exist where the minimum surface temperature falls below 15°C and the salinity ranges lies outside 32–40‰.

It is important to remember that the distribution of warm-water carbonate facies is not only governed by latitude, but also depends to a large extent on ocean

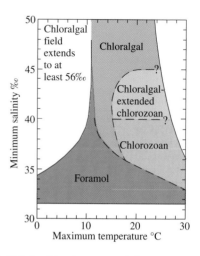

Fig. 5.38 Relationships between salinity-temperature annual ranges and occurrences of skeletal grain associations in modern shelf carbonate environments (modified from Lees 1975)

circulation patterns. On the western side of the Atlantic Ocean, warm-water carbonate facies with reefs and oolites are found up to about 30°N from the Equator, while on the eastern side, off West Africa, the ocean is generally too cold, even close to the Equator. This pattern is due to the east-west equatorial winds producing upwelling on the eastern side of the ocean, and accumulation of warm surface water on the western side.

Hermatypic corals have fairly narrow salinity tolerances lying between 31 and 40‰. Although hermatypic corals cannot live at lowered or elevated salinities, green algae do, giving their name to the chloralgal association.

5.7.2.2 Foramol Association

Where the seawater temperature range is 0°C up to about 15°C, the sediments are dominated by benthic foraminifera and molluscs, together with a somewhat minor contribution from echinoderms, barnacles, bryozoans, calcareous red algae and ostracods. This association is referred to as foramol or heterozoan (James and Clarke 1977). The foramol association thus comprises temperate and cold water carbonates currently being deposited e.g. the Spitsbergen Bank northwest of Bear Island in the Barents Sea. The Spitsbergen Bank is 30–100 m deep and surrounded by deeper channels so receives little clastic matter (Bjørlykke et al. 1975). The area lies on a cold oceanic front where cold currents from the north and east mix with the warmer water of the Atlantic Ocean. This results in powerful currents down to 80–100 m and high productivity of molluscs, benthic foraminifera, bryozoans and barnacles. The seafloor sediments are characterised by a Holocene 80–90% pure carbonate sand deposited across the top of Pleistocene moraines (Bjørlykke et al. 1975). Similar carbonate deposits are encountered on other banks off North America, for example Grand Bank, and along the coasts of western Scotland and Norway.

The biological precipitation of carbonate is not dependent on the water being warm or saturated with calcium carbonate. On the contrary, we often find the highest productivity in cold areas because the water there is richer in nutrients, particularly in areas of upwelling. Cold seawater is undersaturated with respect to carbonate minerals so that the skeletal

material begins to dissolve as soon the organisms die. Net accumulation of skeletal material on the sea bottom is possible because carbonate deposition merely requires that the biological production rate of carbonate skeletons exceeds the rate of dissolution. Skeletal material with a primary aragonite or high-Mg calcite composition will dissolve faster than that composed of low-Mg calcite, because of the lower pH due to increased CO_2 in the cold water.

The accumulation of cold water carbonates also requires a low input of clastic sediments. Such carbonate deposits are well known from many areas of the North Sea because it is presently starved with respect to clastic sediment input.

In the interpretation of ancient limestones we should not always assume that we can use the well-known warm water carbonate models, but also consider the possibility that we are dealing with cold water carbonate sediments at higher latitude.

5.7.3 Modern Environments of Carbonate Sedimentation

5.7.3.1 Reefs

From a sedimentological point of view a reef may be defined as a laterally restricted body of carbonate rock whose composition and relationship with the surrounding sediments suggest that the bulk of its biota were bound together as a framework during deposition, maintaining and developing a positive topographic structure on the sea bottom. Modern reef structures are dominated by hermatypic corals and calcareous red algae (Fig. 5.39), but the biotic composition of the reef structures through the Phanerozoic has been fairly variable (Fig. 5.40). Generally, all these different reef types show similar facies patterns to modern coralgal reefs. Recent reef formation, and the shape of the reef complexes, is essentially a function of the ecological environments of the reef-building organisms:

1. Reef-building corals require warm surface water and their distribution today is therefore limited to the lower latitudes. However, palaeogeographical reconstructions must take into account that surface water temperature is not a simple function of latitude. This is illustrated by the coasts of West Africa

Fig. 5.39 Corals from the modern Great Barrier Reef. Mainly brain corals and stag corals

and along the western side of the American continent. Where there is upwelling of cold water the surface layers are in most cases too cold for reefs to form, even near the equator. Off the coast of East Africa and in the Caribbean, on the other hand, the water is warm because of prevailing onshore winds, and coral reefs are abundant.

2. The hermatypic corals which build reefs live in symbiosis with algae and require sunlight. The reefs are therefore sensitive to changes in sea level since reef organisms tolerate neither exposure nor "drowning" below the photic zone. The Holocene transgression about 10,000 years ago raised sea level about 100 m in a few thousand years, but most reefs managed to grow quickly enough to keep pace with the rising sea level.

3. If sea level rises faster than the coral reef can grow, the reef may drop below the photic zone and "drown", but this is very rare because they can gain height rapidly.

4. Reefs provide a favourable ecological environment for animals which are not part of the reef structure itself, but which live on other organisms.

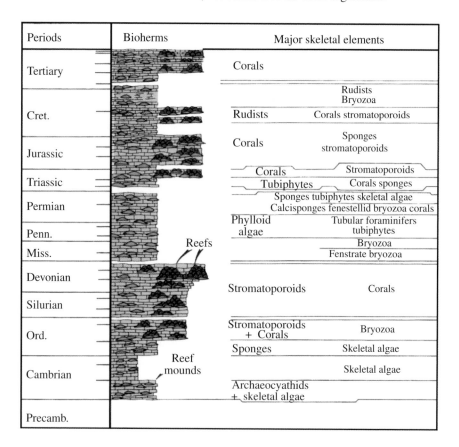

Fig. 5.40 Simplified stratigraphic column showing how the dominating biota in reefs and reef mounds has changed through the Phanerozoic. Gaps indicate times when there appear to be no reefs or reef mounds (modified from James 1983)

5. Many organisms in coral reefs live by filtering water to trap planktonic organisms and organic material. If the water contains too much siliciclastic mud, the clay minerals can choke the filtering organs so that the organisms die. Corals are particularly sensitive to the clay content in the water and can only live in clear water. The addition of clay, for example from a delta, will kill a coral reef. Pollution will have the same effect.

6. Clear sea water, however, is usually very poor in nutrients, so the organisms in a reef are dependent on good water circulation to obtain enough food. Consequently, reefs tend to grow on the edge of ocean basins, or as structures projecting high up from deeper waters. In this way, high water temperature is combined with low mud content and good water circulation.

7. The reef facies is built up as a wave-breaking structure, with powerful networks of corals which are braced by an encrustation of coralline algae. Red algae are massive and strong and have an important function in supporting the reef structure so that it can withstand the strong waves. However, parts of the most exposed parts of the reef structures can be mechanically destroyed by wave action during heavy storms. In this way reef fragments of various sizes may fall down the fore-reef slope, be transported up onto the reef flat or be deposited in gaps and cavities in the reef framework. This mechanical destruction is aided by the boring and grazing action of different bioeroding organisms. The bio-erosion can be so extensive that many of the primary structures of fossil reefs are destroyed. Boring mussels are particularly effective, excavating elongated cavities about 1 cm in diameter in corals, algae etc. Boring sponges, worms and algae are also important reef-damaging organisms, typically forming minutely thin but densely spaced holes which destroy the skeletal material. Coral reefs provide a very favourable environment for a variety of fish, and the cavities in the structure offer protection against predators, but some species of fish also live by literally eating the reef. They take small bites, and obtain nutrients from the surface. In this manner they contribute to the bio-erosion of the reef and the deposition of lime sediment on the sea bed.

A reef grows by virtue of the ability of its reef-building organisms to keep pace with the biological and mechanical processes which are continually breaking it down. Reefs will grow fastest on the outside, where wave energy and nutrient supply are greatest. In the case of barrier reefs, lagoons form on the landward side and contain a mixture of carbonate sand and fragments which have been transported from the reef facies during storms, together with skeletal material derived from organisms living in the lagoon (Fig. 5.41). The lagoons are usually only 2–6 m deep and protected from the highest waves by the reefs (Fig. 5.42) and contain a rich flora of green algae such as the calcareous *Halimeda* and *Penicillus*. These are low, bush-like forms which cannot withstand high energy waves and require the protected environment of the lagoon.

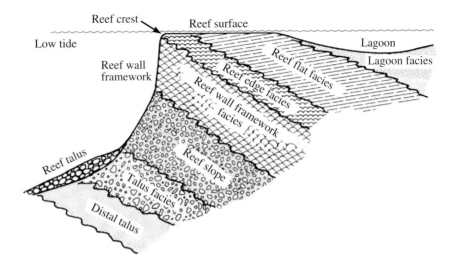

Fig. 5.41 Section through a reef which is developing out into a basin. Note the distribution of various reef facies and their diachronous nature

Fig. 5.42 View of reef facing the deep ocean and the protected lagoon (from Bahamas (Photo G.M. Friedman))

photic zone (<100 m) and most of it occurs within the upper 20 m.

Also the Maldive Islands in South Asia are carbonate platforms. They have grown to form a submarine mountain 3 km above the surrounding seafloor and its growth has kept pace with the subsidence and changes in sea level.

The Bahamas Bank is one of the best researched areas of modern carbonate banks in the world and covers a very large area, about 700 km N–S and 200–300 km wide (Fig. 5.44). The greater part of this area is less than 10 m deep. There is deep water on the west side towards the Mexican Gulf, so there is no clastic sediment supply from land in that direction. Sediments on the Bank are therefore pure carbonate, and the carbonate-producing organisms are not subjected to pollution from clay minerals produced by weathering. Most of the small amounts of clastic sediments deposited on Bahamas are aeolian dust blown all the way from the Sahara region in Africa.

Carbonate platforms like the Bahamas represent the more or less continuous build-up of carbonate on the seafloor from Jurassic, Cretaceous or Tertiary times right up to the present. As the seabed has subsided, carbonate sedimentation has built up the area so that it has remained in the photic zone. This has led to it being surrounded by deeper areas of sea which have acted as traps for clastic sediments from neighbouring continents. East of Andros Island the edge of the carbonate platform gives way to a very steep submarine slope (20–30° in many places). At the Tongue of the Ocean it slopes down to 2,500 m, and east of the little Bahamas Bank that lies to the north, it descends to 4,500 m in the Atlantic Ocean. Throughout its existence, the eastern side of the Bank has been reinforced by reef structures which are solid carbonate rock capable of forming steep submarine slopes without submarine slides. Diving with submersibles has, however, revealed great blocks of limestone at the foot of the slope. The stresses in the rock in a steep slope such as this may have contributed to fracturing, releasing the blocks from the rock wall. Near the base of the slopes there are very high differential stresses due to the high vertical stress from the overlying carbonate rocks compared to the water pressure in the horizontal direction. This may cause fracturing and the release of carbonate slabs from these submarine slopes. A 2 km thick carbonate sequence represents a load (stress)

However, they secrete small (c. 1 μm) needle-like crystals of aragonite. When the algae die, the aragonite needles are released and form carbonate ooze. Green algae are therefore considered to be the most important source of carbonate ooze on the Bahama Bank; chemical precipitation is of minor importance. Other characteristic elements in the lagoon fauna are bivalves, gastropods, echinoderms and annelids. Of particular sedimentological importance are crabs and shrimps (*Callianassa*) which dig tunnels and churn up the sediments, destroying the primary structures. Their tube-like trace fossils (e.g. *Ophiomorpha*) are typical of shallow marine sediments. The stability of the bottom sediments in the lagoons is enhanced by the presence of plants, e.g. *Thallassia* grass which covers much of the sea bottom on the Bahama Bank.

5.7.3.2 Reef and Carbonate Platforms Surrounded by Deep Oceans

Volcanic sea mounts reaching up to sea level are ideal places for carbonate sedimentation since all the clastic sediment will be trapped in the surrounding deeper water (Fig. 5.43). Smaller sea mounts may host circular atolls while larger areas of volcanic rocks may form platforms. Carbonate production may then keep pace with the seafloor subsidence over long periods of geological time. Even during the Pleistocene most of the carbonate platforms have been able to survive sea level changes of more than 100 m without drowning. Nearly all the carbonate production occurs in the

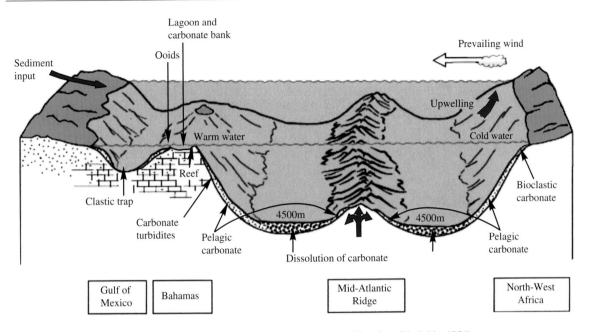

Fig. 5.43 Distribution of carbonate sediments across the Atlantic Ocean (modified from Bjørlykke 1984)

of about 40–50 MPa assuming an average density of 2.0–2.5 g/cm. In the water column at the same depth the water load is only 20 MPa and the differential stress is then 25 MPa.

When the Bahamas Bank, and other carbonate banks in the world, was exposed by the >100 m sea level drop during the glacial periods, it became lithified by percolating meteoric water which dissolved the aragonite and precipitated it as calcite cement. The loose carbonate sediments deposited over the last 10–12,000 years are not more than 3–4 m thick and rest unconformably on well-cemented, Pleistocene carbonate rocks.

Sedimentation on the Bahamas Bank reflects the climatic and bathymetric conditions. The temperature of the surface water varies from about 20–22°C in winter to 30–32°C in summer. Because of the limited exchange between water overlying the Bahamas platform and the surrounding ocean, the water in the interior of the Bank, particularly in summer, has a higher salinity than normal – up to 40‰. In winter the salinity is reduced by increased water circulation with the surrounding ocean, by rainwater and by runoff from Andros Island. Since the prevailing winds are from the east, where we also have deep water, wave energy is strongest on the east side of the platform. We therefore find reef facies along the east side of the

platform, but not along the west side. However, corals and algae (coralalgal facies) are also present on the west side of the bank, but they do not form proper reef structures there.

Oolite banks form in shallow areas with constant wave agitation and strong tidal currents. In water deeper than the normal wave base, we find large bedforms of oolites and other carbonate sand which are only mobilised by major hurricanes. Dunes or sandwaves with a wavelength of about 50–100 m, mapped from aerial photograph surveys, show no sign of having moved in 20–30 years. Extensive areas of seabed where the wave energy is less than on the oolite banks are covered with algal mats. The algal threads forming this mat help to protect the sediments from erosion, inhibiting bed transport and current ripple development. They also contribute, through photosynthesis, to the creation of a chemical environment with low CO_2 concentrations which may cause local precipitation of carbonate within the mat.

In the middle areas, sedimentation of carbonate mud facies prevails. This is because wave energy is reduced along the edge of the Bank, and the whole of the shallow area within is a low-energy environment. Tidal currents are also reduced in the shallow water because most of the tidal energy is dissipated in overcoming the friction against the bottom and the

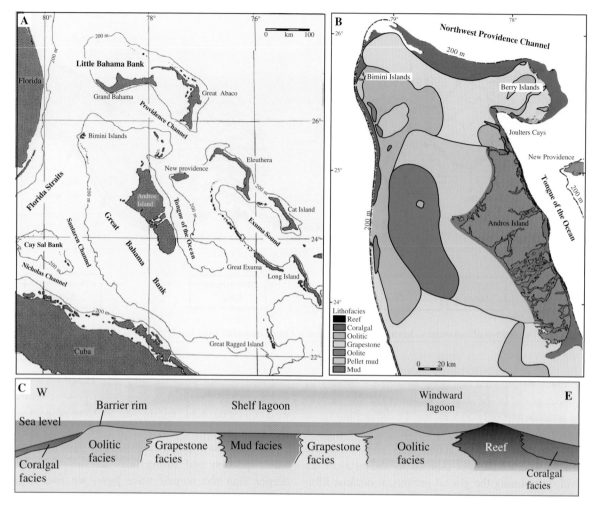

Fig. 5.44 (**a**) Map of the Bahamas Platform with surrounding areas. Note the deep ocean basin on the landward side. (**b**) Part of the Bahamas platform showing the distribution of recent carbonate sediments. (**c**) Simplified E–W section across the Bahamas platform. Note that the coral and oolite facies, which require high energy, are distributed along the edge of the basin, where wave power is greatest. Reef facies, which require the highest wave energy, only develop on the east side against the prevailing winds (modified from Bathurst 1971)

tidal range drops from about 0.7–0.8 m along the edge to practically zero in the centre. The most important sediment transport mechanism is probably hurricanes. These can mobilise normally stable sediments, forming large dunes or sandwaves. Large oolite banks may migrate along the edge of the platform, and storms also inflict some erosion of the reefs. Towards the centre of the platform wind stress can cause changes in sea level of up to 3 m, and create great turbulence which brings large quantities of mud and sand into suspension. Some of this material will be transported off the platform and down the submarine slopes.

The areas west of Andros Island, which are the shallowest and best-protected against storms from the east, consist mainly of lime mud (Fig. 5.45). Over wide areas the carbonate mud is dominated by pellets. They form small grains (0.1 mm) consisting of clay and silt particles so that the "recycled" lime mud now behaves like fine sand when transported by currents. The faecal pellets consist of mud which has passed through the alimentary canal of marine organisms such as gastropods, bivalves and annelids. Crustacea, such as the shrimp *Callianassa*, are also important pellet producers.

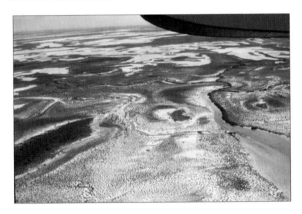

Fig. 5.45 Arial view of supratidal environments with tidal channels. Bahamas. The tidal range is reduced in the inner part of the platform

5.7.3.3 Carbonate Platforms and Reef Environments Along the Coast; the Persian Gulf

Attached carbonate platforms can only form if there is little sediment supply from land. In dry areas, in particular, there will be very little runoff from land which could add clastic sediments. The drainage pattern also plays a major role. During transgressive periods a shallow sea will invade the low-lying land, raising the base level of the rivers. In this way the clastic sediment supply is trapped in the lower valley reaches and much of the shelf will have clear water dominated by carbonate sedimentation.

The Persian Gulf (Fig. 5.46) provides an important present day model of a carbonate-producing environment which is fundamentally different from the Bahamas Bank. Whereas the Bahamas Bank is surrounded by deeper water, and therefore represents a pure carbonate environment, the Persian Gulf lies in a fold zone between the alpine mountain chain of Iran to the north and the stable Arabian shield to the south and southwest. The Gulf is at the most only 80–90 m deep, and a delta is being built out into the northwestern end by the Euphrates and Tigris. Clastic siliceous sediments are also being added from the north, from the mountains in Iran with abundant limestones and dolomites. Much of the sediment therefore consists of impure carbonates (marl). Only on the south (Trucial Coast) side are there purer carbonate sediments, because there is very little runoff from the deserts on the Arabian shield. Here too, though, there

is some supply of clastic material, particularly through aeolian transport from the desert. Sedimentation has been proceeding in a marine basin/embayment in this area since Mesozoic times.

The sea temperature in the Persian Gulf varies from 20°C in winter to 34°C in summer, and in the shallow areas it can be even hotter. The salinity in the open water is around 39–42‰, and may be higher due to evaporation in the lagoons. The tidal range varies from 0.5 to 2 m. The innermost supratidal area consists of a 10–15 km broad marginal belt covered by cyanobacteria and salt deposits. This belt is called *sabkha* in Arabic, and the word has become a geological term.

Because of the high temperature and salinity few organisms can live on the sabkha, which is dominated by algal mats (stromatolites) that form a crust of precipitated carbonate and gypsum. The stromatolites may grow in layers parallel to bedding or develop dome-shaped, columnar or irregular structures. The stromatolite structures are often broken into polygons which are separated from each other by sediment-filled cracks. The cracks are the posthumous reflection of shrinkage cracks as a result of drying.

The slope of the sediment surface here is only 0.4 m/km and during powerful storms water is driven in over the sabkha and later evaporates. Anhydrite, dolomite, magnesite and halite may be precipitated, though the halite is easily dissolved again. The precipitation of gypsum and anhydrite leads to the water developing a higher Mg^{2+}/Ca^{2+} ratio, which favours the formation of dolomite and magnesite.

In the Persian Gulf green calcareous algae such as *Halimeda* and *Penicillus*, which are important producers of lime mud in the Bahamas, are for the most part absent. Chemical precipitation is therefore probably the most important process, particularly during periods when diatoms proliferate and raise the pH by consuming CO_2 and thereby reducing the solubility of calcite. On the continental shelf, biogenic carbonate accumulates in the form of shells of foraminifera, molluscs, gastropods, ostracods, bryozoans and echinoderms. Especially in shallow water environments the skeletal material can be heavily affected by post-mortem micritisation if exposed on the sea bottom for some time. Along the outer edge of the shelf are reefs, built of coral species that are particularly well adapted to the high salinity.

Tidal channels connect the lagoons with the gulf. At the mouths of the channels there are ebb-tidal

Fig. 5.46 (**a**) The Persian
Gulf showing the facies
distribution of Recent bottom
sediments. (**b**) Map of the
carbonate facies in the Abu
Dhabi area (modified from
Wagner and van der Togt
1973, Purser and Evans 1973)

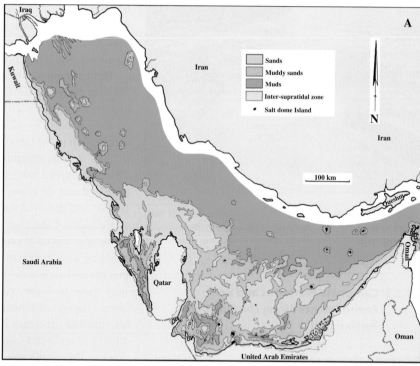

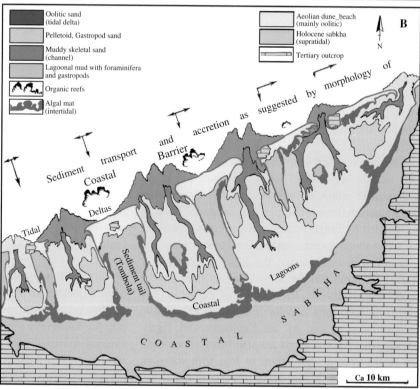

deltas with oolite banks in the shallowest parts (<2 m). In addition to oolites there are grains of bioclastic material. Tidal deltas of this sort would form good oil reservoirs because of the high primary porosity, and in the Mesozoic series we find similar reservoir rocks. Shoreface deposits are not as well sorted as we would normally expect as they consist of fine sand and lime mud. This is possibly because the shoreline is stabilised by plants which bind the sediments.

Gastropods are particularly important in the lagoons and the intertidal environment. On the beach, shrimps and crabs burrow and disturb the lamination in the sediments. The lagoons are surrounded by swamps with bushes and mangroves, or algal flats. Landward of the algal mat facies, which is most typically developed at the high water mark, the sabkha facies continues. The degree of evaporation is higher here in the supratidal zone and dilution with seawater rarer. The sulphates, especially gypsum or anhydrite, are precipitated within the sediment near the groundwater table. The sabkha flats pass landwards into drier areas dominated by aeolian sand.

The facies distribution we have just described, from the open marine lagoon to the supratidal environment, will form a characteristic vertical sequence in the event of relative sea level fall and progradation. Periods with transgressions and subsequent regressive outbuilding will result in a series of sequences (cycles), starting with shallow marine (subtidal) sediments and culminating with evaporites (anhydrite) at the top. These cycles are also found in the Mesozoic and the impervious anhydrite beds forms an ideal cap rock above oil reservoir rocks. Carbonate sand facies, pellet facies, and particularly oolites, make good reservoir rocks, while the marine muds are source rocks.

5.7.3.4 Evaporite Basins

In ocean areas with normal salinity, practically all carbonate deposition is through biological precipitation. This is because the biological precipitation is rather efficient so the sea water does not become saturated with respect to the most common carbonate minerals like aragonite and calcite. In areas with somewhat higher salinity, however, for example in the Persian Gulf, there may also be chemical precipitation of calcium carbonate. This is because there are fewer organisms to precipitate carbonate and because of evaporation concentrating the sea water. Here too this precipitation may nevertheless be linked to biological factors. Periods with algal blooms in the surface waters involve photosynthesis and consumption of CO_2. This raises the pH, creating oversaturation and thus favourable conditions for chemical precipitation, i.e. a high degree of oversaturation.

Carbonates make up only a small percentage of the salt precipitated when seawater evaporates to dryness. However, they are among the least soluble of the common salts in restricted ocean basins where the salinity is too low for the more soluble salts to precipitate (e.g. NaCl). Carbonates together with sulphates like gypsum or anhydrite often form thick evaporite sequences. In highly saline basins with more intense evaporation and restricted circulation, the water will become enriched in Mg^{2+} and dolomite will form, perhaps also magnesite.

5.7.3.5 Carbonate Turbidites on Slopes Skirting Carbonate Platforms

Carbonate sand and mud which is stirred up during storms over shallow water may be transported over the shelf edge to continue downslope as turbidites or debris flows. Carbonate platforms may rise 2–3 km above the surrounding seafloor and the slope may be very steep (20–30°) because of the hard and stable carbonate rocks.

5.7.3.6 Pelagic Carbonate

Pelagic carbonate deposits consist largely of planktonic organisms (coccolithophores, foraminifera, pteropods etc.) which live in the upper part of the water column, sinking to the bottom when they die. How clean the carbonate deposits are depends on how much other biological sedimentation there is, e.g. from siliceous organisms like diatoms and radiolaria, and how rapidly clastic sedimentation takes place. Therefore, carbonate sediments commonly become concentrated on top of submarine highs where the siliciclastic sedimentation rates are low. This is often observed on seismic lines where the reflections become stronger on the top of submarine positive tectonic structures.

In the deep ocean much of the pelagic carbonate production is dissolved as the particles settle through the water column. The sedimentation rate is a function of productivity in the upper water layers minus solution as the dead organisms sink towards the seafloor. The dissolution of skeletal material is due to undersaturation of $CaCO_3$ in deep ocean water because this cold water can dissolve more CO_2 than warm surface water. The combination of low temperatures and increasing hydrostatic pressure with depth involves an increase in the pCO_2 and decrease in the pH. The CO_2 is produced by respiration and decay of pelagic organisms in the deep ocean. Below the depths where the rate of dissolution is equal to the rate of sedimentation of carbonate, no carbonate sediments accumulate. This is called the carbonate compensation depth (CCD) and varies from 1–2 km in cold water at higher latitude to 4–5 km in the warm water equatorial regions.

Since the planktonic carbonate organisms are very small, pelagic carbonate deposits form a fine-grained ooze of clay- and silt-size particles, with occasional larger fossil fragments. Large areas of the South Atlantic and Pacific are covered by sediment containing more than 50% $CaCO_3$ from planktonic organisms. Foraminifera and coccolithophores form the most important deep-sea carbonate deposits. They are also important as an indication of environment, e.g. water temperature (Fig. 5.12). Coccolithophores live mainly in the photic zone. In areas of high productivity, for example in the fjords of Norway, the concentration of coccolithophores may be several millions per litre, but 50,000–500,000 is a more normal level. Although they consist of low-Mg calcite, their size makes them relatively soluble in cold water. In consequence, although production is greatest at high latitudes, it is only at lower latitudes that large quantities of coccolithophores are able to accumulate on the seafloor.

Shallow, warm seas with little other carbonate production provide particularly favourable conditions for the deposition of purer coccolith deposits. The seas of northwest Europe in Cretaceous times were a good example. The climate in the Mesozoic was undoubtedly considerably warmer than today, and northwest Europe also lay further south. Chalk forms a characteristic rock which is exposed in Denmark, South England and France, and continues under the southern and middle sections of the North Sea. It is missing in the north, possibly for palaeoclimatic reasons, although

it is found in northeast Ireland. Chalk sediments were probably deposited in water depths of not more than 1–300 m, mostly below the photic zone. Since chalk is a micritic limestone, one would not expect it to form a suitable reservoir rock. The Ekofisk and associated fields are in fact the world's only major oilfield in such rocks, and the low permeability of this fine-grained lithology creates problems for production.

Pelagic calcareous algae such as coccolithophores did not become common until the late Jurassic and early Cretaceous. Consequently we do not have chalk deposits from older periods. This is an example of where the rock type is totally dependent on which organisms were precipitating carbonate. During the Palaeozoic most carbonate production took place in shallow water, as there were no planktonic calcareous algae to form deepwater pelagic carbonates.

5.7.3.7 Lakes and Inland Seas

Carbonate sedimentation in lakes depends on the rate of weathering and supply of Ca^{2+} from older calcareous sediments and calcium-bearing silicate rocks in the drainage basin. If there is an ample supply of Ca^{2+} and CO_3^{2-} also lacustrine sediments may contain considerable amounts of carbonate, and particularly at lower latitudes we find pure carbonate deposits. In cold lakes the solubility of carbonate is high and the carbonate content in the bottom sediments will mostly be limited to a few species of bivalves and gastropods. Temperate lakes often accumulate calcareous muds called marls, which in some lake sediments contain dolomite. Algae and certain higher plants often play an important role in carbonate production in lakes.

The Dead Sea is a good example of an inland sea where carbonate is precipitated chemically due to strong evaporation. In Africa and other tropical areas, lakes will be subject to seasonal evaporation to dryness, and we may find alternating layers of biogenic carbonate and chemically precipitated carbonate. Here, too, algal blooms in the surface water layer play a major role in precipitating carbonate. Thicker beds of carbonate below lake floors can be associated with longer term climate variation. In Lake Victoria, the sediments older than about 12,000 years have high carbonate content. This is because the lake nearly dried up during the last glacial period, when rainfall in the region was lower than now.

5.7.3.8 Calcareous Tufa Deposits and Travertine

Tufa is the name of a porous and spongy calcareous deposit common in limestone areas, usually at the base of slopes where groundwater emerges at the ground surface. Accreted in thin layers which often incorporate vegetation, it can build up into deposits several metres thick. Travertine is a more massive, relatively dense and sometimes finely banded and laminated carbonate deposit associated with freshwater springs or precipitated as speleothems in caves. Both varieties are precipitated from freshwater supersaturated with respect to calcium carbonate. Groundwater flowing through carbonate rock contains CO_2 at a higher partial pressure than at the surface because of the lower ambient temperature and the higher pressure that is due to the weight of the overlying water column. The increased partial pressure of CO_2 promotes increased dissolution of calcium carbonate from the host rock. When the water flows out of the rock at the surface, or reaches a cave, it will regain equilibrium with atmospheric pressure, causing degassing of CO_2. The loss of CO_2 entails that the water may become supersaturated with regards to calcite, followed by precipitation of this mineral. This is particularly effective in summer when the water will quickly warm up once it emerges. Exposure to light will also cause biogenic precipitation (photosynthesis) by algae. Evaporation may also concentrate the carbonate sedimentation in lakes, though this depends on the rate of weathering and supply of Ca^{2+} from older calcareous sediments and calcium-bearing silicate rocks. Evaporation may also concentrate the water and lead to increased precipitation.

5.7.4 Meteoric Water Flow and Diagenesis

The introduction of meteoric water has a profound effect on carbonate sediments and their potential as reservoir rocks.

A proportion of the rainwater (meteoric water) falling on land infiltrates into the groundwater. The flow of groundwater ultimately is limited by the rate of recharge by rainwater, which determines the water table gradient. As long as the water table is above sea level the groundwater has the hydrodynamic potential to flow beneath the beach and out into the basin beneath the seafloor, floating on top of the more saline basin porewater (Fig. 5.47). The freshwater lens is floating like an iceberg in the sea. With a groundwater density of 1.00 g/cm³ and the more saline water 1.025 g/cm³, the ratio between the groundwater head and the depth of freshwater penetration is theoretically $1/(1.025 - 1.00) = 1/40$. A groundwater head of just 10 m can drive freshwater to a depth of up to 400 m below sea level. Shallow water carbonates deposited in coastal environments and around islands will thus in

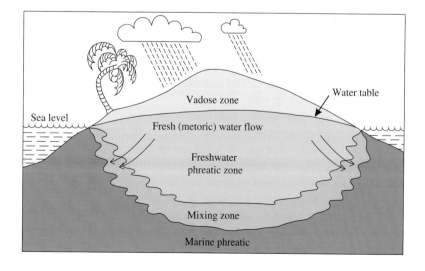

Fig. 5.47 Cross-section of an ideal permeable carbonate sand island showing the distribution of major diagenetic environments in the shallow subsurface. The vadose zone is situated near the surface above the water table, and the pore spaces are occupied by water and air. The pore spaces below the groundwater table are permanently water-saturated. The fresh interstitial waters float on the denser saline waters beneath. There is a mixing zone between the two water phases

most cases be flushed by fresh groundwater after deposition. The diagenetic effects are greatest at shallow depth where the flow rates are highest.

However, coastal carbonate environments are usually rather dry and while carbonate platforms may have more rainfall, the islands on them may be small compared to the size of the platform. Both these factors tend to reduce the flux of meteoric water into marine carbonate sediments, although it may still be very significant, particularly when the sedimentation rate is low. More distal and pelagic facies may avoid this flushing altogether. On land and in the nearshore parts of the basin meteoric water may also be undersaturated with respect to calcite, in which case caverns are likely to develop.

When meteoric water flows through recent carbonate sediments it will be undersaturated with respect to aragonite but become rapidly supersaturated with respect to calcite. Aragonite will therefore dissolve first and calcite will precipitate. Gradually the meteoric water will reach equilibrium with low-Mg calcite, and calcite cement in the form of large crystals (block-shaped cement) may be precipitated (Fig. 5.48). This cement is very different from marine cements precipitated from modified seawater (without sulphate).

On land, the sediments above the water table are located in the *vadose* zone, where the pores are alternately filled with water and air as a consequence of intermittent meteoric water percolation. Partial desiccation results in an unequal distribution of the porewater with it primarily held near grain contacts, by capillary forces; as a result there will be a preferential cementation of pore throats giving a rounded pore geometry. This cement type is called meniscus cement. Porewater will also collect on the underside of grains as pendant droplets and precipitate cement in this form, called pendant cement. Both meniscus cement and pendant cement are characteristic of partial cementation in the vadose zone.

At sea level lowstands, particularly in the Quaternary, marine carbonate sediments were directly exposed to freshwater that caused rapid cementation and hardening. The sea level drop of more than 100 m during the glaciations exposed and cemented all carbonate sediments that had been in the photic zone during the preceding interglacial (highstand) periods. On modern carbonate banks soft sediments are therefore limited to the Holocene (postglacial) deposits, which normally are less than 2–3 m thick.

5.7.5 Shallow-Marine Diagenesis

5.7.5.1 Introduction

Lithification is the process which transforms loose sediment into solid rock. It occurs through new minerals (cement) being precipitated in the pore spaces binding together the primary particles. To cause carbonate cement to be precipitated, we must have porewater which is oversaturated with respect to a carbonate phase.

In general diagenetic processes are driven by a progression towards more mechanically stable grain packings and more thermodynamically stable mineral assemblages. This is also the case for carbonate sediments. During progressive burial the increasing overburden stress causes denser packing of grains so that the porosity is reduced. The reduction in porosity is then a function of the effective stress that may be expressed as the compressibility of the rock. Mechanical compaction is in principle instantaneous but there is usually some additional compaction with time at the same effective stress. This is called creep.

The mechanical compaction of carbonate sand and mud follows the same principles as for terrigenous sand and clay. In the case of carbonate sediments however, chemical processes involving dissolution and cement precipitation are much more important at low temperatures. This is because the kinetics of carbonate reactions are much faster at low temperature than is the case for silicate reactions. The prediction of porosity and permeability in carbonate rocks therefore also depends to a very large extent on chemical diagenesis at shallow depth. Porosity reduction in carbonates can therefore not be predicted solely on the basis of effective stress because it also depends on the primary mineralogical composition of the grains and textural relationships.

Cement formed in a marine environment is aragonite or high-Mg calcite, which forms needle-shaped crystals. High-Mg calcite can also precipitate as micritic cement. Early marine aragonite cement may grow as evenly distributed layers of aragonite needles perpendicular to the surface of the grains. This is called isopachous fibrous cement because a layer of uniform thickness is formed (Fig. 5.49). Isopachous calcite cement may also be precipitated in meteoric (phreatic) porewater.

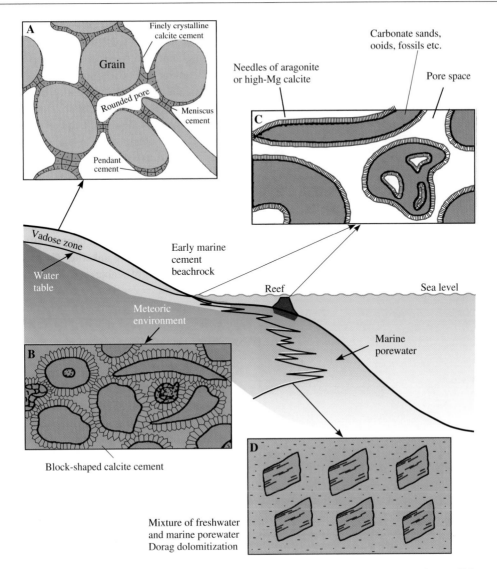

Fig. 5.48 Types of carbonate cement as a function of diagenetic environment. (**a**) Water is held near grain contacts in the vadose zone by capillary forces, and as a result cements form in these places (meniscus cement). In this way the pores will become more rounded due to the preferential cementation of the pore throats. Water will also collect at the underside of grains as pendant droplets, and after multiple phases of drainage and precipitation a cement with pendant form will be precipitated (pendant cement). (**b**) Phreatic calcite cement is typically clear and blocky shaped. (**c**) Early marine cements commonly consist of an isopachous fringe of acicular aragonite or high-Mg calcite. (**d**) Dolomite can be precipitated from a mixture of fresh and marine porewater

The relative growth rate of the cement will depend on the chemical composition of the substrate and its microstructure. The greater the accordance between the substrate and the cement, the higher the growth rate will be. Calcite cement that grows out from echinoderm fragments will have a high growth rate because both will have an identical lattice structure. The cement precipitates as syntaxial rims in optical continuity with the single calcite crystals of echinoderm plates, such that twin lamellae and cleavage planes transgress skeletal plates and cement overgrowths (Fig. 5.50). This style of cementation, called syntaxial or epitaxial cementation, often completely obliterates porosity in echinoderm-rich rocks. The volume of syntaxial overgrowth on echinoderm fragments is generally greater than the volume of cement crystals that have started

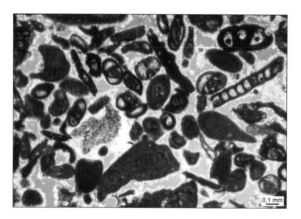

Fig. 5.49 Beach rock dominated by foraminifera and lithoclasts cemented with an isopachous rim of acicular aragonite cement. The pink areas are voids filled with stained epoxy. Thin section as viewed in plane polarized light. Recent Bahamas

Fig. 5.50 Crinoidal limestone. Dusty looking crinoid ossicles with syntaxial calcite cement. The cement is in optical continuity with the echinoderm plates, with twin lamellae transgressing skeletal plates and cement (modified from Greensmith 1978)

their growth on multicrystalline fossil fragments or terrigenous material. In extreme cases, so much syntaxial cement may surround the echinoderm fragments that other grains will also become incorporated (poikilotopic crystal growth).

When calcite replaces earlier aragonite or high-Mg calcite by neomorphism, we sometimes see "ghosts" of the earlier crystals. Radiaxial fibrous cement mosaics frequently form as a result of replacement of aragonite cement but also may form by direct precipitation of calcite. These are mosaics of calcite crystals with optic axes that converge away from the substratum (cavity

wall) on which they grew. They contain twin laminae which are convex towards the substratum.

Calcite cement formed early in oxidised porewater will contain practically no iron, since Fe^{3+} is not soluble in the oxidised state. Only Fe^{2+} can substitute for Ca^{2+} in the calcite structure, and the formation of ferroan calcite therefore requires reducing conditions. Calcite precipitated in the sulphate-reducing zone is free of iron (non-ferroan), however, since all the available Fe^{2+} will form sulphides. Calcite formed at greater depths under reducing conditions will normally contain some iron and manganese, depending on the availability of such ions in the porewater at the time of formation. At temperatures of about 100°C and above, iron-rich carbonates like ankerite ($Ca(Mg,Fe)(CO_3)_2$) become increasingly stable and are often found in minor quantities.

When a particle, e.g. a fossil, which consists of aragonite, is dissolved and replaced by calcite, this can occur by means of two different processes (Fig. 5.51):

A. By complete dissolution of the particle leaving a mould (secondary cavity) showing the outline of the original grain. The mould can subsequently become filled with low-Mg calcite, either entirely or partially depending on the supply of pore fluid supersaturated with calcite. This secondary precipitate has a highly characteristic texture. A narrow zone of small prismatic crystals develops along the periphery of the cavity. The crystals become larger and more equidimensional towards the centre of the cavity; this is called a "drusy mosaic" (Bathurst 1975).

B. Through gradual dissolution of aragonite and simultaneous precipitation of low-Mg calcite. This reaction mechanism, *neomorphism*, will to some extent preserve primary structures.

5.7.5.2 Beach Rock

The tidal zone on sandy beaches in tropical areas is commonly cemented by aragonite or less commonly by high-Mg calcite to produce "beach rock". On modern beaches the lithified zone may be about 0.5 m in thickness containing embedded bottles or other artifacts indicating that cementation has been very rapid by geological standards. Cementation takes place in situ beneath a thin sediment cover and is due to a

Fig. 5.51 Two types of conversion of aragonite to calcite. *Left*: neomorphic replacement of the grain. *Right*: complete solution of the grain and later precipitation of sparry calcite. If oil or gas is introduced before the moulds are cemented up the porosity may be preserved (modified from Bjørlykke 1989)

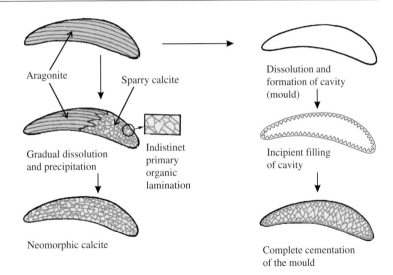

Aragonite

Sparry calcite

Dissolution and formation of cavity (mould)

Gradual dissolution and precipitation

Indistinet primary organic lamination

Incipient filling of cavity

Neomorphic calcite

Complete cementation of the mould

combination of high flux of water due to wave action and rise in ionic concentration due to evaporation of seawater as it drains through the beach at low tide.

5.7.5.3 Hardground

A combination of dissolution and precipitation of carbonate sediments close to the seabed may produce a well-cemented surface called a "hardground". This can be recognised by evidence of borings and/or encrustations of hard bottom faunal elements such as calcareous algae, sponges, corals, serpulids, crinoids and oysters. Hard grounds are best developed in areas of slow sedimentation and high current activity. Long exposure on the seafloor can also lead to impregnation with minerals such as iron hydroxides, phosphorite and glauconite.

5.7.5.4 Cementation in Modern Reefs

Modern reefs have a very high porosity, made up of everything from small cavities to large caverns. It was previously believed that this primary porosity was preserved in older reefs. Boring and blasting into modern reefs, however, has revealed that the primary porosity is rapidly reduced because the hollows are rapidly filled up with fossil fragments and lime mud, and/or become more or less cemented by early marine cement. Reefs can therefore only make good reservoir rocks if secondary porosity develops through the dissolution

of fossils or cement. This may happen by the reef being exposed above sea level to groundwater percolation and also by flow beneath the seabed of meteoric water from land. Aragonite or high-Mg calcite will then dissolve particularly easily and low-Mg calcite will precipitate. This diagenetic process may not significantly increase the overall porosity, but merely redistribute it.

Marine cementation commences early in reefs. The parts most exposed to waves, where water flux is greatest, will undergo rapid marine cementation. In the interior and landward parts there are less marine cement and more secondary porosity due to subaerial exposure and freshwater flushing. Skeletal material of primary aragonite may dissolve as calcite precipitates in the open framework porosity. Thus the porosity to a large extent becomes mouldic due to dissolution of many of the reef-building organisms.

Early cementation of aragonite or high-Mg calcite reduces porosity but the cementation produces a mechanically much stronger rock. The potential for preserving the remaining porosity is therefore higher than in uncemented carbonate sand. Furthermore, pressure solution may be reduced because the cement increases the contact area which reduces the stress per grain contact.

5.7.5.5 Reefs as Reservoir Rocks

The properties of reefs as reservoir rocks vary greatly and depend very much on the type of reef building organisms. These have varied through time and depend also on the environment. The fossils determine the initial mineralogical composition of the reef structures and hence the burial diagenesis.

The coarse carbonate block deposits (talus) down the slope in front of the reef (the fore-reef facies) and bioclastic sand behind the reef experience less water flow and tend to develop less marine cement. These facies constitute better reservoir rocks in terms of primary porosity.

Reefs are often surrounded by organic-rich terrigenous mud which may form a good source rock during burial. The reefs themselves may be good hydrocarbon traps, because they rise up from the seabed and thereby constitute structures which may become sealed if the reef drowns and is covered by terrigenous mud (Fig. 5.52).

5.7.6 Burial Diagenesis

Early cementation by calcite cements reduces the primary porosity but the cement may strengthen the grain framework and thus reduce compaction. Statistically, however, porosity clearly decreases as a function of burial depth (Schmoker and Halley 1982), although the porosity/depth function varies greatly with the lithology. The loss of porosity in carbonate sediments is partly a result of mechanical compaction of the carbonate mud and grains, just as in terrigenous mud and sand, but with an important difference. Unlike silicate reactions, carbonate reactions are relatively fast even at low temperatures and chemical compaction is therefore important both at shallow depth and low temperature, as well as at deeper burial depth and higher temperature.

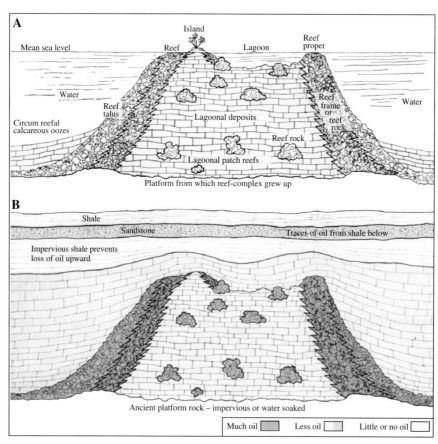

Fig. 5.52 (a) Vertical cross-section through a carbonate platform from which a reef complex grew up. (b) Micritic limestones in the lagoon have low porosity and permeability and oil and/or gas will mainly be present in reef and reef talus facies. The reef has been building into the deep ocean producing a steep submarine slope. The red algae is a strong reefbuilding organism while the more delicate green algae live in the protected lagoonal environments. In the supratidal environment we find blue-green algae and offshore carbonate is precipitated by planktonic algae (coccolithophores)

Aragonite and high-Mg calcite are metastable minerals in the marine environment. They precipitate because low-Mg calcite, which is the thermodynamically stable calcium carbonate phase, fails to precipitate. This is probably due to the poisoning (inhibiting) effect of sulphate on calcite precipitation (Kastner 1984). During burial both aragonite and high-Mg calcite dissolve and are replaced by low-Mg calcite, but we do not know how fast this reaction is in marine sediments. This thermodynamic drive for dissolution and compaction is almost independent of stress. The dissolution of aragonite fossils or cements may cause grain framework to collapse, whereas high-Mg calcite is replaced by calcite that retains much of the original texture so that the grain framework is conserved.

The solubility of carbonate grains is also a function of effective stress and thus burial depth. Pressure solution will often be concentrated at particular horizons where there has been some enrichment of clay particles which increases the rate of solution. Stylolites, which are surfaces where a considerable amount of solution has taken place, will then form. The horizon will be enriched with finely divided silicate minerals and other insoluble material in the limestones (Fig. 5.53). Where carbonate, usually calcite, is dissolved due to pressure solution, the dissolved material tends to be precipitated in the neighbouring pore spaces. When the pores between stylolites are filled with cement, dissolution along the stylolite ceases. Stylolites may have a relief from 1–2 mm up to several tens of centimetres and a spacing of typically 1–5 cm. As the rate of solution along the surface varies, a very irregular relief develops which may be taken as a minimum measure of the thickness of the carbonate layer dissolved.

The carbonate dissolved at grain contacts and along stylolites is transported to the adjacent sediments by diffusion. Here the carbonate will usually re-precipitate in the pores, with a possible reduction in porosity of up to about 20%. Investigation of interbedded layers of limestone and dolostone show that the former is more susceptible to pressure dissolution than the latter. Dolostone usually has a strong framework of interconnected dolomite rhombs. Stylolites are therefore especially well developed in carbonates that are almost exclusively calcitic. Some late diagenetic dolomite rhombs may precipitate along developing stylolite surfaces as a result of a dynamic system with pressure gradients on a microscopic scale.

CO_2 and organic acids generated from decomposing organic matter are rapidly neutralised. The pH decreases with depth because of the increasing amounts of CO_2 that can be dissolved in the porewater as the pressure increases. The solubility of calcite increases with increasing pressure but at normal hydrostatic pressure gradients temperature is the overriding factor determining the solubility gradients. Calcite precipitates only very locally during upwards flow, where there are abrupt pressure drops.

Compaction-driven flow is normally directed upwards and since the solubility then is reduced, dissolution rather than precipitation will occur. The capacity of compaction-driven porewater to transport carbonate is in any case rather limited because of the low solubility gradients and moderate fluid fluxes at greater depth. During burial diagenesis the loss of porosity depends both on the mechanical compaction and on pressure solution. Both are functions of the effective stress.

5.7.7 Classification of Carbonate Rocks

Early petrologists often subdivided limestones according to the size of the dominant mechanically deposited grains: Calcilutite (grains <63 μm), calcarenite (grains between 63 μm and 2 mm) and calcirudite (grains >2 mm). Later Folk's (1959, 1962) classification was widely accepted because of its applicability to a wide range of carbonate rock types and the ease with which its terms could be utilised and understood. Folk's

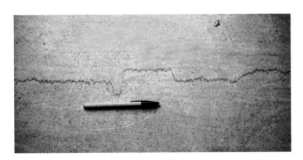

Fig. 5.53 Cross-section of a stylolite. The *dark colour* is due to insoluble material within the limestone (such as clay minerals, pyrite and organic matter) which is concentrated in the seam. The amplitude (the distance between top and bottom) may be taken as a minimum measure of the thickness of the dissolved carbonate layer

classification is based on the idea that, in principle, the sedimentation of carbonate sediments is comparable to that of terrigenous material. However, today most workers prefer to use the classification by Dunham (1962) because it is not based on the composition of the matrix but on the nature of the framework, which is more applicable in revealing the depositional processes.

5.7.8 Folk (1959, 1962) Classification

Disregarding admixture of terrigenous material, Folk (1959, 1962) distinguished between three basic components of limestones:

1. Sediment grains (allochems). The principal allochems are: skeletal grains, ooids, peloids and fragments of carbonate rocks (intraclasts or extraclasts).
2. Microcrystalline lime mud (micrite) comprising clay-size particles (grain size <4 µm). In modern carbonate environments such as the Bahamas most of the mud consists of micron-sized needles of aragonite produced by green algae like *Halimeda* and *Penicillus*. These are transported as clay fraction material.
3. Sparry calcite cement (sparite) which is carbonate crystals that have been diagenetically precipitated in the pore space after deposition. While the carbonate mud and some of the grains have a brown stain due to organic material, the cement stands out as clear and transparent in thin section.

Almost all carbonate deposits contain more than one type of material, and Folk's classification is based on the relative proportions of the three endmembers: allochems, microcrystalline lime mud and sparry calcite cement (Fig. 5.54). Allochems represent the framework of the rock making up the bulk of most limestones. The matrix between the allochems may consist of lime mud if there is little bottom current. In limestones this mud may have recrystallised into small calcite crystals (microcrystalline mud), which is called *micrite* and is thus an indicator of a low-energy environment.

Deposits consisting of well-sorted allochems have primary porosity which may be filled with cement

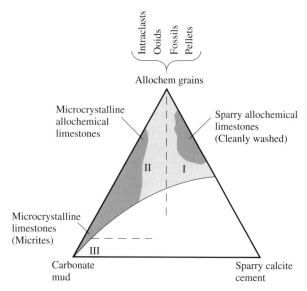

Fig. 5.54 Triangular diagram showing the three major textural types of limestones. *Shaded areas* depict the most commonly encountered limestones. *I.* The limestones consist chiefly of allochemical constituents cemented by sparry calcite. The relative proportions of sparry calcite cement and allochems varies within rather restricted limits depending on the packing and shape of the allochems, which are important factors for the pore volume. *II.* These limestones also consist of a considerable proportion of allochems, but in these cases the currents were not strong enough or persistent enough to winnow away the micrite. In these deposits the restriction of packing imposes a certain maximum on the amount of allochems, but there is no minimum. Therefore the content of allochems may vary from about 80% down to almost nothing in clean carbonate mud. *III.* Limestones consisting almost entirely of micrite with little or no allochems or sparry calcite (modified from Folk 1962)

during diagenesis. The cement is precipitated from aqueous solutions and consists of clear, transparent crystals (spar) which are easy to distinguish from micrite, which tends to be brownish because of the organic content. *Sparite* is thus a term for well-sorted carbonate sand, originally with high primary porosity, which has later been filled with calcite cement (spar). As such it normally represents a high-energy environment. Thus the relative proportions of micrite and sparry calcite cement indicate the degree of "sorting" or current strength of the environment during deposition.

As illustrated in Fig. 5.55 there are two parts to the rock name classification. The first part is contributed by the abbreviated allochems' name based on the dominating type of allochem. The second part reflects the void-filling material (micrite or sparry

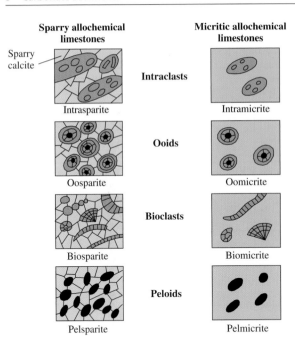

Sparry allochemical limestones

Micritic allochemical limestones

Sparry calcite

Intraclasts

Intrasparite — Intramicrite

Ooids

Oosparite — Oomicrite

Bioclasts

Biosparite — Biomicrite

Peloids

Pelsparite — Pelmicrite

Fig. 5.55 Somewhat simplified schematic representation of the constituents that form the basis for Folk's classification of carbonate rocks (modified from Tucker and Wright 1990)

calcite). Biosparite, for example, means that the carbonate rock is dominated by fossil fragments without any matrix, but cemented with sparry calcite cement. Oomicrites and pelmicrites are sediments consisting of oolites or pellets with a carbonate mud matrix. Intrasparite consists of carbonate fragments eroded inside a basin to form well-sorted carbonate deposits. Sediments with a considerable percentage of grains larger than 1.0 mm are called rudites. A coarse-grained bioclastic sparite would thus be classified as biosparrudite.

If the allochems are found in a matrix of carbonate mud (micrite) the rock is termed biomicrite, oomicrite, intramicrite or pelmicrite. These sediments represent low-energy environments where neither currents nor wave action separate the mud from the grains. Micrites normally have to have low porosity and permeability to be reservoir rocks. While the carbonate mud may initially have high porosity this is strongly reduced by mechanical and chemical compaction. Micrites may, however, become tectonically fractured, thus increasing their porosity and particularly their permeability. Secondary porosity due to dissolution of aragonite and high-Mg calcite fossils may also improve the reservoir quality of micrites.

Carbonate sediments deposited in environments with sufficient wave or current energy to separate the grains from the muddy matrix can be well-sorted with an initial pore space of about 40% between the grains (or even more if the skeletal material shows irregular growth forms). These sparry deposits are potential reservoir rocks if they only become partly cemented with calcite spar. The presence of petroleum may retard the precipitation of carbonate cement.

5.7.9 Dunham (1962) Classification

Working for Shell Oil, Dunham (1962) developed a classification which was more oriented to the needs of the oil industry, and this simple classification has become the most widely used today. His classification is also based on the depositional texture of the limestone, but the fundamental criterion is not the composition on the matrix as in the Folk (1959, 1962) classification, but the nature of the framework (whether it is mud-supported or grain-supported), see Fig. 5.56. Mud is defined as carbonate particles with a grain size <20 μm.

Mud-supported deposits imply deposition in a low energy environment. Limestones with very few grains (<10%) floating in a mud matrix are classified as mudstones. Mudstones with more than 10% grains, but still mud-supported, are classified as wackestones. Grain-supported limestones indicate deposition in a high energy environment. They are classified as packstones or grainstones, the former having a mud matrix, the latter a greater or lesser amount of calcite cement in the intergranular pores. The term boundstone is used where the fabric indicates that the original components are bound together during deposition (e.g. as in reefs). It is equivalent to the term biolithite in the classification of Folk (1959).

The textures of carbonate deposits correspond in many ways to those we observe in clastic terrigenous sediments. Mudstone corresponds to clay, and wackestone and packstone correspond to greywackes. Grainstones are well-sorted sandstones with little matrix, and have many of the same properties as well-sorted quartz sandstones, both as regards primary porosity and diagenetic transformation. We distinguish between grain-supported and matrix- or mud-supported sandstones. Mud-supported sandstones will

Fig. 5.56 Dunham (1962) classified carbonate rocks according to depositional texture (modified from Tucker and Wright 1990)

Original components not bound together during deposition				Original components bound together	Depositional texture not recognizable
Contains lime mud			Lacks mud and is grain-supported		
Mud-supported		Grain-supported			
Less than 10% grains	More than 10% grains				Crystalline carbonate
Mudstone	Wackestone	Packstone	Grainstone	Boundstone	Crystalline

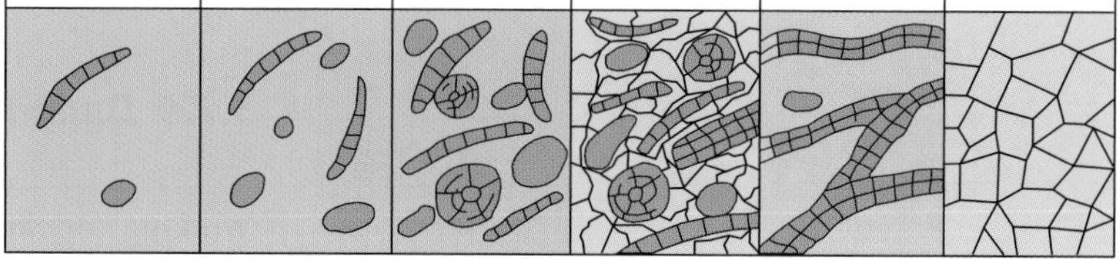

be subject to compaction of the matrix between the grains, and have many of the plastic properties of fine-grained sediments (muds). The degree of compaction will depend on the proportion of grains.

Grain-supported carbonates have a framework of grains which rest upon one another. Compaction cannot take place without the grain framework being deformed. This may take place by the grains being packed more tightly through mechanically crushing due to the force exerted by the overburden, or being chemically dissolved, particularly at the contacts between grains (pressure solution). The mechanical strength of carbonate fragments from fossils may be relatively low, but that of thick-shelled fossils and ooids is high. The contacts between grains will initially be very small, for example round ooids will only have very limited areas of contact. As the overburden increases, the pressure per unit area at the contact points will be very great. Pressure solution will then occur at the contact points, so that the contact area expands and the pressure per unit area decreases.

The distinction between grain-supported and mud-supported rocks is not simply a function of the ratio of grains to mud, because carbonate grains have widely different and often highly irregular shapes. Rocks composed of spherical grains (e.g. ooids) may need a grain content of about 60% to achieve grain support, whereas rocks with highly irregular grain shapes may form a self-supporting framework with a grain content of only 20–25%.

Mudstones, wackestones and packstones would all be poor reservoir rocks because of the carbonate mud in the matrix. In the case of packstones the grains form a grain-supported fabric but the matrix is still carbonate mud. Grainstones represents well-sorted carbonate deposits with good porosity and permeability, which may serve as a good reservoir if cement precipitation is not too advanced. This is the carbonate equivalent of well-sorted siliceous sand. Boundstones are rocks bound together by organisms (fossils) as in reefs, and they may also have high porosity and permeability.

5.7.10 Dolomitisation

The term "dolomite" is used to designate both a mineral (Fig. 5.57) and rocks in which this mineral is the main constituent. To avoid confusion, the term "dolostone" has been introduced for the rock, but has not been widely adopted.

The mineral dolomite ($CaMg(CO_3)_2$), consists of layers of CO_3^{2-} groups separated by alternating layers of Mg^{2+} and Ca^{2+}. This is a highly organised structure (trigonal rhombohedral) and the organisation of more or less pure layers of Mg^{2+} and Ca^{2+} leads to high kinetic energy being required for the crystallisation of dolomite. This is particularly true at low temperatures and so far it has not been possible to synthesise low temperature dolomite ($<100°C$) in the laboratory.

Dolomite is in most cases not formed directly, but as a secondary mineral as a result of reactions between different forms of $CaCO_3$ and Mg^{2+}. The reaction

$$2CaCO_3 + Mg^{2+} = CaMg(CO_3)_2 + Ca^{2+}$$

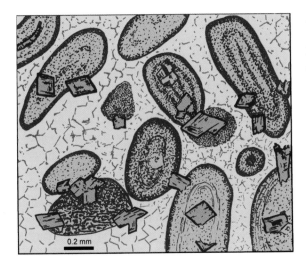

Fig. 5.57 Partly dolomitised oolite. The dolomite rhombs (*dark brown*) transect the oolite grains (*light brown*) and the surrounding spar crystals (*blue*), indicating a late diagenetic dolomitisation (modified from Greensmith 1978)

is dependent on the Mg^{2+}/Ca^{2+} ratio. The dolomitisation process will only proceed with a supply of magnesium maintaining a high Mg^{2+}/Ca^{2+} ratio.

Mg^{2+} is more strongly hydrated than Ca^{2+} in seawater. The Mg^{2+} ion together with its surrounding water molecules, $Mg(H_2O)_6^{2+}$, cannot easily enter a crystal position at surface temperatures. However, the hydration decreases with increasing temperature and this is a factor often cited in favour of explaining late burial dolomitisation.

Experiments show that in the absence of sulphate, dolomite forms rapidly in solutions with $MgCl_2 + NaCl + CaCl_2$. The main reason that dolomite is not common in modern marine environments is that the sulphate ion (SO_4^{2-}) is very efficient at preventing dolomitisation. Consequently dolomitisation takes place more easily when there are few sulphate ions (Baker and Kastner 1981). For this reason dolomite is formed in a number of lakes and in brackish zones with mixing of salt- and freshwater (Dorag model). In a microbial sulphate reduction zone, the SO_4^{2-} concentration will be lower and ammonium (NH_4^+) can be formed by nitrate-reducing bacteria. The sulphate is reduced to hydrogen sulphide (H_2S), which may react with dissolved reduced iron (Fe^{2+}) to form metal sulphides such as pyrite (FeS_2). NH_4^+ can replace Mg^{2+} adsorbed on clay minerals. In this way magnesium is liberated for dolomitisation.

5.7.10.1 Models

Dolomitisation means that $CaCO_3$ is dissolved and dolomite precipitated. The conditions for this are:

1. That calcium carbonate becomes unstable, and that the solution is supersaturated with respect to dolomite.
2. That Mg^{2+} is added to the solution so that dolomitisation can continue.
3. That an inhibitor such as sulphate is absent or at least in low concentration.

Solution of calcium carbonate takes place most easily if we have aragonite at the outset. Fine-grained carbonate mud has a large specific surface which enables it to react more rapidly than massive carbonate.

The dolomitisation process is accelerated if the sediment concerned has a high permeability and a high rate of percolating porewater containing magnesium, though if carbonate sediments already contain a good deal of magnesium, dolomitisation will be able to proceed without any addition of Mg^{2+}. This applies, for example, to carbonate sediments rich in high-Mg calcite as in reefs. We often find thin dolomite beds or finely divided dolomite in shales and this may be due to a supply of Mg^{2+} from clay minerals.

Seawater is a highly complex solution. We cannot simply predict the way in which it will react from the concentrations found through chemical analyses. Some of the Mg^{2+} and Ca^{2+} is associated with Cl^- through ion pairing, and must be excluded from calculations regarding activities involving carbonates.

Theoretical calculations indicate that dolomitisation should occur when the Mg^{2+}/Ca^{2+} activity ratio is about 0.6. Although these figures are somewhat uncertain, it is clear that seawater, in which the Mg^{2+}/Ca^{2+} ratio is 5.6, is oversaturated with respect to dolomite. The fact that dolomite does not form is ascribable to kinetic reasons, one probably being the high SO_4^{2-} concentration, and only when the ratio is over about 7 will dolomitisation take place in seawater. In freshwater the ion strength is lower, and here dolomite can be formed at lower Mg^{2+}/Ca^{2+} ratios because there is very little sulphate. However, freshwater contains little magnesium, so Mg^{2+} must be supplied by mixing with seawater and large amounts of seawater must circulate through the limestone. A steady percolation of freshwater alone will not lead to any great degree of dolomitisation, because freshwater will displace the magnesium-rich salt water, and the mixing zone will be too small.

Seawater is the only source of magnesium for large scale dolomitisation. The seawater which circulates inside atolls and reefs is reducing and therefore low in sulphate, at the same time having abundant magnesium. Reef and atoll facies therefore offer favourable conditions for dolomitisation.

5.7.10.2 Evaporite Model for Dolomitisation

Dolomite is often associated with evaporite environments. The first definite example of dolomite being formed today was found in evaporating sediments in a supratidal environment in the Bahamas in about 1960. It is clear that when seawater evaporates, and aragonite and also gypsum ($CaSO_4 \cdot 2H_2O$) are precipitated, the composition of the fraction still in solution will become increasingly enriched in magnesium, and dolomite will only be precipitated at high Mg^{2+}/Ca^{2+} ratios. In addition magnesite ($MgCO_3$) may form. This explains why dolomite is important in most evaporite sequences. However, evaporite minerals are very soluble, particularly chlorides, and often are not preserved; gypsum may also dissolve and be replaced by carbonate. Dolomite might therefore have been deposited in an evaporite environment despite the fact that we do not find any of the original highly soluble salts preserved. For this reason it is important to look for indirect evidence of evaporite conditions and solution, including replacement of evaporite minerals.

The most important indicators of evaporite conditions are:

1. Absence of ordinary marine fossils, apart from stromatolites which can tolerate high salinity. In Palaeozoic and younger deposits stromatolites are typical of evaporites, because under normal marine conditions cyanobacteria have too much competition from other organisms.
2. Breccias which may have been formed through solution of underlying salt deposits so that beds collapse and form a *collapse breccia*. Such breccias are characterised by angular fragments from an overlying bed, for example of carbonate, which have fallen down into a solution cavity.

3. Pseudomorphosis (replacement) of evaporite minerals, e.g. halite (NaCl) and gypsum ($CaSO_4 \cdot 2H_2O$). Evaporite minerals, which are disseminated through a matrix of less soluble minerals such as carbonates, are often replaced through pseudomorphosis so that the crystal form may reveal the original mineral. The cubic halite crystals are typical, and the characteristic swallow-tail twins of gypsum crystals are easily recognised even if they have been replaced by other minerals.
4. *Chickenwire structure*. Anhydrite often forms very characteristic nodules or continuous layers which look like chicken wire. Even if the anhydrite layers are converted to calcite, these structures may be preserved.
5. Authigenic quartz and feldspar. Evaporites are often associated with microcrystalline quartz (chert) and chalcedony. A high content of authigenic feldspar and zeolites is also typical of many evaporites. With low-grade metamorphism (200–300°C) zeolites will dissolve and be replaced by feldspar.

5.7.10.3 Late Diagenetic Dolomite

While early diagenetic dolomite is normally relatively fine-grained, late-diagenetic dolomite usually forms larger crystals, often well-defined dolomite rhombs. The smaller crystals are formed by rapid crystal growth under conditions of high supersaturation. The larger crystals are formed by slow crystal growth from a few nucleation centres at a very low degree of supersaturation combined with deep burial.

As mentioned previously, hydration of Mg^{2+} decreases with increasing temperature, making Mg^{2+} more readily available for the dolomite structure. Dolomite might then be formed at very low Mg^{2+}/Ca^{2+} ratios; at about 80°C the ratio can be as low as 0.1. Although dolomite can precipitate in solutions with low Mg^{2+}/Ca^{2+} ratios, the question remains of how sufficient magnesium is added for the dolomitisation to take place. The amount of magnesium in solution in the deeper part (>2–3 km burial depth) of sedimentary basins is in most cases very low and is far from adequate as a source for large scale dolomitisation. Water from compaction of mudstones and shales is probably also insufficient to supply much magnesium for dolomitisation. Primary high-Mg calcite will be a source of Mg to form dolomite at depth.

There is only a little magnesium in the porewater of sedimentary basins apart from in the vicinity of evaporites. Probably most of the dolomitisation occurs near the surface where the magnesium comes from seawater. Fine-grained dolomite, formed at an early stage, may be dissolved at depth and recrystallise as coarsergrained dolomite, in which case there is no need to postulate a magnesium supply deep in the basin.

We often find dolomite enriched along stylolites, probably because dolomite is less soluble than calcite and the solution and precipitation round a stylolite will concentrate clay minerals which, in turn, may release some magnesium.

If the composition of the porewater later shifts towards a low Mg^{2+}/Ca^{2+} ratio, dolomite may dissolve and calcite reprecipitate. There are a number of cases where distinctly dolomite-type rhombs are found to consist of calcite. One common cause of reversed dolomitisation – often called *dedolomitisation* – is porewater coming from gypsum which is dissolving. This gives the porewater a high Ca^{2+} concentration. However, many people have recommended that the term "dedolomitisation" should be dropped, and the positive term "calcitisation" be used instead.

5.7.10.4 The Significance of Dolomitisation

For many years there has been intensive research into the processes which lead to dolomitisation. A great deal remains to be learnt, however, before we really understand the precise conditions for dolomitisation so that we can predict the extent of dolomite in sedimentary basins. The reason for this great interest is that dolomitic carbonate rocks are very important reservoirs for oil and gas. The dolomitisation process may create secondary porosity because dolomite has a greater density than calcite so if an identical number of mol dolomite is precipitated as in the original calcite, we would get approximately 12% smaller volume and an equivalent increase in porosity. The dolomitisation process assumes, however, that calcium is removed and magnesium introduced, and there is then no reason why there should be an increase in porosity proportional to the difference in density, since there is no reason why the same number of mol dolomite should be precipitated as were removed by the calcite dissolving. Since dolomitisation involves large-scale percolation of porewater, we may also

have net increase in the porosity associated with this process.

Micritic limestones have in most cases too low porosity and permeability to be regarded as reservoir rocks except when fractured. Tectonically fractured limestones may become cemented in relatively short geologic time before the migration of petroleum. Fractured dolomites are, however, more stable and likely to remain open longer because of the lower solubility of dolomite.

5.7.11 Formation of Carbonate Sediments Rich in Siderite and Chamosite (Ironstones)

Iron carbonates like siderite ($FeCO_3$) are stable carbonate phases with rather low solubility (Berner 1981). In the presence of sulphides, however, most of the iron will be precipitated as iron sulphides (i.e. pyrite or marcasite). Precipitation of siderite and other iron carbonates is therefore restricted to settings where the porewater has a low content of reduced sulphur (sulphides), as in freshwater and below the sulphate-reducing zone.

There is almost no iron in solution in oxidised water and iron can therefore not be taken from seawater. During weathering on land, however, large amounts of iron may be released and transported as small particles of iron oxides, hydroxides or adsorbed on clay and organic matter. These are often concentrated in the distal parts of deltas where clastic sedimentation is low (Fig. 5.58).

Most of the iron supplied by rivers will be reduced just below the seafloor by small amounts of organic matter. It may precipitate at the redox boundary as iron oxides or as glauconite or chamosite. In carbonate sediments, aragonite may be partly replaced by siderite instead of calcite if iron is present. Because of the lower solubility of siderite all the available iron will be exhausted and precipitated as siderite before calcite can begin to precipitate.

Iron-rich sediments are typical of mixed carbonate and clastic sedimentary sequences. On carbonate platforms like the Bahamas there is little supply of iron because it is not connected to a source on land. All clastic iron-rich sediments are trapped in the deep water around the carbonate platform and there

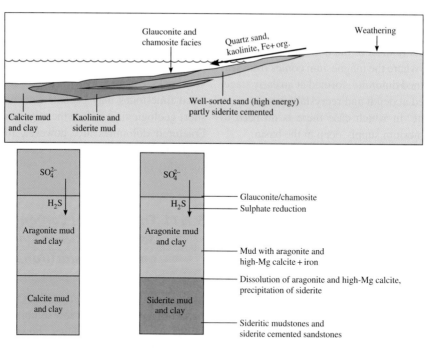

Fig. 5.58 Formation of sideritic limestones. During weathering large amounts of iron are released and transported as red fine-grained iron oxides. Reacting with organic matter iron oxide is reduced to Fe^{2+} and when aragonite becomes unstable, siderite ($FeCO_3$) is formed instead of calcite because siderite is most stable (lower solubility)

is practically no iron in the seawater covering the platform. Only small amounts of aeolian dust with some iron are transported to the Bahamas from Africa (Sahara).

5.8 Carbonate Reservoir Rocks

5.8.1 Introduction

Carbonate reservoir rocks are mostly limestones and dolomites and these rock types contain almost 40% of the oil reserves in the world. They are particularly common in the Middle East.

Carbonate reservoir rocks differ in several important aspects from sandstone reservoirs:

1. The sediment particles are in most cases produced locally within the basin by biological production (fossils) or by chemical precipitation.
2. The mineralogy and the textures depend very much on the organisms precipitating the carbonates.
3. Carbonate minerals have higher solubility than silicate minerals in porewater, and also higher reaction rates at low temperature. Dissolution of grains and

precipitation of cements may therefore be important also at shallow depth (<2 km).

4. The reservoir quality is highly dependent on the porosity, which may be of different types. The porosity in carbonate rocks may be pore space between grains (*intergranular porosity*) or porosity within grains (*intragranular porosity*), commonly fossils. These primary porosity types are usually strongly modified even at shallow depth during burial. Well-cemented carbonate rocks may be brittle, particularly during uplift, and tectonic fracturing may produce important *fracture porosity*.

The most important porosity types in carbonate rocks are the following:

A. *Primary porosity* is the pore space that existed in the sediment just after deposition prior to diagenetic alteration. During burial the primary porosity is reduced by compaction and cementation.

   1. *Primary intergranular porosity* (pore space between carbonate grains)
      Intergranular porosity (inter-particle porosity) is the preserved primary pore space between grains. Pore shapes may vary considerably and

are homogeneous only when the enclosing particles are of uniform size and shape (e.g. well-sorted ooids). The porosity increases with better sorting and more irregular grain shapes, attaining values up to approximately 60%. Also carbonate mud, just after deposition, has high intergranular porosity (70–80%) between the small mud particles, but the permeability is low and in most cases this porosity is rapidly reduced by mechanical compaction during early burial.

2. *Intragranular porosity* (porosity inside grains)
Intragranular porosity (intra-particle porosity) consists of pore space that occurs within grains. It is commonly formed by the decay of organic material within carbonate skeletons (e.g. foraminifera, corals and bryozoans).

3. *Growth-framework porosity* (porosity inside a rigid framework produced by fossils)
Growth-framework porosity is developed between and within organisms during the in-place growth of a carbonate framework. Typical examples are reef corals with uneven and patchily developed pores between the branches or between different colonial organisms.

4. *Shelter porosity*
Shelter porosity is often found below large plate-like grains which have acted as umbrellas protecting the pore space beneath from being filled with finer material. Typically porosity may be preserved below the concave side of molluscs and brachiopod shells. Organic material which is later decomposed may have produced shelter porosity which can be preserved if there is early cementation.

B. *Secondary porosity* is pore space produced by dissolution of grains or cement after deposition. Secondary porosity may therefore be regarded as an addition to the primary porosity. This requires a net transport of carbonate in solution out of the rock. If aragonite dissolves and the same amount of calcite precipitates inside the rock, no porosity is gained, though the distribution of pore space and the permeability may change drastically.

1. *Mouldic porosity*
Mouldic porosity is formed by the selective dissolution of grains, particularly skeletal material with a primary aragonitic composition. Whole fossils or grains like ooids may dissolve after the precipitation of cement in the primary pores. These secondary pores thus become moulds of the dissolved structures. Mouldic porosity may also form by dissolution of fossils from a matrix of carbonate mud which has been lithified. In the Jurassic Smackover Formation of Arkansas, USA, oomouldic porosity was created by the dissolution of ooids during freshwater flushing (More 1989).

2. *Fenestral porosity*
Fenestral porosity consists generally of small elongate to equant pores which are typically 1–10 mm in diameter. These characteristic pores are often arranged in layers within the sediment and are most commonly encountered in algal mats facies where they have been produced by decaying organic matter or by desiccation.

3. *Breccia porosity*
Breccia porosity may form by the collapse of a rock due to dissolution (e.g. dissolution of evaporites or limestone dissolution during karst weathering) or tectonic deformation.

4. *Fracture porosity*
Fracture porosity commonly forms due to folding, faulting, salt doming, differential compaction, salt dissolution or fluid overpressure. Carbonate sediments often become cemented at shallow depth and may fracture and dilate so that fracture porosity is produced. Open fractures will gradually be filled with cement but fractures in dolomite may remain open longer than those in limestones.
Shrinkage porosity is a special type of fracture porosity formed during early diagenesis when sediments may shrink i.e. due to dewatering. Septarian fractures in limestone concretions are also a kind of shrinkage fracture.

5. *Vuggy, channel and cavern porosity*
Vuggy, channel and cavern porosity are not fabric selective, i.e. they cut across grains and/or cement boundaries. The pores are of irregular size and shape and may or may not be interconnected. Many vugs are solution-enlarged moulds where evidence of the precursor grain has been destroyed by dissolution of the neighbouring matrix. Vuggy pores are commonly 1 mm–1 m in diameter. Cavern porosity commonly relates to meteoric (karstic) leaching and is differentiated from vuggy porosity by the larger pore size (man-sized or larger).

5.8.2 Some Examples of Carbonate Reservoirs

5.8.2.1 Grainstones

Well-sorted carbonate sands (grainstones) are good reservoirs if the porosity is not reduced too much by cementation. The sand grains are usually ooids or fossil fragments. If the grains are mainly of low-Mg calcite, most of the porosity tends to be primary porosity which gradually fills with cement during burial. The source of most of the cement is then normally pressure solution along stylolites or grain-grain contacts.

Precipitation of early carbonate cement reduces the stresses at grain contacts and strengthens the grain framework. The mechanical compaction may therefore be reduced. When many of the grains consist of aragonite they may dissolve, producing secondary porosity while the dissolved carbonate fills the primary porosity. Dissolution of aragonitic ooids and fossils produces moulds which have very little communication between them and such mouldic reservoirs are characterized by rather high porosity but low permeability.

5.8.2.2 Fractured Reservoirs

The term fracture is used for any break in a rock and includes cracks, joints and faults. In many cases the fracture is less permeable than the matrix, but some may also be partly or totally open. Open fractures formed by extension are common during uplift and folding. Well-cemented limestones and dolomites tend to have brittle properties and faulting may produce a breccia with good permeability.

The rock fragments produced by the brecciation take the stress and prevent the fault plane from closing. Movements (off-set) along the fault plane have a similar effect. The porosity and permeability produced in carbonates is, however, temporary because it may relatively rapidly be filled with carbonate cement. Filling of open fractures by oil or gas will retard or inhibit calcite cementation and thus contribute to the conservation of the porosity. This is particularly true if the rocks are relatively oil-wet.

Fractures in limestones are commonly cemented by diffusion of carbonate from the adjacent matrix. Flow of water upward along the fault plane will cause dissolution of calcite rather than precipitation because of the retrograde solubility with respect to temperature. In carbonates where the fractures are mostly open and forming a three-dimensional fracture network, the effective permeability can be very good. However, fractures are often formed during several phases of tectonic deformation and then only some of the fractures may remain open. The overall drainage during production is then more difficult to predict.

Fractured carbonate reservoirs often consist of dolomite. Well-cemented massive dolomite is rather brittle forming large fracture networks when subjected to tectonic forces.

Fractures in dolomite may be preserved longer than in limestones. This is because dolomite is both mechanically stronger and less soluble than calcite so that it will take longer for the fractures to be cemented up. For this reason we may find dolomite reservoirs at great depth (5–6 km) and also in uplifted basins. Fractured reservoirs may have very high permeability so that the reservoirs can be produced even if the average porosity is low.

5.8.2.3 Chalk

Chalk is a pelagic sediment where calcareous algae (coccolithophores) are dominant. Coccolithophores consist of low-Mg calcite and are therefore mineralogically stable during early diagenesis. The presence of aragonitic skeletal material is fairly meagre, so these mineralogically unstable fossils are not an important calcite source for early cementation. During periods with little pelagic sedimentation seafloor cementation may form hardgrounds. These are often encrusted by bivalves, boring organisms, bryozoa and sponges and may be mineralised by glauconite and phosphate.

For the most part the chalk is rather pure $CaCO_3$ but more clay-rich intervals like the Plenus Marl occur in the North Sea reservoirs, forming a tight low-permeability zone. Thin clay laminae enhance pressure solution and often develop into stylolites. The dissolution of calcite by pressure solution is the most important source of carbonate cement.

During diapirism of the underlying Zechstein salt (Upper Permian) the chalk became unstable and was reworked by gravity flows (turbidites and debris flows). The eroded sediments were already somewhat cemented and the debris flow units have proved to be better reservoir rocks than the primary chalk deposits. The early cementation may have strengthened the grains or aggregates of grains so that compaction during burial was reduced. In addition, the salt doming has produced fracturing in the chalk and these fractures have been essential in increasing the overall reservoir permeability.

Early filling of oil and gas in carbonate rocks like the chalk will retard calcite cementation and tend to preserved porosity compared to water-saturated chalk.

5.9 Summary

Limestones are mostly organically precipitated except in evaporitic environments. The evolution of calcareous organisms determines to a very large extent the initial grain size and mineralogy of the sediments, which in turn strongly influences the properties of carbonate reservoirs. The depositional environment is very important in controlling primary sorting of carbonate sand and the distribution of framework builders like reefs.

Limestones compact mechanically as a function of effective stress, but mineral dissolution and precipitation (chemical compaction) may also be important at shallow depth (0–1 km). Thermodynamically unstable aragonite dissolves at rather shallow depth causing formation of secondary mouldic porosity and early cement. Sediments consisting of mostly calcite will compact mechanically and preserve their porosity until pressure solution becomes an effective compaction process. Grain-to-grain dissolution and stylolites then provide the sources of cement reducing the primary porosity.

Dolomite forms in most cases near the surface under evaporitic conditions, and under reducing condition in contact with seawater (as in reefs and atolls) where the sulphate content is low. Dolomitisation of calcite requires dissolution and reprecipitation and a large scale supply of magnesium and there is no reason that this in itself should cause increased porosity even if dolomite is denser than calcite.

Dolomite is both mechanically and chemically more stable (less soluble) than calcite and therefore preserves more of its porosity during burial. Extensional fractures formed tectonically will tend to stay open longer in dolomite than in limestones and may therefore form important reservoirs.

Carbonate sediments may compact both mechanically and chemically also at shallow depth and low temperatures. Early cementation due to dissolution of aragonite and precipitation of calcite cement will increase rock strength and prevent further mechanical compaction.

Further Reading

Audet, M.D. 1995. Modelling of porosity evolution and mechanical compaction of calcareous sediments. Sedimentology 42, 355–374.

Bathurst, R.G.C. 1971. Carbonate sediments and their diagenesis. Developments in Sedimentology 12, 620 pp.

Berner, R.A. 1981. A new geochemical classification of sedimentary environments. Journal of Sedimentary Research 51, 359–365.

Buchen, van F.S.P., Gerdes, K.D. and Estebahn, M. 2010. Mesozoic and Cenozoic carbonate systems of the Mediterranean and Middle East: Stratigraphic and Diagenetic Models. Geological Society Special Publication 329, 424 pp.

Dunham, R.J. 1962. Classification of carbonate rocks according to depositional texture. In: Ham, W.E. (ed.), Classification of Carbonate Rocks. American Association of Petroleum Geologists Memoir 1. Tulsa, OK, pp. 108–121.

Folk, R.L. 1959. Practical petrographic classification of limestones. American Association of Petroleum Geologists Bulletin 43, 1–38.

Folk, R.L. 1962. Spectral subdivision of limestone types. In: Ham, W.E. (ed.), Classification of Carbonate Rocks. American Association of Petroleum Geologists Memoir 1. AAPG, Tulsa, OK, 62–84.

James, N.P. and Clarke, J.A.D. 1997. Cool-water carbonates. SEPM Special Publication 56, 440 pp.

Kastner, M. 1984. Control of dolomite formation. Nature 311, 410–411.

Lucia, F.J. 1999. Carbonate Reservoir Characterization. Springer, Berlin, 226 pp.

Moore, C.H. 1980. Porosity in carbonate rock sequences. AAPG Continuing Education Course Note Series from the Fall Education Conference 1980. AAPG Tulsa, OK, pp. 124.

Moore, C.H. 1989. Carbonate Diagenesis and Porosity. Elsevier Science Publishers, Amsterdam, 338 pp.

More, C.H. 2004. Carbonate Reservoirs. Porosity Evolution and Diagenesis in a Stratigraphic Framework. Developments in Sedimentology 46. Elsevier, Amsterdam, 444 pp.

Nielsen, J.K. and Hanken, N.-M. 2002. Late Permian carbonate concretions in the marine siliciclastic sediments of the Ravnefjeld Formation, East Greenland. Geology of Greenland Survey Bulletin 191, 126–132.

Purser, B.H. and Evans, G. 1973. Regional sedimentation along the Trucial Coast, SE Persian Gulf. In: Purser, B.H. (ed.), The Persian Gulf. Springer, Berlin, pp. 211–232.

Purser, B., Tucker, M. and Zenger, D. 1994. Dolomites. Special Publication 21, International Association of Sedimentologists. Blackwell, Oxford, 451 pp.

Roehl, P.O. and Choquette, P.W. 1985. Carbonate Petroleum Reservoirs. Springer, New York, 622 pp.

Schmoker, J.W. and Halley, R.B. 1982. Carbonate porosity versus depth: A predictable relation for south Florida. AAPG Bulletin 66, 2561–2570.

Scholle, P.A. 1978. A color illustrated guide to carbonate rock, constituents, textures, cements, and porosities. AAPG Memoir 27, 241 pp.

Scholle, P.A., Bebout, D.G. and Moore, C. 1983. Carbonate depositional environments. AAPG Memoir 33, 708 pp.

Scholle, P.A. and Ulmer-Scholle, D.S. 2003. Color guide to petrography of carbonate rocks. AAPG Memoir 77, 474 pp (2nd printing).

Swart, P.K. Eberli, G. and Mckensie, J. 2009. Perspectives on Carbonate Sedimentology. International Association of Sedimentologists. Wiley-Blackwell, Hoboken, NJ, 400 pp.

Tucker, M.E. and Wright, V. P. 1990. Carbonate Sedimentology. Blackwell, Oxford, 482 pp.

Zenger, D.H. and Dunham, J. B. 1988. Dolomitization of Siluro-Devonian limestones in a deep core (5,350 m), southeastern New Mexico. In: Shukla, V. and Baker, P.A. (eds.), Sedimentology and Geochemistry of Dolostones. SEPM Special Publication 43, 161–173.

Chapter 6

Shales, Silica Deposits and Evaporites

Knut Bjørlykke

6.1 Mudrocks and Shales

Mudrocks and shales are the most abundant lithologies in most sedimentary basins. They are important because shales include source rocks for oil and gas, and recently large reserves of gas have been found in shales. Shales may therefore be reservoir rocks because they may have significant porosity and some (although small) permeability to flow gas. The seismic image for sandstones also depends on the properties of adjacent shales.

There is however no precise definition for mud or mudrocks. It is used to describe fine-grained rocks with a relatively high content of clay-sized particles, mostly clay minerals but also other minerals. Carbonate mud is discussed under carbonate sediments. The upper limit for clay particles is 0.004 mm in the geological literature, but in the engineering literature (soil science) 0.002 mm is commonly used. Because clay mineral grains are essentially flat flakes, they have large surface areas, some like smectite having several hundred m²/g. There is a cohesion between small particles, and clay minerals also have a surface charge due to broken bonds in the mineral structure. This cohesion plays an important role in sedimentary processes of erosion, transport and deposition since most clastic sediments contain significant amounts of clay. The properties of clays are not only controlled by the mechanical strength of the grains but also by the composition of the pore fluid. This is also true during sediment compaction.

Mudrocks and shales are often treated as one lithology, but they vary greatly as a function of both mineral composition and grain-size distribution. A relatively large fraction of grains may be larger than clay size, but as long as the larger particles are floating in a finer matrix the properties are dominated by the clay-sized particles.

Here we will discuss siliceous (i.e. non-carbonate) mudrocks and clay. The clay minerals in mudrocks may have different origins:

(1) Clay minerals formed by weathering of igneous and metamorphic rocks.
(2) By erosion of older shales and mudrocks.
(3) From volcanic ash.
(4) By diagenesis on the seafloor and during burial.

The clay mineral assemblage produced by weathering depends on the composition of the rocks that are being weathered, and the climate. A humid climate will favour the formation of kaolinite. In a granite or gneiss, feldspar, mica and most other silicate minerals will dissolve and the aluminum and silica will precipitate as kaolinite. The quartz grains will be weathered out as sand and their grain size will reflect the quartz crystal size range in the parent rock (see Chap. 3). The result is a bimodal distribution of sand and kaolinitic clays.

In areas with mostly basic rocks like anorthosite, gabbro and basalts, weathering will only produce clay minerals, since there is no quartz which in more acid rocks forms most of the sand fraction. Minerals like basic plagioclases, pyroxenes and hornblende will quickly break down to clay minerals such as kaolinite.

K. Bjørlykke (✉)
Department of Geosciences, University of Oslo, Oslo, Norway
e-mail: knut.bjorlykke@geo.uio.no

K. Bjørlykke (ed.), *Petroleum Geoscience: From Sedimentary Environments to Rock Physics*,
DOI 10.1007/978-3-642-02332-3_6, © Springer-Verlag Berlin Heidelberg 2010

6.2 Supply of Clay Minerals to Sedimentary Basins

While kaolinite is derived from humid climate weathering, smectite and illite are more typical of deserts because there is less flow of fresh (meteoric) groundwater. Chlorite is mostly derived from erosion of metamorphic rocks in relatively cold climates where weathering is slow. In warmer and wetter climates chlorite will break down, but may occur near basalts and basic volcanic rocks, particularly in the marine environment where chlorite is more stable. Volcanic ash consisting of glass and unstable volcanic mineral assemblages may alter to smectite, both on land and on the seafloor. In deep sea sediments, zeolites like phillipsite are common.

Erosion of older mudrocks and shales can produce nearly all types of clay minerals. Glacial clays are essentially mechanically ground down sedimentary, metamorphic or igneous rocks. Since there is very little chemical weathering, their chemical composition is nearly the same as the rocks eroded. Chlorite and illite are formed by disintegration of mica and metamorphic chlorite, and most of the quartz and feldspar is preserved. We find such clays accumulating in front of modern glaciers that terminate in lakes or the ocean. When the continental ice sheet withdrew from Scandinavia 10,000–9,000 years ago, thick glaciomarine clays were deposited in the fjords, the inland portions of which became valleys through postglacial isostatic uplift.

Clay transported by rivers into lakes will remain suspended for some time and very finely laminated clayey sediments will be deposited. In lakes there is usually less bioturbation to destroy the lamination than in marine environments. Freshwater sediments tend to be finely laminated compared to most marine sediments.

Clay from rivers is not transported far offshore and we see from satellite pictures that there is clear water not so far from the delta front (Fig. 6.1). When the clay minerals come into contact with seawater, the salt content in the water will cause the clay particles to flocculate, which makes them sink faster to the bottom near to the river mouth. This is because clay minerals have a negative charge which prevents them from sticking together in freshwater. In seawater these negative charges are neutralised by the cations in seawater, such as Na^+, K^+, Mg^{++} and Ca^{++}. K^+ is most effective

Fig. 6.1 Satellite photograph showing the distribution of clay *outside* the Mississippi delta. The limited extent of the delta mud is due to flocculation of the clay particles when freshwater is mixed with ocean water

because it is not so strongly hydrated (surrounded by water molecules) as Mg^{++}, Na^+ and Ca^{++} (Fig. 6.2).

Clay minerals transported into marine sedimentary basins will also be subjected to sorting by grain size. Kaolinite is the coarsest grained of the clay minerals and will therefore be deposited in the most proximal parts of deltas and shorefaces while illite and smectite will be transported further out into the more distal parts (Fig. 6.1).

From deltas and the shelf edge, clayey sediments as well as sand may be transported down the slope by turbidity currents or debris flows. In the most distal shelf or deep-water facies, sedimentation rates can be very low and aeolian dust may make up much of the sediment deposited. This is particularly true offshore dry areas like the Sahara in West Africa. This mud is rich in

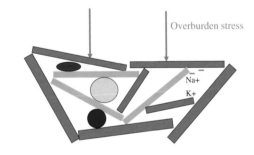

Fig. 6.2 Clay minerals have negative charges due to broken chemical bonds in their silicate structure. In freshwater there is repulsion between the clay minerals, keeping them in suspension. In seawater, cations like sodium and potassium help to neutralise these charges and the clays will form denser aggregates (flocculation)

fine-grained smectite and illite, often with iron oxides which may serve as an important nutrient for organic production in the South Atlantic Ocean. Volcanic ash may be deposited over large areas. During transgressions, extensive layers of mud may be deposited on shelves along the basin margin. This mud may later be eroded and supplied to the basin during periods of uplift, so that the prograding regressive sequences may be rich in clay.

6.3 Silica (SiO_2) Deposits

Silica which is liberated through weathering goes into solution as silicic acid (H_4SiO_4). Even if quartz and also feldspar have relatively low solubility, large quantities of silica are transported by rivers out into the sea. The total amount of dissolved silica added annually to the sea is estimated to be about 4×10^8 tonnes. Some silica is also introduced from the mid-oceanic ridges, but is probably of only very modest significance compared with the fluvial input. An equal amount of silica must be removed for the seawater composition to remain constant. Silica is removed biologically as biogenic silica, and through the formation of silicate minerals in the sea. Radiolaria and diatoms are particularly efficient at removing silica. As a result, even though quartz is only very slightly soluble in seawater (3–6 ppm SiO_2), surface water (seawater) is usually undersaturated with respect to quartz. This is because the organisms can precipitate silica from seawater even if the concentration of SiO_2 is far less than 1 ppm. While amorphous silica has a solubility of about 150 ppm, cristobalite and tridymite have a solubility of 6–15 ppm, depending on the degree of order in the crystals. The most important silica-producing organisms are:

Phytoplankton: diatoms and silicoflagellates
Zooplankton: radiolaria
Silica sponges

These organisms are built up of amorphous silica (opal A). The total production of organic silica in the oceans has been estimated to be from 2×10^{10} up to 10^{11} tonnes/year. The largest contribution comes from diatoms, and also a very large percentage (50–70%) of the total primary production of carbon (2×10^{10} tonnes/year) is ascribable to diatoms consisting of about 60% silica and 40% carbon.

The flux of silica into the oceans from the continents is far less than the organic production of siliceous organisms. The approximately 4×10^8 tonnes/year brought by rivers only represents about 1% of all organic silica precipitation. Some silica is added to the sea through submarine volcanism along the oceanic spreading ridges. Seawater is being circulated (convected) through hot basalt, dissolving silica and precipitating sulphides from the sulphate in the seawater. Even with this source of silica, the supply to the ocean water can not balance the amounts precipitated biologically. Since we must assume that the composition of seawater has been relatively constant, this means that only about 1% of the overall organic silica production is retained in sedimentary deposits. Most of the silica from plankton redissolves in the undersaturated seawater before it reaches the bottom, and some also dissolves on the seafloor and diffuses up into the water column. Consequently it is only when the rate of organic precipitation is higher than the rate of solution that we find deposition of organic silica.

Diatoms dissolve because seawater is undersaturated with respect to silica, and large quantities of organic material can thereby be released without oxidation taking place. Organic matter produced by the solution of diatoms thus constitutes a large part of the total organic matter accumulated. Phosphates, nitrogen and various trace metals are also released through the disintegration of plankton as they sink through the water column. These recycled nutrients can once more provide the basis for organic production when water wells up to the photic zone (Fig. 6.3). The upwelling currents also prevent supply of clastic sediments from land, so that nearly pure silica can be deposited.

Opal A will decompose to opal CT which may consist of bladed crystals forming small spheres called lepispheres. Because more energy (temperature) is required to precipitate quartz, minerals like cristobalite and tridymite are formed. These are minerals which are stable at very high temperatures (1,000–1,500°C), but precipitate out instead of quartz at low temperatures, even though quartz is thermodynamically more stable.

This phase is called opal CT, sometimes also porcellanite. Opal CT will, when subjected to higher temperatures, slowly dissolve and the silica will be precipitated as quartz.

Amorphous silica (opal A) dissolves and is replaced by opal CT, usually at a temperature of around 50–70°C which corresponds to about 1.5–2 km of overburden at average geothermal gradients. The

Fig. 6.3 Upwelling of water rich in nutrients causes biological precipitation of silica and phosphates. The silica deposits (opal A) will, when buried, be altered to opal CT and then to quartz

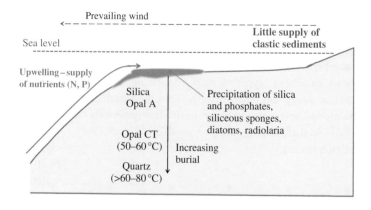

reason why amorphous silica (opal A) can exist so long despite being thermodynamically unstable, is that quartz does not crystallise at low temperatures. Opal CT is transformed into quartz at temperatures around 60–80°C. The change in acoustic impedence which accompanies the transition from opal A to opal CT and then to quartz (chert) may produce a significant seismic reflection. Because these reactions are controlled by temperature they tend to occur as horizontal zones that may be mistaken for a fluid contact (gas/water or oil/water).

Vast amounts of diatoms are found today round Antarctica, and the thick sediment accumulations there have a very high content of amorphous silica. In the North Sea too there are now large amounts of silica but there is little net accumulation. In the Tertiary sequence in the North Sea there are major silica beds. Those with the greatest extent are associated with ash layers from Eocene volcanicity related to the opening of the Norwegian–Greenland Sea and consist of radiolarians and diatoms together with altered volcanic sediments (ash) with abundant smectite. Well-cemented silica beds of Eocene age are called Moler in Denmark, and Balder Formation in the North Sea where it generates a very prominent seismic reflector.

In the Oligocene there are also nearly pure silica beds, and where they are buried to less than 1,500–1,600 m, fossils of opal A such as diatoms are exceptionally well preserved (Fig. 6.4). The silica is still opal A. When altered to opal CT most of the primary structures are gone.

When opal A and opal CT dissolve, the water becomes oversaturated with respect to quartz which then begins to crystallise at a large number of points (nucleii). It thus forms microcrystalline quartz, known as "chert". The dark colouration is due to organic material. In Europe, flint is the best known variety of chert which is abundant as concretions in the Upper

Fig. 6.4 Diatoms and radiolaria from Oligocene siliceous sediments from the North Sea basin (1,430 m depth). From Thyberg et al. (1999)

Cretaceous Chalk. They are formed of silica which has been finely distributed in the sediment, largely as sponges and radiolaria often concentrated along special horizons. Because small particles with a large specific surface are highly unstable, they will dissolve and silica will be precipitated in nodules which are massive structures with a small specific surface and in consequence thermodynamically more stable.

Novaculite is more or less synonymous with light, laminated chert which contains less organic material. Porcellanite is a term applied to more contaminated silica (chert) which has the appearance of unglazed porcelain. Laminated chert is very common in Palaeozoic and Mesozoic sequences immediately above oceanic seafloor (ophiolites) near spreading ridges, sourced from water circulating through the basalts. This chert may be white, grey or dark, depending on the content of organic matter or traces of iron, magnesium etc.

Chert containing oxidised iron has a reddish colouration and is called jasper. Jasper may form on the seafloor near the spreading ridges. When water circulates through the spreading ridges it carries away iron which then oxidises in the seawater and produces the red colour. When mixed with silica which is precipitated mostly by organisms, chert is formed. Jasper is found on top of obducted ocean floor basalts in nappes of Ordovician age in western Norway.

Chert has also been found during ocean drilling, overlying present ocean floor of Cretaceous or Tertiary age. Although all amorphous silica has been converted to quartz, it is often possible in chert to detect the remains of organic particles, e.g. radiolaria, especially in Palaeozoic chert deposits. Younger silica deposits are chiefly precipitated by diatoms. In areas with upwelling, silica may also be deposited in relatively shallow water. The Monterey Formation in California is a well-known example of chert formed in an upwelling zone. Around the Pacific Ocean similar chert of Upper Miocene age is found. The Monterey Formation is both an important source and reservoir rock for oil in California, especially in the Ventura Basin. It is a source rock on account of the organic matter produced by the diatoms. After the transformation to opal C and quartz, the rock fractured due to tectonic folding, so that it became a fractured reservoir (Fig. 6.5).

Subjected to 1–2 km of overburden, amorphous silica is converted to quartz and the rock becomes brittle and fractures during folding and faulting to form good fractured reservoirs.

The Monterey Formation contains organic-rich shales and phosphate deposits. This is a very common association when sediments have been deposited in upwelling areas with a limited clastic supply. Transformation of amorphous silica to quartz appears to depend on the chemical environment (temperature, pH, ionic strength, Mg concentration). Clay minerals reduce the conversion rate.

In soils, however, quartz may precipitate at low temperature near the surface, forming *silcrete* resulting in very hard ground.

Precambrian chert deposits typically occur in conjunction with banded iron formations (BIFs). It is uncertain whether there were organisms which could precipitate silica in Precambrian times, or whether Precambrian chert was chemically deposited. There are no definite indications of biological precipitation of Precambrian chert, and since silica must also have been added to the oceans, we must assume that the ocean was saturated with respect to silica, and that there may have been inorganic precipitation of silica.

Silica deposits may also be formed in lakes by freshwater diatoms. In lakes along the rift systems of East Africa, there are thick deposits of diatomite which are exploited for use in insulating materials. A great deal of CO_2 is taken from the lake water by diatoms and other algae for photosynthesis, and the water therefore becomes strongly basic (pH 9–10). This increases the solubility of silica, thus increasing the corrosion of silicate minerals. Examples of this are found, for instance, at Lake Turkana in Kenya. Volcanic rocks and water from hot springs are also important sources of silica.

Chemical precipitation of silica may also take place in evaporite basins, and in ephemeral lakes which dry up between rainy seasons.

Oil seep in chert. Santa Maria, north of Santa Barbara

Fig. 6.5 Monterey chert with oil seeps, California

6.4 Evaporites

Evaporites consist of minerals which have crystallised out through evaporation of water. This can happen in many ways:

1. Evaporation of seawater in completely or partly cut-off marine basins.
2. In lakes which have little or no outlet and a high evaporation rate.
3. Through evaporation of seasonal precipitation which collects in topographical depressions without outlets (playas).

4. In soil profiles or sandy sediments, through evaporation of groundwater.
5. In arctic areas, the sublimation of ice and the freezing of seawater to ice both increase the salt concentration of sea water, and evaporite minerals such as gypsum may be precipitated.
6. Through solution and precipitation of salts from older evaporite deposits.

The formation of evaporites is therefore not an unambiguous indication of a high temperature, but most evaporite deposits (salt) form in the climatically dry belts about 20–30° from the equator. Evaporites contain a number of salt minerals which are too soluble to be precipitated in normal marine or continental environments. The most important are:

Chlorides
Sulphates
Alkaline carbonates
Ca-Mg carbonates
Borates
Nitrates
Silica deposits
Iron deposits

6.4.1 Marine Evaporite Environments

Although the salt content of seawater varies somewhat in the different parts of the world's oceans, the composition of seawater is relatively constant. The table below shows the percentage composition of dissolved salts in seawater which add up to a salinity of 35‰. To the right are the percentages of the various salts obtained through evaporation.

Percentage by weight of salts in seawater

Ion	Percentage in seawater	Salt	Percentage by weight of common salts after evaporation
Na	30.64	NaCl	77.76
Mg	3.76	$MgCl_2$	10.86
Ca	1.20	$MgSO_4$	4.74
K	1.09	$CaSO_4$	3.60
Cl	55.21	K_2SO_4	2.47
SO	7.70	$MgBr_2$	0.22
CO	0.21	$CaCO_3$	0.35
Br	0.19		
	100.00		100.00

By evaporating seawater one can therefore determine the relative amounts of different salts. An evaporite basin will often have some supply of seawater, such that the evaporation is not total. As a result, evaporite deposits may accumulate carbonates and sulphates, while the chlorides remain in solution.

When seawater evaporates, carbonates are among the first salts to precipitate, but the amount of carbonate in solution is very small. When the volume of seawater is reduced to 1/3–1/5, both $CaCO_3$ (aragonite) and $CaSO_4 \cdot 2H_2O$ (gypsum) have precipitated. Only when the volume is down to 1/10 will NaCl (halite), quantitatively the main constituent, be precipitated (Fig. 6.6). $MgSO_4$ and $MgCl_2$ will be precipitated at the same time. Polyhalite $\left(Ca_2K_2Mg(SO_4)_5 \cdot 2H_2O\right)$ commonly

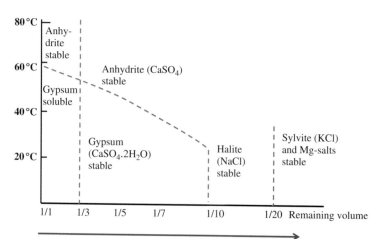

Fig. 6.6 Stability of salt minerals as a function of progressive evaporation of seawater and temperature

Fig. 6.7 Layers of gypsum in sabkha

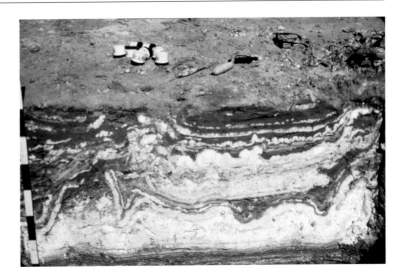

precipitates when the seawater has been reduced to 1/20. KCl (sylvite) and bromides are among the most soluble salts, and are hence the last to precipitate. If the salt concentration increases with evaporation, this is the sequence of salt deposition. During periods with greater circulation and supply of normal seawater the salt concentration will fall, and there will be cyclic salt deposition and solution. Chlorides (NaCl, KCl) will dissolve in moisture from the air unless the humidity is very low.

Periods of increased evaporation and hence high salt concentration may alternate with influxes of seawater with a more normal salt concentration, producing cycles representing changes in salinity. These are called evaporite cycles. In many evaporite basins the salt concentration has been sufficiently high for gypsum to be precipitated, but not high enough for chlorides (see Fig. 6.7).

The stability of the various salts during evaporation can be determined experimentally, or estimated through physical chemistry calculations. However, it is clear that certain metastable mineral phases can also be formed. Calcium sulphate may precipitate both as a hydrated mineral, $CaSO_4 \cdot 2H_2O$ (gypsum), and as a non-hydrated mineral $CaSO_4$ (anhydrite) (see Fig. 6.8). Which of these two phases forms as a result of oversaturation of calcium sulphate depends on the temperature, salinity and water vapour pressure. In a solution of $CaSO_4$ alone, anhydrite forms only at temperatures of over 60°C, but as the concentration of other salts increases, anhydrite may form at temperatures down to 25–30°C (Fig. 6.6).

Fig. 6.8 Gypsum crystals growing on the sabkha surface

Gypsum is the mineral which normally forms in marine evaporites, but anhydrite is also observed in modern evaporites in supratidal zones. This is true, for example, of the sabkha deposits in the Persian Gulf. These are black algal muds in the supratidal zone, where temperatures may be up to 80°C, and anhydrite is deposited in the sediment. In patches of open water where the temperature is lower, gypsum forms instead.

Evaporite sediments have formed throughout geological history, but appear to have formed more abundantly during certain periods, particularly the Permian. In Northern Europe and the North Sea we find thick evaporites from this period in what is called the Zechstein Sea, which serve as a cap rock for the gas in the underlying Permian aeolian sandstones.

The lighter salt layers will flow upward and gradually form large mushroom-shaped or columnar structures. The prerequisite for initiating this salt dome formation, however, is that the salt beds must be at least 100–200 m thick. Salt domes are common in Northern Germany, Denmark and the southern part of the North Sea, where they may form oil or gas traps. During the Permian, NW Europe was in the dry belt 20–30°N, similar to the Sahara today.

Surveys of the bottom of the Mediterranean Sea have revealed the existence of considerable evaporite deposits in Upper Miocene sequences. Adjacent land areas also have numerous localities where salt beds of this age (Messinian salt) have been preserved. This can indicate that the Straits of Gibraltar were closed for large parts of the Tertiary and that the Mediterranean Sea was a vast desert inland sea 2,000–3,000 m below sea level at that time.

6.5 The Sabkha Model

"Sabkha" is the Arabic term for the large flat areas around the Persian Gulf. This area extends from the tidal zone, forming an 8–10 km wide belt with a very low seaward gradient of only c.0.4 m/km. The sabkha flats were created by carbonate sediments infilling lagoons, chiefly during the Holocene. For the last 4,000 years there has been a gradual regression in the area. The climate is very dry (30–100 mm annual precipitation), but this is nevertheless sufficient to cause groundwater from surrounding areas to flow out to the sea. The groundwater flow is extremely slow, only a few centimetres a year, but is nevertheless responsible for the diagenetic environment containing porewater of continental origin. During storms and high tides, marine water floods the sabkha. Some runs off, some evaporates and some filters down into the sediment and mixes with the groundwater. Relatively large amounts of gypsum are then formed. The remainder of the water is highly enriched in magnesium, and dolomite is formed. Evaporation will increase the salt concentration of the porewater, and if it is intense enough, chlorides (halite) are precipitated in addition to gypsum. However, halites will dissolve easily the next time salt water flows in over the sabkha. Moisture in the air

will also help to dissolve halite on the surface, so that it is not preserved in the bedding series.

The sediments which are deposited alternate between gypsum-carbonate and some clastics. A conspicuous lamination develops because of algal growth (cyanobacteria), which does not get destroyed because there is no bioturbation in this salt environment.

Although it is hot, up to 50–80°C on the sabkha, evaporation is limited where there is no open water because of the low permeability within the sediments. When the water table sinks, the evaporation rate rapidly declines. Thick layers of sediments with gypsum make up much of the Sabkha sediments (Fig. 6.8). Whereas the rate of evaporation from open water in the Persian Gulf is about 124 cm/year, it is only about 6 cm/year in the sabkha. We therefore have only moderate "evaporite pumping" where chlorides are precipitated above the water table and sulphates below the water table. The chlorides deposited will, however, redissolve as the overburden increases and the relative position of the water table rises. In some semi-arid regions, e.g. the Coorong region of Australia, dolomite lakes form which evaporate to dryness in the summer (ephemeral lakes). Evaporite minerals deposited in summer will redissolve during the winter rains, and will not be preserved in the bedding series.

6.6 Marine Evaporites

Evaporation from the surface of the sea will cause the salinity of the water in a basin to increase. If there is little wave or current action, the warm surface water will not mix quickly with the underlying, colder water. If the salt concentration increases, the density will increase and the surface water will sink to greater depths and mix with the water there, despite being warmer. To create a high salinity basin we must have physical barriers reducing or totally blocking the connection to the open ocean, and the evaporation must be greater than the total amount of freshwater added to the basin by precipitation, rivers and groundwater. If the evaporation is lower than the supply of freshwater it will develop into a freshwater basin.

An evaporite basin with a limited connection to the open sea may not dry out; it can then accumulate large amounts of gypsum, but not more soluble salt like halite (NaCl). The Mediterranean has enough seawater exchange through the Straits of Gibraltar to prevent the formation of evaporites in the Mediterranean today.

6.7 Tectonic Control of the Formation of Evaporite Basins

Partial or complete severing of marine basins occurs as a result of tectonic uplift of barriers. Rifting and incipient spreading of the ocean floor provide ideal conditions for the formation of evaporite basins, as marine basins may be partly or entirely cut off by horsts or by volcanoes and lavas. The Red Sea used to be an active evaporite basin but gradually the connection with the Indian Ocean has become too large.

Rifting, and spreading of the ocean floor in connection with the formation of the Atlantic Ocean, led to the accumulation of thick evaporite series. In Jurassic and Lower Cretaceous times early seafloor spreading resulted in the formation of a series of evaporite basins in the area between Africa and South America south of the equator, and in the Gulf of Mexico and North Africa north of the palaeo-equator. Where they were sufficiently thick, they formed diapirs which greatly influenced further sedimentation and the structural development of these parts of the continental shelf. As seafloor spreading continued, such that the Atlantic Ocean widened and the ocean floor basalts cooled and subsided more rapidly, the opportunities for forming closed basins diminished. After the mid-Cretaceous no major evaporite basins formed in this region.

6.8 Evaporites in Lakes and Inland Seas

Basins with no outlet are formed particularly in tectonically active areas. In East Africa we find numerous freshwater evaporites in the rift basins where the climate is sufficiently arid, and at the end of the Cainozoic during the Mio-Pliocene there were many landlocked lakes and inland seas in connection with rift valleys. Since the chemistry of river water is quite different from that of seawater, such "continental" evaporites are quite different from marine series. The composition

will vary according to the types of rocks and the weathering in the drainage area around the lake. Lake evaporites normally contain large amounts of carbonate, particularly sodium carbonate and a number of other salts, hence the name "soda lakes". The mineralogical composition of these soda deposits is very complex. Two important minerals are trona $Na_2CO_3 \cdot NaHCO_3 \cdot 2H_2O$ and gaylussite, $CaCO_3 \cdot Na_2CO_3 \cdot 5H_2O$. If there is volcanism in the area, as is often the case around rift valleys, this will modify the composition of the water, both by weathering of ash and lava and by the direct addition of volcanic water (springs) rich in dissolved salts.

In playa lakes there may be total evaporation of seasonal rainfall resulting in thin layers of carbonates, including dolomite and sulphates. Signs of dessication and wind reworking will be common. Salt and clay particles in the dried-out lakes will tend to be transported by the wind, to form small dunes. Water from occasional rains in deserts will often collect between large aeolian dunes and form interdune lakes and sabkhas.

6.9 Evaporation of Groundwater

Where groundwater evaporation exceeds rainfall there is net upward transport to the surface of the soil and salts have to precipitate out in the soil profile. This can happen in the vadose zone near the sea, or inland where evaporation is high and the groundwater has flowed from another area. Gypsum and chlorides exert great crystallisation power, and can push aside the sediment matrix so that large euhedral crystals form. In brown soil types such as prairie soils, where the rainfall is high enough to prevent the accumulation of more soluble salts, a layer of carbonate (caliche) is commonly formed and in drier areas also gypsum. When such soils are eroded the broken up caliche will form pebbly conglomerates.

Evaporites from older geological periods are easily dissolved by groundwater or surface water. The Dead Sea is an example of an evaporite basin where water flowing into the basin is already rich in dissolved salts. This is because the Jordan river runs through evaporite sediments of Cretaceous and Tertiary age. In addition, water of volcanic origin enters the rift valley through faults and fractures.

6.10 The Stability of Gypsum and Anhydrite During Burial Diagenesis

Higher pressure and temperature will favour the stability of anhydrite, which has a more compact structure than gypsum. Gypsum formed in evaporite environments will therefore turn into anhydrite when there is sufficient overlying sediment, usually at depths ranging from just under 1,000 to 3,000 m. When anhydrite-bearing sediments are uplifted and come into contact with groundwater due to erosion of overlying sediments they will gradually, depending on the water circulation, hydrate to gypsum which is then the stable phase. The loss of water resulting from the transition from gypsum to anhydrite leads to a volume reduction (compaction) of 38%, and the transition from anhydrite to gypsum leads to a corresponding increase (expansion) of 60%. This compaction and expansion creates serious geotechnical problems in those areas of Europe, e.g. Switzerland and Germany, where evaporite deposits are common.

6.11 Iron- and Manganese-Rich Sediments

Iron and manganese share many similarities in their geochemical behaviour in sedimentary environments. Both elements are poorly soluble in the oxidised state because they form hydroxides and oxides: $Fe(OH)_3$, $Mn(OH)_4$, Fe_2O_3 and MnO_2. In the reduced state they occur as Fe^{2+} and Mn^{2+} and are then much more soluble. Both iron and manganese can be precipitated as carbonate ($FeCO_3$, $MnCO_3$), either as separate minerals, or as part of calcite, dolomite or ankerite. They are therefore not very soluble in basic solutions with low redox potentials, but are quite soluble at low pH values and in the reduced state. Consequently we can precipitate iron and manganese in two ways: (1) through oxidation, (2) by changing the solution from acid to basic.

Fe^{2+} forms iron sulphides with very low solubility while manganese sulphides are far more soluble and therefore more soluble in porewater which is reducing and acid. Based on these geochemical considerations, we can predict a great deal about the deposition of iron and manganese in sediments. As silicates and other minerals dissolve in the oxidising environment during weathering, the concentration of insoluble iron oxides

($Fe(OH)_3$) and manganese ($Mn(OH)_4$) will increase, along with aluminium ($Al(OH)_3$), and constitute a major part of laterite.

Bogs, such as we find in the Scandinavian highlands, have conditions ideal for such precipitating iron ("bog iron"). Water seeping through bogs will have low Eh and pH, and iron-containing minerals from underlying rocks will be dissolved by humic acids, transported in the reduced state and precipitated through oxidation where the water flows out of the bog.

Porewater in sediments and sedimentary rocks is normally reducing. This is because most sediments contain reducing agents, particularly organic material. Only the upper 0–20 cm below the sea (lake) floor are oxidised. The depth to which the groundwater will be oxidising depends on the supply of oxygen, i.e. the oxygen content and rate of flow of groundwater compared to the consumption of oxygen due to oxidation. Desert sediments contain little organic material which can function as a reducing agent, and the groundwater will therefore remain oxidising longer. This can explain the red colouration that is trivalent iron.

The redox boundary, which normally lies just below the sediment/water boundary on the ocean bottom, represents an important geochemical trap. The concentration of elements on each side of this boundary is very different because of the different solubility of elements in the two separate chemical environments. A high concentration gradient of Fe^{2+} and Mn^{2+} on the reducing side of the redox boundary leads to a diffusion and precipitation of oxidised iron and manganese immediately above the boundary. At a depth where there is little oxygen in the porewater, sulphate-reducing bacteria consume the oxygen in the sulphate ions to form H_2S. The sulphate concentration in the porewater declines rapidly downwards causing a downwards diffusion. H_2S is an acid which can release iron bound in various clastic minerals to form FeS (machinawite), which is black and easily oxidised, and then pyrite (FeS_2) which is a little more stable. If the sediments have a high silica content, chamosite may also be formed. Glauconite must be formed right at the redox boundary, since it contains both bivalent (Fe^{2+}) and trivalent (Fe^{3+}) iron. Uranium and vanadium, which have low solubility in the reduced state, will be able to diffuse downwards and be precipitated in the reducing zone below the redox boundary.

Manganese will not be as easily trapped in the sulphate-reducing zone, as Mn^{2+} does not form such stable sulphides as Fe^{2+}, and it will therefore have a greater tendency to be precipitated in the oxidised zone.

During breaks in sedimentation, or slow sedimentation, porewater expelled by compaction will cross the redox boundary and this may cause the precipitation of iron and manganese on the seafloor. *Manganese nodules* are concretions of manganese and iron hydroxides and oxides found on the seafloor, particularly in the ocean basins (South Atlantic and Pacific). They also contain relatively high concentrations of metals like Ni, Cu, Zn and Co. The nodules grow by very slow concentric accretion in a pelagic ooze. Plankton is capable of accumulating very high concentrations of metals from seawater and when the organisms dissolve, the planktonic ooze becomes very rich in these metals, which will be precipitated together with the manganese hydroxides in the concretions.

Iron and manganese are virtually insoluble in oxygenated sea water, and consequently cannot be transported in ordinary solution. However, rivers can carry a good deal of iron and manganese adsorbed onto organic particles or as clay-sized iron oxides which in many cases may produce a red colour.

Iron can be precipitated not only by oxidation, but also in the reduced state in a basic environment. Thin carbonate laminations in shales represent a local high-pH environment, where reduced iron which is normally soluble can be precipitated:

$$Fe^{2+} + CaCO_3 = FeCO_3 + Ca^{2+}$$

This reaction occurs at relatively low Fe^{2+}/Ca^{2+} ratios because iron carbonate (siderite) is less soluble than calcite. If there is sulphur present, iron will first form the sulphide, however, and iron carbonate will not be stable. Siderite is therefore most typically formed in freshwater basins, and not directly in the sulphate-reducing zone in marine sediments. Siderite can also be formed below the sulphate-reducing zone, for example by reaction between aragonite and iron in the sediment.

6.12 Phosphorite (or "Phosphatic Deposits")

Phosphorites are sedimentary layers in which phosphate minerals are the main components, mainly

apatite $Ca_5(PO_4)_3(F,Cl,OH)$. These crystals may contain fluor (F^-), chlorine (Cl^-) or hydroxyl (OH^-). Fluorapatite $Ca_5(PO_4)_3F$ is the most important in marine sediments. We also have francolite $(Ca,Na)_5(F,OH)(PO_4,CO_3)_3$, where some carbonate has substituted for phosphate and where sodium substitutes for calcium. Iron apatites such as strengite ($FePO_4 \cdot 2H_2O$) are also relatively common minerals in secondary (weathered) phosphatic deposits.

Phosphorite rocks are an important source of fertilizer. Large deposits such as those found in Morocco, Spanish Sahara and Senegal in West Africa, and in Florida, are of great commercial value. Understanding how phosphate deposits form is therefore a matter of considerable economic interest. In most sedimentary rocks phosphorus is a trace element, and very special conditions are necessary for phosphate enrichment to take place. These deposits therefore tell us something important about the environment of deposition.

Guano deposits are formed on land from bird or bat excrement in which the phosphate gradually becomes concentrated as the other organic components are leached out. For the phosphates to be preserved, the rainfall must not be too high, because it will dissolve phosphate sediments on land. We also have freshwater phosphatic deposits, but it is the marine ones that are the most important.

The first prerequisite for marine phosphate deposits is that sedimentation must proceed very slowly, i.e. there must be virtually no clastic sedimentation. In consequence we find phosphate beds associated with major or minor breaks in sedimentation, or with periods of very slow sedimentation. It has long been known that phosphate is formed in areas with strong upwelling and high organic productivity. We have good examples of this along the edge of the continental slope off Chile and Peru. Water welling up from great depths brings with it nutrients which are liberated when marine organisms disintegrate in deeper water. When the water flows up to the surface the nutrients are consumed by organisms, mostly plant plankton, which provide a high primary production. These organisms contain about 1% P (dry weight), which is an enrichment of 140,000 compared to dissolved phosphorus in ordinary seawater.

Organisms with an amorphous silica skeleton (e.g. diatoms and radiolaria) will readily dissolve in seawater, and in some cases carbonate (calcite and particularly aragonite) will also dissolve, resulting in further phosphate enrichment. Phosphate minerals then crystallise out of the phosphate-rich sediments and often replace other minerals, e.g. carbonate. Apatite is a heavy mineral (specific weight 3.18) and may also be enriched mechanically by weak traction currents.

Apatite will often crystallise out as concretions in bottom ooze, and erosion by traction currents may concentrate these nodules into a conglomerate. On the continental shelf, phosphate forms at depths of between 100 and 400 m, i.e. below the photic zone. However, phosphate nodules and massive beds of fine-grained phosphate mud (phosphate micrite) may also form in lagoons where the water is less clear. These phosphate deposits will be easily eroded even as a result of minor regressions, and form conglomerates of phosphate mudstone (micrite). Today phosphate deposits are forming only in a few areas with strong upwelling, but in previous geological periods we find very extensive phosphate beds, often associated with transgressions. The transgressions will hold clastic sediment back for a while, enabling biogenic matter to be concentrated. Phosphate beds are often associated with other authigenic (formed *in situ*) minerals which take a long time to form, particularly glauconite and also manganese deposits.

In Florida there are large phosphate deposits of mid-Tertiary age, and the same beds are exposed on the floor of the continental shelf off South Carolina (Blake Plateau), where the manganese-rich phosphatic deposits total 10^9 tonnes.

Marine phosphate deposits may also be formed by fossils with a phosphatic skeleton, for example fish. These deposits may also be extensive, and are often called "bone beds".

Phosphate minerals such as apatite may contain considerable amounts of uranium and rare-earth metals which substitute for calcium. Weathering of phosphate deposits will lead to oxidation of uranium to UO_2^{2+}, which is soluble in the form of uranyl ions, and uranium may be precipitated again in the reduced state (UO_2) when it comes into contact with organic matter.

Weathering of phosphate sediments will initially cause solution of carbonate and carbonate-containing apatite so that the sediments become enriched in fluorapatite. As calcium is removed through solution by rainwater, the sediment will become richer in iron and aluminium, and iron and aluminium phosphates will form. Phosphate deposits are also formed through weathering of phosphate-bearing igneous rocks. Carbonatites from the East African rift system contain a high percentage of apatite, and when carbonate is dissolved during weathering, apatite becomes concentrated and forms phosphate deposits which are also rich in iron minerals.

Further Reading

Aplin, A.C., Fleet, A.J. and Macquaker, J.H.S. 1999. Muds and mudstones: Physical and fluid-flow properties. Geological Society Special Publication 158, 190 pp.

Chamley, H. 1989. Clay Sedimentology. Springer, New York, 623 pp.

Schieber, J., Zimmerle, W. and Sethi, P. 1998. Shales and Mudstones. Schweizerbartsche Verlagsbuchhandlung, Stuttgart, 384 pp.

Thyberg, B.I., Stabell, B., Faleide, J.I. and Bjørlykke, K. 1999. Upper Oligocene diatomaceous deposits in the northern North Sea – Silica diagenesis and paleogeographic implications. Norsk. Geol. Tidskrift 79, 3–18.

Weaver, C.E. 1989. Clays, mudrocks and shales. Developments in Sedimentology 44, 818 pp.

Chapter 7

Stratigraphy

Jenø Nagy and Knut Bjørlykke

7.1 Introduction and Concepts

Stratigraphy is the study of the succession and time-related architecture of rock strata. Stratification is not limited to sedimentary rocks, but is also found in igneous rocks, particularly volcanic rocks, and in certain plutonic rocks. All bedded rocks can be treated stratigraphically, i.e. to establish age relations between beds. However, the term "stratigraphy" is used primarily in the context of sedimentary research.

Stratigraphy involves the study, subdivision and documentation of sedimentary successions, and on this basis to interpret the geological history they represent. In order to reconstruct an environment or special events in the geological past, regionally or even globally, it is necessary to correlate sedimentary horizons from different areas. It is important to establish which sedimentary units were deposited at the same time or by the same or similar sedimentological or biological processes, even if they are not quite contemporary. Correlation is generally a question of what is possible to achieve with the available amount and quality of data.

We usually have no possibility of determining definitely which beds were deposited at the same time, but attempt to use all the information available in the rocks. This information falls into five main categories: (1) Rock composition and structures resulting from sedimentological processes. (2) Fossil content, which is a result of biological, environmental and ecological evolution throughout geological history. (3) Content of radioactive fission products in minerals or rocks which may be used for age dating. (4) Magnetic properties of rock strata. (5) Geochemical features of sediments. These correlation methods are so different that it has been found useful to work with three forms of stratigraphy which can be used in parallel:

1. *Lithostratigraphy:* Classification of sedimentary rock types on the basis of their composition, appearance and sedimentary structures.
2. *Biostratigraphy:* Classification of sedimentary rocks according to their fossil content.
3. *Chronostratigraphy:* Classification of rocks on the basis of geological time.

Geochronology is the actual subdivision of geological time.

The first two are thus based on rock relationships which can be described, and are often collectively referred to as rock stratigraphy.

Despite radiometric dating, geological time is not absolute. Even this dating method gives different ages, depending on which half-life is used for calculating radioactive decay, and is encumbered with many other uncertainty factors. Chronostratigraphy is therefore a theoretical and abstract concept which describes a time scale we cannot measure exactly. Rock stratigraphy, on the other hand, takes the rocks themselves as its starting point and is based on boundaries which are identified as potentially suitable for correlation across greater or lesser areas. For rules for stratigraphic nomenclature, see stratigraphy.org. The different types of stratigraphic units are listed below in hierarchic arrangement. See also Stratigraphy.com.

J. Nagy (✉)
Department of Geosciences, University of Oslo, Oslo, Norway
e-mail: jeno.nagy@geo.uio.no

K. Bjørlykke (ed.), *Petroleum Geoscience: From Sedimentary Environments to Rock Physics*,
DOI 10.1007/978-3-642-02332-3_7, © Springer-Verlag Berlin Heidelberg 2010

Lithostratigraphic	Biostratigraphic	Chronostratigraphic	Geochronological
Supergroup	Assemblage zone	Eonothem	Eon
Group	Range zone	Erathem	Era
Subgroup	Acme zone	System	Period
Formation	Interval zone	Series	Epoch
Member	Biozone	Stage	Age
Bed		Chronozone	Chron

7.2 Lithostratigraphy

The *formation* is the fundamental lithostratigraphic unit defined in a bedded succession. An important property of a formation is that it should be easily recognised in the field or a borehole due to its composition (lithology). The formation is thus a mappable unit, which can be visualised on an ordinary geological map (e.g. 1:50,000) or recognised in the description of a bedded succession. There is in principle no limit to the thickness of a formation, but it usually varies from a few tens to several hundred metres. A 50–300 m thick sandstone, over- and underlain by quite different rocks such as shale or limestone, would constitute a natural formation. However, a formation will seldom be homogeneous, and for more detailed mapping it will be useful to divide the formation into smaller units. Parts of a sandstone formation containing shale or conglomerate beds, for example, can be defined as *members* of the formation.

The smallest unit in the lithostratigraphic classification is a *bed* which is assumed to have been deposited by a single depositional process without a break in sedimentation, and is distinguishable from the beds above and below. When sections are logged, beds are often recorded as "units". Beds which are <1 cm thick are called *laminae*, but this is a purely size-descriptive term. Most beds/units contain, or may even be entirely comprised of, laminae. Only when a lamina lies isolated between significantly different sediments will it be named, e.g. if it is a thin tephra (ash bed) or bentonite in a shale succession.

For some purposes it is expedient to group several formations together into a larger unit called a *group*. A group normally consists of three to six formations and can be divided into two or more *subgroups*. The largest lithostratigraphic unit is a *supergroup*, which consists of two or more groups. This unit is used when a common name is needed to encompass a thick sedimentary package.

In sedimentary basins which are partly or entirely buried in the subsurface, for example the North Sea basin, formations and groups are defined on the basis of records from wells, such as well logs. In offshore areas where it is difficult to find enough geographical names to give the stratigraphic units, other names are then also used, including historical names, names of animals etc.

7.2.1 Lithostratigraphic Terminology

As research into sedimentary successions expanded, it required a standardised lithostratigraphic nomenclature that could be applied anywhere. Therefore, a set of international rules was established for naming lithostratigraphic units. This is summarised below:

1. A stratigraphic unit should preferably be defined with reference to a type section (stratotype) present in a good exposure, or in a well where the unit is adequately represented.
2. Each stratigraphic unit ought to be named after a geographical site, located preferably near the type section if exposed on land. Offshore stratigraphic units may be named after marine features (e.g. fishing banks), although in the North Sea basin other names have also been used.
3. The same name ought not to be used for more than one stratigraphic unit. The unit which is first defined has priority.
4. Stratigraphic names may consist of a name, i.e. from a locality, and a stratigraphic unit, e.g. Kimmeridge Formation or Kimmeridge Clay. In the US many stratigraphic names are well established even if they do not strictly follow the international rules for stratigraphic nomenclature.

7.3 Biostratigraphy

Biostratigraphy is based on the fossils in sedimentary rocks. Its tasks are to group strata into units based on fossil content, and apply these to correlate sedimentary successions. Biostratigraphy has made it possible to correlate sedimentary rocks on a global basis, and forms the foundation for global stratigraphic classification of sedimentary successions in combination with radiometric and other age dating methods. The biostratigraphic application of fossils is based on the fact that during the course of geological time, biological evolution took place whereby some species died out and new ones appeared. Any particular species will therefore be represented only in sediments deposited within a limited time span, and can be used to recognise this time span or part of it.

7.3.1 Nature of the Fossil Record

7.3.1.1 Fields of Application

Industrial application of biostratigraphy is mainly based on microfossils, owing to the fact that these microscopic remains commonly occur in large amounts, and can be extracted easily both from drill cuttings, sidewall cores and conventional cores. The rapid evolution of many microfossil groups makes them valuable tools in subsurface stratigraphic work (Fig. 7.1). In addition to the general tasks of age determination and correlation, their applications include: unconformity identification, characterisation and correlation of seismic (depositional) sequences, fingerprinting of formations, reservoir zonation and palaeoenvironmental modelling. These applications are crucial in petroleum exploration, contributing significantly to assessments of reservoir distribution, source rock evaluation, trap evaluation, estimation of reserves, field development studies and impacts on drilling problems. The fast evaluation potential of microfossil samples is also an important factor in any subsurface work.

7.3.1.2 Distribution of Microfossils

Practically all rocks of sedimentary origin contain microfossils, although their abundance, diversity and state of preservation are highly variable. These features are strongly influenced by the age, depositional

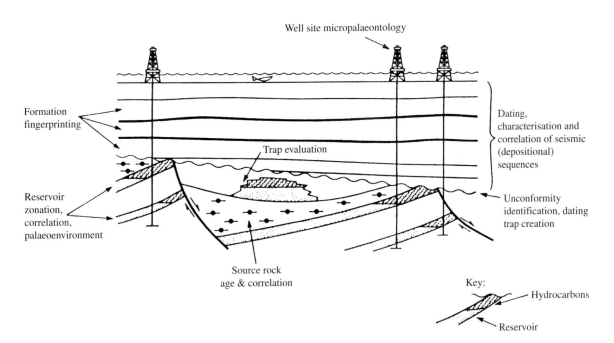

Fig. 7.1 Application fields of micropalaeontological methods in petroleum exploration and appraisal activities (from Copestake 1993)

Rocks \ Microfossils	Spores, pollen	Dinoflagellates	Acritarchs	Foraminifera	Conodonts	Ostracods	Prasynophyceans	Calpionellids	Chitinozoans	Botryococcaceans	Coccolithophores	Radiolarians	Silicoflagellates	Diatoms
Clays and shales	●	●	●	●	◉	●	◉	◉	●	⊗	●	●	⊗	⊗
Limestones and marls	◉	◉	◉	●	●	●	⊗	●	◉		●	⊗	⊗	
Flints and cherts	⊗	◉	⊗	◉	⊗	⊗					⊗	●	◉	◉
Coal, lignite, peat	●	◉	◉				○			◉				
Sands and sandstones	◉	⊗	⊗	◉	⊗	○	○		⊗					
Dolomites, ankerites	⊗	⊗	⊗	◉	◉	○	○	○						
Evaporites: gypsum, halite	⊗	○	○											
Metamorphic rocks: slates, phyllites, marbles	○	○	○	○	○									

Legend: ● Abundant, ◉ Common, ⊗ Rare, ○ Sporadic

Fig. 7.2 Distribution trends and importance of major microfossil groups in different types of rocks

environment, composition and diagenetic history of the sediments. Marine mudstones, marls and limestones usually have a particularly rich microfossil content of high diversity (Fig. 7.2). Well sorted sandstones have a generally low microfossil content, as most microfossils are both lighter and smaller than the average sand grains. Dolomites are also poor in microfossils because most of the fossil organisms with primary calcite have been dissolved during dolomitisation, or because the beds represent an evaporite environment. Coals and related organic sediments are rich in terrestrial microfossils (spores and pollen). Evaporites are characterised by small quantities of microfossils, mainly of terrestrial origin.

Microfossils occur in Precambrian rocks but rapid development of many central groups started in the Cambrian (Fig. 7.3). Every part of the stratigraphic column from Cambrian to recent time contains one or several microfossil groups which are potentially useful for biostratigraphical or palaeoecological purposes. According to skeletal composition, microfossils can be arranged into five groups, which comprise several taxonomic divisions:

(1) Calcareous microfossils: calcitic and aragonitic foraminifera, single-celled heterotrophs; ostracods, microscopic crustaceans; calpionellids,

uncertain origin; carophytes, algal reproductive organs; calcareous nannoplankton, single-celled algae.

(2) Arenaceous microfossils: agglutinated foraminifera, single-celled heterotrophs.

(3) Siliceous microfossils: radiolarians, single-celled heterotrophs; diatoms, single-celled algae; silicoflagellates, single-celled algae; ebridians, single-celled heterotrophs.

(4) Phosphatic microfossils: conodonts, primitive vertebrates.

(5) Organic-walled microfossils: acritarchs, uncertain origin; prasinophytes, cysts of green algae; chitinozoa, uncertain origin; chlorophytes, freshwater algae, algal colonies; dinoflagellates, single-celled algae.

7.3.2 Factors Controlling Stratigraphic Application

There are several factors influencing the applicability of microfossils to stratigraphic purposes. The most important ones having a positive effect on the application potential are: (1) Presence of hard parts with high probability of preservation. (2) High evolution rate of taxa. (3) Extensive regional distribution, or

Fig. 7.3 Outline of the stratigraphic distribution of microfossil groups from Precambrian to present time, arranged according to skeletal composition (modified after Haq and Boersma 1978)

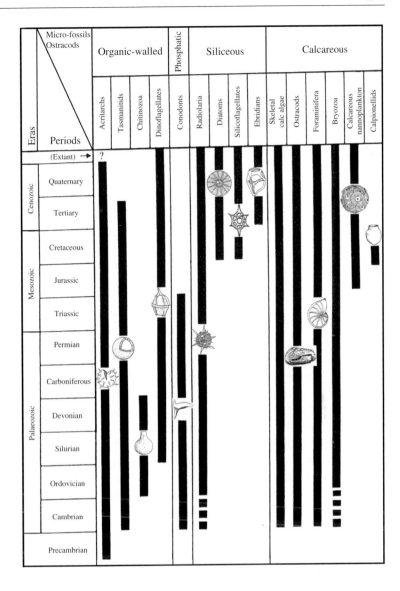

high potential for dispersal. (4) High degree of facies independence. (5) Small redepositional potential. (6) Presence of many distinct morphological features, increasing the precision level of taxon identification.

7.3.2.1 Mode of Life

The mode of life (*ethology*) of organisms is crucial for the regional distribution of their fossil remains. Living and fossil biota can be arranged into two ethological groups (Fig. 7.4). *Pelagic* organisms live freely in the water column and comprise the *planktonic* forms (passively floating) and *nektonic* forms (actively swimming). The *benthos* comprises bottom dwelling organisms belonging to *epibenthos* (living at the sediment/water interface) and *endobenthos* (living buried in the sediment). Foraminifera as a group reveal the widest environmental range by being represented in all the ethological categories except nektonic. Ostracods and diatoms also have a relatively wide range as they occur both in the benthic and pelagic habitats. Conodonts are regarded to have a nektonic origin. Acritarchs, dinoflagellates, calcareous nannoplankton, radiolarians and several minor groups are exclusively planktonic. Spores and pollen occupy a special position because they are the reproductive organs of terrestrial plants, but are dispersed principally as plankton (by air or water).

Fig. 7.4 Mode of life and habitats of major microfossil groups in aquatic environments

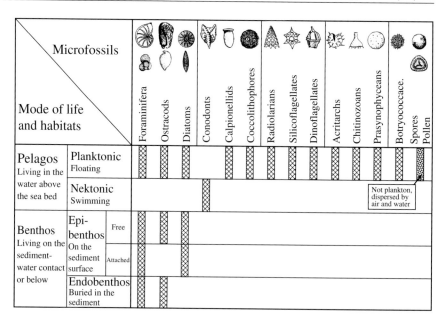

It is obvious that the pelagic group contains microfossils with particularly high stratigraphic applicability, owing to their relative independence of facies, e.g. planktonic foraminifera, calcareous nannoplankton, dinoflagellates and the nektonic conodonts. Benthic biota usually have less stratigraphic potential, although they are also commonly used for this purpose, particularly in marginal marine deposits where planktonic biota are rare. On the other hand, benthic forms are commonly excellent facies indicators, e.g. benthic foraminifera and ostracods.

7.3.2.2 Environmental Distribution

The life habitat of biota is of prime importance both for stratigraphical purposes and biofacies analysis. In non-marine subaquatic environments important fossil groups include diatoms and ostracods (Fig. 7.4). In marginal marine brackish environments low diversity assemblages of foraminifera, ostracods, and diatoms are dominant. Shallow shelf to bathyal environments are typified by their high diversity assemblages of benthic and pelagic biota. In abyssal environments, below the compensation depth of calcium carbonate, microfossils are exclusively siliceous, arenaceous or organic-walled.

The members of the ubiquitous group of organic-walled microfossils are produced in a wide range of contrasting habitats (Fig. 7.4). Pollen and spores are produced by land and freshwater flora. They are

transported far out into the marine realm but show decreasing frequency with increasing distance from source. They are commonly used in the stratigraphy of terrestrial and marginal marine deposits in the absence of other fossils, although they provide a low age resolution. Chlorophyte freshwater algae (as Bothryococcus) are also prone to be transported to marine areas. Prasionophyte marine algae and acritarchs are common in brackish marginal marine and shallow shelf settings. Dinoflagellates are most common in marine shelf and oceanic waters, and are widely used for stratigraphical purposes.

Foraminifera occur in a wide variety of environments ranging from open ocean to estuarine waters, and are extensively used in stratigraphic and facies analyses. Marginal marine environments (such as tidal marshes, estuaries and lagoons) are typified by low diversity benthic assemblages composed of agglutinated and mixed calcareous-agglutinated components. The low species diversity restricted nature of these assemblages reduces their stratigraphical applicability. Normal marine shelf areas are typified by high diversity calcareous faunas and increasing frequency of planktonic species toward the shelf break.

The planktonic component reaches its maximum in bathyal areas. Shelf and bathyal assemblages form an excellent basis for age determination and stratigraphic correlation. The amount of calcareous benthic and planktonic foraminifera decreases downslope through the lysocline and they disappear below the compensation depth of calcium carbonate. Below this level the sea bed is populated by agglutinated species, usually of a cosmopolitan nature, which provide an adequate basis for stratigraphic application.

The environmental distribution of the various microfossil groups in the North Sea basin Mesozoic and Cenozoic succession is well known owing to the prevailing oil and gas exploration (Fig. 7.5). The dominance pattern in this area reflects the general bathymetric distribution pattern of the various groups. Abyssal deposits are dominated by radiolarians and cosmopolitan agglutinated foraminifera. Bathyal and outer shelf sediments are characterised by high dominance of planktonic foraminifera and calcareous nannoplankton. Mid-shelf sediments contain abundant calcareous benthic foraminifera and dinoflagellates. Deltaic deposits are typified by spores,

pollen and agglutinated foraminifera. In non-marine deposits the most common fossils are spores and pollen.

7.3.2.3 Redeposition

Erosion of fossil-bearing sediments with subsequent transport and redeposition of microfossils can lead to serious problems for dating and correlation. Palynomorphs are particularly liable to redeposition owing to their small size and weight. Conodonts also have a high redeposition potential because their calcium phosphate skeletons are strongly resistant to corrosion by weathering and transport. Redeposition is suspected if microfossils appear at higher stratigraphic

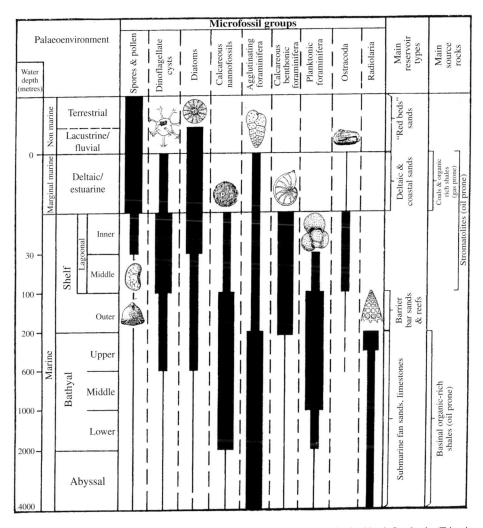

Fig. 7.5 Depth-related palaeoenvironmental distribution of major microfossil groups in the North Sea basin (Triassic to Tertiary), and position of the main reservoir and source rock facies (from Copestake 1993)

levels than expected. Reworked microfossils can be distinguished by changes in colour and traces of wear.

7.3.2.4 Facies Barriers and Biostratigraphy

Sediment cores from deep oceanic settings usually provide undisturbed sections containing a continuous fossil record, commonly of high stratigraphic resolution. Based on this type of material, biostratigraphy combined with magnetostratigraphy and chemostratigraphy contributed heavily to development of a global biochronozonal framework. This framework has been extended to formations deposited in marine shelf, deltaic and coastal areas, where it forms an important basis for dating and correlation.

The basic unit of biostratigraphic correlation is the *taxon chronozone*, also called *biochronozone*. As defined below, a taxon chronozone represents all sediments deposited between the evolutionary appearance and extinction (total range) of its marker species (Fig. 7.8). In contrast, the *biozone* represents only those sediments at a given place that contain the one or several species. In the case of a single marker, the biozone can be defined by the local appearance and disappearance (local range) of the species. The time interval represented by the biozone may differ from place to place.

In deep oceanic regimes, the latitudinal distribution of each species is generally defined by the water mass temperature and chemistry. Therefore, the regional (lateral) extent of biozones and taxon chronozones is of restricted nature. Outside its optimal stratigraphic distribution area, the range of the marker species can vary considerably owing to environmental factors. At such sites the chronozone cannot be defined and the term biozone is used. Correlation between tropical and temperate oceanic regions is difficult, because of the low number of biota that include reliable zonal indicator species common to both areas. Interfingering of biozones (local ranges) and integration with magnetostratigraphy and chemostratigraphy is a useful approach to this type of correlation.

7.3.3 Biozones

The basic unit of biostratigraphy is the biozone which is a body of strata defined by its fossil content. During

the history of biostratigraphy, a variety of names and definitions of units have been proposed, but there are only three basic types of zones: interval zones, assemblage zones and abundance zones. Zones defined in sedimentary successions are usually arranged into a zonal scheme, which can be developed for an extensive region and even for global application.

7.3.3.1 Interval Zones

These are the most commonly used type of zones, with boundaries defined by particular fossil events, specifically the first and last occurrence of taxa (usually species). Thus, the interval zone comprises the strata deposited between these two events. There are five types of interval zones (Fig. 7.6).

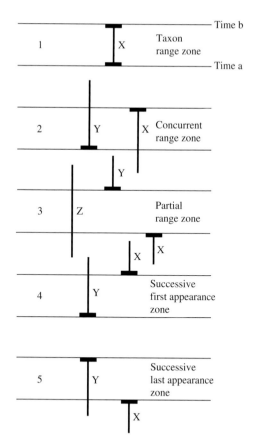

Fig. 7.6 Outline of the five types of biostratigraphic interval zones shown by occurrences of microfossils X, Y, and Z. *Vertical lines* are taxon ranges, *horizontal bars* are first and last occurrences

1. *Taxon range zone* is defined by the first and last occurrence of a particular taxon.
2. *Concurrent range zone* is the interval between the first occurrence of a taxon and the last occurrence of another taxon.
3. *Partial range zone* is the interval between the last occurrence of a taxon and the first occurrence of another taxon partitioning the range of a third taxon.
4. *Successive first appearance zone* represents the interval between the successive first appearance of two taxa.
5. *Successive last appearance zone* is the interval between the successive last appearance of two taxa.

The successive last appearance zone is the type most commonly used in industrial biostratigraphy, because the samples are mainly derived from well drilling cuttings. In such sample sets, the actual first appearances are unreliable owing to downhole (caving) contamination. Precise positions of first appearances are required in order to recognise the first four interval zone types.

7.3.3.2 Other Biozones

1. *Assemblage zone* is a biozone characterised by the association of three or more taxa (Fig. 7.7). This

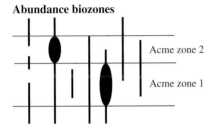

Fig. 7.7 Outline of the biostratigraphic assemblage and abundance zones

zone is recognised without regard to range limits of taxa, and is usually of regional importance.
2. *Acme zone* is based on the numerical maximum of one or several taxa, and is also termed *abundance zone*. Taxon maxima are strongly influenced by local environmental conditions, and therefore this zone is usually of local stratigraphic importance.
3. *Taxon chronozone* represents the interval between two unique bioevents: the first evolutionary appearance of a taxon, and its extinction. Thus, this zone represents the total range of the taxon, and is of global nature (Fig. 7.8). The distribution of taxa, however, is to a varying degree facies-controlled, which leads to the development of geographically restricted taxon range zones. One or several of such local taxon range zones can be included in the total range of the taxon.

7.3.4 Biostratigraphic Correlation

Biostratigraphic correlation is based on the fact that during the course of geological time, biological evolution has taken place, whereby some species and higher taxa have died out and new ones have appeared (Fig. 7.3). Any particular species will therefore be represented only in sediments deposited within a limited time span. However, this method also has many limitations. Some fossil groups developed rapidly, with individual species only existing for a short period. Other groups evolved slowly and persisted throughout long periods of geological time. Some fossil organisms are found only in rocks deposited in particular environments or facies. Many were dependant on a certain salinity, oxygen content and water depth. Bottom conditions played a major role for infaunal (burrowing) animals.

Biozones of sedimentary successions are arranged into zonal schemes, which are the main tools of fossil-based correlation. The ideal index fossil for defining a biozone ought to belong to a fossil group which underwent rapid biological evolution and had a global distribution. It should not have had special environmental requirements, and should have been cosmopolitan with respect to habitat. One such fossil group is the graptolites, which are common in Cambro-Silurian shales. They were planktonic and drifted in the surface seawater. However, even this type of plankton may show

Fig. 7.8 Facies-related distribution of an imaginary fossil taxon exemplifying the difference between a taxon chronozone (global) and a biozone (local)

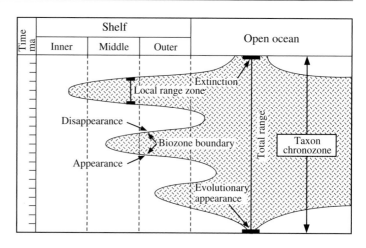

a distribution pattern which reflects ocean currents and temperatures etc. At the present microfossils are being used more and more for biostratigraphic correlation. Several animal and plant groups (e.g. conodonts, planktonic foraminifera, radiolarians and spores and pollen) have wide distributions and have the advantage that even small samples, e.g. from boreholes, provide adequate material for statistical treatment.

Although fossils are always helpful in stratigraphic correlations, we now avoid defining important geological boundaries, for example that between the Ordovician and the Silurian, by means of fossil occurrences. Such a boundary would have to be moved whenever one found a new occurrence of a certain fossil. The geological periods are now defined by international committees which select a type section with continuous sedimentation and preferably fossiliferous facies, and the boundary is physically marked as a fixed point in the section. The boundary is then unambiguously defined. All available means can then be used, including fossils, to correlate the boundary with other areas. This is the principle of the arbitrary boundary. In reality it is not entirely arbitrary. We try to put it on a section with optimal potential for correlation with other areas and the global time scale (Fig. 7.9).

Geologists used to define a stratigraphic boundary at a break (hiatus) in the sedimentary record. However, sediments later found elsewhere that had been deposited in the period represented by the hiatus, could not be assigned to either unit. For example, the boundary between the Tertiary and Cretaceous

was defined in England where a hiatus is developed between these systems. As a result it was difficult to reach agreement on whether sediments which were deposited for example in Denmark during this period (Danian), should belong to the Cretaceous or the Tertiary.

7.4 Time Stratigraphy

Chronostratigraphy is an attempt to correlate rocks deposited at the same time, across larger areas. The accuracy achievable with chronostratigraphic correlation depends on whether the sediments contain evidence of well-defined geological events which were simultaneous across the region. These events may be biological (e.g. appearance of new species), sedimentological (e.g. deposition of ash layers) or geophysical (e.g. reversals of the Earth's magnetic field). Geological research has a long tradition of correlating synchronous events in geological history. Such correlations are independent of an absolute timescale. Only after the development of radiometric dating methods did it become possible to set up a series of datings approaching an absolute timescale, but radiometric datings are not absolute with respect to time.

We thus distinguish between two types of time stratigraphy: (1) Geochronology, which is the subdivision of the Earth's history into finite time units. A geochronological unit is a specific interval of geological time. (2) Chronostratigraphy or time-rock

Fig. 7.9   Geological time scale for the Phanerozioc Aeon (from Gradstein and Ogg et al. 2009)

stratigraphy is the subdivision of sedimentary successions, and their correlation, on the basis of time. Chronostratigraphic units are by definition synchronous.

A geochronological unit represents a specific interval in the geological time scale. For example, the Jurassic *period* is a geochronological unit. A geochronological unit may define the time span between two specific geological events. We can say, for example, that some of the rocks in the North Sea were deposited during the Jurassic period. The corresponding chronostratigraphic unit (system) signifies the rocks which were formed during the same period. We therefore say that some of the rocks in the North Sea belong to the Jurassic *system*.

The basic chronostratigraphic unit is a *chronozone*. A chronozone includes all the deposits formed during a particular, relatively short, geologic time interval and which are defined by a geological phenomenon or by a particular interval of a rock succession. In most cases a chronozone is a *taxon chronozone*, which is defined as the period between the first appearance and last occurrence of a particular fossil taxon (Fig. 7.8). Whereas a biozone can only be defined where the fossil is present, a chronozone represents all the rocks that were formed during this period, regardless of whether they contain fossils. A *chron* is the interval of time during which the rock in a chronozone was formed. It is thus the geochronological equivalent of a chronozone.

A chronozone may be named after a biostratigraphic unit, e.g. a *Didymograptus extensus* chronozone, or after a lithostratigraphic unit, which can be recognised over large areas. Such characteristic strata are often called *marker beds, key beds*, or *datum beds*. Chronostratigraphic horizons are important for correlation within sedimentary basins, and form the framework for all facies reconstructions. If we have two *chronohorizons* or datum beds, we can measure the variation in thickness and composition of the sediments which were deposited during a particular time period. This may make it possible to map variations in sedimentation rate and the ratio of sandstone to shale. Maps showing the sediment thickness between two marker beds are called isopach maps. This is very useful for reconstructing sedimentary facies on the basis of borehole data, a standard method in connection with oil prospecting.

The best chronostratigraphic horizons (marker beds) are bentonite (ash) layers; on a smaller scale also coal beds, phosphate beds, thin limestone or sandstone beds, or particular fossil horizons. When analysing stratigraphic records from wells (logs) one tends to use beds which produce a distinctive log pattern. Seismic reflectors can also in certain instances be used as datum beds (horizons).

A *stage* is a chronostratigraphic unit which includes one or more chronozones, but which nevertheless covers a limited period of time, usually 3–10 million years (Fig. 7.9). This is the smallest unit in the chronostratigraphic hierarchy which is used for correlation all over the world. A stage is defined in a type section and usually designated by a geographical name near the type profile. For example, the Kimmeridge stage is well exposed on the Dorset coast at Kimmeridge. The correlation of a stage is usually based on biostratigraphy. An *age* is the period of time (geochronological unit) which corresponds to a stage.

A *series* is a chronostratigraphic unit larger than a stage. For example, the Late Jurassic is a series constituting part of the Jurassic system. The geochronological unit which corresponds to a series is an *epoch*. We can say that a certain limestone was deposited during the Late Jurassic epoch.

A geochronological *period* varies in duration from about 20–30 million years (Silurian) to about 60–70 million years (Cretaceous). The Quaternary period, however, is much shorter, only about 2.5 million years. The rocks formed during a period constitute a *system*. An *era* is comprised of two or more periods. The Palaeozoic era had a duration of about 300 million years, but the Cenozoic era did not last longer than the longest Palaeozoic periods (65 million years).

The largest units in the chronostratigraphic scale, *erathem* and *enothem*, are not used much, since it is seldom relevant to group rocks which were deposited over such long periods of time. However, when we discuss geological history in relation to the biological development on the earth, for example, it can be useful to speak about the Phanerozoic *aeon* which covers the Palaeozoic, Mesozoic and Cenozoic eras. The Precambrian is divided into two aeons: the Proterozoic (542–2,500 million years ago) and the Archean (2,500–4,000 million years ago).

7.5 The Relation Between Lithostratigraphy, Biostratigraphy and Chronostratigraphy

These three types of stratigraphy are based on different criteria, but geological experience has shown that it is useful to maintain this tripartite division. Stratigraphical boundaries defined in these three ways may sometimes nearly coincide, but usually show considerable divergences particularly over larger distances. This can be demonstrated schematically by the occurrence of a sandstone formation between two shale units, Fig. 7.10. Lithostratigraphically, this sandstone formation is unambiguously defined by the boundaries between shale and sandstone. However, if we find good index fossils, X and Y, in both the shale and the sandstone, it may turn out that the top of the sandstone in area B corresponds chronostratigraphically to the bottom of the sandstone in area A. Sedimentation of sand has thus moved from B to A during the course of a measurable period of geological time. Sandstone sedimentation along coasts and on deltas will tend to shift over long periods of time, and the sandstones deposited will therefore have an upper and a lower boundary which are not parallel to a theoretical time plane, i.e. they are time-transgressive.

In practical geological work (mapping, well and borehole studies, etc.) one is primarily interested in correlating rock types, in other words lithostratigraphy. To construct a facies map that gives an overview of the distribution of sediments at a particular time, it is necessary to establish a time correlation. Good index fossils or marker beds will provide important information about the depositional conditions, for example during regressive or transgressive phases. If we study a modern coastal area, we find different fauna in different environments. We can easily see that the distribution of fauna within a limited geological time period is not controlled by stratigraphic time, but by facies. Many animal groups provide good indices for sedimentary facies (Figs. 7.4 and 7.5).

7.5.1 Correlation

Rocks from the Phanerozoic Aeon have been classified on the basis of fossils. Committees for each period have been set up, responsible for selecting type profiles where the lower and upper boundaries are physically defined by a bolt ("golden spike"). Correlation with other areas can then be carried out with the aid of fossils or other age indicators, anywhere in the world.

Although biostratigraphic correlation is still the most important method, magnetostratigraphy and radiometric dating methods are gaining importance. Within the confines of a sedimentary basin, however, certain types of lithostratigraphic correlation can turn out to be the most accurate. The various types of well logs provide good opportunities for lithostratigraphic correlation. Seismic profiles have perhaps to an even greater degree made it possible to correlate lithostratigraphy over great distances (see next chapter).

7.6 Radiometric Dating Methods

7.6.1 Concept of Dating

Our geological time scale is founded on age determinations based on radioactive processes. There are a number of other indirect methods of measuring geological time, but only radiometric datings give satisfactory quantitative results. The measurements are based on the fact that radioactive nuclides undergo fission at a certain rate. The rate of fission can be determined with some accuracy, and there is no reason to believe that it has varied through geological time, even though this is difficult to prove. The rate at which radioactive decay proceeds is proportional

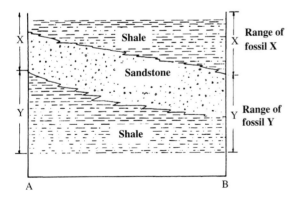

Fig. 7.10 Occurrence of a time-trasgressive sandstone formation between shale units. X and Y are good index fossils which are little affected by facies. Consequently, they indicate that the sandstone is younger in area A than in B, and has been deposited by progradation from B to A

to the assumed number of fissionable atoms present. The number of atoms remaining which can undergo fission decreases according to the following formula, however:

$$dN/dt = -\lambda N - \text{which is the rate of decay at}$$
$$\text{any time } t.$$
$$N = N_o e^{\lambda t}$$

where

N_o = number of atoms at time T_o

N = number of atoms at time T

λ = decay (disintegration) constant, which is the fraction of the total number of atoms which will decay in a given time.

A convenient measure of the rate of fission is the half-life $(T_{1/2})$ i.e. the time it takes for half of the original atoms to decay. That is to say

$$N/N_o = 1/2 = e_{1/2}^{-\lambda T}$$

The half-life is then:

$$T_{1/2} = \ln 2/\lambda = 0.693/\lambda$$

Radioactive decay processes result in new nuclides called daughter nuclides. In order to date geological material we measure the ratio between the fissionable nuclide (parent nuclide) and the daughter nuclide.

7.6.2 Potassium-Argon Method

One of the most commonly used dating methods is the potassium-argon method. ^{40}K is an unstable nuclide which decays mainly to ^{40}Ca with the emission of beta particles, but about 12% is transformed into argon-40 through capture of electrons and emission of X-rays (gamma-rays). Any mineral which contains potassium, for example biotite, will also contain some ^{40}K which decays to ^{40}Ar. The amount of argon in the mineral is thus an expression of the mineral's age, which can be calculated if we know the half-life $(T^{1/2})$ or the decay rate (λ) (half-life is 1.31×10^9 years).

The amount of argon gas in the minerals is analysed with a sensitive mass spectrometer. Argon is a very volatile gas, however, and when minerals containing argon are heated up or deformed, for example during metamorphism or folding, the argon will escape and the radioactive "clock" will indicate an age which corresponds more to the time of metamorphism than to the formation of the mineral. Therefore, a radioactive age determination does not necessarily give a figure for the age of the rock, but will tell us something about the geological processes which have affected the rock and thus help us to reconstruct its geological history.

7.6.3 Rubidium-Strontium Method

Another commonly used dating method is the rubidium-strontium rato, ^{87}Rb to ^{87}Sr. Rubidium has a radioactive isotope, ^{87}Rb, which decays to ^{87}Sr with a half life of 4.89×10^{10} years. The amount of ^{87}Sr is then a function of the time since the formation of the mineral or rock. Instead of measuring absolute amounts of ^{87}Sr, it is simpler to measure the $^{87}Sr/^{86}Sr$ ratio on the mass spectrometer. Then the $^{87}Rb/^{86}Sr$ and $^{87}Sr/^{86}Sr$ ratios are plotted as axes on a graph. Analyses of minerals or rocks which have the same age will then form points on a straight line (isochron). The slope of this line will be an expression of the age of the rock.

Using this method we can analyse both minerals and whole rocks, since we assume that the ^{87}Sr formed does not escape from the rock as easily as argon (in the K-Ar method). When dating sediments one should choose the finest-grained clay sediments with low permeability. It can then be assumed that after deposition the sediment was homogeneous with respect to strontium isotopes, and will provide an isochron which dates any diagenesis occurring a relatively short time after deposition. Larger clastic fragments, however, will contain a strontium isotope ratio which corresponds to the age of the source rock, and the obtained date will be intermediate between the age of the source rock and the time of deposition of the sediments.

7.6.4 Carbon 14 Method

The carbon 14 method is the one most commonly used for dating the youngest sediments, from about 50,000 years old up to the present. ^{14}C is formed in the atmosphere by cosmic rays when a ^{14}N atom absorbs a neutron and gives off a proton. ^{14}C is unstable, and decays to ^{14}N. The production of ^{14}C thus occurs only in the atmosphere, at altitudes over 10,000 m. The ^{14}C then mixes with the lower air layers and the seawater.

The ^{14}C method is based on the assumption (discussed below) that the ^{14}C content of the atmosphere has been constant for a long time, due to an equilibrium between the ^{14}C added from the atmosphere and the ^{14}C which decays to ^{14}N. The half-life of ^{14}C is 5,730 years. This means that after 5,730 years half of the ^{14}C atoms will have changed to ^{14}N. ^{14}C enters the carbon dioxide (CO_2) in the air and is taken up by plants through photosynthesis. If we assume that the carbon dioxide in the air in the past had as much ^{14}C as now, the ^{14}C content of older plants or plant remains is an expression of their age. This age can be determined analytically with relatively great accuracy, for example 10,200 ± 100 years, depending on the nature of the sample. When 50–60,000 years has elapsed since plant material formed (i.e. since it ceased to take up CO_2), the ^{14}C content will be so small that we will be approaching the limit of detection. This is then the upper limit to the age of material we can analyse. We now know that the concentration of ^{14}C has varied in the last 10,000 years and more due to variation in cosmic radiation.

^{14}C from the atmosphere also becomes mixed with sea water and freshwater. Carbonate-secreting organisms which live in the sea and in lakes will take up CO_2 from the water, and the amount of ^{14}C in their $CaCO_3$ skeletons can be measured. Mollusc and foraminifera shells are particularly suitable for age dating. Chemically precipitated carbonates can also be dated in this way. Living organisms have a tendency to fractionate the lightest isotopes from the heaviest. The ratio between two stable carbon isotopes, ^{12}C and ^{13}C, can therefore be used to correct ^{14}C determinations.

7.6.5 Other Radiometric Methods

There are numerous other radiometric dating methods, of which the uranium-lead method and the lead-lead method (relationship between different lead isotopes) are the most important.

The methods based on fission processes mentioned above, involve long half-lives, 10^9–10^{10} years, so for young sediments the quantity of decay products available to be analysed will be small and the accuracy poor. An exception is the protactinium method (^{231}Pa/^{230}Th), which has given age determinations on younger sediments showing good agreement with the ^{14}C method.

The fission-track method is especially well suited for younger rocks. It is applied to glass or minerals which contain a sufficient amount of uranium 238. The material will show tracks from fission products which are observed as deformations. By counting the number of such tracks, an expression of the age is obtained. In apatite the tracks disappear (aneal) at about 100°C, which is called the blocking temperature. The frequency of tracks is therefore an expression of the time elapsed since rocks have been uplifted and cooled to below the blocking temperature.

7.7 Chemostratigraphy

7.7.1 The ^{87}Sr/^{86}Sr Method

The ^{87}Sr/^{86}Sr ratio is particularly useful for carbonates and other minerals precipitated in the ocean. In the ocean water the ^{87}Sr/^{86}Sr ratio has varied greatly over geologic time (Fig. 3.7). By analysing this ratio in minerals precipitated in marine environments the ages can be constrained. During Tertiary time, the ^{87}Sr/^{86}Sr ratio increased and analyses of calcareous organisms will give a unique age, in some cases with a resolution of 1–2 million years. This method can be applied also to calcareous microfossils such as foraminifera and coccoliths. In carbonate the Sr/Ca ratio may be useful also in detecting depositional facies, because the data will reflect the primary aragonite/calcite ratio.

7.7.2 Bulk Chemical Composition Analyses

Chemical analyses of bulk samples of cuttings and cores may be useful in correlation between wells, because the chemical composition reflects changes in the composition of sediments supplied to the basin. Particularly good results are expected in correlation of reservoirs. For this purpose, both major elements, trace elements and isotopic composition can be used. The samples can be analysed by XRF, energy dispersive systems in a SEM, or in mass-spectrographs. Analyses can be performed for isotopes, bulk chemical composition and trace elements.

The bulk chemical composition varies as a function of grain size but the ratio between major elements such as Na/Al and K/Al may be characteristic for the provenance rocks of sediments. Both sandstones and shales may have very different ratios between Na-feldspar and K-feldspar. Trace elements may be useful also in provenance studies.

7.8 Magnetostratigraphy

7.8.1 Palaeomagnetism

When clastic sediments are deposited or when volcanic rocks solidify, minerals orientate themselves in the prevailing magnetic field. Magnetic minerals in clastic sediments can orient themselves according to the magnetic field during deposition, while diagenetic minerals will become oriented in the magnetic field during diagenesis. The magnetic minerals in rocks thus define a magnetic vector, which indicates the direction and strength of the magnetic field during formation. By compensating for later magnetic effects, the orientation of this *remanent magnetism* can be measured. The data will reveal the position of the geographic pole, and indicate the palaeogeographical latitude and longitude.

Palaeomagnetic measurements are of great help in reconstructing the positions of the continents during the geological past, and are important for plate tectonic and palaeoclimatic reconstructions. From the Palaeozoic onwards there is quite good agreement between palaeomagnetic determinations of latitude and palaeoclimatic indications like biogeography, glacial deposits and evaporites, but this is not the case when it comes to Precambrian deposits.

7.8.2 Magnetic Field Polarity Changes

It has become evident that the polarity of the Earth's magnetic field has been reversed for time periods of varying duration. A number of measurements of magnetism in rocks of known age have provided us

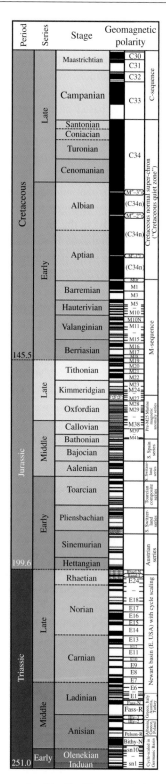

Fig. 7.11 Reversals of the Earth's magnetic field from Triassic to Cretaceous time. Normal polarity: black bands, reversed polarity: white bands (From Gabi Ogg 2010, http://stratigraphy. science.purdue.edu/charts/educational.html)

with a time scale with periods of normal and reversed magnetic field. We thus find that we can divide geological time into long periods with dominantly normal or reversed polarity (Fig. 7.11). Within these we also find several shorter intervals when the magnetic field switched between the two polarities. Since we must assume that the switching between normal and reversed magnetism has taken place simultaneously and suddenly all over the world, such physical changes offer an ideal basis for correlation.

There are often major practical problems, however, since many periods of the Earth's history are characterised by a predominance of successions with rapidly changing reversals. Where we have many measurements and continuous profiles, e.g. in deep-sea cores, we will be able to correlate with relative certainty on the basis of the longer periods of magnetic field stability. Volcanic rocks and sediments deposited in fluvial or shallow-water environments, however, will have numerous hiatuses between beds, hampering registration of continuous variations in the residual magnetism. During the last 700,000 years we have had apparently normal polarity, possibly with the exception of a short period about 200,000–300,000 years ago. If we find sediments or volcanic rocks with reversed magnetism, we know that they are highly probably more than 700,000 years old.

7.9 Sequence Stratigraphy

7.9.1 General Aspects

Sequence stratigraphy is defined as the subdivision of sedimentary basin fills into genetically related stratal packages bounded by unconformities and their correlative conformities. Thus, the concept of sequence stratigraphy is closely related to that of allostratigraphy. Allostratigraphic analysis applies bounding discontinuities, e.g. erosional surfaces and marine flooding surfaces, for recognition of sedimentary entities independent of any genetic model. Sequence stratigraphy applies allostratigraphic features to interpret the depositional origin of sedimentary packages by analysing these in a framework of transgressive-regressive developments producing base level changes.

The infilling of a basin is controlled by the interaction of tectonics and eustasy, producing base level changes which define the accommodation space for sediments to be deposited. This is the space or height between the seafloor and the sea level. Climate is an important factor determining the production and supply of sediments. The evolution of basin architecture is controlled by the balance between accommodation space and sediment supply. If sediment supply exceeds the accommodation space available, *prograditional geometries* will be the result. If accommodation space and sediment supply are roughly balanced, *aggradational geometries* result. When sediment supply is less than the creation of accommodation space, *retrogradational geometries* are formed.

The major bounding and subdividing surfaces of a depositional sequence are commonly represented by: the lower and upper sequence boundaries, transgressive surfaces and maximum flooding surface (Fig. 7.12). These surfaces principally define three sediment bodies: lowstand systems tract, transgressive systems tract and regressive systems tract. The building blocks of the systems tracts are parasequences, forming parasequence sets.

7.9.2 Accommodation Space

Local accommodation space for sediments represents simply the water depth in the depositional area. A global factor influencing the accommodation space is the eustatic sea level stand which is defined by a combination of several factors. One of these is the volume of the ocean basins, which is believed to be mainly controlled by the rate of seafloor spreading. Young lithosphere is warm and buoyant, therefore it will rise and decrease the volume of the world ocean. On the contrary, old lithosphere is colder and denser, and sinks. Another factor contributing to ocean volume changes is the sedimentation rate.

The volume of water in the ocean basins is also a primary factor influencing eustatic sea level changes. Continental and mountain glaciations and deglaciations decrease and increase, respectively, the volume of ocean water. These developments are exemplified by the well-documented regressions and transgressions accompanying Quaternary glacial and interglacial periods (Fig. 7.13). Other factors influencing the volume of oceanic waters are the expansion of water with increasing temperature and the amount of continental groundwater and surface water.

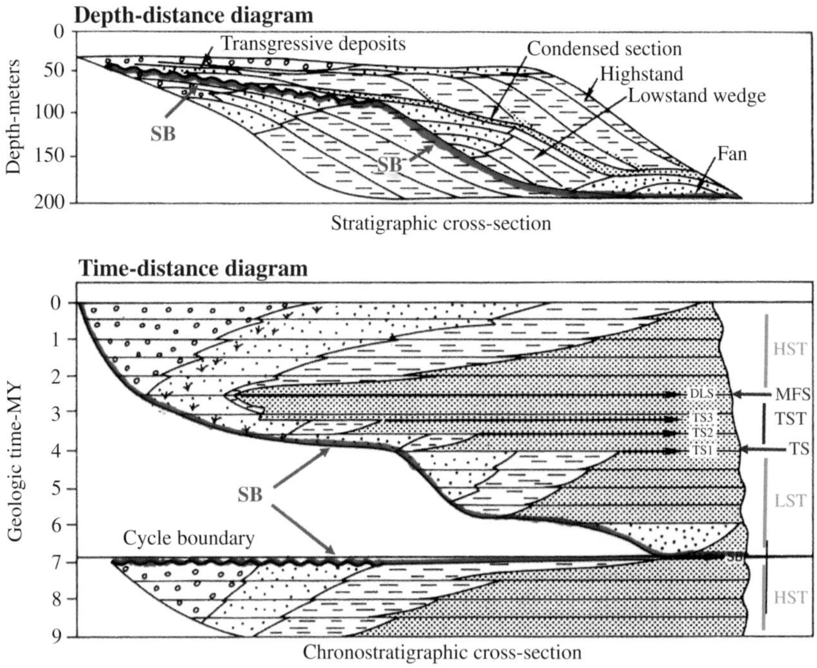

Fig. 7.12 Depositional
sequence portrayed in a
depth-distance and a
time-distance diagram. It
shows the regional extent of
the condensed section from
deep to shallow water, with its
time span strongly expanding
from coastal to basinal areas.
The time span of the sequence
boundary unconformity
decreases from marginal to
basinal areas. Abbreviations
of sequence elements:
SB = sequence boundary,
LST = low stand systems
tract, TS = transgressive
surface, MFS = maximum
flooding surface, HST = high
stand systems tract (modified
after Loutit et al. 1988)

Depth-distance diagram

Transgressive deposits · Condensed section · Highstand · Lowstand wedge · SB · SB · Fan

Stratigraphic cross-section

Time-distance diagram

Geologic time-MY

HST · DLS · MFS · TS3 · TST · TS2 · TS · TS1 · LST · SB · Cycle boundary · HST

Chronostratigraphic cross-section

Fig. 7.13 Schematic
illustration of
transgressive-regressive
cycles in the Quarternary,
resulting from
glacially-induced changes of
the volume of ocean water.
During a complete cycle,
sedimentation takes place
only during short episodes
(note the three types of
depositional architecture).
During most of the time there
is non-deposition or erosion
(hiatus)

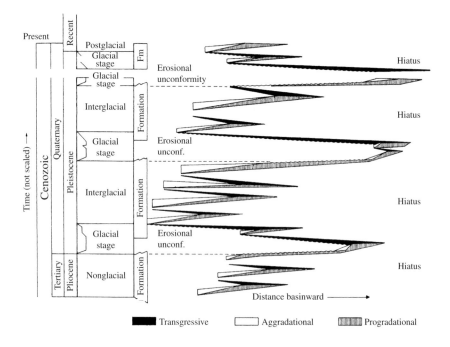

Transgressive · Aggradational · Progradational

Tectonic movements result in transgressions and regressions of local to regional nature. These movements are driven by factors that affect how continental lithosphere floats on the asthenosphere, which is controlled by three main mechanisms: (1) Stretching of continental lithosphere resulting in replacement of continental lithosphere by denser and thinner asthenosphere which sinks. (2) During stretching, continental lithosphere is heated, becomes less dense and tends to uplift. Subsequent cooling results in subsidence. (3) Weight of tectonic load added to lithosphere can produce subsidence e.g. wedges of fold and thrust belts pushed into foreland basins.

When analysing sedimentary successions, it is usually extremely difficulty to distinguish between the effects of eustatic sea level changes and tectonic movements. Therefore, in sequence stratigraphic studies, the phrase "relative sea level" is used to express the local sum of global sea level and tectonic movements.

7.9.3 Parasequences

Parasequences are sediment packages of genetically related beds representing a single minor transgression event followed by sediment progradation. In the marine realm parasequences are bounded by flooding surfaces. Most of the marine parasequences are formed in offshore shelf to shoreface environments, and show an upwards-coarsening development (Fig. 7.14). In coastal areas upwards-fining parasequences are formed during transgressions, encroaching over the supratidal zone, and contain shallowing-up subtidal, intertidal to supratidal strata (Fig. 7.15).

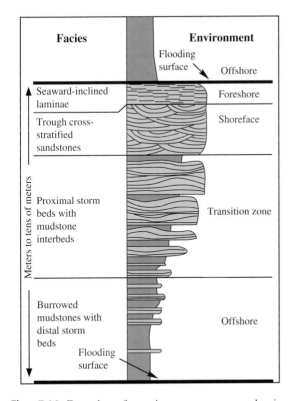

Fig. 7.14 Example of marine parasequence showing coarsening- and shallowing-upwards from offshore shales to foreshore sandstones (from University of Georgia web site, 2009)

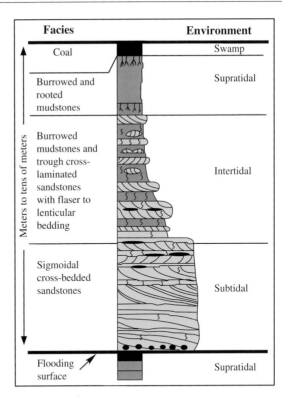

Fig. 7.15 Example of coastal parasequence showing fining- and shallowing-upwards from subtidal sandstones to supratidal swamp deposits (from University of Georgia web site, 2009)

Stacked parasequences form parasequence sets, which show stacking patterns according to their position in the sequence stratigraphic architecture (Fig. 7.16). If sediment supply exceeds the available accommodation space, progradational stacking geometry will be the result. If accommodation space and sediment supply are roughly balanced, aggradational geometries occur. When sediment supply is less than the creation of accommodation space, retrogradational stacking geometries are formed.

7.9.4 System Tracts

The lowstand systems tract is bounded by the sequence base and the transgressive surface (Fig. 7.12). It was deposited when the sea level was located at or below the shelf break. The shelf was exposed to erosion and crossed by rivers eroding incised valleys and transporting sediments to the shelf margin. The lowermost part of the lowstand systems tract is the basin floor fan,

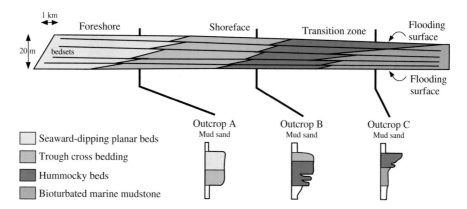

Fig. 7.16 Lateral and vertical relationships between parasequences forming a parasequence set, which progrades from foreshore to offshore shelf (from University of Georgia web site, 2009)

dominated by gravity-driven deposition of turbidite lobes and in feeder channels. The upper part of the lowstand systems tract is the lowstand wedge, which consists of progradational parasequences. Uppermost, estuarine sediments are deposited in drowned incised valleys (Fig. 7.17).

The transgressive systems tract is bounded by the transgressive surface and the maximum flooding surface. It is composed of retrogradational parasequence sets showing a generally upward-fining and upward-deepening development. This systems tract is generally thin because much sediment is trapped in estuaries, and the advancing transgression successively creates new accommodation space.

The highstand systems tract is developed between the maximum flooding surface and the upper sequence boundary. This systems tract represents a relatively thick succession of progradational parasequences showing an upward-coarsening and upward-shallowing development (Fig. 7.17). The parasequences downlap to the maximum flooding surface. The estuaries are now filled up by sediments, therefore the systems tract is characterised by outbuilding of deltas and shorelines.

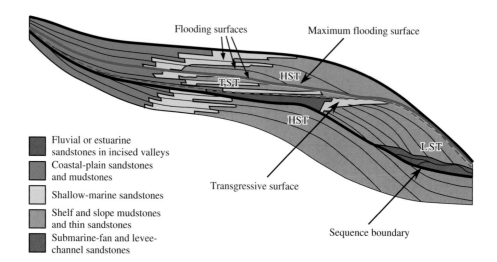

Fig. 7.17 Depositional sequence (type 1) portrayed as a down-dip section across an idealised passive continental margin with shelf break. The diagram shows sequence boundaries, systems tracts and parasequences. LST = low stand systems tract, TST = transgressive systems tract, HST = high stand systems tract (from University of Georgia web site, 2009)

7.9.5 Condensed Sections

The transgressive surface and maximum flooding surface were formed during periods of transgression characterised by sediment starvation, which leads to development of condensed intervals forming important marker horizons in the sequence stratigraphic architecture of a basin (Fig. 7.12). Condensed sections are thin marine horizons commonly showing several of the following features: pelagic to hemipelagic sedimentation; presence of apparent hiatuses; borrowed horizons and hardgrounds; presence of authigenic minerals such as glauconite, phosphate, siderite; high faunal abundance (if not anoxic); high faunal diversity (if oxic), low diversity (if hypoxic). Deposition of organic-rich black shales with high gamma readings is typical of many condensed sections.

Condensed sections have a large regional extent, and show an expanding time span from coastal to basinal areas (Fig. 7.12). They are particularly well developed at the maximum flooding surface and help to identify this horizon. Thus, condensed sections represent important horizons of correlation within sedimentary basins. They are also applicable to recognition of global transgressive-regressive cycles.

7.9.6 Sequence Types

During development of the sequence boundary the relative sea level fall is of varying magnitude. The sea level fall can continue to a position below the shelf-slope break, or stop on the shelf. According to these developments, two types of sequences are distinguished.

Type 1 sequence: This is formed if the sea level falls below the shelf break at the sequence boundary (Fig. 7.17). The shelf is exposed for erosion and a low stand systems tract is formed. The boundary reveals a sea level fall below the previous shoreline. A type 1 sequence can be formed in two basinal developments: by deposition in a basin having a shelf-slope break; or by deposition in a basin that lacks a shelf-slope break but has an evenly sloping ramp-like margin.

Type 2 sequence: The sea level position above the sequence boundary corresponds to the position of the previous shoreline below the boundary. Consequently, the boundary reveals an aggradational development. During the sea level low stand a shelf margin systems tract is deposited.

7.9.7 Carbonate Sequence Stratigraphy

Sequence stratigraphy was originally developed for siliciclastic systems, but the same principles can be applied to carbonate systems as well. However, the special features of carbonates make the appearance of stratigraphic elements somewhat different in the two system types.

Carbonate production takes place mainly in situ, and is particularly high during moderate sea level rise, which typically results in extremely thick transgressive systems tracts. Parasequences formed in peritidal environments are particularly thick. The highstand systems tracts in carbonate successions are much thinner than in siliciclastic sequences. During transgression, extremely rapid sea level rise stops in situ carbonate production, which leads to the formation of condensed sections. Carbonates are prone to dissolution, leading to sequence boundaries commonly marked by solution surfaces, karst development and palaeosols.

Further Reading

Armstrong, H.A. and Brasier, M.D. 2005. Microfossils, 2nd edition, Blackwell Publishing, Oxford, 296 pp.

Copestake, P. 1993. Application of micropaleontology to hydrocarbon exploration in the North Sea. In: Jenkins, D.G. (ed.), Applied Micropaleontology. Kluwer Academic Press, Dordrecht, pp. 93–152.

Gradstein, F.M. and Ogg, J.G. 2004. Geological time scale 2004 – Why, how and where next! Lethaia, 37, 157–181.

Hallam, A. 1992. Phanerozoic Sea-level Changes. Columbia University Press, New York, NY, 266 pp.

Haq, B.U. and Boersma, A. 1978. Introduction to Micropaleontology. Elsevier, North-Holland, Inc. 376 pp.

Hedberg, H.D. (ed.). 1976. International Stratigraphic Code Guide. Wiley, New York, 200 pp.

Jones, R.W. 1996. Micropaleontology in petroleum exploration. Oxford University Press, Oxford.

Jones, R.W. 2006. Applied Palaeontology. Cambridge University Press, Cambridge, 432 pp.

Kautsakos, E.A.M. (ed.). 2005. Applied Stratigraphy. Springer, Dordrecht, 488 pp.

Lindberg, D.R., Lipps, H.J. and Hazel, J.E. 1993. Micropaleontology. In: Lipps, H.J. (ed.), Fossil Prokaryotes and Protists. Blackwell Scientific Publications, Boston, MA, pp. 31–50.

Lipps, J.H. (ed.). 1993. Fossil Prokariotes and Protists. Blackwell Scientific Publications, Boston, MA, 342 pp.

Loutit, T.S., Hardenbol, J., Vail, P.R. and Baum, G.R. 1988. Condensed sections: The key to age determination and correlation of continental margin sequences. In: Wilgus, C.K., Hastings, B.S., Kendall, C.G.St.C., Posamentier, H.V., Ross, C.A. and Van Wagoner, J.C. (eds.), Sea Level Changes – An Integrated Approach. SEPM Special Publication 42, pp. 183–213.

Nystuen, J.P. 1998. History and development of sequence stratigraphy. In: Gradstein, F.M., Sandvik, K.O. and Milton, N.J. (eds.) Norwegian Petroleum Society, Elsevier, Amsterdam, Special Publication 8, pp. 31–116.

Ogg, J.G. 1995. Magnetic polarity time scale of the Phanerozoic. Global Earth Physics – A Handbook of Physical Constants. American Geophysical Union, Washington, DC, pp. 247–270.

Van Wagoner, J.C., Posamentier, H.V., Mitchum, R.M., Vail, P.R., Sarg, J.F., Loutit, T.S. and Hardenbol, J. 1888. An overview of the fundamentals of sequence stratigraphy and key definitions. In: Wilgus, C.K., Hastings, B.S., Kendall, C.G.St.C., Posamentier, H.V., Ross, C.A. and Van Wagoner, J.C. (eds.), Sea Level Changes – An Integrated Approach. SEPM Special Publication 42, pp. 39–45.

Chapter 8

Sequence Stratigraphy, Seismic Stratigraphy and Basin Analysis

Knut Bjørlykke

8.1 Seismic Stratigraphy

Seismic records are based on measurements of the time sound waves (seismic waves) take to travel through rock. The sound or signal is produced by explosives or compressed air (air guns). Rock is an elastic medium and the velocity of sound conveys a lot of information about the properties of the rock. Normal sound waves (P-waves) travel through both the solid phase, which for the most part consists of minerals or rock fragments, and the liquid or gas in the pores. Shear waves (S-waves) on the other hand can only go through the solid phase.

The most important parameters influencing the velocity of sound are: porosity, mineral composition, and the degree of cementation. The velocity of sound waves in water is about 1,500 m/s, but depends on temperature and salt concentration. Sound passes through unconsolidated sediments at velocities which are only slightly higher than the velocity in water (1,500–2,000 m/s, and sometimes even lower) because they have high water content and because the framework on which the sediment grains are based does not offer any real strength (stiffness) as a medium for the seismic waves.

Cementation of sand with carbonate or siliceous cement will bind the grains together in a framework which will increase the stiffness and velocity considerably even if the porosity is relatively high. Compaction due to overlying sediments which causes water to be expelled will also give higher velocities, not only because the water content decreases, but because more numerous and larger contacts are formed between the clastic grains. Velocities in moderately consolidated sediments, such as the Tertiary sediments of the North Sea, are 2–3 km/s. In more consolidated (compacted and cemented) sedimentary rocks which have not been subjected to metamorphosis, velocities are mostly between 3 and 5 km/s. This is the case for many of the Mesozoic sediments in the North Sea. Metamorphic and eruptive rocks will have velocities of about 5–6 km/s. Limestones will often have higher velocities than sandstones at the same depth because they often are more cemented and because carbonate cement has a high degree of rigidity and low compressibility. Carbonate reefs may be strongly cemented and have high velocities at shallow depth. Sandstone in turn provides a more rigid medium for sound waves than shale at the same depth, because of its grain-supported structure.

If the rocks do not contain oil and gas we can assume that their porosity is identical with the water content in the rock. Velocity will then be a function of porosity (φ), and if we know the velocity of sound in the rock matrix, we can calculate the porosity using Wyllie's equation:

$$1/V_r = (1 - \varphi)/V_m + \varphi/V_f$$

where

V_r = velocity in rock when saturated with liquid, i.e. the measured velocity

V_f = velocity in the fluid

V_m = velocity in the rock matrix.

K. Bjørlykke (✉)
Department of Geosciences, University of Oslo, Oslo, Norway
e-mail: iknut.bjorlykke@geo.uio.no

K. Bjørlykke (ed.), *Petroleum Geoscience: From Sedimentary Environments to Rock Physics*,
DOI 10.1007/978-3-642-02332-3_8, © Springer-Verlag Berlin Heidelberg 2010

The inverse values of the velocities are expressions of the time the signals take to travel through a layer of certain thickness.

If the pores are filled with gas instead of liquid (water or oil), the velocity reduction will be even greater because the sound travels much more slowly through gas than through a liquid. Whyllie's equation is however not an accurate description of the relation between porosity and velocity.

V_r thus approaches V_m at zero porosity, and for sandstone V_m is about 5.5–6 m/s.

If we know the velocity of the rock matrix and the fluid we should be able to calculate the porosity as a function of velocity, but the Wyllie equation is very much a simplification. Rocks with the same porosities can have rather different velocities depending on the type of grain contacts and the distribution of cement.

When sound waves move between sedimentary beds with different velocities, they will be refracted according to Snell's law (see Fig. 8.1):

$$\sin x_1 / \sin x_2 = v_1 / v_2$$

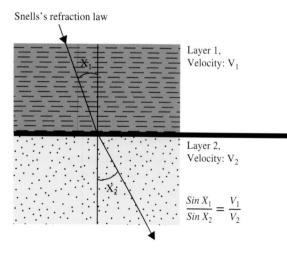

Fig. 8.1 Snell's law for the refraction of sound waves

Here x_1 is the angle of incidence of the waves where they meet the boundary plane between two strata, and x_2 is the angle of refraction of the emergent waves. v_1 and v_2 are the velocities through the respective rock strata.

If the two beds have different velocities, they will as a rule also have different densities, and part of the

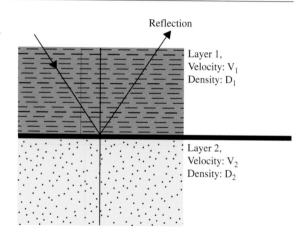

Fig. 8.2 Diagram for the reflection of waves in a layered sedimentary sequence. The amount of energy reflected is a function of acoustic impedance, which is the product of the density of the beds (ρ) and the velocity of the sound waves (v)

acoustic energy will not be refracted, but reflected. How much of the energy is reflected depends on the difference in the *acoustic impedance*, which is the product of velocity and density (Fig. 8.2).

The coefficient for reflection (R) is then:

$$R = (\rho_2 \cdot v_2 - \rho_1 \cdot v_1) / (\rho_2 \cdot v_2 + \rho_1 \cdot v_1)$$

where ρ_1 and ρ_2 are the densities of the two rocks, and v_1 and v_2 their respective velocities (Fig. 8.2). We see that the greater the difference in density and velocity, the greater the amount of energy which will be reflected. Sandstone will often have significantly different acoustic impedance from shale, and a considerable amount of sound energy will be reflected from the boundary between a sandstone bed and a shale bed.

This is however not always true and the contrast depends on the type of clays and their clay mineral composition. Limestones will tend to have both high velocities and high densities. The result will be even greater contrast in acoustic impedance between limestones and, for example, shales. However, this contrast will always depend on the porosity of the limestone in question.

On a seismic section, beds which have greatly contrasting acoustic impedances stand out as strong reflectors. This makes it possible to map characteristic rock boundaries, e.g. the top of a limestone or the boundary between shales and sandstones, using seismic sections (Fig. 8.3).

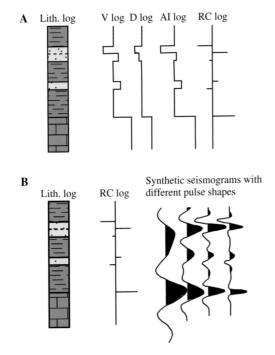

Fig. 8.3 (a) Illustration of a reflection coefficient log based on velocity (v) and density (ρ) (from Anstey 1982). (b) Synthetic seismogram of the different reflection coefficients (RC log) in a bedded sequence. The resolution of the seismogram varies with the width of the seismic pulse (from Anstey 1982). Normally, sediments have to exceed 20–30 m in thickness to be distinctly recorded on a seismic section, depending on the wavelength of the seismic signal. The upper sand contains some gas in the upper part, causing lower velocities

As we have seen, the critical parameters determining the reflection coefficient are the velocities and densities of the different lithological units. Using a well, we may measure a velocity log (sonic log) and a density (ρ) log which record how these properties change through the sequence (see Fig. 8.3a). The product of velocity and density may then be computed and presented as an acoustic impedance (p·v) log. Note that at the boundary between limestones and shales there is a very significant drop in both velocity and density, resulting in a marked change on the acoustic impedance log. Sandstones usually have higher velocities than shales, but they may not differ much in density, so the difference in acoustic impedance will be small. The reflection coefficient, which is an expression of differences in acoustic impedance, is a synthetic seismic trace such as we would have seen on a seismic cross-section through the sequence.

If we have gas instead of water in a rock, the velocity will be considerably reduced. The velocity of sound in gas is much lower than it is in liquid, depending on composition, temperature and pressure. The boundary between gas-bearing and water-bearing rocks may produce a strong reflection because there is a large difference in impedance between the two layers. For this reason the boundary between gas and oil is often revealed as a strong reflector because it is horizontal and does not always follow the other rock strata. This is called a "flat spot" and exemplifies direct indication of hydrocarbons (usually gas) through seismic methods. At greater depth and higher pressure the contrast between gas and oil and also oil and water will be lower.

Reflections which are multiples of a relatively flat sea-bottom reflection are also near-horizontal and may be confused with "flat spots", but these may be removed by filtering the data during processing. Temperature-dependent diagenetic reactions may also produce horizontal reflections, e.g. the transformation of amorpheous silica (Opal A) and Opal CT to quartz which produce a strong increase in velocity and density. If the geothermal gradients are rather uniform (horizontal isotherms) this diagenetic transition will form horizontal reflections which may crosscut the bedding.

It may also be possible using AVO (see Chap. 17) to detect contacts between oil and water. Oil has a lower velocity than water and if the oil has a high gas content, the difference is even greater.

Changes in the oil/water contact during production can be monitored. By shooting seismic in an oilfield at intervals of several years, changes in the oil/water contact during production can be monitored (see 4D Seismics). In deeper reservoirs with lower porosities it is more difficult to detect fluid contacts.

High pore-pressure causing reduced effective stress and stiffness (elasticity) results in lower seismic velocities in sandstones and clays, particularly at depths less than 3 km. Limestones can have relatively high velocities even at shallow depth.

At greater depths chemical compaction is the most important factor, and in clastic sediments it is generally a function of temperature and less dependent on the effective stresses. Nevertheless, seismic velocities are often observed to fall in overpressured rocks even if the porosity is not significantly reduced. This may be related to reduced stress at grain contacts.

As the quality of seismic data has improved, one has been able to use seismic profiles for detailed interpretation of stratigraphic relations and even depositional environments. The basis for this is that seismic reflections usually follow time lines in a sedimentary sequence. In other words, seismic reflections follow surfaces which constituted the seafloor surface at the time when the sediments were deposited.

Seismic reflections can be followed from a sandy facies into a clay/siltstone shale facies. We can, for example, follow seismic reflections from the fluvial part of a delta out into the pro-delta muds. Relatively thin transgressive sandstones may be deposited on the delta top, and are useful for correlation. This is because the relief on land may be very low so that a few metres sea level rise causes a large transgression. Transgressions will then result in a wide shelf with little clastic supply, causing deposition of carbonates. Thin calcareous sediments and limestones may therefore also be relatively close to time lines. Limestones typically have rather high velocities and so we find strong reflectors. On the delta slope, however, fluvial sediments and marine delta front sand may prograde into a marine basin over considerable geological time, depending on the depth of the basin and the sediment supply. The seismic reflections will follow the surface from delta front sand to delta slope, where we have sand and mud (shale) beds lying parallel with the slope. Even though we can often follow the reflector a little further into the basin, it will be less marked there because there is less contrast in lithology and hence in acoustic impedance. Different types of clay may however also produce differences in seismic response and particularly smectitic clays are characterised by low density and velocity.

The shifting of sedimentation input from one part of the delta to another through channel switching (distributary abandonment as part of delta-lobe shifting), also contributes to the formation of lithological contrasts on the delta slope. Progradation of new delta lobes results in deposition of sheets of sand over mud near time-stratigraphic boundaries. The inactive delta lobes will be compacted and often develop a thin carbonate or transgressive sandstone layer, while sedimentation takes place in the active lobe. The small unconformities produced in this way also tend to produce lithological contrasts which may be recorded on the seismic record.

Seismic sections through prograding deltas provide information about the water depth, the rate of sediment input and the wave energy in the basin.

8.2 Different Types of Seismic Signatures

A stratigraphic unit which is composed of a conformable bedding series, genetically linked together at the top and bottom by unconformities, is called a *depositional sequence*. A depositional sequence is thus a package of sediments deposited during a definite period of time, defined by unconformities above and below.

The unconformities may be due to a break in sedimentation due to relative changes in sea level or other causes such as changes in sediment supply.

A seismic profile through a sedimentary succession has reflectors that show layers of contemporary deposits. Terminolgy has been established to describe the geometry of seismic reflections (Fig. 8.4).

Baselap is the term for gradual deposition above the lower boundary of a depositional series and represents a small unconformity.

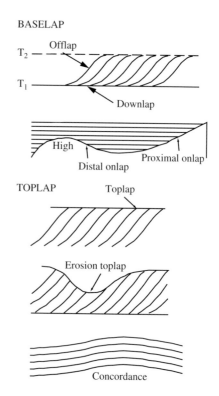

Fig. 8.4 Different types of seismic stratigraphic relations. The seismic reflectors represent time lines as a rule, i.e. rocks deposited at the same time

Fig. 8.5 Coastal onlap followed by a sea level drop and a renewed onlap. The coastal onlap indicates a relative sea level rise of H metres but this may be due to local tectonic movements and loading by water and sediments

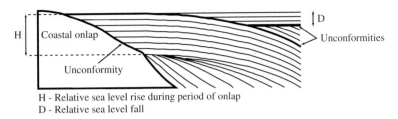

H - Relative sea level rise during period of onlap
D - Relative sea level fall

If a sequence progrades out across an unconformity, depositing successively younger beds basinwards, we call this type of contact *downlap*. The building of beds out into the basin like this is called *offlap*. We are thus dealing with a bed which has a primary depositional slope with respect to the unconformity surface.

Onlap is the term for primarily almost horizontal beds against a sloping unconformity which may be a submarine or a subarial slope. It occurs most commonly as a result of sedimentation gradually burying an unconformity during a transgression onto a land surface that provides the unconformity (Fig. 8.5). We call this *proximal onlap* or *coastal onlap*. If sedimentation covers a positive relief structure in the basin, for example a horst or a salt dome, we get *distal onlap* (Fig. 8.4).

Toplap is the contact between the seismic reflectors and an upper unconformity. An erosion surface will truncate the reflections sharply, forming an erosional truncation (*erosional toplap*).

Transgression may form a coastal onlap across an unconformity (Fig. 8.5). Later the relative sea level fell, and a prograding offlap sequence formed. Finally the sea level rose again and an onlap sequence formed. H and D are not to be regarded as absolute values for sea level changes. They must be adjusted for isostatic responses to loading and unloading and to tectonic uplift or subsidence. H is the relative rise in sea level during the period with onlap. D is the relative fall in sea level which led to a downward shift in the prograding downlap.

8.3 Interpretation of Lithology and Sedimentary Facies by Means of Seismic Profiles

In addition to structural data, a seismic profile provides us with information about the properties of

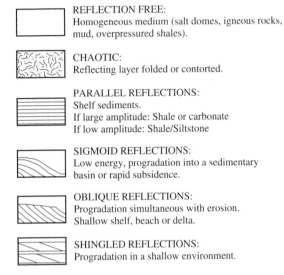

Fig. 8.6 Classification of internal structures in seismic units. Layering in the sedimentary sequences causes changes in the acoustic impedance

sedimentary rocks. The internal properties of seismic units provide important information about the lithology (Fig. 8.6).

Because seismic reflections mainly represent time lines, i.e. sedimentary beds which were deposited contemporaneously, it is also possible to a certain extent to interpret the depositional environment. The most important parameters we use are:

1. *Reflection amplitudes* – the strength of the reflections. As we saw above, the proportion of the energy reflected at the boundary between two beds is a function of the difference in the acoustic impedances (velocity multiplied by density). If we have an alternating series of different beds, the distance between the bed boundaries in relation to the wavelength of the transmitted seismic waves will play a major part (Fig. 8.3).

2. *Reflector frequency*. The distance between the reflectors will indicate the thickness of the bed, but there will be a lower limit to the thickness that can be detected, which should correspond to about the half wavelength of the seismic waves.

3. *Internal velocity in the beds*. The internal velocity of the bed can provide information about lithology and porosity.

4. *Reflector continuity*. The continuity of reflectors will be a function of how continuous the sediment beds are, information which is essential for reconstructing the environment.

5. *Reflector configuration*. If we take the compaction effect into account, the shape of the reflecting beds gives us a picture of the sedimentation surface as it was during deposition. The slope of the reflectors, for example, represents the slope of prograding beds in a delta sequence with later differential compaction and tilting superimposed. Erosion boundaries with unconformities will in the same way show the palaeo-topography during erosion.

Seismic profiles provide information about the filling of the basin in response to the type of subsidence and sedimentary facies.

When interpreting lithology and depositional environment from seismic profiles, it is important to use sedimentological models as aids. If there are well data on lithologies, these must also be integrated. The information we gain from seismic profiles is often not sufficient by itself for an unambiguous interpretation. There may be several lithological compositions and environments which could give similar seismic signatures. Only by looking at the whole basin in a sedimentological context do we have a good basis for selecting the interpretation which seems most reasonable.

8.4 Seismic Interpretation of Sedimentary Basins

Seismic profiles present a picture of the way the basin has been filled in. This is a result of an interaction between the rate of subsidence, rate of deposition and the energy of the depositional environment. The seismic signatures can be used to interpret facies and basin infilling.

8.5 Faults and Tectonic Boundaries

It must be remembered that the principle we use for calculating the depth to a reflecting boundary assumes that the layering is relatively horizontal.

Primary seismic reflections will be deformed through tectonic deformation so that they become tilted or folded. Folded beds will only be realistically depicted if the folds are sufficiently gentle that the beds have a low angle of dip.

We can distinguish faults where good reflections suddenly stop, suggesting an abrupt lateral change in lithology. Faults are generally too steep to reflect the sound wave straight back again, and the fault plane itself will not appear as a reflector on the seismic profile. Because of the special "edge" effects near faults, the ends of the reflecting layers which should define faults will not be quite correctly located on the seismic profile, and it may therefore be difficult to trace the fault entirely accurately. The termination of beds against faults may produce diffraction from a point source, giving a curved alignment. Special treatment of seismic data (migration) will rectify a fair number of these errors and give a more correct picture. In recent years seismic lines have been shot with smaller and smaller grid spacings to obtain a better map of the reservoir structure. A three-dimensional seismic data set is then produced and seismic sections can be constructed at any angle relative to the grid. This method also allows us to construct horizontal time-slices through the structure. This is almost like a topographic or geological map which is a horizontal projection of the geology. It is also a very powerful method of delineating faults and other important structural elements.

Another relatively new development is *borehole seismics*, particularly vertical seismic profiles (VSPs). This method involves firing shots near the seafloor close to a well and recording signals at regular depth intervals in the well. The main advantage of VSPs is that they produce a very good profile of the seismic velocity as a function of depth, better than a synthetic seismic log.

8.6 Changes in Sea Level

It has been clear for a long time that there are unconformities in sedimentary sequences which can be correlated over long distances, and that there were periods in geological history with a high sea level and others when it was low.

Proximal onlaps are due to sedimentation moving landwards over an unconformity surface. If we are dealing with a coastal deposit, a proximal onlap will mean that the sea level has risen in relation to the land surface which forms the top of the unconformity. On seismic profiles we can see onlaps onto the land, measure the height range between the lowest and uppermost onlaps, and calculate the difference in seismic time, and convert this into approximate thickness.

However, we must remember that the thickness of the sediments deposited is due not only to a rise in eustatic sea level, but also to local subsidence of this part of the basin. The weight due to increased water depth will cause further subsidence, and sedimentation will increase the load, resulting in further subsidence to attain isostatic equilibrium. Local tectonic subsidence may produce a relative change in coastal onlap in a seismic profile. Regressions are defined as the boundary between land and sea being displaced out into the basin. They may be caused by a fall in sea level which will shift the coastline to further out on the shelf or to the edge of the continental slope. Here unloading of some of the water plus erosion of sediments leads to isostatic uplift of the area landward of the coast line, so that the measured regression is greater than the real lowering of the sea level.

The definition of a transgression is that the sea encroaches over what was previously land, while a regression is a situation where the boundary between sea and land ("shoreline") moves seaward so that seabed becomes land.

Transgressions and regressions are not always directly related to sea level changes. When a delta builds out into the sea, there is a local regression on the delta even if the sea level has not fallen. If sedimentation is sufficiently rapid, we can have a local regression on a delta even with a rising sea level.

Along a coastline we can have transgressions in some areas and regressions in others at the same time, depending on the rate of sedimentation or erosion with respect to sea level change.

The term "forced regression" is used to indicate that there is a primary lowering of the sea level.

Changes in sea level can be due to:

a. Local tectonic movements, for example uplift of a horst or subsidence of a graben structure.
b. Plate-tectonic movements which can be of great extent, but are not global.
c. Sea level changes. These are global and are called eustatic sea level changes.

During the Quaternary, cyclic changes in sea level of up to 120 m accompanied the growth and decay of continental ice sheets. These changes were rapid and of large magnitude and can be traced throughout much of the world. Nevertheless it is often the local conditions which count most. In the areas which had supported ice sheets, such as Scandinavia, the melting of the ice led to uplift due to unloading, and this exceeded the rise in sea level so that there was a regression. A 2,000 m thick ice sheet will cause about 600–700 m of isostatic depression of the crust under the ice because the density of ice ($0.9 \ g/cm^3$) is about one third of that of rocks ($2.7 \ g/cm^3$).

When seismic stratigraphy was established (Vail et al. 1976), it was assumed that most of the variations in sea level that could be interpreted from seismic records, were attributable to eustatic changes and thus could be employed for correlation across great distances. There was a problem in that one did not know of any other processes than the accumulation of ice on the continents capable of producing large and rapid global sea level changes. However we only know of such ice ages from the Quaternary and late Tertiary, the Carboniferous-Permian, late Ordovician and the end of the Precambrian.

The general consensus now is that the rapid sea level changes outside the ice ages were mainly caused by tectonic activity. Nevertheless, transgressions and regressions can be correlated over greater or smaller distances, depending on the type of tectonic movements. In particular large scale plate tectonic movements involving changes in seafloor spreading rates and subduction rates cause global sea level changes.

In order to prove that a transgression is eustatic, we need rather accurate age determinations of the transgressive deposits from many parts of the world, but it is difficult to get high resolution datings. However seismic reflectors correlated with well data can be used to obtain good stratigraphic control and a picture of onlap and offlap can be interpreted in terms of sea level variation. They may represent relative sea level changes for a part of a sedimentary basin, or eustatic (global) transgressions or regressions.

Although one might perhaps expect that the sedimentation within a basin would primarily be characterised by local tectonic conditions, drainage, depositional conditions etc., studies of thousands of seismic profiles and wells from many sedimentary basins indicate that there have been simultaneous transgressions and regressions in completely different parts of the world. This has been documented by correlating seismic profiles with oil wells in the same area where the age of seismic unconformities and depositional sequences can be dated by means of biostratigraphy. It turns out that characteristic seismic reflectors which represent falls in sea level are of approximately the same age, for example in the North Sea, South China Sea, Mexican Gulf and Alaska. It has become clear that many areas, especially continental margins, have a rather similar tectonic history which is ascribable to global seafloor spreading.

Much of geological time is not represented by deposits, though; there are usually greater or lesser breaks in deposition (hiatuses). This becomes very clear when one looks closely at the continental sequences. Ocean floor sequences are more continuous, but also there one finds clear breaks in deposition. The time represented by breaks (hiatuses) varies greatly because there will always be some sedimentation somewhere, and correlation with sequences with continuous sedimentation entails trying to find signs of rapid changes in facies and deposition depth. Microfossils can be helpful in indicating water depth, even though they are not always reliable.

Mapping unconformities in the field can be difficult in the absence of abundant exposures. Many of the best examples have therefore been found from desert areas in the USA and elsewhere.

8.7 Changes in the Volume of the Ocean Basins

As we have seen, large scale ocean floor topography is a function of the age of the seafloor, i.e. how long it has had to cool down since it was formed, and of the overlying sediment thickness. Periods with rapid seafloor spreading will result in relatively broad spreading ridges which cause the volume of the ocean basins to decrease, and seawater then will spread further onto continent margins. If all seafloor spreading ceased, the spreading ridges would slowly sink and within about 100 million years would have disappeared almost completely, The volume of the oceans basins would then be greater, as there would be a sea level drop corresponding to the volume of the ocean ridges. This mechanism can explain the great fluctuations in sea level through geological time. Note too that periods with major transgressions can be correlated with periods of rapid seafloor spreading, for example in the Cretaceous and Carboniferous periods. The deep channels formed in connection with subduction are, in fact, small in relation to the width of the spreading ridge. In Permian and Triassic times we had one big supercontinent and little seafloor spreading. This was a regressive period with a large land area.

Drying out of cut-off ocean basins may also lead to eustatic changes in sea level. There is much to indicate that the Mediterranean Sea was cut off from the Atlantic and dried up in Upper Miocene (Messinian) times. This reduced the world's total volume of ocean basins, increasing the sea level by about 5–6 m.

Crustal thickening and thinning. An increase in the depth to the *Moho* (seismic discontinuity separating the earth's crust and mantle) will lead to elevation of the land. The greatest land elevation results from continental collision, when the thickness of the continental crust may be doubled (to 70–80 km), as in the Himalayas.

Stretching and thinning of the continental crust will move heavy mantle rocks upwards and increase the average density of the rocks down to a compensation depth of 100 km. This will lead to subsidence and more low density sediments can be accumulated. We see this at the transition between continental and oceanic crust, and where we have rift formation the continental crust thins below the rift, causing graben formation.

Variations in the temperature gradient affect the density of the rocks and thereby the isostatic equilibrium. Rifting causes elevation of the areas along the margin of the rift where the crust is not thinned (e.g. East Africa) and subsidence of the whole area when the rifting ceases and the crust cools (e.g. the North Sea).

The major transgressions in the Cambrian, Ordovician and Cretaceous, can be explained fairly satisfactorily by means of plate tectonic models. A relatively low sea level at the end of the Palaeozoic (Carboniferous-Permian) and the beginning of the Mesozoic (Triassic) can be explained as being due to limited seafloor spreading, for example along the Atlantic Ocean. With fewer and smaller spreading ridges the oceans could accommodate more water and consequently less seawater would flood the continents. On the continents there was active rifting without seafloor spreading and this increased the geothermal gradient and elevated areas which were previously covered by shallow epicontinental seas, to above sea level.

The short-term variations in sea level are more difficult to explain, however. For geological periods which experienced major continental glaciations (Quaternary-Upper Miocene, Permian-Carboniferous, Upper Ordovician and late Precambrian) we can resort to glacioeustatic changes in sea level (changes in sea level due to glaciation), which are very rapid in geological terms. Widespread sea level changes in the Lower Tertiary and Mesozoic must be due to plate tectonic factors.

Subsidence of the seafloor and continental crust of various thicknesses is a function of time after the rifting stage. The spreading ridges with hot basalt (oceanic crust) often lie at about 2 km water depth or above sea level (island). Seafloor basalt gradually subsides to almost 6 km due to cooling over 100 ma without sediment loading (Fig. 8.7.)

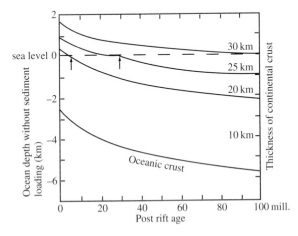

Fig. 8.7 Subsidence of the sea bed and continental crust of various thicknesses as a function of post-rifting age (after Kinsman 1975). Basaltic spreading ridges (oceanic crust) are often at 2 km depth, then the seafloor sinks to almost 6 km depth after 100 million years

8.8 Sedimentation and Isostatic Equilibrium

8.8.1 Why Do Sedimentary Basins Subside?

Sedimentary basins can be assumed to be in isostatic equilibrium in relation to the earth's crust. However, the crust has a certain rigidity, and it takes time before equilibrium is attained after loading. Studies of uplift curves, for example from Scandinavia, show that during the course of 10,000 years the crust has undergone major adjustments in the form of uplift to compensate for the unloading of ice after the last glaciation. This is geologically a very fast response and we can assume that major sedimentary basins are in approximate isostatic equilibrium with regard to most geological processes. Nevertheless, even the filling of water reservoirs for hydroelectric plants causes local subsidence.

For the most part, then, we can apply the classical Airy isostasy model to sedimentary basins, which enables us to draw a number of interesting conclusions.

Uplift and subsidence are also linked to variations in geothermal gradients and heat flow in sedimentary basins.

Movement of rocks in relation to the surface affects the geothermal gradients. Erosion removes the uppermost, colder strata so that warmer strata come closer to the surface, and the geothermal gradient increases. As a result of subsidence of a sediment basin and sedimentation, heat flow upward will be partly offset by rock subsidence, giving lower geothermal gradients. Sedimentary basins with high rates of sedimentation are therefore often characterised as "cold basins". Tectonic elevation and erosion will increase geothermal gradients.

The geothermal gradient in seafloor rocks is consistently greater than it is over the continents, and on the continents it is highest in areas of volcanic activity.

Areas with a high geothermal gradient due to volcanism will slowly cool down to normal gradients when volcanic activity ceases. It may take about 100 million years before a normal geothermal gradient is re-established, due to cooling and contraction in accordance with the crust's coefficient of expansion and isostatic subsidence due to increased density. Seafloor basalt is hot and flows at relatively shallow depths in the earth's crust, and the spreading oceanic ridges are only about 2.2 km below the surface of the sea. After about 180 million years of cooling, the water depth at isostatic equilibrium is about 5.7 km without sediment loading. With sediment loading the oceanic crust may subside to about 17 km (Fig. 8.7). This is the theoretical maximum thickness of a sedimentary sequence overlying oceanic crust. Sediment basins on the continental crust have a sedimentary thickness which is a function of the thickness of the crust and the density of the sediments and the basement. The thinner the continental crust beneath a sedimentary basin, the more sediments can accumulate while maintaining isostatic equilibrium (Fig. 8.8). Along continental margins sedimentary basins have formed on thin (stretched) continental and oceanic crust.

The formation of sedimentary basins clearly requires that the density of the rocks below the basin is greater than that of the rocks which surround it. The sediment and water which fill the basin are lighter and have to be compensated for with heavier rocks below the basin. We can assume a depth of compensation of about 100 km. This means that the weight of a column of rock plus any water present down to a depth of 100 km must be the same everywhere. Subsidence of a sedimentary basin may be due to several crustal processes:

1. Stretching and thinning of the continental crust. Heavier mantle rocks then make up a greater percentage of the rock column down to a compensation depth of about 100 km.
2. Cooling, i.e. lower geothermal gradients in the crust, lead to contraction and a higher rock density (thermal contraction), and this will result in subsidence.
3. Increased loading by water, sediments or other rocks will cause subsidence. Water loading could be due to a transgression increasing the weight of the water column. Sediment loading occurs with basin infilling.
4. Subsidence along subduction zones. This results in lower geothermal gradients in the downward-deflected crust, so that the density also increases.
5. Tectonic loading. Thrusting of tectonic plates leads to increased loading on the part of the earth's crust in question and we have subsidence, especially in front of nappes (foreland basins).

8.8.2 Changes in Sea Level and Sedimentation and Isostatic Compensation

Variations in sea level due to eustatic transgressions or tectonic subsidence will represent loading or unloading of the crust which will reinforce the primary change in sea level. If the sea rises 100 m, for example due to ice sheet melting, this will increase the isostatic loading on the seafloor. We can calculate that there will be a further 43.5 m of subsidence, so that the total increase in the depth of water at equilibrium will be 143.5 m.

If a sedimentary basin with this depth of water is filled with sediment, there will be further isostatic subsidence because of the weight of sediments, providing accommodation for deposition totalling 250–300 m

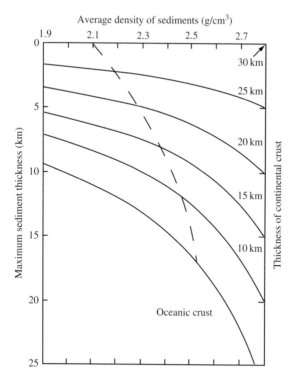

Fig. 8.8 Potential loading capacity to isostatic equilibrium on oceanic crust and of various thicknesses of continental crust as a function of sediment density (after Kinsman 1975). If the average sediment density is 2.5 g/cm³, the maximum thickness of sediments can be 17 km on top of the oceanic crust. If there is 20 km of continental crust and the sediment density is 2.3 g/cm³, there is room for 4 km of sediments

depending on the density of the sediments. A 100 m rise in sea level will thus lead to deposition of almost 300 m of sediment. In the same way, primary tectonic subsidence due to cooling of the ocean floor will lead to further subsidence due to increased water and sediment loading.

Using stratigraphic data in the form of measured profiles or oil wells as our starting point, we can calculate backwards to the primary tectonic or eustatic changes in sea level. This method is called "backstripping".

$$Z = Y((\rho_m - \rho_s)/(\rho_m - \rho_w))\Delta H \rho_w/(\rho_m - \rho_w) + (H - \Delta H)$$

where Z is the primary tectonic subsidence, F is a factor which is a function of the rigidity of the earth's crust, ΔH is the change in sea level and H is the water depth (Watts 1983). Y is the sediment thickness compensated for compaction, i.e. the sediment thickness (and density ρ), shortly after deposition (Stickler and Watts 1978b, Bally et al. 1981). If we know the variations in sea level and the depth of the water from environmental interpretations, for example, these can be inserted into the equation so that the primary tectonic movement can be worked out.

The backstripping technique which is used to reconstruct the subsidence history of different parts of a sedimentary basin lends itself very well to computer modelling. The data obtained on the depth and temperature history of the source rock in particular has allowed much better assessments of kerogen maturity and the times of oil expulsion and migration. One parameter which is crucial but difficult to estimate is the variation of the geothermal gradient as a function of geological time. It is also often difficult to estimate the palaeodepth during the deposition of different sedimentary formations. Whether a formation was deposited at a depth of 200 or 1,000 m will make a very significant difference, and it is often difficult to make accurate palaeoenvironmental estimates of the palaeobathymetry.

To form a sedimentary basin with a thick infill of sediments requires a crustal depression large enough to provide the *accommodation space* for the deposits. This may be a result of large-scale crustal movements like seafloor spreading and crustal thinning (extension). The supply of sediments is also critical and

clastic sediments have to be supplied from adjacent land areas which are being uplifted and eroded.

Chemical and biogenic sediments are formed locally from seawater and thus are not reliant on sediment supply from land areas. This typically applies to limestones, which accumulate where there is little clastic sedimentation. Sedimentary basins that are more or less isolated from the sea in arid regions can be filled up with evaporites at a rather high sedimentation rate.

8.9 Continental Rifting

Stretching and thinning of the continental crust takes place by tensional tectonics in connection with rifting. Crustal stretching may be a result of tensional forces and uplift due to the high geothermal gradients associated with rifting. This leads to a thinning of the continental crust, causing isostatic subsidence.

Fracture zones in the continental crust (rift valleys) produce subsidence which often is located in a sedimentary basin because the continental crust is thin and because heavy rocks from the mantle push their way up, thus increasing the average density of the rocks. Those parts of the rift valley system which have much volcanism will have less space for the sediments. Along the margins of a rift system, where the continental crust is not stretched, uplift occurs due to the higher geothermal gradient. This causes the basement rock at the surface to slope away from the rift valley as the geothermal gradients are reduced. This will to some extent be compensated for by the formation of erosional valleys which cut backwards into the raised shoulders on the sides of the rift valley.

Because of the high relief around these basins and the short transport distance for the sediment eroded from the bedrock, these basins will be characterised by mineralogically very immature sediments, largely arkoses and conglomerates deposited in fan deltas along the active faults. In the central and deeper parts of the basins we find deposition of finer-grained sandstones and clayey sediments. Rift valley basins may be continental lacustrine basins as in East Africa, or marine as those offshore East Africa and the Jurassic basins of the North Sea. Horsts, which are unstretched (thick) pieces of continent crust, may become topographically very high due to high heat flow. The Ruwenzori Mountains of East Africa, reaching more than 5,000 m, are one example. Both marine and lacustrine rift basins will tend to have reducing conditions in the deeper part due to limited circulation of oxygenated water.

Rift basins formed in areas with wet climates will be occupied by large lakes. Lacustrine basins often have an even better potential for producing source rocks than marine rift basins, because water stratification (density stratification) is usually more marked in lakes. We will therefore often find black, organic-rich shales in these basins.

Rift basins formed in arid zones are characterised by evaporite deposits. Block faulting will readily lead to isolated basins or horsts which cut off contact with the open sea. Evaporites are typical of rift deposits today, e.g. in East Africa, and were widespread in Europe and North America during the Permo-Triassic, before the ocean-floor spreading which created the Atlantic Ocean started in the Mid-Jurassic. The Zechstein Salt deposits in Germany and the North Sea are typical examples. Jurassic and early Cretaceous rifting during the early phase of the opening of the South Atlantic Ocean evaporite basins was located in the arid regions of the time. They can form almost perfect cap rocks that are not likely to leak. If they are thick enough they may form salt domes which produce structural traps in overlying rocks, as at Ekofisk and in the Gulf Coast basin. Rising salt domes also strongly influence the clastic sediment distribution in a basin.

Salt has high conductivity causing the temperatures above the salt to be higher than normal and the sediments below the salt to be cooler than normal. This must be taken into account when modelling the maturation of source rocks associated with the salt. These temperature anomalies may also be important when modelling diagenesis and reservoir properties. Large subsalt discoveries have been made in recent years both in the Gulf Coast and offshore Brazil.

The subsidence due to rifting renders the adjacent rocks unstable and promotes gravitational sliding of blocks in the crust, in towards the rift structure. There is a tendency for listric faults to form, i.e. parallel, curved fault planes which start as normal faults and curve round with depth until they are almost horizontal. The blocks then become rotated so that they turn over and slope away from the rift. During the initial part of the spreading phase, basins with limited circulation will be formed so that evaporites and carbonates are often deposited. Upper Jurassic and Lower

Cretaceous deposits of this sort are found extensively along the margin of the Atlantic Ocean.

8.10 Subsidence Along Passive Margins

Passive margins develop from a rift phase to a spreading phase. The transition between continental crust and oceanic crust therefore consists of a thinned continental crust with listric faults and horsts formed during the rifting phase (Fig. 8.9). As ocean-floor spreading progresses, the geothermal gradients in this part of the continental shelf will decline, resulting in cooling and thermal subsidence of the continental margins. The oceanic crust will also experience thermal subsidence, as a function of the age of the seafloor. Subsidence flexure will develop where the subsidence is most rapid, on the outer parts of the continental shelf and slope nearest the oceanic crust. Further in from the passive margin the subsidence rate will be slower. Eventually sedimentation fills the prism between the oceanic and continental crust. With cooling, the rigidity of the crust increases, so that the bending of the continental crust near the continental margin broadens and there is overall subsidence of the continental shelf, and in consequence transgression and onlap. It will subside isostatically and make accommodation space. During high sea level the clastic supply is pushed back onto the continent and during low stand a progradational sequence is formed (Fig. 8.10).

In other places the drainage discharge from land focuses the sediment in large delta areas. This is the situation with the Mississippi delta in the Mexican Gulf, which has been receiving sediment from large expanses of the North American continent since Mesozoic times.

The sediment supply is controlled by the drainage system on the continents, which may follow old rift systems because most of the sediment is produced on the continents and has plenty of space to be deposited along the margin of the deep ocean.

The Niger delta represents a similar focusing of sedimentation on the African side of the Atlantic. Sediment basins along passive margins are deposited above old rift basins which have cooled. If basin subsidence is rapid, this will also contribute to a low geothermal gradient because the sediments have to be heated during burial (20–30°C/km). A relatively

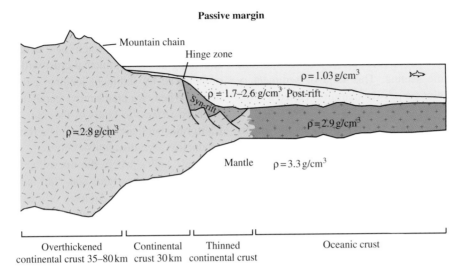

Fig. 8.9 Simplified cross-section of a passive continental margin. The sediments that were deposited during the initial phase of rifting lie beneath the younger sequence which was deposited along the passive margin. The thinner the crust is, the more sediments can accumulate and the maximum thickness is reached when the progradation reaches cold oceanic crust

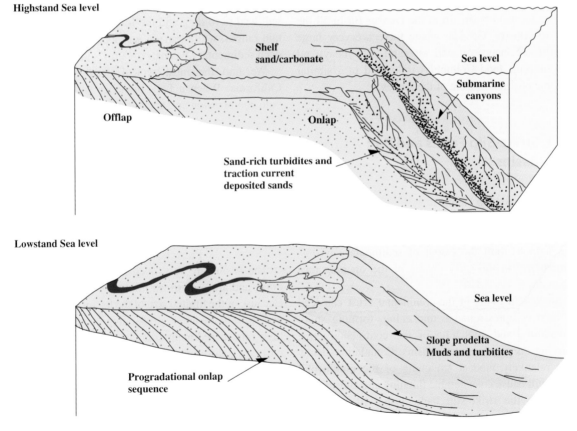

Fig. 8.10 A modal for transgression with sediment starvation and renewed progradation across the shelf onlapping (downlapping) onto an unconformity

great thickness of sediment (4–5 km) must therefore be deposited for the underlying source rocks to achieve maturation. That is to say, attain 100–150°C, depending on the subsidence rate and hence the time available for the heating. This kind of sedimentary basin situated along a passive margin is called a "cold basin".

Foreland basins form in front of mountain chains when they are uplifted and eroded. The advancing nappes help to depress the continental crust and provide accommodation space for the sediments shed from the mountains. Foreland basins tend to be broad and gently folded into giant structures in the distal parts. The Middle East is a large foreland basin in front of the alpine mountain chain running along Iran and Turkey. The Persian Gulf is a continuation of this basin collecting sediments coming from the mountains of Iran. The Jurassic and Cretaceous sequences in Iraq, Saudi Arabia, Kuwait and The Emirates form some of the largest oil fields in the world.

On the east side of the Rocky Mountains we have similar foreland basins from the Denver Basin all the way to Alberta, Canada, where there are very large deposits of heavy oil and tar sand. The heavy oil in Venezuela (and Columbia) is located in a similar tectonic position in front of the Andes.

8.11 Strike-Slip Faults and Pull-Apart Basins

We find "strike slip" faults in both the oceanic and the continental crust. Transform faults in the oceanic crust result in ridges of younger basaltic material which can help to limit the extent of sedimentary basins, particularly in the early phases of the opening (see Chap. 6).

Strike-slip faults in the continental crust can lead to both compression, i.e. thickening, forming small mountains, and stretching of the continental crust forming deep sedimentary basins. Calculations show that relatively modest stretching of the continental crust will cause considerable thinning and subsidence. Bends in the strike-slip fault system like the San Andreas Fault open up deep holes in the continental crust which can be filled in with very thick sequences of sediments like in the Ventura and Los Angeles Basins. The sedimentation rate may be very high

because of the large uplifted land areas and the relatively small basins.

Here, too, relief which has developed through stretching of the continental crust will be made more pronounced by sediment loading and thermal subsidence. When two plates move parallel to one another, we have *strike-slip* faults of the San Andreas type in California, but there is no new plate formation, nor any subduction. We distinguish between "right-lateral" or "dextral" faults, where the opposite side of the fault has moved to the right, and "left-lateral" or "sinistral" faults. The San Andreas is a dextral fault, and the western part of California (Salina block) has moved northward in relation to the North American continent. If the fault plane follows a completely straight structure in the rock parallel with the direction of movement there will be neither tension nor compression along the fault plane. However, faults usually follow older structures in the basement rocks which may be curved and strike slip movements may then produce both compression and tension along the fault plane. Compression will lead to folding and the formation of small mountain ranges, while tension will lead to the opening of deep sedimentary basins.

We find the most typical examples of this in California. When a fault branches we may also see both compression and tension, with elevation and subsidence respectively of blocks, depending on their orientation. If the fault shifts to another parallel fault plane, we have crustal tension in the area in between, and often also basalt flows. A so-called "pull-apart" basin forms, which is a "hole" in the continental crust formed by the strike slip movement. The Salten Trough in Southern California is a good example and much of the Los Angeles and Ventura basins originated in this way during the Pliocene.

The basins are very deep holes in the continental crust, and are sometimes floored with volcanic rocks. Because of the limited size of the basins and the continued uplift of the areas surrounding them, we get a very high rate of sedimentation. In the Miocene-Pliocene basins in California, 5–8 km or more of sediments have been deposited in roughly as many million years. This corresponds to an average of 1 mm/year, which is a very high figure in this context. The sediments consist largely of proximal turbidites along the edge of the basin and distal turbidites and fine-grained clay sediments in the more central, deeper parts. In the Ventura Basin which continues into the Santa Barbara

Channel, there are about 15 km of young sediments (<4–5 million years) which have been folded and which contain large amounts of oil.

Basins of this type provide almost ideal conditions for oil accumulation and are amongst the richest oil regions in the USA. Very few basins can compete with the California strike-slip related basins in terms of barrels of oil per square kilometre. This is because these deep basins often have reducing conditions in their bottom water due to limited circulation of oxidised water. We therefore get organic-rich sediments deposited. Turbidite sandstones, particularly the more proximal parts, often have sufficiently high porosity to become reservoir rocks. Even if the porosity is not especially high, turbidite deposits often form thick sequences that extend over large areas. The reservoir quality can also be controlled by turbidite lobes or channels with levees.

Sediments deposited in basins of this type tend to undergo folding, which will result in good structural traps. Geothermal gradients will be relatively high, and allow thorough maturation.

8.12 Converging Plate Boundaries

When plates collide (converge), there can be boundaries between:

- Oceanic crust and continental crust, such as along the west side of South America (Andes type).
- Continental crust against continental crust (Alpine or Himalayan type mountain chain).
- Oceanic crust against oceanic crust, with formation of volcanic islands.

The oceanic plate has a higher density than the continental crust, so in a subduction zone the oceanic plate will be carried downwards under the continental one. The oceanic plate is relatively cold and its geothermal gradient is low because since it is descending, the upward heat flux is slower. Where there is active subduction, ocean trenches are formed, such as the 10–11 km deep ones outside the Philippines and Japan.

In subduction zones the downward-moving part of the oceanic crust is characterised by extremely low geothermal gradients. This is because the heat flow must move counter to the direction of movement of the

plate undergoing subduction which is therefore colder and denser than the average old ocean floor crust. The subduction zones form the deepest parts of the ocean (up to 10–11 km deep) where there is not a great deal of sedimentation. Ordinary, cold, 100–150 million year old ocean floor is in isostatic equilibrium at a depth of 5–7 km without sediment loading.

Where we have subduction of oceanic crust beneath continents we find the following types of basin:

- Trench basin (oceanic sedimentation within the trench).
- Accretionary prism.
- Fore-arc basin (relatively shallow basin in front of island arcs).
- Inter-arc basin (sedimentary basin between island arcs of continental crust).
- Back-arc basin (sedimentary basin on the oceanic crust).
- Retro-arc basin (formed if continental blocks are upthrust and faulted, creating a secondary basin in the continental crust).

Trench basins seldom contain large amounts of sediment if separated from the continent by an island arc. One of the reasons why they are the deepest oceanic areas is because they do not fill up with sediment. This is because the ocean floor beneath trench basins is part of a dynamic system and is being consumed through subduction as sedimentation occurs. The downward movement of the oceanic crust helps to maintain a low geothermal gradient in the trench since heat must flow up through rocks that are descending. Because of the continual lateral displacement of the rock floor under the trenches, the sediment thickness will be limited. The island arcs are the main source of sediment supply to the trench basins.

Most of the sediment coming from the continents will be trapped in back-arc basins. The sediments in trench basins consist of deep-water conglomerates, turbidites and pelagic sediments. Trench basins lie mostly below the carbonate compensation depth and therefore contain little carbonate. Sedimentary transport of very fine-grained pelagic sediments tends to occur along the axis of the trench. The inner slope up to the island arcs is steeper than the outer one, and the subduction zone (Benioff zone) is at the foot of this slope (Fig. 2.48).

When the oceanic plate descends into the Benioff zone, sediments deposited on the sea bed may be scraped off, forming wedges of sediment. This is called an *accretionary prism* and consists of a complicated pattern of folded and brecciated sedimentary rocks exhibiting overthrusts and underthrusts (Fig. 2.49). The sedimentary sequence is the right way up within each thrusted unit, but the sequence of thrusted units is inverted. The youngest sediment wedges are at the bottom because they are the last sediments to have been scraped up from the seafloor.

On the inside of the trench, slope basins are formed with sediment supplied by the island arcs. These deposits are folded, deformed and reworked by the subduction processes.

Although sediments in trench basins may be rich in organic material, their oil potential is poor due to a lack of reservoir rocks and the intense tectonic deformation. In addition the geothermal gradient will also be lower than in other basins, at least at the time of deposition.

Fore-arc basins form between the actual island arcs with volcanoes, and the subduction trench. They are not subjected to the intense tectonic deformation which occurs in trench basins. Fore-arc basins often transgress onto the island arcs as the belt of volcanic activity gradually withdraws towards the continent. These basins will contain fluvial and deltaic sediments closest to the island arcs, then shallow marine continental shelf sediments. In the outer parts sediments are deposited in deeper water towards the oceanic slope. There is good potential here for organic-rich sediments to accumulate, but the geothermal gradient may be low despite the proximity to a volcanic island chain, and maturation consequently slow.

Large parts of California in front of the Rocky Mountains (Sierra Nevada) are a fore-arc basin (San Joaquin Basin) and there are numerous oil fields here, not least in the area around Bakersfield.

Intra-arc basins are a result of tensional tectonics in the island arcs within an area otherwise characterised by convergent plate movement. Here a system of fault-controlled basins (graben) is formed. Sediments deposited in these basins will for the most part be immature, and volcanic material will be common. Volcanic sandstones will often lose their primary porosity rapidly due to unstable minerals and diagenetic transformation. The geothermal gradient is high, but conditions for formation of source and reservoir rocks are not very good.

Intra-arc basin sediments will frequently be subjected to intense tectonic deformation which may deform potential reservoirs.

Back-arc basins develop on the oceanic crust due to seafloor spreading behind the island arcs. There is likely to be an abundant supply of sediment from the continent, and back-arc basins may also fill with deltaic and shallow marine sediments. Because of later tectonic deformation, back-arc sedimentary basins are not the most promising oil prospects.

The Jurassic and Cretaceous seaway through North America east of the Rocky Mountains was a back-arc basin during subduction of the Pacific plate. The area is a good oil region but Tertiary uplift resulted in a very strong meteoric water drive which flushed out many of the oil fields.

8.13 Conclusions

Stratigraphy is a response to changes (movements) in the Earth's crust and sediment supply. Changes in the environments and the biological production are also important.

Geophysical methods, mainly seismics, are used to map out sedimentary basins on a regional scale and also on a small scale for detailed exploration and production. It is also important to reconstruct the evolution of sedimentary basins though time and the changes with respect to the sediment supply and the environment.

This must to a large extent be based on reconstruction of plate tectonic movements and basin modelling.

Further Reading

Berg, O.R. and Woolverston, D.G. 1985. Seismic stratigraphy II. An integrated approach to hydrocarbon exploration. AAPG Memoir 39, 276 pp.
Brown, A.R. 1996. Interpretation of three-dimensional seismic data. AAPG Memoir 42, 424 pp.
Sclater, J.G. and Christie, P.A.F. 1980. Continental stretching: an explanation of the post-mid Cretaceous subsidence of the

Central North Sea Basin. Journal of Geophysical Research, 85, 3711–3739 [Appendix A, 3730–3735].

Tearpock, D.J. and Bischke, R.E. 1991. Applied Subsurface Geological Mapping. Prentice Hall, Englewood Cliffs, NJ, 648 pp.

Vail, P.R. et al. 1977. Seismic stratigraphy and global changes of sea level. I. In: Payton, C.E. (ed.), Seismic Stratigraphy – applications to hydrocarbon exploration. AAPG Memoir 26, 49–212.

VanWagoner, J.C. et al. 1990. Siliciclastic sequence stratigraphy in well logs, cores and outcrops. AAPG Methods in Exploration Series 7, 55 pp.

Payton, C. 1977. Seismic stratigraphy – applications to hydrocarbon exploration. AAPG Memoir 26, 516 pp.

Suppe, J. 1985. Principles of Structural Geology. Prentice-Hall, Inc., Englewood Cliffs, NJ, Chapter 1, 17–24.

http://strata.geo.SC.edu

Chapter 9

Heat Transport in Sedimentary Basins

Knut Bjørlykke

The temperature increases downwards in the crust and there is therefore a transport of heat upwards, referred to as the heat flow. Most of the flow is by conduction (thermal diffusion). Flow of porewater will also transport heat in the subsurface but the flow rates in sedimentary basins are normally so small that we can ignore the contribution from fluid flow (advection). Around igneous intrusions there is usually thermal convection with high flow rates and heat transport. In shallow areas with high flow rates of meteoric water, advective heat transport is also significant.

The source of the heat is mainly radioactive processes which are particularly important in the continental crust due to the enrichment of uranium, thorium and potassium in granitic rocks.

Heat is transported through sedimentary basins mostly by conduction following the heat flow equation:

$$Q = c \times \frac{dT}{dz} \qquad (9.1)$$

The thermal conductivity (c) is expressed as $Wm^{-1}{}^{\circ}C^{-1}$ or $Wm^{-1}{}^{\circ}K^{-1}$ which is the heat (W) transported over a given distance (m) with a certain drop in temperature ($^{\circ}C$). Temperature is expressed as Celsius ($^{\circ}C$) or Kelvin (K) but Fahrenheit was and still is commonly used in the USA. Conductivity may also be expressed in terms of calories (cal), which is an alternative unit for energy/heat. 1 W equals 1 J/s or 0.239 cal/s.

The heat flow (Q) is most commonly expressed by W/m^2 but can also be expressed as cal/cm^2s or joule/cm^2s. A heat flow of 70 mW/m^2 corresponds to 1.4 $\mu cal/cm^2$s. This is the heat flow unit (mW/m^2) which may be referred to as HFU. The temperature of a volume of rock is a function of the heat flux and the conductivity of rocks and fluids. The increase in temperature with depth (temperature gradient) is called the geothermal gradient (dT/dz). Typical geothermal gradients may also be written as ∇T and in sedimentary basins are usually 25–45°C/km.

$$Joule = \frac{kg \times m^2}{s^2}$$

$$Watt = \frac{kg \times m^2}{s^3}$$

$$1\ Joule = 0.239 cal$$

$$1\ Watt = 1\ joule/s$$

In a sedimentary basin there is the background heat flux from the underlying basement, and granitic rocks have higher heat production and temperatures than basic rocks. The sedimentary sequences overlying the basement also produce heat by radioactive reaction and black shales with a high content of organic matter often have a relatively high uranium content. This additional heat source may be significant in terms of increasing the heat flux and the geothermal gradients in sedimentary basins.

The temperature distribution (geothermal gradients) in sedimentary basins can vary regionally and over geologic time. This determines both the generation and expulsion of petroleum, and it also strongly influences the reservoir quality. It is therefore important to understand the processes that control heat transport in sedimentary basins.

K. Bjørlykke (✉)
Department of Geosciences, University of Oslo, Oslo, Norway
e-mail: knut.bjorlykke@geo.uio.no

K. Bjørlykke (ed.), *Petroleum Geoscience: From Sedimentary Environments to Rock Physics*,
DOI 10.1007/978-3-642-02332-3_9, © Springer-Verlag Berlin Heidelberg 2010

The temperature gradient ∇T (dT/dz)

$$= \text{Heat flow } (Q)/\text{Conductivity } (c). \quad (9.2)$$

In rocks with low conductivity (like mudstones and shales) the geothermal gradients will be high and in highly conductive rocks (like salt) the geothermal gradients will be low (Fig. 9.1). The thermal conductivity of salt (halite and anydrite) is 5.5 $Wm^{-1}°C^{-1}$, while shales may have conductivities between 1.0 and 2.5 $Wm^{-1}°C^{-1}$. Sandstones and limestones have values between shales and salt.

In a situation where a rock is filled with stationary porewater the total heat flux (Q) is the sum of the heat conducted through both the rock's matrix and pores (porosity φ filled with fluids):

$$Q = C_r (1 - \varphi) + \varphi C_f \quad (9.3)$$

C_r is the conductivity of the solid rock and C_f is the conductivity of the fluids (usually water) in the pore space. Water (fresh) at 20°C has a conductivity of 0.6 W/mK while seawater and saline brines are much more conductive. The conductivity of common sedimentary minerals ranges from 7.7 $Wm^{-1}°C^{-1}$ for quartz to 1.8 for illite and smectite. The conductivity is therefore to a large extent a function of the quartz content and the water content (porosity).

The conductivity of shales from the North Sea ranges from about 0.8 to 1.1 $Wm^{-1}°C^{-1}$ (Midttømme et al. 1997) so they are not very much more conductive than water. The conductivity parallel to bedding may be up to 70% higher than perpendicular to bedding.

Most of the heat transport is vertical except around hydrothermal or igneous intrusions, but in the case of steeply dipping beds the conductivity would be higher.

Over a limited vertical interval of the sedimentary section the heat flow may be relatively constant and we see from Eq. (2) that the geothermal gradient is inversely related to the conductivity.

When there are rocks with low conductivity near the surface the geothermal gradient will be higher so that the underlying sediments will be warmer. This is called a blanketing effect. Mudstones with low thermal conductivity on top of granites or older sedimentary rocks will have this effect.

Salt with high conductivity has the opposite effect. Because the temperature gradient through salt is low, the temperature will be relatively high at the top of the salt and low at the bottom. This has consequences for maturation of source rocks. This is a very important effect for petroleum prospects below thick salt layers, i.e. in the Gulf of Mexico, offshore Brazil, West Africa and the North Sea. The temperatures below the salt will be significantly lower than normal at this depth. This means that the reservoir quality of sandstones reservoirs will be better due to less quartz cement. Lower temperatures will also preserve more petroleum as oil or condensate as there will be less cracking to gas.

The heat flux is only constant in an equilibrium situation. When sediments subside they are heated and a part of the background heat flux is used to heat the subsiding rocks (Fig. 9.2). This is equal to the heat capacity of the rocks and the subsidence rate. In basins with high sedimentation rates the heat flow is strongly reduced and in the Plio-Pleistocene depocentres the geothermal gradients are down to 20–25°C/km (Harrison and Summa 1991).

During subsidence and sedimentation the sediments must be heated, and this heat is taken from the background heat flow and the geothermal gradient is reduced.

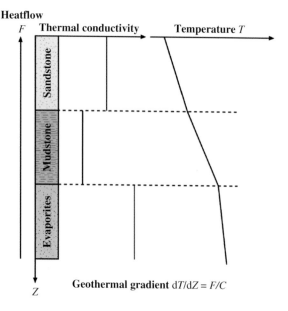

Fig. 9.1 Relationship between heat flow, conductivity and geothermal gradients. The geothermal gradients have an inverse relation with the conductivity for the same heat flux. Thick salt layers or domes cause higher geothermal gradients above the salt and low temperatures below the salt

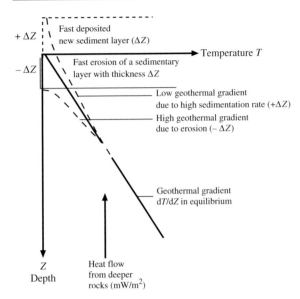

Fig. 9.2 Geothermal gradients as a function of rapid uplift (erosion) or subsidence (sedimentation). During subsidence some of the heat flow is used to heat the subsiding sediments and underlying basement and this will reduce the geothermal gradients, forming cold basins. During uplift the heat from cooling rocks will add to the heat flux, producing steeper geothermal gradients

During subsidence the heat flux is:

$$Q_s = Q_b - F_s \cdot C_{hw} \cdot dT/dZ$$

The heat capacity of water $(C_{hw}) = 4{,}200 \, J/kgK$.

Here Q_b is the background heat flux from the basement, F_s is the subsidence rate (downward flux of rocks) and C_{hw} is the heat capacity of the rocks. When rocks are uplifted and cooled the heat given off from the cooling rocks adds to the background heat flow: During uplift the heat flux becomes:

$$Q_s = Q_b + F_u \cdot C_{hw}$$

Here F_u is the rate of uplift.

The heat capacity of the mineral matrix may be estimated at about 8–900 J/kg K.

In terms of volume the heat capacity of rocks is however close to 2,500 J/dm^2/K.

This means that at 20% porosity about 2.5 times as much heat is stored in the mineral matrix (density 2.7) as in the water phase. During the first period of advective flow along a fault or through permeable sandstone beds a high percentage of the advected heat will be lost by conduction to the mineral matrix.

9.1 Heat Transport by Fluid Flow

When fluid, usually water, is transported in a sedimentary basin there is also heat transport unless the transport is parallel to the isotherm (Fig. 9.3).

The advective heat transport Q_t is proportional to the flux of water (Darcy velocity $F = m^3/m^2$), the heat capacity of water (C_{hw}), and the geothermal gradient (∇T).

$$Q_t = F \cdot C_{hw} \nabla T \cdot \sin \alpha$$

Here α is the angle between the direction of fluid flow and the isotherms which are lines with equal temperature.

During compaction-driven flow the flow rates are in most cases too small for this heat transport to be significant. Focused compaction-driven flow may cause a significant heat flow by advection, but only if the rate of porewater flow is very high. The average water flow upwards relative to the sediments is very low, but relative to the seafloor the porewater is in most cases sinking.

Meteoric water fluxes along aquifers into sedimentary basins are many orders of magnitude faster than in compaction-driven flow; in some cases the downwards flow of cool meteoric water from mountains into sedimentary basins may cause a significant reduction in the geothermal gradients.

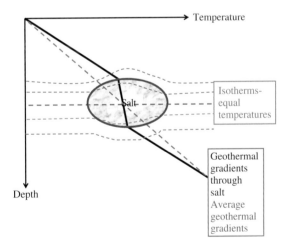

Fig. 9.3 Geothermal gradients are strongly influenced by layers of salt or salt domes. Since the heat flow is relatively constant the geothermal gradient must be very low through the highly conductive salt. As a result the overlying sediments will be warmer than normal while the underlying sediments will be cooler

9.2 Heat Transported by Conduction and by Fluid Flow (Advection)

The relative contribution from these types of heat transport can be expressed by the Peclet number (Pe):

$$\mathrm{Pe} = \rho_f\, C_f\, Q_z\, L / C_r\, (1 - \varphi) + \varphi C_f$$

Here ρ_f is the fluid density, C_f the heat capacity of the fluid, Q_z the vertical component of the Darcy velocity, L the length of the flow path, and C_r and C_f the respective thermal conductivities of the solid phases (minerals) and the fluids (water) (Person and Garven 1992).

The conductivity of water depends on temperature and salinity but is much lower than that of the matrix (0.6 W/m°C and 2.5–3.5 W/m°C, respectively). Hot porewater therefore rapidly looses its heat to the mineral matrix. Convection is driven by the primary temperature gradients. Porewater convection does change the temperature field but temperature perturbations due to this flow are not very large (Ludvigsen 1992).

Numerical calculations of fluid flow in modern sedimentary basins like the Gulf of Mexico basin show that compaction-driven porewater flow is quite insignificant in terms of advective transport of heat (Harrison and Summa 1991). In the North Sea basin, too, the geothermal gradients only vary within rather narrow limits (35–40°C/km). The occurrence of locally higher values in the offshore Western Norway has been attributed to the effect of recent glacial erosion producing transient thermal heat flows (Hermanrud et al. 1991).

In the Mississippi Valley, USA, compaction-driven flow has been shown to be quite insufficient to generate hot fluids capable of precipitating ores (Bethke 1986). Compaction-driven flow from thrust belts may produce significant thermal perturbations on a relatively local scale, but modelling suggests that such flow is insufficient to cause large-scale thermal anomalies in the adjacent foreland (Deming et al. 1990). In continental rifts like the Rhine Graben, where the rift margins are exposed and elevated topographically, groundwater flow can to a large extent explain the observed thermal anomalies (Person and Garven 1992).

9.3 Importance of Heat Flow and Geothermal Gradients

Heat flow is a very important parameter, which strongly influences geothermal gradients and rates of petroleum generation. It also strongly influences rates of quartz cementation and other types of chemical compaction in siliceous sediments.

Heat flow and geothermal gradients are also very important for the utilisation of geothermal energy and heat pumps in the ground or in rocks.

In sedimentary basins we have a heat flow from the basement into the overlying sedimentary sequence. The composition of the basement rock determines the rate. Granitic rocks with high potassium and uranium content will produce more heat than rocks like anorthosites and gabbros which are very low in potassium. Offshore Norway the background heat flow varies significantly depending on the basement rocks. Organic-rich shales like the Upper Jurassic source rocks from the North Sea basin may also contribute significant heat because of the radioactivity (high uranium content).

As we have seen above, high sedimentation rates will reduce the geothermal gradient because some of the heat flux is used to heat the subsiding sediments. Cold basins with rapid subsidence and low geothermal gradients (<20–25°C/km) require deep burial of the source rocks before they can generate petroleum, both because of low temperature and the short geologic time (<2–3 million years) for petroleum generation. The small time/temperature integral will also result in little quartz cement in reservoir sandstones. The amount of quartz cement can also be used as a measure of the time/temperature index. The temperature history is an important parameter in basin modelling because it influences the sediment density and therefore the rate of subsidence and the generation of hydrocarbons, and also the reservoir quality.

The heat flow and the geothermal gradients may change over geologic time and that complicates basin modelling. In subsiding basins, however, it is the geothermal gradients during the last 20–30% of the subsidence which are most important. In the North Sea Basin we may have had relatively high geothermal gradients in late Jurassic times, but both the source rocks and the reservoir rocks were then only buried to rather shallow depths and the temperatures were still relatively low, except close to volcanic intrusions and hydrothermal activity.

Further Reading

Bethke, C. 1985. A numerical model of compaction-driven groundwater flow and heat transfer and its application to the paleohydrology of intracratonic sedimentary basins. Journal of Geophysical Research 90, 6817–6828.

Bethke, C.M., Deming, D., Nunn, J.A. and Evans, D.G. 1990. Thermal effects of compaction-driven ground water flow from overthrust belts. Journal of Geophysical Research 95, 6669–6683.

Demongodin, L., Pinoteau, B., Vasseur, G. and Gable, R. 1991. Thermal conductivity and well logs: A case study in the Paris Basin. Geophysical Journal International 105, 675–691.

Harrison, W.J. and Summa, L.L. 1991. Paleohydrology of the Gulf of Mexico Basin. American Journal of Science 291, 109–176.

Hermanrud, C., Eggen, S. and Larsen, R.M. 1991. Investigations of the thermal regime of the Horda Platform by basin modelling: Implication for the hydrocarbon potential of the Stord basin, northern North Sea. In: Spencer, A.M. (ed.), Generation, Accumulation and Production of Europe's Hydrocarbon. European Association of Petroleum Geoscientists, Special Publication 1, Oxford University Press, Oxford, 65–73.

Midttømme, K., Roaldset, E. and Aagaard, P. 1997. Thermal conductivities of argillaceous sediments. In: Mcann, D.M., Eddleston, M., Fenning, P.J. and Reves, G.M. (eds.), Modern Geophysics in Engineering Geology. Geological Society Special Publication 12, 355–363.

Person, M. and Garven, G. 1992. Hydrologic constraints of petroleum generation within continental rift basins: Theory and application to the Rhine Graben. AAPG Bulletin 76, 466–488.

Chapter 10

Subsurface Water and Fluid Flow in Sedimentary Basins

Knut Bjørlykke

The pore spaces in sedimentary basins are mostly filled with water. Oil and gas are the exceptions and most of the information we have about fluid flow in sedimentary basins is derived from the composition of water and the pressure gradients in the water phase. It is therefore important to characterise and understand the variations in the composition of these waters. All the porewater may be referred to as *subsurface water* but the water that is analysed from exploration wells or is produced during oil production is usually called *formation water* or *oil field brines*.

The composition of subsurface water can provide vital information for several different practical purposes:

(1) The water composition may give information about the origin of the water and the pattern of fluid flow in the basin.
(2) Differences in water composition with respect to salinity or isotopic content may provide important information about communication (permeability) across barriers like shale layers or faults.
(3) The composition of water produced together with oil may give information about the rock units from which the water is being drained. When water injection is applied to a reservoir, the composition of the injected seawater is different from that of the formation water and this

can be used to determine if there is a breakthrough of the injected water to the production well.
(4) The water composition determines the density of the water column which is required to interpret and calibrate the fluid pressure data from wells.
(5) Water composition and its resistivity are critical for the calibration of well logs. The resistivity is crucial for calculating the formation factor F, which is the resistivity of the formation water relative to the resistivity of the whole fluid-saturated rock.

Water buried with the sediments has long been referred to as *connate* water and thought to represent the original seawater. This is an unfortunate term because the origin of such water is very complex and meteoric water has been found to have a much stronger influence in sedimentary basins than earlier assumed. In addition, seawater changes its composition significantly as soon as it is buried; the first major change is the removal of sulphate ions in the sulphate-reducing zone (Figs. 10.1 and 10.2).

Subsurface waters are mainly derived from:

(1) Seawater buried with the sediments.
(2) Meteoric water, which is groundwater (originally rainwater) that can flow from land to far offshore.
(3) Water released by dehydration of minerals i.e. from gypsum, or clay minerals like smectite and kaolinite.
(4) Hydrothermal water introduced by igneous activity. This is also referred to as juvenile water.

K. Bjørlykke (✉)
Department of Geosciences, University of Oslo, Oslo, Norway
e-mail: knut.bjorlykke@geo.uio.no

K. Bjørlykke (ed.), *Petroleum Geoscience: From Sedimentary Environments to Rock Physics*,
DOI 10.1007/978-3-642-02332-3_10, © Springer-Verlag Berlin Heidelberg 2010

Fig. 10.1 Simplified illustration of how the composition of porewater is influenced by reaction near the seafloor and supply of freshwater into sedimentary basins from land areas

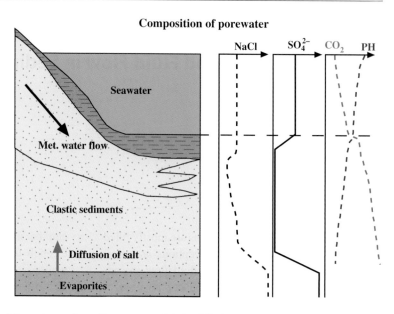

Composition of porewater

Transport and precipitation of elements near the seafloor occurs mainly by diffusion across the redox boundary.

Evolution of the concentrations in water

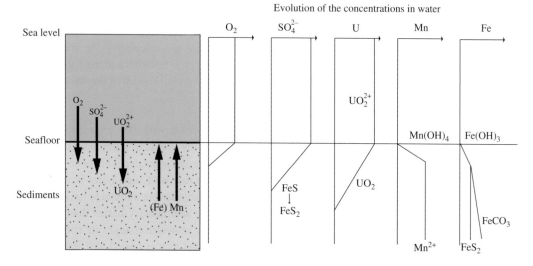

The redox boundary is a function of the supply of free oxygen or sulphate from the seawater and the consumption of oxygen mainly by oxidation of organic matter.

In the absence of H_2S from sulphur reduction, Fe^{2+} is more soluble but may precipitate as $FeCO_3$.

Fig. 10.2 Only a few centimetres below the seafloor there are important reactions between the seawater, which is normally oxidising, and reducing porewater. Elements that are most soluble in the oxidised state (sulphur and uranium) are concentrated below the redox boundary while elements like iron and manganese are precipitated above the redox boundary on the seafloor

10.1 Composition of Formation Water

In most cases the formation water is saline and Cl^- is by far the dominant anion, so that the Na^+/Cl^- ratio is less than 1. The rest of the positive charge is mostly made up of Mg^{2+} and Ca^{2+}. The composition of porewater is a function of its primary origin, modified by the mineral composition of the sediments, the temperature and the quantity of dissolved gases, particularly CO_2.

Pure water of meteoric origin derived from fresh groundwater may gradually mix with more saline water. Meteoric water usually has less than 10,000 ppm dissolved material compared to seawater, which has 35,000 ppm and contains more bicarbonate (HCO_3^-) and small amounts of magnesium, sodium and calcium. Because meteoric water comes from the surface, it brings with it oxygen and bacteria which can break down hydrocarbons (see "Biodegradation"). In sediments with even small amounts of organic matter the porewater quickly becomes reducing. The composition of meteoric water alters as it reacts with the more readily soluble minerals which it passes through. Small amounts of carbonate makes the meteoric water basic and will act as a buffer with respect to pH.

Drilling on the continental shelves, for example off the east coast of the USA and also offshore Africa, shows that meteoric water aquifers are widespread and that water which is virtually fresh is sometimes found below the seabed as far out as 100 km from the coast. In the northern North Sea there is also isotopic evidence that the formation water in shallow reservoirs is partly of meteoric origin.

Marine porewater, which is present in sediments when they are deposited, will initially have an approximately normal salinity but sulphate is removed by sulphate reduction just below the seafloor. Clay minerals act as ion exchangers, absorbing cations like K^+ and Mg^{2+} from the porewater. Compacted clay and mud may function as a membrane. Clay minerals are normally negatively charged on the surface and particularly at the ends, where there are broken silicate bonds. Negatively charged ions are repulsed and held back by the membrane due to the negative charges of the clay minerals. In order for the charges on both sides to equalise, the small ions, particularly H^+, must move in the opposite direction, and as a result there is a higher H^+ concentration (lower pH). This process, which will concentrate salt, is called *salt sieving*. Membranes also discriminate selectively amongst different cations, depending on their size and charge. Those alkali ions which are strongly hydrated (Na^+, Li^+) and also Mg^{2+}, will to a lesser degree be adsorbed onto the surface of the clays and will be more mobile than for example K^+ and Rb^+, which are less hydrated. At higher temperatures the hydration becomes less effective and ions like Mg^{2+} are more available to be adsobed on clay minerals and form carbonate minerals like dolomite.

At $70-100°C$, smectite will dissolve and form illite and water. At $120–140°C$ kaolinite will become unstable and form illite, quartz and water. This process binds cations, particularly K^+, and releases pure water, thus reducing the salinity of the porewater. As a result the porewater in shales often has a lower salinity than that in sandstones at the same depth. The composition of porewater in sandstones varies greatly, depending on whether they contain meteoric water or not.

The quantity of dissolved solids in porewater increases as a function of depth in most cases. Concentrations of 100,000–300,000 ppm of dissolved matter (total dissolved solids, TDS) are typical for basins with evaporite beds. With increasing temperature, the kinetic obstacles to mineral solution and precipitation reactions are reduced. At temperatures above about $80°C$ the porewater will tend to be in equilibrium with most of the minerals present. In sedimentary basins with evaporites, these will greatly influence the composition of the porewater in overlying formations. Dissolved salts move upwards through compaction-driven porewater flow; diffusion due to high concentration gradients also plays a major role. Sedimentary basins along the Atlantic Coast in areas where Mesozoic evaporites are deposited have porewater compositions which are essentially different from those in areas north of this palaeoclimatic belt. The South American continental shelf, the Gulf Coast and the area off the East Coast of the USA are characterised by Jurassic/Cretaceous evaporites. In the North Sea the extension of Zechstein evaporites forms an important boundary which is reflected in the porewater composition of overlying Mesozoic sediments. In addition to chemical analysis, isotope composition ($^{13}C/^{12}C$ and $^{18}O/^{16}O$) can provide valuable information concerning the origin of the porewater.

Many of the dissolved ions in porewater are in equilibrium with the minerals present. They are then not very useful as indicators of the origin of the water. Cl$-$ and Br$-$ are better tracers for fluid flow as they do not react very much with the minerals present.

The water produced from drill stem tests may be strongly contaminated by the drilling mud filtrate and the composition of the mud should be considered when using the analyses of such waters. Water produced during production is less likely to be strongly contaminated because of the larger volume involved. The results of the analyses may be expressed as % or ppm.

$$mg/l = ppm/density of water$$

$$meq/l = mg/l \times valence/mol. weight$$

The main anions in subsurface waters are Cl−, HCO_3^- and SO_4^{2-}, and the main cations are Na^+, K^+ and Ca^{++}.

Meteoric water is characterised initially by low ionic strength but it then reacts with minerals and also amorphous phases like opal A. Unless meteoric water flows through evaporites the chlorinity will remain very low and the main anions will be bicarbonate (HCO_3) or carbonate $\left(CO_3^{2-}\right)$. Sulphate $\left(SO_4^{2-}\right)$ may form in meteoric water due to oxidation of sulphides in rocks and of sulphur in organic matter, but the sulphate will tend to be reduced to sulphides by sulphate-reducing bacteria.

Porewater of marine origin naturally starts with the composition of seawater but only a few centimetres below the seafloor most of the free oxygen is removed from the porewater due to oxidation of organic matter in the sediments. A few metres below the seabed nearly all the sulphate which was in the seawater has been consumed by sulphate reducing bacteria. It is therefore a characteristic of so-called connate water that it has very low sulphate content. The chlorinity, however, remains practically unchanged because Cl− is not consumed by any significant diagenetic process.

Steep concentration gradients and a strong drive for transport by diffusion exist across the redox boundary because of the difference in solubility of many ions between the seawater, which is normally oxidised, and the porewater below. Sulphate $\left(SO_4^{2-}\right)$ is transported downwards and is reduced to sulphides below the redox boundary. Reduced sulphur reacts with iron-bearing minerals including iron oxides (haematite) to form pyrite (FeS_2). Reduced manganese and iron (Mn^{2+}, Fe^{2+}) will be transported upwards and precipitated above the redox boundary. While much of the iron will be trapped in the reduced state as sulphides, manganese sulphides are rather soluble and very little Mn will therefore be trapped below the redox boundary. Ferrous iron (Fe^{2+}) may also be trapped as carbonate such as siderite ($FeCO_3$). Manganese is therefore transported upwards more efficiently than iron and

may form large deposits (as manganese nodules) in deepwater environments where the sedimentation rate is low.

The composition of subsurface water is strongly influenced by the dissolution of evaporites where these are present but the effect can often be shown to be rather local (radius <1 km). Meteoric water can dilute the chlorinity of porewater but mixing of porewater is not very efficient in sedimentary basins. This is because the flow is slow and laminar and pore waters with different salinities have different densities, which also tends to inhibit mixing. Transport by diffusion from high to lower salinity may nevertheless be significant over distances of a few hundred metres, depending on the diffusion constant of the sediment matrix. Low permeability shales also have low diffusion constants. Dehydration of minerals such as gypsum or clay minerals like smectite, kaolinite and gibbsite also causes reductions in salinity because pure crystal-bound water is released into the porewater.

Higher salinities than seawater are in most cases due to the dissolution of evaporites.

High salinity porewater is found in the central North Sea above the Permian evaporites in the Central Graben, while in the northern North Sea the porewater is of normal salinity or brackish composition. In the northern North Sea where there are no evaporites, formation water salinity is in most cases close to that of seawater, or less saline (brackish) due probably to a component of meteoric water. Near evaporites the porewater is also often rich in calcium due to gypsum or anhydrite, with a corresponding reduction in sodium, so that $CaCl_2$ is an important dissolved salt. Dissolved NaCl may be transported away from the evaporites by diffusion or by porewater flow (advection). In both cases the salinity will be reduced away from the salt deposit and it is unlikely that the dissolved salt should become sufficiently concentrated for halite to be precipitated again.

The halite and also gypsum commonly observed in small amounts in sandstone cores is due to evaporation in the core store; they are not diagenetic minerals as has sometimes been reported. During tectonic uplift and erosion, hydration of minerals like anhydrite causes increased salinity but this is normally diluted with meteoric water. Near the surface the salinity can of course increase by evaporation. In the seawater at shallow water depth the pH is

Fig. 10.3 Illustration of fluid potential which is the difference between the fluid pressure at a certain depth and the weight of the overlying column of porewater and seawater

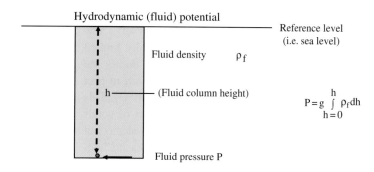

Fluid potential $F_p = P - \rho_f gh$. When $F_p = 0$, the pressure is hydrostatic

The fluid density (ρ_f) is a function of both temperature and salinity and we should try to estimate an average density over the depth interval (h).

relatively high because the water is often at least nearly saturated with respect to calcite and the pCO_2 is low. At greater water depth in the ocean and also below the seafloor the solubility of CO_2 increases, lowering the pH. Carbon dioxide is consumed near the ocean surface by photosynthesis and released by oxidation of organic matter sinking towards the ocean floor. Just below the seafloor in the sulphate reduction zone more CO_2 is produced during fomentation of organic matter. Then at greater burial depth, thermal maturation of kerogen releases CO_2. Organic acids are also generated from the source rocks and probably also from the oil. Porewater is, however, like seawater a buffered solution and organic acids are weak acids. Compared to the total buffering system of, firstly, the silicate mineral system and, secondly, the carbonate system, the addition of relatively small amounts of comparatively weak acids (organic acids) will not change the pH of the porewater significantly (Hutcheon 1989). These different types of CO_2 have characteristic ranges in $\delta^{13}C$ composition.

Formation analyses from sedimentary basins show that the pore waters are frequently crudely stratified with respect to salinity and this puts constraints on porewater flow. Also the oxygen isotope compositions may vary with depth, probably at least partly due to diagenetic reactions. At shallow depth and low temperature the porewater is normally out of equilibrium with respect to the silicate minerals because of the slow reaction rates. Meteoric water is usually highly supersaturated with respect to quartz because it does not precipitate at low temperatures (<70–80°C). Seawater,

on the other hand, is usually very much undersaturated with respect to quartz because silica is taken out of seawater by siliceous organisms, mainly diatoms.

Amorphous silica from organisms (opal A) may survive during burial down to 1.5–2 km (60–80°C) before dissolving. Opal A is replaced first by opal CT, which is unstable and will be replaced by quartz. As long as opal A and opal CT exist in the sediments the porewater is supersaturated with respect to quartz. At higher temperatures the porewater becomes more and more in equilibrium with quartz and other minerals present.

The isotopic composition of subsurface water reflects to a large extent the initial composition. Porewater with a marine origin has characteristic values close to Standard Mean Ocean Water (SMOW). More negative values may indicate the introduction of meteoric water into a marine basin. At greater burial depth the composition of porewater is more influenced by the reactions of the minerals. As a general rule dissolution of minerals formed at low temperature, i.e. during weathering (kaolinite, smectite), causes the pore waters to become more positive when they dissolve at greater depth (higher temperatures). This is because the minerals that precipitated at a higher temperature, in this case illite, will contain less ^{18}O. Dissolution and precipitation of clastic quartz, which was originally precipitated at high temperature, will shift the porewater in a negative direction when low temperature quartz, with more ^{18}O, is precipitated.

We now have considerable data on the composition of porewater in sedimentary basins from exploration and production wells. It is clear that the porewater composition varies greatly over distances of a few hundred metres and this is evidence of very limited mixing by advection. Except around salt domes the porewater seems to be crudely stratified with respect to both salinity and isotopic composition $(\delta^{18}O)$. This puts important constraints on the transport of solids in solution by fluid flow. The saline porewater is not transported very far from the salt. With increasing temperature the density of porewater is reduced, while it is increased by increasing salinity.

10.2 Composition of Porewater in an Oil Field

Analyses of porewater that is present in the oil field along with the oil provide information about fluid flow and migration of oil. Differences in salinity and chemical composition between parts of a reservoir may indicate lack of communication by fluid flow and also diffusion between compartments. The isotopic composition of the formation water may also help to indicate degrees of communication in a reservoir in addition to changes in the composition of oil. Strontium isotopes have been used for this purpose. If there are primary differences in the composition of oil across an oil field this can be used to monitor the contribution from the different parts during production. During onshore production freshwater may be used for water injection into the reservoir. Offshore, seawater is injected, but this has a composition which is distinctly different from the formation water.

At any given time the amounts of solids in solution are very small except in highly saline porewater near evaporites. The solubility of most silicates and also carbonate minerals is sufficiently low that the porewater composition of sedimentary basins is almost totally controlled by the solid phases. Porewater in sedimentary basins often shows evidence of stratification with respect to both salinity and isotopic composition, which precludes large scale mixing of porewater. The composition of the formation water in reservoirs may provide useful information about the source of the water and communication within the reservoir, in the same way as the composition of the oil gives information about the source of the oil.

10.3 Fluid Flow in Sedimentary Basins

Fluid flow in sedimentary basins is important because it determines the distribution of pore pressures in the water phase, and also in oil and gas. High pore pressures may also be a hazard when drilling wells. The fluid phases have the capacity to transport solids in solution, and heat by fluid flow (advection) and by diffusion. The flow of fluids may be through the pore network in the rock matrix or along fractures, and this is a principal difference between these two types of flow. This part of the chapter will discuss the factors controlling the flow of fluids in sedimentary basins.

The fluids are mostly water, but may also be oil and gases including air, filling the pores in the sediments between the grains. The grains are in most cases minerals but some may be amorphous (e.g. silica – Opal A, or organic matter and kerogen). *Porosity* is the percentage (or fraction) of the rock volume which is filled with fluids. This may be also expressed by the *void ratio* which is the ratio between the volume of voids (porosity) and solids. What is called *void* is not strictly void but filled with fluids and is the same as porosity. The relation between porosity (φ) and void ratio (V_r) is thus:

$$V_r = 1 / (1 - \varphi)$$

In the oil industry porosity is mostly used while rock and soil mechanics literature tends to use void ratio.

Porewater flow in sedimentary basins can be classified according to the origin of the water and the driving mechanism for the flow:

(1) Meteoric water flow is sourced by groundwater (originally rainwater) and in most cases the groundwater table is above sea level, providing a drive downwards into the basin. This water is normally fresh (low salinity) except in very arid regions where meteoric water may dissolve evaporites.

(2) Compaction-driven water is driven by the effective stress and chemical compaction which causes a reduction in available pore space. The upward component of this flow is a function of the rate of porosity reduction (compaction) in the underlying sediments.

(3) Density-driven flow is driven by gradients in the fluid density due to differences in salinity or temperature. Thermal convection is driven by the thermal expansion of water. As the temperature increases downwards in sedimentary basins the density is reduced, creating a density inversion with depth. Flow driven by thermal convection differs from the two other types of flow in that the same water is used over again and is not dependant on an external supply of porewater.

10.4 Fluid Potentials

The flow of fluids in sedimentary basins follows simple fluid dynamics laws. Fluids do not necessarily flow from higher to lower pressures, but from higher to lower fluid potentials. The fluid potential is defined as:

$$F_p = P - \rho gh.$$

Here P is the fluid pressure, ρ is the density of the fluid, g the acceleration of gravity and h is the distance up to some reference level, which in a sedimentary basin could be the sea level or the water table (Fig. 10.3). The fluid potential is thus the potential for fluid flow, so if the fluid potential is zero there can not be any flow.

The fluid potential is an expression of the *deviation* from the pressure gradient due to the density of the fluid column. In the case of water the hydrodynamic potential expresses the deviation from the hydrostatic pressure gradient ($\rho_w g$), which is defined by the weight of the water column. In the case of groundwater flow the fluid potential is usually referred to as the hydraulic potential, which is the mechanical energy per unit volume of groundwater. The lithostatic stress, which may also be referred to as the total stress, is the weight of the rock column saturated with fluids (ρ_r), for the most part water. The difference between the total stress ($\rho_r gh$) and the pore pressure (P) is the effective stress (σ_{ev}) which is transmitted by the sediment particles or the rock:

$$\sigma_{ev} = (\rho_r gh - P)$$

Differences in hydrodynamic potentials can also be expressed in terms of potentiometric (or piezometric) surfaces which are the heights to which water would rise above sea level (or some other reference datum like the groundwater level) in an open pipe from a rock in the subsurface. Pore waters with a higher potentiometric surface than sea level are defined as *overpressured* while those that have a potentiometric (piezometric) surface close to sea level or the groundwater table are normally pressured.

It is often stated that fluids flow from high to lower pressure but this is obviously not always true. In the ocean the pressure increases from the surface down towards the bottom but water does not flow from the bottom to the surface because there is in most cases no potentiometric head. The pressure gradient in the ocean water is close to the density gradient ($\rho_r gh$) and ocean currents are driven by very small differences in fluid potential due to changes in temperature.

To maintain significant pressure (potentiometric) gradients, there must be a resistance to flow; in the case of flow in porous rocks this is measured as *permeability*. Fluid flow is a function of permeability (k) and viscosity (μ), and the flow of water and other fluids can be described by the Darcy equation:

$$F = \nabla P \cdot k / \mu.$$

The flux F can be expressed as volumes of fluids passing through a certain cross-section in a given time, i.e. $cm^{-2} \cdot cm^{-2} s^{-2}$ (Fig. 10.4). This is also called the *Darcy velocity*. The *absolute velocity* of the flow is v (cm/s) = flux $(cm^3 \cdot cm^{-2} s^-)$ divided by the porosity (φ). So if the porosity is 0.1 (10%) the velocity is 10 times the flux. In reality the velocity is a little higher because the fluid pathway is not along a straight line. Since the porosity varies greatly along the flow path it is more useful to use the flux (cm^3/cm^2) or the Darcy velocity (cm/s) as a measure of fluid flow rates. ∇P is the potentiometric gradient, i.e. the change in potential over a certain distance. In the horizontal direction this is the same as the pressure gradient.

Given a pressure P_1 in a point X_1 at a depth h_1, and a pressure P_2 in a point X_2 at the depth h_2, the *potentiometric gradient* (∇P) is:

$$\nabla P = ((P_1 - \rho gh_1) - (P_2 - \rho gh_2)) / X_2 - X.$$

$$(X_2 - X_1 \text{ is the distance between } P_1 \text{ and } P_2.)$$

The *permeability k* is an expression of the resistance to flow and is a constant in the Darcy equation which

Fig. 10.4 Illustration of the Darcy equation for fluid flow in sedimentary basins

Permeability

The permeability is the resistance to flow in any material.

If the permeability is 1 Darcy (1 D) the fluid flux is 1 $cm^3 \cdot cm^{-2} \cdot s^{-1}$, when the pressure gradient is 1 $Atm.cm^{-1}$ (100 $kPa.cm^{-1}$) in the direction of flow and the viscosity of the fluids is 1 centipoise (as for water at 20°C).

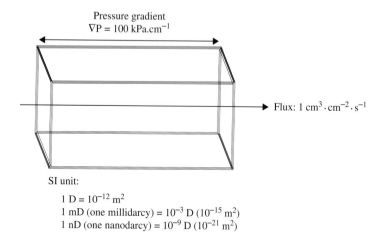

Pressure gradient
$\nabla P = 100 \ kPa.cm^{-1}$

Flux: 1 $cm^3 \cdot cm^{-2} \cdot s^{-1}$

SI unit:

1 D = 10^{-12} m^2
1 mD (one millidarcy) = 10^{-3} D (10^{-15} m^2)
1 nD (one nanodarcy) = 10^{-9} D (10^{-21} m^2)

relates solely to the properties of the rock. Permeability has the dimension of m^2 $\left(1 \text{ Darcy} = 10^{-12} m^2\right)$. The permeability of a rock may be referred to as the *absolute or intrinsic permeability* to make it clear that it only relates to the characteristics of the rock, as opposed to the *relative permeability* which is the permeability of one immiscible fluid in the presence of another fluid, compared to the permeability with 100% saturation of one fluid.

The *hydraulic conductivity (K)* or *transmissibility* is an expression of the ability of the rocks to conduct or transmit fluids of a certain viscosity (μ). The conductivity has the dimension of m/s or ft/s and can be calculated from the permeability if the viscosity is known ($K = k \cdot g/\mu$). At about 20°C, the kinematic viscosity of water is $1:10^{-6}$ m^2/s and at 100°C it is 0.210^{-6} m^2/s which is one centipoise. 1 Darcy (permeability) $(k) = 10^{-5}$ K (conductivity) for water (strictly 9.66×10^{-6} K), 1 Darcy = 10^{-12} m^2 or 10^{-8} cm^2. Well sorted and poorly cemented sand may have permeabilities between 1 and 10 Darcy. In tight shales the permeability is typically below 1 nanodarcy (10^{-9} Darcy) or even much lower (Fig. 10.4).

Water is not very compressible. The compressibility is close to $4.4 \times 10^{-10}/Pa^{-1}$ (Freeze and Cherry 1979). The pressure of a 1 km water column (10 MPa) causes a compression of water of about 0.4% or 4 m. The expansion of water during tectonic uplift or release

of overpressure is therefore relatively small. Usually the cooling of water associated with uplift will cause a thermal contraction which is greater than the expansion due to pressure reduction, so that the porewater becomes denser.

Porewater flow is oriented perpendicular to points of equal hydrodynamic potential (isopotential lines) because that will represent the steepest potentiometric gradient. That will be the case in relatively homogeneous sediments but generally the flow is very much controlled by the distribution of highly permeable *aquifers* such as poorly cemented sandstone layers, and of shales which serve as low permeability *aquitards* (fluid barriers) (Fig. 10.5).

The distribution of very high permeability aquifer sandstones and low permeability aquicludes (mostly shales, salts and cemented layers in salts) that control most of the flow in sedimentary basins is closely linked to primary sedimentary facies, tectonic developments and diagenesis. These depositional and diagenetic processes determine the "plumbing system" in the basin and this must be the starting point for fluid flow modelling.

Aquicludes are beds of such low permeability that they can not transmit significant quantities of fluids under normal hydraulic gradients. These are groundwater hydraulics terms but over geological time even low permeability shale will transmit some fluid. Some

Fig. 10.5 Fluid flow along a sandstone which is a confined aquifer overlain by a shale which is an aquitard

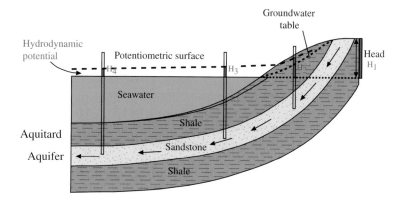

shales may have extremely low permeability and be practically impermeable, thus capable of maintaining overpressures on this timescale. The term *seal* is often used as equivalent to aquiclude in the oil industry, and may describe a seal for both oil and water. Shales which are not completely sealing with respect to water may nevertheless prevent oil from entering into the small pores, due to capillary forces. Fractures may be open and provide highly permeable pathways, but others may be almost impermeable barriers to fluid flow because they are closed or cemented up. In a massive sand the flow is very different and controlled by the orientation of the isopotential lines which may be controlled by the groundwater table (Fig. 10.6).

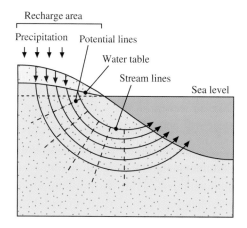

Fig. 10.6 Flow of meteoric water in massive sand, perpendicular to the pressure distribution (potential lines)

Fluid flow modelling that uncritically applies the Darcy equations to fluid flow in sedimentary basins can lead to quite unrealistic results. Frequently the porewater flux is calculated from an observed pressure gradient and from assumed or modelled permeabilities for the different lithologies. One of the main problems with such calculations is that the permeability can not be determined very accurately. It typically varies by several orders of magnitude from bed to bed up through a sequence and there may also be large variations along the bedding due to sedimentary structures.

In the case of flow *perpendicular* to the bedding the average permeability is the *harmonic mean* of the permeability of the individual beds. The harmonic mean of the permeabilities of a number of beds (n) is defined by:

$$k_\mathrm{h} = \left[i/n \sum_{i=1}^{n} 1/k_i \right]^{-1}$$

The harmonic mean is strongly influenced by the bed with the lowest permeabilities such as a tight shale or a carbonate-cemented interval. For flow *parallel* to bedding the arithmetic mean is relevant. Even if we measure the permeability at relatively short intervals in a cored section it is very difficult to come up with a good average permeability, partly because relatively thin, low permeability, layers affect the value to such a degree.

In exploration we want to make predictions ahead of drilling, but without core data it is very difficult to provide assumptions about the permeability distribution with the degree of confidence needed for modelling fluid flow. In most cases, however, the fluid flux (*F*) is primarily constrained by the supply of fluids. The flux of meteoric water (groundwater) flow is limited by the infiltration of rainwater into the ground. The potentiometric gradient near the surface is the slope of the groundwater table (or the potentiometric surface).

If the rainfall is 1 m/year the infiltration may be 0.3 m/year and this is the initial flux in the recharge area. If this flux is continuous for 1 million years the total flow is then $3 \times 10^5 \, m^3 \cdot m^{-2}$ which is very significant, several orders of magnitude larger than the average compaction-driven flow. However the meteoric water flux is greatest near the surface and decreases rapidly with depth. We must remember that the meteoric water flowing into the basin also must flow up to the surface. Sandstones that pinch out in mudstones or shales will support only a very low flux despite the high permeability of the sandstones.

Fluid flow resulting from differences in hydrodynamic potential can be treated mathematically using Darcy's equation and mass conserving equations during flow (continuity equations). Modelling flow for whole basins is very complex. The orientation and distribution of permeable sediments (usually sandstones and limestones) will to a large extent dominate the pattern of fluid flow. At depths greater than 3–4 km (100°C) quartz cementation may be more extensive along the fault planes than in the adjacent sandstones.

The importance of fault planes as conduits for fluid flow is greatest in well-cemented uplifted sedimentary rocks or in metamorphic rocks, because the matrix permeability is so low. These are rocks that have been subject to unloading and usually some extension (fracturing), and they possess high rock strengths which can prevent fractures from closing. Deformation of well-cemented rocks may produce rock fragments (brecciation) which by wedging the faults may help to resist horizontal stress, keeping the fractures open. In more porous sedimentary rocks like sandstones and limestones, which are not well cemented and have higher matrix permeabilities, fractures are less critical. In subsiding sedimentary basins the sediments do not usually have sufficient strength to resist the horizontal stress that is trying to close any open faults and fractures. Faults in this setting are therefore more likely to be barriers than conduits for fluid flow (see Chap. 11 on rock mechanics).

Prediction of fluid pressure ahead of drilling in the basin depends on the permeability distribution in three dimensions over distances of several kilometres. Even if the geology is known in great detail, it would still be difficult to specify sufficient details about the permeability to obtain a realistic fluid flow model based on sedimentology and structural geology. During exploration we normally have insufficient data

to model fluid pressure. During production much more data is available on the distribution of permeabilities and the model can be constrained by the pressure response to production, and to water injection.

10.5 Meteoric Water Flow

If the permeability is homogeneous in all directions, the porewater flow can be calculated from the elevation of the groundwater table and the fluid densities alone. The flow is perpendicular to lines with equal potentials. In an isotropic rock matrix with constant fluid density, the flow of meteoric porewater follows a curved pattern perpendicular to the isopotential lines (Fig. 10.6). Sedimentary rocks are generally very inhomogeneous and the contrasts in permeability caused by clay layers and sand or gravel beds will generally totally dominate the flow of water (Fig. 10.5). While sand and gravel beds may have permeabilities of 1–10 Darcy, the permeability of poorly-compacted mud may be 0.01 mD or lower. In such cases, mathematical modelling is of little value if the stratigraphy, sedimentology and structural deformation of the sedimentary sequences are not interpreted correctly. Modelling can nevertheless help to constrain some of the interpretations by calculating the consequences of different alternatives.

Meteoric water (freshwater) has a potentiometric head defined by the groundwater table, which is normally higher than sea level. Fresh porewater will thus flow into marine sedimentary basins from the coastlines (Fig. 10.7). Along the coast and underneath islands, lenses of freshwater float on more saline porewater like an iceberg in the sea. The depth to which freshwater will penetrate is a function of the density difference between the freshwater and the saline water:

$$D = H\rho_{mw} / (\rho_{sw} - \rho_{mw})$$

Here, D is the depth of penetration below sea level, H is the height of the groundwater table above sea level and ρ_{sw} and ρ_{mw} are the densities of saline water and meteoric pore waters, respectively. For $\rho_{sw} = 1.025 \, g/cm^3$ and $\rho_{mw} = 1.0 \, g/cm^3$. $D = 40\,H$, meaning that the depth of the fresh water wedge is 40 times the height of the groundwater table underneath an island or within a confined aquifer along the coast. If the meteoric water becomes brackish by mixing

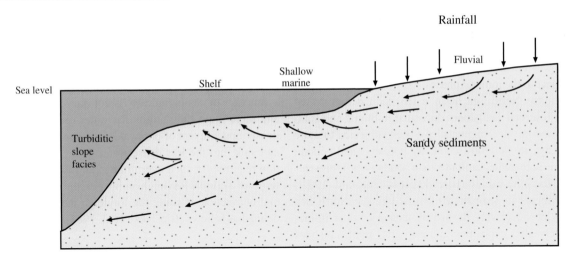

Fig. 10.7 Flow of meteoric water into sedimentary basins. The fluid flux will decrease away from the coastline

with saline waters, the density difference is reduced and the depth of penetration can be deeper because $(\rho_{sw} - \rho_{mw})$ becomes smaller. The depth of meteoric water flow into highly saline porewater is very much less. Relative sea level changes serve as a pumping mechanism, driving meteoric water into sedimentary basins at low sea level stands due to the increase in hydrodynamic head. In this way early diagenesis may be linked to sequence stratigraphy.

During the last glaciations sea level was lowered more than 100 m several times. This increased the head of the groundwater table on the land area, pushing meteoric water deep into sedimentary basins. Under completely hydrostatic conditions, a 100 m head (10 MPa pressure) should theoretically correspond to a penetration down to about 4 km following the above calculations. However, the flow will mostly follow permeable beds (aquifers) and may extend a great distance out from the coastline. Well log analyses from offshore Georgia, USA, suggest that freshwater extended for up to 100 km offshore just a few metres below the seafloor (Manheim and Paull 1981). In the modern Gulf of Mexico Basin, the depth of meteoric water penetration is estimated to be about 2 km (Harrison and Summa 1991). The depth of penetration does not depend only on the head of the meteoric water but also on compaction processes in the sediments, which can generate overpressures that may exceed the meteoric water head. Even slight overpressures due to compaction will strongly reduce the depth of meteoric water penetration.

The rainfall, the catchment area and the percentage of infiltration into the groundwater determine the upper limit of meteoric water flow. Meteoric water which flows down into a sedimentary basin must eventually flow up to the surface, in order to maintain a continuous flow. Permeable sandstones are hydraulically almost dead ends if they pinch out into very low permeability mudstones. Pressure will then build up in the aquifer and reduce the meteoric water inflow, forcing water to flow through the overlying mudstones. We must remember that even if the fluid flux is small, the large area of contact between a sand layer which serves as aquifer and the overlying mud will allow relatively high volumes of porewater to escape upwards through the mud. Even if the flow per area (flux) is small through the mudstones the area for vertical flow is large compared to the vertical cross-section. At shallow depths (<500 m) prior to severe compaction, overpressure is rarely developed as the permeability is much higher than in compacted mudstones and shales.

The degree of meteoric water flushing is highly dependent on climate and facies. The land surface and vegetation determine the percentage of rainfall which infiltrates down to the groundwater. Fluvial, deltaic and nearshore shallow marine sediments will be flushed by meteoric water shortly after deposition. The flux is then likely to be high and the porewater is still very much undersaturated with respect to feldspar and mica. The total volume of water flowing through each volume of sediment is inversely related to sedimentation rates. At high sedimentation rates the sediments spend less time in the zone of meteoric water flushing. Sands deposited in more distal shelf facies (nearer the shelf edge) and turbidites (on the slopes) are normally less well connected to the main groundwater wedge, so that the flux is lower and the porewater is closer to equilibrium with respect to the mineral phases.

In the North Sea basin it has been demonstrated that reservoir sandstones deposited in fluvial and shallow marine environments have been subjected to more feldspar dissolution (secondary porosity) and contain more authigenic kaolinite than sandstones representing turbidite facies (Bjørlykke and Aagaard 1992). Sediments in sedimentary basins like the North Sea may be intensively flushed by meteoric water immediately after deposition and also after uplift episodes and erosion. When tectonic uplift results in subaerial exposure and the formation of islands, meteoric water is collected on land and driven into the subsurface around the islands and adjacent to other land areas. The sediments most strongly affected by meteoric water leaching are constantly being removed by erosion, however. This may explain why there is not always much evidence of feldspar leaching and high kaolinite contents immediately below unconformities.

Good examples of clay mineral diagenesis related to modern groundwater systems have been observed down to 3–400 m depth in the Mississippi Gulf coastal plain (Hanor and Mcmanus 1988). In Canada there is isotopic evidence of recent meteoric water diagenesis extending several hundred metres below the land surface (Longstaffe 1984). It must be stressed that the isotopic composition of the porewater acquires a meteoric signature as soon as a volume of meteoric water has displaced the marine (connate) porewater. New minerals (like calcite, kaolinite and quartz) precipitated in this porewater will reflect that isotopic signature.

10.6 Porewater Flow Driven by Thermal Convection

Thermal convection is an effective mechanism for the mass transfer of dissolved material in sedimentary basins, because the same water can be used over and over again (Cassan et al. 1981, Wood and Hewett 1982, Davis et al. 1985). The limitation of fluid (water) supply, which constrains compaction-driven flow, is then eliminated. Thermal convection may occur because the

density of water decreases with depth in a sedimentary basin as the temperature increases, due to the thermal expansion of water. This creates an inverse density gradient which may be unstable. If the isotherms (lines of equal temperature) are horizontal, the density as a function of temperature will not vary horizontally and there is no flow unless the water overturns. The denser upper layers of porewater may start to overturn and sink into the less dense water below. The condition required for such overturning can be expressed in terms of a critical Rayleigh number. In the case of thermal convection of water the Rayleigh number can be defined as follows (Bjørlykke et al. 1988):

$$R = g\beta\Delta THk/\kappa\mu$$

Here, g is acceleration due to gravity, β is the coefficient of thermal expansion of the fluid (water), ΔT is the temperature difference between the upper and lower boundaries of the convection cell, H is the thickness of the layers, k is the permeability, κ is the thermal diffusivity and μ is the viscosity. Assuming reasonable values for the properties of water the equation can be expressed in a simpler form (Bjørlykke et al. 1988):

$$R = 1.2 \times 10^{-2} k\nabla TH$$

The critical Rayleigh number which must be exceeded for Rayleigh convection to occur is about 40. We see that the most critical factors are the height of the water column and the permeability of the rocks. If the permeability is 1 Darcy and the geothermal gradient 30°C/km, the thickness (H) of the permeable layer (sandstone) must exceed about 300 m for the critical Rayleigh number to be exceeded so that thermal convection can occur. Sedimentary rocks, though, are rarely uniform and the vertical permeability typically changes abruptly in a sequence of sandstones and shales. Thin layers (0.1 m) of low permeability shales or cemented layers in sandstones may cause almost complete flow separation, producing smaller convection cells instead of potentially larger ones (Fig. 10.8) (Bjørlykke et al. 1988). Each of the convection cells may then have insufficient height (small H) to exceed the critical Rayleigh number. The low vertical permeability in layered sequences suggests that Rayleigh convection is probably not very important in sedimentary basins. Several hundred metre thick sandstones

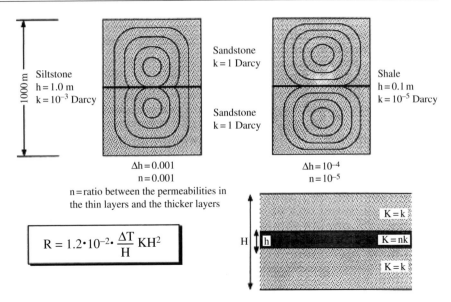

Fig. 10.8 A modelling of the low permeability layers on vertical Rayleigh convection in sedimentary basins (Bjørlykke et al. 1988)

$$R = 1.2 \cdot 10^{-2} \cdot \frac{\Delta T}{H} KH^2$$

with no thin shales or cemented intervals are rarely encountered.

Non-Rayleigh convection will always take place when the isotherms are not horizontal. This is because the temperature and the fluid density are then not constant in the horizontal direction. This situation is always unstable because there is a potentiometric drive for the waters to overturn, which will produce some fluid flow without the need to exceed a critical Rayleigh number. The velocity for non-Rayleigh convection is:

$$v = g \cdot k \cdot \beta \sin \alpha \nabla T / \mu$$

When the geothermal gradients vary only moderately within a basin, the slope of the isotherms (α) will be small and the flow velocity very low. Sloping isotherms will also result from sloping beds because the heat flux is reflected when the conductivity of the beds varies, but this effect is also normally quite small. From the equation above we see that the flow rates are a function of the isotherm slope, as well as a function of the height of the convection cell which often is equal to the thickness of a sandstone bed. The thickness of the sandstone beds (H) or the distance between the low permeability shales, will normally define the height of the convection cells and in most sedimentary sequences the distance between thin shales or even clay laminae is only a few metres or less. Although non-Rayleigh convection nearly always occurs to some

degree it is probably rather insignificant in most cases in terms of the transport of solids in solution in sedimentary basins. This is because of the low slopes of the isotherms and of the low vertical permeability, resulting in very low flow rates mostly inside rather thin sandstones separated by shales. These are probably not very significant in terms of solid transport in connection with diagenetic process. Around igneous or hydrothermal intrusions, however, the lateral change in geothermal gradients and the slope of the isotherms may be very high and then thermal convection is very important. Also around salt domes geothermal gradients may change due to the higher conductivity. Inverse salinity gradients will increase the drive for thermal convection, while normal salinity gradients will make the porewater more stable. Even moderate salinity gradients strongly influence fluid flow in sedimentary basins (Fig. 10.9).

The increase in density due to the salinity may totally or partly offset the density reduction due to the thermal expansion of water. At a salinity gradient of 30 ppm/m, and an average geothermal gradient, the effects of the thermal expansion of water are more than offset, so that the water becomes denser with depth (Fig. 10.9). This effectively removes any drive for convective flow (Bjørlykke et al. 1988). When such trends are recorded in formation water analyses or well logs it provides strong evidence that vertical mixing is not taking place, because convection would have destroyed the salinity gradients (Gran et al. 1992). Around salt

Fig. 10.9 Density of water as a function of salinity and temperature (from Bjørlykke et al. 1988). The increase in temperature causes a thermal expansion and a density inversion while salinity gradients may have the opposite effect

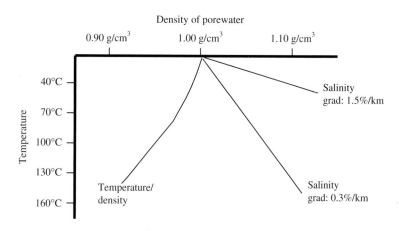

domes the permeabilities are often rather low, further reducing the potential for fluid flow.

The salinity in sediments surrounding salt diapirs can be used to trace fluid flow, since Cl⁻ is not consumed to any significant degree by diagenetic reactions. Analyses of the salinity distribution around salt domes from offshore Louisiana show some evidence of convection, but the observed salinity stratification and the lack of more mixing and dilution of the saline porewater suggest this convection is very slow (Ranganathan and Hanor 1988, Evans and Nunn 1989). Inverted salinity gradients are only likely to develop around salt diapers or underneath salt layers. The inverse salinity gradients can only sustain flow on the down-going limb of a convection cell. Unlike thermal convection there is no mechanism to make the water flow up again and the water would also gradually homogenise with respect to salinity. In basins like the North Sea, porewater analyses demonstrate that the porewater is at least stratified in a crude way with respect to its composition. The salinity increases downwards towards the Permian salts, effectively ruling out large scale convection and excursions of compaction-driven flow from the deeper parts of the basin into the overlying sequence (Gran et al. 1992).

There is also a trend towards more positive $\delta^{18}O$ values with depth, which again confirms some degree of porewater stratification (Moss et al. 2003). The vertical salinity gradients and the isotopic composition of the porewater confirm that the porewater is not undergoing convection on a large scale and that there is no large flow of porewater from the deeper part of the basin to shallower depths. This has important consequences for diagenetic models in connection with fluid transport of solids in solution.

10.7 Compaction-Driven Porewater Flow

As sediments compact they lose porosity and the excess porewater has to be expelled. This is the driving force for compaction-driven flow. The rate of porosity loss is a function of effective stress, lithology, temperature and time. The porosity/depth functions observed in sedimentary basins may be very complex, depending on the lithologies. At the transition between two lithologies the porosity may increase with depth but for a uniform lithology the porosity will decrease with depth. If we integrate the porosity/depth function from the seafloor though the sedimentary sequences to the underlying basement we obtain an area A below the porosity/depth curve. This is an expression of the total volume of water in the basin per unit area (Fig. 10.10). The total compaction-driven flow of water in the basin is a function of the changes in the porosity/depth curve (Fig. 10.10). As new layers of sediment are deposited the underlying sediments compact and the porewater is forced upwards.

It is possible to show that at a constant sedimentation rate the average upward component of the compaction-driven flow is always equal to or lower than the subsidence rate (Caritat 1989, Bjørlykke 1989). The porewater is therefore moving upwards through the sedimentary sequence but nearly always downwards relative to the seafloor. We may say that the sediments are sinking through a column of porewater. Typical sedimentation rates in sedimentary basins are 0.1–0.01 mm/year and the average rates of upwards porewater flow are lower than these values. Assuming there is no flow of water from the basement there is practically no upward flow in the basal layer of

Fig. 10.10 Illustration of a porosity/depth function in a sedimentary basin. The integrated area defined by this *curve* is an expression of the total volume of water in the basin and the *slope* (the derivative) is an expression of the compaction-driven water flux from each layer

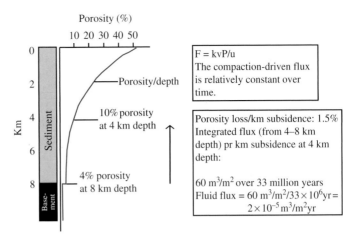

Porosity (%)

F = kvP/u
The compaction-driven flux is relatively constant over time.

Porosity loss/km subsidence: 1.5%
Integrated flux (from 4–8 km depth) pr km subsidence at 4 km depth:

60 m³/m² over 33 million years
Fluid flux = 60 m³/m²/33 × 10⁶yr = 2 × 10⁻⁵ m³/m²yr

sediments. The porewater in this layer subsides at almost the same rate as the basement. Higher up in the sequence the compaction-driven flow receives contributions from more and more layers. In the uppermost layer the porewater flux is equal to the total integrated porosity loss over time in the underlying sequence. The upwards-flowing porewater is filling the pore space of new layers deposited on the seafloor, and during continued subsidence there is normally no porewater flow up though the seafloor into the water column except when there is local focusing of the flow. During periods with no sedimentation (hiatus), the deposited sediments continue to compact and porewater flows across the redox boundary just below the seabed and dissolved ions like Fe^{2+} and Mn^{2+} may precipitate as $Fe(OH)_3$ and $Mn(OH)_4$.

Very high degrees of focusing are required to obtain high flow rates. It is then easy to understand why compaction-driven flow is several orders of magnitude lower than meteoric water flow, at least in the shallow parts of the basin near land. The slow porewater flow rates also imply that this water has time to approach thermal and chemical equilibrium with the minerals. Porewater will always transport some heat but, compared to the background heat flow by conduction, this is very small and in most cases can be ignored (Bethke 1985, Ungerer 1989, Ludvigsen et al. 1992).

Modelling compaction-driven flow in the Gulf Basin, Harrison and Summa (1991) found that the maximum rate of the vertical component is 2 mm/year (2 km/million years), which is approximately equivalent to the maximum sedimentation rate. This flow rate is too slow to contribute significantly to the heat flow.

The temperature distribution in the Gulf Basin at the present day does not reflect compaction-driven porewater flow. Most of the heat transport is by conduction. The average compaction-driven flow within a basin is more or less independent of the permeability because it is a function of the rate of loss of porosity in the underlying sediments. During mechanical compaction, low permeability sediments may result in overpressure and a certain reduction of fluid flow, but at greater depth where the compaction is mostly chemical, the fluid flux is independent of the permeability although variations in permeability may focus the flow.

10.8 Constraints on Water Flow in Sedimentary Basins by Porewater Chemistry

Many sedimentary basins contain evaporites, often occupying the basal part, having formed during the initial rifting. Very high salinity is then typically found in a zone of a few hundred metres adjacent to the salt. This is the case with the Zechstein salt in the North Sea basin. The shallower parts of a basin and in particular near tectonically uplifted areas, may contain porewater of meteoric origin with very low salinity. The isotopic composition of the porewater often shows a very clear stratification. Oxygen isotopes are often negative in the shallow parts of the basin due to the inflow of meteoric water, while they may be positive at greater depth due to diagenetic reactions.

During subsidence there are generally no open fractures because of the ductile properties of subsiding sediments (Bjørlykke and Høeg 1997). The permeabilities in shales are very low, probably less than a nanodarcy (Leonard 1993). Samples measured in the laboratory may show erroneously high values because of fracturing resulting from unloading during core retrieval. Increasing the effective stress during laboratory measurements to compensate for this has been shown to lower the permeabilities by two orders of magnitude (Katsube et al. 1991). In sediments which have been subject to uplift, however, fracture permeability may be important. Laboratory measurements of shale permeabilities may be several orders of magnitude larger than the results of well tests in the field (Oelkers 1996).

In subsiding basins the effective permeabilities on a large scale must be low, not above about 10^{-9} Darcy, to maintain overpressures over time. Calculating the porewater flux due to compaction and using observed pressure gradients, the effective permeability of thick shale sequences can be calculated. Using data from Haltenbanken offshore Norway this method gave permeabilities between 10^{-9} and 10^{-10} Darcy (Olstad et al. 1997). These calculations are based on one dimensional models and the results are most probably minimum values since lateral drainage is not included.

10.9 The Importance of Faults

Faults may greatly affect fluid flow in sedimentary basins and may serve either as conduits or barriers, depending on the situation. A clear distinction must be made between flow along, and across, the fault plane. Faults are also important because they may offset porous sandstones (aquifers) against shales, rendering permeable sandstones a dead-end in terms of fluid flow. Faults that cut through sandstones may have a clay smear from adjacent shales or from mica or authigenic kaolinite inside the sandstone, and this may significantly reduce the permeability, in some cases sufficiently to form an oil trap.

Flow along fault planes requires that they are kept open to some degree. A force equivalent to the horizontal stress acts on the fault plane, trying to close it; an equivalent overpressure is required to counteract this in order for the fracture to remain open (Fig. 10.11). However, this corresponds closely to the fracture pressure and even without the presence of a fault the rocks would fracture. At such high overpressure there are very low effective stresses and the sediments are unable to compact mechanically. We must also consider the source of the fluids. When there is no compaction (porosity reduction) in the adjacent sediments, water can not flow from the rock matrix into the fractures. Well-cemented sedimentary rocks

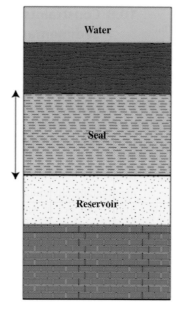

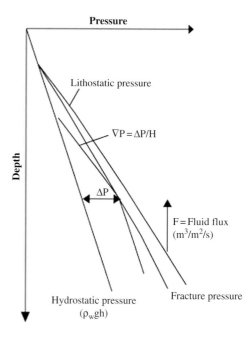

Fig. 10.11 Illustration of fluid pressure and rock pressure (lithostatic). It the permeability in shales (seals) is low enough overpressure will build up. The pressure can not exceed the fracture pressure which is equal to the horizontal stress, as the seal will then leak

and basement rocks have high shear strength and may produce rock fragments (brecciation) during faulting. These rock fragments may wedge the fault plane, helping to resist the horizontal stress. Brecciated fault planes may therefore be important conduits for fluid flow but over time the permeability will gradually be reduced by cementation.

The cements will in most cases not be due to precipitation from advective flow but form by diffusion from the adjacent rock matrix. Carbonate and silicate minerals next to the faults are under lithostatic stress and therefore more readily soluble than when unstressed. In the case of carbonate cement this is fairly clear because upwards (cooling) flow will dissolve calcite rather than precipitate it, due to its retrograde solubility. Brecciated faults contain broken rock fragments and quartz grains that are good nucleation sites for quartz cement formation. A thermodynamic drive towards dissolution of the minerals under stress is therefore likely, with ensuing precipitation in the fault plane. Renewed fracturing is therefore required for the faults to remain permeable.

10.10 Seismic Pumping

In crystalline and well-cemented sedimentary rocks tectonic shear may result in an extensional shear failure which opens up fractures. Increases in the tectonic stress may reduce the opening of such fractures and produce fluid flow that is often referred to as seismic pumping (Sibson 1990). This mechanism may work in metamorphic and well-cemented sedimentary rocks where the rock strength is high enough to keep fractures from being closed by tectonic stress. In these types of brittle rocks most of the water is present in the fractures and little in the rock matrix. Softer sediments are by contrast more ductile in their response to tectonic stress and fractures will normally not stay sufficiently open to transmit fluid rapidly. In compacting (normally consolidated) sediments most of the water is in the sediment matrix and even if fractures should remain open the rate-limiting step is the flow of porewater from the matrix, which often consists of low permeability mudstones and shales.

If the low permeability seal overlying or surrounding an overpressured part of the basin is broken, i.e. through faulting, a rapid pulse of porewater flow

upwards may follow. However due to the low compressibility of water ($4.3 \times 10^{-10}\mathrm{Pa}^{-1}$) the upward flow of porewater necessary to reduce the pressure is relatively small. For an overpressure of 10^7 Pa (potentiometric surface 1 km above sea level), the average expansion of the water would be 4.3×10^{-3} (Leonard 1993). In a vertical column through an overpressured sequence this would result in an average upwards flow of 4.3 m for each km of sequence, if all the overpressure was released at one time. If the flow was focused through a smaller cross-section, this figure would increase proportionally.

It is very unlikely that large volumes of shales would reduce their overpressure over a very short time, given their low permeabilities, so the potential for episodes of rapid flow of compaction water is limited. If the water is saturated with respect to gases like carbon dioxide or methane the compressibility of the pore fluid will increase significantly. A reduction in the pressure then will cause gas to come out of solution and form a separate phase, which has a high compressibility. The degree of overpressure is reduced by porewater flow through a leaking seal, but only by a small amount. The pressure will not drop much below fracture pressure before the fine fractures close. In the relatively shallow section this small increase in effective stress may result in some mechanical compaction. Chemical compaction (>100°C), however, occurs at a very slow rate which is mostly a function of temperature and time and is relatively independent of the pressure changes related to fracturing. Compaction will therefore slowly build up the pore pressure again unless there is continued flow from the overpressured section.

As we saw from the calculations, the flow resulting directly from the pressure release is rather limited. The main expulsion of porewater is due to sediment compaction, which is an indirect consequence of the reduction of overpressure and increase in effective stress (effective stress = overburden stress minus pore pressure). Compaction and expulsion of porewater resulting from increased effective stress is a gradual and rather slow process and this strongly influences the rate of water supply to the faults from mudstones. The permeability of the surrounding mudstones further limits the rate of flow into the fault zone. If a fault plane does not extend up to the surface (or seafloor), the upwards-moving porewater will have to be accommodated in shallower strata. Large volumes of porewater can not suddenly be injected into shallower sandstones even if the latter are normally pressured.

Flow into shallow aquifers of limited extent will result in a temporary pressure build-up before the water can be displaced, and this will reduce the flow rate. The highest flow rates can be expected when the fault extends all the way to the surface so that the porewater can escape into the water column or onto the land surface. In many sedimentary basins like the North Sea basin, most of the faults extend only into the Cretaceous or lower Tertiary section. These faults were therefore not very active during the early Tertiary, and certainly not during the later Tertiary and Quaternary. The timing of faulting must be taken into account when faults are called upon to explain fluid flow and diagenetic reactions in sedimentary basins. In a sequence with a high clay/sand ratio, faults may be sealing between two sandstones due to the clay smear on the fault plane.

10.11 Episodic Flow

When the source of the fluids is hydrothermal the flow may be episodic, at least when considering individual fractures or a limited area, because new fractures develop and close. Igneous intrusions (sills and dykes) may cause boiling and episodic flow. For a cooling batholith as a whole the fluid flow may be more uniform, depending on the rate of cooling and on the circulation of groundwater by thermal convection. It has also been argued that compaction-driven flow may be episodic when the fracture pressure is reached. Fluids flow through the fractures produced by hydrofracturing but as they escape, pressure will be reduced and the fractures will be reduced or close. It is not so clear if the fracture will stay closed for some time and then leak again or whether it will continue to transmit fluids at a variable rate corresponding to the rate of compaction. At depths greater than about 3 km where most of the compaction is chemical, the rate of compaction will be slow and relatively independent of changes in the stress field. When considering larger volumes of rocks they can not be heated or cooled rapidly, at least not in a non-hydrothermal environment, because of the high specific heat capacity of the rocks. The rate of compaction and the resulting fluid flow will then be relatively uniform over a limited time.

10.12 Formation of Overpressure (Abnormal Pressure)

Overpressure is a term used for subsurface pressures that significantly exceed the hydrostatic pressure. This implies that the flow of porewater to the surface during compaction is resisted to a considerable degree, so that the pressure gradients are increased in the least permeable part of the sediments. Overpressure may be produced by different mechanisms. The simplest type of overpressure is due to the pressure (head) of an elevated groundwater table connected to the basin through an aquifer. Rainwater infiltration into the ground will then help to maintain the pressure even if the aquitards (seals) do not have very low permeability. The overpressure will nevertheless decrease away from the area of recharge. This is also expressed by decreasing piezometric surfaces along the direction of flow. For overpressure to develop due to compaction the permeability in the seal must be many orders of magnitude lower because the fluid flux is very much lower than in the case of meteoric water flow (Bjørlykke 1993). In a meteoric water aquifer the flux may be up to 0.1–1.0 m/year, while the compaction-driven flux is usually less than the sedimentarion rate which may typically be 0.1 mm/year.

During compaction of sediments there must always be a slight overpressure because there must be sufficient pressure gradients for the excess porewater to flow out so that the porosity can be reduced. The Darcy equation shows that the pressure gradient must be an inverse function of the effective permeability of the rocks forming the seal. When we have very low permeabilities a high pressure gradient will build up, which will drive the water out. The term *disequilibrium compaction* has been used to describe the development of overpressure because the permeability is too low for the water to be expelled at lower pressure gradients. A disequilibrieum is always required for compaction to take place, but if the permeablities are not very low, only a slight overpressure is required for the explusion of porewater during compaction. The build-up of overpressure reduces the effective stress with the effect of stopping or at least reducing mechanical compaction (Fig. 10.11). Overpressure thus provides a negative feedback on mechanical compaction. Chemical compaction in siliceous rocks involving quartz cementation will still continue as a function of temperature even if the effective stress does not increase. The pore pressure can then build up to fracture pressure.

From the Darcy equation we see that the pressure gradient (∇P) is:

$$\nabla P = F \cdot \mu / k$$

It is clear that increases in the fluid flux (F) could cause high overpressure, but the pressure gradient is very sensitive to variation in permeability.

High sedimentation rates and basin subsidence will increase the compaction-driven fluid flux (F) and will contribute to the build-up of overpressure if the permeability is considered to be constant. As shown above, the average upwards flow of porewater is equal to the integrated change in the porosity/depth curve in the underlying sediments.

In addition there is a fluid flux driven by the release of crystal-bound water which in sediments with high contents of water-bearing minerals may be significant. When porewater is heated the thermal expansion of water will also add to the fluid flux but calculations show that this is not very significant.

This is partly because the porewater is not heated very much during basin subsidence since it is moving upwards as the basin and the sediments are sinking.

When solids (e.g. kerogen) are transformed into fluids like oil and gas, a volume expansion may occur. Cracking of oil and the formation of gas also involves a phase change, which will cause increased pressure because of the expansion of gas.

Generation of oil from kerogen gives an increase in volume but calculations suggest that it is not very large because this leaves some solid material (coke) remaining. Generation of gas will cause a higher volume increase, especially at relatively shallow depths.

If the total increase in volume is moderate the convertion of solid kerogen to fluid petroleum is a very efficient mechanism for increasing the pore pressure. This is because the fluid/solid ratio is changed and this may cause source rocks to hydrofracture so that the petroleum is expelled. An illustration of such a phase change can be observed in the spring when lenses of ice frozen in the ground melt. The conversion from solid ice to water represents a reduction in overall volume but may generate overpressure because the fluid/solid ratio has been very much increased.

When considering larger compartments in sedimentary basins, however, the pressure is mainly controlled by the water phase which is much more abundant.

Development of overpressure depends on the fluid flux in relation to the permeability of the rocks. The permeability of shales which may serve as seals for overpressure compartments varies greatly and is difficult to predict. A change from 1.0 nD (10^{-9}D) to 0.1 nD will increase the pressure gradient 10 times and correspond to an increase in the fluid flux by a factor of 10. We must also remember that for vertical flow perpendicular to bedding the effective permeability is the harmonic average of the permeabilities in the different layers. Thin layers with very low permeability may therefore control the flow rate and the build-up of overpressure. It is therefore very difficult to model and predict overpressure. High overpressures can not be due to mechanical compaction, at least not of sediments at the same depth range. Fluid transfer from greater depth may cause higher overpressure at shallower levels, though. The onset of chemical compaction, and in particular quartz cementation at temperatures higher than 80–100°C, will reduce the porosity and permeability not only in the sandstones but also in the shales. High sedimentation and subsidence rates will thus reduce the rate of compaction-driven flux and also the rate of permeability reduction in the shales (seals). Many shales are almost impermeable.

Extremely low permeabilites are required to maintain overpressures in basins that are uplifted and no longer undergo compaction. Overpressured reservoirs in onshore basins in North America have, for the most part, not subsided since the early Tertiary and it is remarkable that the overpressures have been retained without more recent compaction. In the Anadarko Basin the Missisippian and Pennsylvanian sequence is overpressured close to fracture pressure, while the underlying Ordovician rocks are normally pressured (Al-Shaieb et al. 1994). In the Powder River Basin, Cretaceous shales are highly overpressured but not to fracture pressure (Maucione et al. 1994). Widely distributed free gas (Surdam et al. 1994) may also reduce the permeability for water in the shales. It is not clear if these pressures may be maintained by active gas generation at depth. Reservoirs flanking mountain chains like the Rocky Mountains may also be overpressured, here due to meteoric water flow from the mountains.

In sedimentary basins the permeabilities are very much higher parallel to bedding than perpendicular to bedding, and overpressure usually depends more on the lateral drainage than on variations in the vertical permeabilities. Modelling 1 D vertical flow is therefore not very realistic. Faulting that offsets permeable sandstones against tight shales may contribute to the development of overpressure. Synsedimentary growth faulting in particular is very common in basins with high sedimentation rates like the Gulf Coast basins. This is one of the reasons for the widespread overpressuring in such basins.

Models for the prediction of overpressure are often based on changes in fluid flux with less emphasis on the permeability. If the permeability is kept constant, it is possible to model overpressure as a function of other variables such as rates of compaction, hydrocarbon generation and thermal expansion of the porewater. These variables must, however, be compared with the range of permeability values that are likely to exist in a sedimentary basin. Modelling of fluid pressures and the build-up of pore pressure are often based on permeability distributions derived from porosity distributions, which are also poorly constrained. Smectitic mudstones have very low permeabilities even at shallow depth and are particularly effective seals and may be a significant factor causing overpressure. In the North Sea smectite-rich Eocene and Oligocene mudstones and associated sandstones are frequently overpressured at just 1–2 km depth.

The permeability (k) of fine-grained sediments may be related to porosity (φ) and will therefore decrease during compaction. This relationship may be expressed as $k = c\varphi^5$ (Rieke and Chillingarian 1974). It is clear that rather small variations in porosity can produce large variations in permeability. The most important factor in the sediment composition is the surface area (s), which is closely linked to grain size (d). This is expressed in the Konzeny-Carman equation:

$$k = c\varphi^3 / (1 - \varphi)^2 \, s^2$$

The effect of the surface area can also be expressed in terms of tortuosity (t):

$$k = \varphi^3 d^2 / 72 \, t(1 - \varphi).$$

In the case of smectitic clays the specific surface may be several hundred m^2/g while kaolinite and illite typically have about 10 m^2/g (Skjeveland and Kleppe 1992). The specific surface of mudstones rich in smectite may be more that 10 times that of mudstones with mostly kaolinite, chlorite and illite. According to the Konzeny-Carman equation, the permeability in smectite-rich layers may thus be lower by a factor of 10^{-2} compared to other mudstones.

10.13 Summary

The origin of porewater in sedimentary basins may be seawater, meteoric water (freshwater) or water released from minerals by dehydration. Pore waters change their composition by reacting with minerals and amorphous phases and approach equilibrium with the mineral phases present at a rate which is kinetically controlled. Highly soluble ions like chlorides, however, are not in equilibrium with the minerals except within evaporite deposits with halite (NaCl). Fluid flow in sedimentary basins is constrained both by the pressure gradients and the supply of fluids. Meteoric water is the most important supply of fluids and because it is renewed by rainfall the flow can be maintained for a very long time.

Compaction-driven flow is limited by the volume of water buried in the basin and fluids produced in situ by mineral dehydration and petroleum generation. The upwards flow of porewater is usually lower than the sedimentation rate so that the porewater is moving downward relative to sea level. High flow rates can therefore not be sustained except by extreme focusing of the flow.

Further Reading

Audet, D.M. and McConnelli, J.D.C. 1992. Foreward modelling of porosity and pore pressure evolution in sedimentary basins. Basin Research 4, 147–162.

Berner, B.A. 1980. Early Diagnosis, A Theoretical Approach. Princeton University Press, Princeton, NJ,141 pp.

Bethke, C.M. 1989. Modelling subsurface flow in sedimentary basins. Geologische Rundschau 78, 129–154.

Bethke, C.M. 1985. A numerical model of compaction-driven groundwater flow and heat transfer and its application to the paleohydrology of intracratonic sedimentary basins. Journal of Geophysical Research 90, 6817–6828.

Bethke, C. 1986. Hydrothermal constraints on the genesis of the Upper Mississippi valley mineral district from Illinois Basin brines. Economic Geology 81, 233–249.

Bethke, C.M., Harrison, W.J., Upson, C. and Altaner, S.P. 1988. Supercomputer analysis of sedimentary basins. Nature 239, 261–267.

Bjørlykke, K., Mo, A. and Palm, E. 1988. Modelling of thermal convection in sedimentary basins and its relevance to diagentic reactions. Marine and Petroleum Geology 5, 338–351.

Bjørlykke, K. 1993. Fluid flow in sedimentary basins. Sedimentary Geology 86, 137–158.

Bjørlykke, K. and Aagaard, P. 1992. Clay minerals in North Sea sandstones. In: Houseknecht, D.W. and Pittman, E.D. (eds.) Origin, Diagenesis, and Petrophysics of Clay Minerals in Sandstones. SEPM, Tulsa, OK (Special Publication) 47, 65–80.

Bjørlykke, K. and Høeg, K. 1997. Effects of burial diagenesis on stresses, compaction and fluid flow in sedimentary basins. Marine and Petroleum Geology 14, 267–276.

Buhrig, C. 1989. Geopressured Jurassic reservoirs in the Viking Graben: Modelling and geological significance. Marine and Petroleum Geology 6, 31–48.

Caritat, P. de. 1989. Note on the maximum upward migration of pore water in response to sediment compaction. Sedimentary Geology 65, 371–377.

Cassan, J.P., Palacios, M.C.G., Fritz, B. and Tardy, Y. 1981. Diagenesis of sandstone reservoirs as shown by petrographical and geochemical analyses of oil bearing formations in the Gabon Basin. Bulletin des Centres de Recherches Exploration-Production Elf-Aquitaine 5, 113–135.

Cathles, L.M. and Smith, A.T. 1983. Thermal constraints on the formation of Mississippi Valley-Type Lead -Zinc Deposits and their implications for episodic basin dewatering and deposit genesis. Economic Geology 78, 983–1002.

Chapman, R.E. 1987. Fluid flow in sedimentary basins: A geologist's perspective. Geological Society Special Publication 34, 3–18.

Chester, R. 1990. Marine Geochemistry. Unwin Hyman, London, 698p.

Davis, S.H., Rosenblat, S., Wood, J.R. and Hewett, T.A. 1985. Convective fluid flow and diagenetic patterns in domed sheets. American Journal of Science 285, 207–223.

Evans, D.G. and Nunn, J.A. 1989. Free thermohaline convection in sediments surrounding a salt column. Journal of Geophysical Research 94, 12413–12422.

Evans, D. et al. 2001. The Millennium Atlas. Petroleum Geology of the Central and Northern North Sea. Geological Society.

Giles, M.R. 1987. Mass transfer and problems of secondary porosity creation in deeply buried hydrocarbon reservoirs. Marine and Petroleum Geology 4, 188–201.

Gran, K., Bjørlykke, K. and Aagaard, P. 1992. Fluid salinity and dynamics in the North Sea and Haltenbanken basins derived from well log data. In: Hurst, A., Griffiths, C.M. and Worthington, P.F. (eds.) Geological Application of Wireline Logs II. Geological Society Special Publication 66, 327–338.

Hanor, J.S. and Mcmanus, K.M. 1988. Sediment alteration and clay mineral diagenesis in a regional ground water system, Mississippi Gulf Coastal plain. Transactions-Gulf Coast Association of Geological Society 38, 495–502.

Harrison, W.J. and Summa, L.L. 1991. Paleohydrology of the Gulf of Mexico Basin. American Journal of Science 291, 109–176.

Hautshel T. and Karuerauf, A.I. 2009. Fundademntals of Basin and Petroleums Systems. Springer, New York, NY, 476 pp.

Hutcheon, I.E. 1989. Application of chemical and isotopic analyses of fluids to problems in sandstone diagenesis. In: Hutcheon, I.E. (ed.), Short Course in Burial Diagenesis. Mineral Association of Canada, Québec, 270–310.

Katsube, T.J., Mudford, B.S. and Best, M.E. 1991.Petrophysical characteristics of shales from the Scotian Shelf. Geophysics 56, 1681–1689.

Leonard, R.C. 1993. Distribution of subsurface pressure in Norwegian Central Graben. In: Parker, J.R. (ed.), Petroleum Geology of N.W. Europe. Proceedings of the 4th Conference. Geological Society, pp. 1295–1393.

Longstaffe, F.J. 1984. The role of meteoric water in diagenesis of shallow sandstones: Stable isotope studies of the Milk River Aquifer and Gas Pool, southeastern Alberta. In: McDonald, D.A. and Surdam, R.C. (eds.), Clastic Diagenesis. American Association of Petroleum Geologists Memoir 37. AAPG, Tulsa, OK, pp. 81–98.

Ludvigsen, A. 1992. Thermal convection and diagenetic processes in sedimentary basins. PhD Thesis, University of Oslo.

Manheim, F.T. and Paull, C.K. 1981. Patterns of ground water salinity changes in a deep continental-oceanic transect off the Southeastern Atlantic coast of the USA. Journal of Hydrology 54, 95–105.

Mudford, B.S., Gradstein, F.M., Katsube, T.J. and Best, M.E. 1991. Modelling 1D compaction driven flow in sedimentary basins: A comparison of the Scotian Shelf, North Sea and Gulf Coast. In: England, W.A. and Fleet, A.J. (eds.), Petroleum Migration. Geological Society Special Publication 59, 65–85.

Oelkers, E.-H. 1996. Physical and chemical properies of rocks and fluids for chemical mass transport calculations. In: Lichtner, P.C., Stefel, C.I and Oelkers, E.H. (eds.) Reactive transport in porous media. Reviews in Mineralogy 34, 131–191.

Olstad, R., Bjørlykke, K. and Karlsen, D.K. 1997. Pore water flow and petroleum migration in the Smørbukk Field area, offshore Norway. In: Møller-Pedersen, P. and Koestler, A.G. (eds.), Hydrocarbon Seals – Importance for Exploration and Production. Norwegian Petroleum Society 7. Elsevier, Amsterdam, pp. 201–216.

Ortoleva, P. 1994. Basin Compartments and Seals. AAPG memoir 61. AAPG, Tulsa, OK, 459p.

Person, M. and Garven, G. 1992. Hydrologic constraints of petroleum generation within continental rift basins: Theory and application to the Rhine Graben. AAPG Bulletin 76, 466–488.

Ranganathan, V. and Hanor, J.S. 1988. Density driven ground water flow near salt domes. Chemical Geology 74, 173–188.

Rieke, H.H. and Chillingarian, G.V. 1974. Compaction of Argillaceous Sediments. Elsevier, New York, NY, 424p.

Warren, E.A. and Smally, P.C. 1994. North Sea Formation Water Atlas. Geological Society Memoir 15, 208p.

Wood, J.R. and Hewett, T.A. 1982. Fluid convection and mass transfer in porous sandstones – A theoretical model. Geochimica et Cosmochimica Acta 46, 1707–1713.

Wood, J.R. 1986. Thermal transfer in systems containing quartz and calcite. In: Gautier, D.L. (ed.), Roles of Organic matter in Sediment Diagenesis. SEPM, Tulsa, OK, Special Publication 38, 181–189.

Chapter 11

Introduction to Geomechanics: Stress and Strain in Sedimentary Basins

Knut Bjørlykke, Kaare Høeg, and Nazmul Haque Mondol

At shallow depths in sedimentary basins there are soft clays and loose silts and sands, while diagenetic processes have transformed these sediments to claystones, shales, silt- and sandstones at greater depths. Sedimentary rocks continuously undergo physical and chemical changes as a function of burial depth, temperature and time, and important hydro-mechanical parameters change during burial, erosion and uplift. An understanding of these processes is important in order to predict the magnitude and distribution of sediment properties and stresses in the basin. The in-situ stress condition affects the rock response to changes in the stress field due to drilling and petroleum production.

Soil and rock mechanics (geomechanics) have mainly been developed to solve engineering problems in relation to landslides and surface and underground construction. These are usually at very shallow depths compared to that of a petroleum reservoir. We will here focus on some aspects of geomechanics of particular relevance for the petroleum geologist.

Porosity loss (volumetric compaction) with time due to increased effective stress is referred to in the engineering literature as *consolidation*, while compaction at constant effective stress is usually called *secondary compression* or *creep*. Compaction in deep sedimentary basins has occurred over geologic time scales at very low strain rates, and at higher temperatures than shallow sediment compaction. Therefore, in addition to mechanical compaction, there are important effects of mineral grain dissolution, precipitation and cementation. This process is called chemical compaction.

Compaction determines the porosity, density and permeability of the sediments which are essential input for basin modelling; petroleum reservoir quality dependent on the porosity and permeability. The processes of mechanical and chemical sediment compaction (diagenetic processes) determine the physical properties and are also important for understanding seismic velocity records and seismic attributes in sedimentary basins.

11.1 Subsurface Fluid Pressure and Effective Stress Condition

A distinction should be made between total stress, effective stress and fluid pore pressure. This is not always done in technical reports and publications related to petroleum geology.

11.1.1 Total and Effective Stress

In general, stress (σ) is defined as force per unit area. The overburden weight of the sediment including the weight of the fluid in the pore space produces a vertical stress (σ_v). For a sedimentary basin with a fairly horizontal surface, and without major lateral variations in the sediment compressibility, the vertical stress at any point can simply be computed as:

$$\sigma_v = \rho_b g h \tag{11.1}$$

where ρ_b is the average sediment bulk density of the overlying sequence, h is the sediment thickness and g is the acceleration of gravity. This is the *vertical*

K. Bjørlykke (✉)
Department of Geosciences, University of Oslo, Oslo, Norway
e-mail: knut.bjorlykke@geo.uio.no

K. Bjørlykke (ed.), *Petroleum Geoscience: From Sedimentary Environments to Rock Physics*,
DOI 10.1007/978-3-642-02332-3_11, © Springer-Verlag Berlin Heidelberg 2010

total stress or the *lithostatic stress*. It may be calculated more accurately by integrating the varying density over the depth of the sediment column. The effective vertical stress (σ'_v) is defined as the difference between the vertical total stress (σ_v) and the pore pressure (u):

$$\sigma'_v = \sigma_v - u \qquad (11.2)$$

This is the *effective stress* which is sometimes called the average intergranular stresses because it is transmitted through the grain framework. It is the effective stress that governs the mechanical compaction of sediments where little chemical compaction (cementation) has taken place. It should be noted that the local intergranular particle-to-particle contact stress is many times higher than the effective stress as defined here, due to the small area of contact. The total overburden weight is carried by the mineral grain framework and the pore pressure (Fig. 11.1).

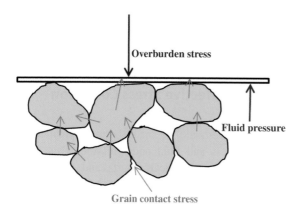

Fig. 11.1 The effective stress from the overburden (σ'_v) is carried by the mineral grain framework (solid phase) and the pore pressure (fluid phase). The effective stress is defined as the overburden vertical stress minus the pore pressure

The effective stress in the horizontal direction is defined as total horizontal stress minus the pore pressure. The horizontal stress will in general not be equal to the vertical stress as discussed in Sect. 11.3 below. However, the pressure in the pore fluid (pore pressure) is the same in all directions.

The bulk density (ρ_b) of sedimentary rocks varies as a function of the porosity (ϕ), the density of the fluid (ρ_f) in the pore space, and the density of the solid phase (ρ_m) which is comprised mainly of minerals:

$$\rho_b = \phi\rho_f + (1 - \phi)\rho_m \qquad (11.3)$$

The solid phase may also have variable density due to different mineral composition, and in some cases amorphous phases also play a role. Usually the density of the mineral matrix in sandstones and shales is close to $2.65-2.70$ g/cm³. If there are significant contents of denser minerals such as siderite or pyrite the bulk density will be higher. Smectite and mixed-layer minerals have variable but generally lower densities. The fluid density also varies with the composition of water and petroleum. In the case of gas-saturated rocks the bulk density becomes significantly lower. The increase in total vertical stress per metre of depth is commonly called the *lithostatic stress gradient* (Fig. 11.2). At about 10% porosity (and assuming pores filled with water) the lithostatic stress gradient is typically 25 kPa/m (25 MPa/km) corresponding to a mineral density of about 2.66 g/cm³ as in quartz and illite. At 30% porosity the bulk density of sediments is typically 2.1 g/cm³.

The rock density is critical for modelling isostasy and backstripping and it is mostly a function of

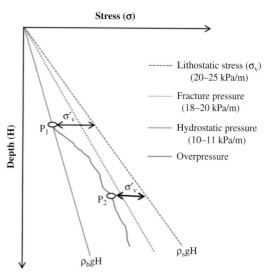

At hydrostatic pressure (P_1) the effective stress at a depth H is $\sigma'_v = Hg(\rho_b-\rho_w)$ but at overpressure (P_2) the effective stress is $\sigma'_v = Hg(\rho_b-P_2)$.

Fig. 11.2 Simplified diagram showing the increase in vertical total stress (lithostatic) and hydrostatic pressure as a function of depth. In reality these lines are not strictly straight because the total vertical stress varies as a function of the sediment bulk density (ρ_b) which tends to increase with depth. The hydrostatic pressure curve is a function of the density of the formation water (ρ_w) which varies with temperature and salinity

the degree of compaction since mineral densities normally do not vary greatly even if the mineral composition does. Carbonates, particularly dolomite, are however significantly denser than shales and sandstones.

11.1.2 Fluid Pressure

In general, the pressure in the pore fluid at any given point may be equal to the weight of the water column to sea level or groundwater table. The porewater may then be said to be in a hydrostatic state. There may also be a pressure gradient in the porewater which is different from the hydrostatic, and then there will be a fluid flow which is a function of the permeability and the viscosity. The flow may be relatively constant over long time, but if the pore pressure changes over a relatively short time the flow is said to be transient. This would be the case with fluid flow related to earthquakes.

In offshore sedimentary basins the pore fluid pressure (u) measured in oil or water is a result of several contributions:

$$u = \rho_p g H_p + \rho_w g H_w + \rho_{sw} g H_d + \Delta u \qquad (11.4)$$

where $\rho_p g H_p$ is the pressure contribution from a petroleum column of height H_p with a density ρ_p, $\rho_w g H_w$ is the pressure due to a water-saturated sequence (H_w), and $\rho_{sw} g H_d$ is the pressure contribution of the seawater column (H_{sw}) with density (ρ_{sw}). Δu is the *overpressure* (or sometimes underpressure) which is any deviation from the *hydrostatic pressure* (Fig. 11.2). Overpressure is sometimes also called abnormal pressure. If u is equal to the hydrostatic pore pressure, then Δu is zero, and the sediment sequence is said to be normally pressured. There is then no tendency to fluid flow.

The overpressure (Δu) can be expressed as the equilibrium height (ΔH) of a hypothetical water column above the level corresponding to hydrostatic pressure. This is the level that water would rise to in a pipe from the formation to the surface. It is called the piezometric or potentiometric surface level. As discussed in Chap. 10 on fluid flow, differences in fluid potentials or piezometric surfaces are expressions of the driving forces for fluid flow in sedimentary basins.

The fluid densities referred to in the equations above are the average densities for a certain fluid column. The density of water decreases with depth due to the temperature increase, but near evaporites the salinity gradient can offset the thermal expansion so that the water becomes denser with depth. The fluid pressure can be calculated more accurately by integrating the fluid density over the height of the fluid column. If the water density is constant and equal to 1.0 g/cm³, the hydrostatic pressure gradient is 10 kPa/m (0.1 bar/m). Expressed in *psi* (pounds per square inch) the equivalent gradient for freshwater is 0.434 psi/ft. In basins like the North Sea and the Gulf Coast the water density varies significantly, and typical Gulf Coast pressure gradients are 0.465 psi/ft or 10.71 kPa/m (Dickey 1979).

If a standpipe (well) is installed in a normally pressured sediment (no overpressure) the water would rise in the pipe to the sea level or groundwater table. Artesian overpressures may be due to meteoric water flow from, for instance, a mountain lake and into a sedimentary basin. If fluid pressure is higher than the weight of the fluid, the water in the standpipe would rise ΔH m above the local water table depending on the degree of overpressure.

During burial and basin subsidence (compaction, compression) the pore pressure is above hydrostatic (transient overpressure) and the water flows out of the sediments as they compact. Unless the permeability in the sediments is very low only very small overpressures are required for the expulsion of water during compaction. If there are low-permeability barriers to flow in all directions, high overpressures may develop during burial because it takes a long time for the pore pressures caused by the added overburden to dissipate/drain. There are also other processes that may lead to overpressure. High sedimentation rates will cause higher rates of compaction and compaction-driven flux. Overpressure will retard mechanical compaction because the effective stress is reduced. The porosity reduction is however very much a function of time in the case of chemical compaction. Since chemical compaction in siliceous sediments is mainly a function of temperature, compaction will continue even at high overpressures and reduced effective stress.

Lower than hydrostatic pressures can also develop but are less common and are usually formed during uplift in the sedimentary basin. Below are listed three ways in which underpressure may develop:

(1) Tectonic extension may slightly increase porosity and create some fractures which need to be filled with fluids, thus lowering the fluid pressure. As water with no gas bubbles has low bulk compressibility (4.10^{-4} MPa^{-1}), a small increase in the porosity caused by the creation of new fractures will produce a significant lowering of pressure. During uplift the sediments no longer compact and water can therefore not flow in from the rock matrix to fill the fractures without lowering the pressure. Extension during uplift will thus tend to draw in meteoric water from above, but if the fractures are not connected so that the water can flow up to the surface, the flow will be rather limited. Compressional tectonics or strike slip tectonics may produce episodes of rapid fluid flow along fractures, often referred to as *seismic pumping*.

(2) Condensation of gas to fluid petroleum may cause reduced fluid volume and lower pore pressure. This may, however, often be compensated for by the expansion of dry gas.

(3) Cooling and contraction of water (the opposite of aquathermal pressuring) may cause lowering of fluid pressure below hydrostatic.

11.2 Normally Consolidated Versus Overconsolidated Sediments

A layer in a sediment sequence that never before in its geological history has been subjected to higher vertical effective stress than at present, is called normally consolidated (NC). If, on the other hand, the sediment has been subjected to higher effective stresses, e.g. by previous glacial loading, by higher overburden that subsequently has been eroded, and/or by pore pressures in the past that were lower than at present, the sediment is called overconsolidated (OC) as it has been preloaded. The ratio between the past maximum effective vertical stress and the present stress is commonly called the overconsolidation ratio (OCR).

At relatively shallow depths in a sedimentary basin (less than $2-3$ km, $<70-90°$C), the mechanical

compaction processes dominate over the chemical compaction in siliceous sediments. At higher temperature (deeper burial) chemical compaction processes become dominant in controlling the rate of compaction. Carbonate sediments may, however, become cemented and highly overconsolidated at shallow depth.

The hydro-mechanical properties at shallow depths may be very different for a normally consolidated sediment sequence compared with an overconsolidated one, depending on the magnitude of the OCR. For the overconsolidated sediment, the compressibility and permeability are usually much lower and the shear strength significantly higher. As discussed below, the lateral stresses in overconsolidated sediments may be higher than in normally consolidated sediments.

11.3 Horizontal Stresses in Sedimentary Basins

Knowledge of the magnitude and distribution of horizontal stresses in sedimentary basins is important in relation to petroleum exploration, drilling and production. Their magnitude is also important in the interpretation of seismic signals used in field exploration and in reservoir production management. In a sedimentary basin the geomechanical properties vary from those of loose cohesionless sediments at shallow depths to dense and cemented sedimentary rocks at greater depth. This affects the horizontal (lateral) stress distribution with depth.

While the vertical stresses are determined by vertical equilibrium (Eq. 11.1), the magnitude of lateral stresses cannot be determined by equilibrium equations and is statically indeterminate. Their magnitudes are governed by a number of factors, including the overburden/erosion (loading/unloading) and uplift history of the basin and the deformation characteristics of the sedimentary rocks. These are a result of gravitational and tectonic forces, and also of stress changes caused by chemical compaction and the accompanying volume change. Their magnitude is determined based on an understanding of the geological history, theoretical and semi-empirical relationships, and field measurements.

11.3.1 Theoretical and Semi-empirical Relationships

In a basin which is wide compared to its thickness, the compaction process due to added overburden may be modelled as a one-dimensional deformation situation (i.e. only strain in the vertical direction, no strain horizontally). This is often denoted as a uniaxial strain compaction situation. If the sediment mineral skeleton (framework) may be assumed to behave in a linearly elastic and isotropic manner (see Sect. 11.4), the horizontal stress which is built up as the vertical overburden is increased, is defined by:

$$\sigma'_H = \frac{\nu}{1 - \nu} \sigma'_\nu \qquad (11.5)$$

where ν is the Poisson's ratio for the sediment mineral skeleton. The same relationship would hold for unloading if the material really exhibits linearly elastic behaviour. Furthermore, for isotropic material, the magnitude of horizontal stress would be the same in all directions. If one assumes anisotropic behaviour, the equations corresponding to Eq. (11.5) would be somewhat more complicated, and the horizontal stresses would be different in the different directions. The horizontal stress coefficient for a uniaxial deformation situation is commonly called K_0 in geomechanics. For an assumed Poisson's ratio $\nu = 1/3$, K_0 becomes 0.5 from Eq. (11.5).

As discussed in Sect. 11.4, linearly elastic behaviour may be an acceptable approximation for a cemented sediment (sedimentary rock) undergoing minor deformation. However, during the initial gradual build-up of loose sediments in a basin, it is not realistic to assume linear elastic behaviour of the sediment framework. Its behaviour is very non-linear and inelastic, undergoing mainly permanent deformation. In soil mechanics one uses the following semi-empirical relationship for normally consolidated (NC) sediments. It is based on idealised theoretical considerations and on laboratory and field measurements:

$$K_{0nc} = 1 - \sin\varphi' \qquad (11.6)$$

where φ' is the angle of shearing resistance (friction angle) used in the Mohr-Coulomb failure criterion expressed in terms of effective stresses, and it is assumed that there is no cementation (cohesion

$c = 0$). The K_0 value defined this way refers to the ratio of effective horizontal and vertical stresses (not total stresses). For $\varphi' = 30$, K_{0nc} becomes 1/2.

When such a sediment is unloaded (erosion) and thus becomes overconsolidated (OC), the horizontal stress does not decrease proportionally with the reduction in vertical effective stress because the sediment skeleton does not exhibit elastic behaviour. Horizontal stresses are "locked in" in the sediment, and the K_0-value increases. The following semi-empirical relationship is used, based on laboratory experiments and field measurements:

$$K_{0oc} = K_{0nc}(OCR)^n \qquad (11.7)$$

where OCR is the overconsolidation ratio and n is a coefficient experimentally determined to usually be between 0.6 and 0.8, depending on the sediment properties. K_0 may well reach values above 1 (2–3 have been measured), which means that the horizontal stress is significantly larger than the vertical stress.

During burial and compaction and erosion (uplift), the horizontal stresses may change due to tectonic movements, and with time due to combinations of mechanical loading and unloading and also chemical compaction which may be independent of stress.

If extension occurs in a sediment with friction angle (φ') but no cohesion intercept (c), the effective horizontal stress coefficient would decrease from K_0 to a lower limiting value of:

$$K_{ext} = 1 - \sin\varphi'/1 + \sin\varphi' \qquad (11.8)$$

At this low lateral effective stress, shear failure would occur and a shear plane (normal fault) would form. Such extension may occur due to general basin extension, or more locally over a salt dome, over an elevated fault block of sedimentary rock with softer sediments on either side, or at the top of a slope.

If on the other hand, lateral compression occurs in the same sediment, the lateral effective stress coefficient would increase to a limiting value (reverse faulting):

$$K_{com} = 1 + \sin\varphi'/1 - \sin\varphi' \qquad (11.9)$$

Assuming $\varphi' = 30$, K_{ext} and K_{com} would be 1/3 and 3, respectively. If a cohesion intercept (c) is included, the value for K_{ext} would be smaller and K_{com} higher than given by Eqs. (11.8) and (11.9), respectively. For the case of horizontal compression the maximum horizontal effective stress is:

$$\sigma_H' = \frac{1 + \sin \varphi'}{1 - \sin \varphi'} \sigma_v' + 2c' \sqrt{\frac{1 + \sin \varphi'}{1 - \sin \varphi'}} \quad (11.10)$$

This is the linear Mohr-Coulomb failure criterion expressed in terms of the major and minor principal stresses. In this case σ_H is the major principal stress and σ_v the minor principal stress.

Zoback et al. (1985) and other investigators have measured high horizontal stresses in basement rocks. These stresses probably reflect compressional plate tectonic movements. However, basin sediments are much more compressible than the basement rocks (uncemented sands have been found as deep as 1.5–2 km in the North Sea). Therefore, the external plate tectonic and regional tectonic stresses that are transmitted through the underlying basement and the deeper well-cemented sedimentary rocks will have little effect on the horizontal stresses in the compressible sedimentary basin above, unless the lateral tectonic movements (compressive strains) are very large (Bjørlykke and Høeg 1997, Bjørlykke et al. 2005, 2006).

In the North Sea the horizontal stress has been found to increase with depth faster than the vertical stress, and at 4 km and deeper the total horizontal stress is usually 0.8−0.95 of the total vertical stress, approaching unity. The magnitude of horizontal stress at these depths is influenced by the effects of chemical compaction and creep. It should be noted that the ratio between the total horizontal and vertical stresses is not the same as the ratio between the effective stresses at the same location. In a sediment with high pore pressures (overpressure), and a ratio between total stresses of 0.9, the corresponding ratio between effective stresses may be about half that value, depending on the magnitude of overpressure. It is the ratio between effective stresses that indicates how close the sediment may be to a local shear failure, and it is the magnitude of the minimum effective stress that governs whether a tension fracture may occur.

The strain rates are important for the laboratory determination of K_0 because deformation by creep is a function of time. Some rocks behave very differently

at low strain rates and at high strain rates. For further discussion on brittle and ductile behaviour see Sect. 11.4.4.

When the effective stress in the reservoir is increased due to reduced fluid pressure during petroleum production, the strain rates are fairly high. Therefore, the ratio between the horizontal and vertical stress will be controlled by mechanical compaction, and the K_0 values determined in laboratory tests may be applied.

11.3.2 Field Measurements of Horizontal Stress

As it is difficult to predict the horizontal in-situ stress condition in the basin, and as the horizontal stresses may be very different in two orthogonal directions, it is common to resort to field measurements. The maximum horizontal stress is termed σ_H while the minimum horizontal stress is called σ_h. If the vertical stress is the major principal stress σ_1, the two horizontal principal stresses are σ_2 and σ_3, respectively.

To measure these two horizontal stresses and their orientation one may use so-called hydrofracturing tests in a borehole (e.g. Goodman 1989, Fjaer et al. 1992). For a typical situation where the major principal stress in the sediment is vertical, a vertical radial fracture will open in the wall of a vertical borehole when the fluid pressure in the borehole is increased. This is because the fracture will be oriented perpendicular to the direction of lowest effective stress which is in the horizontal direction. If the maximum and minimum horizontal stresses are different, the tangential stresses around the borehole vary. When the fluid pressure is equal to the minimum effective stress plus the rock tensile strength (τ) at the most critical location around the borehole, i.e. the location with the smallest initial tangential compression stress, a fracture opens in the wall of the borehole. This fluid pressure level is called the fracture pressure. By lowering the fluid pressure after the crack has opened, and then increasing the fluid pressure again, one may determine the tensile strength of the rock as the difference between the fracture pressure during the first and second load cycles. The tensile strength of a sedimentary rock is only a small fraction of the compressive strength, and tensile strength can

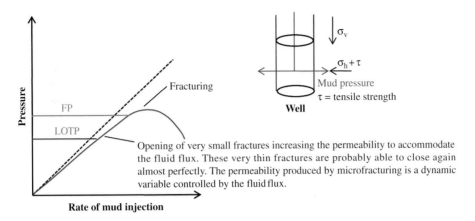

Opening of very small fractures increasing the permeability to accommodate the fluid flux. These very thin fractures are probably able to close again almost perfectly. The permeability produced by microfracturing is a dynamic variable controlled by the fluid flux.

Fig. 11.3 Principle of leak-off test (LOT). The leak-off pressure must be higher than the lowest stress; fractures develop perpendicular to the direction of shear stress. In sedimentary basins the horizontal stress is lowest and the fractures will be vertical

often be neglected. Thus the fracture from the first load cycle may be used to determine the horizontal stresses.

The best way to measure the fracture pressure during the drilling phase of exploration is to perform a series of "minifrac" tests. However, this is not normally done, and it has become common practice in the petroleum industry to perform a simpler measurement that is called a "leak-off" test (Fig. 11.3). This is a pressure test in the well which is closed using the blow-out preventer valves. After the string of drill casing is set and cemented in the well, a leak-off test is normally run after a few metres of hole are drilled below the drill shoe. Mud is pumped into the well through the string using the cement pump of the drill rig. Return flow is prevented by cementing the casing, and the mud pressure is recorded as a function of time and injected volume. Before any fracture is opened, little or no mud is leaked into the sediment formation and the pressure simply increases as a linear function of the volume of mud injected. When the first fracture is produced the pressure increase is reduced relative to the injected volume of mud, thus changing the slope of the curve. This leakage of mud into the formation is an indication that a thin fracture(s) has formed and that the fracture pressure has been reached (hence the name leak-off test). When repeated, the leak-off test fracturing starts a little earlier because now $\tau = 0$ across the fracture that was created during the first load cycle.

The results are often very reproducible which indicates that the rock was not seriously damaged during the first test, and that the small fractures produced probably closed again rather efficiently. The location of the fracture(s) may be determined by modern formation imaging tools called FMI. The location of the radial fracture gives the direction of the minimum principal stress as the latter is normal to the fracture (tangential to borehole wall at that location). The small fractures produced during a leak-off test may resemble the fractures formed during natural hydrofracturing at the top of an overpressure compartment in a sedimentary basin.

The directions of the maximum and minimum horizontal stresses may also be determined from borehole break-outs as shown in Fig. 11.4. The maximum compressive stress in the wall of the borehole occurs at

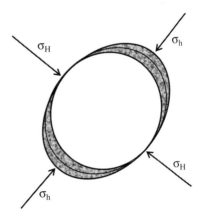

Fig. 11.4 Borehole break-out and the orientation of principal stresses. σ_H is the highest horizontal stress and σ_h is the lowest horizontal stress

each end of a diameter normal to the maximum horizontal stress direction. If the tangential stress is high enough to cause a compressive failure, a break-out occurs. By locating the break-out, one can determine the direction of the maximum horizontal stress. In practice the location may be determined by calliper measurements. The calliper measures the width of the borehole by recognising an oval shape. Modern formation imaging tools (FMI) can also be used to see the borehole break-outs.

11.4 Deformation Properties of Sedimentary Rocks

The fluid in the pores of the sediment compresses, but if there is no gas in the pore fluid, this effect is very small and insignificant when computing volumetric compaction in sedimentary basins. It is the compression bulk modulus of the grain structure that will govern the volumetric deformations, and the permeability and neighbouring drainage boundary conditions that will govern how quickly the fluid may escape from the pores and allow the volume change to occur.

11.4.1 Concepts from the Theory of Elasticity

Elastic material behaviour, linear or non-linear, means that all strains (volume change and shear distortion) caused by a stress change are recovered when the stresses return to their original condition. If the grain skeleton (framework) of a sedimentary rock may be considered linearly elastic and isotropic for very small deformations, the deformational characteristics may be defined by the theory of elasticity.

Youngs modulus (E) is the ratio between the increase in normal stress and the resulting strain in the stress direction, when there is no change in the orthogonal normal stresses:

$$E = \sigma_z/\varepsilon_z \qquad (11.11)$$

where ε_z is the compressive strain in the z-direction, and $\Delta\sigma_x = \Delta\sigma_y = 0$. Poisson's ratio ($v$) is defined as $\varepsilon_x/\varepsilon_z$ in this situation. ε_x is the strain in the x-direction

and is equal to ε_y if the material is isotropic. For a linearly elastic, isotropic material, only two constants (for instance E and v) are required to fully define all the deformation characteristics.

The bulk modulus (K) is defined as the ratio between the increase in equal-all-round stress and the resulting volumetric compression (compaction):

$$K = \Delta\sigma/\varepsilon_{vol} \qquad (11.12)$$

For an isotropic material the bulk modulus is $K = E/3(1 - 2v)$. The shear modulus (G) is the ratio between the increase in shear stress and the resulting shear strain (angular change due to deformation). For an isotropic material it may be shown that $G = E/2(1 + v)$.

For a uniaxial strain compaction situation with strain only in the vertical direction and no lateral strain allowed, the compaction (compression) modulus (M) may be expressed as $M = E(1 - v)/(1 + v)(1 - 2v)$. As stated in Sect. 11.3.1, the compaction process in a fairly homogeneous and wide sedimentary basin may be considered uniaxial. This is not the case in a narrow basin or where there are abrupt changes in depth of basin, lithology and material compressibility, for instance due to faulting and block rotations.

The strains induced by the transmission of seismic signals are so small that linear elastic behaviour may be assumed. However, anisotropic deformation characteristics should be allowed for when the seismic signals are used to derive deformation characteristics of the sedimentary rocks. A special type of anisotropy, which is a useful extension of the isotropic material theory, assumes the sedimentary rock to be transversely isotropic. This implies that the elastic properties are equal for all directions within a plane, but different in the other directions. Transverse isotropy may be considered to be a representative symmetry for horizontally layered sedimentary rocks. The properties are assumed isotropic, but different, in the vertical and horizontal planes. Five elastic constants fully define all the deformation characteristics for such a material.

In general, sedimentary rocks cannot be treated as linearly elastic materials because the normal and shear strains are not recovered when the element is unloaded. However, the modulus concepts from elasticity theory, as outlined above, are very useful even when analysing the more realistic non-linear behaviour of such rocks, including permanent (plastic) deformation.

It should be pointed out that for a linearly elastic ideal material like the one described above, there is no coupling between the effects of normal stresses and shear stresses and between normal strains and shear strains. For instance, if an element is only subjected to shear stresses, there will not be any normal strains. As discussed below, the behaviour of sediments and sedimentary rocks is not that simple, and there can be significant volume expansion (dilatancy) or contraction even if only shear stresses are applied. This phenomenon is also related to the degree of overconsolidation, as the higher the overconsolidation ratio, the more significant is the degree of shear dilatancy.

11.4.2 Non-linear, Inelastic Behaviour in Uniaxial Strain Compression

The stress–strain results from a saturated sediment tested in uniaxial strain compression are shown in Fig. 11.5a. The starting point (I) for the curve represents the initial state of stress, and it is assumed that initially the specimen is normally consolidated. The uniaxial compression modulus increases with the level of applied vertical effective stress. We therefore use the term tangent modulus (M_t) and/or secant modulus (M_s) to represent the behaviour. The tangent modulus gives the slope of the curve at any specified stress level,

while the secant modulus gives the slope of the secant between two stress levels, commonly between the initial point I and point A. At stress level A the specimen is unloaded to the initial stress level (point B). As may be seen from the figure, a significant irrecoverable strain has been accumulated, given by the horizontal distance IB. The specimen is then reloaded back to A and up to a higher level, point C. It is found that the curve from A to C is a natural elongation of the curve from I to A. From I to A and A to C the specimen is normally consolidated, while during the unloading/reloading sequence it is overconsolidated. It should be noted that linearly elastic loading and unloading behaviour would be represented by only one common straight line in this diagram, from I to C.

Extensive laboratory testing of different types of sediments has shown that the tangent modulus may be determined by the following general expression:

$$M_t = m\,p_0 \left(\frac{\sigma'_v}{p_0}\right)^{1-a} \tag{11.13}$$

where p_0 is a reference stress to make the ratio inside the parenthesis non-dimensional (often p_0 is set equal to 0.1 MPa in the published literature), and "m" and "a" are non-dimensional coefficients depending on the type of sediment and its geological history. To fit the different experimental curves, the exponent "a" is

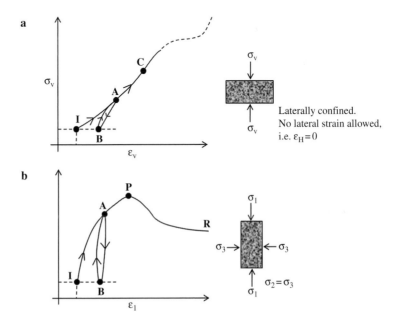

Fig. 11.5 Stress–strain behaviour of sedimentary rocks. (**a**) Stress–strain behaviour of sediment during uniaxial loading, unloading and reloading. (**b**) Stress–strain behaviour of sediment under triaxial compression with no restraint of lateral strains; P is the peak strength and R is the residual strength

found to lie between 0 and 1. For a normally consoli-dated clay sediment $a=1$ fits quite well, and this gives an M_t value which is directly proportional to σ_v. For overconsolidated clay sediments (weak claystones) an a-value of 0 gives a fair approximation, and that corresponds to an M_t which is constant as given by the almost straight lines for the unloading and reloading stress–strain curves in Fig. 11.5a.

When the effective vertical stress is increased much beyond the C-level, one may find that the stress–strain curve bends rather sharply over to the right (see dotted curve in Fig. 11.5a) before it again starts to rise for a further increase in the vertical effective stress. In a sand this is caused by crushing of the coarser sand grains because the intergranular contact stresses become so high (Chuhan et al. 2002, 2003); this is further discussed in Sect. 11.5. A similar phenomenon happens in sediments with a cemented but open and porous structure. This was clearly demonstrated for the reservoir chalk at the Ekofisk Field in the North Sea. When the effective vertical stress was increased by reducing the fluid pressure in the reservoir, it reached a level at which the coccolith structure (framework) of the chalk collapsed and caused large vertical strains, reservoir compaction and seafloor subsidence. Subsequent injection of seawater maintained the fluid pressure in the Ekofisk reservoir and avoided further increase of effective stresses. The rate of compaction was reduced, but not stopped, because it was later found that the seawater weakened the chalk framework and increased the compressibility.

11.4.3 Non-linear, Inelastic Behaviour When Approaching Shear Failure

An element in a state of true uniaxial strain compression (Sect. 11.4.2) can undergo large vertical compression (compaction), but it cannot fail in shear by creating failure planes and fractures. This is prevented by the lateral confinement of the element (one-dimensional compression).

However, consider now a cylindrical specimen of a saturated sediment/sedimentary rock as shown in Fig. 11.5b. The specimen is first subjected to an axial stress (σ_1) and a radial stress (σ_3) representing the initial in-situ stress condition. The initial pore pressure in the specimen is also the same as the one in-situ and

thus the initial effective stresses. Then the axial stress is increased while the lateral stress is kept constant. The loading is performed so slowly that any tendency to overpressure in the pore fluid is avoided by allowing the fluid to drain out of the specimen (dissipate), so that the pore pressure remains at its initial value. In geomechanics this is called a drained test, as compared to an undrained test in which the fluid is not allowed to drain and overpressures (positive or negative) build up during the loading.

The recorded stress–strain curve for this axial loading is shown in Fig. 11.5b. Loading occurs from the initial point I to point A. The initial section of the curve is fairly straight (linear), but as the axial stress increases the curve starts to bend. The slope of the curve at any point is called the tangent Young's modulus (E_t). The secant from point I to A gives the secant modulus (E_s) up to that stress level. If the axial stress at point A is reduced down to the original axial stress level, the unloading curve goes down to point B and an irrecoverable strain given by the distance IB has been accumulated. Upon reloading the curve climbs back up to point A. The slopes of the unloading-reloading curves are very similar and close to the initial slope of the curve I to A. When the axial stress is increased beyond point A, the curve continues to bend as the specimen approaches a shear failure condition. Both the tangent and secant modulus decrease significantly. As the stress difference ($\sigma_1 - \sigma_3$) becomes even larger, so does the maximum shear stress in the specimen. The stress difference cannot exceed a certain level (strength), as the specimen is not able to carry any more load, and large axial strains (and shear strains) ensue.

If one were to start the test described above with a higher horizontal effective stress level, the stress–strain curve would be steeper and climb higher. For loose sediments the moduli and shear strength are strong functions of the horizontal effective stress level, therefore it is so important to be able to estimate the effective stresses. For sedimentary rocks (shales and sandstones) with strong cementation caused by chemical processes, the modulus and strength are also influenced by the horizontal effective stress, though to a much smaller extent. Note that the triaxial test shown in Fig. 11.5b may also be run as a uniaxial strain test. This is done by adjusting the horizontal stress at all stages of the test such that no horizontal strain is allowed to occur. This is called a K_0 test.

11.4.4 Brittle Versus Ductile Stress–Strain Behaviour

For some sedimentary rocks, and depending on the magnitude of the lateral effective stress, the stress–strain curve in Fig. 11.5b may drop abruptly after it has reached a peak at point P (the peak strength). This is accompanied by a drop in shear resistance with further strain down to a level R which is called the residual strength after the failure. The behaviour from the peak down to residual is termed strain-softening or strain-weakening.

For other sedimentary rocks there is no strain-softening, and the stress–strain curve is fairly horizontal or even slightly climbing as the strain increases. This is called ductile behaviour. A sedimentary rock found to behave as a brittle material at low horizontal effective stress, may well behave as a ductile material when the effective horizontal stress becomes sufficiently high, i.e. the horizontal stress has reached the brittle-to-ductile transition stress level (e.g. Goodman 1989).

Other factors also affect the degree of brittleness. The higher the temperature and the lower the stain rate, the lower the less tendency for brittle behaviour. In this connection it should be pointed out that the rate of stress change in and around a reservoir during petroleum production is orders of magnitude higher than the geological stress changes, except for earthquake occurrences. Some rocks behave very differently at low strain rates compared with at high strain rates. Rock salt is brittle when loaded at a very high strain rate, but flows like a viscous fluid over geological time. Also carbonate sandstones and shales yield to stress in a ductile manner by mechanical as well as chemical compaction if the strain rate is low enough.

11.5 Compaction in Sedimentary Basins

The deformation characteristics and typical values of compression modulus for sedimentary rocks are presented in Sect. 11.4.

In sedimentary basins down to depths of 1.5–2 km mechanical compaction caused by the increase of effective vertical stresses is the dominating compaction process and has the greatest influence on the sediment's hydro-mechanical properties. However,

at greater depths where temperatures are higher (>70–80°C), it is mainly chemical compaction that contributes to the volume change and to the hydro-mechanical properties through the effects of dissolution, precipitation and cementation.

11.5.1 Sands and Sandstones

The effective stress of the overburden is transmitted through a framework of load-bearing grains. These grain-to-grain contact stresses may become very much higher than the average effective stress. Not all grains will be subjected to high stresses because they may be shielded by other grains within the load-bearing grain framework. The stress from the overburden will thus be concentrated on these other grains, which then may fracture. The small areas of grain contact make the stress at these contacts very high, even at moderate burial depth.

Natural sand grains of quartz and feldspar are mostly blocky rather than spherical and have an irregular surface. This means that the area of contact is likely to be very small even for the larger grains, resulting in much higher contact stresses than for smaller grains. If sand grains had been perfectly spherical the contact would be controlled by the elasticity, and the stress would then be independent of the grain size.

Experimental compaction of well sorted sand reveals that coarse-grained sand aggregates are subjected to significant grain fracturing and compaction at 20–30 MPa effective stresses while fine-grained sand does not fracture, and compacts much less at the same stress levels (Fig. 11.6) (Chuhan et al. 2002, 2003, Bjørlykke et al. 2004).

Well sorted sand (e.g. beach sand) with an intitial porosity of 40–45% may compact mechanically to 30–38% porosity, depending on the stress level and the grain size. Poorly sorted sand and sand with high mud contents will compact much more at even lower stresses.

With increasing compaction due to grain rearrangement or breakage, and cementation, the rock becomes less porous, less compressible and stronger. This process increases both the number and area of grain contacts.

Loose uncemented sands subjected to shear deformation may develop thin deformation bands. Such shear bands may be composed of densely packed

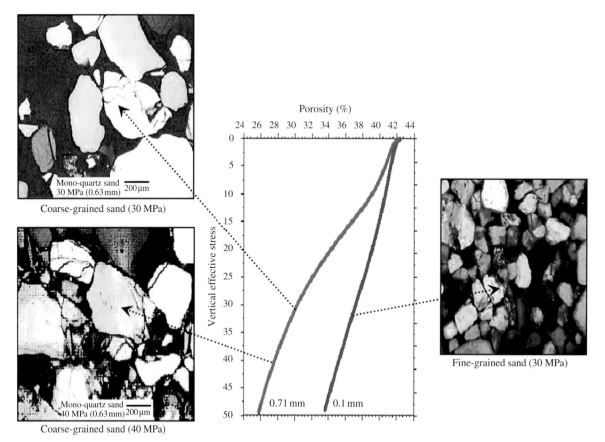

Fig. 11.6 Experimental compaction of loose sand grains (after Chuhan et al. 2002). Coarse-grained sand is more compressible than fine-grained sand. This is because there are fewer grain contacts and more stress per grain contact in coarse-grained sand, resulting in more grain fracturing

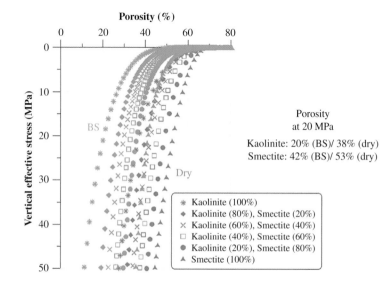

Fig. 11.7 Experimental mechanical compaction of dry (in *grey*) and brine-saturated (in *color*) clay aggregates under uniaxial compression strain (after Mondol et al. 2007). Porosity at 20 MPa effective stress of dry and brine-saturated pure smectite and kaolinite mixtures are shown

grains where smaller silt-sized grains have been packed between larger grains during the shearing. If there is little clay the shear strength of the shear band will exceed the shear strength of the matrix and the shear deformation will shift laterally to an area where there has been no strain (deformation). This is called strain-hardening and results in a network of shear bands which have only been subjected to small offsets. Large displacements recorded on seismics may in reality consist of a broad zone of deformation bands.

Shear deformation in sand may also result in grain crushing. Experimental deformation of sand suggests, however, that the effective stress normal to the shear band must be at least 10 MPa for coarse sand to be crushed. In the case of normal faults the horizontal stress must have been 10 MPa and the vertical stress 20–25% higher, which corresponds to burial depths of 1–1.5 km.

Precipitation of quartz or other cements increases the stiffness of sand and reduces its compressibility, transforming loose sand into indurated sandstones. Only relatively small amounts of quartz cement, probably only 2–4%, are required to effectively stop the mechanical compaction that is due to rearrangement of grains. As a result the velocity, and particularly the shear velocity, will increase sharply for a modest reduction in porosity.

Sandstones may behave as if they were overconsolidated due to cementation (chemical compaction) and will only compact following the stress-strain curve for overconsolidated rocks. We must distinguish between overconsolidation due to previously higher effective stresses and "pseudo overconsolidation" caused by cementation and chemical compaction. This because the highest effective stress must be estimated from the burial curve and the pore pressure while the chemical compaction may be rather insensitive to changes in stress.

Chemical compaction involving the dissolution and precipitation of quartz is controlled mainly by temperature because the rate of quartz cementation seems to be the rate-limiting step. This implies that the effective stress plays only a minor role. The mineral cement growing as overgrowth on detrital quartz grains into the pore space between the grains is not subjected to stress at the time of cement precipitation and is not influenced by the stress. However, after further compaction it may become a part of the load-bearing structure. Therefore the grain-to-grain contact stresses in a sandstone may

be lower at depths of 3–4 km than at 1–2 km because the stress is distributed over a larger grain contact area and is also supported by quartz cement. Most mechanical grain crushing therefore occurs at depths shallower than about 2–3 km where there is little quartz cement present. In the case where quartz cementation (overgrowth) is prevented by coatings (e.g. chlorite), grain fracturing may occur at 3–4 km at about 35–40 MPa effective stresses. Grain fracturing will then expose uncoated fresh quartz surfaces for quartz cementation which gradually will reduce the porosity (Fig. 11.6).

During production of a reservoir, reduced pore pressure and higher effective stress may cause compaction of the reservoir rocks. In the case of well-cemented reservoir sandstones this effect is very small, but in shallow reservoirs with loose sand such compaction can be significant.

11.5.2 Clays and Mudstones

11.5.2.1 Clays and Mudstones Are Poorly Defined but Mudstones May Have a High Content of Silt and Sand

Clays, mudstones and shales have very different physical properties compared to coarser-grained sediments like siltstones and sandstones. The physical properties of clays depend not only on the strength of the sediment particles (mostly clay minerals), but also on their surface properties and chemical bonds which are controlled by the composition of the pore fluids. Clays, mudrocks and shales may vary greatly and have very different physical properties, depending on the clay mineral composition and on the content of silt and sand.

The most common clay minerals are smectite, illite, chlorite and kaolinite. When the silt and sand content exceeds 40–50% there may be a grain-supported structure, and this marks the transition into clay-rich siltstones and sandstones. A sediment of sand and silt grains floating in a matrix of clay has for many purposes the same properties as the clay fraction, but the density is higher and so is the seismic velocity. This is because the stiffness is determined by the load bearing grains.

Poorly sorted clays like glacial clays compact readily at relatively low effective stresses because the clay particles are packed in between the silt and sand grains. Data on compaction of clay and mudstones may be obtained from measuring the degree of compaction where the burial history and maximum effective stress can be estimated, or by experimental compaction in the laboratory.

In fine-grained sediments like clays and mudstones the total overburden stress is distributed over a very large number of grain contacts and the stress per grain contact may be quite low. Relatively small amounts of minerals precipitated as cement between the primary grains can then cause a very significant increase in stiffness and seismic velocity even at shallow burial. Most commonly this involves carbonate cement, which makes soft clay grade into marls and calcareous mudstones with higher bulk moduli. Quartz cementation requires higher temperatures (>80°C) corresponding to 2–2.5 km in basins with normal geothermal gradients.

11.5.2.2 Experimental Compaction of Clays

Experimental compaction of clays is difficult. It requires very careful sample preparation, and compaction tests up to 50 MPa stress may take 5–6 weeks for smectite-rich clays. This is because the permeability is so low that it takes a long time for the excess water to drain. Time is also required to allow for the slight compaction at constant stress (creep) which may also be referred to as secondary compaction.

Compaction of mud to mudstones and shales is the result of natural processes during burial, usually over several million years. We can determine the resultant rock properties by analysing natural rock samples in the laboratory. It is nevertheless still difficult to estimate the effective stress and temperatures to which these rocks have been subjected. This is particularly true in the case of samples exposed on land after substantial uplift. Samples from offshore wells in subsiding basins are much better constrained with respect to the burial history but representative mudstones are rarely cored. Cuttings can be analysed mineralogically, but it is difficult to test their mechanical properties without reconstituting the samples.

There is often a need to predict the compaction of sediments (soils) including clays in an engineering context and they are then tested in the laboratory to measure the strain (compaction) as a function of effective stress and other soil and rock mechanical

parameters. To simulate natural burial in sedimentary basins we use rather high stresses, up to 50 MPa or more, corresponding to 4–5 km of overburden. In most cases, though, chemical compaction becomes dominant at shallower depth.

By testing artificial mixtures of clays we can measure their physical properties as a function of clay mineralogy and silt and sand content. Kaolinitic clays compact much more readily than smectite, which is the most fine-grained clay mineral and has very low compressibility (Mondol et al. 2007). At about 20 MPa effective stress, corresponding to about 2 km of burial, pure smectite has more than 40% porosity while kaolinite has less than 20% (Fig. 11.7). Even at 50 MPa corresponding to 4–5 km burial depth at hydrostatic pressure the porosity may still exceed 40%. Clay minerals compact more when wet than dry (Mondol et al. 2007). This is probably because the friction between the grains is higher in dry clays (Fig. 11.7).

In the case of smectite the large surface area and the water which is bound to these clay surfaces make it difficult to define the proportion of free water, and hence determine the exact porosity, which will also depend on the composition (electrolytic strength) of the porewater. Clay minerals tend to have a negative charge, causing repulsion between the clay particles. However, the negative charges will adsorb cations like Na^+ and K^+, thus neutralising this repulsion. This is what causes flocculation when river-borne clays enter the sea. Marine clays therefore have a more stable clay mineral fabric and higher shear strength than freshwater clays. Slow weathering and leaching of saltwater (Na^+, K^+) from marine clays uplifted by glacial unloading in Scandinavia may therefore lead to slope instability and "quick clay" slides. Experimental compaction of clays confirms that their compressibility and shear strength are a function of the salinity of the porewater.

Smectitic clays are less compressible than kaolinitic clays and this may be partly due to chemical bonds and partly because of the fine grain size. The negative charge of the clay minerals also causes water to be bound to the mineral surfaces by the positive charge of the dipole of the water molecule. This is important in the case of smectite which has a surface area of several hundred m^2/g but is not so significant for coarser clay minerals like illite, chlorite and kaolinite which have much lower surface area bound water. The total stress is divided by the number of grain contacts and

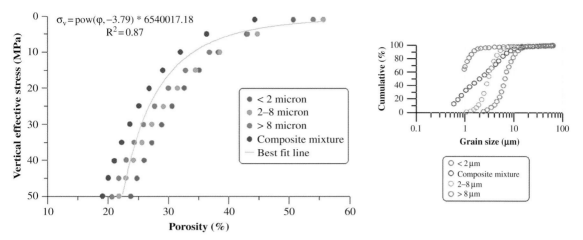

Fig. 11.8 Experimental mechanical compaction of brine-saturated kaolinite aggregates sorted by grain size (after Mondol et al. 2008). The sample containing less than 2 μm sized kaolinite aggregates retained higher porosity compared to all the other mixtures. The maximum porosity reduction is observed in the composite mixture containing all the grain sizes, demonstrating the importance of both grain size and sorting for the rock properties

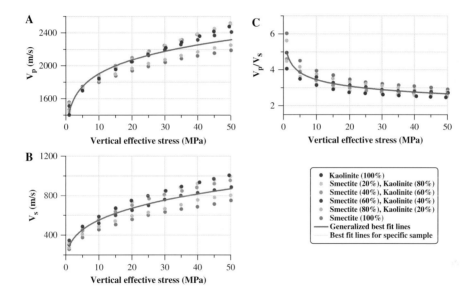

Fig. 11.9 Crossplots of ultrasonic velocities vs. vertical effective stress of brine-saturated smectite-kaolinite (modified after Mondol et al. 2008b). Pure smectite has lower V_p (**a**) and V_s (**b**) compared to pure kaolinite. The V_p/V_s ratio is higher in pure smectite than in pure kaolinite (**c**)

the stress per grain is then less in smectitic clays. It has been shown experimentally (Mondol et al. 2008a) that fine-grained kaolinite is less compressible than coarse-grained kaolinite (Fig. 11.8). Similarly, well sorted fine-grained sand is less compressible than coarse-grained sand (Chuhan et al. 2003). Smectitic clays are characterised by low velocities and low density because they have high porosity (Fig. 11.9) (Mondol et al. 2008b).

11.5.2.3 Chemical Compaction of Clays and Mudstones

Clays compact mechanically at shallow depth and at temperatures below 70–80°C, but at higher temperatures compaction may be controlled by chemical reactions. Chemical compaction must have thermodynamic drive so that less stable minerals dissolve and more stable minerals precipitate. Clay minerals like smectite become unstable and are replaced by mixed-layer minerals and illite. The silica released by this process must be precipitated as quartz cement for this reaction to proceed and this causes marked stiffening and higher velocities. Compaction is then mainly controlled by temperature rather than effective stress.

Amorphous silica from volcanic sediments, and from amorphous silica (opal A) from fossils like diatoms and siliceous sponges, will be a silica source for precipitation of quartz cement even at low temperature. Carbonate cements from calcareous fossils will also cause a marked increase in the stiffness and velocity. In the absence of thermodynamically unstable minerals like smectite, mudstones may remain nearly uncemented to greater depth. At about 130°C kaolinite becomes unstable in the presence of K-feldspar and causes precipitation of illite and quartz. Gradually, however, mudstones become harder and develop a schistocity, becoming a shale. The cleavage is produced by pressure solution of quartz and other minerals in contact with layers enriched in clay minerals.

The transition from mudstones to shales does not only involve a marked increase in stiffness and velocity but also an increase in the anisotropy. This is due to the reorientation of clay minerals during diagenesis so that the velocity parallel to bedding will be higher than perpendicular to bedding. In addition the resistivity will to a large extent be controlled by the orientation of the clay minerals and the quartz cementation in mudrocks and shales.

11.6 Summary

In subsiding sedimentary basins, mechanical compaction caused by the increase of effective vertical stresses is the dominating compaction process and has the greatest influence on the sediment's porosity and hydro-mechanical properties down to depths of 2–2.5 km. At greater depths where temperatures are higher (>70°C), it is mainly the chemical compaction that contributes to the volume change and to the hydro-mechanical properties due to the effects of dissolution, precipitation and cementation. Carbonate rocks, though, may undergo chemical compaction at shallower depths.

The magnitude and distribution of stresses in sedimentary basins are important in relation to petroleum exploration, production and reservoir management. Knowledge of in-situ stress is also important in connection with drilling, particularly during deviation and horizontal drilling. The magnitude and orientation of stresses affect the propagation and interpretation of seismic signals, particularly the S-waves, through sedimentary rocks.

The total vertical stress (σ_v) of a rock sequence is carried partly by transmission of stress in the solid grain framework (effective stress σ'_v) and partly by the pressure in the fluid phase (porewater or petroleum).

Determination of effective stresses depends on reliable estimates of the fluid pore pressure which often is in excess of hydrostatic (overpressure). There exist semi-empirical relationships for the ratio between horizontal and vertical effective stresses, but reliable estimates of horizontal stresses depend on field measurements like hydraulic fracturing tests. It is common practice in the petroleum industry to use a simplified procedure (leak-off tests) to determine the magnitude and orientation of the minimum horizontal stress.

The virgin (in-situ) distribution of stresses in sedimentary basins is the result of both mechanical and chemical compaction, usually over geological time. Changes in stresses during petroleum production from a reservoir are much more short term and are mainly mechanical. In a carbonate reservoir the chemical processes may be so fast that also chemical compaction may become significant at that time scale.

The effective stresses in sand may cause grain-to-grain contact stresses which are so large that compression (compaction) may occur due to crushing and fracturing of grains. This is more pronounced in coarse-grained rather than fine-grained sands and may account for a significant component of the porosity reduction. After the grain crushing and permanent collapse deformations have occurred, the grain size is reduced and the reservoir regains stiffness. With increasing depth such grain crushing is less likely due to increased cementation of the grain structure caused by chemical processes. In most sandstone reservoirs quartz cementation starting at 2–2.5 km (70–80°C) will stabilise the grain framework and prevent further mechanical compaction. Sandstone reservoirs with a critical content of quartz cement (>2–3%) will therefore experience very little compaction even if the effective stress is increased during production. Similar compaction by grain breakage may also occur for high effective stresses in a reservoir where the framework of the sedimentary rock is very porous (e.g. the chalk in the Ekofisk reservoir, North Sea). This framework collapse led to very large reservoir compaction and subsequent seafloor subsidence (c. 10 m).

Smectitic clays are characterised by very high V_p/V_s ratios when compared with other clays (Fig. 11.9c). This implies that in a sequence of mudstones, smectitic clays will stand out with very different characteristics compared with kaolinite and probably also illite-rich sequences, which have much lower velocities and V_p/V_s ratios. At temperatures above 70–80°C smectite will no longer be stable and mudstones will be more influenced by chemical compaction and cementation. In cold basins, however, mechanical compaction can be dominant down to 4–5 km burial depth. Every mudstone has a unique compaction curve which will depend on a number of factors such as mineralogy, grain size, pore fluids, pore pressure, pore aspect ratio, etc.

The effect of time on compaction will always be difficult to evaluate in the laboratory but prolonged compaction tests suggest that compaction due to creep is rather small. At depths shallower than about 2 km, high porosity (>30%) is preserved in Upper Palaeozoic sandstones despite long burial time.

In spite of the limitations, laboratory porosity/density/velocity-stress relations and their comparison with data found in well logs will provide important constraints on evaluation of burial depths, pore pressure prediction and/or the amount of uplift and erosion found in sedimentary sequences (Mondol et al. 2007, 2008b, c,).

Further Reading

Barton, N. 2007. Rock quality, seismic velocity, attenuation and anisotrophy. Taylor and Francis/Balkema, London, 729 pp.

Bjørlykke, K. and Høeg, K. 1997. Effects of burial diagenesis on stresses, compaction and fluid flow in sedimentary basins. Marine and Petroleum Geology 14, 267–276.

Bjørlykke, K. 2006. Effects of compaction processes on stress, faults, and fluid flow in sedimentary basins. In: Buitersand,

S.H.J. and Schreurs, G. (eds.), Analogue and Numerical Modelling of Crustal-Scale Processes. Geological Society Special Publication 253, 359–379.

Bjørlykke, K., Chuhan, F., Kjeldstad, A., Gundersen, E., Lauvrak, O. and Høeg, K. 2004. Modelling of sediment compaction during burial in sedimentary basins In: Stephansson, O., Hudson, O. and King, L. (eds.), Coupled Thermo-Hydro-Mechanical – Chemical Processes in Geo-systems. Fundamentals, Modelling, Experiments and Applications. Geo-Engineering Book Series, vol. 2, Elsevier, London, pp. 699–708.

Bjørlykke, K., Høeg, K., Faleide, J.I. and Jahren, J. 2005. When do faults in sedimentary basins leak? Stress and deformation in sedimentary basins: Examples from the North Sea and Haltenbanken Offshore Norway – A discussion. AAPG Bulletin 89, 1019–1031.

Bjørlykke, K. 2003. Compaction (consolidation) of sediments. In: Middleton, G.V. (ed.), Encyclopedia of Sediments and Sedimentary Rocks, Kluwer Academic Publishers, Dordrecht, pp. 161–168.

Chuhan, F.A., Kjeldstad, A., Bjørlykke, K. and Høeg, K. 2002. Porosity loss in sand by grain crushing. Experimental evidence and relevance to reservoir quality. Marine and Petroleum Geology 19, 39–53.

Chuhan, F.A., Kjeldstad, A., Bjørlykke, K. and Høeg, K. 2003. Experimental compression of loose sands: Relevance to porosity reduction during burial in sedimentary basins. Canadian Geotechnical Journal 40, 995–1011.

Dickey, P.A. 1979. Petroleum Development Geology. Petroleum Publishing Co., Tulsa, OK, 398 pp.

Fjær, E., Holt, R.M., Horsrud, P., Raaen, X. and Risnes, R. 2008. Petroleum-related rock mechanics 2nd ed. Developments in Petroleum Science 53, 491 pp.

Gueguen, Y. and Palciauskas, X. 1994. Introduction to Physics of Rocks. Princeton University Press, Princeton, 294 pp.

Goodman, R.E. 1989. Introduction to Rock Mechanics. Wiley, New York, 562 pp.

Mavko, G., Mukerji, T. and Dvorkin, J. 1998. The Rock Physics Handbook. Cambridge University Press, Cambridge, 327 pp.

Mondol, N.H., Bjørlykke, K., Jahren, J. and Høeg, K. 2007. Experimental mechanical compaction of clay mineral aggregates – Changes in physical properties of mudstones during burial. Marine and Petroleum Geology 24(5), 289–311.

Mondol, N.H., Bjørlykke, K. and Jahren, J. 2008a. Experimental Compaction of Kaolinite Aggregates: Effect of Grain Size on Mudrock Properties. 70th EAGE Conference & Exhibition, Extended Abstract I037, June 2008.

Mondol, N.H., Bjørlykke, K. and Jahren, J. 2008b. Experimental compaction of clays. Relationships between permeability and petrophysical properties in mudstones. Petroleum Geoscience 14, 319–337.

Mondol, N.H., Fawad, M., Jahren, J. and Bjørlykke, K. 2008c. Synthetic mudstone compaction trends and their use in pore pressure prediction. First Break 26, 43–51.

Mondol, N., Jahren, J., Bjørlykke, K. and Brevik, I. 2008d. Elastic properties of clay minerals. The Leading Edge 27, 758–770.

Peltonen, C., Marcussen, Ø., Bjørlykke, K. and Jahren, J. 2008. Mineralogical control on mudstone compaction; a study of Late Cretaceous to Early Tertiary mudstones of the Vøring and Møre basins, Norwegian Sea. Petroleum Geoscience 14, 127–138.

Voltolini, M., Wenk, H.-R., Mondol, N.H., Bjørlykke, K. and Jahren, J. 2009. Anisotropy of experimentally compressed kaolinite-illite-quartz mixtures. Geophysics 74(1), 1–11.

Wiprut, D. and Zoback, M.D. 2000. Constraining the stress tensor in the Visund field, Norwegian North Sea: Application to wellbore stability and sand production. International Journal of Rock Mechanics and Mining Sciences 37, 217–336.

Zoback, M.D., Moos, D., Mastin, L. and Andersen, R.N. 1985. Well bore breakouts and in situ stress. Journal of Geophysicsl Research 90, 5523–5530.

Chapter 12

The Structure and Hydrocarbon Traps of Sedimentary Basins

Roy H. Gabrielsen

12.1 Tectonic Regimes and Stress

12.1.1 Principal Stress Regimes and Types of Stress

The concept of plate tectonics offers a useful framework for structural geological analysis on all relevant scales in petroleum geology, from regional in the exploration stage, to local in the reservoir evaluation and production stages. This is natural, because the principal geological stress systems are ruled by processes in the deep earth like mantle convection and lithosphere subduction, the secondary effects of which are manifested at the base of the lithosphere and along plate margins. Based on these concepts, the basic dynamics of the lithosphere can be quantified, which is a prerequisite for the evaluation and calculation of the state of stress at any point. As seen in the perspective of the petroleum structural geologist, understanding and quantifying the stress situation at the plate margins is a prerequisite for understanding the state of stress in any basin system and in any reservoir.

We term principal stresses originated at plate margins *far-field* or *contemporary* stresses. The stress situation in a basin or a reservoir may be a sum of several far-field stresses combined with a *local* stress, which may be related to burial, erosion, geothermal gradients, topography, basement relief and structural inhomogeneities in the substratum. In other words, the plate tectonic framework provides a basic and general concept on which any structural geological analysis of a sedimentary basin rests, but it must be supplied with information on the local stress system that is superimposed on it. In the context of the far-field plate tectonic stress, we distinguish between the *plate boundary* and the *intra-plate* component.

The plate boundary stress is subdivided into three basic plate margin settings, namely *constructive boundaries* where adjoining plates are moving away from each other and new crust is formed by magmatic activity, *destructive boundaries* where lithosphere is consumed by subduction or obduction, and *conservative boundaries* where plates are moving past each other in a strike-slip sense and where lithosphere is neither created nor consumed (Fig. 12.1). It is important to note that plate boundaries have been generated and destroyed throughout large parts of the history of our planet, so that many may be preserved inside the present plates and hence do not coincide with present continent/ocean margins. The three basic types of plate margin coincide with the three principal stress configurations, namely the tensional, the compressional and the strike-slip regimes. To describe these regimes, we rely on the concept of principal stress configurations and their definition in the context of the principal axes of stress.

12.1.2 Tectonic Stress in the Earth's Lithosphere

As we have seen already, in addition to stresses generated at tectonic plate contacts, several other processes contribute to the generation of stresses. We may, for example, have *residual stress*, which is inherited from previous deformation, where a rock body may have

R.H. Gabrielsen (✉)
Department of Geosciences, University of Oslo, Oslo, Norway
e-mail: roy.gabrielsen@geo.uio.no

K. Bjørlykke (ed.), *Petroleum Geoscience: From Sedimentary Environments to Rock Physics*,
DOI 10.1007/978-3-642-02332-3_12, © Springer-Verlag Berlin Heidelberg 2010

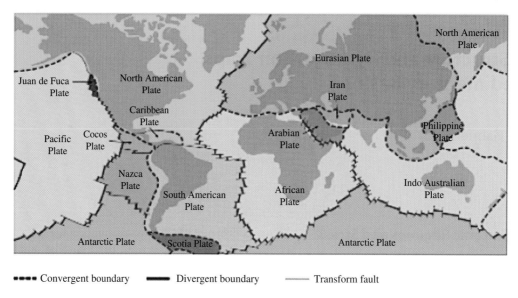

Convergent boundary ● ● ● ●   **Divergent boundary** ▬▬▬   **Transform fault** ─────

Fig. 12.1 The major tectonic plates and their boundaries (from Nystuen in Ramberg et al. 2008). Reprinted with permission from the Norwegian Geological Society and the author

become bent with the elastic stress component remaining unreleased, and *thermal stress* which is related to expansion of rocks during heating or contraction during cooling and stress related to local gravitational gradients. Accordingly, the *total stress* consists of several components:

$$\text{Total stress} = \text{reference stress} + \text{residual stress} + \text{thermal stress} + \text{tectonic stress}$$

where the reference stress refers to the stress inside the plate, devoid of plate tectonic stresses, and thus the

$$\text{Tectonic stress} = \text{contemporary stress} + \text{local stress}.$$

We will not go further into the analysis of residual and thermal stress here, but concentrate on stress generated by primary interactions at plate margins or forces derived thereof.

Taking into consideration the reference stress conditions, one can describe the stress situations for the three principal conditions of deformation, namely extension, contraction and strike-slip. This was done by Anderson in two influential works in 1934 and 1951, in which the framework for all modern tectonic and structural geological analysis is defined. Anderson used the vertical lithostatic stress (σ_v) as a reference:

$$\sigma_v = \rho g z \qquad (12.1)$$

(where ρ is the specific weight, g is the constant of gravity and z is the height of the rock column). This applies for the three principal stress conditions because σ_v can be considered similar for one rock type for a constant h, assuming that the burial history has been similar. Thereby the three principal stress systems can be defined, each corresponding to a regime of deformation (Fig. 12.2):

$$\sigma_v > \sigma_H > \sigma_h; \text{ extension,} \qquad (12.2)$$

$$\sigma_H > \sigma_v > \sigma_h; \text{ strike-slip,} \qquad (12.3)$$

$$\sigma_H > \sigma_h > \sigma_v; \text{ contraction,} \qquad (12.4)$$

where σ_H and σ_h are the maximum and minimum horizontal stresses, respectively. By using σ_v as a reference, further calculations become dependent on the reference system applied. In the contractional regime there will be a tectonic component (σ_t^*) in addition to the reference stress, so that the greatest (horizontal) stress is:

$$\sigma_H = \rho g z + \sigma_t^* \qquad (12.5)$$

Assuming uniaxial stress, which implies that the reference stress is a function of the elastic properties

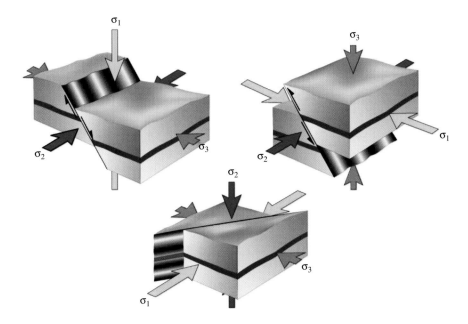

Fig. 12.2   Principal stress configurations (from Fossen and Gabrielsen 2005)

of the rock, and ignoring the thermal expansion, we can describe the stress as:

$$\sigma_H = \left[\frac{\upsilon}{1-\upsilon}\right]\rho g z + \sigma_t \qquad (12.6)$$

where υ is Young's modulus (see Chap. 11). Because $(\upsilon/1 - \upsilon) < 0$, the tectonic stress is $\sigma_t > \sigma_t^*$. Accordingly, the reference stress condition selected also influences the calculated total tectonic stress as long as the buried rock is compressive. For a non-compressive rock ($\upsilon = 0.5$), lithostatic and uniaxial reference systems are equal.

12.1.3 Deformation Mechanisms and Analogue Models

Our daily contact with the physical world tells us that materials deform in many ways, depending on the type of material and its physical state. Thus, a liquid reacts to outer stress very differently from a piece of rock, and one type of rock like chalk has very different physical properties as compared to another like granite. Furthermore, one material may change its mechanical properties dramatically by change of temperature and pressure. These contrasts are founded on

processes occurring on the scales of the grains (in a rock), the molecule and the atom. We have given these processes and their associated meso- and macroscopic physical expressions names like brittle, elastic, plastic, viscous and ductile, and sometimes combinations like elastico-plastic.

Unfortunately, these terms are not always used in a consequent manner and are therefore liable to cause confusion when taken out of context or not precisely defined. This particularly concerns the term "brittle", because it is used in a double sense, namely as a deformation mechanism and a deformation style.

When applied in the context of *deformation mechanisms*, *brittle deformation* implies that existing bonds are physically broken between mineral grains, or that fracturing of the individual grains themselves takes place. As a consequence, the rock loses its cohesion and (potentially) physically falls apart. In contrast, the *plastic deformation* mechanism implies that deformation takes place by the transfer of dislocations on the atomic scale. This means that the mineral can change its shape without loss of cohesion. Generally, plastic deformation occurs at higher p,T-conditions than those accompanying brittle deformation.

Concerning *deformation style*, the term *brittle* is used about localised strain, like that associated with jointing and faulting, and particularly in cases when the rock loses its cohesion and where the deformation occurs at lower p,T-conditions (though not necessarily so). The *ductile* deformation style characterises strain which is distributed over a wider area, as commonly observed in connection with folding and meso- and mega-scale shear-zones. A ductile style of deformation is predominant at high p,T-conditions, but may also occur under very low p,T-conditions, if it involves weak materials like sand and clay. In such cases, however, displacement takes place along grain boundaries or along borders between rock bodies and not by dislocation creep or other atomic-scale mechanisms that characterise the plastic deformation mechanism.

12.2 Petroleum Systems in Extensional Regimes

Areas of extension are affiliated with horizontal divergent stress and are found in association with constructive or passive plate boundaries and in intra-plate settings. Thus, extensional stress regimes either are associated with subsidence and basin formation (in intra-plate settings) or characterise active break-up of continents (along constructive or passive margins).

Although the conditions for development of petroleum systems in areas of active spreading may be meagre, the remnants of the earlier stages of break-up, now situated in passive margins settings, fulfil all the requirements that characterise productive petroleum provinces. This is because such tectonic regimes have undergone crustal thinning and associated subsidence, which involves all the processes essential for petroleum to be generated, trapped and accumulated in sufficient volumes and concentrations for petroleum fields to be commercially interesting. Accordingly, such settings frequently display an attractive combination and distribution of source, reservoir and cap rocks, structural and stratigraphic traps and the conditions for maturation, expulsion, migration and accumulation of hydrocarbons.

12.2.1 Extensional Basins

The formation of extensional basins may be seen as the first stage of the *Wilson Cycle*, which begins with thinning, stretching and rifting of the continental crust followed by continental break-up and mid-oceanic

spreading. The concept of the Wilson cycle predicts that this process sometimes becomes reversed, causing closure of the ocean, collision between the adjacent continental plates, and hence the construction of a mountain chain along the zone of collision (Fig. 12.3). The junction between the continental plates defines the *suture* between the two.

If we use the present North Atlantic as one example, the highly hydrocarbon-rich northern North Sea basin system is situated in a passive continental margin configuration, where the extensional basin system developed during continental break-up. In contrast, Iceland, where petroleum resources are less abundant, is situated on the top of the mid-oceanic spreading ridge. However, if one looks more closely at the structural configuration at depth, one finds that the northern North Sea basin system, which includes the Viking Graben that developed in Jurassic-Cretaceous times, is underlain by an older (Permo-Triassic) basin system (Fig. 12.4). The Permo-Triassic basin system is in turn superimposed on the even older Caledonian suture, which was subsequently affected by gravitational collapse in Devonian times, representing the last stage in a previous Wilson Cycle.

There are several models for the lithospheric configurations that accompany extensional crustal thinning, the end members of which are the "pure shear" (symmetrical) and "simple shear" (asymmetrical) models (Fig. 12.5). It should be noticed that these models are not necessarily mutually exclusive; we can find basin systems that display elements from more than one model, such as the "delamination model" (Fig. 12.6).

The "pure-shear model" for extensional crustal thinning was suggested by Dan McKenzie in 1978, in a paper that has become the most frequently cited in geosciences in modern times. This model assumes thinning of the weak lower crust/lower lithosphere by pure shear, and hence is characterised by the development of a symmetrical configuration (Fig. 12.5a). The pure-shear extension of the ductile lower crust is accompanied by thinning of the upper crust by brittle faulting and subsequent development and rotation of fault blocks.

In this context it is possible to separate the active stretching stage, which is associated with fault-controlled thinning of the upper crust, and later subsidence controlled by thermal processes. As a response to extension, the crust and upper mantle lithosphere becomes thinned and, promoted by extensional faulting, the basin floor will subside quickly.

Fig. 12.3 The major stages in the Wilson Cycle (from Nystuen in Ramberg et al. 2008). Reprinted with permission from the Norwegian Geological Society and the author

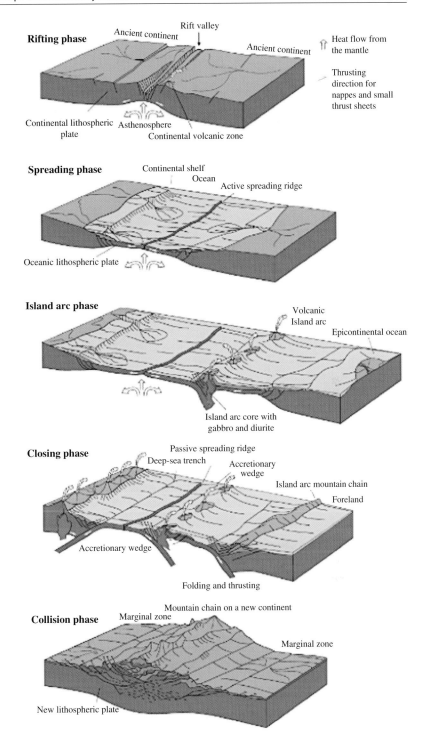

This implies that deeply seated warm rocks are transferred upwards in the lithosphere so that the isotherms in the thinned area become elevated and the thermal gradients become steepened accordingly. These deep processes influence the relief of the basin floor because heating causes rock volumes to expand and elastic, quasi-plastic and isostatic adjustments to occur simultaneously at lithospheric, basin (e.g. by uplift of the basin margins) and fault block scales. In the next stage of development (the post-rift stage), the basin

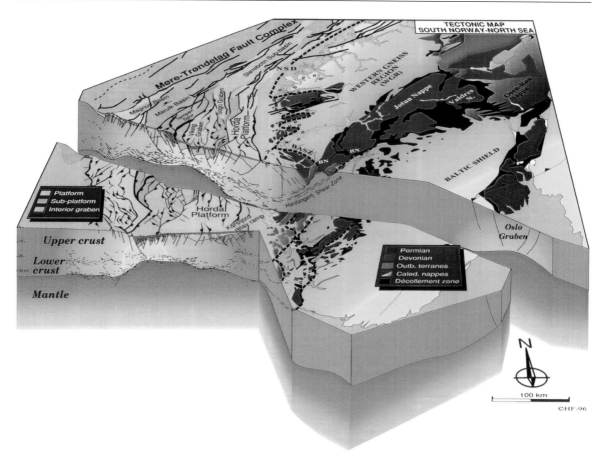

Fig. 12.4 The architecture of the North Sea continental shelf. Note the deep structure, showing a Permo-Triassic rift system buried beneath the younger Jurassic – Cretaceous Viking Graben (from Fossen 2002)

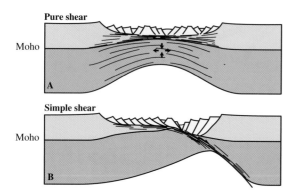

Fig. 12.5 Basic configuration of (**a**) pure shear and (**b**) simple shear extensional basins (modified from Fossen and Gabrielsen 2005)

will continue to subside due to a combination of thermal contraction, sediment compaction and sediment loading. This sounds complex, but luckily these processes are well understood and can be modelled with good accuracy on the basis of the algorithms proposed by McKenzie and supplied with additional modelling tools, developed particularly in the late 1990s.

For modelling purposes and for the analysis of extensional basins with respect to petroleum exploration, three stages of development can be distinguished (Fig. 12.7).

The pre-rift stage is characterised by gentle flexuring and fracturing of the lithosphere. In some rifts we see the development of a gentle bulge, caused by mantle doming and associated warming – and hence expansion – of the lithosphere. In other cases, a gentle subsidence, defining a broad, shallow basin is seen, caused by mild extension of the cold (not-yet-heated) lithosphere. In both cases, the lithosphere is prone to develop steep fractures on a crustal or even lithospheric scale. These fractures have the capacity to accommodate magma, generating dikes. Regarding hydrocarbon reservoir potential characterising the pre-rift stage,

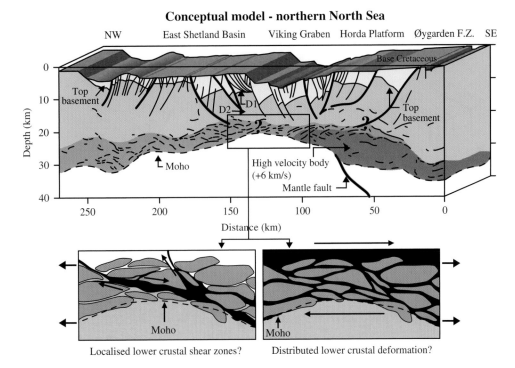

Conceptual model - northern North Sea

Fig. 12.6   Model of the Viking Graben, displaying elements of pure and simple shear. Modified after Odinsen et al. (2000)

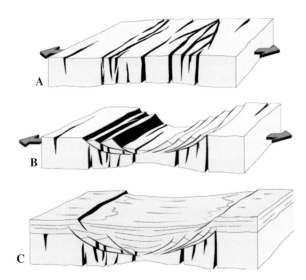

Fig. 12.7   Three major stages in the devlopment of extensional basins. After Gabrielsen (1986)

sand deposits are likely to be sheet-like and relatively thin, with few structural traps developing at this stage. Sediment transport is mainly transverse to the basin axis, but quite homogeneous due to lack of pronounced gradients in the basin. The marginal sediment transport

system is prone to act in concert with the axial transport system, feeding the latter with sediments. This may consist of braided or meandering river systems, depending on factors like axial basin gradient and climate. Since most rifts are generated by break-up of continents, a terrestrial depositional environment would be most common for the pre-rift stage, so that source rocks and cap rocks, which are mostly of marine depositional origin, may be scarce (Fig. 12.8a). There are, however, numerous examples of both source rocks and cap rocks of terrestrial origin.

In the *active stretching stage* extension, and hence also subsidence, accelerate. Simultaneously, heat input increases due to upheaval of hot layers of the mantle lithosphere. The steep fractures generated in the pre-rift stage will not be able to accommodate the extension and a new set of low-angle planar or listric faults will be activated, separating fault blocks that are detached from the lower crust by a subhorizontal zone of weakness. Gliding on the system of detachments, the fault blocks and their internal beds will rotate away from the basin axis (Fig. 12.9a). From the view of the petroleum explorationist, the active stretching stage deserves particular attention because of the

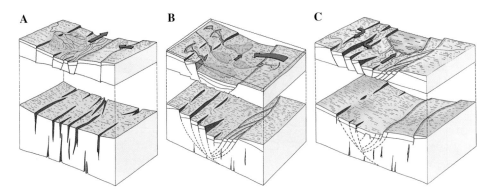

Fig. 12.8 Principal sedimentary transport systems associated with the three mains stages in the development of extensional basins. After Gabrielsen et al. (1995)

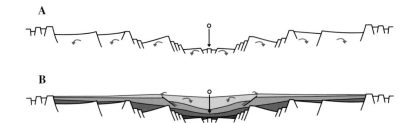

Fig. 12.9 Pattern of rotation of sedimentary units in the (**a**) syn- and (**b**) post-rift stages. After Gabrielsen et al. (1995)

variety of structural and stratigraphic traps that may develop. This stage is also characterised by a complex sediment distribution system that is likely to produce a variety of lithofacies due to the increasing topographic relief associated with high fault activity. The marine transgression that commonly follows the increased subsidence of the basin floor also contributes to this variety in sedimentary facies. Sand that is eroded from the high-standing parts of the basin (e.g. basin shoulders and crests of rotated fault blocks) may be trapped in lows in various structural positions and these units are likely later to be covered by transgressive marine sediment accumulations. The sediment transport system in the active stretching stage is likely to be dominated by complex transverse and locally bi-directional fluvial systems that are strongly influenced by the elongated, rotated fault blocks, generating axis-parallel transport in segments along the basin margin. The central part of the basin may be less complex and axial-parallel sediment transport would prevail there (Fig. 12.8b).

In *the thermal subsidence stage*, thermal contraction of the lithosphere dominates the basin subsidence pattern. Because solids typically contract during cooling, the parts of the basin that have experienced the strongest extension (i.e. those that have been thinned the most and hence heated the most) will contract and subside more than other parts. In a pure-shear configuration this is most likely to be the central segment running along the basin axis. This means that the rotation of strata upwards away from the basin axis becomes reversed so that strata begin to rotate downwards towards the basin axis (Fig. 12.9b). This rotation is strengthened by sediment loading and compaction (thickest sequence in the central part of the basin).

The transverse sediment transport will persist during the thermal subsidence stage, while the basin floor becomes gradually smoothed. An axial transport system may also still be active, but is likely to become less pronounced through this stage of development (Fig. 12.8c). Depending on the balance between subsidence and sediment input, the water depth will vary from one basin to another, but the depositional environment is likely to be marine and the central part of the basin may attain great water depth (thousands of metres). The fault systems that dominated the basin floor geometry during the active stretching stage are now quiescent, and stratigraphic hydrocarbon traps rather than structural ones are likely to be the most common.

Syn-rift to post-rift transition. The *pure-shear model* predicts that a simple geometrical change of the outline of extensional basins will accompany the transition from the syn- to the post-rift stage. In this model the margins of the relatively narrow, steep-walled rift, which traps the syn-rift sediments, become overstepped at the syn- to post-rift transition. This implies that the basin becomes wider and the rate of subsidence decreases asymptotically during the following post-rift stage. Thus, one defines the beginning of the post-rift development as the stage by which the syn-rift faults become inactive and subsidence becomes controlled dominantly by thermal contraction and sediment loading.

In practical terms, the identification of this stage in the basin development is not trivial, because the transition is frequently not synchronous all over the basin, and the criteria for identifying the transition in reflection seismic data are not always well constrained. To overcome this problem, the syn- to post-rift transition should be defined more precisely as the point in time when net heat out of the system is greater than net heat into the system. It is recognised that a lateral heat flow gradient commonly exists perpendicular to the basin axis. This implies that the area closest to the basin axis, which coincides with the area of greatest thinning, is also the part of the basin displaying the highest heat flux at the end of the syn-rift stage. The lithosphere beneath the central part of the basin will accordingly undergo the greatest vertical contraction during the post-rift stage. The enhanced subsidence at the basin axis is further enhanced in cases where the basin is filled by sediments, creating an extra load and also a greater total compaction. Hence, the syn- to post-rift transition coincides with a regional shift in tilt from fault block rotation *away* from the graben axis during the syn-rift stage to tilting directed *towards* the basin axis during the post-rift development (Fig. 12.9). This change is due to a shift from bulk thermal expansion to bulk thermal contraction of the lithosphere and is in most cases clearly distinguishable in reflection seismic data.

It needs to be emphasised that the syn- to post-rift transition is unlikely to occur simultaneously throughout the entire basin. This is due to differences in structural configurations, e.g. the existence of *graben units*, and thermal inhomogeneities associated with variable stretching both along and transverse to the basin axis. For reasons discussed below (Chap. 22), the entire Cretaceous sequence of the northern North Sea is included in the post-rift development *sensu stricto*. Furthermore, analysis of the basin topography permits three sub-stages to be identified within the framework of the post-rift development: the incipient, the middle and the mature post-rift stages. The configuration at the syn-rift/post-rift transition is treated separately in the present analysis (Section 12.3.1).

In the analysis of basin subsidence it is important to remember that in addition to the effects of fault-related subsidence and thermal expansion and contraction, the basin's subsidence is affected by elastic deformation and isostasy, and in many cases also by extra-basinal stress.

The simple-shear model for extensional basins is in considerable geometrical and mechanical contrast to the pure-shear model for extensional basins in that the simple-shear model assumes that extension is concentrated along one or several inclined fault zone(s) affecting the entire crust (Fig. 12.5b). Still, when thermo-tectonic and isostatic responses are concerned the principles are similar to those of the pure-shear model. The simple-shear model is based on observations in the Basin-and-Range of North America and was formulated by Brian Wernicke in 1981. The Basin-and-Range basin system displays a particular geometry in that the lithosphere is extended to the degree that the lower crust, described as a metamorphic core complex, has become uplifted and exposed in the central part of the basin. The asymmetrical configuration of the basin particularly influences the pattern of isostatic response to extension. An important factor is the relative thickness of the upper mantle/lithosphere. This is because the lower crust commonly is denser than the upper astenosphere, causing large-scale contrasts in differential subsidence and uplift across the basin. Superimposed on this are more local isostatic effects, associated with contrasting thicknesses of layers with different densities and the topography of the basin.

Since the same tectono-thermal principles that apply for the pure-shear basin also are valid for simple-shear basins, the main basin stages and the conditions for hydrocarbon generation and entrapment are also the same. Even though the simple-shear model was inspired by analysis of the Basin-and-Range basin system it has proved relevant for many other basins too, suggesting that simple shear is a common component in the formation of basins.

The delamination model can be seen as a combination of the simple- and pure-shear models. In this case the upper and middle crust extends by simple shear. At depth, the master fault flattens and merges with the lower crust, which becomes thinned by pure shear.

The Viking Graben of the northern North Sea seems to have a configuration that fits the delamination model (Fig. 12.6). Also in this case, the thermo-mechanical pure-shear model applies and, with some modifications, can be used to model the basin development. However, the delamination model makes it necessary to take into account an additional variable parameter, namely that the two parts of the lithosphere situated above and beneath the delamination surface have undergone different amounts of extension.

12.3 The Structural Architecture of Extensional Basins

Comparison of many extensional basins reveals that they have many architectural elements in common. These include the position and geometry of the dominant fault systems, the position of the most prominent terraces or platforms, and the position of structural highs. This does not imply that all basins are similar or that all contain the same types of structures. Nevertheless, a systematisation suggested in the following, is useful in the analysis of extensional basins and their exploration for hydrocarbons. Before looking in detail at the exploration leads that are typical for extensional basins, it is therefore instructive to examine the principal structural building blocks of the extensional basin (Fig. 12.10a, b).

The extra-marginal fault complex. The distal part of the basin is separated from the foreland by an enhanced fault frequency (as compared to the foreland of the basin) and a set of planar to listric faults commonly arranged in an *en echelon* geometry and sometimes generating a horst-and-graben-system with a moderate relief. In some cases dike systems affiliated with the initial stage of extension are filling in some of the faults of the extra-marginal fault complex. In many cases, a horst is seen to separate the extra-marginal fault complex from the platform, defining a topographic threshold between the foreland and the basin itself.

The platform is a relatively flat tectonic unit that is less intensely faulted. The extension may be concentrated on a limited number of faults, but even those display only moderate throws as compared to the major fault complexes that delineate the interior basin. Hence, the platform is characterised by a few broad, and only slightly tilted, structural traps. Where age determinations are possible, the fault systems of the platform area are seen to have been activated at an early stage of the basin development, but the activity slowed down or became arrested when the subsidence accelerated in the inner part of the basin system. The platform may also be delineated on its basinward side by a horst.

A *platform marginal horst* is sometimes developed at the basinward side of the platform, defining the transition from the tectonically quiet platform to the much more heavily faulted sub-platform or the inner marginal fault complex. The platform marginal horst is asymmetrical in outline in that its platform-facing margin is defined by a few faults with moderate throws, whereas its basinward border consists of a complex system of faults with great throws, sometimes in the order of several 100 m. The fact that the marginal platform horst is such a common feature in many mature extensional basins suggests that it has an important mechanical significance, the position of its distal (antithetic) fault perhaps being determined by the mechanical strength of the platform rocks.

The *sub-platform* is delineated by the marginal platform high on its distal side and the interior graben on the other. The *inner marginal fault system* that separates the subplatform from the interior graben, together with the extra-marginal fault complex, are the most profound fault zones of the basin, and the two are likely to be linked along the principal detachment found within the lower crust. The subplatform is heavily faulted and encompasses a number of secondary rotated fault blocks that again may be criss-crossed by a third order of faults. There are examples that the second order faults flatten along local detachments at shallower levels than both the primary master faults of the marginal fault complex and the inner and outer marginal fault systems. The *inner marginal fault system* also coincides with the axis of basinward rotation as activated during the post-rift stage.

The interior basin is the unit of the basin which is underlain by the most extensively thinned crust and where the maximum post-rift subsidence occurs. In the case of a symmetrical (pure-shear) basin, it is delineated on both margins by inner margin fault systems,

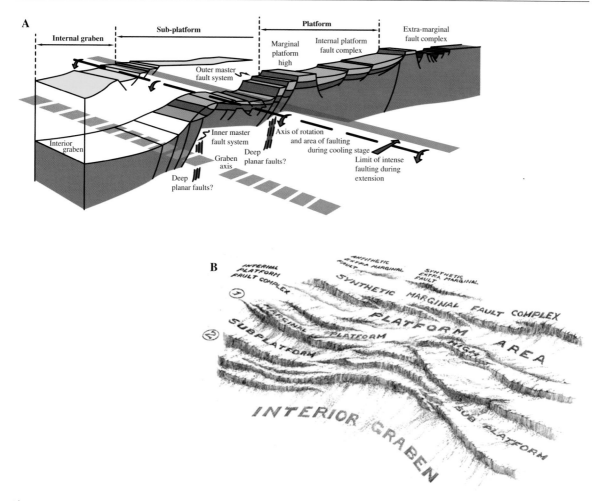

Fig. 12.10 (**a**) Principal sketch of major structural elements in graben systems. (**b**) Graben margin with its main structural elements as seen in three dimensions. Note the change of configuration as seen along the strike of the margin. Redrawn from Gabrielsen (1986)

whereas in the case of an asymmetrical basin (simple shear) only one of the margins has this status. Even the deepest part of the basin is underlain by rotated fault blocks, initiated during the active stretching stage of the basin formation.

It should be noted that the structural elements described above are not likely to be present along the entire basin margin. Thus, in some segments, the platform may be in direct contact with the interior basin, whereas in other segments the platform or the platform marginal high may be missing. This inconsistency may reflect the influence of structural or lithological inhomogeneities in the basement, varying strain rates or uneven bulk extension along the basin axis. Indeed, it

is common for large rifts that the basin is divided into several sub-basins or basin units, each distinguished by its particular geometry and even polarity.

12.3.1 The Structural Influence on Reservoir and Source Rock Distribution in Extensional Basins

The types of traps related to the different stages in graben formation are illustrated in Fig. 12.11. The numbered trap types mentioned in the following sections refer to this figure.

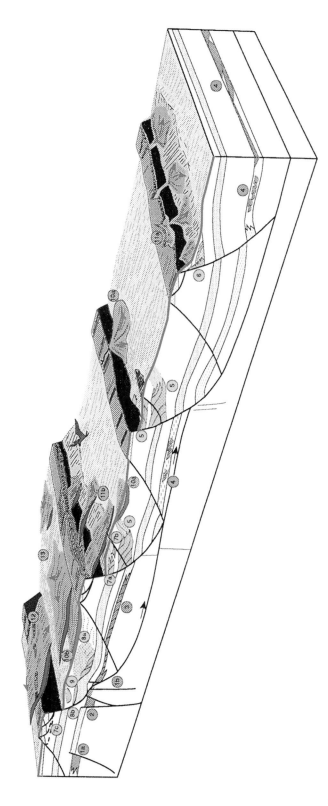

Fig. 12.11 Variations in trap types associated with the pre-rift (*orange*), syn-rift (*yellow*) and post-rift (*green*) stages. The trap types associated with the syn-rift/post-rift transition are marked in *red*. See text for detailed explanation. From Gabrielsen et al. (1995)

12.3.1.1 The Initial Stage

According to the model presented above, sedimentation during the initial stage will occur in a broad, shallow basin with moderate surface gradients. Due to the moderate stretching at this stage, sedimentation will generally keep pace with subsidence. Local depocentres would be related to few, steep normal faults with attached accommodation structures and shallow catchments. Because of the moderate surface gradients and only few and minor fault escarpments, it is reasonable to assume that source areas for sand must be sought outside the marginal platform fault system.

Traps generated at this stage can be buried to very great depths (several kilometres) during the total graben subsidence. In addition to potential overmaturation due to deep burial, the scarcity of source rocks may be a general problem for prospectivity of traps related to this stage, because terrestrial conditions are likely to prevail. However, exceptions to this are for example the West African rift basin lacustrine shales.

In cases where later extension has caused significant rotation of the pre-rift succession, the burial problem may be locally avoided, with pre-rift strata riding structurally high, for example along the marginal areas of the Viking Graben tilt blocks. As a consequence of their early establishment in the subsidence history, traps of this type may become faulted and fragmented by later movements. Even though the topography is influenced by active faults to a limited degree, traps related to such structures still may occur.

The general three-stage model suggests that axial transport dominates in the initial graben stage and stratigraphic traps might be generated by the axial fluvial system (Fig. 12.11, trap type 4). The modest tectonic subsidence characteristic of this stage would favour large lateral extension and good continuity of the sand sheet. The stratigraphic traps would be related to meandering or braided river systems and shallow lakes, and the geometry of the system would to a large extent be ruled by the subsidence rates in the incipient central graben. Examples of axially transported sandstones of this type in the primitive Viking Graben include the Statfjord Formation and Lomvi Formation. These sheet sands contrast with the more lenticular sand bodies of the Lunde and Teist formations, deposited under the control of greater subsidence rates.

In the evolving graben which should be characterised by increasing fault activity and eventual magmatism, the continuity of those sand bodies would be later broken.

Units related to transverse sediment transport encompass both pure stratigraphic and mixed traps (Fig. 12.11, trap type 5). Alluvial fan systems, which offer pure stratigraphic traps with potentially great lateral extension and considerable thickness, might represent the most important type. The general model for this stage in the graben development suggests that axial transport systems dominate, but that most of the axial transport comes ultimately from transverse systems upstream.

Clastic fans related to primary synthetic fault-growth represent a mixed trap type, which depends on proximal sealing faults and a distal pinch-out (Fig. 12.11, trap types 1 and 2). Taking the likely moderate relief, low surface gradient and simple fault geometry into consideration, it is probable that the reservoirs in this type of trap are characterised by good lateral continuity in the dip direction. The thickness of the source rock may be considerable in stable basins.

Clastic fans related to accommodation structures along master faults (Fig. 12.11, trap type 1b) classify as mixed palaeomorphic/structural traps. Because of the steep geometries of the master faults in the initial rifting stage, the accommodation structures will be laterally restricted transverse strike, and to constitute a trap of some size accordingly demands good continuity along strike. This is especially the case in late stages of development where channel amalgamation and progradation lead to a sheet-like geometry.

12.3.1.2 The Active Stretching Stage

During this stage the area of subsidence narrows, and the structural elements typical for mature grabens start to appear. The subsidence, and hence the sedimentation pattern, which to a large extent will be influenced by the faulting within the graben area, is characterised by numerous fault-bounded depocentres that are only partly interconnected. The geometry of faults will change from steep planar to low-angle, either by fault-plane rotation or by development of listric faults. Both low-angle rotational and listric faulting will enhance the internal graben relief by rotation and upheaval of fault-block crests and by isostatic and flexural adjustments of the edges of the fault blocks.

The fault pattern, which is initially relatively simple and stable and consists of isolated fault strands, develops into linked structures, and the relief is enhanced. This gives the possibility of accumulation of considerable thicknesses of sediments, and with potentially good strike-continuity in the hangingwall fault blocks of the master faults. Such catchments may be fed with sediments which are transported across the fault scarp from the foot-wall hinterland, and also from the eroded crests of the neighbouring hangingwall dip slope. With continued subsidence, the initiation of accommodation structures (hangingwall anticlines, antithetic faults and forced folds) will restrict both the area of deposition and the continuity of sands. Simultaneously, fault complexity will increase with the development of different types of fault-related transfer zones and steps, which in turn may act as foci for drainage systems.

Finer-grained pelagic and other, mature sediments derived from the hinterlands would dominate over the coarser sandy or conglomeratic deposits eroded from local structural highs, although the latter may dominate in isolated basins. Initially, the sediment transport is still essentially axial, but due to enhanced relief in the central parts of the basin will be branched to a lesser extent. A growing local influence from relief caused by rotated fault blocks is expected. This may cause sediment transport parallel to the graben axis also in the more distal parts of the graben system. Locally this pattern will be broken by transverse transport systems cutting across highs related to rotated fault blocks, or in connection with relay ramps and bridges between fault blocks. In broader tilted fault-blocks or in grabens immediately adjacent to the hinterlands, the volume of transversely-transported coarse sediment may be great.

Compared to the initial stage, the final burial depth attained by traps generated in the active stretching stage will of course be shallower. Due to the likely marine influx and sub-basinal restricted conditions, the chances for generation of source rocks will also be higher, and semi-regional or local areas of enhanced subsidence may promote the presence of local pockets of source rocks in addition to those of regional significance.

Fans related primarily to synthetic growth along faults give potentially a mixed trap type, but may also include stratigraphic trap elements (Fig. 12.11, trap type 5). The trap depends on the sealing of the fault zone. This trap type is initiated before accommodation structures are activated, and may therefore be characterised by wide extension towards the basin centre. The trap size and complexity depend upon sediment influx relative to basin floor subsidence rate and the stability (potential for reactivation) of the master fault.

In these syn-rift traps the width of the belt of reservoir sand/conglomerate out from the master fault zone is directly related to the subsidence rate and the size of the crestal area being drained. Abundant sediment supply and moderate subsidence rates will cause larger radius fans to develop, whereas high subsidence rates cause all clastic sediment to be trapped in much narrower belts in the immediate vicinity of the fault.

This trap type is well known from the late Jurassic interval in the North Sea, where the most prominent examples are the Brae Field and the Magnus Field. Because the coarse-sediment gravity-flow sands and conglomerates are restricted to narrow fault-parallel belts, they commonly have problematic low continuity in this same direction (Fig. 12.12).

Fig. 12.12 Drainage systems associated with rotated fault block and associated fault scarps. Note that the fan systems associated with the fault scarps may be discontinuous and that sand is also derived from the rotated hanging wall fault block. Modified from Nøttvedt, Gabrielsen and Steel (1995)

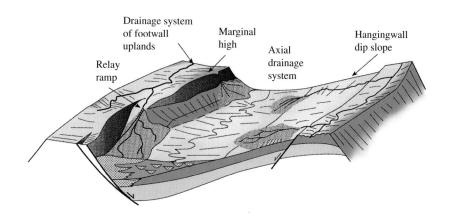

Fans associated with accommodation structures may develop where extended fault activity takes place. Such structures are likely to develop in the hanging-wall close to the master fault. The type and geometry of the accommodation structure depend on the geometry of the master fault as well as on the amount of subsidence. Because of development of a fault scarp and the instability associated with such scarps, the foot-wall fault block is likely to be the major source for the hangingwall catchment, even though the hanging-wall dip slope may represent the more extensive surface of erosion. Examples of stratigraphic traps in the hangingwall dip-slope position are documented several places in the Norwegian shelf, and are well described from the western margin of the Viking Graben.

The first stage of hangingwall deformation in listric faulting may be normal drag, followed by development of a hangingwall anticline ("roll-over") which mirrors the curvature of the master fault. A structural low is defined on the proximal side of the hanging-wall anticline, and opens for entrapment of sediments (Fig. 12.11, trap type 6). Traps in this position are primarily mixed stratigraphic/palaeotopographic and may be related to a local unconformity. Since displacement rates along faults are likely to increase with progressive displacement, it is possible also locally for subsidence to outpace sedimentation with time, to create considerable space for deposition along active faults.

A special type of palaeotopographic trap related to hangingwall anticlines may be expected where forced folds or fault-bend folds above flats in an irregular fault plane have caused surface deformation. As in regular hangingwall anticlines, there is room for sediment entrapment between the master fault and the forced-fold anticline (Fig. 12.11, trap type 9). This trap type is in principle equivalent to that of the conventional roll-over described above.

In addition, sediments may be trapped along the flanks of the forced fold (Fig. 12.11, trap types 8a and 8b). Where fault-bend folds are associated with a large flat with a significant inclination to the regional dip of the master fault, and where sediment supply is good, the trapped volumes might be considerable. In fact, the total volume may further be increased since the depocentre will move as the hangingwall is transported across the flat, causing stacking of the units. The major disadvantage foreseen for this type of stratigraphic trap – especially in view of its distal position – would

be the difficulty for sediments to bypass the hanging-wall anticline. On the other hand, the model allows the possibility of eroding the top of the anticline, thus utilising a local source area. Large-scale fault-bend folds are not uncommon (e.g. Njord Field), but we are not aware that stratigraphic traps related to such features have been reported in the literature from the North Sea.

By continued subsidence along the master fault, antithetic faults are likely to form on the distal flank of the hangingwall fold, and a graben develops parallel to the master fault. If the sedimentation rate does not keep up with subsidence the graben may influence the drainage pattern, and generate a sediment trap as well. This eventually heralds the shift from an open to a closed system. At starved basin margins the accommodation graben may further restrict development of the primary fan, and in systems where the graben floor is close to the marine level, the top of the accommodation structure may be eroded, and there may be local sediment transport towards the master fault (Fig. 12.11, trap types 7a, 7b).

Examples of this type of trap are found in the "Ula trend", which locally defines the margin of the Central Graben. In block 2/2 a system of antithetic faults, partly triggered by halokinesis, has developed a local high, contributing to the initiation of a graben where the sands of the Ula Formation have been trapped.

Gravity slides across fault escarpments may occur along the crestal areas of the rotated fault blocks (Fig. 12.11, trap type 7c). These are the most unstable areas of the graben system and have their instability enhanced by the flexural cantilever effect, which predicts accentuation of uplift in this area. Eventually, slight inversion may further contribute to the uplift of fault-block crests along basin margins.

In reflection seismic data, erosion of uplifted and rotated fault block crests may be the most easily detectable effect, but deformation of the escarpment by gravity sliding should not be neglected. The basic transport mechanisms would be block sliding, rock fall, or mass flow. In all these cases large volumes of reservoir rocks may become resident in the hangingwall in proximal or distal positions relative to the master fault.

Different types of gravity slides have been reported in this position in front of rotated fault-blocks from the North Sea, but have not so far been deliberately drilled.

12.3.1.3 Unconformity Traps Related to Transition from Active Stretching to Thermal Cooling

The transition from active stretching to thermal cooling is manifested by a change in style of subsidence pattern. In the active stretching stage, fault blocks rotate away from the graben axis, causing similar tilting in the sedimentary cover. The thermal cooling stage, however, causes tilting of strata towards the graben axis, mainly because the most rapid subsidence takes place along the graben centre (Fig. 12.9). In strongly asymmetric grabens, this picture will of course be modified accordingly.

Overall transgression and onlap of the crestal areas may be expected towards the end of the active stretching stage because of the overall subsidence of the rifted area, and because sediment starvation is not uncommon when subsidence outpaces sediment yield. On a regional scale relief may be enhanced by upheaval of graben shoulders due to isostacy and elastic response to faulting. The spatial distribution and the magnitude of elevated (possibly eroded) and subsiding areas is influenced by the geometry and depth of the detachments, and by whether the crustal thinning happens during simple or pure shear. Accordingly, these factors will be of importance to the development, distribution and preservation of stratigraphic traps.

The accelerated axial subsidence, which is likely to occur during the early stages of cooling because of the exponential nature of the thermal decay, will normally be associated with marine transgression, and at the stage where earlier graben walls are onlapped and drowned, a break-up unconformity will develop. Depending on the nature of the subsidence, the unconformity will be diachronously onlapped both along the axis and transversely in the graben system. Depending upon the relation between subsidence and sedimentation rate, isostatic stability of the graben margins and individual fault blocks, the unconformity may be very complex. In this situation a number of unconformity traps may develop.

Truncational unconformity traps are formed by erosion of units deposited during the active stretching stage. Fault block rotation and incipient compaction of the sediment package are likely to cause slightly tilted units which may be eroded and sealed by the thermal subsidence shales (Fig. 12.11, trap type 10a). In principle, this trap type may form in all parts of the graben system, but it is most likely to develop along the graben shoulders.

On-lapping unconformity traps are found above the unconformity, and are dependent upon whether erosion on nearby highs has taken place or not (Fig. 12.11, trap type 10b). The relief across master faults may be considerable at this stage, and it is likely that there is significant erosion and reworking of sands from the marginal highs at this stage. Since the fault activity is retarded, the graben relief will diminish during this process, and the traps may be of considerable lateral extent and contain large volumes of sand.

12.3.1.4 Stratigraphic Traps Related to Thermal Subsidence and Sediment Loading

At this stage sediment transport may be both axial and transverse within the graben system, but it is likely that the transverse systems will dominate in the basin margin areas close to major hinterland relief. Minor continued fault activity should still be expected along the master faults of the graben margins due to isostatic adjustments. These areas will also act as pivots during the shift in subsidence pattern. Altogether this implies relatively smooth graben slopes, with an increased possibilty to develop thick and extensive sandsheets with axes oriented transversely to the graben axis.

Because of the high rate of subsidence of the graben floor at this stage, deposition is likely to take place in a marine environment and the axial basin may be starved of sediment. As the thermal gradients of the system approach equilibrium, the basin will fill in and finally level out the relief completely.

Basin-margin fans (Fig. 12.11, trap type 11a) represent a well-described trap type. These are true stratigraphic or palaeotopographic in type, and will normally have great lateral extent. Their thickness will depend upon the degree to which the graben relief was levelled out when deposition took place.

As the graben is expected to be filled by water, transport agents may be gravity mass flows and turbidity currents (Fig. 12.11, trap type 11b). Large transport distances are therefore possible and the submarine fans may be completely separated from the delta systems along basin margins, particularly in periods with lowstand of sea level (Fig. 12.11, trap type 12). Examples here include some of the main reservoirs in the North Sea, like the Frigg, Forties and Bruce fields.

The platform-vergent fans are closely related to the graben-vergent fans, but occur between crests of rotated fault-blocks (Fig. 12.11, trap type 13). During infilling of the graben, sedimentary packages in areas with the thicker sedimentary fill will tend to suffer a stronger compaction than areas with thin packages. This results in development of a hangingwall compaction syncline, which, if it has surface expression, may act as a local sediment trap.

The platform-vergent fans will, however, be more restricted than the basin-vergent ones, and are more dependent on a local sediment source. These circumstances make this trap type less attractive due to small potential sediment volumes.

12.4 Strike-Slip Systems

The strike-slip structural regime is characterised by horizontal orientations of σ_1 and σ_3, whereas σ_2 is vertical (Fig. 12.2). Hence, the orientation of the plane of τ_{max} is vertical, which is also the orientation of the master faults. The displacement along the master fault will be in the horizontal plane (parallel to strike) and the faulting includes initiation of a complex system of secondary fractures.

The general development of shear systems can conveniently be analysed by the use of analogue mechanical experiments. Strike-slip-systems are highly dynamic, and the geometry of the initial stages is very different from that of the mature stages of development. Figure 12.13a shows the relation between the structural elements at the initial stage of strain for a right-lateral (dextral) shear system. To analyse the shear system, one decomposes the shear-forces into compressional and tensional vectors by constructing a vector parallelogram. The dominant features define a system of conjugate fractures (Riedel- and Riedel'-shears) that are related to the compressive component of the shear. Of these, the Riedel-shears are synthetic to the major shear (meaning that they are sub-parallel and have similar shear-sense), whereas the Riedel'-shears are oriented at a large angle to the regional orientation of the fault zone and also display an opposite (antithetic) shear-sense. In addition to the Riedel- and Riedel'-shears, contractional structures (reverse fault, thrusts and folds) with their axes oriented 90° to the compressional stress component

may develop. Due to their favourable orientation and synthetic sense of shear, with continued movement the Riedel-shears tend to become dominant at the expense of the antithetic Riedel'-shears. In accordance with the orientation of the tensional vectors, a set of tensional fractures (T-fractures) may be initiated with orientation parallel to the compressional component of the system.

By continued displacement there will be a tendency for both Riedel- and Riedel'-shears to rotate and for the tips of Riedel-shears to become joined to constitute a system of linked fractures striking parallel to the main shear trend. These Y-shears require that some displacement has taken place and are accordingly not present at the initial state of shear. Another set of shear fractures, P-shears, also occurs at a more advanced stage of development. These are oriented at an angle of 60° to the contractional component (Fig. 12.13a) and are accordingly symmetrical with the Riedel-shears, with the Y-shear-direction as the plane of symmetry. Dynamically, the P-shears nucleate at the profound, principal fault trace and develop up-section to create an array of fractures with an *en echelon* geometry.

In total, the interaction between the sets of secondary fractures (R, R', T, P and Y) contributes to the complex geometry of the strike-slip fault and generates an uneven and step-like morphology. The arrangements of the steps in left-stepping and right-stepping arrays generate contrasting stress-configurations along the strike of the strike-slip fault, depending on the relative shear-sense (Fig. 12.13b). Thus a right-lateral (*dextral*) shear that affects a right-stepping system of strike-slip fault branches causes extension in the overlap-zones (ramps or bridges) between the individual fault branches, whereas a left-stepping system generates overlap zones of contraction for the dextral system. For a left-lateral (*sinistral*) shear-sense, the relations are opposite.

This implies that a variety of structures, and hence a variety of hydrocarbon trap types, are likely to develop along a strike-slip fault. In the ideal case, where the fault trace is planar and the movements of the opposing fault blocks are absolutely parallel, the trace would be one vertical plane. But since this is the case only for very restricted segments of strike-slip faults, there will be segments where material is squeezed up and out of the fault zone, and cases where slivers of the footwall and hangingwall fall into the fault zone. In both cases the faults are likely to have

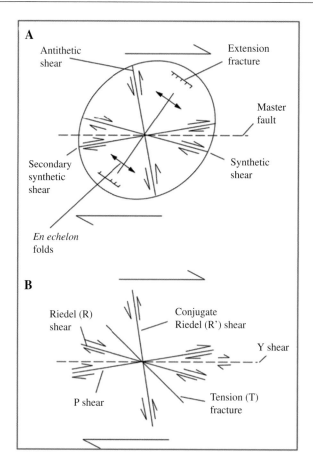

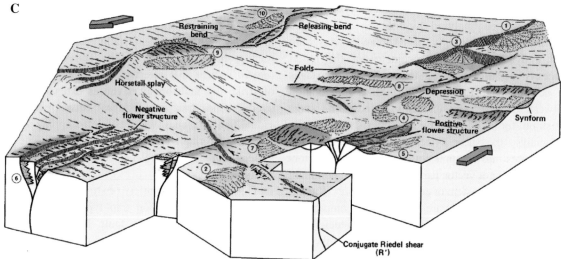

Fig. 12.13 (**a–b**) Fracture types typical for a dextral strike-slip system. By advanced stages of strain, synthetic shears (Riedel shears) and fractures associated with the master fault (Y-shears) tend to dominate the geometry of the fault zone. (**c**) Most common hydrocarbon traps in strike-slip systems. Types of structures are indicated in the figure. (**a**) and (**b**) are redrafted from Crowell (1974)

a steeply dipping root, creating diagnostic geometries for strike-slip faults called (positive and negative) flower-structures (Fig. 12.13c). In cases where distinct fault segments overlap, but are not in direct contact, zones with pull-apart basins or turtle-back structures will occur, whereas extensional and contractional duplexes will develop where the fault-segments are in contact in zones of releasing or restraining bends.

Movements in shear-zones are in many cases not entirely parallel, so that a contractional or extensional component adds to the shear. These situations are called transpressional and transtensional, respectively, and contribute to exaggerating the morphological expressions of the structures described above. In such cases there is a tendency for forces to decompose along weak beds in the deforming units so that strain is taken up in separate systems. Thus a transpressional stress can be decomposed into a pure contractional regime and a pure strike-slip regime. The process is called *strain partitioning* and is well known for e.g. the transform delineating the western Barents shelf, where a dextral transpression is decomposed into collision in the West Spitsbergen Fold- and Thrust Belt and shear along the Hornsund Fault Zone.

12.4.1 Hydrocarbon Prospectivity in Strike-Slip Regimes

Large-scale strike-slip systems may be highly dynamic depositional systems for sediments and also offer a great variety of structural and sedimentary traps. However, compared to extensional basins, there are two significant differences. Firstly, the thermal development is different in that the steep dips and deep roots of the master faults are likely to cause very significant and fast thinning. This involves the substratum of the basin down to the level where the faults detach, which may be top of the lowermost crust or even the Moho. This implies that the thermal gradient may very rapidly steepen during early stages of the basin formation and that significant leakage of heat may start before the syn-rift stage is passed. This may be accompanied by fast burial and maturation (and perhaps over-maturation) of the source rock, and also make basin modelling difficult. Secondly, the geometry and tectonic position of strike-slip systems are such

that the likelihood for accumulation of marine source rocks is less than for extensional basin systems now situated in passive margin settings.

On the other hand, the structural complexity and variability may develop structural and stratigraphic trap types that are not found in extensional basin systems. Examples are flower-structures and arrays of anticlines that may be found along the strike-slip fault at regular intervals.

12.5 Contractional Regimes

When reading the following, it is important to remember that *compression* characterises the stress system, whereas *contraction* describes the physical process of shortening. Contractional regimes are associated with large-scale orogenic processes (formation of mountain chains), but areas of shortening also may occur on local scales within regional extensional or strike-slip realms. Compression is characterised by the principal stresses being oriented so that $\sigma_1 > \sigma_2 > \sigma_3 = \sigma_{hmax} > \sigma_{hmin} > \rho gz$. In other words, the smallest stress acts in the vertical plane so that the most energy-efficient way of shortening is by transferring excess mass towards the surface. Also, due to the orientation of the principal stresses, the dip of the plane of maximum shear is 30°, promoting the development of thrust faults (Fig. 12.2).

Orogens are extremely mobile tectonic zones and such systems may accommodate displacements that are several orders of magnitude greater than that which is typical for extensional basins. This implies that the pattern of deformation may be very complex and involve a number of stages or "phases", each phase representing a unique set of stress conditions and p,T-relations. Still, even the most intricate pattern can be analysed by the utilisation of relatively simple geometrical methods and modelling techniques.

The major strain in a large-scale contractional system like an orogen is generally associated with plate margins. In the deep and central parts of the oregen, deformation takes place under high to extremely high p,T-conditions. Such settings are not optimal for the generation and accumulation of petroleum resources and will therefore not be considered further here. However, in the upper (shallow) parts of an orogen as well as along its frontal parts, sedimentation and structuring take place during p,T-conditions that are compatible with the generation and accumulation of hydrocarbons. Here basins related to the interior development and collapse of the central part of the orogen will occur, particularly in the later stages of the mountain building, whereas foreland basins may be active throughout the entire life of the orogen.

12.5.1 The Architecture of Thrust Systems

The most important building blocks of orogens are folds and contractional faults. Both these types of structures affect rocks volumes, so that they may be constrained from their surroundings by a certain geometry and intrinsic style of deformation. Thrust faults tend to climb up-section, because this is the direction of σ_3. Still, contractional faults are characterised by shallow dip and a tendency to flatten over greater distances, particularly where they follow beds of low mechanical strength. They are also frequently seen to merge with other faults along horizontal fault strands and surfaces between mechanically weak beds, and they regularly intimately affiliate with folds. This is so because the folds and faults frequently are seen to develop in concert. One example can be initial buckling of a bed followed by a fault breakthrough along the fold hinge. In other cases folds can develop in front of an advancing fault (fault-propagation folding). In both these examples, the folds would be asymmetrical with the longer fold limb dipping at a shallow angle away from the transport direction, and the shorter fold limb at a steeper angle. The fold axes would be oriented transverse to the direction of transport and may constitute structural traps of considerable magnitude. By continued shortening the faults may link up, trapping isolated, lens-shaped rock bodies, which commonly incorporate folded beds. The lenses are referred to as *horses* and where they are stacked between a horizontal floor fault and a roof fault, they constitute a *duplex* (Fig. 12.14a).

In cases where the faults propagate systematically in the direction of the front of the contractional system, detachments become linked by climbing fault branches. Together these faults may generate duplexes developed by foreland-directed in-sequence thrusting (Fig. 12.14b). In cases where the system halts e.g. due to increasing friction, the younger horses may pile up on top of the older ones, and an antiformal stack is generated. Changes in overburden and friction may also cause fault activity to switch from one place to another in an unsystematic way. This is termed out-of-sequence thrusting (Fig. 12.14b). Thus, it is not uncommon that shortening is accommodated by hinterland-directed thrusting. In cases where the single faults do not become joined along a roof fault, a system of "blind" faults may develop, terminating at a tip-line. Alternatively, the faults may break the surface. In rarer cases the faults climb down-section in the direction of transport. This is an indication that the strain rate is greater along the roof-fault than it is along the floor fault.

The regions where contractional faults climb up-section are termed ramps and the total geometry of the fault is that of a ramp-flat ramp. In such cases, the pure geometry of the fault planes forces the strata inside the horses to become folded. The folds reflect the geometry and steepness of the fault plane because the front of the horse depends on the cut-out angle of the original ramp. Ramps parallel to the transport direction may also influence the development and the geometry of the thrust system. Such features may potentially separate subunits of contrasting deformational style. In cases where strain rates are not similar across the ramp, shear and strong rotation occur, and when the structures propagate to affect the surface topography they may strongly influence the depositional systems associated with the mountain chain.

It is obvious that the tectonic processes described above produce a variety of structural traps, among which anticlines with along-strike closure and horses delineated by sealing faults may be the most obvious. Because the subsurface structuring also per definition affects the topography during mountain building, different types of subtle and stratigraphic traps are also likely to be generated (Fig. 12.14a).

12.5.2 Hydrocarbon Prospectivity of Contractional Regimes

The very dynamic character of contractional systems obviously produces a variety of sedimentary systems and structural and stratigraphic traps. The relief associated with orogens normally is measured in

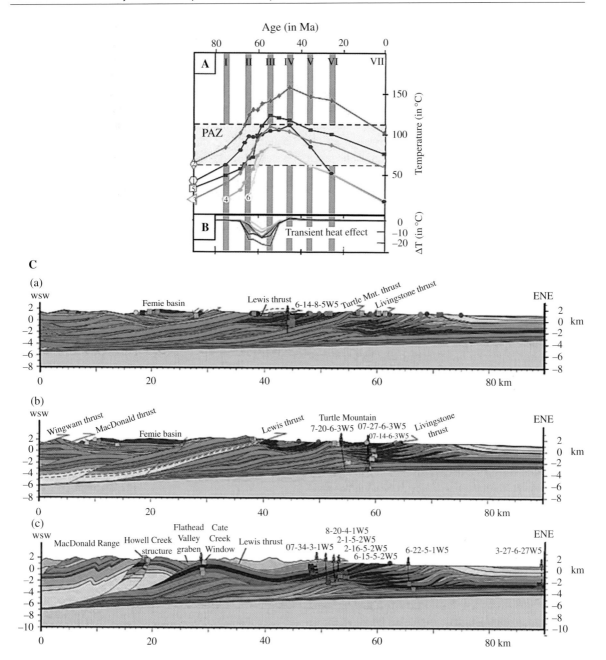

Fig. 12.14 (**a–b**) Temperature-history curves for different parts of the Canadian Cordillera at different positions relative to the apatite annealing zone (PAZ). (**c**) Structural cross-sections with corresponding organic maturity indications (*blue*: low maturation, *dark red*: over-maturation). *Light blue, green* and *orange* colours indicate oil – gas maturation for the same profile. From Hardebol et al. (2009). Reproduced by permission of the American Geophysical Union

kilometres. Strong erosional forces, gravitational instability and climatic influences are important parameters in the development of mountain chains. The system is flooded with a variety of clastic erosional products, the mineralogical composition of which reflects the types of rocks that are involved in the orogen in the first place.

This implies that reservoir rocks and hydrocarbon traps of all kinds are abundant. Because the mountain chain necessarily is uplifted, however, organic-rich marine deposits of the kind that would produce the source rock, would be rare. An exception to this would be cases where a source rock deposited before the contraction started becomes involved in the orogen. In such cases, the critical factors would be the depth of tectonic burial of the source rock, the geothermal gradient of the greater orogeny and the positioning of the source rock relative to the reservoirs. In the dynamic environment of a nappe pile, it must be taken into consideration that units now separated by tens of kilometres may have been juxtaposed at the time of maturation and migration.

Three principally different basin settings can be distinguished. *Intramontane basins* are collapse structures or structural lows generated by folding and thrusting inside the realm of the mountain chain. Those active at the peak tectonic activity are likely to trap large amounts of coarse clastics over a short period of time. They are in most cases of restricted size and source rocks are rarely associated with them. *Foreland basins* are far more interesting from a hydrocarbon exploration point of view. These are stabilised as accommodation areas for sediments due to the gravity load of the progressing orogen, and may trap the bulk of the sediments eroded from the rising mountains and transported towards the orogenic front. The central parts of foreland basins may reach thousands of metres in depth and constitute deep marine depositional systems. Thus, the foreland basin may offer a whole range of sedimentary environments from fluvial, via shallow marine to deep marine. The structuring in the foreland basin position is moderate, but increasing during the progressive development of the orogen. Thus, the basin will eventually become overrun by the advancing deformation front and cannibalised. In this process, a variety of structures are developed from gravitational extensional to contractional complexes of folds, thrust sheets and duplexes. In *basins related to subduction zones*, different types of accretionary prisms may provide source rocks and reservoir rocks, as well as traps, both stratigraphic and structutral. However, these are very dynamic systems, sometimes too dynamic to provide low-risk exploration targets. Also, the sediments in such systems are likely to be too fine-grained to provide good reservoirs.

Although perhaps more rare than in extensional regimes, source rocks may be involved in orogens, such as in the Canadian Cordillera. Detailed studies here performed by Hardebol and his co-workers

reveal a complex maturation history, where both tectonic and seminary burial have to be taken into account. Although some mountain chains may reveal surprisingly homogeneous bulk geothermal gradient patterns, the individual tectonic units may have undergone contrasting histories of burial and uplift, making a full tectonic restoration necessary in the evaluation of the hydrocarbon maturation of the system (Fig. 12.14).

12.6 Structural Inversion

By the term "structural inversion", or simply "inversion", we generally mean a system of extensional structures that has subsequently undergone contraction. This implies that the principal axes of stress have been changed from being orientated such that

$$\sigma_1 > \sigma_2 > \sigma_3$$
$$=$$
$$\sigma_v > \sigma_{hmax} > \sigma_{hmin}$$
$$=$$
$$\rho g z > \sigma_{hmax} > \sigma_{hmin}$$

switches to

$$\sigma_{hmax} > \rho g z > \sigma_{hmin}$$

This implies that the dip of the plane of maximum shear will switch from 60° to 30° (Fig. 12.15). Thus, although an established zone of weakness will represent a potential zone of reactivation when structural inversion occurs, it is unlikely that the already established faults will be able to accommodate much strain, meaning that new faults with lower angles of inclination will be initiated. The most common characteristics of an inverted system are:

- Reverse reactivation of (extensional) faults
- Generation of new, low-angle fault traces
- Development of secondary contractional structures (folds, reverse faults, thrusts)
- Uplift of basin margins
- Uplift of central parts of basins.

The most common configuration at an early stage of inversion of a fault is shown in Fig. 12.15a. In this case, the accommodation space generated in the hangingwall during extension is completely filled by

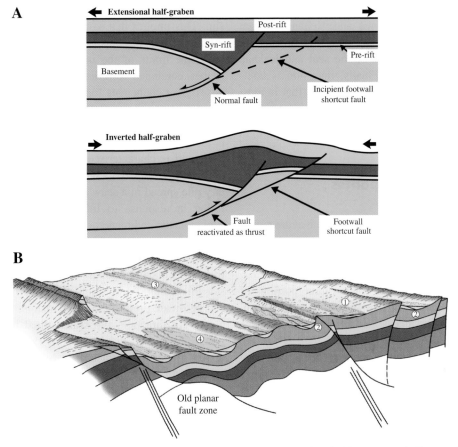

Fig. 12.15 (**a**) Typical geometry of an inverted extensional fault. Note the low-angle footwall shortcut fault (modified from Cooper et al. 1989). (**b**) Surface expressions of deformation in an inverted terrane. Elevated areas, prone to yield erosional products, are marked in *red*, sand accumulations *yellow*.

1: Topographic low inside a pop-up. 2: Topographic low in the hangingwall of an inverted fault. 3: Incipient syncline between a train of emergent anticlines. 4: Syncline between two emergent and eroded anticlines

sediments. During inversion of the master fault, these sediments will be squeezed out of their position in the hangingwall and onto the footwall, accompanied by uplift and folding. For inversion without any oblique component, the fold axes will be oriented parallel to the strike of the extensional fault and, accordingly, orthogonal to the new σ_{hmax}. By continued deformation, the pre-existing fault may be squeezed against the hangingwall and become steepened as a consequence, whereas new, low-angle faults generated in the footwall may create local thrusts (Fig. 12.15b).

If one looks at the entire basin, the response on inversion will depend on the geometry of the basin and the mechanical properties of the crust and lithosphere. In the case of a basin that has already been affected by thinning and thermal weakening, the central part may become overdeepened and the basin shoulders uplifted. In contrast, in the case of a mechanically strong basin fill, the central basin may be uplifted, forming an inverted eye-shaped basin geometry. Alternatively, the basin fill may be folded and squeezed out of the basin, as described above for faults.

12.6.1 Hydrocarbon Prospectivity in Basins with Structural Inversion

From a petroleum exploration point of view, structural inversion is an effect that comes on top of and subsequent to the development of a regular extensional basin, and particularly affects the basin margins. Inversion structures may provide additional structural traps as very well exemplified in the mid-Norwegian margin by the Ormen Lange and Helland Hansen structures. On the other hand, inversion invokes an additional risk for breaking of the seal and leakage though reactivated faults. Finally, inversion is commonly associated with uplift and erosion, that in most cases add to the complication of the geological history and reservoir pressure.

12.7 Basins with Evaporites

Rock salt is strictly speaking a crystalline aggregate of the mineral halite (NaCl), which is one of more than 20 evaporite minerals formed by precipitaion from saturated brines, most commonly from solar evaporation. Although salt deposits may contain large portions of other evaporite minerals like anhydrite and gypsum, most studies of the mechanical properties, and hence the dynamics, of evaporites, consider rock salt as the dominant mineral. Although the very strong influence of water on the mechanical strength and flow properties of halite is well established (the deformation mechanism changes from dislocation creep to diffusion creep at a water content as low as 0.05%), there is still not enough data available on the rheological properties of evaporites to predict the detailed strain path and geometry that large evaporite bodies develop when exposed to gravity and loading from a clastic sedimentary overburden. In addition, many evaporite sequences contain a high proportion of clastic material that may be involved in the deformation, and the rheological effect of these clastic "contaminations" is not easily predictable.

Although one tends to associate the problem of seismic imaging in areas with extensive salt deposits to be affiliated with complex salt structures, one should not overlook that many basins contain evaporites that are tabular, flat-lying and stable, and where the problem of seismic imaging is restricted to seeing through the salt as such. It is still fair to say that the structural geology of evaporites has attracted much attention, partly because the high mobility of such deposits poses intriguing structural geological problems, and even more so because salt structures are associated with a variety of structural and stratigraphic traps of significance for the petroleum industry. The structural geology of salt very much reflects the local tectonic environment, be it extensional, contractional or strike-slip. This implies that salt bodies come in an almost infinite variety of shapes, some of which even challenge the limits of the imagination. This variety of shapes, combined with the acoustic properties of salt, poses great challenges for reflection seismic imaging of salt bodies and the strata beneath and close to them.

Gravity-driven deformation at continental margins is generated by the regional gradient of the margin itself and is characterised by upslope extension and downslope contraction. Salt and its overlying sedimentary pile spread in a seaward direction due to regional tilting in response to lithosphere cooling, whereas synkinematic sedimentation induces loading instabilities. At basin scale, thin-skinned deformation may induce extreme upslope salt thinning, leading to the formation of salt welds, as well as massive downslope salt thickening. Most often, the extensional domain can be divided into three sub-domains, which are, in a seaward direction, the sealed tilted block, growth fault/rollover, and diapir domains (Fig. 12.16a). The upper domain is characterised by tilted blocks that are sealed early by synkinematic sedimentation. The rollover domain displays a large amount of extension, whereas the domain of diapirs is generally considered as gently translating, accommodating small amounts of extension. Diapirs correspond to weak zones and are easily and often squeezed. Downslope of the margin, contractional structures balance the amount of upslope stretching.

The domain of shortening is also divided into three sub-domains. In a seaward direction they are composed of diapirs squeezed at late stage, polyharmonic folds and thrust faults developed at early stage, and folds and thrusts developed at late stage. Contractional structures are initiated in a domain located at a distance from the initial salt edge. Compression remains localised in this domain during the initial stages of evolution by continued deformation. The upslope migration of contraction can then reach the extensional domain and squeeze the diapirs.

Analogue experiments show that the overall structural zoning is mainly controlled by the initial condition (salt basin) and the basal slope angle, whereas the type of structures in the structural domains strongly depends on sedimentation rate (Fig. 12.16b).

An early attempt to classify salt structures systematically and to set this into a dynamic context was made by Trusheim in 1960. He suggested that salt

Fig. 12.16 (**a**) Analogue experiment showing salt structures and related structuring of sediments between salt ridges in an extensional, inclined slope (from Brun and Fort 2008). (**b**) Surface expressions of deformation in a terrane affected by halokinesis. Elevated areas, prone to yield erosional products, are marked in *red*, sand accumulations given in *yellow* for different structural positions. 1: Rim syncline. 2: Rim syncline between a stock and a pillow. 3: Rim syncline along a salt wall. 4: Salt-induced graben. 5: Graben on *top* of collapsed salt pillow. 6: Salt-induced rotated fault block. 7: Stratigraphic trap covered by overhang in salt pillow

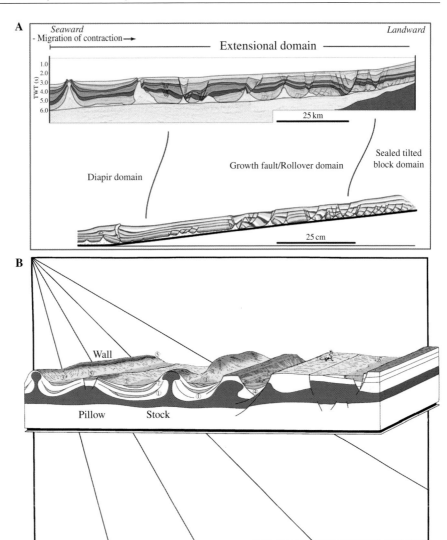

impiercements grow from elongated low-profile ridges (anticlines and rollers) triggered by gravitational contrasts, developing into rows of pillows, diapirs and eventually into walls and sheets of salt. The diapirs come in a variety of shapes from regular massive stocks, via irregular masses to elegant mushrooms. This geometric classification is undoubtedly valid for a tectonically stable, evenly subsiding basin. But even this relatively predictable kinematic growth pattern of salt structures causes great problems in seismic imaging due to the complex pattern of internal flow in the salt structure itself, including horizontal displacement, affiliated with the development of overhanging or even horizontal walls.

12.7.1 Hydrocarbon Prospectivity in Salt-filled Basins

Deep parts of basins where salt structures tend to be situated may be excellent sediment traps. The growth of diapirs contributes to the development of local depocentres and the areas around salt diapirs may accumulate large volumes of reservoir rocks of good quality and be associated with excellent structural and stratigraphic traps. However, due to the capacity of salt to flow horizontally at shallow levels and develop overhanging bulges and sheets, and even to become detached from its deeper sources, the detailed geometric configuration around the stem of the salt structure, including its diameter, is commonly disguised and the diameter of the stem itself may be impossible to determine from reflection seismic data.

In addition to the bulge above the salt empiercement itself, which will reflect the geometry of the upper layers of the empiercement, the main types of features that may constitute structural hydrocarbon traps adjacent to salt diapirs are the rim syncline system (sometimes several generations), anticlines associated with the rim syncline system, faults generated due to volume reduction during vertical transport of salt, and drag structures close to the stem of the diapir (Fig. 12.16b). Due to the circular nature of the diapir, all these structures are likely to be closed when seen in three dimensions. In addition, numerous types of stratigraphic trap may be related to any of these structural features. For salt anticlines and simple walls, which have not developed overhangs, seismic imaging of the structures is usually relatively straightforward. For mushroom-shaped diapirs, however, this is much more challenging and several parameters have to be taken into consideration in the structural analysis: the distance of the rim syncline system from the centre of the diapir and its amplitude and wavelength depend on the thickness of the original salt sequence, on the diameter of the diapir, and the salt flow rate relative to sedimentation rate. In many cases however, this structure is situated sufficiently far away from the diapir for seismic imaging to be unproblematic. When it comes to the fault systems and the drag structures, these occur close to the stem of the diapir and are likely to be covered and completely obscured by the overhanging diapir bulb.

One particular difficulty in seismic interpretation may occur in cases where dynamic salt interacts with faulting. At the crests of salt diapirs and stocks a combination of ring-shaped and radial fault systems is commonly found, but these are unproblematic in seismic imaging and interpretation. In addition, numerous examples exist of fault activity promoted by the underlying active salt, providing a substratum for detachment. Again, such structural relations are also clearly displayed in reflection seismic data. Finally, though, due to transfer of larger volumes of salt towards the basin axis, smaller pillow-shaped volumes of salt are frequently left and trapped along basin margins, where they interact with the basin margin fault system. In such cases different configurations are sometimes developed in the hangingwall and the footwall of the salt-involved fault, causing complex and contrasting sedimentary conditions across the fault so that significant problems arise in sequence correlation. Where salt has intruded along the fault-plane, interpretation of seismic data may be hampered by reduction of the general data quality.

In more complex tectonic environments (strike-slip and contraction), the complexity and variety of salt body configuration is commonly much greater, because the final geometry of the salt body will be determined by directed flow reflecting varying differential stress and strain. For example, the importance of evaporite sequences in the development of many thrust belts like the Pyrenees and the West Spitsbergen thrust-and-fold belt is well documented. In such settings seismic imaging may be complicated by salt being involved as an extensive, continuous or disrupted unit during the thrusting, and also because it may have accumulated unevenly and become integrated in contraction structures like fold cores and duplexes, and as intrusions along fault planes.

The quality of seismic imaging performance in areas of salt has been greatly improved in recent years. Still, the days of surprises are not yet over when results from drilling become available. The effort in improving seismic imaging techniques must therefore continue. And it should go hand in hand with field study and analogue modelling.

12.8 Faults and Fault Architecture

Fractures (faults and joints) are found in practically all hydrocarbon reservoirs and are crucial elements because they both influence the migration of hydrocarbons within the reservoir and contribute to the entrapment of fluids. Due to great variation in fault rocks and fracture types and their distribution, the influence of fractures on reservoir communication is not easily predictable. It is therefore natural that the analysis of single faults and fracture systems is receiving increasing attention.

For the assessment of the architecture of faults and fracture systems one can, in principle, chose between stochastic and deterministic or a combination of the two. Given the complexity and generalised architecture of larger faults, however, stochastic methods are less favourable in analysis of such features. This also seems to be the case for fracture systems generated in stress situations where σ_1 is distinctly different from σ_3. To provide input to a deterministic reservoir model that aims at taking the complexity of larger faults into consideration, we have performed field studies in order to constrain realistic characteristics for the units that commonly can be defined within the realm of a fault zone.

Since mesoscopic and macroscopic faults affect volumes of rock, and accordingly should be described as composite rock bodies that include a complex system of structures (fault rock, folds and fractures), the term "fault zone" is applied here. In the following we use "fracture" as a general term that includes faults, deformation bands and joints, and we distinguish between shear fractures (microscale) and faults (meso- and megascale). Also, we use the term "high-strain zone" for the parts of the fault core where shear is concentrated.

12.8.1 The Structural Elements of the Fault Zone

It is well established that faults are commonly zoned and composed of several units with distinct deformation styles (Fig. 12.17). These include the fault core,

where most of the displacement is accommodated, and its associated damage zones that are geometrically and mechanically related to the development of the fault.

The fault core is in general separated from the footwall and hangingwall damage zones by distinct fault-branches. Lozenge-shaped rock bodies frequently dominate the cores of extensional faults. These are commonly referred to as fault lenses or horses, which may occur in isolation, as *en echelon* trains, or be stacked to constitute duplexes. The fault-rock lenses may consist of relatively undeformed country rock derived from the footwall or the hangingwall of the fault core. In faults with greater displacement, the fault lenses may represent lithologies exotic to that of the observable footwall and hangingwall or be completely reworked fault rocks like cataclasites and breccias. The geometry of the lenses, their relative arrangement and their relation to intervening high-strain zones are important for the fluid communication along and across faults in cases where contacts between units of high or low permeability control the fluid flow.

Field study of the shape of fault core lenses suggests that such features have relatively regular shapes and that the a:c-ratio (relation between length measured in the dip-direction and maximum thickness) in extensional faults is in the order of 10:1 and that the b:c-ratio (relation between length measured in the strike-direction and maximum thickness) is somewhat

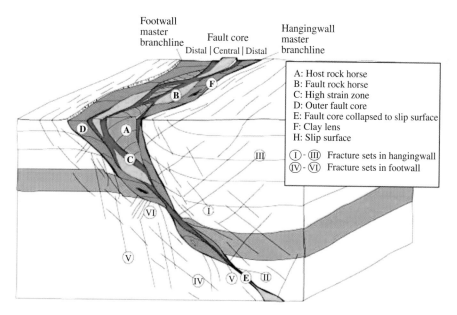

Fig. 12.17 The structural elements of an extensional fault

less than this, perhaps 9:1 or 8:1. It is also suggested that the lenses are close to symmetrical with reference to both the central a- and b-axes. The available data also suggest that these relations are roughly valid also for the higher-order lenses (2nd, 3rd, 4th order etc.), although there is a tendency for the a:c- and b:c-values to become slightly reduced for the higher-order lenses.

The high-strain zones separating individual or groups of fault lenses may include deformed units that can be recognised as country rock in the footwall and hangingwall, as well as several types of fault rocks, the host rock of which cannot be determined. For example the most intensely deformed zone of the fault core is easily distinguishable and this zone may represent the latest area of deformation, indicating that strain softening has occurred.

The damage zones define a halo of fractures on both sides of the fault core. The fractures of the damage zones are associated with the dynamic development of the fault and may encompass remnants of the propagation of the incipient fracture, commonly termed the process zone. The strain intensity in the damage zones is generally modest compared to that of the fault core, and in sedimentary rocks bedding and other primary features can commonly be recognised. The fracture distribution, frequency and orientation in the hangingwall and the footwall are generally different, and it is therefore convenient to distinguish the footwall damage zone from that of the hangingwall.

Field investigations show that damage zones are not symmetrically distributed around the fault core. Accordingly, it is commonly observed that the fracture frequency curves build up towards their maximum values more steeply in the footwall than in the hangingwall. Also, there are differences where orientation of the major fracture populations is concerned. The footwall damage zone is characterised by a set of fractures oriented subparallel to the footwall master fault-branch. This fracture population may interfere with fracture sets that dip both more and less steeply than the master fault-branch. The hangingwall damage zone is influenced to a greater extent by antithetic fractures to the master fault and a wider area is affected by these fractures. The fractures of the damage zones may be of different kinds, depending on lithology, depth of burial at the time of deformation, and strain intensity.

Faults that are in their early stage of development do not possess the complexity described above. In sandstones they develop from single or arrays of deformation bands that with increasing strain coalesce into one zone of focused shear. This is generally also the case for incipient faults in carbonates.

Further Reading

Allen, M.R., Goffey, G.P., Morgan, R.K. and Walker, I.M. (eds.). 2006. The deliberate search for the subtle trap. Geological Society Special Publication 254, 304 pp.

Biddle, K.T. and Christie-Blick, N. (eds.). 1985. Strike-slip deformation, basin formation and sedimentation. Society of Economic Palaeontologists and Mineralogists (Tulsa OK), Special Publication 37, 386 pp.

Boyer, S.E. and Elliot, D. 1982. Thrust systems. American Association of Petroleum Geologists Bulletin 66, 1196–1230.

Brun, J.-P. and Fort, X. 2009. Entre Sel et Terre. Structures et Méchanismes de la Tectonique Salifière. Société Géologique de France, Paris, 154 pp.

Butler, B.W.H. 1982. Terminology of structures in thrust belts. Journal of Structural Geology 4, 239–245.

Cooper, M.A., Williams, G.D., de Graciansky, P.C., Murphy, R.W., Needham, T., de Paor, D., Stoneley, R., Todd, S.P., Turner, J.P. and Ziegler, P.A. 1989. Inversion tectonics – A discussion. Geological Society, Special Publications 44, 335–347.

Coward, M. 1994. Continental collision. In: Hancock, P.L. (ed.), Continental Deformation. Pergamon Press, New York, pp. 55–100.

Crowell, J.C. 1974. Origin of late Cenozoic basins in southern California. In: Dorr, R.H. and Shaver, R.H. (eds.), Modern and Ancient Geosynclinal Sedimentation. SEPM Special Publication 19, 292–303.

Cunningham, W.D. and Mann, P. 2007. Tectonics of Strike-Slip Restraining and Releasing Bends. Geological Society Special Publication 290, 482 pp.

Dahlstrom, C.D.A. 1970. Structural geology in the eastern margin of the Canadian Rocky Mountains. Bulletin Canadian Petroleum Geology 18, 332–406.

Fossen, H. and Gabrielsen, R.H. 2005. Strukturgeolgi. Fagbokforlaget Vigmostad & Bjørke AS, 375 pp (in Norwegian).

Gabrielsen, R.H. 1986. Structural elements in graben systems and their influence on hydrocarbon trap types. In: Spencer, A.M. et al. (eds.), Habitat of Hydrocarbons on the Norwegian Continental Shelf. Norwegian Petroleum Society. Graham & Trotman, London, pp. 55–60.

Gabrielsen, R.H., Steel, R.J. and Nøttvedt, A. 1995. Subtle traps in extensional terranes: A model with reference to the North Sea. Petroleum Geoscience 1, 223–235.

Hancock, P.L. (ed.). Continental Deformation. Pergamon Press, Oxford, 421 pp.

Hardebol, N.J., Callot, J.P., Bertotti, G. and Faure, J.L. 2009. Burial and temperature evolution in thrust belt systems: Sedimentary and thrust sheet loading in the SE

Canadian Cordillera. Tectonics 28, TC3003, 1–28, doi: 10.1029/2008TC002335.

Jackson, M.P.A. and Talbot, C.J. 1991. A glossary of salt tectonics. Geological Circular 91-4, University of Austin, Texas, Bureau of Economic Geology, 44 pp.

Kearey, P., Klepeis, K.A. and Vine, F.J. 2009. Global Tectonics, 3rd edition, Wiley-Blackwell, New York, NY, 482 pp.

Lister, G.S., Etheridge, M.A. and Symonds, P.A. 1986. Detachment faulting and the evolution of passive continental margins. Geology 14, 246–250.

McKenzie, D.P. 1978. Some remarks on the development of sedimentary basins. Earth and Planetary Science Letters 40, 25–32.

Nøttvedt, A., Gabrielsen, R.H. and Steel, R.J. 1995. Tectonostratigraphy and sedimentary architecture of rift basins, with reference to the northern North Sea. Marine and Petroleum Geology 12, 881–901.

Nøttvedt, A. (ed.). Dynamics of the Norwegian Margin. Geological Society Special Publication 167, 472 pp.

Nøttvedt, A., Larsen, B.T., Gabrielsen, R.H., Olaussen, S., Brekke, H., Tørudbakken, B., Birkeland, Ø. and Skogseid, J. (eds.), Dynamics of the Norwegian margin. Geological Society Special Publication 167, 41–57.

Nystuen, J.P. 2008. The changing face of the Earth, Chapter 2 in Ramberg, I.B., Bryhni, I., Nøttvedt, A. and Rangnes, K. (eds.) The Making of a Land – Geology of Norway, Norsk geologisk forening. The Norwegian Geological Association, Oslo, pp. 20–61.

Odinsen, T., Christiansson, P., Gabrielsen, R.H., Faleide, J.I. and Berge, A.M. 2000. The geometries and deep structure of the northern North Sea rift system. In: Nøttvedt, A. (ed.), Dynamics of the Norwegian Margin. Geological Society Special Publication 167, 41–57.

Ramberg, I., Bryhni, I., Nøttvedt, A. and Ragnes, K. 2008. The making of a land. Geology of Norway. Published by the Norwegian Geological Associations, 624p

Ranalli, G. 1995. Rheology of the Earth, 2nd ed. Chapman & Hall, London, 413 pp.

Tchalenko, J.S. 1970. Similarities between shear zones of different magnitudes. Geological Society of America Bulletin 81, 1625–1640.

Twiss, R.J. and Moores, E.M. 1992. Structural Geology. W.H. Freeman & Co., New York, NY, 532 pp.

Wernicke, B. 1981. Low-angle normal faults in the Basin and Range province: Nappe tectonics in an extending orogen. Nature 291, 645–647.

Wernicke, B. 1985. Uniform-sense normal simple-shear of the continental lithosphere. Canadian Journal of Earth Science 22, 108–125.

Chapter 13

Compaction of Sedimentary Rocks Including Shales, Sandstones and Carbonates

Knut Bjørlykke

The physical properties of sedimentary rocks change continuously during burial as a response to increasing stress and temperature; they also change to a certain extent during uplift and cooling. There is an overall drive towards lower porosity with depth, which increases the density and velocity.

Increased effective stress from the overburden or from tectonic stress will always cause some compaction (strain), expressed by the compressibility and the bulk modulus (See Chap. 11) which can be measured in the laboratory. During mechanical compaction the solids, mainly minerals, remain constant so that the reduction in bulk volume is equal to the porosity loss.

Chemically, the mineral assemblage will be driven towards higher thermodynamic stability (Lower Gibbs Free Energy) (Fig. 13.1a). These reactions involve the dissolution of minerals or mineral assemblages that are unstable, and precipitation of minerals that are thermodynamically more stable with respect to the composition of the porewater and the temperature. Higher temperatures will favour minerals with lower water content, for example by dissolving smectite and kaolinite and precipitating illite (see Chap. 4). The rates of these reactions are controlled by the kinetic parameters such as the activation energy and thereby the temperature.

The main lithologies in sedimentary basins are shales, sandstones and carbonates, and they respond very differently to increased stress and temperature during burial.

This is important both for basin modelling and in seismic data interpretation.

There are no precise definitions for mud, mudrock and shale. The term mud is used to describe fine-grained sediment with a relatively high content of clay-sized particles, chiefly clay minerals. Carbonate mud will be discussed separately, under carbonate compaction.

The compaction (porosity loss) as a function of burial depth varies greatly because each primary lithology has a different compaction curve. While porosity may increase with depth through an interval due to changes in lithology, for each individual lithology the porosity will nearly always be reduced with depth (Fig. 13.1b).

13.1 Compaction of Mudrocks and Shales

Mudrocks and shales are often treated as one lithology in connection with basin analyses, seismic interpretation and well log analyses, but in reality they span a wide range of properties determined by the diversity of mineral composition and grain-size distribution. Furthermore, the composition of mudstones and shales changes during progressive burial due to diagenesis, which includes both mechanical and chemical compaction.

Just after deposition the porosity of the mud near the sea or lake bottom may be extremely high, up to 70–80%. After about 1,000–2,000 m burial depth much of the mechanical compaction has taken place even if the porosity may still be relatively high (20–40%). Muddy

K. Bjørlykke (✉)
Department of Geosciences, University of Oslo, Oslo, Norway
e-mail: knut.bjorlykke@geo.uio.no

K. Bjørlykke (ed.), *Petroleum Geoscience: From Sedimentary Environments to Rock Physics*,
DOI 10.1007/978-3-642-02332-3_13, © Springer-Verlag Berlin Heidelberg 2010

Fig. 13.1 (a) Principal aspects of sediment compaction (burial diagenesis). During burial, sediments are subjected to changes in physical properties as a function of increasing stress and temperature. From an initial sediment composition the porosity is reduced and the density and velocity are increased. Mechanically the compaction is a strain due to effective stress. Chemical compaction resulting from dissolution and precipitation of minerals is controlled in siliceous rocks by thermodynamics and kinetics and is therefore a function of temperature and time. The strain (compaction) is here independent of stress. (b) The porosity/depth trends will be different for different lithologies (primary mineralogical and textural composition). A simple exponential function may be rather far off from the real porosity/depth function

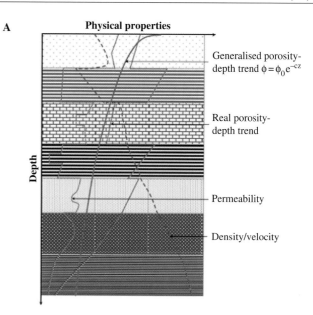

A **Physical properties**

Generalised porosity-depth trend $\phi = \phi_0 e^{-cz}$

Real porosity-depth trend

Permeability

Density/velocity

Depth

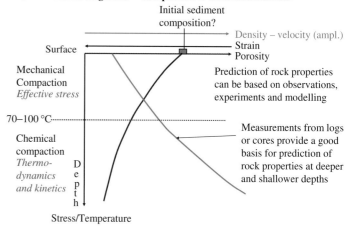

B **Burial diagenesis – Compaction of siliceous sediments**

Initial sediment composition?

Surface

Density – velocity (ampl.)
Strain
Porosity

Mechanical Compaction
Effective stress

Prediction of rock properties can be based on observations, experiments and modelling

70–100 °C

Chemical compaction
Thermo-dynamics and kinetics

Depth

Measurements from logs or cores provide a good basis for prediction of rock properties at deeper and shallower depths

Stress/Temperature

sediments can become very compact because silt and clay can occupy much of the pore space between the larger ones, resulting in a densely packed mass.

Clay minerals, which usually account for the bulk of the finest fractions, have an impressive size range. Kaolinite particles are sheets where the longest dimension is typically 1–20 μm, while smectite particles may be smaller by a factor of 1,000 (only a few nm). Illite, chlorite and most other clay minerals have grain sizes that are intermediate between these end members. Smectite has a very high specific surface area (several hundred m²/g) because of the small grain size and is very sensitive to the chemical composition of the porewater. Additions of salt (NaCl or KCl) are used to stabilise soft clays for engineering purposes

(construction), increasing their compressive and shear strengths. Addition of KCl in the drilling mud is also used to stabilise clays when drilling.

Sediments are often highly anisotropic and parameters like velocity and resistivity can vary greatly with the orientation of the measurement relative to the bedding. Mudstones may also become increasingly anisotropic with burial depth, giving higher velocity parallel to the bedding than in the vertical direction. Experimental compaction shows, however, that the degree of grain reorientation varies markedly with the clay mineralogy and the content of sand and silt (Voltolini et al. 2009).

The composition of mudstones and shales with respect to their clay mineralogy and their content of silt and sand can provide not only important information about the environment both in and around the basin, but also about the rock properties controlling compressibility, density, seismic velocity and resistivity. These parameters are important in the interpretation of seismic data and also electromagnetic surveys. This is clearly seen in some of the silty Jurassic shales from the North Sea basin (Fig. 13.2).

In the North Sea basin and Haltenbanken, poorly sorted, clayey, partly glacial sediments of Pleistocene and Pliocene age fall on a nearly linear compaction trend reaching velocities up to 2.8 km/s near the base of this sequence (Fig. 13.3). Glaciomarine clays overrun by glaciers can become very hard and compact; in the Peon gas field in the North Sea they have developed sufficiently low permeability to trap gas at just 160 m below the seafloor.

Eocene and Oligocene smectitic clays of volcanic origin have much lower velocities (<2 km/s) and densities (Fig. 13.3). Kaolinitic clay is far more compressible because it is very much coarser-grained so that the stress per grain contact is higher. Experimental compaction also shows that fine-grained kaolinite

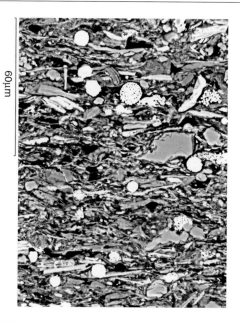

Fig. 13.2 Jurassic mudstone from the North Sea basin buried to 2.5 km depth. Note that many of the grains are of silt-sized quartz and that mica grains have a parallel orientation (scale = 0.06 mm). The white spherical structures are framboidal pyrite. The velocity in this shale is about 3 km/s (V_p 3,019–V_s 1,665 m/s)

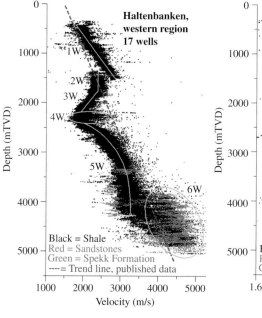

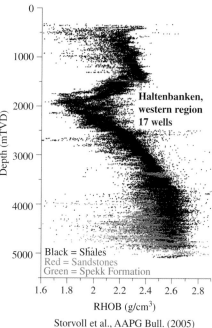

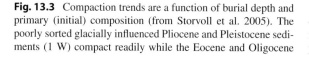

Storvoll et al., AAPG Bull. (2005)

Fig. 13.3 Compaction trends are a function of burial depth and primary (initial) composition (from Storvoll et al. 2005). The poorly sorted glacially influenced Pliocene and Pleistocene sediments (1 W) compact readily while the Eocene and Oligocene smectite-rich sediments of volcanic origin (4–3 W) have low compressibility. The underlying Cretaceous and Jurassic sediments (5 W) show increases in density and velocity which probably are caused mostly by chemical compaction

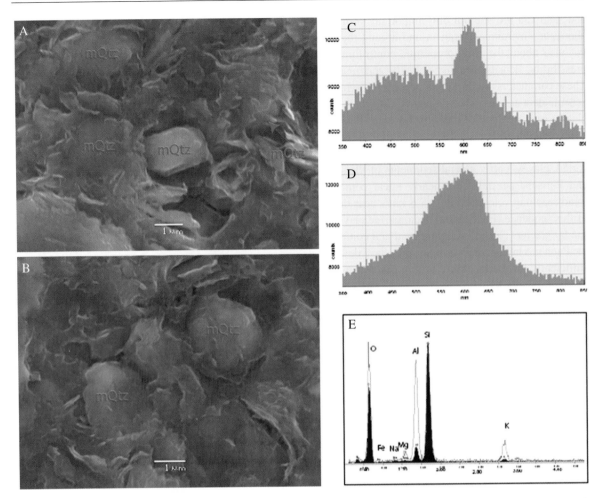

Fig. 13.4 (**a** and **b**) Authigenic micro-quartz precipitated in smectite-rich Late Cretaceous mudstones from the northern North Sea (from Thyberg et al. 2009b). They can be distinguished from clastic quartz grains by their cathode luminescence responses (**c** and **d**) and their chemical composition (**e**). They are found in mudstones which have been buried deeply enough to reach temperatures (>70–80°C) which make illite replace smectitic, providing excess silica which is then precipitated as micro-sized quartz crystals

is less compressible than coarse-grained kaolinite (Fig. 11.7). Illite and chlorite are much more difficult to characterise. The clay mineral illite as determined by XRD includes both relatively coarse-grained detrital mica and very much finer-grained diagenetic ilitte, i.e. formed from smectite. Chlorite also varies considerably, from detrital chlorite from metamorphic rocks to authigenic, usually Fe-rich, diagenetic chlorites.

In the laboratory the velocites (V_p and V_s) can be measured as a function of stress for mixtures of different clay minerals (see Chap. 11). Smectitic clays have very much lower velocities than kaolinitic clays but additions of silt increase the velocity. The V_s/V_p ratio also varies as a function of clay mineralogy. This is very important since this ratio is used to determine the fluid content in sand and siltstones. Mudstones and shales also may have a significant content of gas which changes the V_s/V_p ratio.

The primary composition of the mud deposited on the seafloor depends on the clay mineralogical composition and the amount of silt- and sand-sized grains. Carbonate and silica from biogenic debris are critical components with respect to burial diagenesis. Relatively moderate amounts of carbonate cement in mudstones result in high velocity at shallow depth. The source of the carbonate cement is in most cases biogenic carbonate. Fossils composed of aragonite are particularly important because they dissolve and become a source of calcite cement.

Mudstones may contain significant amounts of organic (amorphous) silica (Opal A), particularly radiolarian and diatoms, in areas with high organic productivity. Siliceous sponges may also be an important source of silica in fine-grained siltstones and sandstones.

Biogenic silica will react to form opal CT and microcrystalline quartz at about 60–80°C and this will also produce a strong stiffening of the mudstones (Thyberg et al. 2009a, Peltonen et al. 2008, Marcussen et al. 2009). Smectite becomes unstable and dissolves at temperatures above 70–100°C and mixed layer minerals and illite precipitate.

$$\text{Smectite} + K^+ = \text{illite} + \text{quartz}$$

In the case of iron-rich smectite, chlorite may form as well. This reaction releases excess silica which must precipitate as quartz. Smectite is only stable when the concentration (activity) of silica is high. The precipitation of quartz provides a sink for the silica and lowers the silica concentration so that the reaction can continue. Therefore the rate of quartz cementation, which is a function of temperature, is controlling this reaction.

It has been suggested that the silica released from the above reaction could be transported by diffusion into adjacent sandstones and precipitated as quartz cement there. Recently, small authigenic (grown in place) quartz crystals have been identified in smectite-rich mudstones which have been heated to more than about 80–85°C (Fig. 13.4). This shows that the silica is conserved locally in the mudstones. Even if the silica concentration in the mudstones were higher than in sandstones, diffusive transport in shales would be very inefficient. In mudstones without smectite or amorphous silica there are no obvious sources of silica to be precipitated as early quartz cement. At greater depth most of the quartz cement is probably derived from pressure solution of detrital quartz.

In sandstones the quartz cement is sourced by pressure solution of quartz grains, but it is not clear to what extent silt and sand grains floating in a matrix of clay will dissolve and cause precipitation of quartz as cement or as overgrowth on the grains. While quartz grains dispersed in a clay matrix may dissolve in contact with clay minerals, the surrounding clay may prevent or retard overgrowth.

At greater burial and temperatures (>130°C) kaolinite becomes unstable in the presence of K-feldspar and releases silica which is precipitated as quartz (Bjørlykke 1983, Bjørlykke et al. 1986):

$$Al_2Si_2O_5OH_4 + KAlSi_3O_8 = KAl_3Si_3O_{10}(OH)_2 + SiO_2 + 2H_2O$$
$$\text{Kaolinite} + \text{K-feldspar} = \text{illite} + \text{quartz} + \text{water}$$

This reaction is driven towards increased density (lower water content).

Kaolinite is however stable up to more that 200°C if there is no K-feldspar or other source of potassium available locally in the rock. It may then be replaced by pyrophyllite ($AlSi_2O_5(OH)$) which contains less water.

Mud containing mostly quartz, illite and chlorite will be chemically stable up to high temperatures because these are metamorphic minerals. With increasing overburden and temperature, massive mudstones develop a more pronounced cleavage typical of shales.

This is due to a higher degree of parallel orientation of the sheet silicate minerals, particularly mica, illite and chlorite.

During folding, high horizontal stresses may produce an axial plane cleavage. This is controlled by a reorientation of clay minerals (sheet silicates) and also a flattening (elongation) of quartz grains by stress-driven dissolution and precipitation.

Mudstones and shales which have a high organic content contain kerogen which often occurs as thin lamina. Before the source rock becomes mature the kerogen is a part of the solid phase and can carry some of the overburden stress. When most of the kerogen is altered to oil and gas it becomes part of the fluid phase, thus changing the fluid/solid ratio so that the pore pressure reaches fracture pressure which makes expulsion more efficient. Shales with a lower organic content also will generate petroleum and gas which may migrate into the most porous and permeable layers. Some of the oil and gas, though, will be retained in the small pores as *shale gas*. There is now considerable interest in shale gas, particularly in onshore basins where drilling costs are moderate.

13.2 Sandstones

Compaction of sand and sandstones has been discussed in Chap. 4. Dissolution at grain contacts (pressure dissolution) is driven by the increased solubility due to stress causing a slight supersaturation of silica with respect to quartz and a precipitation of new authigenic quartz (overgrowth). It is now generally assumed that the precipitation, which is a function of temperature, is the rate-limiting step and that this chemical compaction is therefore rather insensitive to the stress. Dissolution at grain contacts occurs preferentially in contacts with mica and clay minerals which favour the development of stylolites. Transport distance between the dissolution and precipitation sites is very short and is driven by diffusion, and will be limited by the distance between the stylolites.

Compaction-driven porewater can not explain significant transport of silica. At normal geothermal gradients $3 \cdot 10^9$ volumes of water are required to precipitate one volume of quartz. In addition, porewater is generally not moving upwards in relation to the seafloor so there is little cooling of the porewater (see Chap. 4).

As in mudstones, chemical compaction in sandstones is mostly controlled by temperature and both sandstones and mudstones compact chemically during burial. Overpressure reduces the effective stress and therefore has little effect.

The loss of porosity results in higher density and a reduction in rock volume or shrinkage (Fig. 13.5). Even a very small loss of porosity (strain) by chemical compaction will reduce the bulk volume so that the stress is reduced.

This shrinkage will contribute to a reduction in horizontal stress because some of the compaction may occur in the horizontal direction. This is indicated by the leak-off pressures at greater depth (Fig. 13.6). This may reduce horizontal tectonic stresses.

Sediment compaction - rock shrinkage

- ☐ Bulk modulus = Stress/strain(ΔV)
- ☐ If the strain ΔV is 0.001 or 0.1% and the bulk modulus is 50G Pa the effective stress is reduced by 50 MPa
- ☐ k volume (V_R) = Solids (V) + Fluids (porosity)
- ☐ Void ratio = $V_S / V_f = \varphi/(1-\varphi)$
- ☐ For isochemical reactions V_S = const.
- ☐ $\Delta V = \Delta \varphi$, $dV/dt = d\varphi/dt$

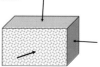

Fig. 13.5 Some definitions related to sediment compaction. During mechanical compaction the strain is produced by an increase in the effective stress. Chemical compaction in sandstones and other siliceous sediments produces strain without stress. The strain will however reduce differential stresses

In the upper parts of sedimentary basins (<70–80°C) the compaction of siliceous sediments follows the laws of soil and rock mechanics. At greater depth compaction is mainly chemical and controlled by temperature (Fig. 13.7).

When the compaction is mechanical any reduction in the effective stress due to uplift or the build-up of overpressure will cause the sedimentary rocks to become overconsolidated, and the deformation does not follow the virgin loading curve (see Chap. 11).

Chemical compaction in siliceous sediments will continue during uplift as long as the temperature is higher than 70–80°C. The mechanical extension due to unloading will then at least partly be compensated for by chemical compaction, and open fractures will gradually be healed by quartz cement.

13.3 Carbonate Compaction

Compaction of carbonates is controlled by principally very different processes than in siliceous sediments. Because the kinetics of carbonate dissolution and precipitation are so much faster than for siliceous rocks (mudstones and shales), temperature is not the main control on carbonate compaction. Cementation of carbonate sediments into hard solid rocks may occur right near the surface. In addition the presence of aragonite which is thermodynamically less stable than calcite provides a strong drive for cementation. The sediments

Fig. 13.6 Leak-off pressure is an indication of the horizontal stress and at 3–4 km this is nearly equal to the vertical stress. This suggests that during chemical compaction the rocks compact both *vertically* and *horizontally*, thus reducing differential stress. At very slow strain rates a sandstone may respond nearly as a fluid where the stress is equal in all directions. The horizontal stress is close to the vertical overburden stress. From The Millennium Atlas. Geological Society of London, 2003

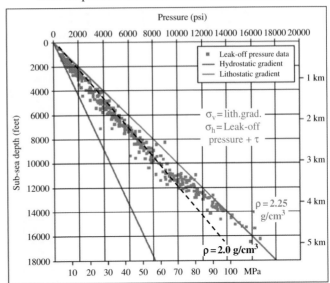

Leak-off pressure data from Central Graben, North Sea

Stress in passive margin basins with mostly siliceous sediments

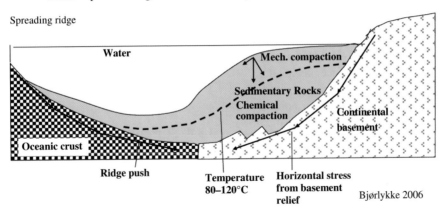

Bjørlykke 2006

Fig. 13.7 Simplified cross-section through a sedimentary basin on a passive margin. Most of the tectonic stress is transmitted through the basement and the well-cemented sedimentary rocks. In the case of ice loading, the strain rates are relatively high and the response in the sediments will be mostly mechanical compaction. Gravitational stress may also be important

then become mechanically overconsolidated and may be unable to undergo further mechanical compaction even when subjected to 40–50 MPa (4–5 km depth). Carbonate sediments like the Chalk, composed almost entirely of low-Mg calcite, undergo little cementation by pressure solution along stylolites at depths exceeding 1–1.5 km depth. Overpressure is very important in reducing both mechanical compaction and pressure dissolution. In the Ekofisk Field, Chalk may have porosities exceeding 30% at nearly 3 km burial depth due to high overpressure, because the effective stress only corresponds to about 1 km without overpressure.

The processes controlling porosity loss in carbonate sediments are still poorly understood.

The dissolution rate may be more important compared to sandstones. At the contact between two calcite grains there is probably only a very thin layer of water, while clay minerals have a double layer due to the negative surface charges. The transport of calcium along the calcite grain contacts may also be rate-limiting.

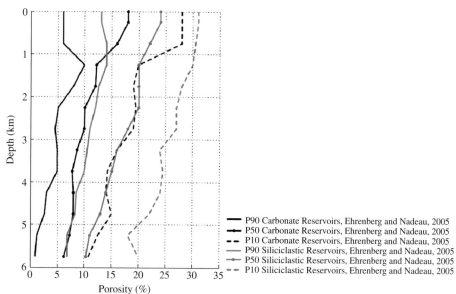

Fig. 13.8 Compaction trends for carbonates and sandstones (from Ehrenberg and Nadeau 2005). Average porosity versus top depth for global petroleum reservoirs. P90, P50, and P10 indicate that 90, 50 and 10% of the reservoirs' values have higher porosity than this value

Carbonate grains, particularly of fossils, may have an organic coating which may influence precipitation.

Early porosity reduction in carbonate at shallow depth may help to preserve the resultant porosity during deeper burial. Carbonates have generally lower porosity than sandstones at the same depth (Fig. 13.8) but there is a wide range of porosity/depth values, particularly for carbonates.

Near the surface where there may be meteoric water flow the system is relatively open and net porosity may be created by dissolution. There is, however, limited potential for mass transport of carbonate in solution during burial and the reactions must be nearly isochemical. This is because the porewater will always be closely in equilibrium with the carbonate minerals that are present. This leaves little potential for transport by diffusion, or by advection, because the flow rates are so small, particularly in relation to the isotherms. Focused flow as along faults will cause some dissolution because of the retrograde solubility of carbonates like calcite. The solubility is also a function of pressure but flow across pressure barriers is rather limited.

Much of the compaction of carbonate rocks occurs along stylolites because the dissolution and transport along grain contacts are enhanced by the presence of sheet silicates (see Chap. 5).

13.4 Summary

Shales, sandstones and carbonates follow different compaction trends and they are controlled by principally different processes.

Both shales and sandstones compact mechanically as a function of effective stress until chemical compaction takes over and further compaction is mainly a function of temperature and time. The initial mineralogical and textural composition is very important both for sandstones and mudstones (shales).

Carbonate sediments may compact chemically at very shallow depth and low temperature and the compaction process is driven by a complex interaction between stress and chemical compaction, but the temperature is less important. One of the main factors controlling compaction and rock properties in carbonates is the primary content and distribution of aragonite, causing early cementation.

Further Reading

Bjørlykke, K. 2003. Compaction (consolidation) of sediments. In: Middleton, G.V. (ed.), Encyclopedia of Sediments and Sedimentary Rocks. Kluwer Academic Publ., Dordrecht, pp. 161–168.

Bjørlykke, K. 2006. Effects of compaction processes on stress, faults, and fluid flow in sedimentary basins. In: Buiters, S.H.J. and Schreurs G. (eds.), Analogue and Numerical Modelling of Crustal-Scale Processes. Geological Society Special Publication 253, 359–379.

Bjørlykke, K., Aagaard, P., Dypvik, H., Hastings, D.S. and Harper, A.S. 1986. Diagenesis and reservoir properties of Jurassic sandstones from the Haltenbanken area, offshore mid – Norway. In: Spencer, A.M. et al. (eds.), Habitat of Hydrocarbons on the Norwegian Continental Shelf. Norwegian Petrolium Society, Graham & Trotman, London, pp. 275–286.

Bjørlykke, K., Chuhan, F., Kjeldstad, A., Gundersen, E., Lauvrak, O. and Høeg, K. 2004. Modelling of sediment compaction during burial in sedimentary basins. In: O. Stephansson, J. Hudson and L. King (eds.), Coupled Thermo-Hydro-Mechanical-Chemical Processes in Geosystems. Elsevier, London, pp. 699–708.

Chuhan, F.A., Kjeldstad, A., Bjørlykke, K. and Høeg, K. 2002. Porosity loss in sand by grain crushing. Experimental evidence and relevance to reservoir quality. Marine and Petroleum Geology 19, 39–53.

Chuhan, F.A., Kjeldstad, A., Bjørlykke, K. and Høeg, K. 2003. Experimental compression of loose sands: Relevance to porosity reduction during burial in sedimentary basins. Canadian Geotechnical Journal 40, 995–1011.

Ehrenberg, S.N. and Nadeau, P.H. 2005. Sandstone vs. carbonate petroleum reservoirs; a global perspective on porosity-depth and porosity-permeability relationships. AAPG Bulletin 89(4), 435–445.

Ehrenberg, S.N., McArthur, J.M. and Thirlwall, M.F. 2006. Growth, demise and dolomitization of Miocene carbonate platforms on the Marion Plateau, offshore NE Australia. Journal of Sedimentary Research 76, 91–116.

Ehrenberg, S.N, Nadeau, P.H. and Steen, Ø. 2008. A megascale view of reservoir quality in producing sandstones from the offshore Gulf of Mexico. AAPG Bulletin 92, 145–164.

Hesthammer, J., Bjørkum, P.A. and Watts, L. 2002. The effect of temperature on sealing capacity of faults in sandstone reservoirs – Examples from the Gullfaks and Gullfaks Sør Fields, North Sea. AAPG Bulletin 86(10), 1733–1751.

Hovland, M., Bjørkum, P.A., Gudemestad, O.T. and Orange, D. 2001. Gas hydrate and seeps – Effects on slope stability: The "hydraulic model". ISOPE Conference proceedings, Stavanger, pp. 471–476, ISOPE (International Society for Offshore and Polar Engineering), New York.

Marcussen, Ø., Thyberg, B.I., Peltonen, C., Jahren, J., Bjørlykke, K. and Faleide, J.I. 2009a. Physical properties of Cenozoic mudstones from the northern North Sea: Impact of clay mineralogy on compaction trends. AAPG Bulletin 93(1), 127–150.

Marcussen, Ø., Thyberg, B.I., Peltonen, C., Jahren, J., Bjørlykke, K. and Faleide, J.I. 2009b. Physical properties of Cenozoic mudstones offshore Norway: Controlling factors on sediment compaction and implications for basin modeling and seismic interpretation. AAPG Bulletin 93(2), 1–24.

Mondol, N.H., Bjørlykke, K. and Jahren, J. 2008. Experimental compaction of clays. Relationships between permeability and petrophysical properties in mudstones. Petroleum Geoscience 14, 319–337.

Mondol, N.H., Bjorlykke, K., Jahren J. and Høeg, K. 2007. Experimental mechanical compaction of clay aggregates – Changes in physical properties of mudstones during burial. Marine and Petroleum Geology 89, 289–311.

Peltonen, C., Marcussen, Ø., Bjørlykke, K. and Jahren, J. 2008. Mineralogical control on mudstone compaction; a study of late Cretaceous to early Tertiary mudstones of the Vøring and Møre basins, Norwegian Sea. Petroleum Geoscience 14, 127–138.

Storvoll, V., Bjørlykke, K. and Mondul, N.H. 2005. Velocity-depth trends in Mesozoic and Cenozoic sediments from the Norwegian Shelf. AAPG Bulletin 89, 359–381.

Teige, G.M.G., Hermanrud, C., Wensås, L. and Nordgård Bolås, H.M. 1999. Lack of relationship between overpressure and porosity in North Sea and Haltenbanken shales. Marine and Petroleum Geology 16, 321–335.

Thyberg, B., Jahren, J., Winje, T., Bjørlykke, K. and Faleide, J.I. February 2009a. From mud to shale: Rock stiffening by micro-quartz cementation. First Break 27, 27–33.

Thyberg, B., Jahren, J., Winje, T., Bjørlykke, K., Faleide, J.I. and Marcussen, Ø. 2009b. Quartz cementation in Late Cretaceous mudstones, northern North Sea: Changes in rock properties due to dissolution of smectite and precipitation of micro-quartz crystals. Marine and Petroleum Geology In press.

Voltolini, M., Wenk, H.-R., Mondol, N.H., Bjørlykke, K. and Jahren, J. 2009. Anisotropy of experimentally compressed kaolinite-illite-quartz mixtures. Geophysics 74, D13–D23.

Chapter 14

Source Rocks and Petroleum Geochemistry

Knut Bjørlykke

As discussed in Chap. 1, petroleum is generated from organic matter which accumulates in sedimentary basins. Only a small fraction of the organic matter produced in the photic zone in the ocean becomes trapped in sediments (Fig. 14.1). Most of the organic matter is oxidised in the water column or on the seafloor and the nutrients are released into the water and become available for new organic production near the surface during upwelling. Most source rocks are black shales like the Upper Jurassic Kimmeridge Clay and its equivalents in the North Sea basin (Fig. 14.2).

The organic matter is transformed into kerogen which consists of very large and complex molecules. We do not usually apply the term kerogen to fresh organic material, but to material which is somewhat dehydrated after burial to about 100 m or more. Kerogen is formed gradually within the upper few hundred metres of the sediment column after deposition from precursor products like humus, and humic and fulvic acids. The organic matter may be derived from marine organisms, mostly algae, or from plants derived from land.

The transformation of amino acids, carbohydrates, humic acids and other compounds into kerogen is achieved by the removal of functional groups such as acid groups, aldehydes and ketones. This involves a loss of oxygen from the organic material, also of nitrogen, water and CO_2.

Kerogen therefore has higher H/C, and lower O/C, ratios than the initial compounds (Fig. 14.3).

Kerogen may also include organic particles of morphologically recognisable biological origin such as vitrinite (derived from woody tissues and liptinite materials, e.g. algae spores, cuticles, etc.). Because of its resistance to strong oxidising acids kerogen can be recovered from sedimentary rocks by dissolving most of the rock away with HCl or HF.

It is also possible to separate kerogen by a density method, using heavy liquids, because kerogen is lighter than minerals. The resulting concentrate of kerogen can be studied microscopically using transmitted and reflected normal light, to identify the biological origin and the degree of thermal alteration. These phases of altered organic material are called macerals. Algal material has a dull appearance, while wood and material from higher plants is called vitrinite. Vitrinite becomes increasingly shiny when exposed to higher temperatures and by measuring the amount of reflected light in the microscope we obtain an expression for the degree of thermal alteration (Vitrinite index).

It is also useful to use ultraviolet light microscopy, since certain components, i.e. liptinites, display characteristic fluorescence colours. Infra-red spectroscopy (IR) or nuclear magnetic resonance (NMR) spectroscopy can be used to investigate the chemical composition and structure of kerogen.

Being a complex of very large molecules (polymer), kerogen is difficult to analyse, but upon heating to 350–450°C in an inert atmosphere (pyrolysis) it will break down into smaller components which can then be analysed by means of gas chromatography and mass spectrometry.

Kerogen has a wide range of compositions, dependant on the original organic composition, but may be classified into 3 main types which may be plotted as a function of the H/C ratio and the O/C ratio (Fig. 14.2)

K. Bjørlykke (✉)
Department of Geosciences, University of Oslo, Oslo, Norway
e-mail: knut.bjorlykke@geo.uio.no

K. Bjørlykke (ed.), *Petroleum Geoscience: From Sedimentary Environments to Rock Physics*,
DOI 10.1007/978-3-642-02332-3_14, © Springer-Verlag Berlin Heidelberg 2010

Fig. 14.1 Formation of source rocks. Only a small fraction of the organic matter is preserved. The formation or organic-rich source rocks requires restricted water circulation and oxygen supply

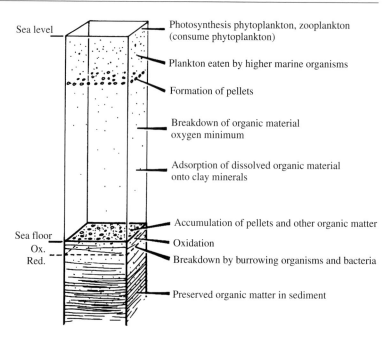

Type I sapropelic kerogen is formed from organic material with a high content of lipids with long aliphatic chains. It consists of spores and planktonic algae, as well as animal matter, which have been broken down microbially after deposition in the sediment. Saprolitic material which consists of fats, oils, waxes, etc., has a high H/C ratio, usually between 1.3 and 1.7. This kind of kerogen is often called Type I, and contains little oxygen (O/C <0.1). It will provide mainly oil, with less gas (CH_4 and CO_2). Type I kerogen is typical of oil shales, especially in freshwater basins

Fig. 14.2 Draupne shale (Kimmeridge shale) cores from northern North Sea (Block 34/7)

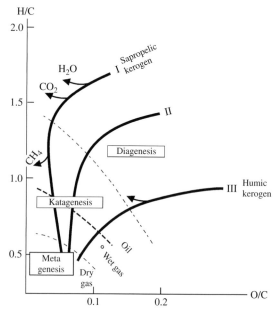

Fig. 14.3 Diagram (Van Krevelen diagram) showing the primary composition of the different types of kerogen and the changes as a function of heating (maturation) during progressive burial

like the Green River basin in Colorado, Wyoming and Utah, but is also found in marine basins.

Type II kerogen is a mechanically and chemically complex mixture of algae and other marine organisms and plant debris. The composition varies considerably, depending on the initial organic precursor materials which again may be linked to depositional facies. Type II kerogen represents a composition midway between types I and III but it does not represent a mixture of these end members. It has relatively high H/C, and low O/C, ratios, but contains more oxygen-containing compounds (ketones and carboxyl acid groups) than Type I. Esters and aliphatic chains are also common. This is the usual type of kerogen found in marine basins where mixtures of phytoplankton, zooplankton and micro-organisms have accumulated under reducing conditions, sometimes along with land-derived plant material. This type of kerogen is the most common source of oil (Fig. 14.3).

Type III humic kerogen is derived from organic matter from land plants, such as lignin, tannins and cellulose. This type has a low initial H/C ratio, and a high initial O/C ratio, reflecting the composition of the precursor plant matter. In maturing (through the effect of temperature) this kerogen, which is often called Type III (Tissot and Welte 1978), generates abundant water, CO_2 and methane (CH_4). Most coals have a composition and structure similar to Type III kerogens. Coal generates mostly gas but some coals may also generate some oil.

14.1 Transformation of Kerogen with Burial and Temperature Increase

With increasing temperature the chemical bonds in these large molecules (kerogen) are broken and kerogen is transformed into smaller molecules which make up oil and gas. This requires that the temperature must be 80–150°C over long geological time (typically 1–100 million years).

The conversion of kerogen to oil and gas is thus a process which requires both higher temperatures than one finds at the surface of the earth and a long period of geological time. Only when temperatures of about 80–90°C are reached, i.e. at 2–3 km depth, does the conversion of organic plant and animal matter to hydrocarbons very slowly begin to take place. About 100–150°C is the ideal temperature range for this conversion of kerogen to oil, which is called maturation. This corresponds to a depth of 3–4 km with a normal geothermal gradient (about 30–40°C/km). In volcanic regions organic matter may mature at much lesser depths due to high geothermal gradients (e.g. high heat-flow areas). In large intracratonic sedimentary basins or along passive margins, however, the geothermal gradient may be only 20–25°C/km and the minimum overburden required to initiate petroleum generation will be correspondingly greater (4–6 km).

In general one can say that petroleum can not be generated near the surface except locally through the influence of hydrothermal and igneous activity. Shallow deposits of oil and gas which we find today were actually formed at great depths and either the overburden has been removed by erosion or the hydrocarbons have migrated upwards considerable distances. As we have already seen, however, large amounts of natural gas, chiefly methane (CH_4), may be formed near the surface by biochemical processes.

Temperature increases with increasing overburden, causing the carbon-carbon bonds of the organic molecules in the kerogen to rupture. This results in smaller hydrocarbon molecules. When kerogen maturation reactions are completed, the kerogen's "organic" components, which may be derived from lipids, fatty acids and proteins, have been converted into hydrocarbons.

As the temperature rises, more and more of the bonds are broken, both in the kerogen and in the hydrocarbon molecules which have already been formed. This "cracking" leads to the formation of lighter hydrocarbons from the long hydrocarbon chains and from the kerogen. The removal of gas, mainly CH_4, leaves the residual kerogen relatively enriched in carbon. At the outset kerogen (Type I and II) has an H/C ratio of 1.3–1.7. Humic kerogen (Type III), which has high initial oxygen contents, gives off mostly CO_2 gas and so its oxygen/carbon ratio gradually diminishes. This diagenetic alteration begins at 70–80°C and as water and CH_4 are removed the H/C ratio will fall to about 0.6 and the O/C ratio will become less than 0.1 at about 150–180°C.

In the North Sea basin, most of the oil is generated at temperatures around 130–140°C, which equates with a depth of about 3.5 km. If temperatures higher than 170–180°C persist for a few million years, all the longer hydrocarbon chains will already have been broken (cracked), leaving us only with gas – mainly methane (dry gas). The kerogen composition will gradually be depleted in hydrogen and move towards pure carbon (graphite) (H/C→0).

14.2 What Factors Influence the Maturation of Kerogen?

The term "maturity" refers here to the degree of thermal transformation of kerogen into hydrocarbons and ultimately into gas and graphite. The conversion of kerogen into hydrocarbons is a chemical process which takes place with activation energies of around 50–60 kcal/mol. This energy is required to break chemical bonds in the kerogen which consists of very large molecules (polymers) so that smaller hydrocarbon molecules can be formed.

It has been assumed that formation of oil is a first order reaction, the rate of which is an exponential function of time. Understanding the factors which influence the rate of this reaction is of great interest. Four factors are thought to contribute:

1. Temperature
2. Pressure
3. Time
4. Minerals or other substances which increase the rate of reaction (catalysts) or which inhibit reactions (inhibitors).

Temperature is clearly the most important factor, and hydrocarbons can be produced experimentally from kerogen by heating it (pyrolysis). This reaction is time-dependent and in laboratory experiments, where time is more limited than it is in nature, fairly high temperatures (350–550°C) have to be used in pyrolysis. Pressure appears to play a minor role but increasing pressure should reduce the rate of the reaction because of the increase in volume involved in the formation of hydrocarbons (Le Chatelier's rule).

There is a relatively small volume increase when kerogen becomes oil, even though oil is lighter than kerogen. This is due to the residual (coke) which remains unaltered.

When kerogen is converted directly into gas, or from oil which has been formed first, there is a marked volume increase. This should lead to slower reaction rates under high pressure in a closed system and retard generation of gas. Generation of petroleum, particularly gas, may contribute to the formation of overpressure but in a sedimentary basin the pressure will for the most part be controlled by the flow of water which is the dominant fluid phase. In the source rock however overpressure is likely to develop, causing hydrofracturing which helps to expel the generated petroleum. The main cause for overpressuring is, however, not only the increase in fluid volume but the transformation of solid into fluids. When solid kerogen is transformed into fluid oil or gas the ratio between the solid phase and the fluid phase is changed, as expressed by the porosity and the void ratio.

Temperature is however the most important factor controlling petroleum generation.

It has long been suspected that minerals, particularly clay minerals, might affect the rate of hydrocarbon generation. A number of laboratory experiments have been carried out in which kerogen is mixed with various minerals but the results have not been conclusive.

The conversion of organic matter begins at 70–80°C, given long geological time. Between 60 and 90°C the transformation of kerogen proceeds very slowly, and it is only in ancient, organic-rich sediments that significant amounts are formed. Most of the maturation process occurs between 100 and 150°C. Here the degree of kerogen transformation is also a function of time. This means that rocks which have been subjected to 100°C for 50 million years are more mature than rocks which have been exposed to this temperature for 10 million years. As the organic-rich sediment (source rock) is buried in a sedimentary basin, it will normally be subjected to increasing temperature as a function of increasing burial depth. If we know the stratigraphy of the overlying sediment sequence and the geothermal gradient and the subsidence curve, we can calculate the temperature as a function of time.

At low degrees of maturity we find more of the alkenes (olefins) and cykloalkenes (naphtenes), which have high H/C ratios, while with greater maturity there is an increase in the proportion of aromates and polyaromates (low H/C ratio). Oil thus acquires increasing gas content with increasing maturity.

During this transformation of organic matter, water and oxygen-rich compounds are liberated first, then compounds which are rich in hydrogen. This conversion results in enrichment of carbon and the colour of the residual kerogen changes from light yellow to orange, brown and finally black. These gradations can best be registered by measuring light absorption of fossil pollen and spores (palynomorphs). It is also possible to analyse colour changes in other kinds of fossils, for example conodonts.

For application in exploration a rapid semi-quantitative method has been developed whereby these colour changes are estimated from smooth spores examined under transmitted light and compared with a standard colour scale. This parameter is called the "Thermal alteration index" (TAI) and will give a rough idea of the thermal maturity of the sediments and their temperature history.

Another way of assessing palaeotemperatures at which alteration took place in sedimentary rocks is to record the degree of carbonisation of other plant remains which are usually present. Vitrinite, which originally was fragments of woody tissue, is a common component of coal but is also found in smaller amounts in marine source rocks. This material is analysed by measuring the amount of light it reflects. It becomes shinier and reflects light better as the degree of carbonisation increases. By measuring the reflectivity of vitrinite particles under a reflected light microscope an exact value is obtained for this maturity parameter, expressed by the reflectivity coefficient R_0 (% vitrinite reflectance). If R_0 is less than 0.5% in a shale it can not have generated much oil and is classed as immature. Shales with $R_0 = 0.9$–1.0 have been exposed to temperatures corresponding to maximum oil generation. $R_0 = 1.3$ represents the upper limit for oil generation, above which the shale will only produce condensate (light oils) or gas.

For certain source rocks the ratio between extractable alkenes (paraffins) with an even number of carbon atoms per molecule and those with an odd number may also be an expression of maturity. In plant material and in marine algae, one finds a higher abundance of alkenes with an odd number of carbon atoms than in transformed organic matter like waxes and fatty acids. The decrease in this predominance of odd over even in source rocks with increasing maturity is due to the dilution of the original biologically derived n-alkanet mixture with a newly generated mixture which has a regular carbon number distribution. This odd/even ratio is normally expressed by means of an index called the Carbon Preference Index (CPI):

$$\text{CPI} = \text{n-alkenes (odd)} / \text{n-alkenes (even)}$$

It is based on analyses of alkenes with carbon number between 25 and 33 (C_{25} and C_{33}) in oils.

However, organisms also start off with different CPI index values, and land plants have high ratios between odd and even carbon numbers. Bacteria have a predominance of even carbon numbers.

The maturation process will cause a shift in the carbon number distribution towards smaller molecules, particularly in the range C_{13}–C_{18}.

Oil that comes from carbonate source rocks often has a low CPI index, while oil derived from plants has a high index value. With increasing temperature the CPI index goes towards 1, that is to say, equal even and odd carbon numbers.

The isotopic ratio also changes because the bonding between hydrogen and ^{13}C, and between hydrogen and ^{12}C, are not equally stable. Light gases such as methane, are enriched in the light isotope ^{12}C, and the hydrocarbons that remain will therefore have an increasing $^{13}C/^{12}C$ ratio with increasing temperature.

There is isotopic fractionation of carbon, and when kerosene is releasing petroleum, this phase is somewhat enriched in ^{11}C corresponding to the precursor kerosene. Gases, particularly methane, normally have lighter carbon isotopes than kerosene and oil. When methane is formed from larger hydrocarbon molecules by thermal cracking, the ^{12}C–^{12}C bond is less stable than the ^{13}C–^{12}C bond and the product becomes enriched in ^{12}C (low $\delta^{13}C$).

14.3 Modelling of Petroleum Generation

The rate of petroleum generation can be calculated and modelled.

We assume that the rate (k_i) of petroleum generation follows an Arrhenius function:

$$(k_i) = A \exp\left(-E_i/RT\right)$$

where R is the gas constant, T is the temperature and E_i is the activation energy. The activation energy mostly varies between 50 and 80 kcal/mol or about 200 kJ/mol. A is an exponential constant dependant on the type of kerogen.

Since the temperature changes during the burial history the effect of temperature has to be integrated over the range of temperatures that the kerogen is exposed to. This may be expressed by the Time Temperature Index (TTI)

$$(\text{TTI}) = \int_{t_0}^{t_x} 2^{F(T)} dt$$

This is the integrated temperature (T) over time from the time of deposition (t_0) to the present day (t_x). F is a factor dependant on the activation energy for the reactions. The reaction rates may approximately double for each 10°C increment. The maturation of kerogen to petroleum can be successfully modelled (Makhous and Galushkin 2005) but the results depend very much on the input data with respect to the temperature history. The activation energy may also vary significantly for different types of source rocks. If the geothermal gradient can be assumed to have been constant throughout the relevant period, this is relatively simple. However, if the geothermal gradient has varied considerably through time it is a lot more complicated.

A theoretical maturity parameter (P) can be calculated by integrating temperature with respect to time:

$$P = \ln \int_0^t 2^{T/10} \cdot dT,$$

t, geological time (million years); T, temperature (°C).

We see that a doubling of the reaction rate for every 10°C is built into this expression (Geoff 1983). This is an expression which is very similar to Lopatin's Time-Temperature Index (TTI) (Waples 1980) which integrates the temperature the source rock is subjected to with respect to the burial time.

When the temperature rises above about 130–140°C, maturation proceeds very rapidly, and then the time factor is less crucial.

There are differing views as to how much emphasis should be placed on time in relation to temperature in the matter of maturation. Oil companies use different formulae for calculating these temperature factors and the 10-degree rule is now found not always to be valid, particularly for very young sedimentary basins with high geothermal gradients.

The rate of sediment heating, i.e. temperature increase per unit time, can be expressed as dT/dt. The rate of subsidence is dZ/dt, and the geometrical gradient dT/dZ. We then get:

Rate of heating (dT/dt) = rate of subsidence $(dZ/dt) \cdot$ geothermal gradient (dT/dZ).

In basins like the North Sea the rate of heating is typically only 1–2°C/million years but during rapid subsidence and sedimentation in the late Cenozoic the heating rate was higher. During the deposition of

1–1.5 km of upper Pliocene and Pleistocene sediment in 2–3 million years gradient was lowered but the heating rate must have been about 10°C/million years.

The maturity of source rocks can now be calculated with the help of basin modelling integrating temperature over time. The subsidence curve for the source rock is determined from the stratigraphic age and thickness of the overlying sequence.

By estimating a geothermal gradient, depth can be converted to temperature. This way we obtain a curve that shows the temperature history of the kerogen through geological time. Integration of this curve enables the maturity to be calculated (Fig. 14.4).

If we know the stratigraphy of the overlying sediments and the geothermal gradient, temperature can be calculated as a function of time.

Calculating the maturity is important not only for predicting where the source rocks are sufficiently mature to produce oil and gas (or perhaps over mature). It is also important for estimating the timing of the generation and migration of oil and this type of basin modelling has become an important part of oil exploration. The most important input into the

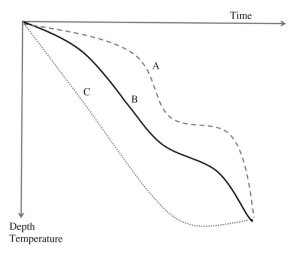

Fig. 14.4 The maturity of a source rock is a function of the time – temperature index (TTI). This can be calculated from the burial curve if the geothermal gradients are known. The geothermal gradients are most critical during the deepest burial because the maturation is an exponential function of the temperature. At a certain depth and temperature the maturity may vary greatly and burial curve C will produce the highest maturity at this depth. Source rocks buried following curve A and B are less mature because their exposure to greater burial depth (higher temperatures) has been much shorter

calculation is the subsidence history as derived from the stratigraphic record, and the estimated geothermal gradient as a function of geological time. The success of the basin modelling depends as always very much on the quality of the input data.

14.4 Rock-Eval Analyses

Rock-Eval is a standard routine analysis of source rocks, usually shales, to establish how much of the kerogen has been transformed into petroleum and how much can be transformed at a higher temperature.

The sample of shale is crushed and heated to 300°C, at which point one measures the amount of hydrocarbons that are already formed in the source rock but have not migrated out. The content of hydrocarbon with carbon numbers between C_1 and C_{25}, is called S_1. It is measured as the area beneath the peak S_1 (Fig. 14.5).

On further heating from 300 to 550–600°C, new petroleum is formed in the laboratory from the kerogen by heating (pyrolysis), and this amount is called S_2.

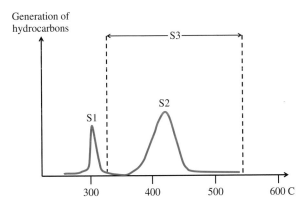

Fig. 14.5 Rock-Eval analyses. A rock sample representing a possible source rock is heated gradually to about 550°C while the amount of hydrocarbons generated is measured. At about 300°C oil and gas which has already been generated in the source rock is expelled and measured as the S_1 peak. The peak at about 400–460°C represents the amount of hydrocarbons generated from the kerogen in the sample. The temperature of peak HC generation is called the Tmax. The Hydrogen Index (HI) = S_2 (S)/TOC (total organic carbon) is a measure of the potential of the source rock to generate petroleum. The total amount of CO_2 generated is measured as the S_2 peak. The Oxygen Index (OI) = P3/TOC

This is a measure of how much oil and gas could have been generated if the source rock and been buried deeper. The reason it requires such high temperatures is that the heating in the laboratory lasts just a few minutes or hours, instead of some millions of years.

During heating from c. 300 to 550°C, CO_2 is also formed and is collected and measured separately as the S_3 peak (Fig. 14.5). Most of the CO_2 groups dissociate from the kerogen between 300 and 390°C. The generation of petroleum varies with temperature and reaches a peak corresponding to S_2 (Fig. 14.5). This temperature, which gives the maximum petroleum generation, is called T_{maks} and is typically in the range 420–460°C.

The ratio between the amount of petroleum generated (S_2) and the total content of organic material (TOC) is known as the Hydrogen Index (HI).

The quantity of CO_2 which is formed (S_3), is limited by the oxygen content of the kerogen. The S_3/TOC ratio is the Oxygen Index (OI).

The ratio between the quantity of free oil already formed (S_1) and the total amount of petroleum (S_1 + S_2), is an expression of how much petroleum is still left in relation to how much has already been generated. The $S_1/(S_1 + S_2)$ ratio is the Production Index (PI). Good source rocks have a high production index.

When we analyse source rocks, we must however take into account that some of the oil which has been generated has migrated out of the source rock: the S_1 peak represents the remaining petroleum.

Good source rocks are the first prerequisite for finding oil and gas in a sedimentary basin. If they are not present, one can save oneself the trouble of further prospecting. However the quality of the source rock can vary through the basin and the type of source rock determines the composition of the oil.

The timing of the oil and gas generation is also very important. It also determines the timing of oil migration with respect to the formation of traps. Once we know with reasonable certainty when oil generation and migration took place, we can attempt to construct maps to show the structures and faults at that time. For example, if a trap was formed by folding after the oil had migrated, it can not have captured any of the oil, but perhaps the gas which is formed later.

The generation of oil and gas lends itself to mathematical modelling, and there are good programs

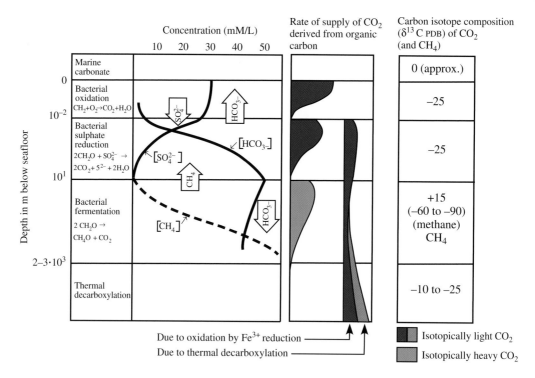

Fig. 14.6 Natural fractionation of carbon isotopes in CO_2 generated in sediments during burial. During oxidation by oxygen or sulphate reduction the $\delta^{13}C$ values are highly negative while bacterial fermentation produces positive values. Thermal decarboxylation also produces negative $\delta^{13}C$ values

which make this easier. The most important data which have to be fed in are the subsidence rate, the geothermal gradient and the activation energy of the kerogen.

The activation energy varies according to the type of kerogen and is an important factor when it comes to calculating the timing of the oil generation.

In nature natural processes cause fractionation of carbon which is reflected in the isotopic composition of CO_2 (Fig. 14.6). Note that the CO_2 released by bacterial fermentation has a positive $\delta^{13}C$ while the carbon from maturation of kerogen (thermal decarboxylation) has negative values.

14.5 Composition of Petroleum

Naturally occurring petroleum has a very complicated chemical composition. The most important hydrocarbons occurring in oil are alkanes (paraffins, C_nH_{2n+2}). Important gases are methane (CH_4), ethane (C_2H_6), propane (C_3H_8) and butane (C_4H_{10}). Paraffins from

carbon numbers 5, pentane (C_5H_{12}) to 15, pentadecane ($C_{15}H_{32}$) occur mainly as liquids at room temperature. These paraffins are an important part of the gasoline fraction of oil. Higher paraffins are waxy and almost a solid phase at room temperature but become less viscous at higher temperature.

Naphthenes (cycloparaffins) are a series of cyclic, saturated hydrocarbons with the general formula $(CH_2)_n$. Cyclopentane (C_5H_{10}) and cyclohexane (C_6H_{12}) are important members of the naphthene series which are found in all types of petroleum. Naphthenes with lower carbon numbers, C_3H_6 and C_4H_8, are gases. The percentage of naphthenes in petroleum varies from 7 to 31. Petroleum rich in naphthenes is called asphalt-based crude, because many of the more complex naphthenes have high boiling points and high viscosity.

Olefins (*alkenes*) are hydrocarbons with the same chemical composition as alkanes, but contain one or more double bonds. Hydrocarbons of this type do not normally occur in crude oil. *Aromatic hydrocarbons* are unsaturated, cyclic, and have the general formula C_nH_{2n-6}. For example benzene, C_6H_6, and toluene,

$C_6H_5CH_3$, are important ingredients of car petrol on account of their high octane numbers. Aromatic hydrocarbons have a strong odour and a lower boiling point than aliphatic hydrocarbons with the same number of carbon atoms. Aromatic compounds make up 10–39% of crude oil. Refinement of crude oil with a high content of aromatic compounds results in a high octane product.

Asphalt is brown to black, solid or highly viscose petroleum consisting of hydrocarbons with high molecular weights. Asphalt is rich in aromatic compounds and naphthenes, and is enriched in sulphur, nitrogen and oxygen.

The sulphur content of crude varies from 1 to 5–6% and occurs as pure sulphur, hydrogen sulphide (H_2S) and organic sulphur compounds. Components in oil that contain sulphur, nitrogen and oxygen, are called NSO compounds.

The density of crude oil is measured using the unit API gravity where "gravity" is expressed as an inverse value in relation to density as defined by the American Petroleum Institute.

$$API\ gravity = 141.5/\rho - 131.5$$

Here ρ is the density of the petroleum at 60°F (15°C).

We see from the formula that a light oil with a density (ρ) of 0.8 g/cm^3 has a gravity close to 45, whereas a heavy oil with a density of 1.0 g/cm^3 has a gravity of 10. Oil with API gravity of about 15–20 or lower is considered heavy oil.

Crude with low API gravity, contains more sulphur than the light crudes.

Pure (native) sulphur also occurs in great quantities in some oil fields, and most of the world's sulphur production comes from sources of this type. In most oil reservoirs, however, sulphur is an unwelcome component, since it causes corrosion and problems in the refining process. Burning oil of this type releases SO_2 into the atmosphere. Much of it can be removed during refining, but that increases the cost.

Elemental composition of hydrocarbons in percent (after Hunt 1979)

	C	H	S	N	O
Oil	84.5	13	1.5	0.5	0.5
Asphalt	84	10	3	1	2
Kerogen	79	6	5	2	8

The NSO compounds include resins and asphaltenes which are enriched in oxygen, nitrogen and sulphur. In addition hydrocarbons contain a number of trace metals derived from the organic matter in the source rock. The most important are Ni and V.

API gravity and density of some hydrocarbons

Hydrocarbon	Molecule type	Formula	H/C ratio	API	Density gravity g/cm^3
Hexane	Paraffin	C_6H_{14}	2.3	82	0.6594
Cyclohexane	Naphthene	C_6H_{12}	2.0	50	0.7786
Benzene	Aromatic	C_6H_6	1.0	29	0.8790

Hydrocarbons with a high H/C ratio tend to have lower densities.

Classification of crude oil:

1. Low shrinkage oil: Oil which after production and separation at ordinary pressure and temperature contains a high percentage of fluid (>80%) and little gas.
2. High shrinkage oil: Oil which after production and separation at ordinary temperature and pressure contains a smaller percentage of liquid (less than 70%) and a large amount of gas.
3. Condensate: Gas which turns into a liquid with pressure reduction. Condensates have low densities (about 40–60° API).
4. Dry gas: This is gas that does not form a liquid phase at normal temperature. This is especially methane and other molecules with low carbon numbers (less than 6, hexane).
5. Wet gas: This is gas that in addition to methane also contains significant amounts of alkanes with high carbon numbers. It may be in a gaseous phase in the reservoir, but after production and separation may form a good deal of liquid at normal pressure and temperature.
6. Undersaturated crude: Oil which contains less gas than is potentially soluble at reservoir temperature and pressure.
7. Saturated crude: Oil which contains as much gas as can be dissolved at reservoir pressure and temperature.

Oil saturated with gas will form gas bubbles if the pressure is reduced due to tectonic uplift or during production in a reservoir.

14.6 Gas

Methane is a thermodynamically stable compound, even at temperatures of 500–600°C or more. As a result we may find methane gas at very great depths in sedimentary basins and also deep in the Earth's crust. The limiting factors for economic exploitation of gas of this type are the low porosity and permeability we normally find in reservoirs at these depths (6–8 km), and the expense of deep drilling. The main component of natural gas is normally methane (CH_4), but ethane (C_2H_6), propane (C_3H_8) and butane (C_4H_{10}) may also be important. In addition we have varying amounts of CO_2, H_2O, nitrogen, hydrogen and inert gases such as argon and helium.

Methane produced by bacterial breakdown of organic matter at low temperatures (<60°C) is characterised by a light carbon isotope composition. This kind of gas is often referred to as "shallow biogenic gas".

14.7 Biodegradation

Hydrocarbons can be broken down by microorganisms (bacteria, yeast and fungi). This is a form of biological oxidation whereby hydrocarbons are oxidised to alcohols, ketones and various acids. The biological breakdown of hydrocarbons proceeds far more rapidly with smaller molecules (carbon number <20). When it comes to molecules with the same carbon number, n-paraffins will break down first, and then isoparaffins, naphthenes and aromatics. Isoprenoids, steranes and triterpanes show the greatest resistance to biodegradation. Relatively rapid biodegradation depends on a supply of oxygen, usually dissolved in water or in air. If sulphate is present hydrocarbons may also be broken down by sulphate-reducing bacteria which use the oxygen in sulphates (e.g. gypsum) to oxidise some of the oil. Minerals containing trivalent iron (Fe^{3+}), such as haematite, can also contribute to the biodegradation and oxidation of oil deeper in the basin when ferric iron is reduced to ferrous (Fe^{2+}). The content of haematite and other minerals with trivalent iron in sediments can thus also be an oxidant.

Anaerobic biodegradation may occur under reducing conditions deeper in a sedimentary basin and the supply of nutrients such as phosphorus in the sediments may then be critical.

The temperature must however be lower than about 80°C. The bacteria eat the lighter compounds so that the larger, more asphaltic, compounds are enriched. Biodegraded oils (tar) have therefore high viscosity.

Biodegradation occurs as soon as the oil flows out at the surface as oil seepage and the lighter compounds will evaporate. Bacteria can then use oxygen from the air and the biodegradation is relatively fast.

Biodegradation may occur in relatively shallow reservoirs (<1,500–2,000 m, <70–80°C). Very close to the surface and in contact with groundwater flow, the biodegradation may be oxic but for the most part is not possible to supply oxygen from the surface and we must assume that the biodegradation is anoxic. As a result the oil is less valuable and difficult to produce. Injection of steam to heat the oil and reduce its viscosity is the most effective method of increasing production from reservoirs containing heavy, biodegraded oil. Increasing oil prices have however made production of heavy oil and tar sand more profitable. Heating the oil may however require up to 30% of the energy recoved thus increasing the CO_2 emission (see tar sand and oil shales, Chap. 21). Excavation and separation of oil from sand also require much more energy than production of conventional oil.

Reservoirs buried to greater depth (>80–100°C) are sterilised by the heat and may remain without biodegradation also after uplift to lower temperatures. It is difficult to introduce new bacteria.

Further Reading

Geoff, J.C. 1985. Hydrocarbon generation and migration from Jurassic source rocks in the East Shetland Basin and Viking Graben at the North Sea. Journal of the Geological Society 140, 445–474.

Hunt, J.M. 1996. Petroleum Geochemistry and Geology. Freeman and Co., New York, 743 pp.

Karlsen, D.A., Nedkvitne, T., Larter, S.R. and Bjørlykke, K. 1993. Hydrocarbon composition of authigenic inclusions – Application to elucidation of petroleum reservoir filling history. Geochemica et Cosmochemica Acta 57, 3641–3659.

Makhous, M. and Galushkin, Y. 2005. Basin Analysis and Modeling of the Burial, Thermal and Maturation Histories in Sedimentary Basins. Editions Technip, Paris, 379 pp.

Tissot, B.P. and Welte, D.H. 1984. Petroleum Formation and Occurrence. Springer, Berlin, 669 pp.

Wapples, D.W. 1980. Time and temperature in petroleum exploration. AAPG Bulletin 64, 916–926.

Chapter 15

Petroleum Migration

Knut Bjørlykke

The transport of petroleum from the source rock to the reservoir rocks is called *migration*. It is important to understand this process so that the direction of migration and trapping of petroleum can be predicted. Many different theories have been proposed in the past but it is now clear that petroleum is mainly transported as a separated phase and that the process is mainly driven by the buoyancy of petroleum relative to water. The solubility of oil in water is very low for most compounds. The solubility of gas, particularly methane, is much higher both in oil and water and increases with depth (pressure). There is, however, also very limited flow in sedimentary basins to transport petroleum.

Considerable amounts of gas can bubble out of water or oil if the pressure is reduced due to uplift or due to pressure reduction in a reservoir during production.

15.1 Primary Migration

The expulsion of petroleum from a source rock into adjacent rocks is called *primary migration.*

Kerogen is a solid compound of very large molecules (polymers) formed from organic matter and may occur as dispersed particles in the sediments or as laminae in a claystone. Kerogen may be load-bearing and capable of transmitting stress before it generates petroleum. As the kerogen matures, much of this solid matter breaks down to generate oil or gas and is thus transformed into fluid phases. If the fluids are not expelled immediately, this process increases the volume of the fluid phase (porosity) compared to the original volume of the solid phase in the source rock.

The ratio between the volume of the fluid phase (porosity) and the solid phase is often referred to as the void ratio: $V_r = \varphi/(1 - \varphi)$.

Not all kerogen is transformed to fluids during maturation. There is a residue of solids which is called coke.

It has usually been assumed that there is a volume expansion during maturation of kerogen because the density of the oil and gas and the remaining solids in the kerogen may be smaller than the density of the primary kerogen, thus causing a volume expansion. This expansion may not necessarily be very large in the case of oil generation. Even if there was no overall volume expansion the generation of oil would contribute to the build up overpressure, since the main factor is the change in void ratio when solid kerogen is altered to fluid petroleum. To illustrate this point we may make the analogy of looking at frozen ground with lenses of ice formed during the winter. Ice is a solid that can carry the weight of the overburden, but when it melts in the spring the ice becomes fluid and a part of the porosity unless it is expelled. Overpressure develops and sometimes small mud volcanoes may form, despite the fact that there is a reduction in volume from ice to water, rather than an expansion.

Source rocks may include thin layers of siltstones or sandstones that can serve as pathways for the migration of the petroleum fluids generated from the kerogen. If, however, these more permeable layers are absent, the permeability of the shale matrix is in most cases low enough for the fluid pressure to build up

K. Bjørlykke (✉)
Department of Geosciences, University of Oslo, Oslo, Norway
e-mail: knut.bjorlykke@geo.uio.no

K. Bjørlykke (ed.), *Petroleum Geoscience: From Sedimentary Environments to Rock Physics*,
DOI 10.1007/978-3-642-02332-3_15, © Springer-Verlag Berlin Heidelberg 2010

where petroleum is generated until fracture pressure is reached. Should the source rock consist of kerogen in a fine-grained clay-rich matrix, the flow of oil out of the source rock is resisted both by very high capillary pressures and the low permeability. In such cases oil can not migrate out of source rocks through the matrix. Very thin open fractures allowing the expulsion of petroleum will develop when the fluid pressure in the source rock has reached fracture pressure. The fracture pressure is controlled by the horizontal stress (σ_h) which is in most cases lower than the overburden stress (σ_v).

The kerogen is normally not distributed homogeneously in the source rocks. Organic-rich mud deposited under reducing conditions tends to be finely laminated due to a lack of bioturbation, and some laminae may consist of almost pure kerogen (Fig. 15.1).

If these layers of kerogen then matured into petroleum which was not expelled, this fluid phase would have had to support the full overburden stress (σ_v). The fracture pressure corresponding to (σ_h) is exceeded, however, before the overburden stress (σ_v) is reached, allowing petroleum to escape through vertical fractures (perpendicular to the direction of least stress). Even if the kerogen is distributed more evenly in the source rock, the generation of fluid petroleum increases the volume of the fluid phase. Layers of source rock with 10% TOC by weight make up about 20% of the volume. If the water content of the source rock is 10%, maturation and fluidisation of 50% of the kerogen would increase the fluid content (porosity) by 100% if expulsion did not occur. The excess fluid must therefore be expelled during maturation because a shale with such high porosity would compact mechanically and thus squeeze the oil out.

The primary migration is then controlled by the rate of petroleum generation and this process therefore seems fairly unproblematic. Either the source rocks have sufficient permeability for the petroleum to migrate out through the rock matrix or hydrofracturing creates sufficient permeability for the primary expulsion. If the source rock is very lean a significant fraction of the petroleum could be retained by the source rock, but in the case of richer source rocks a relatively high percentage of the oil generated will be expelled. The actual percentage of petroleum expelled from source rocks is not well known, though.

The petroleum remaining in the source rock can not usually be produced by drilling wells because of the low permeabilities. However, it may contain large amounts of gas which will flow.

In recent years there has been a major development of shale gas production, particularly in the Devonian and Carboniferous shales of North America such as the Barnett Shale (Mississippian). Production is enhanced by horizontal drilling and artificial fracturing of the shales. The remaining oil is normally difficult to produce without mining the shale.

Oil shales are mostly source rocks that have not been buried deeply enough to become mature and expel petroleum. If they have been uplifted and exposed, they can be mined and the kerogen heated in ovens to about 500°C to generate the petroleum.

15.2 Secondary Migration of Petroleum

The flow of petroleum from source rock to reservoir rocks is called secondary migration and must be understood in terms of two-phase and in some cases three-phase flow. The relative permeability for oil or gas is then critical. At low oil or gas saturation the hydrocarbons will only occur as small droplets in the water, which will not flow because of the capillary resistance and the fact that the buoyancy effect will be very weak.

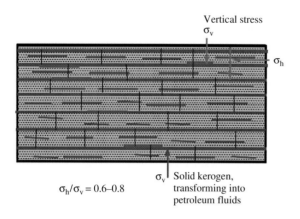

Fig. 15.1 Schematic illustration of a source rock. The kerogen often occurs as distinct laminae which are load-bearing prior to maturation and petroleum generation. The change from solid kerogen to fluid petroleum therefore creates an overpressure which may cause fracturing, helping the primary migration of petroleum out of the source rock

The migration of a separate hydrocarbon phase is limited by the capillary forces resisting flow through small pore throats; the relative permeability for hydrocarbon is then very small so that little migration can occur. If there is a pressure gradient in the fluid phase the water will flow past the petroleum droplets, which are held back by capillary forces. Asphalt-rich oil, on the other hand, which may be formed from biodegradation, sticks to the grain surfaces which are clearly oil-wet and water will flow in the pores.

On a much larger scale, a sedimentary basin will always have low average oil saturation. Regional fluid pressure gradients in sedimentary basins will only move water since at this scale it is the only continuous phase.

Once the primary expulsion from the source rock has been achieved, the oil and gas phases will flow upwards driven by buoyancy, along pathways where the petroleum is concentrated. In this way high oil saturation can be attained locally, increasing the relative permeability. Secondary migration requires that a continuous pathway with high petroleum saturation is established; where flow is prevented by high capillary entry pressure or low permeability, petroleum may be trapped in both small and large scale dead ends (micro-traps).

When oil or gas flows upwards and accumulates in traps, it is normally not accompanied by water flow. The trap is in a way a hydrodynamic "dead end", essentially because of the low permeability of the cap rock, unless this has fractured. As an oil or gas lag starts to accumulate near the top of the structure, the permeability with respect to water flow is reduced further. If permeability to water is nevertheless not zero, while the migration of oil may be prevented by capillary forces water may in some cases slowly seep through the cap rock even if the permeability is very low (Teige et al. 2005).

Reduction in overburden pressure due to uplift and erosion may bring large volumes of gas out of solution in the water or oil phase, to form separate gas accumulations. This gas may then fill the structures and displace the oil out of the traps so that oil migrates up to the surface or into a shallower trap.

Migration of both oil and gas takes place mostly as a separate phase and the driving force is the buoyancy of the hydrocarbon phase in water. The density of the petroleum (ρ_o) varies from 0.5 to 1.0 g/cm^3. The water phase may have densities (ρ_w) of 1.00–1.20 depending on the salinity. The buoyancy force ($F1$) is therefore

$$F1 = (\rho_w - \rho_o) H$$

where H is the height of the continuous petroleum column. For a 100 m high and narrow oil column with a density of 0.8 g/cm^3, the pressure in the water phase is 1 MPa and in the oil phase 0.8 MPa. The pressure difference between the oil phase and the water phase for each 100 m is thus 0.2 MPa.

It is this pressure difference which overcomes the capillary forces resisting the migration of oil and gas. The capillary forces depend on the surface properties of the grains in relation to the fluids. In water-wet sandstone, water is drawn in (imbibed) to replace oil if the oil saturation is too high, and the excess pressure in the petroleum phase helps to resist that. Most sandstones are water-wet in contact with normal oil and gas but the wetting angle may vary.

Assuming water-wet conditions the capillary resistant force is

$$F2 = (2\gamma \cos \phi)/R$$

where γ is the interfacial tension between petroleum and water, ϕ is the wetting angle and R is the radius of the pore throats between the pores which the petroleum has to pass.

In coarse-grained sandstones with relatively large pores the resistance from the capillary forces is lowest and migration will then follow such layers.

The interfacial tension between gas and water may vary from 30 to 70 dynes/cm whereas for oil the range is 5–35 dynes/cm (Schowalter 1979). The capillary force ($F2$) resisting petroleum migration is therefore higher for gas than oil, but the buoyancy force ($F1$) is also higher because of the lower density of gas. The capillary resistance to flow is as we have seen an inverse function of the critical pore radius (R).

Migration depends on the size of the pore throats along a continuous pathway for oil migration. The smallest pore throats are therefore critical for the migration, not the radius of the pores themselves.

The capillary forces can be tested experimentally by measuring the pressure required to displace water and force oil into sandstone. The pressure measured corresponds to oil columns from 0.3 to 3 m (Schowalter 1997). In siltstones the displacement pressure is much higher and in mudstones and shales the displacement pressure corresponds to hundreds of metres or even kilometres of oil column.

It is difficult to control the surface properties of oils in the laboratory. Experimentally it is easier to use mercury as the displacing fluid and mercury capillary-pressure curves (mercury injection curves) are the standard way to analyse the effective pore size distribution in reservoir rocks. Since the surface tension is known the critical pore throat radius (R) can be calculated.

In sandstones the displacement pressure increases with increasing cementation and the resultant reduction in porosity and permeability. Even relatively well cemented sandstones are not normally barriers to oil migration, though. Well-cemented carbonate layers and also thin clay layers and stylolites may serve effectively as seals for further migration. The main problem is usually not the migration through sandstones, but from one sandstone body to the next through shales. Faults in sandstones may have a clay smear, reducing the permeability and increasing the capillary entry pressure.

During progressive burial, faults and fractures do not tend to be open conduits for fluid flow, but during uplift rocks are more brittle (overconsolidated) and fractures and faults may be more open.

Shales and mudstones buried to more than about 3 km have relatively low porosity and permeability and high capillary entry pressure. The vertical extension of oil columns is limited by the depth at which oil is generated (3–5 km) and the reservoir depth. Extractions of fluids from shales in near contact with oil and condensate in reservoirs do not show evidence of oil saturation (Olstad et al. 1997), suggesting that oil has not displaced water in the shales. Gas molecules, and particularly methane, are very much smaller than those composing oils and can probably diffuse through shales, though at relatively slow rates. Both oil and water flow follows the Darcy Law ($F = \nabla P \cdot k/\mu$). The flux F is a function of the permeability k, the potentiometric gradient (∇P) and the viscosity (μ). In the case of a single fluid phase the permeability is a function of the size of the connections between the pores along the flow pathway. When two fluid phases like oil and water are present, the permeability of one fluid phase is also a function of the relative abundance of the two phases. Oil can only flow through the percentage of the fluid phase which is filled with oil

and in a water-wet rock there is a layer of water around each grain reducing the cross-section of the pore throats available for oil flow. The permeability of oil in the presence of water varies as a function of the *oil saturation*, which is the percentage of oil in the pore space, compared to the total fluid volume. Similarly the permeability with respect to water flow depends on the cross-sections of the water-filled parts of the pores between grains. The *relative permeability* is the permeability of a fluid phase in the presence of another fluid phase compared to the permeability in the same rock when only one fluid is present. If the percentage of oil in the pores is less than 20–30% the relative permeability of oil is so low that it will move very slowly or not at all compared to water. In some cases pore networks may be filled with water, oil and gas and then we have to consider 3-phase flow. Shales have normally low permeability for water and even lower relative permeability for oil. Shales are barriers to oil migration both because the capillary forces resist the flow and because of the low permeability.

15.3 Migration in Sandstones

Migration takes place along pathways with the lowest capillary entry pressures. These are in most cases sandstones or open fractures. Mostly the migration follows the upper parts of sandstone beds due to the buoyancy of oil in water. Very little of the migration occurs vertically through sandstones.

The nature of the transition between the permeable sandstones and the overlying shales is therefore very important. In a sandstone which is coarsening-upwards the maximum permeability is near the top of the sequence just below the shale (Fig. 15.2). This is the case in sandstones deposited in shallow marine environments (shoreface, beach and delta-front successions). When they are not well-cemented, these uppermost beds possess high porosity and permeability, and the oil saturation and relative permeability will therefore also be high.

In fining-upward units such as fluvial sandstones and turbidites the permeability and capillary entry pressure decreases upward. Due to its buoyancy, oil will flow along the top of sandstone or siltstone beds

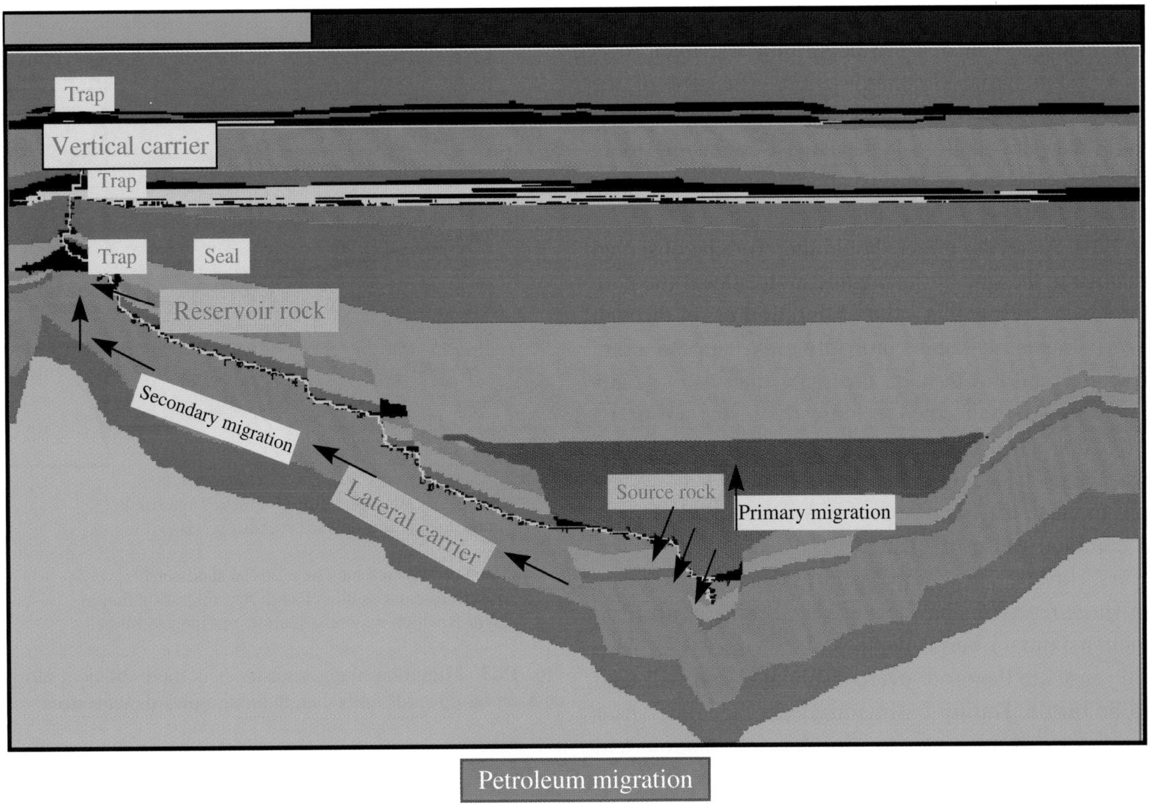

Fig. 15.2 Petroleum migration in a sedimentary basin. From primary migration out of the source rock and secondary migration up to a trap. From Steven Larter (unpublished)

against the overlying mudstones or shales. Here the permeability as well as the oil saturation will be lower in siltstones and poorly sorted sandstones and the flow rate and oil saturation is reduced compared to cleaner sands. Sedimentary structures such as cross-bedding may contain laminae that are barriers for oil because they have very high capillary entry pressure. This may produce micro-traps for the migrating oil and gas, increasing the loss during migration.

The direction of migration is up the maximum slope of sandstone beds in regular planar sand sheets which may be folded, and this may be modelled based on 3D seismic. In elongate sandstone bodies like fluvial channels and certain submarine bars, however, the direction of migration is also controlled by the sandstone geometry. Abandoned channels may become separate traps and synsedimentary faults (i.e. growth faults) may trap migrating oil and gas.

15.4 Migration of Oil Through Shales

There are rarely continuous connecting sandstone bodies from the source rock to the reservoir and petroleum must therefore migrate through shales.

Fluid flow through shales may occur in different ways:

1. Matrix-controlled (intergranular flow)
2. Flow in fractures produced by hydro-fracturing due to overpressure (mostly microfractures)
3. Flow in tectonically induced macrofractures

In the case of matrix-controlled flow the capillary entry pressure and the permeability are a function of the pore size distribution and diagenetic alterations. In shales the typical pore sizes may be a few 100 Å or less. In the Gulf Coast the pore diameter may be less than 25 Å at 3–4 km depth and less than 10 Å at 4–5 km burial (Leonard 1993). The size of the largest pores and their connections will however determine the permeability. It the largest connecting pore size is below 50 Å this is approaching the size of the asphaltenes in crude oil (Tissot and Welte 1978, p. 170). A sieving effect should therefore be observed relative to the size of the organic molecules if the pore throats are below this value. Migration of oil through low permeability shales probably only happens along fractures and not through the shale matrix. Fractures in shales may be formed tectonically during uplift and extension when the shales are brittle. Relatively large fractures may occur because the horizontal stresses trying to close the fractures are generally small. This is a common cause of oil and gas leakage from reservoirs. However, nearly all onshore reservoirs and many offshore reservoirs have experienced some uplift from their maximum burial depth and in a rock mechanical sense are therefore overconsolidated and will tend to be brittle. During basin subsidence, tectonic shear produces fractures during ductile deformation and these fractures are then no more permeable than the matrix.

Oil fields are often highly overpressured, with many of them leaking petroleum at the top of the structure which is often close to the fracture pressure. This means that it is the horizontal stress and the tensional strength of the rocks that control the pressure. When the pressure is close to fracture pressure it implies that faults are no longer the conduits for fluid flow because the rock matrix would then fracture and let the oil through. If faults were a zone of weakness the fluid pressure should have remained below fracture pressure.

In the case of traps formed by rotated fault blocks the top of the structure will usually coincide with a fault. Even if seismic evidence indicates gas leakage from the top of the reservoir, this does not necessarily mean that the leakage is along the fault, because fracturing of the cap rock will occur in approximately the same position (Fig. 15.3).

We must distinguish clearly between migration of oil *along* a fault plane and *across* it. If there is sand on both sides of the fault plane it is difficult to predict if the fault will be a barrier for oil migration. Sealing faults are often critical for the formation of traps in

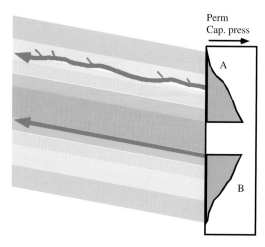

Flow of oil through a fining-upwards sandstone (A) and a coarsening-upwards sandstone (B).

In layer A which may be a fluvial sandstone or turbidite, more oil is lost during migration that in layer B which may be a shallow marine sandstone.

Fig. 15.3 Migration along sandstones is more efficient along coarsening-upwards units than in fining-upwards sequences

rotated fault blocks. Clay smears from adjacent shales can serve as a barrier at shallow depth, while at greater depth intensive quartz cementation may reduce the permeability and the capillary entry pressure for oil. See further discussion on rock mechanics in sedimentary basins (Chap. 11).

In subsiding basins, fault planes are at shallow to moderate burial depth and are subjected to shear deformation which produces clay smearing, so that the permeability along the fault is normally lower than through the rock matrix. After tectonic displacement, faults may also be subject to cementation. During tectonic uplift, however, faults may be extensional and much more permeable.

Migration due to hydrofracturing of shales. If the fluid pressure exceeds the fracture pressure the rock will hydrofracture. The pressure required to fracture the rock can be measured in a well by using a leak-off test (LOT test, see Chap. 11). However, the LOT creates microfractures which develop at a lower pressure than that required to form proper fracturing. These microfractures should theoretically develop when the pressure exceeds the sum of the least horizontal stress and the tensional strength of the rock. Since the pressure required for large scale fractures to form (fracture pressure) is higher than the LOT test value, the tensional strength which allows microfractures to form is lower than when forming proper hydrofracturing.

In the case of microfractures (LOT tests) these probably deform the rock in a different way, so that the rock can heal once the pressure is released.

Fractures developed by hydrofracturing during leakage of oil are likely to be vertical because they develop parallel to the direction of maximum stress which is normally vertical in subsiding basins with little external tectonic stress. As a fracture opens, the permeability along it is increased. This reduces the pressure gradient along the fault plane to less than the fracture gradient. The top of the fracture may therefore be above fracture pressure, while the lowest part is below fracture pressure and subject to effective stress trying to close it. The fractures produced by hydrofracturing must therefore propagate upwards and are of limited vertical extent. They develop first in the least permeable parts of the shale (cap rock) which may only contain water because of low capillary entry pressure. The pressure in these very small water-saturated pores should not be influenced by the pressure in the petroleum phase in the sandstones (Fig. 15.4).

The excess pressure in the petroleum phase compared to the water pressure, is held by the capillary forces and does not influence the pressure causing onset of hydrofracturing (Bjørkum 1998). However once the first fracturing has occurred the petroleum will be the continuous phase along the fracture, and it is the pressure in the hydrocarbon phase which causes leakage when the horizontal stress is exceeded.

In the laboratory, water has been shown to flow though a cap rock while oil has been retained by the capillary pressures (Teige et al. 2005).

If we consider migration in two dimensions it is clear that it is not only the source rock that will fracture. All the overlying shales that could serve as potential cap rocks could reach fracture pressure and leak petroleum if there are no lateral drainage paths. This is because the fracture pressure gradient is steeper than the fluid pressure gradient. Accumulation of petroleum in a trap capped by shale which does not fracture, requires that the pressure be reduced by lateral flow of water to maintain pressure below fracture pressure (Fig. 15.5).

15.5 Migration Through Tectonically Fractured Rocks

A clear distinction must be made between different types of tectonic fracturing and hydrofracturing, although these processes can occur together. It is important also to distinguish between faults formed during subsidence and those formed during uplift.

Some of the fractures observed may have been formed during soft sediment deformation just after deposition or be associated with growth faulting, and thus unrelated to tectonic stress.

During subsidence and uplift the direction of stress may have changed several times, producing fractures with different orientations. We must therefore not assume that all the fractures were open or closed at the same time. Fractures produced during late stages of uplift are most likely to be open, but they have very little relevance to the conditions during oil migration. However, rocks that are now outcropping have usually had a long history of deformation.

Numerous studies have been carried out on the pattern of fracturing in outcrops on land to serve as analogues for subsurface fracture patterns which can not readily be mapped from seismic or cores. This has then served as a basis for sophisticated fluid modelling and its consequence for oil migration, assuming in some cases that the fractures are open and more permeable than the matrix, in others that they are less permeable than the matrix.

15.6 Trapping of Petroleum Below a Cap Rock

A cap rock traps petroleum if the flow into the trap exceeds the flow out of the trap. A trap may leak petroleum through the matrix of the seal or through fractures produced by overpressure or tectonically. If there is leakage through the matrix of a shale it is because the capillary forces are not high enough to resist the buoyancy of the petroleum. As we have seen, the capillary forces are a function of the wetting angle and the pore size, and the buoyancy forces are a function of the density of the petroleum and the thickness of the petroleum column. A shale may then be able to hold a limited thickness of oil or gas column in

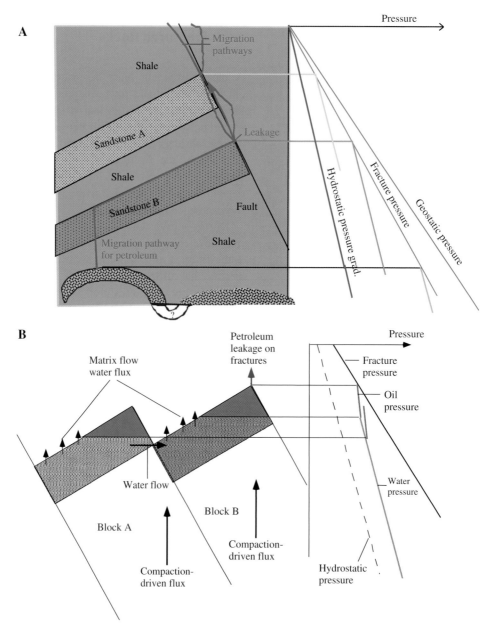

Blocks A and B are connected but separated from the rest
of the basement:

Oil leakage (Fracture flow, F1) = compaction-driven flux (Flux A + Flux B) - matrix flow (F2)

At shallow depth the matrix flow may relativly high and compaction mostly mechanical.
Fracture pressure can then only be reached by fluid supply from deeper
sediments driven by compaction.

Fig. 15.4 (**a**) Fracturing will always occur at the top of the structure. The fluid pressure will intersect the fracture gradient at the top because the pressure gradients inside sandstones are nearly hydrostatic. (**b**) If two rotated fault blocks are in pressure communication the lower block (**A**) will not reach fracture pressure and fracture. The higher block (**B**) then serves as a safety valve for block A. If they represent separate pressure cells they could both fracture and leak

Fig. 15.5 Fracturing in seals is likely to occur where the permeability for water is lowest. The fracture pressure may therefore be controlled by the water-saturated shale even if the pressure is higher in the oil-saturated zone. Water may therefore leak from the structure when petroleum is retained. Based on Bjørkum et al. (1998)

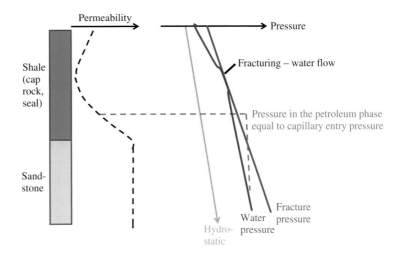

the reservoir (Schowalter 1979). There is nevertheless considerable evidence that fine-grained shales buried to 3–4 km have more than enough capillary force to hold several hundred metres of oil column. In the case of gas the density is lower, producing higher buoyancy, but the interfacial tension is also higher. Gas may also to a much larger extent be dissolved in the water phase and diffuse through the cap rock.

It may seem natural to assume that the extra pressure due to the buoyancy of the petroleum may be critical with respect to the fracturing because the fracture pressure then is reached earlier. However, the extra pressure of the petroleum column is balanced by the capillary forces and does not change the water pressure critical for reaching fracture pressure (Bjørkum et al. 1998). Also, the fracture pressure may be reached in the cap rock some distance above the top of the petroleum column (Fig. 15.4). Once hydrofracturing starts and petroleum becomes the continuous phase in the fractures, the pressure in the petroleum phase will resist the horizontal stress and will determine when the fractures open and close.

For practical purposes this may not play an important role because if there is no adequate horizontal leakage of water or petroleum, the pressure will continuously build up in the structure and the fracture pressure will be reached, causing the structure to leak. A cap rock thus serves as a valve which slowly allows water or petroleum through, maintaining a constant pressure equal to the fracture pressure. The flow is not likely to be very episodic because a high flow rate would lower the pressure and close the fractures.

15.7 The Rate of Oil Migration

Attempts have been made to calculate the rate of petroleum migration from the source rock to the reservoir. Many assumptions then have to be made about the capillary resistance and the permeability of the rocks through which the oil and gas migrate. The oil saturation and the cross-section of the oil-saturated rocks would also play an important role. There are good reasons, though, to think that the oil and gas migration normally is not rate-limiting for the accumulation of a trap.

Modelling of oil migration through sand suggests that migration may occur at very high velocities exceeding 100 km/million years and that the oil column may be thin (Sylta 1997). This modelling is, however, based on migration through sandstones with relatively high permeability (100 mD). The oil-saturated pathway may then be only a few centimetres thick. A 100 million m^3 (600 million bbl) oil field could be filled in 3 million years with oil flowing at just 1 cm^3/s, equivalent to a tiny trickle out of a garden hose. Thicker oil columns will probably build up where the migration has to pass through less permeable sandstones and siltstones which require higher capillary pressures.

The rate of migration is limited by the petroleum generation in the source rocks in the drainage area. The oil and gas can not flow faster than it is generated. If the migration is slower than the generation, a trap has formed somewhere along the migration pathway. It the rate of oil generation increases, both the cross-section of oil-saturated rocks along the pathway will increase and a thicker oil column produce higher oil saturation and relative permeability. These are positive feedbacks that will make it possible for the migration rate to increase in order to adjust to the rate of generation. Low permeability shales along the pathway will either fracture due to the high pressure or form a cap rock. A trap which does not have pressure communication with any higher structures will easily reach fracture pressure and leak. This will not apply to a lower trap if there is continuous permeable sandstone up to a higher trap; the lower trap will not then reach fracture pressure and leak (Fig. 15.3).

15.8 Regional Migration

In a basin with well-defined shales serving as cap rocks the sandstones represent the migration routes.

In the North Sea basin, Cretaceous shales are important cap rocks and the base of these Cretaceous shales can be mapped out in great detail using 3D seismics. The depth contours make the traps stand out like mountains or hills on a topographic map. Using that analogy, the migration of oil follows the highest ridges and fills the structures down to the lowest contour connecting it to the next mountain which is the spill point. The closure of the structure, which defines the maximum thickness of the oil and gas column, is represented by the vertical distance from the valley up to the top of the mountain.

In an overpressured part of a sedimentary basin the highest structures within one pressure compartment are likely to leak. Deeper structures in the same compartment may have a high chance of retaining oil because they will not reach fracture pressure and leak (Fig. 15.3).

The evaluation of a trap depends very much on the contours on this map, which is a function of the accuracy of the depth conversion of the seismic data.

Faults offsetting the carrier bed and the shale seal may obstruct migration but represent at the same time a possible trapping mechanism.

Much of what has been said about oil migration holds true also for gas migration. Gas molecules, particularly methane, are very small and may diffuse through shales and mudrocks in a way which is not possible for oil. The migration of gas above structures is now easily seen on seismic sections, showing that many gas fields do leak but are filled at a rate which compensates for the loss.

15.9 Loss of Petroleum During Migration

If we know the volume of source rocks in the drainage area and how much each volume can yield, the total volume of petroleum generated can be calculated. Assuming a certain expulsion efficiency, the maximum amount of petroleum which could migrate into a structure can be calculated. This has been done for different sub-basins in the North Sea (Geoff 1983).

The percentage of oil and gas which was generated but is not found in the major oil fields, could have been lost through leakage from reservoirs or have accumulated in smaller traps that have not been discovered. Some oil and gas is probably also left behind along the migration route.

The properties of the sandstones which serve as migration routes depend on the depositional environment and subsequent diagenesis. However, shallow marine sandstones are normally coarsening-upwards and have the cleanest and most permeable sand at the top of the sequence. There will then often be a sharp contact to the overlying shale and migration will be very efficient with relatively little loss. In fining-up sequences, commonly found in fluvial sandstones and turbidites, the buoyancy will force oil and gas up into the finer-grained part of the sandstones where much of it may be trapped in small pores.

It has been claimed that organic acids generated in the source rocks could dissolve feldspar and mica in the reservoir sandstones. Many of these acids are rather water-soluble and will diffuse into the water phase in the source rock, which often contains carbonate, and during migration would be neutralised before reaching the reservoirs.

The sandstones adjacent to or interbedded with the main source rock in the North Sea basin (Kimmeridge shale) show little or no evidence of dissolution of feldspar and precipitation of authigenic kaolinite. This suggests that the generation of organic acids in the source rocks has an insignificant effect on the development of secondary porosity (Bjørlykke and Aagaard 1992).

Compared to the total buffering system of, firstly, the silicate mineral system and, secondly, the carbonate system, the addition of relatively small amounts of comparatively weak acids (organic acids) will not change the pH of the porewater significantly.

15.10 Petroleum Seepage and Exploration

Oil exploration started by drilling where there already was evidence of seepage. This was also the case in Oil Creek in Pennsylvania, and in California.

When the reservoir is buried more deeply the amount of leakage is highly variable and the migration pathway to the surface may be very complex. The idea that the presence of petroleum should be detected directly has always seemed attractive and geochemical surveys to detect oil have been carried out on a large scale both on land and on the seafloor.

A leakage of petroleum may be a good sign because it shows that some petroleum is present. On the other hand the fact leakage occurs indicates that large fractions of the petroleum generated may have been lost to the surface. Many large oilfields have no visible seeps directly above the structures (Transher et al. 1996).

A study of the Southeast Asian Basin showed that leakage was concentrated over tectonic structures such as faults and diapirs.

The probability of detecting petroleum on the land surface or on the seabed is highest in areas of active migration. Active seeps with gas may also be detected by side-scan sonar (Hovland et al. 2002). Good examples of active seeps are found in many basins in California, the Gulf of Mexico, North Sea and Indonesia (Abrams 1996).

Passive seeps where there is at present no active seepage are more difficult to detect including by refined geochemical methods.

In overpressured reservoirs the migration is likely to be vertical due to hydrofracturing. However once the petroleum is out of the overpressured compartment further migration will tend to occur laterally, often along gently dipping sandstones in a shaly sequence.

This is well illustrated from Haltenbanken, offshore Norway, where petroleum is mostly leaking vertically from overpressured Middle to Lower Jurassic reservoirs (Karlsen et al. 1995) and migrating up dip through Palaeocene and Eocene sands up to the seafloor near the coast (Bugge et al. 1984, Thrasher et al. 1996).

15.11 Summary

Primary migration of petroleum from a source rock to the reservoir rock may occur through the rock matrix if it contains sediments with sufficiently high permeability and low capillary entry pressure. If petroleum can not flow through rock matrix the generation of petroleum will contribute to the build-up of the pore pressure until fracturing occurs.

Secondary migration has an upward component and is driven by the buoyancy of the petroleum phases in the porewater, and is resisted by the capillary forces. The rate of migration is a function of the rate of generation and primary expulsion. The rate of migration is therefore very low and the migration pathway may have a very small cross-section, but with a high degree of petroleum saturation.

Migration through sandstones occurs mostly along the top of sloping beds. The depositional environments determine the sandstone geometry, and diagenesis determines the internal flow properties. Coarsening-upwards sequences have a higher migration efficiency than fining-upwards sequences or poorly sorted sand.

Migration through shales requires that they fracture due to overpressure or tectonic deformation to overcome the capillary resistance. A cap rock must have a capillary entry pressure which is higher than the pressure in the petroleum phase.

During progressive burial to about 4 km depth (120°C) the shales often become almost impermeable and fracture pressure is likely to be reached unless the fluids can be drained laterally. Fracturing and leakage of petroleum will occur at the top of the highest structure in the pressure compartment where the fracture pressure is lowest (least overburden).

During subsidence, fault zones are not likely to provide open conduits for oil migration unless fracture pressure is reached.

Further Reading

Abrams, M.A. 1996. Distribution of subsurface hydrocarbon seepage in near-surface marine sediments. In: Schumacher, D. and Abrams, M.A. (eds.), Hydrocarbon Migration and its Near-Surface Expression. AAPG Memoir 66, 1–14.

Bjørkum, P.A., Walderhaug, O. and Nadeau, P. 1998. Physical constraints on hydrocarbon leakage and trapping revisited. Petroleum Geoscience 4, 237–239.

Bjørlykke, K. and Aagaard, P. 1992. Clay Minerals in North Sea Sandstones. In: Houseknecht, D.W. and Pittman, E.D. (eds.), Origin, Diagenesis and Petrophysics of Clay Minerals in Sandstones. SEPM Special Publication 47, 65–80.

Bugge, T., Knarud, R. and Mørk, A. 1984. Bed rock geology on the mid-Norwegian continental shelf. In: Spencer, A.M. et al. (eds.), Petroleum Geology of the North European Margin: Norwegian Petroleum Society. Graham and Trotman, London, pp. 549–555.

Geoff, J.C. 1983. Hydrocarbon generation and migration from Jurassic source rocks in the E. Shetland Basin and Viking Graben of the northern North Sea. Journal of Geological Society 140, 445–474.

Hantchel, T. and Kauerauf, A.I. 2009. Fundamental of basin and Petroleum System Modelling. Springer, New York, NY, 476 pp.

Hovland, M., Garder, J.V. and Judd, A.G. 2002. The significance of pockmarks to understanding fluid flow processes and geohazards. Geofluids 2, 127–136.

Karlsen, D., Nylend, B., Flood, B., Ohm, S.E., Brekke, T., Olsen, S. and Backer-Owe, K. 1995. Petroleum migration of the Haltenbanken, Norwegian continental shelf. In: Cubitt, J.M. and England, W.A. (eds.), The Geochemistry of Reservoirs. Geological Society Special Publication 86, 203–256.

Leonard, R.C. 1993. Distribution of subsurface pressure in the Norwegian Central Graben and applications for exploration. In: Parker, J.R. (ed.), Petroleum Geology of Northwest Europe: Proceedings of the 4th Conference. The Geological Society, pp. 1295–1303.

Lindgren, H. 1987. Molecular sieving and primary migration in the Upper Jurassic and Cambrian claystones source rock. In: Brooks, J. and Glennie, K. (eds.), Petroleum Geology of Northwest Europe. Graham and Trotman, London, pp. 357–364.

Nadeau, P.H., Bjørkum, P.A. and Walderhaug, O. 2005. Petroleum system analysis: Impact of shale diagenesis on reservoir fluid pressure, hydrocarbon migration, and biodegradation risks. In: Doré, A.G. and Vining, B.A. (eds.), Petroleum Geology: North-West Europe and Global Perspectives – Proceedings of the 6th Petroleum Geology Conference. The Geological Society, pp. 1267–1274.

Olstad, R., Bjørlykke, K. and Karlsen, D.K. 1997. Pore water flow and petroleum migration in the Smørbukk Field area, offshore Norway. In: Møller-Pedersen, P. and Koestler, A.G. (eds.), Hydrocarbon Seals – Importance for Exploration and Production. Norwegian Petroleum Society, Special Publication 7, 201–216.

Schowalter, T.T. 1979. Mechanism of secondary hydrocarbon migration and entrapment. AAPG Bulletin 63, 723–760.

Sylta, Ø., Pedersen, J.I. and Hamborg, M. 1998. On the Vertical and Lateral Distribution of Hydrocarbon Migration Velocities During Secondary Migration. Geological Society Special Publication 144, pp. 221–232.

Teige, G.M.G., Hermanrud, C., Thomas, W.L.H., Wilson, O.B. and Nordgård Bolås, H.M. 2005. Capillary resistance and trapping of hydrocarbons; A laboratory experiment. Petroleum Geoscience 11, 15–129.

Thrasher, J., Fleet, A.J., Hay, S.J., Hovland, M. and Düppenbecker, S. 1996. Understanding geology as the key to using seepage in exploration: Spectrum of seepage styles. In: Schumacher, D. and Abrams, M.A. (eds.), Hydrocarbon Migration and Its Near-Surface Expression. AAPG Memoir 66, 223–241.

Chapter 16

Well Logs: A Brief Introduction

Knut Bjørlykke

Logging is a way of recording the physical properties of the rocks penetrated by a well.

Logging started with simple electric logs measuring the electrical conductivity of rocks, but it is now a technically advanced and sophisticated method. Here only the basic principles will be introduced but there are several specialised textbooks on well logging.

A drilling mud is used to balance the water pressure in the formation and also any gas or oil which is encountered. It is also used to transport the rock fragments (cuttings) from the drill bit. The drilling mud which is lining the hole (the mudcake) or penetrated into the formation may strongly influence the readings.

One of the advantages is that a continuous downhole record is acquired, providing a detailed picture of both gradual and abrupt changes in physical properties from one bed to the next. Usually only selected parts of the reservoir rocks are cored, and samples of cuttings from the rest of the well give no more than a general idea of the lithology. Only the well logs are able to adequtely reveal the whole of the drilled sequence. Logging has the added advantage that it measures, in situ, rock properties which cannot be measured in a laboratory from either core samples or cuttings.

A borehole is logged by sending a probe with measuring instruments down a well after the drilling tool is pulled up. The measurements from the instruments in the logging tool are recorded digitally at intervals of between 3 and 15 cm and the data is processed near the well on land, or on the platform in the case of offshore wells. Well logging is carried out by specialised companies which work under contract for the oil companies. Some types of logging and seismic analyses may also occur while drilling.

Most logs (except radioactivity logs) are dependent on direct contact with the rock via the walls of the well, and have to be run after successive intervals of the drilling, before each stage of the steel casing is installed in the well.

Modern logging tools make several types of records at the same time, and the instruments are built into a long steel pipe which is only about 10 cm in diameter.

The following are the most important types of log:

1. *Electric logs* – self-potential, resistivity and conductivity logs. Electric logs were the first type to be employed in petroleum exploration, because it was fairly simple to make the measurements. This involved measuring the electric resistance (R) (resistivity) and the current that is set up between the drilling mud and the porewater in the rock (formation), i.e. the self-potential (SP).
2. *Radioactivity logs* – gamma ray and neutron logs. Gamma logs measure the natural emission of gamma rays from rocks in the well. A neutron log is obtained by using a neutron source which sends radiation into the rocks. The absorption, mostly by hydrogen atoms, occurring in water and hydrocarbons is then measured.
3. *Acoustic (sonic) logs* – measure how fast sound travels through rocks, and in particular provide information about porosity. This also indicates whether a liquid or gas phase occupies the pore spaces.

K. Bjørlykke (✉)
Department of Geosciences, University of Oslo, Oslo, Norway
e-mail: knut.bjorlykke@geo.uio.no

K. Bjørlykke (ed.), *Petroleum Geoscience: From Sedimentary Environments to Rock Physics*,
DOI 10.1007/978-3-642-02332-3_16, © Springer-Verlag Berlin Heidelberg 2010

4. *Dipmeter logs* – a type of electric log which measures the slope of beds and laminations in rocks.

Logs which directly measure properties of the well itself:

– *Caliper logs* register variations in the diameter of the well.
– *Temperature logs* – record borehole temperature and can be used to calculate the true formation temperature.
– *Image logs* provide a picture of the well wall and may reveal layering, sedimentary structures and fractures.

Well logs are used both qualitatively and quantitatively. Qualitatively, the characteristic reactions from different types of rocks are used for stratigraphic correlation, identification of sedimentary facies etc. Quantitatively it is possible on the basis of logs to determine porosity and, if relevant, the water and oil saturation of the rock. Well logs are the most important basis for correlating sequences in a sedimentary basin and for evaluating the properties of reservoir rocks and their fluid content for production purposes.

It is very important to know the type of drilling mud used, as it strongly influences what is recorded on the logs. These may be muds with seawater or freshwater and the salinity of the mud will determine the resistivity. Oil-based mud is now also used more frequently. The pressure in the drilling mud must exceed the formation water pressure, which means the mud will therefore be squeezed into the formation. A mudcake will line the borehole wall because the solid particles in the mud are concentrated there as the fluid penetrates the formation (Fig. 16.1). Further into the formation there will be a flushed zone where nearly all the primary pore fluids are replaced by the fluids from the drilling mud. Beyond this there will be a zone where the primary pore fluids are partially replaced by drilling fluids. This is called the invaded zone. In the case of formations with low permeability, such as shales, and tight sandstones and mudstones, flushing and invasion will be rather limited. In porous sandstones, however, the drilling mud and fluid may extend much more deeply into the formation, changing the formation's electrical and other properties (Fig. 16.1).

16.1 Resistivity Logs

Resistivity is a relatively simple property to measure, and resistivity logs were one of the first types to be used by the oil industry.

Resistivity logs are the result of measuring the resistance between 2 and 4 electrodes which are in contact

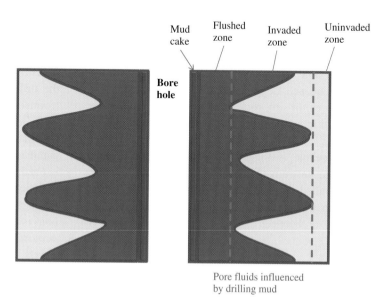

Fig. 16.1 The drilling mud will invade the formation to different degrees. This will influence all recordings from the logging tools

with the rocks in the well wall. Resistivity (in ohms) is measured as a function of the cross-section (m^2) of the rock and the distance (m) between the electrodes. The resistivity $R = ohm \cdot m^2/m$, and the unit is ohm metre.

Resistivity is the inverse of conductivity. Most minerals are very good insulators and it is only the clay minerals and salts such as KCI, NaCI which have a significant conductivity. Almost all conduction takes place through the liquid phase, and the resistance therefore depends primarily on the pore liquid and its salt content. Rocks containing porewater with high salt concentrations have a lower resistivity than rocks with fresh porewater. Conductivity is also a function of the amount of porewater relative to rock volume (hence porosity) and the distribution of pores in the rock (permeability). Each of the electric log measurements depends on the degree to which drilling mud invades the formation, because this will influence the electrical properties.

Since it is not the conductivity of the porewater itself that we are interested in, but the properties of the rocks, we measure what is called the formation factor (F):

$$F = R_o/R_w$$

where R_o is the resistivity of the rock when it is 100% saturated with water with resistance R_w. R_w is thus the resistance of an equivalent volume of just formation water.

The measured resistivity is thus a function of water saturation (S_w) which is the percentage of the porosity filled with water. If there is oil and gas in the formation, this will be detectable on account of higher resistivity.

The formation factor (F) is a function of the porosity and permeability of the rock and is an expression of rock properties independent of the conductivity of the porewater. For sediments with a high primary porosity the formation factor will be an expression of the diagenetic alteration of the rock. The relationship can be expressed:

$$F = a/\phi^m$$

where ϕ is the porosity and m is the cementing exponent which varies from 1 for porous rock to 3 for very well cemented rock (average value is c.2.0). a is a constant (tortuosity factor) which for carbonate

rocks is about 1.0 depending on the permeability (or tortuosity).

Carbonate rocks	$F = 1/\phi^2$
Consolidated sand	$F = 0.81/\phi^2$
Average sand	$F = 1.45/\phi^{1.54}$

Sediments with low salinity porewater (meteoric water) have higher resistivity than sediments with normal marine porewater. Oil, and particularly gas, will greatly increase the resistivity. Limestones are good insulators. If they are well-cemented (have low porosity and permeability), limestones have very high resistivity. This is a characteristic feature which helps to identify thin limestones that may be important for correlations. Coal beds have even higher resistivity because pure coal has virtually zero conductivity. Even relatively thin coal beds show up as strong responses on a resistivity log. The minimum bed thickness which can be registered depends on the distance between the electrodes. Pure, well-cemented sandstones have a higher resistivity than impure, clay-rich sandstones. Evaporites, including salt, are characterised by very good conductivity and low resistivity.

The water saturation (S_w) is determined from the resistivity of the rock partly saturated with formation water (R_t) and the resitivity of the rock fully saturated with water (R_0).

$$S_w = (R_0/R_t)^{1/n}$$

Here n is the saturation exponent which is usually between 1.8 and 2.5.

Since $R_0 = (F \cdot R_w)$ we obtain:

$$S_w = (F \cdot R_w/R_t)^{1/n}$$

This is called the Arches equation and for $n = 2$ this is $S_w^2 = (F \cdot R_w/R_t)$

Subsituting for the formation factor (F) we obtain:

$$S_w \left(a \cdot R_w/\phi^m \cdot R_t \right)^{1/n}$$

The resistivity is strongly influenced by the invasion of drilling mud into the invaded zone. Some logs are designed to measure the resistivity as far into the rock from the well as possible, beyond the zone impregnated with drilling mud. This is known as a *lateral log*. Another version, the *dual induction focused log*, measures the resistivity both adjacent to the well and further into the rock, comparing the paired readings to assess the effect of the drilling mud in the formation. Yet another type, the *micro log (ML)*, is used to determine the thickness of the drilling mud (mudcake) adhering to the well wall.

An induction tool is a means of focusing the induced current used for resistivity measurements. With a distance of 1 m between the coils, resistivity is measured 1–5 m into the formation away from the well.

Once we know the resistivity of the drilling mud, the porosity of the rock can be calculated from the resistivity of that part of the rock which has been invaded by the mud.

Oil and gas have far higher resistivity than water, so resistivity logs may be used to locate the contact between oil and water (OWC) and the gas/water or gas/oil contact (GWC, GOC) in a reservoir. See simplified overview of log responses in Fig. 16.2.

16.2 Spontaneous Self-potential Logs (SP Logs)

These logs measure the currents generated naturally in the rocks by the difference in the composition of the pore fluids and was one of the first measurements that was used in the oil industry.

Self-potential logs measure the electric potential which develops between an electrode which moves up and down the well and a fixed electrode near the surface. An electric current is created due to the difference in the concentrations of electrolytes in the liquid phases. We are chiefly interested in the relative variations in the spontaneous self-potential. The SP log is therefore calibrated so that the self-potential in shale in the sedimentary sequences in question gives the background value which is a straight line called the *shale baseline*. Readings normally cover a range of about 100 mV, and sandstones produce more negative values (to the left) than shale. The readings for porewater with a lower resistivity than the drilling mud will also be negative. The readings are a good indication of how pure a sandstone is, i.e. the degree of sorting, or clay

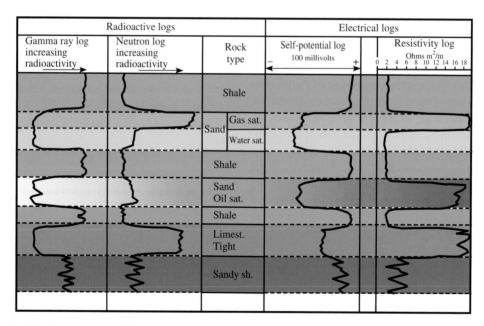

Fig. 16.2 Simplified log response to different lithologies. From Hobson and Tiratsoo (1981). The resistivity is particularly sensitive to the oil saturation and therefore the determination of the gas/water contact (GWC) and the oil/water contact (OWC)

content. SP logs are well suited for interpreting sedimentary facies because coarsening- and fining-upward sequences appear very distinctly on the logs. SP logs are often used as an expression of sorting and permeability in clastic sediments which again controls the conductivity. Sandstones with relatively fresh porewater will give less negative readings than sandstones with marine porewater, because the difference with respect to porewater is less and if the formation water is fresher than the mud filtrate, the values will be positive. When using an oil-based mud, readings can not be obtained with SP logs.

Sandstones with a high oil or gas saturation will have low SP response because of low conductivity in the water phase. The SP log responds to beds which are more permeable than shale, but only if the porewater has a different resistivity to the mud filtrate.

If we have porous sandstones the SP value can be used to determine the resistivity of the water.

In most cases readings from limestones lie between those which correspond to a shale and to a well-sorted sandstone, depending on porosity.

Self-potential is generated from porewater over a certain depth range, and SP logs do not generally give the best bed resolution. Partly for this reason, SP logs are often replaced by gamma-ray logs, which give very much the same information with better bed resolution.

16.3 Radioactive Logs

16.3.1 Gamma Logs

Gamma-ray logs measure the natural radioactivity which is produced in the rock. The elements which produce gamma radiation of significance in ordinary sedimentary rocks are potassium, thorium and uranium. The relative contribution to the total recorded radiation is such that 1 ppm uranium corresponds to 3.65 ppm thorium and 2.70% potassium. Shale normally contains the most of these elements and the gamma reading of shales is almost always higher than that of sandstones. Even though the potassium content of clay minerals varies a good deal, and sandstones also contain potassium, the gamma log will give readings which are basically a function of the sand/shale ratio. In consequence it is very similar to the SP log.

Sandstones with a high content of feldspar and mica will have proportionately greater gamma intensity than purer, quartzose sandstones. The Jurassic sandstones from the North Sea have rather unusually high gamma-ray readings due to their high content of mica and K-feldspar. Glauconite will also produce an intense gamma-ray response due to the potassium content. Some sandstones may have high contents of uranium or thorium. In most cases however the gamma log is a good measure of the sand/shale ratio and will show similar trends to those obtained by SP logs.

Black shales in particular, with their substantial organic content, produce marked reactions on the gamma log because they normally have higher uranium content than other shales. The Kimmeridge (Upper Jurassic) shale in the North Sea contains 2–10 ppm uranium, and the contribution to gamma radiation from the uranium is therefore high. It is therefore often referred to as the "hot shale". Normally shales contain less than 1 ppm/uranium, but 10–12 ppm thorium, which represents a very significant percentage of the total radioactivity (c.50%). Gamma radiation is measured in API units from 1 to 200 and this scale can be calibrated with a standard radiation intensity (micrograms of radium per tonne).

Limestones have very low concentrations of U, Th and K, and give very low gamma-ray responses. In evaporite sequences, however, gamma logs are very sensitive indicators of potassium salts.

In recent years the gamma-ray logging tool has been equipped with a scintillation counter capable of distinguishing different energy levels (expressed in MeV of gamma radiation from rock), so that it can distinguish between the relative contributions of K (1.46 MeV), U (1.76 MeV) and Th (2.62 MeV). Such logs are called spectral gamma-ray logs and by using only the Th log, for example, it is often possible to obtain a better estimate of the shale content, because the concentration of thorium in shales varies less than that of uranium and potassium.

The gamma ray index is calculated from the minimum gamma ray readings in clean sand.

The maximum reading in the shales is recorded. The percentage of shale (Vsh) can then be calculated.

Source rocks usually contain relatively high concentrations of uranium and they are often referred to as "hot shales". This is because uranium dissolved in oxidised seawater is reduced and concentrated on organic matter.

16.4 Neutron Logs

The neutron log method is based on a probe which emits neutrons at high velocity. The neutron rays are absorbed by rock, and particularly by the water in the rock. This is due to collisions with atomic nuclei, and the absorption of the neutron radiation is primarily a function of hydrogen atom concentrations (Hydrogen Index). The reduction in neutron radiation at a specified distance from the neutron source can then be measured. The collision frequency of the neutrons can also be recorded by measuring the secondary X-ray radiation created by the absorption of the neutron radiation. Since most of the hydrogen in rocks is present as water, neutron logs provide an expression of the water content and thereby the porosity of a sediment. Neutron logs are particularly useful for determining the porosity of shales but clay minerals also contain hydrogen in their structure. As opposed to SP logs and resistivity logs, neutron log response is not dependent on permeability. Gas is less dense and has fewer hydrogen atoms per unit volume than water and oil, and therefore has a lower hydrogen index. Neutron logs can therefore be used to detect gas and distinguish it from oil. Rocks with little water (low porosity) will absorb correspondingly less of the radiation and therefore produce a strong response on the neutron log. Porous sandstones by contrast produce little response (Fig. 16.2). Calculation of porosity based on neutron logs (neutron porosity) results in too low porosity values when the pores are filled with oil and gas because they contain less hydrogen per unit volume compared to water. This is called the gas effect.

In shales and sandstones with high clay content the neutron logs record higher porosity because hydrogen is also present in the clay minerals which is the solid phase. This is called the shale effect and is most pronounced in mudstones with high content of smectite and kaolinite compared to those with mostly illite and chlorite. Limestones however give very reliable porosity values because carbonate minerals contain little hydrogen.

Organic matter such as coal and other kerogens also has a high hydrogen index. The hydrogen index of limestones and sandstones can be converted into neutron porosity units, and neutron logs are the best logging tool for determining the porosities of reservoir rocks.

Neutron porosity may be presented as PHIN, NPHI or φ_N.

Compensated Neutron logs (CNT or CN) give the porosities in percent after compensating for borehole irregularities.

Using NMR logs based on nuclear magnetic resonance can distinguish between free water in the pore space and H_2O and OH groups in minerals. Also bound water on mineral surfaces has a different NMR signature.

16.5 Sonic or Acoustic Logs

With this method a probe sends out acoustic pulses which travel through the rock surrounding the well to the other end of the logging tool, and the velocity of sound in the rock is recorded. The velocity is usually presented as the time Δt a signal takes to travel a certain distance, which is the inverse of velocity (slowness) This is called "interval transit time", and is presented on the logs on a scale of 40–140 $\mu s/ft$ $(\mu s = 10^{-6} s)$ or $\mu s/m$. 100 $\mu s/ft$ corresponds to 10,000 ft/s, or 3,048 m/s. The interval transit time (t) is the reciprocal of the sonic transit velocity (v). Since the velocity of sound in water, which here means porewater, is considerably lower than it is in minerals and rocks, the measured velocity will be more or less inversely proportional to the rock porosity.

The velocity is also dependent on how the rock is cemented, i.e. the pore distribution and the nature of the cementing minerals.

Wylli's equation expresses the relationship between the velocity of sound and porosity:

$$1/v_r = (1 - \varphi)/v_m + \varphi/v_f$$

where v_r is the measured velocity recorded on the log, φ is the porosity, v_m is the velocity in the minerals or rock at zero porosity and v_f the velocity in the pore fluid.

Using the interval transit time $(t = 1/v)$ we obtain:

$$t_r = (1 - \varphi)t_m + \varphi t_f$$

and the porosity (φ) can be calculated: $\varphi = (t_r - t_m)/(t_f - t_m)$

The porosity ϕ can be calculated based on the log velocity (t_r) if we have reliable values for the interval transit time for the rock matrix t_m and the interval time (t_f) of the pore fluid (gas, oil or water).

The velocity measured by the velocity log is however not a direct function of the porosity. In sandstones, small amounts of cement (i.e. quartz cement) may produce a grain framework with high stiffness and velocity despite relatively high porosity.

In mudstones and shales the porosity and velocity vary greatly as a function of the clay mineralogy and the presence of carbonate or quartz cement.

In addition to p-wave velocities (v_p), shear-wave velocities (v_s) are now also often recorded.

This makes it possible to analyze the v_p/v_s ratio which is important for the interpretation of rock properties and also the fluid saturation.

16.6 Density Logs

Density logs measure the density of the rocks and their pore fluid by measuring the electrodensity of a rock. Gamma rays from cobolt-60 or cesium-137 are focused on the formation, and their attenuation due to collisions with electrons (Compton scattering) measured by a separate detector.

The electron density is very closely related to the rock density expressed in g/cm^3. Density logs produce important information which helps to identify different lithologies as a function of their densities.

If the density of the minerals (ρ_m), the bulk rock density (ρ_b) and the fluid density (ρ_f) (oil, gas or water) are known, the porosity can be calculated:

$$\text{Porosity } (\varphi) = \frac{\rho_m - \rho_b}{\rho_b - \rho_f}$$

The bulk density of a rock also depends on the pore fluid and in gas reserves the density of gas must be used when calculating porosity. At the same time the gas/oil or gas/water contact can be detected as a change in bulk density if it occurs in an homogeneous part of the reservoir rock.

16.7 Dipmeter Logs

Dipmeter logs are used to measure the dip of beds or laminations, for example cross-bedding within a bed. Three electrodes set at angles of 120° to each other measure the spatial orientation of the electrical properties of the rocks. Dipmeter measurement makes it possible to measure the dip of the beds in a single well. Because the dipmeter measures the spatial orientation of physical properties in rocks, this method is also be used to measure tectonic deformation of rocks, and not least the primary orientation of sediment particles (fabric). This is useful in beach sediments as the long axis of sand grains tends to be parallel with the shoreline, while in fluvial sandstone the sand grain will be oriented parallel with the transport direction. Dipmeter logs can therefore be very useful for reconstructing sedimentary environments, transport direction etc.

There are many uncertainties involved in interpreting dipmeter logs, however, and it is not always possible to find dipmeter patterns that are compatible with sedimentological models of cross-bedded strata.

16.8 Temperature Logs

Geothermal gradients in sedimentary basins are extremely important for the calculations necessary for predicting kerogen maturation, and also for the general modelling of basin subsidence. When logging a well, the temperature of the mud may be recorded as one of the physical variables. The temperature is measured during logging but it is influenced by the temperature of the drilling mud.

Because the mud is circulating, the drilling mud will not be in thermal equilibrium with the formation water. When the circulation of drilling mud is stopped, it will have risen in temperature, slowly approaching that of the formation water. By plotting the increase in temperature against time after circulation has stopped, the correct formation temperature can be estimated. However, it takes several days or even weeks before the temperature of the drilling mud approaches that of the formation water, and there is therefore considerable uncertainty involved in such calculations. The temperatures measured during logging are called "bottom-hole temperature" (BHT). Before using such data one should see if the proper corrections have been made.

. If the total heat flow (Q) in the basin is known from other measurements, the conductivity (k) of the sediment can be calculated from the temperature log data which records the geothermal gradient (∇T)

$$\text{Conductivity } (k) = Q/\nabla T$$

16.9 Caliper Logs

- Caliper logs measure the diameter of the borehole. Some also record the three-dimensional shape of the borehole walls. Normally caliper logs are presented as hole diameter in inches, and thus record cavities where the well has caved in, and also the hardness of the rocks cut during drilling. Holes in clay and poorly compacted shale will have a larger diameter than the drill bit due to caving in after drilling and erosion by the drilling mud. Where we have porous sandstone, a mud cake may develop, causing the hole diameter to become smaller.
- Variations in diameter of the hole and cave-ins do influence the records of the different logs described above. Therefore it is important to consult the caliper log to identify any artifacts due to variation in the geometry of the hole which might appear on the other types of logs.

Well logging normally takes place before the steel casing is set, to obtain contact with the formation, but gamma rays will penetrate the steel so that logging can be performed after setting the casing. It is now also possible to have sensors combined with the drill string so that we may obtain logging and other data while drilling (LWD). This is called logging while drilling (LWD) and measurements while drilling (MWD).

16.10 NMR Logs (Nuclear Magnetic Resonance)

This method can characterise the fluids as well as the solid components in a rock.

It will measure porewater also in isolated pores. Adsorbed water on clay minerals etc. produces a different signal compared to free water and therefore distinguishes between irreducible and moveable fluids.

It can identify and type hydrocarbons and estimate the HC saturation.

16.11 Image Logs

Bore hole images are pictures of the rocks from inside the well and require a good light source. Video cameras may record the lamination and bedding, and because they are oriented, palaeocurrent direction may be inferred. This method may be very useful recording fractures and large pores (vugs). It is important to detect open fractures and also partly healed fractures which can not be closed by increased stress. Electrical images are similar to those obtained by dipmeter logs. Acoustic images recording the sound waves which are bounced back from the borehole wall may also be used.

Modern image logs provide a good overview of the lamination and also fracture pattern.

This may be compared with sedimentological logging of cores (Fig. 16.3).

16.12 Horizontal and Deviated Wells

In vertical wells the hole has in most cases a high angle relative to bedding. This means that the lateral logs record changes in the properties of the same beds away from the hole. In the case of horizontal wells however the properties of *different* layers are recorded away from the well. The rock mechanical stability of horizontal wells is more critical because the layering has a low angle to the well.

In vertical wells the boundaries between different beds may be relatively sharp but when the deviation angle is high so that the wells penetrate the sedimentary layers at low angle the logs will measure a gradual transition between a sandstone and shale even if the contact is sharp.

Clastic core description

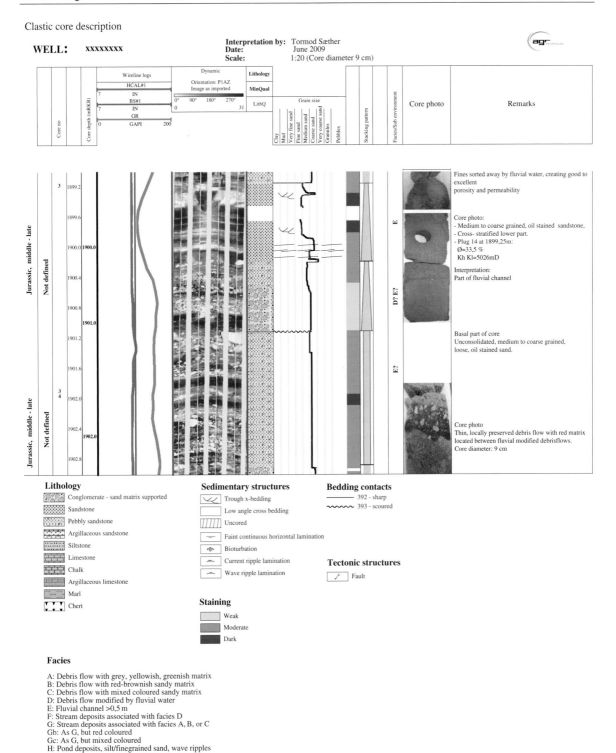

Fig. 16.3 Clastic core description from a North Sea well with image logs through a coarse clastic section. Courtesy from Tormod Sæter and Lundin Petroleum

16.13 Interpretation of Depositional Environment by Means of Well Logs

In addition to providing information about the porosity and permeability of rock types, well logs can be used directly for interpreting depositional environments. Gamma logs and self-potential logs record characteristic coarsening-upward and fining-upward sequences very efficiently. We can therefore often recognise fluvial channels (fining-upward) and shallow marine deposits (coarsening-upward). Turbidities give a characteristic alternation between coarse and fine material.

In delta sequences coal or lignite beds give characteristically strong deflections on the resistivity log.

It is important, however, to bear in mind that the response of the logs is not due to sedimentological parameters such as grain size and sorting, but can be indirectly correlated with these parameters. The gamma-ray log does not measure the clay content, but the content of radioactive gamma-ray emitters in the sediments. Sand frequently also has a high concentration of such elements (uranium, potassium, and thorium). Fluvial channels should have a nice "bell-shaped" curve due to a sharp erosional base and a fining-upward trend which results in increasing clay content and consequently an increasing gamma response upwards. However, fluvial channels often have clay or shale clasts near the base which will tend to destroy this simple pattern.

Sandstones with a high content of feldspar and mica will have a relatively high gamma response due to high potassium contents and may be interpreted as sand rich in clay.

Where core are available the well logs' responses may be combined with the core logs in the interpretation of facies and reservoir properties (Fig. 16.4).

Log data and other types of information are available on the web from many areas.

For the Norwegian offshore see www.npd.no (Fig. 16.5).

16.14 Summary

Well logs may be used both as a visual qualitative tool to identify the main rock types and also as a basis for sedimentological and stratigraphic interpretation.

More quantitative calculations of porosity, density and oil and gas saturation can also be made based on log data. Crossplots which are calibrated for different types of lithologies are then required. These methods will not be included here, but there are several good textbooks on well logging. In oil companies special programs are used for calculations.

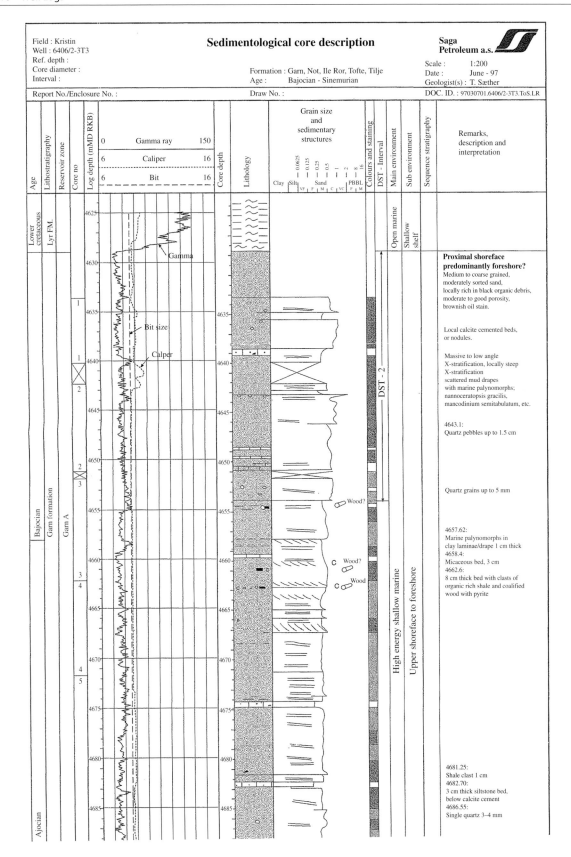

Fig. 16.4 Sedimentological log from Haltenbanken, Offshore Norway, produced by Tormod Sæter

Fig. 16.5 Log from a North Sea well. The remaining part of the log can be obtained from http://www.npd.no/engelsk/cwi/pbl/ wellbore composite_logs/4054.pdf along with a very large data base for logs, seismics, and data on exploration and production

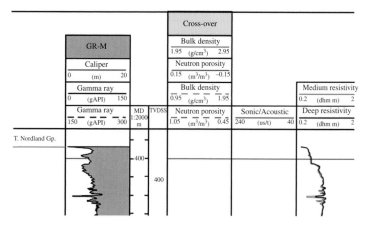

Wellbore 35/11–12

RKB 24 m

For more information see the NPD factpages at www.npd.no.

Find more on www.npd.no

SP (gamma ray) and resistivity response to sedimentary environments

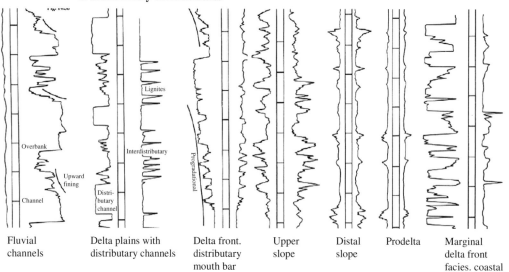

| Fluvial channels | Delta plains with distributary channels | Delta front. distributary mouth bar | Upper slope | Distal slope | Prodelta | Marginal delta front facies. coastal |

Fig. 16.6 Log response as a function of sedimentary environments in a deltaic setting (Missisippi delta). Based on work by W. Fisher. These are SP (on the *left side*) and Resistivity on the *right side*. The Gammay log will have a similar response as the SP log

Further Reading

Asquith, G. and Krygowski, D. 2004. Basic well log analysis. AAPG Methods in Exploration Series 16, 244 pp.

Hearst, J.R., Nelson, P.H. and Paillett, F.L. 2000. Well Logging for Physical Properties. Wiley, New York, 483 pp.

Hurst, A. Griffits, C.M. and Worthington, P.F. 1992. Geological applications of wireline logs.II. Geological Society Special Publication 65, 406 pp.

Kulander, B.R., Dean, S.L. and Ward, B.J., Jr. 1990. Fractured core analyses. AAPG Methods in Exploration Series 8, 88 pp.

Luthi, S.M. 2000. Geological Well Logs: Use in Reservoir Modeling. Springer, Berlin, 271 pp.

Rider, M.H. 2004. The Geological Interpretation of Well Logs. Blackie, Glasgow, 280 pp.

Serra, O. 2007. Well Logging and Reservoir Evaluation. Editions Technip, Paris, 250 pp.

Chapter 17

Seismic Exploration

Nazmul Haque Mondol

17.1 Introduction

Of all the geophysical exploration methods, seismic surveying is unequivocally the most important, primarily because it is capable of detecting large-scale to small-scale subsurface features. Simply stated, seismic methods involve estimation of the shapes and physical properties of Earth's subsurface layers from the returns of sound waves that are propagated through the Earth. Early wildcatters found oil by drilling natural oil seeps and large folds (anticlines) in exposed rocks. These easy oil prospects were all quickly discovered and drilled, and geologists then turned to seismic surveys to find less obvious oil and gas traps. Seismic technology had been used since the early 1900s to measure water depths and detect icebergs, and by 1924, seismic data were first used in the discovery of a Texan oil field (Milligan 2004). Several introductory and advanced textbooks (e.g., Telford et al. 1990, Sheriff and Geldart 1995, Yilmaz 2001) describe the principles of acquisition, processing and interpretation of seismic data. This chapter reviews the fundamental concepts employed in seismic exploration.

In general, two types of seismic method (reflection and refraction) are common, with reflection seismic the most widely used technique in hydrocarbon exploration. This technique provides an image of the subsurface in two or three dimensions (2D or 3D) (Fig. 17.1). The subsurface seismic images are produced by generating, recording and analysing sound waves that travel through the Earth (such waves are also called seismic waves). The density and velocity changes between rocks reflect the waves back to the surface, and how quickly and strongly the waves are reflected back indicates what lies below. Seismic pulses for exploration surveys are generated in one of three ways, employing an air-gun, vibrator or dynamite. An air-gun source is used for marine acquisition whereas vibrator and dynamite are the common sources for land seismic surveys. The strength of pulses associated with different seismic surveys varies, depending on site-specific factors such as rock types, how deep the survey needs to image and the required source.

17.2 Basic Principles

The first step in seismic exploration is to acquire data, which in most cases is carried out from the surface. To understand the seismic data, a review of the physical principles that govern the movement of seismic waves through layered media is necessary. A seismic source at any point on the Earth generates four types of seismic waves: compressional (P-wave), shear (S-wave), Rayleigh (ground roll) and Love, that travel through the layers (Fig. 17.2). Each layer will have a specific density and velocity. Rayleigh and Love waves are surface waves and propagate approximately parallel to the Earth's surface. Although surface waves penetrate to significant depth in the Earth, these types of waves do not propagate directly through the Earth's interior and have limited significance in oil and gas exploration. On the other hand, P- and S-waves are often

N.H. Mondol (✉)
Department of Geosciences, University of Oslo;
Norwegian Geotechnical Institute (NGI), Oslo, Norway
e-mail: nazmul.haque@geo.uio.no

K. Bjørlykke (ed.), *Petroleum Geoscience: From Sedimentary Environments to Rock Physics*,
DOI 10.1007/978-3-642-02332-3_17, © Springer-Verlag Berlin Heidelberg 2010

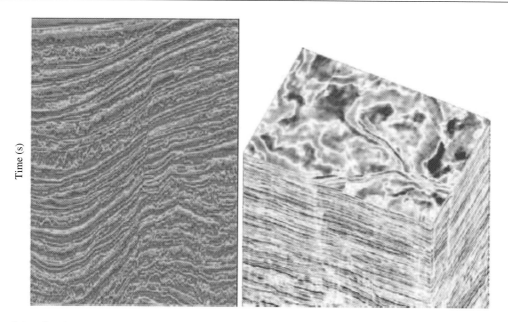

Fig. 17.1 Subsurface imaging by 2D (*left*) and 3D (*right*) seismic reflection data

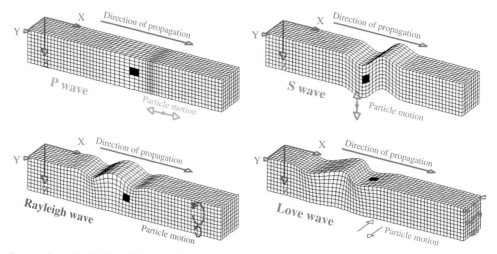

Fig. 17.2 Propagation of body (P and S) and surface (Rayleigh and Love) waves as a function of particle motions. P-waves shake the ground in the direction they are propagating, while S-waves shake perpendicularly to the direction of propagation. Rayleigh waves shake the ground both in the direction of propagation and perpendicular (in a vertical plane) so that the motion is generally elliptical – either prograde or retrograde. Love waves shake the ground perpendicular to the direction of propagation and generally parallel to the Earth's surface (Source: Braile 2000)

called body waves because they propagate outward in all directions from the source and travel through the interior of the Earth and have great significance in seismic exploration. P-waves move faster than S-waves. The P-wave is a longitudinal wave, the force applied in the direction that the P-wave is travelling. The ground must move in that direction. The ground or Earth is incompressible, so the energy is transferred pretty quickly. In the S-wave, the medium is displaced in a transverse way (up and down – compared to the line of travel), and the medium must move away from the material right next to it to cause the shear and transmit the wave. This takes more time, which is why the S-wave moves more slowly than the P-wave in seismic events. S-waves do not travel through fluids as fluid has no shearing capacity.

The basic principle of seismic survey is to initiate a seismic pulse from a *seismic source* at or near the Earth's surface and record the amplitudes and travel times of waves returning to the surface after being reflected or refracted from the interface(s) of one or more layers. When a seismic source emits a pulse that propagates through the sedimentary layers, the sound waves travel between the layers with different velocities and will be refracted according to Snell's law:

$$\sin\theta_1 / \sin\theta_2 = V_2/V_1 \tag{17.1}$$

where V_1 and V_2 are the velocities of the first and second media, $\sin\theta_1$ and $\sin\theta_2$ are the sines of the incidence and refracted angles, and θ_3 is the reflected angle (Fig. 17.3). Snell's law describes the changes in the direction of a wavefront as it travels in media of different velocities. If the seismic wave is incident at an angle, both reflected and refracted P- and S-waves will be generated at an interface between two media. However, at a fluid-solid interface like the seafloor, S-waves will not exist in the fluid part.

A line or grid of geophones or hydrophones called *seismic receivers* records the reflected and refracted seismic signals. Reflections from the layer interfaces

in the subsurface are then measured at receivers (time measurements). If the two layers have different velocities, they will as a rule also have different densities, and part of the acoustic energy will not be refracted, but reflected. How much of the energy is reflected depends on the difference in the *impedance* [P-impedance (Z_p) or S-impedance (Z_s)], which are the product of P-wave (V_p) or S-wave (V_s) velocities and density (ρ). The *reflection coefficient* (R) of a normally incident P-wave on a boundary is given by:

$$R = \frac{\rho_2 V_2 - \rho_1 V_1}{\rho_2 V_2 + \rho_1 V_1} \tag{17.2}$$

where ρ_1 and ρ_2 are the densities of the upper and lower layers, V_1 and V_2 are their respective P-wave velocities, and $\rho_1 V_1$ and $\rho_2 V_2$ are the P-impedances of the upper and lower layers respectively. Therefore, anything that causes a large contrast in impedance in the target zone can cause a strong reflection (Fig. 17.4). The possible candidates include changes in *lithology, porosity, pore fluid, degree of saturation* and *diagenesis*. We see that the greater the difference in density and velocity of two layers, the greater the amount of energy which will be reflected. Sandstone will often have significantly different acoustic impedance from shale, and a considerable amount of sound energy will be reflected from the boundary between a sandstone bed and a shale bed. Limestones will tend to have both high velocities and high densities. The result will be even greater contrast in acoustic impedance between limestones and, for example, shales. However, this contrast will always depend on the porosity of the limestone, though even rather porous limestones have relatively high velocities because they are usually well cemented.

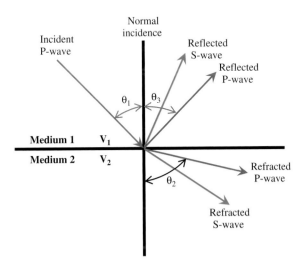

Fig. 17.3 A schematic diagram of reflected and refracted waves generated from an incident P-wave. The angle between the normal to the interface of two media and an incident P-wave is the angle of incidence (θ_1), and is equal to the angle of reflection (θ_3) in isotropic media. The angle of refraction (θ_2) depends on the velocity of the wave in that medium

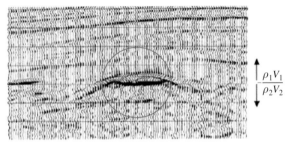

Fig. 17.4 A seismic section showing a large impedance contrast within the target zone (marked by *red circle*)

17.3 Seismic Sources and Reservoirs

It is essential at this point to understand the principle of seismology. Seismology is based on the transmission of sound waves by the rocks of the crust. Strong earthquakes create pressure waves (natural sources of seismic waves) that are transmitted through the entire Earth and detected by seismographs (receivers) on the other side (Fig. 17.5). Seismic exploration, however, as employed by the petroleum geologist, makes use of artificially generated pulses *(seismic sources)*. The principle is simple; an impulse source sends acoustic energy into the Earth. This energy propagates in many directions and is reflected and refracted when it encounters boundaries between two layers. Sensors *(seismic receivers)* placed on the surface measure the reflected or refracted acoustic energy. These artificial sources are much weaker than the natural seismic source (earthquake) but are more focused towards areas of specific stratigraphic interest.

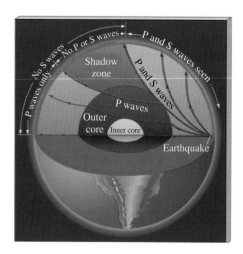

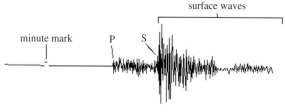

Fig. 17.5 A cross-section of the Earth with earthquake wave paths defined and their shadow zones highlighted (*top*). A typical seismogram (*bottom*) shows the fastest P-waves. The next set of seismic waves is the S-waves. The surface waves (Love and Rayleigh waves) travel a little slower than S-waves (which, in turn, are slower than P-waves) and have lower frequency

17.3.1 Seismic Sources

Different seismic sources are usually used in land and marine acquisitions. In marine environments seismic energy is normally generated using arrays of air-guns, whereas in land seismic one often uses explosives or vibrators. An *air-gun* is a device that releases highly compressed air (at typically 2,000–5,000 psi) into the water surrounding the gun (Fig. 17.6b). A *vibrator* is an adjustable mechanical source that delivers vibratory seismic energy into the ground (Fig. 17.6a). A vibrator source sends a controlled-frequency sweep into the ground. The recorded data are then convolved with the original sweep to produce a usable signal. *Dynamite* – a combination of explosive and detonator, is used as a seismic source. The detonator helps to ignite the explosives. When dynamite ignites, a shock wave propagates with a speed of 3,000–10,000 m/s. It provides an impulsive energy that can be converted into ground motion. It is customary to drill a hole to load dynamite and fill it with heavy mud before shooting. Dynamite can generate usable signal strengths and a bandwidth that covers a wide spectrum of seismic energy. It includes a variety of energy sources based on varying explosive output parameters to meet geological and climatic conditions.

17.3.2 Seismic Receivers

Hydrophones and *geophones* serve as receivers for seismic signals. The *hydrophone* is a device designed for use in detecting seismic energy in the form of pressure changes in water during marine seismic acquisition (Fig. 17.7a). It measures pressure variations with the aid of piezoelectric material, which generates a voltage upon deformation. The two piezoelectric elements in one hydrophone are connected and polarised so that voltages due to pressure waves (returning signal) add and voltages due to one-directional acceleration will cancel. In this way the influence of movements due to currents, wave action and so on will be minimised. Hydrophones are combined to form streamers that are towed by seismic vessels or deployed in a borehole. A typical length of a streamer is about 4–6 km where a single receiver section is typically 75 m long and contains 96 hydrophones which

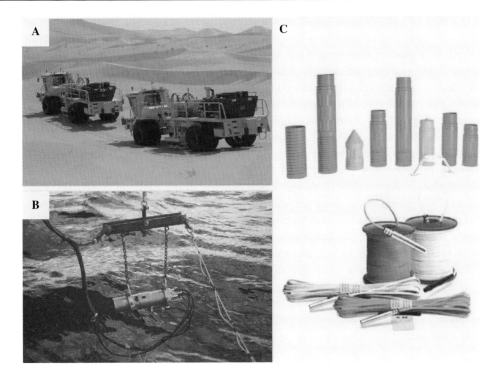

Fig. 17.6 (a) Seismic acquisition in a desert where vibrator is the most common source of seismic energy. (b) An air-gun before deployment in the water. It releases compressed air into the water during marine seismic survey. (c) Explosives (*top*) combined with detonators (*bottom*) form dynamite which is a common source of seismic energy in land acquisition where vibratory trucks can not be used due to rugged topography

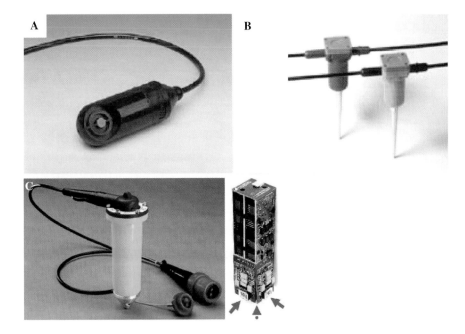

Fig. 17.7 (a) Hydrophone, (b) geophone and (c) multi-component geophone (Courtesy ION Geophysical)

are grouped in arrays of a pre-defined length, mostly 12.5 or 25 m.

The *geophone* is a device used in surface seismic acquisition, both onshore and on the seabed offshore, that detects ground velocity produced by seismic waves and transforms the motion into electrical impulses (Fig. 17.7b). Geophones, unlike hydrophones, detect motion rather than pressure. Conventional seismic surveys on land use one geophone or a group of geophones per receiver location to detect motion in the vertical direction. The three-component (3C) geophone is used for direct measurements of shear waves at the seafloor. An essential feature of a seafloor seismic acquisition is the four-component (4C) detector unit, which includes a hydrophone and a three-component (3C) geophone (Fig. 17.7c). The hydrophone and the vertical geophone measure pressure waves, whereas the two extra horizontal geophones measure particle velocity associated with shear wave energy.

17.4 Seismic Acquisition

The seismic survey is an essential part of the whole cycle of petroleum exploration and production. Seismic surveys are carried out on land and in transition zone, shallow marine and marine environments in different ways. The basic principle is an impulse source such as dynamite, air-gun or vibrator that sends acoustic energy into the Earth. This energy propagates in many directions. Downward travelling energy reflects and refracts when it encounters boundaries between two layers with different acoustic properties (Fig. 17.3). Sensors or geophones placed on the surface measure the reflected acoustic energy, converting it into an electrical signal that is displayed as a seismic trace (Fig. 17.8). The typical recorded seismic frequencies are in the range of 5–100 Hz. P-waves are the waves generally studied in conventional seismic data. P-waves incident on an interface at other than normal incidence angle can produce reflected and transmitted S-waves. S-waves travel through the Earth at about half the speed of P-waves and respond differently to fluid-filled rocks, and so can provide different additional information about lithology and fluid content of hydrocarbon-bearing reservoirs.

The recorded *seismic trace* is a *convolution* (∗) of the *source signal* and the *reflectivity sequence* of the Earth plus *noise* (Fig. 17.9). A seismic trace can simply be expressed by the Eq. (17.3) where multiples are not considered. Transmission losses and geometric spreading are not included and the frequency-dependent absorptions are also ignored in the equation.

$$S = W * R + \text{Noises} \qquad (17.3)$$

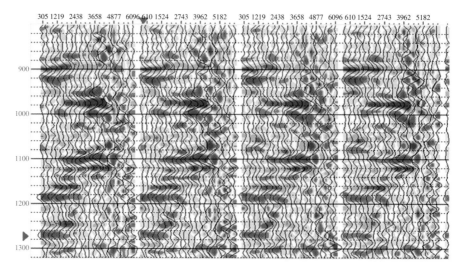

Fig. 17.8 Recorded seismic traces. Each trace consists of one recording corresponding to a single source-receiver pair. Seismic energy recorded at the receivers arrives at different times because of distance of receivers from source. In conventional acquisition, strings of geophones hard-wired together average the individual sensor measurements and deliver one output trace

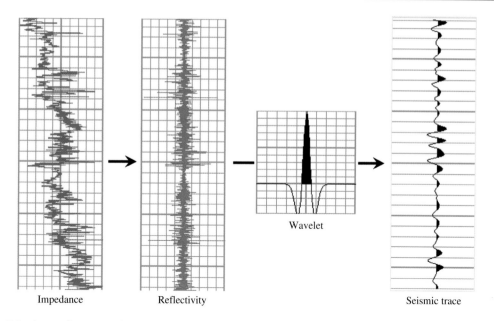

Wavelet

Impedance   Reflectivity   Seismic trace

Fig. 17.9 Seismic trace is a result of convolution of a wavelet and the reflectivity series plus noises

where S is the recorded seismic trace, R is the reflectivity and W is the wavelet. A *wavelet* is a kind of mathematical function used to divide a given function into different frequency components and study each component with a resolution that matches its scale. Accurate wavelet estimation is absolutely critical to the success of any seismic inversion. The inferred shape of the seismic wavelet may strongly influence the seismic inversion results and therefore subsequent assessments of the reservoir quality. *Attenuation* (amplitude loss) of seismic waves is an important phenomenon and caused by three major factors: (a) Geometric spreading: progressive diminution of amplitude (proportional to the inverse of propagation distance) caused by the increase in wavefront area, (b) Intrinsic attenuation: energy losses due to internal friction and (c) Transmission losses: reduction in wave amplitude due to reflection at interfaces.

17.4.1 Terminology Used in Seismic Acquisition

A *Trace* is a seismic time measurement corresponding to one source-receiver pair, and *offset* is the distance between source and receiver for a given trace. In practice, traces from one *source* are simultaneously recorded at several *receivers*. Then, sources and receivers are moved along the survey line and another set of recordings is made. When a seismic wave travels from a source to a reflector and then back to receiver, the elapsed time is called the *two-way traveltime*. The *common depth point* (CDP) is the halfway point of the path only where the Earth is horizontally layered; it is situated vertically below the *common midpoint* (CMP). A *gather* is a family of traces (e.g. *shot-point gather* is the family of all traces corresponding to the same source firing). *Sorting* of traces by collecting traces that have the same *midpoint* (CMP) is called a *common midpoint gather* (CMP-gather). The number of traces summed or stacked is called a *fold*. For instance, in *24-fold* data, every *stacked trace* represents the average of 24 *traces*. The nominal *fold (F)* is the maximum number of *traces* in a *CMP-gather*. For a standard marine seismic acquisition *folds (F)* is given by the formula:

$$F = \frac{N\Delta g}{2\Delta s} \qquad (17.4)$$

where N is the number of channels, Δg is the group interval and Δs is the source interval. The rate of repetition of complete wavelengths of seismic waves is referred to as *frequency (f)* and is measured in cycles per second or *hertz*.

17.4.2 Marine Acquisition

Marine seismic acquisition is generally accomplished using large ships with one or multiple air-gun arrays for sources (Fig. 17.10). Air-guns are deployed behind the seismic vessel and generate a seismic signal by forcing highly pressurised air into the water at a given interval. Receivers are towed behind the ship in one or several long streamer(s) that are several kilometres long. Marine receivers are composed of piezoelectric hydrophones, which respond to changes in water pressure. In marine acquisition, seismic vessels sail along predetermined patterns of parallel circuits (Fig. 17.11). The length of straight segments is calculated from fold plots, and must include additional length – run in and run out – to allow the cable to straighten after each turn. Marine seismic acquisition is faster than conventional land seismic because it does not require jug hustlers to lay and pick up geophones (Rygg et al. 1992).

The advance to 3D seismic acquisition and imaging of the subsurface, introduced in the 1980s, was perhaps the most important step in seismic exploration (Beckett et al. 1995). The 3D seismic images began to resolve the detailed subsurface structural and stratigraphic conditions that were missing or not discernable from 2D seismic data. With 3D seismic acquisition potential reservoirs are imaged in three dimensions, which allows seismic interpreters to view the data in cross-sections along 360° of azimuth, in depth slices parallel to the ground surface, and along planes that cut arbitrarily through the data volume. Information such

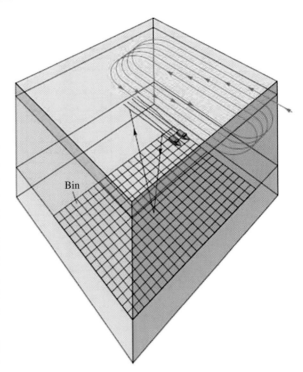

Fig. 17.11 Marine acquisition geometry showing seismic vessels looping in oblong circuits (Source: Ashton et al. 1994)

as faulting and fracturing, bedding plane direction, the presence of pore fluids, complex geological structure, and detailed stratigraphy are now commonly interpreted from 3D seismic data sets.

In 2D marine data acquisition a single streamer is deployed, whereas in 3D acquisition multiple

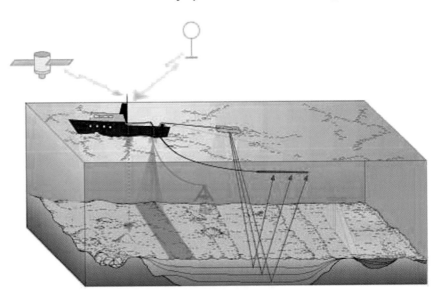

Fig. 17.10 A 2D marine seismic acquisition

Fig. 17.12 Showing a towed streamer for 3D seismic acquisition (Courtesy ION Geophysical)

streamers are towed behind the boat (Fig. 17.12). 3D acquired data can be processed in a more consistent manner and be further manipulated using modern visualisation tools. In 2D data acquisition the data collection occurs along a line of receivers. The resultant image represents only a section below the line. Unfortunately, this method does not always produce a clear subsurface image. 2D data can often be distorted with diffractions and events produced from offline geologic structures, making accurate interpretation difficult. Because seismic waves travel along expanding wavefronts they have a surface area. A truly representative image of the subsurface is only obtained when the entire wavefield is sampled.

A 3D seismic survey where the vessel is towing many streamers (up to 20) at the same time with multiple arrays of airguns is more capable of accurately imaging reflected waves because it utilises multiple points of observations (Fig. 17.12). In the case of 3D survey we have seen that seismic data are sampled from a range of different angles (azimuth) and source-receiver distances (offsets). After seismic processing the data can then be represented as 3D volume images of the subsurface. In 2D processing, traces are collected into *CMP gathers*, while in 3D, traces are collected into *common-cell gathers (binning)*. To perform 3D binning, a grid is first superimposed on the survey area. This grid consists of cells with dimensions of half the receiver group spacing

in the inline direction, equivalent to the CMP spacing in 2D processing, and the line spacing in the crossline direction (Fig. 17.11). In reality, midpoint distributions within a cell are not necessarily uniform since cable shape varies from shot to shot and line to line. Such side drift of the cables is called *feathering*.

Another advanced technique of marine seismic is ocean bottom seismic (OBS) acquisition. It gives the possibility of direct measurement of S-wave data in addition to P-wave data by using ocean bottom cables that have three component geophones (3C) and a hydrophone in addition (thus 4C in total). The 4C cable can be up to 6 km long with 240 stations (i.e. 960 channels since 4C). A typical OBS layout involves 4 or more cables (Fig. 17.13). The optimal choice of acquisition geometry for a 4C survey hinges on both geophysical and financial considerations. Most designs can be classified as either *patch* or *swath*. In *swath* designs, the source lines are parallel to receiver lines, while in *patch* designs, source lines are perpendicular to receiver lines. *Patch* design produces seismic data of a relatively wide range of azimuths, whereas the *swath* design produces data of a limited or narrow range. Moreover, the *swath* design offers a more uniform sampling of offsets with better near-offset coverage. The use of 4C OBS recording has several advantages over conventional towed streamer technology, which includes:

Fig. 17.13 An Ocean
Bottom Seismic (OBS)
marine acquisition technique
(Courtesy ION Geophysical)

(1) Dual-sensor summation (3C geophone + hydrophone) for the suppression of receiver-side multiples.
(2) Utilising P–S wave conversions for enhanced imaging (Fig. 17.14).
(3) Attenuation of free surface multiples when combined with towed streamer recording.

A comparison of the migrated P–S stack versus the P–P stack is shown in Fig. 17.15. The P–S stack is produced from OBS converted wave data whereas the P–P stack is produced from 3D towed streamer P-wave data. From this comparison it is clear that OBS data can be used to successfully image through a gas chimney.

17.4.3 Land Acquisition

A complication in land acquisition is that, unlike marine data, a seismic line is rarely shot in a straight line because of the presence of natural and man-made obstructions such as lakes, buildings and roads (Fig. 17.16). The shot points and the receivers may be arranged in many ways. Many groups of geophones are commonly used on a line with shot points at the end or in the middle of the receiver array. The shot points are gradually moved along a line of geophones. The variations in ground elevation in land acquisition causes sound waves to reach the recording geophones with different traveltime. The Earth's near-surface layer may also vary greatly in composition, from soft alluvial sediments to hard rocks. This means that the velocity of sound waves transmitted through this surface layer may be highly variable. *Static corrections* – a bulk time shift applied to a seismic trace – are typically used in seismic processing to compensate for these differences in elevations of sources and receivers and near-surface velocity variations (Ongkiehong and Askin 1988).

17.4.4 Vertical Seismic Profile (VSP)

VSP is another technique of seismic acquisition, used for correlation with conventional seismic data (land or marine seismic). The defining characteristic of a VSP (of which there are many types) is that either the energy source, or the receivers (or sometimes both) are in a borehole. In the most common type of VSP, hydrophones, or more often geophones in the borehole record reflected seismic energy originating from a seismic source at the surface (Fig. 17.17).

The VSPs vary in the well configuration, the number and location of sources and geophones, and how they are deployed. VSP uses the reflected energy contained in the recorded trace at each receiver position as well as the first direct path from source to receiver. VSPs include the zero-offset VSP, offset VSP,

Fig. 17.14 A comparison of seismic data acquired by the towed streamer (*top*) and OBS (*bottom*) techniques. The OBS survey significantly improves the subsurface image (Source: Thompson et al. 2007)

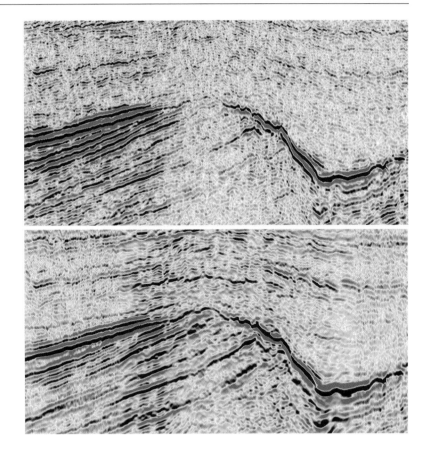

Fig. 17.15 Comparison of P–P stack of conventional 3D streamer data (*top*) and P–S stack of OBS data (*bottom*). Note how the OBS data produces a much better deeper image in the presence of gas versus 3D streamer data (Source: Granli et al. 1999)

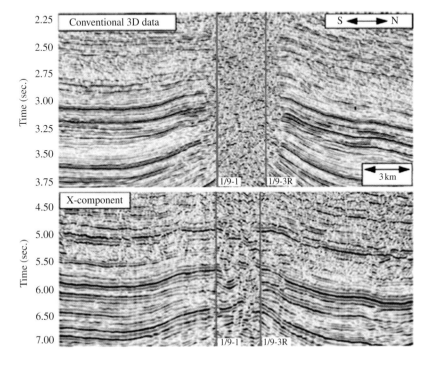

Fig. 17.16 Land seismic
acquisition

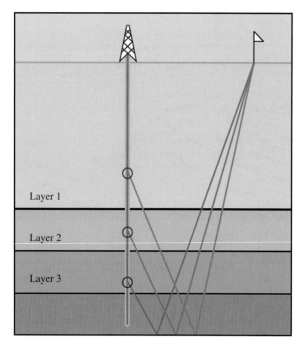

Fig. 17.17 Acquisition of VSP. The downhole geophones
record important structural and stratigraphic data generated by
a surface energy source

walkaway VSP, walk-above VSP, salt-proximity VSP,
shear-wave VSP, and drill-noise or seismic-while-
drilling VSP.

17.4.5 4D Seismic

The acquisition of *4D* or *time-lapse seismic* has opened
new horizons for monitoring reservoir properties such
as fluids, temperature, saturation and pressure changes
during the productive life of a field (Aronsen et al.

2004). 4D seismic is based on the analysis of *repeated
3D* seismic data. The differences in seismic attributes
over time are caused by changes in pore fluid and pore
pressure associated with the drainage of a reservoir
under production. Detection of areas with significant
changes or with virtually unchanged hydrocarbon-
indicating attributes helps to determine new drilling
sites in an already existing production field. For this
method it is critical that the observed seismic changes
can be related to the fluid flow. Differences in data
acquisition, survey orientation, processing and data
quality can introduce significant noise in a 4D analysis.
Hence, such differences must be corrected for as best
as possible. Further details of 4D seismic are discussed
in Chap. 19. The known applications of 4D seismic can
be summarised as:

(1) Monitoring the spatial extent of steam injection
    used for thermal recovery.
(2) Monitoring the spatial extent of the injected water
    front used for secondary recovery.
(3) Imaging bypassed oil or gas.
(4) Determining the flow properties of sealing or leak-
    ing faults.
(5) Detecting changes in oil-water contact.

17.4.6 Permanent Seismic Monitoring

Permanent seismic monitoring is becoming an important tool in the reservoir management toolkit. It is a
4C fiber-optic advanced seismic acquisition technology that is installed permanently on the seabed over
a producing field (Fig. 17.18). It reduces acquisition
time and cost. Permanent seismic monitoring helps
to improve data quality by employing more accurate

Fig. 17.18 Permanent installation of 4C cables at the sea *bottom* over a producing field. It improves data quality by ensuring more accurate receiver locations within the repeated 3D surveys over a period of time

survey orientation and acquisition geometry (receiver locations) within the repeated 3D seismic surveys compared to conventional OBS 4D survey. Such a method is important in monitoring a reservoir injection process employed to enhance recovery from a producing reservoir.

17.5 Seismic Processing

Seismic technology has achieved amazing feats in exploration and production activities in the past few decades. What we record in the acquisition stage is called *raw seismic data* which contains real signals together with *noise and multiples*. This raw data must then be processed by employing advanced methods within signal processing and wave-theory to get better images of the subsurface. The prime objective in the processing stage is to enhance the signal and suppress the coherent and noncoherent noises and multiples (Fig. 17.19). *Coherent noise* is unwanted seismic energy that shows consistent phase from one seismic trace to another. This may consist of waves that travel through the air at very low velocities such as airwaves or air blast, and ground roll that travels

through the top of the surface layer, also known as the *weathering layer*. The energy trapped within a layer known as multiples which is another form of coherent energy. *Multiples* are internal reflections in a layer, which occur when exceptionally large reflection coefficients are present. In marine seismic the water-bottom multiples normally dominate. *Noncoherent energy* is typically nonseismic-generated noise, such as noise from wind, moving vehicles, overhead power line or high-voltage pickup, gas flares and water injection plants.

It has been stated earlier that seismic processing is the alteration of seismic data to suppress noise, enhance signal and migrate seismic events to the appropriate location in space. Seismic processing facilitates better interpretation because subsurface structures and reflection geometries are more apparent. The typical *sampling rate* of seismic acquisition is 2 ms. Digital recording of the incoming wavefield at densely spaced receiver positions ensures that the recorded signal and noise are properly *sampled* and are therefore unaliased. *Aliasing* is the ambiguity that arises because of insufficient sampling. It occurs when the signal is sampled less than twice the *cycle*. The highest *frequency* defined by a sampling interval is termed the *Nyquist frequency* and is equal to the inverse of $2\Delta t$, where Δt is the sampling interval. Frequencies higher than the Nyquist frequency will be folded back. In the noise-free case, aliasing can be avoided by a finer spatial sampling that is at least twice the Nyquist frequency of the *waveform*.

It is important to note that no standard processing sequence exists which can be routinely applied to all types of raw seismic data. The actual sequence will be determined by (a) the purpose of the investigation, (b) extensive testing on selected parts of the dataset and (c) a trade-off between quality and cost. The 2D seismic processing steps typically include static corrections, deconvolution, velocity analysis, normal and dip moveout, stacking and migration. The following routines are generally applied to raw 2D seismic data in different processing stages:

- True Amplitude Recovery (TAR)
- Autocorrelation
- CMP sorting
- Deconvolution
- Trace Muting
- Velocity picking or velocity analysis

Fig. 17.19 Raw seismic data with coherent and noncoherent noise (*top*). Noise attenuation image after autocorrelation, deconvolution and trace muting (*bottom*) (Courtesy Western Geco)

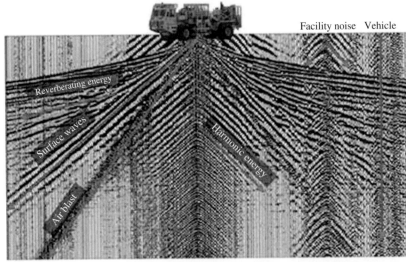

Before DGF

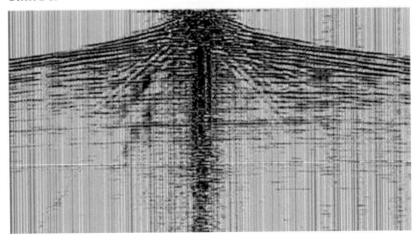

- NMO correction
- DMO correction
- Filtering (F-K and Bandpass filtering)
- Stacking and
- Migration

In the processing stage, bad measurements are edited, datuming applied and corrections of wave-energy decay introduced. The *true amplitude recovery* is applied to increase the amplitude at large travel times. The *autocorrelation* and *deconvolution* are done to compress the wavelet and to attenuate multiples. *Deconvolution* – a technique that can compress the source signature and eliminate multiples – is applied after sorting the data into CMP gathers. The *trace muting* is applied to get rid of unwanted energy. Contributions from the direct waves and possible head

waves are removed by trace muting. *NMO correction* and *F-K filtering* are usually applied to attenuate multiples. Linear coherent noises are also removed by employing F-K filtering (Fig. 17.20).

NMO correction is applied from a space-variant velocity field assuming a horizontal reflector and hyperbolic normal moveout algorithm. The NMO is the difference between the travel time for a certain offset (X) and the vertical (zero-offset) traveltime $T(0)$. Velocities are interpolated for each CDP. Normal moveout is applied according to the following formula:

$$T(X) = \sqrt{\left[T^2(0) + \left(\frac{X}{V} \right)^2 \right]} \qquad (17.5)$$

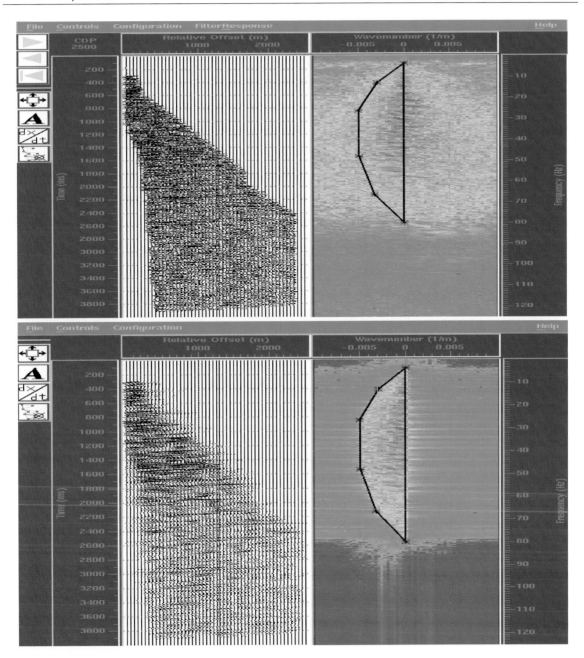

Fig. 17.20 (**a**) A CMP-gather before the F-K filtering: the primaries dipping up and the multiples dipping down in a time-distance display. The F-K domain (*top, right*) shows energy distributions of both primary and multiples energy, respectively. (**b**) The same CMP gather after F-K filtering. The F-K filtering accepted only primary energy (within *polygon*) and filtered out multiples energy (*bottom, right*)

where $T(X)$ is the two-way travel time for a seismic event, X is the actual source-receiver offset distance, V is the NMO or stacking velocity for this reflection event and $T(0)$ is the two-way travel time for zero offset. A sample-by-sample velocity is built at each of the locations where time-velocity pairs are defined. For any point before the first velocity location or beyond the last location, the first or last velocity function is used. Once the correct velocity function has been interpolated, the exact moveout at each sample is

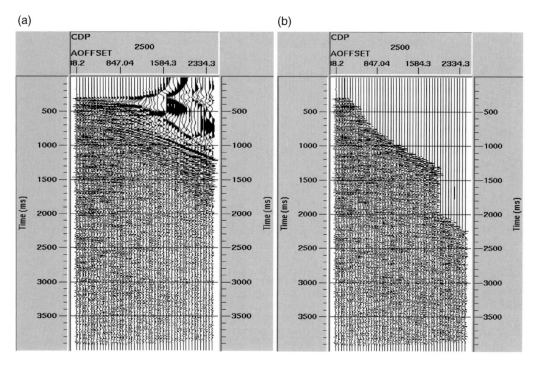

Fig. 17.21 NMO corrected CDP gathers show NMO stretch (**a**) and stretch muting (**b**) at the far offsets. Muting to remove NMO stretch may destroy far offsets information

computed based on the actual source-to-receiver off-set and velocity at that time sample. *NMO stretch* is a fundamental and long-standing problem in seismic processing. After normal moveout correction the early events are stretched at the far offsets (Fig. 17.21a). If we stack this unmuted gather, the early events suffer a severe loss of high-frequency energy, and thus resolution. This can appreciably reduce the interpretability of the section. There have been many attempts to solve the NMO stretch problem. The most universal is front-end or stretch muting, where samples at the beginning of a trace that have suffered severe NMO stretch are zeroed out (Fig. 17.21b). Stretch muting may leave very little fold at early times, reducing the noise suppression provided by stacking.

In the case of dipping beds, there is no common depth point shared by multiple sources and receivers, so *dip-moveout* (DMO) processing becomes necessary to reduce smearing or inappropriate mixing of data (Fig. 17.22). DMO is not routinely employed; however, it can be useful to solve the problem of conflicting dip.

In practice, a *velocity analysis* has to be carried out, where the main purpose is to determine the velocity distribution as a function of time which will give the most accurate NMO correction. The velocity analysis

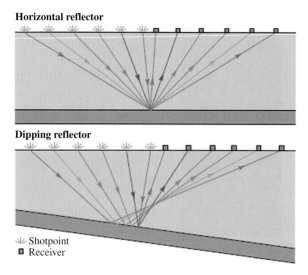

Fig. 17.22 Effect of reflector dip on the reflection point. When the reflector is flat (*top*) the CMP is a common reflection point. When the reflector dips (*bottom*) there is no CMP. A dipping reflector may require changes in survey parameters, because reflections may involve more distant sources and receivers than reflection from a flat layer

is used to compute corrections in traveltime that will be applied to all the traces belonging to a CMP-gather. The most common type of velocity analysis is to repeat the procedure of correcting and stacking CMP data for

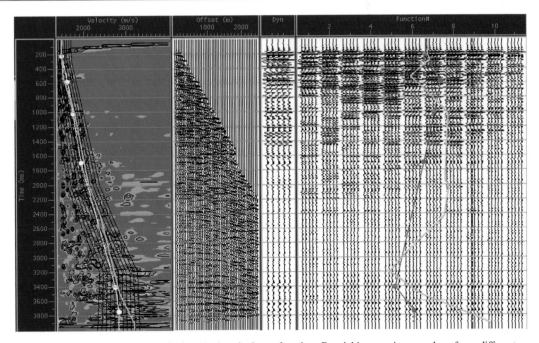

Fig. 17.23 An example of velocity analysis. (**a**) A velocity spectrum or the semblance histogram. (**b**) Plot of velocity analysed CMP data and (**c**) mini-stacks based on picked velocity function. By picking maximum values from different panels a time-velocity function can be constructed

many different velocities and within a discrete time window. By measuring the average absolute value of the data or more precisely the *semblance* within time-windows of different test velocities, and plotting these results in a time versus velocity histogram, it is possible to interpret the velocity information. This type of plot is denoted a *velocity spectrum*. In general, the interpreted velocity function is picked so that it goes through areas with highest values. The basic assumptions in the CMP method velocity analysis is that the geological model corresponds to a slowly varying velocity as a function of depth. In addition, it is assumed that drastic lateral velocity changes do not appear. Then we can interpret high semblance values to be usually related to *primaries* and low semblance values to be related to *multiples* (Fig. 17.23). If the velocity does not increase with depth in some areas it is much more complicated to distinguish between primaries and multiples. If these basic assumptions of velocity analysis are violated both the velocity analysis and the stacked section can be distorted. Such velocity anomalies can be caused by:

- Diffractions (edges, faults)
- Shallow gas pockets
- Multiples

- Complicated reflection sequence (interference)
- Dip
- Reflection from the side (out-of-plane).

Stacking is an important step in seismic processing. Stacking represents summation of *NMO*-corrected traces in a CMP family. The collection of stacked traces forms a seismic section which gives an image (slice) of the subsurface (Fig. 17.24b). The stacking process has two major advantages: (a) it increases the signal-to-noise (S/N) ratio and (b) it amplifies primary energy relative to multiple energy. This second point depends on a good velocity analysis. In the case of an accurate velocity model (Fig. 17.24a), stacking is the most efficient multiple removal method.

Migration is the process that reverses wave propagation effects to get clear images of the subsurface. Migration processing is extremely important in geologically complex areas. The term *migration* came about because, compared to stack sections, the echoes "migrate" to their true subsurface position. Migration is used for several reasons; the most important one is to move reflectors from seismic "apparent" position to their geological "true" position. Another reason for doing migration is to collapse and focus diffractions. Since we cannot measure the direction from where the

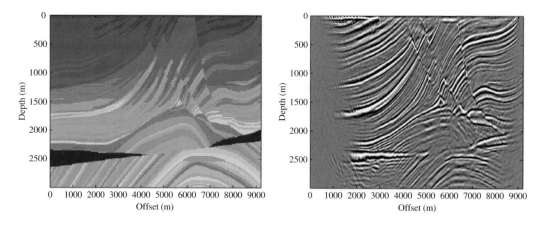

Fig. 17.24 A velocity model (*left*) and a stacked seismic section (*right*) (Source: Versteeg 1994)

signal has been reflected we assume zero offset. It is not correct to assume zero offset where the underlying earth layer is not plane. Before doing the migration a good velocity model is needed as an input.

The migration procedure consists of two well-defined steps: *downward extrapolation* and *imaging*. Downward extrapolation is a simulation process (based on the one-way wave-equation), where the receivers are moved from the surface down to an arbitrary depth. For each downward extrapolated depth a new seismic section can be formed (*imaging*). From each of these sections only the part close to $t = 0$ is retained. A *migrated section* is now formed by combining these strips. The exploding reflector model or ER-model (Lowenthal et al. 1977) has also been used for migration of seismic data. According to this model, a zero offset seismic section can be produced, where all the seismic reflectors "explode" simultaneously at time zero, and subsequently the data is recorded at the surface. However, the exploding reflector model does not account for reflections for which the corresponding down-going ray path differs from the up-going ray path. In order to produce correct travel times with this concept, all the velocities in the subsurface need to be halved with respect to their actual values. The advantages of seismic migration are:

- Diffractions are collapsed to points
- Deeping reflections move up dip and become steeper
- Triplications (bow ties) associated with synforms are unwrapped
- Crossing reflectors avoided.

Seismic migrations are of four types: Pre-stack time and Pre-stack depth migration and Post-stack time or Post-stack depth migration (Fig. 17.25). In *time migration* the images are displayed in two-way travel times, and wave field extrapolation is done in a time-stepping way (Fig. 17.26a). In *depth migration* the wave-stepping is done with respect to depth, and the images can be represented in a true vertical depth (Fig. 17.26b). Depth migration can handle strong lateral velocity variations. Time migration is most common in practice.

17.6 Seismic Resolution

Seismic resolution is the ability to distinguish separate features; the minimum distance between 2 features so that the two can be defined separately rather than as one. The limit of seismic resolution usually makes us wonder, how thin a bed can we see? Normally we think of resolution in the *vertical* sense, but there is also a limit to the *horizontal* width of an object that we can interpret from seismic data.

17.6.1 Horizontal Resolution

The horizontal dimension of seismic resolution is described by the *Fresnel zone*. The Fresnel zone is a frequency and range dependent area of a reflector from which most of the energy of a reflection is returned and arrival times differ by less than half a period from

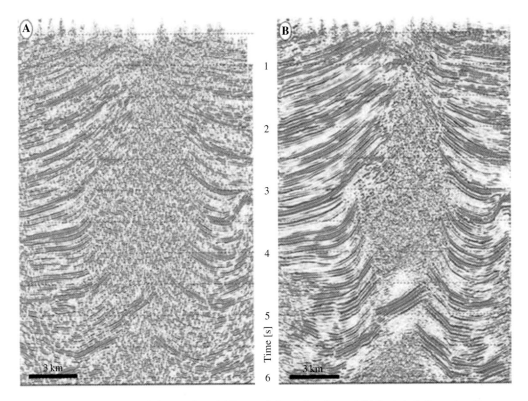

Fig. 17.25 Comparison of time domain images from (**a**) Pre-stack time migration and (**b**) Post-stack time migration

the first break (Fig. 17.27). Waves with such arrival times will interfere constructively and so be detected as a single arrival. Subsurface features smaller than the *Fresnel zone* usually cannot be detected using seismic waves. At spacing greater than one-quarter of the wavelength, the event begins to be resolvable as two separate events. *Migration* can improve lateral resolution by reducing the size of the *Fresnel zone*. For a plane reflecting interface and coincident source and receiver, the Fresnel zone will be circular and with a radius of R_f is

$$R_f = \sqrt{\frac{\lambda Z}{2}} \qquad (17.6)$$

where λ is the dominant wavelength and Z is the depth down to the target surface. Horizontal resolution depends on the frequency and velocity of seismic waves. If we introduce the centre frequency f_c of the pulse (i.e. representing the most energetic part), we

have $\lambda \approx V/f_c$, with V being the wave velocity. Hence, we can rewrite the formula for the Fresnel zone as

$$R_f = \sqrt{\frac{VZ}{2f_c}} \qquad (17.7)$$

17.6.2 Vertical Resolution

Vertical resolution is the ability to separate two features that are close together. A seismic wave can be considered as a propagating energy pulse. If such a wave is being reflected from the top and the bottom of a bed, the result will depend on the interaction of closely spaced pulses. In order for two nearby reflective interfaces to be distinguished well, they have to be about $\lambda/4$ in thickness which is called the *tuning thickness*. This is also the thickness where interpretation criteria change. For smaller thickness, the limit of visibility is reached and positional uncertainties are introduced.

Fig. 17.26 A comparison of Pre-stack time migrated (*top*) and Post-stack depth migrated (*bottom*) images

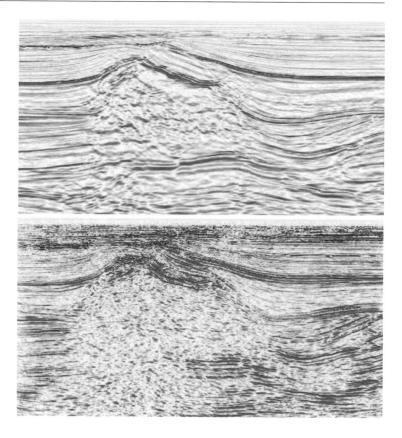

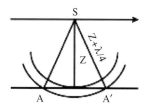

Fig. 17.27 A Fresnel zone in 3D seismic is circular and has diameter A–A′ where S is the source position, Z is the depth down to the target and λ is the wavelength. The size of the Fresnel zone helps to determine the minimum size feature that can be seen in a seismic section

The typical recorded seismic frequencies are in the range of 5–100 Hz. High frequency and short wavelengths provide better vertical and lateral resolution. One could argue that we could simply increase the power of our source so that high frequencies could travel farther without being attenuated. However, there is a practical limitation in generating high frequencies that can penetrate large depths. The Earth acts as a natural filter removing the higher frequencies more readily than the lower frequencies (absorption effect). This means the deeper the source of reflections,

the lower the frequencies we can receive from those depths and therefore the lower resolution we appear to have from great depths (Fig. 17.28). The vertical resolution decreases with the distance travelled (hence depth) by the ray because attenuation preferentially robs the signal of the higher frequency components. *Deconvolution* can improve vertical resolution by producing a broad bandwidth with high frequencies and a relatively compressed wavelet.

As an example, if we introduce the centre frequency f_c of the energy pulse (disturbance) we obtain the following simple relationship between the dominant wavelength (λ), the wave velocity (V) and the centre frequency (f_c):

$$\lambda \cong \frac{V}{f_c}. \qquad (17.8)$$

The typical values for the dominant wavelength are then (a) $\lambda = 40$ m at shallow depth (upper 300–500 m depth), where $V = 2{,}000$ m/s and $f = 50$ Hz, (b) $\lambda = 100$ m at intermediate depths (about 3,500 m), where $V = 3{,}500$ m/s and $f = 35$ Hz

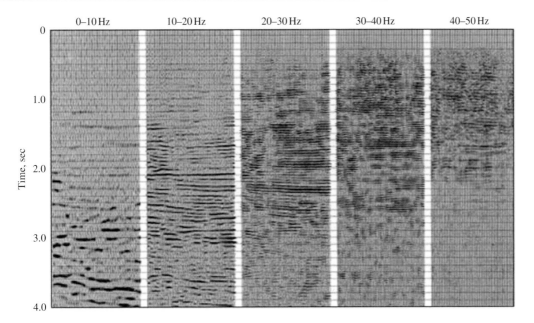

Fig. 17.28 Filtered seismic data showing frequency content variation with depth. Each panel has been filtered to allow a different band of frequencies. As the bandpass rises, the maximum depth of penetration of seismic energy decreases. Lower frequencies (*left*) penetrate deeper. Higher frequencies (*right*) do not penetrate to deeper levels (Source: Ashton et al. 1994)

and (c) $\lambda = 250$ m at depths of about 5,000 m), where $V = 5,000$ m/s and $f = 20$ Hz. For smaller thicknesses than $\lambda/4$ we rely on the *amplitude* to judge the bed thickness. For thicknesses larger than $\lambda/4$ we can use the *waveform*.

17.7 Seismic Interpretation

Seismic data are studied by geoscientists to interpret the composition, fluid content, extent and geometry of rocks in the subsurface. Interpretation of seismic data will be based on an integrated use of seismic inlines, crosslines, time slices and horizon attributes (Dalley et al. 1989, Hesthammer et al. 2001). The seismic sections or images represent slices through the geological model, which can be input to advanced workstations where the actual interpretation can take place. Seismic data can be used in many ways such as regional mapping, prospect mapping, reservoir delineation, seismic modelling, direct hydrocarbon detection and the monitoring of producing reservoirs. Based on the seismic interpretation one will decide if an area is a possible prospect for hydrocarbon (oil or gas). If the answer is positive, an exploration well will be drilled. The ultimate goal will be the drilling of production wells if the target area proves to be a commercial reservoir. Seismic data contain a mixture of signal and noise. It is therefore crucial to understand the signature of the noise, whether it is systematic or random, dipping or flat-lying, planar or non-planar. It is also necessary to investigate the origin of the noise. The challenge of seismic interpretation is then to fully utilise all the information contained in the seismic data. *Systematic noise* can be related to acquisition procedures, processing artefacts, water-layer multiples, faults, complex stratigraphy and shallow gas. *Random noise* includes natural noise (e.g. wind and wave motion), incoherent seismic interface and imperfect static corrections. Without a sound understanding of these factors as well as knowledge of the limitation of seismic resolution, there is a danger of misinterpreting noise as real features.

In terms of the parameters that are analysed and the interpretation that may be drawn from the analyses, a fourfold hierarchy of seismic interpretation can be achieved. These are *seismic facies analysis*, *seismic structural analysis*, *seismic attribute analysis* and *seismic sequence analysis*. The most important parameters used for interpretation of seismic data are:

Reflection amplitudes: The strength of the reflections. As we discussed above, the proportion of the energy reflected at the boundary between two beds is a function of the difference in the acoustic impedances. If we have an alternating series of different beds, the distance between the bed boundaries in relation to the wavelength of the transmitted seismic signals will play a major part.

Reflector spacing: The distance between the reflectors will indicate the thickness of the bed, but there will be a lower limit to the thickness that can be detected, which will depend on the wavelength.

Interval velocity: The interval velocity of a sequence can provide information about lithology and porosity but this will depend on the stacking velocity and will not be very accurate.

Reflector continuity: The continuity of reflectors will be a function of how continuous the sediment beds are, information which is essential for reconstructing the environment.

Reflector configuration: If we take the compaction effect into account, the shape of the reflecting beds gives us a picture of the sedimentation surface as it was during deposition. The slope of the reflectors, for example, represents the slope of prograding beds in a delta sequence with later differential compaction and tilting superimposed. Erosion boundaries with unconformities will in the same way show the palaeo-topography during erosion.

Instantaneous phase: A seismic trace can be considered an analytical signal where the real part is the recorded seismic signal itself. Mathematically we can compute the complex seismic trace (imaginary parts of the signal) and the instantaneous attributes. The Instantaneous phase is a measure of the continuity of the events on a seismic section. The Instantaneous phase is on a scale of $+180°$ to $-180°$. The temporal rate of change of the instantaneous phase is the instantaneous frequency.

17.7.1 Seismic Facies Analysis

A seismic profile provides information about the properties of sedimentary rocks. Seismic facies analysis is the description and geological interpretation of seismic reflectors representing sequence boundaries. It includes the analysis of such parameters as the configuration, continuity, amplitude, phase, frequency and interval velocity. These variables give an indication of the lithology and sedimentary environment of the facies (Selley 1998) (Fig. 17.29). Because

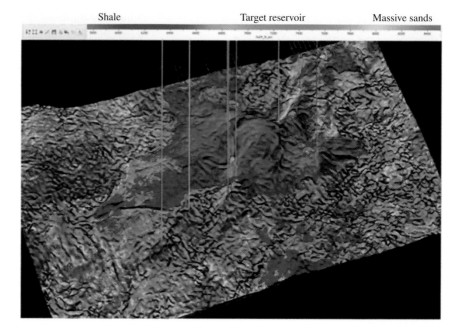

Fig. 17.29 A time slice of a special seismic attribute (S-impedance co-rendered with K_{max} curvature) that is constructed for lithology determination. Fractured, interbedded sandstones (*green*) juxtaposed against the organic-rich shales (*purple*) are the most productive reservoirs in this horizon (Courtesy ION Geophysical)

seismic reflections mainly represent time lines, i.e. sedimentary beds which were deposited contemporaneously, it is also possible to a certain extent to interpret the depositional environment. Large scale sedimentary features that may be recognised by the configuration of seismic reflectors include prograding deltas, carbonate shelf margins and submarine fans (Sherif 1976).

17.7.2 Structural Analysis

Interpreters can draw horizons and faults on in-lines, cross-lines and arbitrary lines, as well as slices. Horizons can be automatically tracked on vertical seismic displays and horizontal slice displays. Improved tracking algorithms for horizon interpretation are combined with user-interpreted faults and fault polygons can produce seismic based interpretation maps (Fig. 17.30). Faults on seismic sections are typically expressed by a loss of reflection amplitude. It must be remembered that the principle we use for calculating the depth to a reflecting boundary assumes that the layering is not too far from horizontal. High-angle faults do not reflect the sound wave back to the receivers to give a signal and faults can therefore not be seen directly on seismic sections. It is the truncations and offset of good reflectors which usually allows us to identify faults. Due to special "edge effects" that we observe near a fault, the reflector terminations which should define the fault are not located in quite their true position on the seismic profile, making it difficult sometimes to define them accurately. Migration of the seismic data goes some way to remedying the problem. Folded beds will only be realistically depicted if the folds are sufficiently gentle that the beds have a low angle of dip.

17.7.3 Seismic Attribute Analysis

A seismic attribute is a quantitative measure of a seismic characteristic of interest. Seismic attribute analysis is concerned with the study of amplitude, polarity, continuity and wave shape. One of the goals of seismic attributes is to somehow capture maximum information by quantifying the amplitude and morphological features seen in the seismic data through a suite of

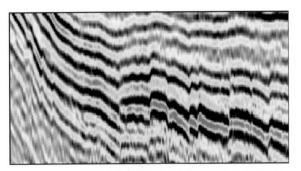

Fig. 17.30 A seismic section showing abundant faulting. Horizons can be interpreted by variable amplitude and pick trough-pick trace shape (*top*). A plan view of a horizon (*bottom*) which can be useful to identify fault patterns from the missing picks and discontinuities in colour (Source: James 2003)

deterministic calculations performed on a computer. The extraction of seismic attributes, such as amplitude envelope, dominant frequency, apparent polarity and instantaneous phase can produce remarkable 3D images of subsurface rock formations (Fig. 17.31). Such analysis may give an indication of the thickness and nature of the upper and lower contacts of a sand body (Selley 1998). Comparison of observed seismic waves with synthetic traces computed from a geological model give some insight into the depositional environment of the sand, and hence help to predict its geometry and internal reservoir characteristics (Fig. 17.32).

There are now more than 100 distinct seismic attributes calculated from seismic data that can be applied to the interpretation of geological structures, stratigraphy, and rock/pore fluid properties. Taner et al. (1994) divide attributes into two general categories: *geometrical* and *physical*. The objective of *geometrical attributes* is to enhance the visibility of the geometrical characteristics of seismic data; they include dip, azimuth, and continuity. *Physical attributes* have to

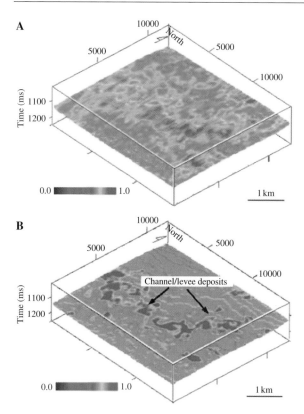

Fig. 17.31 A comparison between average absolute amplitude (**a**) and energy (**b**) in a horizon slice at the same stratigraphic level. Notice that the channel/levee deposits can be recognised, mapped and detected more efficiently from the energy volume than from the amplitude volume (after Gao 2003)

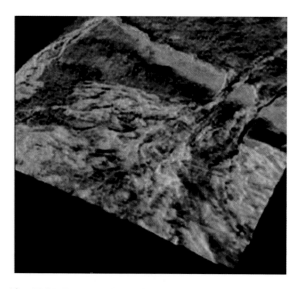

Fig. 17.32 Interpreted 3D seismic volume showing a channel fan complex with a shelf edge (Source: James 2009)

do with the physical parameters of the subsurface and so relate to lithology. These include amplitude, phase and frequency. Liner et al. (2004) classified attributes into general and specific categories. The general attributes are measures of geometric, kinematic, dynamic or statistical features derived from seismic data. They include reflector amplitude, reflector time, reflector dip and azimuth, complex amplitude and frequency, generalised Hilbert attributes, illumination, edge detection/coherence, AVO and spectral decomposition. *General attributes* are based on either the physical or morphological character of the data tied to lithology or geology and are therefore generally applicable from basin to basin around the world. In contrast, *specific attributes* have a less well-defined basis in physics or geology. While a given specific attribute may be well correlated to a geological feature or to reservoir productivity within a given basin, these correlations do not in general carry over to a different basin.

Over the past decades, we have witnessed attribute developments tracking the breakthroughs in reflector acquisition and mapping, fault identification, bright-spot identification, frequency loss, thin-bed tuning, seismic stratigraphy and geomorphology (Fig. 17.33). More recently, interpreters have used cross-plotting to identify clusters of attributes associated with either stratigraphic or hydrocarbon anomalies. Today, very powerful computer workstations capable of integrating large volumes of diverse data and calculating numerous seismic attributes are a routine tool used by seismic interpreters seeking geological and reservoir engineering information from seismic data.

17.7.4 Seismic Sequence Analysis

Apart from structural and facies analysis, seismic signatures can be used to interpret the way the basin has been filled in. This is a result of an interaction between the rate of subsidence, rate of deposition and the energy of the depositional environment (Fig. 17.34).

On seismic profiles we can see onlaps onto the land, measure the height range between the lowest and uppermost onlaps, calculate the difference in seismic time, and convert this into approximate thickness. However, we must remember that the thickness of the sediments deposited is due not only to a rise in eustatic

(a)

(b)

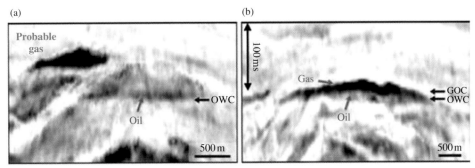

Fig. 17.33 (**a**) A strong soft amplitude anomaly (oil column) at *top* reservoir corresponds to a clear flat event within the reservoir. (**b**) This section shows an acoustically soft body (gas column) within the reservoir with a sharp, flat base (after Blom and Bacon 2009)

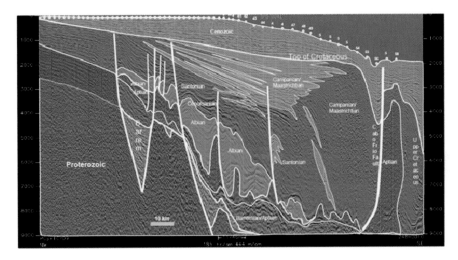

Fig. 17.34 Geological interpretation of a seismic section

sea level, but also to local subsidence of this part of the basin. The weight due to increased water depth will cause further subsidence, and sedimentation will increase the load, resulting in further subsidence to attain isostatic equilibrium. Local tectonic subsidence may produce a relative change in coastal onlap in a seismic profile. *Regressions* are defined as the boundary between land and sea being displaced out into the basin. They may be caused by a fall in sea level which will shift the coastline to further out on the shelf or to the edge of the continental slope, or progradation of a coastline or a delta front. Here unloading of some of the water load and erosion of sediments led to isostatic uplift of the area landward of the coastline, so that the measured regression is greater than the real lowering of the sea level. Here, too, local tectonic movements will play a part. *Transgressions* (geological events during which sea level rises relative to the land and the shoreline moves toward higher ground,

resulting in flooding) and *regressions* are not always directly related to sea level changes. When a delta builds out into the sea, there is a local regression on the delta even if the sea level has not fallen. If sedimentation is sufficiently rapid, we can have a local regression on a delta even with a rising sea level. Along a coastline we can in fact have transgressions in some areas and regressions in others, at the same time, depending on the rate of sedimentation or erosion with respect to sea level change. A regression caused by a fall in sea level is called a forced regression. Changes in sea level can be due to:

• Local tectonic movements, for example uplift of a horst or subsidence of a graben structure.
• Plate-tectonic movements which can be of great extent, but are not global.
• Sea level changes. These are global and are called eustatic sea level changes.

During the Quaternary, cyclic changes in sea level of up to 120 m accompanied the growth and decay of continental ice sheets. These changes were rapid and of large magnitude and can be traced throughout much of the world. In the areas which had supported ice sheets, such as Scandinavia, the melting of the ice led to isostatic uplift due to unloading, and this exceeded the rise in sea level so that there was a regression. When seismic stratigraphy was established (Vail et al. 1977), it was assumed that most of the variations in sea level that could be interpreted from seismic records were attributable to eustatic changes and thus could be employed for global correlation. However accumulation of ice on the continents is the only known process capable of producing large and rapid global sea level changes. Such ice ages are known from the Quaternary, the Carboniferous-Permian, late Ordovician and the end of the Precambrian.

Nevertheless, transgressions and regressions can be correlated over greater or lesser distances, depending on the type of tectonic displacement. It is often difficult to distinguish between eustatic sea level changes and ones caused by more local or regional conditions and we therefore prefer to use the term *relative sea level change*. On a delta, one delta lobe may prograde and produce a regressive sequence while a transgression occurs with carbonate sedimentation on an adjacent lobe.

17.8 Integration and Visualisation of Seismic Data

In order to establish a reservoir model, different geophysical measurements such as seismic, well logs and rock physics need to be integrated. Since these measurements are carried out employing very different frequency regimes, there is a problem of scaling. Rock physics models of laboratory measurements use ultrasound frequencies (kHz to MHz) whereas sonic tools (well logging) operate with frequencies in the kHz range. Seismic data fall within the range of hertz (Hz). Assuming the velocity of a sedimentary layer is 3,000 m/s, a seismic frequency of 30 Hz will give 100 m wavelength whereas sonic log frequency of 30 kHz will give 10 cm wavelength and the ultrasound frequency of 300 kHz will give 1 cm wavelength. Therefore, the vertical resolutions of seismic, sonic and ultrasonic frequencies are 25 m, 2.5 cm and 2.5 mm, respectively (Fig. 17.35). The major problem is then how to best incorporate all geophysical data into a reservoir model. Rock physics investigations define how pore-scale variations in properties like mineralogy, fluid content and grain geometry affect the acoustic response of a core sample. As the scale of investigations increases from laboratory to well log to seismic, low frequency measurements (seismic or well logs) average over different rock types and

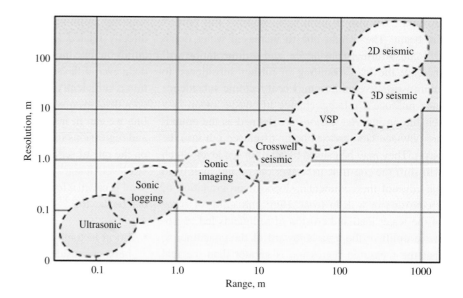

Fig. 17.35 Plot of range versus resolution of various geophysical techniques (Courtesy: Schlumberger Oilfield Glossary)

Fig. 17.36 Shows a 3D volume with fault (*red*), horizons (amplitude maps) and a producing well path through a channel. Once a channel is identified, its shape can be extracted by visualising the 3D volume (Courtesy: Halliburton)

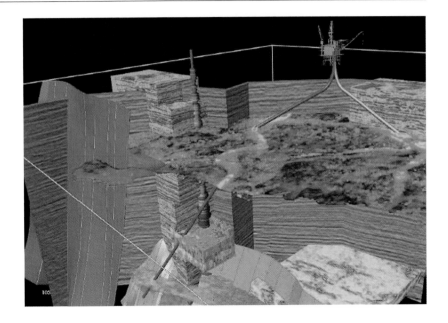

sediments in addition to local pore-scale variations. As a result, spatial heterogeneity and preferential sampling at well logs or seismic scale can cause a shift in the rock physics relationship away from that determined in the lab. Therefore, critical understanding and strategies are necessary to incorporate seismic, well logs and rock physics to find links between qualitative geological parameters and quantitative geophysical measurements. Integration strategies and utilisation of different geophysical data are discussed thoroughly in Chap. 18.

As the use of 3D seismic increasingly becomes an integral part of hydrocarbon exploration, visualisation techniques also continue to evolve as software and hardware improves. Visualisation of target datasets of various sizes and formats in a powerful graphical environment is the answer to efficiently understanding data quality problems, anomalies and trends. Such an environment responds dynamically to manipulation when analysing many datasets on one common canvas, or it can allow travel through time in a 4D visualisation process comparing one survey to another. Advance visualisation techniques allow us to interactively explore regional datasets, integrate live well information into a reservoir model and understand reservoir dynamics in association with a 4D predictive model (Fig. 17.36). 3D visualisation technique is also used to estimate thickness, continuity and lateral extend of a reservoir and its relationship with horizons

and faults to see the finest reservoir details. Visualising combinations of complex seismic attributes like amplitude, continuity and AVO (Amplitude Versus Offset) provide a far more accurate picture of hydrocarbon potential, and in less time, than viewing those attributes discretely. By integration and 3D visualisation of massive amounts of data one can add greater value to decision making to differentiate between economic and sub-economic wells in a marginal area. The ability to understand data, gain useful insight, and communicate these with others is greatly enhanced by the use of interactive 3D visualisation. However, the ability to adjust the parameters and the display interactively is crucial to exploring the data and finding the combinations that highlight specific features and relationships.

Further Reading

Aronses, H.A., Osdal, B., Dahl, T., Eiken, O., Goto, R., Khazanehdari, J., Pickering, S. and Smith, P. 2004. Time will tell: New insights from time-lapse seismic data. Oilfield Review 16(2), 6–15.

Ashton, C.P., Bacon, B., Mann, A., Moldoveanu, N., Déplanté, C., Dickilreson, Sinclair, T. and Redekop, G. 1994. 3D Seismic Survey Design. Oilfield Review 6, 19–32.

Beckett, C., Brooks, T., Parker, G., Bjoroy, R., Pajot, D., Taylor, P., Deitz, D., Flaten, T., Jaarvik, L.J., Jack, I., Nunn, K., Strudley, A. and Walker, R. 1995. Reducing 3D seismic turnaround. Oilfield Review 7(1), 23–37.

Blob, F. and Bacon, M. 2009. Application of direct hydrocarbon indicators for exploration in a Permian-Triassic play, offshore the Netherlands. First Break 27, 37–44.

Braile, L.W. 2000. Seismic Waves and the Slinky: A Guide for Teachers. http://www.gg.uwyo.edu/geol2005a/lectures/seismics/WaveDemo.pdf

Dalley, R.M., Gevers, E.C.A., Stampfli, G.M., Davis, D.J., Gastaldi, C.N., Ruijtenberg, P.A. and Vermeet, G.J.O. 1989. Dip and azimuth displays for 3D seismic interpretation. First Break 7, 86–95.

Gao, D. 2003. Volume texture extraction for 3D seismic visualization and interpretation. Geophysics 68, 1294–1302.

Granli, J.R., Arnsten, B. Anders, S. and Hilda, E. 1999. Imaging through gas-filled sediments using marine shear-wave data. Geophysics 64, 668–667.

Hesthammer, J., Landrø, M. and Fossen, F. 2001. Use and abuse of seismic data in reservoir characterisation. Marine and Petroleum Geology 18, 635–655.

James, H. 2003. Has volume interpretation of structure been cracked at last? First Break 21, 38–40.

James, H. 2009. Visualizing 3D features in 3D seismic data. First Break 27, 57–62.

Liner, C., Li, C.-F., Gersztenkorn, A. and Smythe, J. 2004. SPICE: A new general seismic attribute: 72nd Annual International Meeting, SEG, Expanded Abstracts, pp. 433–436.

Loewenthal, D., Lu, L., Roberson, R. and Sherwood, J. 1976. The wave equation applied to migration. Geophysical Prospecting 24, 380–399.

Milligan, M. 2004. What are seismic surveys and how much shaking do they create? Survey Notes, Utah Geological Survey 36(3), 10–11.

Ongkiehong, L. and Askin, H.J. 1988. Towards the universal seismic acquisition technique. First Break 6(2), 46–63.

Rygg, E., Riste, P., Nottvedt, A., Rod, K. and Kristoffersen, Y. 1992. The Snowstreamer-a new device for acquisition of seismic data on land. In: Vorren, T.O., Bergsager, E., Dahl-Stamnes, O.A., Holter, E., Johanses, B., Lie, E. and Lund, T.B. (eds.), Arctic Geology and Petroleum Potential. NPF Special Publication 2, 703–709.

Selley, R.C. 1998. Elements of Petroleum Geology. Academic Press, San diego, CA, 470 pp.

Sheriff, R.E. 1976. Inferring stratigraphy from seismic data. AAPG Bulletin 60, 528–542.

Sheriff, R.E. and Geldart, L.P. 1995. Exploration Seismology. Cambridge University Press, New York, 592 pp.

Taner, M.T., Schuelke, J.S., O'Doherty, R. and Baysal, E. 1994. Seismic attributes revisited: 64th Annual International Meeting, SEG, Expanded Abstracts, 1104–1106.

Telford, W.M., Geldart, L.P. and Sheriff, R.E. 1990. Applied Geophysics. Cambridge University Press, New York, 770 pp.

Thompson, M., Arntsen, B. and Amundsen, L. 2007. Full azimuth imaging through consistent application of ocean bottom seismic. SEG Expanded Abstract, pp. 936–940.

Tucker, P.M. and Yorston, H.J. 1973. Pitfalls in seismic interpretation. Society of Exploration Geophysicists, Monograph Series, No. 2, 50 pp.

Vail, P.R., Mitchum, R.M. and Thomson, S. 1977. Seismic stratigraphy and global changes of sea level, Part 3: Relative changes of sea level from coastal onlap. In: Payton, C.E. (ed.), Seismic Stratigraphy – Applications to Hydrocarbon Exploration. AAPG Memor 26, 63–97.

Versteeg, R. 1994. The Marmousi experience: Velocity model determination on a synthetic complex dataset. The Leading Edge 13, 927–936.

Yilmaz, O. 2001. Seismic Data Analysis: Processing, Inversion and Interpretation of Seismic Data. SEG Press, Tulsa, 2027 pp.

Chapter 18

Explorational Rock Physics – The Link Between Geological Processes and Geophysical Observables

Per Avseth

The field of rock physics represents the link between qualitative geological parameters and quantitative geophysical measurements. Increasingly over the last decade, rock physics has become an integral part of quantitative seismic interpretation and stands out as a key technology in petroleum geophysics. Ultimately, the application of rock physics tools can reduce exploration risk and improve reservoir forecasting in the petroleum industry.

This chapter covers basic rock physics principles and practical recipes that can be applied in the field. The importance and benefit of linking rock physics to geological processes, including depositional and compactional trends, is demonstrated. It is further documented that lithology substitution can be of equal importance to fluid substitution during seismic reservoir prediction. It is essential in exploration and appraisal to be able to extrapolate away from existing wells, taking into account how the depositional environment changes, together with burial depth trends. In this way rock physics can better constrain the geophysical inversion and classification problem in under-explored marginal fields, surrounding satellite areas, or in new frontiers.

Finally, practical examples and case studies are presented to demonstrate a best-practice workflow and associated limitations and pitfalls. Rock physics models are combined with well log and pre-stack seismic data, sedimentological information, inputs from basin modelling and statistical techniques, to predict reservoir geology and fluids from seismic amplitudes.

18.1 Quantitative Seismic Interpretation Using Rock Physics

The main goal of conventional, *qualitative* seismic interpretation is to recognise and map geological elements and/or stratigraphic patterns from seismic reflection data. Often hydrocarbon prospects have been defined and drilled entirely on the basis of this qualitative information. Today, however, *quantitative* seismic interpretation techniques have become common oil industry tools for prospect evaluation and reservoir characterisation. The most important of these techniques include post-stack amplitude analysis (bright-spot and dim-spot analysis), offset-dependent amplitude analysis (AVO-analysis), acoustic and elastic impedance inversion, and forward seismic modelling. These techniques seek to extract additional information about the subsurface rocks and their pore fluids from the reflection amplitudes and, if used properly, they open up new doors for the seismic interpreter. Seismic amplitudes primarily represent contrasts in elastic properties between individual layers and contain information about lithology, porosity, pore fluid type and saturation, as well as pore pressure – information that cannot be gained from conventional seismic interpretation. Seismic amplitude maps are increasingly important in prospect evaluation and reservoir delineation. As shown in Fig. 18.1, the amplitude patterns often provide a good insight into depositional patterns. Seismic amplitude maps can be very useful in the delineation of subtle traps that are not easily revealed from conventional (i.e. stratigraphic and structural) seismic interpretation.

However, to make sure we understand the meaning of the seismic amplitudes, a quantitative link is

P. Avseth (✉)
Odin Petroleum, Bergen; NTNU, Trondheim, Norway
e-mail: pavseth@yahoo.com

K. Bjørlykke (ed.), *Petroleum Geoscience: From Sedimentary Environments to Rock Physics*,
DOI 10.1007/978-3-642-02332-3_18, © Springer-Verlag Berlin Heidelberg 2010

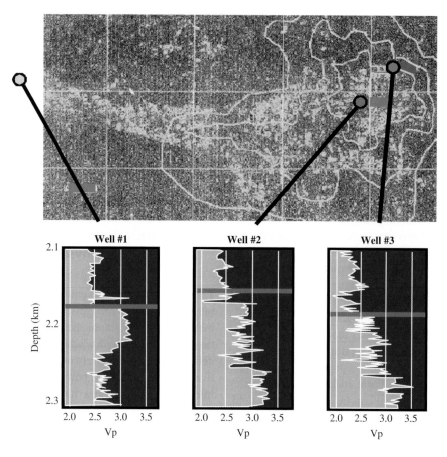

Fig. 18.1 Seismic amplitude map from the Glitne Field, with sonic well log data for three wells penetrating a submarine fan system at different locations. Drastic changes in the seismic velocities and the log patterns are observed as we go from the feeder-channel to the more distal part of the lobe system (Adapted from Avseth et al. 2005)

needed between the geological parameters and the rock physics properties. The seismic reflections are physically explained by contrasts in elastic properties, and rock physics models allow us to link the elastic properties to geological parameters. Hence, the application of rock physics models can guide and improve on the qualitative interpretation (e.g. Mavko et al. 2009, Avseth et al. 2005). Moreover, if we understand the link between geological parameters and rock physics properties, we can avoid certain ambiguities in seismic interpretation, particularly fluid/lithology, sand/shale and porosity/saturation. During fluid substitution, it is very common to assume that the rock type and porosity are constant, neglecting the possibility that lithology can change from the brine zone to the hydrocarbon zone. The link between rock physics and various geological parameters, including cement volume, clay volume and degree of sorting, allow us to perform

lithology substitution from rock types observed at a given well location to rock types assumed to be present nearby. Hence, during quantitative seismic interpretation of a reservoir we can do sensitivity analysis not only of fluid types, but also of the reservoir quality.

The way in which geological trends in an area can be used to constrain rock physics models is also investigated. Geological trends can be split into two categories: compactional trends and depositional trends. If we can predict the expected change in seismic response as a function of depositional environment or burial depth, this will increase our ability to predict hydrocarbons, especially in areas with little or no well log information. Understanding the geological constraints in an area of exploration reduces the range of expected variability in rock properties and hence reduces the uncertainties in seismic reservoir prediction. Figure 18.2 depicts this problem, where the

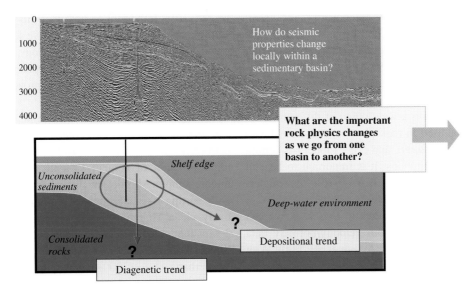

Fig. 18.2 Rock physics properties change with depositional environment and burial depth. These geologic trends must be taken into account during hydrocarbon prediction from seismic data

only well log control we have is in the shallow interval on the shelf edge. Before extending the exploration into more deeply buried zones, or to more distal deep-water environments, it is important to understand the rock physics trends in the area.

18.2 Rock Physics Models for Microstructure Interpretation

Rock physics represents the link between geological reservoir parameters (e.g. porosity, clay content, sorting, lithology, saturation) and seismic properties (e.g. acoustic impedance, V_p/V_s, density, elastic moduli). Rock physics models can either be used to interpret observed sonic and seismic velocities in terms of reservoir parameters, or they can be used to extrapolate beyond an observed range to predict certain "what if" scenarios in terms of fluid or lithology substitution. Rock physics models can also be used to estimate expected seismic properties from observed reservoir properties.

The rock physics link between seismic properties and geological parameters is most commonly displayed through crossplots, and Fig. 18.3 shows an example of rock physics models plotted as porosity versus an elastic modulus (i.e. bulk or shear modulus). Pore fluid and stress sensitivity in reservoir sandstones

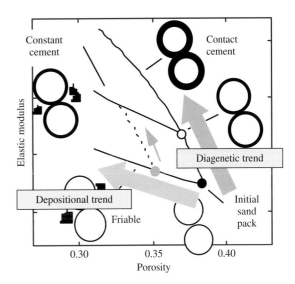

Fig. 18.3 Rock physics models link rock microstructure to elastic properties. Rock physics diagnostics is to apply these models to infer geologic texture from elastic measurements. Rock physics models can also be applied to do forward modelling to better understand the expected elastic response to various geological scenarios. The elastic modulus may be compressional, bulk or shear (from Avseth et al. 2000)

are highly affected by reservoir heterogeneity and sandstone microstructure, and it is therefore important to include these geological factors in the rock physics analysis. Through the rock physics models in Fig. 18.3, we can infer or diagnose the rock texture of a sandstone if we know the porosity and corresponding elastic modulus. The various models include the unconsolidated sand model and the contact cement model (Dvorkin and Nur 1996), together with the constant cement model (Avseth et al. 2000). For more details and equations behind these models, see Avseth et al. (2005). There is a wide range of different models that can be used in rock physics analysis (Mavko et al. 2009, Dræge 2009), and every model has certain advantages and limitations. It is important to bear in mind the famous saying by Box (1976): "All models are wrong, but some are useful".

Using the diagnostic models in Fig. 18.3, we can infer the microstructure from velocity/porosity data (e.g. Dvorkin and Nur 1996, Avseth et al. 2000). With good local validation of the models, we can even quantify the degree of sorting and cement volume from these diagnostic crossplots. Figure 18.4 shows an example from Avseth et al. (2009). Here, the rock physics diagnostics is done in the V_s versus porosity domain, in order to avoid significant pore fluid effects. The cement volume is estimated by interpolating between the constant cement volume trends. For a given constant cement trend, sorting is defined to vary between 1 and 0. Sorting equals 1 for data points that fall right on the contact cement model. Deteriorating sorting implies increasing volume of pore-filling (noncementing) material which will cause lower porosities. The data points will then plot further away from the contact cement model, towards zero porosity. At the zero porosity end member, sorting is also defined to

be zero. It is important to note that this definition of sorting is not the same as the qualitative sorting parameter defined in the field of sedimentology, but they will be highly correlated, with values approaching 0 when sands are poorly sorted and values approaching 1 when sands are well sorted.

Having estimated cement volume and sorting, we can plot these as logs and compare them with other petrophysical logs. Figure 18.5 shows the resulting estimation of cement volume and sorting. The middle subplot shows cement volume as magnitude, with sorting as superimposed colour. For the relatively clean Heimdal sandstones starting at around 2 km depth, we observe a clear depth trend in the cement volume. This matches observations made by Walderhaug et al. (2000) (right subplot). The sorting shows a more erratic pattern, lacking any consistent depth trend, as we would expect since sorting is associated with depositional trends.

It is essential that thin-section observations are used to verify the presence of initial cementation predicted from the rock physics relations. Figure 18.6 shows a thin-section from the relatively clean Heimdal sands, but at first glance the sandstone looks unconsolidated with grains loosely arranged and moderately well sorted. A closer investigation, however, reveals the presence of initial quartz overgrowth covering original grain surfaces, indicated by dust rims (see arrows, Fig. 18.6). This observation confirms what we see in the rock physics crossplots of the well log data. It is interesting that the well log data with 10s of cm

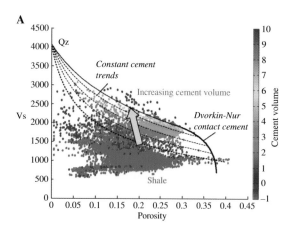

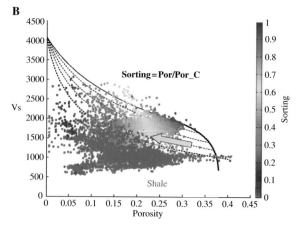

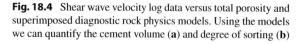

Fig. 18.4 Shear wave velocity log data versus total porosity and superimposed diagnostic rock physics models. Using the models we can quantify the cement volume (**a**) and degree of sorting (**b**)

(*Green data* points are shale data with high GR values, and are for practical reasons given the value –1 in cement volume and 0 in sorting) (Taken from Avseth et al. 2009)

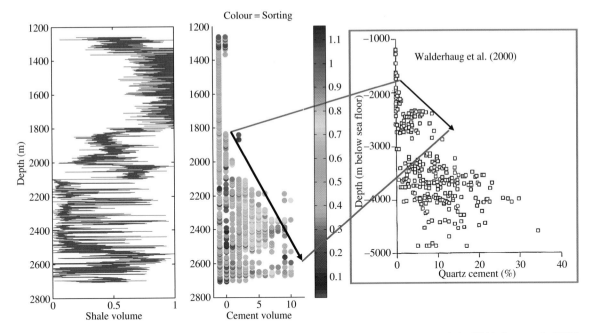

Fig. 18.5 Estimated cement volume and sorting versus depth for a North Sea well. Note the onset of cement starting at around 2,000 m depth (corresponding to c.70°C) and the increasing cement volume with depth (second column). This is in agreement with observations made by Walderhaug et al. (2000). Sorting, however, shows a more erratic pattern with no depth trend (colour in second coloumn; see also Avseth et al. 2009)

resolution reflects what we observe at the microscale. A comparison between cement volume estimated from rock physics models and the cement volume from the thin-section point count analysis (only available in the upper part of the Heimdal sands in the well analysed in Figs 18.4 and 18.5) is shown in Fig. 18.7. The volume of quartz cement from the thin-section analysis is somewhat lower than the volume estimated from the rock physics models. However, if we compare the rock physics estimates with the total cement volume, the match is very good. While most of the cement in the Heimdal Formation is quartz in this well, other types are also present, like feldspar overgrowths and carbonate cement.

18.3 Rock Physics and Depositional Trends

There exist several rock physics models for depositional trends in siliciclastic environments (e.g. Marion 1990, Dvorkin and Gutierrez 2002, Xu and White 1995). Marion (1990) introduced a topological model for sand/shale mixtures to predict the interdependence between velocity, porosity and clay content. When clay content is less than the sand porosity, clay particles are assumed to be located within the pore space of the load-bearing sand. The clay will stiffen the pore-filling material, without affecting the frame properties of the sand. As the clay content increases, so will the stiffness and velocity of the sand/shale mixture, as the elastic moduli of the pore-filling material (fluid and clay) increase. Once the clay content exceeds the sand porosity, the addition of more clay will cause the sand grains to become separated, as we go from grain-supported to clay-supported sediments (i.e. shales).

Marion assumed that, as with fluids, the pore-filling clay would not significantly affect the shear modulus of the rock. This assumption was supported by laboratory measurements on unconsolidated sand/shale mixtures (Yin 1992). The impact on the velocity-porosity relationship of increasing clay content in a sand/shale mixture is depicted in Fig. 18.8. From the measured data we can see that when clay content increases, porosity decreases and velocity increases up to a given point called the *critical clay content*. This point represents the transition from shaly sands to sandy

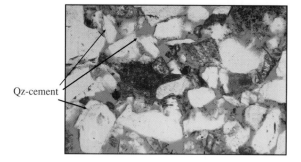

Qz-cement

Fig. 18.6 Thin-sections from Heimdal Formation sands. The *upper image* shows a loosly packed, poorly consolidated sand. Analysis of a zoomed-in image confirms the presence of quartz overgrowths and contact cement. On detrital quartz grains we observe dust rims representing the original grain surfaces that have been covered by quartz cement (*arrows*). Feldspar overgrowth and calcite cement also occur, yet quartz cement is dominating (from Avseth et al. 2009)

shales. After this point, porosity increases with increasing clay content, and velocity decreases. It also has to be mentioned that clay particles can be deposited as laminae between sand grains or intervals of sands, and these will yield a completely different elastic response than pore-filling clays (e.g. Sams and Andrea 2001).

Until recently, shales have often been regarded by geophysicists as a single type of lithology, with little attention given during seismic data analysis to the wide variation in mineralogy, texture and porosity of shales. This is partly because the rock properties of clay minerals are difficult to measure in the laboratory, but also because the acquisition of detailed log data and core samples in shale sequences has been given little priority in the oil industry. Geologists, however, have documented the complexity of shales and there is a vast amount of published literature on the geochemistry and sedimentology of shales (e.g. Bjørlykke 1998, MacQuaker et al. 2007, Peltonen et al. 2009). With increased focus on cross-disciplinary integration, geophysicists are starting to incorporate this geological knowledge into the modelling and analysis of geophysical data (e.g. Dræge et al. 2006, Brevik et al. 2007, Mondol et al. 2007, Marcussen et al. 2009). As with sands, we can distinguish between depositional and diagenetic trends in shales. Depositional trends will affect clay mineralogy, but more particularly, the silt content will have major impact on the seismic properties. Avseth et al. (2005) used simple isotropic

Fig. 18.7 The estimated cement volume based on rock physics models (*black line* in *right-hand subplot*) compared with point-count cement volume in the *upper* 100 m of the Heimdal reservoir sands in the well analysed in Figs. 18.4 and 18.5. The estimated cement volume is slightly larger than the point-counted quartz cement, but for most of the depth-range it matches nicely with the point-counted total cement volume (from Avseth et al. 2009)

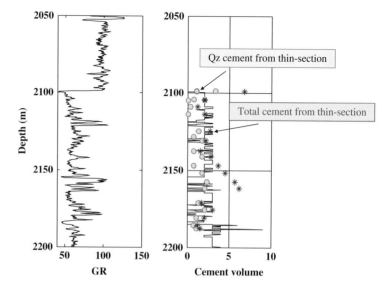

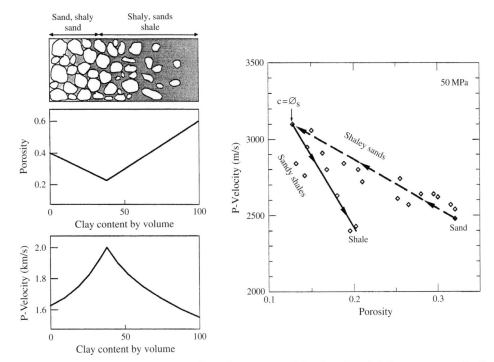

Fig. 18.8 *Left*: The Yin-Marion topological model of porosity and P-wave velocity versus clay content for shaly sands and sandy shales (from Marion 1990). *Right*: Laboratory experiments (Yin 1992) showing P-wave velocity versus porosity for unconsolidated sands and shales at constant effective pressure of 50 MPa. A clear V-shape trend is observed with increasing clay content, where velocity reaches a maximum and porosity a minimum when the clay content equals the sand porosity

lower Reuss bound to model vertical velocities of silty shales. More rigorous anisotropic modelling of shales has been performed by Hornby et al. (1994), Ruud et al. (2003), Johansen et al. (2002) and Dræge et al. (2006), among others.

One of the fundamental aspects of this chapter is to establish a link between rock physics and sedimentology. More specifically, we want to relate lithofacies to rock physics properties. This will improve our ability to use seismic amplitude information to interpret depositional systems, as facies have a major control on depositional geometries and porosity distributions. Furthermore, facies occur in predictable patterns in terms of their lateral and vertical distribution, and can also be linked to sedimentary processes. Hence, facies represent an important parameter in seismic exploration and reservoir characterisation.

Avseth et al. (2001) defined *seismic lithofacies* as a seismic-scale sedimentary unit which is characterised by its lithology (sand, silt and clay), bedding configuration (massive, interbedded or chaotic), petrography (grain size, clay location, and cementation) and seismic properties (P-wave velocity, S-wave velocity, and density). Using the concept that seismic lithofacies represent seismic-scale sedimentary units, we can improve our lateral facies prediction, as we can link facies observed in vertical well logs to seismic attribute maps in accordance with Walther's law of facies (e.g. Middleton 1973).

In the Glitne turbidite system depicted in Fig. 18.1, Avseth et al. (2001) identified various seismic lithofacies (I = conglomerates/gravels, II = massive sandstones, III = interbedded sands/shales, IV = silty shales, V = shales), and these represent a more or less gradual transition from clean sandstone to pure shale. Three subfacies of Facies II are defined based on seismically important petrographic variations within the thick-bedded sand facies. These subfacies were determined from core, thin-section and SEM analyses, in combination with rock physics diagnostics, and include cemented clean sands (Facies IIa), uncemented or friable clean sands (Facies IIb) and plane-laminated sands (Facies IIc). Our seismic lithofacies can be linked to depositional sub-environments and sedimentary processes within a deepwater clastic system (e.g. Walker 1978, Reading and Richards 1994).

Fig. 18.9 Identifying seismic
lithofacies from well log data;
example from Palaeocene
interval in the Glitne Field,
North Sea (from Avseth et al.
2001)

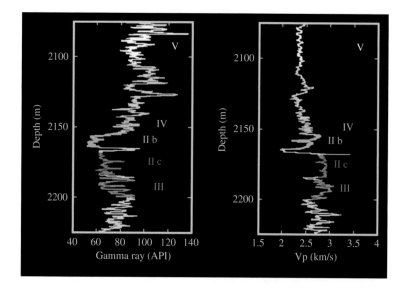

Avseth et al. (2001) selected a type-well for identification of seismic lithofacies from well log data
(Fig. 18.9). Primarily, the gamma ray log was used
to determine the different facies, as it is a good clay
indicator in the quartz-rich sediments of the North
Sea. Density and sonic logs were also used to ensure
that each facies occurs as significant clusters in terms
of rock physics properties. Rock physics analysis can
furthermore be used diagnostically to determine lithofacies when direct core and thin-section data are not
available. This was essential in order to confirm the
presence of cement in Facies IIa. The cementation
in Facies IIa is volumetrically not very significant,
but in terms of elastic properties it has an important

impact. The seismic velocities and impedances are relatively high because of the stiffening effect of initial
cementation.

Figure 18.10 shows the different seismic lithofacies plotted as acoustic impedance versus gamma ray
(left), and V_p/V_s versus gamma ray (right). For acoustic impedance, we observe an overturned V-shape,
and an ambiguity exists between Facies IIb and IV/V.
Cemented sands (IIa) and laminated sands (IIc), as
well as interbedded sands/shales, have relatively high
impedances. The sand/shale ambiguity is not observed
in the V_p/V_s versus gamma ray plot. Here we see a
more linear trend where V_p/V_s increases with increasing gramma ray values (i.e. clay content) as we go from

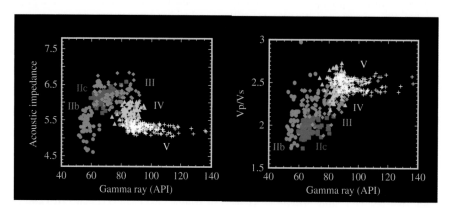

Fig. 18.10 Acoustic impedance and Vp/Vs ratio versus gamma
ray for different seismic lithofacies in the Glitne turbiditic field
depicted in Fig. 18.1. Note the overturned V-shape in the *left
hand subplot*, where clean sands and shales have similar acoustic

impedance values, whereas shaly/interbedded sands have relatively high impedance values. In the Vp/Vs versus gamma ray
plot, we observe a more linear trend as we go from clean sands
to shaly facies (from Avseth et al. 2001)

clean sands (Facies IIa and IIb) to shales (Facies IV and V). The overturned V-shape we observe in acoustic impedance can be explained physically: for grain-supported sediments, increasing clay content tends to reduce porosity (i.e. increase density) and therefore stiffen the rock. However, for clay-supported sediments, porosity will increase with increasing clay content due to the intrinsic porosity of clay, and the rock framework will weaken. Hence, velocity will reach a peak when clay content is approximately 40%, cf. the Yin-Marion model shown in Fig. 18.8. Higher V_p/V_s ratios are expected in shales than sands, since the shear strength in shales tends to be relatively low compared to sands, due to the platy shapes of clay particles.

18.4 Rock Physics and Compactional Trends

Rock physics depth trends are important in seismic exploration and borehole drilling for several reasons. Commonly, overpressured zones can be detected from seismic velocity data, indicated by negative velocity anomalies. Knowing that effective pressure and pore pressure increase linearly with depth, rock physics depth trends can be used to quantify overpressure.

It is extremely important to locate such zones since they can cause hazardous blowouts during drilling. The depth trends for sands and shales can also be used to study the expected seismic signatures of sand/shale interfaces as a function of depth.

In this section, we use existing empirical porosity/depth trends for sands and shales as input to rock physics models of V_p, V_s and density. We can for example use Hertz-Mindlin theory (Mindlin 1949; Chap. 19 by Landrø, this book) to calculate the velocity/depth trends for unconsolidated sands and shales, whereas Dvorkin-Nur's contact-cement model (Dvorkin and Nur 1996) can be used for cemented sandstones. The depth trends allow us to discriminate between pore fluids and lithologies at different depths.

Figure 18.11 shows a schematic representation of shale and sand compaction curves, and a sequence of interbedded turbidite sands and marine shales, typical for the North Sea deep-marine environment of Tertiary age. The depositional porosity in shales is normally much higher (60–80%) than in sands (c. 40%), but we expect a shallow crossover with depth due to the mechanical collapse of the shales. The platy clay fabric in the shales is more prone to compaction than the assemblage of spherically shaped grains in sands; hence the more rapid mechanical porosity reduction in shales than sands. During burial to ~2 km depth,

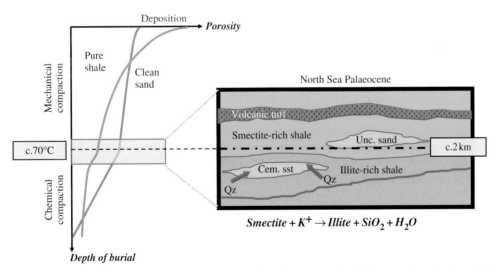

Fig. 18.11 Schematic illustration of sand and shale compaction. At c.70°C it is common to observe a change from mechanical compaction to predominantly chemical compaction in siliciclastic systems. In deep-marine depositional systems, smectite-rich shales will experience illitisation and release of bound water, causing both a porosity reduction and mineralogy change with depth. For quartz-rich sands, initial cementation tends to start at the same depth level. One possible external source of cement is in fact derived from the smectite to illite transition in embedding shales (from Avseth et al. 2008)

both sands and shales are exposed mainly to mechanical compaction. The marine shales in the North Sea Tertiary are very rich in smectite, which gives them very low permeability. In thick smectite-rich shale masses, it is therefore normal to observe undercompaction and associated overpressure even at burial depths of just some 100s of metres. At about 70°C, however, chemical alteration of smectite will commence and we expect a mineral transformation to illite. This is a typical mineral transformation seen in marine shales all over the world (Bjørlykke 1998). Bound water in the smectite layers is released when the temperature reaches this critical temperature, resulting in a porosity decrease. Moreover, the presence of potassium cations (for instance in feldspar or mica) causes quartz to be produced as well. This quartz can precipitate as microcrystalline quartz within the shale matrix (Thyberg et al. 2009), or if connectivity allows, the quartz may precipitate as cement in adjacent sandstones (Peltonen et al. 2009).

Even though there can be a link between the quartz cementation of sandstones and illitisation of smectite-rich shales, it is important to note that the source of quartz cement in sandstones can result from a variety of geological processes (Worden and Morad 2000). Figure 18.11 also includes a volcanic tuff layer that is typically encountered in the Tertiary of the North Sea (i.e. within the Balder Formation). Smectite is often generated from the alteration of volcanic tuff, but tuff also includes amorphous silica that can precipitate as quartz cement even at temperatures below 70°C. Amorphous silica is thus able to dissolve at lower temperatures than crystalline quartz.

There are also potential internal sources of quartz cement in the sandstones, either from authigenic or detrital clays in the sandstone matrix, or from the quartz grains themselves. Moreover, it has been confirmed that clay minerals can both inhibit the precipitation, and catalyse the dissolution, of quartz in sandstones. The presence of oil has also been shown to inhibit quartz cementation in oil-wetting sandstones (e.g. Giles et al. 2000). There is considerable debate among geologists over whether effective pressure at quartz grain contacts in sandstones can cause quartz dissolution and re-precipitation around the grain contacts. More likely, time and temperature control the quartz cementation (e.g. Bjørlykke and Egeberg 1993).

Regardless of the complexity of the clay and quartz diagenesis, there is empirical evidence that both the smectite to illite transformation in shales and the quartz cementation of sandstones are geochemical processes that tend to happen concurrently at around 70–80°C. In the North Sea, this corresponds to a burial depth of about 2 km which is around the target depth of the prolific Palaeocene and Eocene reservoir sands that represent major prospects for the oil industry. Geophysicists should be focused on these geological factors during seismic data analysis of reservoir sands, because dramatic changes in the seismic signatures may reflect not pore fluid changes, but diagenetic alterations in the cap-rock shales and/or reservoir sandstones.

Figure 18.12 shows well log data from a North Sea well that penetrates siliciclastic sediments and rocks of Tertiary age, superimposed with rock physics depth trends for shales and sandstones. We observe a nice match between the calculated velocity depth trends for different lithologies and the well log data. The sandstone rock physics depth trends are modelled by combining Hertz-Mindlin contact theory (see Chap. 19 by Landrø, this book) for unconsolidated sands with the Dvorkin-Nur contact cement model for cemented sandstones (Dvorkin and Nur 1996, Avseth et al. 2003, Avseth et al. 2005, Avseth et al. 2008). The input porosity/depth trends are calibrated with local compaction trends according to empirical relations (e.g. Mondol et al. 2008, Ramm and Bjørlykke 1994). The light blue model trend curves in Fig. 18.12 show how the velocities for sands increase drastically as we go from the unconsolidated regime with only mechanical compaction to the cemented regime with predominantly chemical compaction. The onset of quartz cement happens at about 70°C corresponding to about 2 km burial depth, in accordance with the observations made in Fig. 18.5, and as documented from thin-section analysis (Fig. 18.6).

We apply the Shale Compaction Model (Dræge et al. 2006, Ruud et al. 2003) to estimate the anisotropic effective properties in mechanically compacted shales. The first seismically important mineral reaction in shales is commonly the smectite to illite reaction. This reaction has several implications for the shale; the soft smectite is replaced by stiffer illite which might be distributed differently in the rock, the reaction produces water, the volume of solids is decreased (i.e. illite has a denser mineral structure than smectite), quartz is generated as a by-product, and porosity is reduced by chemical compaction. Where the shales lie within the chemical compaction regime,

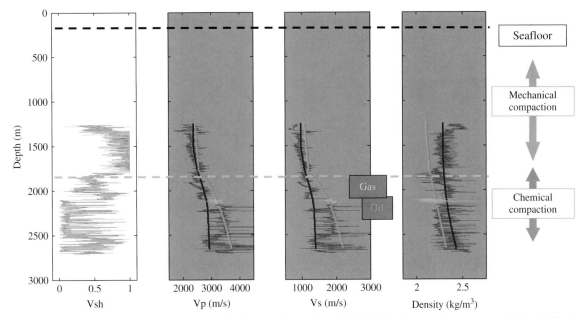

Fig. 18.12 Rock physics depth trends for shales (*dark blue*) and sandstones (*cyan*) juxtaposed on North Sea well log data penetrating a Tertiary sequence of siliciclatic sediments and rocks (same well data as in Fig. 18.5). A gas zone is indicated in *yellow* (2,099–2,151 m), and an oil zone in *red* (2,151–2,168 m). The remaining interval of the Heimdal Formation is brine-filled. The Heimdal Formation is embedded in the Lista Formation shale (Adapted from Avseth et al. 2008)

a new set of rock physics models is applied to estimate the seismic properties. The anisotropic version of a Differential Effective Medium (DEM) model and Self Consistent Approximation (SCA) are used to approximate the elongated pores and grains in shales (Hornby et al. 1994). In the present discussion, pores in chemically compacted shales are considered to be isolated, while the pores in the mechanical compaction regime are connected (cf. Dræge et al. 2006). We define a transition zone, where the properties change from the mechanical to chemical regime. In Fig. 18.13, the initial shale (<1,500 m) is considered to be smectite-rich, while the deeper (>2,200 m) illite-rich shale is somewhat stiffer. There are two counteracting effects on anisotropy. The initial alignment of grains leads to pores becoming more aligned and hence increasing anisotropy. But decreasing porosity leads to decreasing anisotropy, culminating in the anisotropic properties of the solid material at zero porosity. The pores introduce higher anisotropy than the solid, since pores commonly are weaker orthogonal to the longest axis, while the solids are less dependent on direction of wave propagation. In addition to V_p, V_s and density, we estimate the Thomsen parameters of anisotropy, δ and ϵ, which can

be significant during interpretation of angle-dependent seismic reflectivity (AVO analysis).

18.5 Seismic Fluid Sensitivity and Gassmann Theory for Fluid Substitution

Seismic fluid sensitivity is determined by a combination of porosity and pore-space stiffness. A softer rock will have a larger sensitivity to fluids than a stiffer rock at the same porosity. The most common theory for fluid substitution is the so-called Gassmann theory (Gassmann 1951, Mavko et al. 2009, Avseth et al. 2005), which describes the fluid sensitivity of porous, isotropic rocks at seismic frequency (i.e. when no grain-scale viscous flow effects will stiffen the rock frame). The Gassmann theory is very important in all rock physics modelling, therefore both the formulation and workflow of the theory are included in this overview chapter.

For the fluid substitution problem there are two fluid effects that must be considered: the change in rock

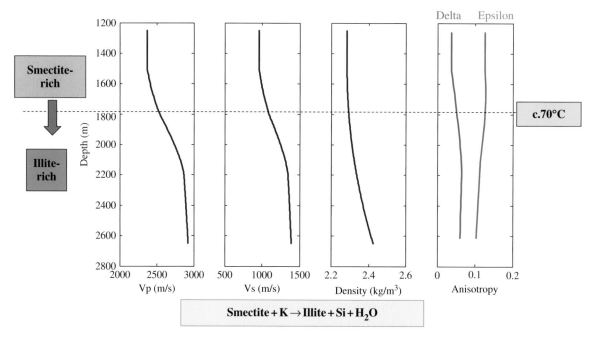

Fig. 18.13 Modelled rock physics depth trends of shales, showing the effect of illitisation of marine smectite-rich shales (from Avseth et al. 2008)

bulk density, and the change in rock compressibility. The compressibility of a dry rock (reciprocal of the rock bulk modulus) can be expressed quite generally as the sum of the mineral compressibility and an extra compressibility due to the pore space:

$$\frac{1}{K_{dry}} = \frac{1}{K_{mineral}} + \frac{\phi}{K_\phi} \qquad (18.1)$$

where ϕ is the porosity, K_{dry} is the dry rock bulk modulus, $K_{mineral}$ is the mineral bulk modulus, and K_ϕ is the pore space stiffness defined by:

$$\frac{1}{K_\phi} = \frac{1}{v_{pore}} \frac{\partial v_{pore}}{\partial \sigma} \qquad (18.2)$$

Here, v_{pore} is the pore volume, and σ is the increment of hydrostatic confining stress from the passing wave. Poorly consolidated rocks, rocks with microcracks, and rocks at low effective pressure are generally soft and compressible and have a small K_ϕ. Stiff rocks that are well cemented, lacking microcracks, or at high effective pressure, have a large K_ϕ. In terms of the popular, but idealised, ellipsoidal crack models, low aspect ratio cracks have small K_ϕ and rounder large aspect ratio pores have large K_ϕ.

Similarly, the compressibility of a *Saturated* rock can be expressed as:

$$\frac{1}{K_{sat}} = \frac{1}{K_{mineral}} + \frac{\phi}{K_\phi + \dfrac{K_{fluid}K_{mineral}}{K_{fluid} + K_{mineral}}} \qquad (18.3)$$

or approximately as:

$$\frac{1}{K_{sat}} \approx \frac{1}{K_{mineral}} + \frac{\phi}{K_\phi + K_{fluid}} \qquad (18.4)$$

where K_{fluid} is the pore fluid bulk modulus. Comparing equations (18.1) and (18.4), we can see that changing the pore fluid will modify the effective pore space stiffness (Avseth et al., 2005). From equation (18.4) we see also the well-known result that a stiff rock, with large pore space stiffness K_ϕ, will have a small sensitivity to fluids, and a soft rock, with small K_ϕ, will have a larger sensitivity to fluids.

Equations (18.1) and (18.3) together are equivalent to Gassmann's (1951) relations. We can algebraically eliminate K_ϕ from Eqs. (18.1) and (18.3) and write Gassmann's relations in one of the more familiar, but less intuitive, forms:

$$\frac{K_{sat}}{K_{mineral} - K_{sat}} = \frac{K_{dry}}{K_{mineral} - K_{dry}} + \frac{K_{fluid}}{\phi(K_{mineral} - K_{fluid})} \qquad (18.5)$$

and the companion result:

$$\mu_{\text{sat}} = \mu_{\text{dry}} \qquad (18.6)$$

Gassmann's equations (18.5) and (18.6) predict that for an isotropic rock, the rock bulk modulus will change if the fluid changes, but the rock shear modulus will not.

These dry and saturated moduli, in turn, are related to P-wave velocity $V_p = \sqrt{(K + (4/3)\mu)/\rho}$ and S-wave velocity $V_s = \sqrt{\mu/\rho}$, where ρ is the bulk density given by:

$$\rho = \phi\rho_{\text{fluid}} + (1 - \phi)\rho_{\text{mineral}} \qquad (18.7)$$

In the equations above, ϕ is normally interpreted as the total porosity, though in shaly sands the better choice is to use effective porosity during fluid substitution (Dvorkin et al. 2007). Another uncertainty stems from Gassmann's assumption that the rock is monomineralic. Clay-rich sandstones actually violate the monomineralic assumption, but an effective mineral modulus can be approximated via the Hill's average or Hashin-Shtrikman modelling (see Mavko et al. 2009). Another issue is whether the clay should be considered only as part of the mineral frame, or should the bound water that constitutes part of the clay mineral be considered part of the fluid when we do Gassmann fluid substitution? If the latter, then the functional Gassmann porosity is actually larger than the total porosity, but the pore fluid should be considered a muddy suspension containing clay particles.

Gassmann's relations were originally derived to describe the change in rock modulus from one pure saturation to another – from dry to fully brine-saturated, from fully brine-saturated to fully oil-saturated, etc. Domenico (1976) suggested that mixed gas-oil-brine saturations can also be modelled with Gassmann's relations, if the mixture of phases is replaced with an effective fluid with bulk modulus $\bar{K}_{\text{fluid}}$ and density $\bar{\rho}_{\text{fluid}}$ given by

$$\frac{1}{\bar{K}_{\text{fluid}}} = \frac{S_{\text{gas}}}{K_{\text{gas}}} + \frac{S_{\text{oil}}}{K_{\text{oil}}} + \frac{S_{\text{br}}}{K_{\text{br}}} = \left\langle \frac{1}{K_{\text{fluid}}(x,y,z)} \right\rangle \qquad (18.8)$$

$$\bar{\rho}_{\text{fluid}} = S_{\text{gas}}\rho_{\text{gas}} + S_{\text{oil}}\rho_{\text{oil}} + S_{\text{br}}\rho_{\text{br}} = \langle \rho_{\text{fluid}}(x,y,z) \rangle \qquad (18.9)$$

where $S_{\text{gas, oil, br}}$, $K_{\text{gas, oil, br}}$, and $\rho_{\text{gas, oil, br}}$ are the saturations, bulk moduli and densities of the gas, oil and brine phases. The operator $\langle \cdot \rangle$ refers to a volume

average and allows for more compact expressions, where $K_{\text{fluid}}(x,y,z)$ and $\rho_{\text{fluid}}(x,y,z)$ are the spatially varying pore fluid modulus and density.

Substituting Eq. (18.8) into Gassmann's relation is the procedure most widely used today to model fluid effects on seismic velocity and impedance for low frequency field applications. The fluid properties are obtained through the Batzle and Wang (1992) equations, and these normally comprise brine salinity, gas gravity, oil reference density, gas-oil-ratio (GOR), temperature and pore pressure.

A problem with mixed fluid phases is that velocities depend not only on saturations but also on the spatial distributions of the phases within the rock (Fig. 18.14). Equation (18.8) is applicable only if the gas, oil and brine phases are mixed uniformly at a very small scale, so that the different wave-induced increments of pore pressure in each phase have time to diffuse and equilibrate during a seismic period (Mavko et al. 2009). Equation (18.8) is the Reuss (1929) average or "isostress" average, and it yields an appropriate equivalent fluid when all pore phases have the same wave-induced pore pressure. A simple dimensional analysis suggests that during a seismic period pore pressures can equilibrate over spatial scales smaller than $L_c \approx \sqrt{\kappa K_{\text{fluid}}/f\eta}$, where f is the seismic frequency, κ is the permeability, and η and K_{fluid} are the viscosity and bulk modulus of the most viscous fluid phase. We refer to this state of fine-scale, uniformly mixed fluids as "uniform saturation."

In contrast, saturations that are heterogeneous over scales larger than $\sim L_c$ will have wave-induced pore

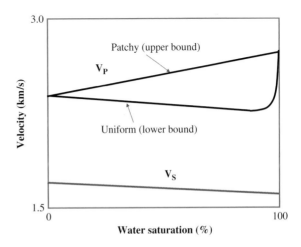

Fig. 18.14 Uniform versus patchy saturation, mixing gas and water in a porous rock according to Gassmann theory

pressure gradients that cannot equilibrate during the seismic period, and Eq. (18.8) will fail. We refer to this state as "patchy saturation." Patchy saturation, or "fingering" of pore fluids, can easily be caused by spatial variations in wettability, permeability or shaliness. The rock modulus with patchy saturation can be approximated by Gassmann's relation, with the mixture of phases replaced by the "isostrain" (Voigt, 1910) average effective fluid (Mavko et al. 2009):

$$\bar{K}_{\text{fluid}} = S_{\text{gas}}K_{\text{gas}} + S_{\text{oil}}K_{\text{oil}} + S_{\text{br}}K_{\text{br}} \qquad (18.10)$$

This implies that patchy saturation will cause higher velocities and impedances than when the same fluids are mixed at a fine scale. The fact that the fluids can not move freely and equilibrate during a seismic wave period results in a stiffer effective rock than is the case with uniform saturation where the fluids have time to equilibrate.

Equation (18.10) appears to be an upper bound, and data seldom fall on it, except at very low gas saturations.

18.6 Rock Physics Templates (RPTs)

We can combine the depositional and diagenetic trend models presented above with Gassmann fluid substitution, and make charts or templates of rock physics models for predicting lithology and presence of hydrocarbons. We refer to these locally constrained charts as *Rock Physics Templates* (RPTs), a methodology first presented by Ødegaard and Avseth (2004). Furthermore, we expand on the rock physics diagnostics presented earlier as we create RPTs of seismic parameters, in our case acoustic impedance versus V_p/V_s ratios (Fig. 18.15). This will allow us to perform rock physics analysis not only of well log data, but also of seismic data (e.g. elastic inversion results).

The motivation behind RPTs is to generate an atlas or collection of relevant rock physics models for different basins and areas. Then, the ideal interpretation workflow becomes a fairly simple 2-step procedure: (1) Select the appropriate RPT for the area and depth under investigation, using well log data to verify the validity of the selected RPT(s). (2) Use the selected and verified RPT(s) to interpret elastic inversion results. RPT interpretation of well log data may also be an important stand-alone exercise both for the interpretation and quality control of well log data, and in order to assess seismic detectability of different fluid and lithology scenarios.

The RPTs are site (basin) specific and are constrained by local geological factors, including lithology, mineralogy, burial depth, diagenesis, pressure and temperature. All these factors must be considered when generating RPTs for a given basin. In particular, it is essential to include only the expected lithologies for the area under investigation when generating the rock physics templates. The water depth and the burial depth determine the effective pressure, pore pressure and lithostatic pressure. The pore pressure is important for the calculation of fluid properties, and for determining the effective stress on the grain contacts of the rock frame carrying the overburden.

In modelling RPTs we also need to know the acoustic properties of mud-filtrate, formation water and hydrocarbons in the area of investigation. Required input parameters include temperature, pressure, brine salinity, gas gravity, oil reference density and GOR. In areas where hydrocarbons have yet to be encountered, gas gravity can be assumed (normally 0.6–0.8). However, oil reference density is more uncertain. Also, the seismic response of oil can be difficult to distinguish from that of brine. One should, however, expect oil to show values similar to low gas saturation in AI versus V_p/V_s ratio crossplots. Regarding saturation distribution, we often assume uniform distribution in the template modelling (Ødegaard and Avseth, 2004), which gives the famous effect where residual amounts of gas will produce almost the same seismic properties as commercial amounts of gas. However, a patchy distribution of gas will show a more linear change in seismic properties with increasing gas saturation (cf. Fig. 18.14).

Pore fluid sensitivity in reservoir sandstones is highly affected by reservoir heterogeneity and sandstone microstructure, and it is therefore important to include these geological factors in the rock physics analysis. As already indicated, initial cement reduces the pressure and fluid sensitivity of the sandstones. Figure 18.15 shows schematically the outline of a rock physics template, where calibrated rock physics models have been selected that fit local data observations (well log data or seismic inversion data) of various lithologies and pore fluids. The presence of diagenetic quartz cement will move brine-saturated sandstone

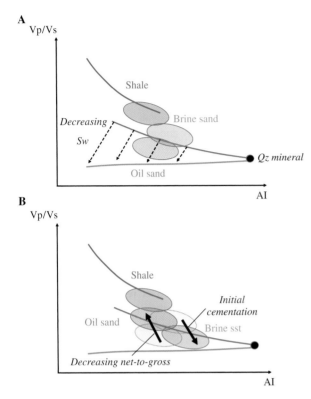

Fig. 18.15 Rock physics templates (RPTs) can be made from the rock physics models in Fig. 18.3 combined with Gassmann theory, to define regions where expected facies and fluids will plot. In particular, the Vp/Vs ratio is known to be a great fluid discriminator in siliciclastic environments. Homogeneous, unconsolidated sands filled with oil are normally well separated from ditto brine filled sands in an RPT of Vp/Vs versus acoustic impedance (AI), c.f. schematic illustration in the upper cross-plot. However, the effect of initial cement will reduce the fluid sensitivity of sandstones and the Vp/Vs ratio of cemented brine sandstones can be similar to the Vp/Vs ratio of unconsolidated sands filled with oil. The effect of net-to-gross will normally move data in the opposite direction in a Vp/Vs-AI cross-plot. Hence, oil sands with relatively low net-to-gross can have higher Vp/Vs ratio than homogeneous brine sands with initial cement, c.f. schematic illustration in lower cross-plot (Adapted from Avseth et al. 2009)

data in an AI versus V_p/V_s crossplot to an area of very low V_p/V_s, where we would expect hydrocarbon-saturated sandstones to plot. By contrast, reservoir heterogeneity and decreasing net-to-gross associated with interbedded sands and shales tend to move data points in the direction of the shale cluster. The cement effect is a microstructural effect, whereas the net-to-gross is related to the scale of shale/sand interbedding and is frequency dependent. In the case where the interbedded shale is relatively soft compared to the

sand, the net-to-gross effect will counteract the effect of cement on effective rock stiffness, hence these two effects will have opposite trends in the AI-V_p/V_s cross-plot. The schematic, but qualitatively correct illustration in Fig. 18.15 demonstrates why it is important to include these geological factors when analysing the rock physics properties and seismic fluid sensitivity in reservoir sandstones.

Figure 18.16 shows an RPT including data from two neighboring wells penetrating Palaeocene sands in the North Sea (Avseth et al. 2009). One well penetrated a thick, turbiditic gas sand with a thin oil-leg, whereas the adjacent well penetrated a turbidite sand filled with oil. Furthermore, one of the wells is the same one as was used in the rock physics diagnostics in Figs. 18.4 and 18.5, and in the depth trend modelling above (Fig. 18.12). It turns out that the sandstone quality changes from the one well to the other, and this drastically distorts the fluid sensitivity to hydrocarbons. The gas-saturated Heimdal sands in well 1 show a small increase in acoustic impedance relative to the cap rock shale, while the oil-saturated sands in well 2 show a significant drop in acoustic impedance. This drastic change in sandstone quality over a short distance will also yield a corresponding change in seismic signatures, see next section below.

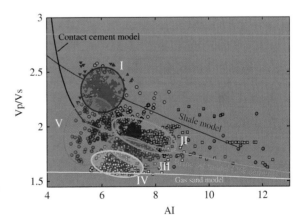

Fig. 18.16 RPT of Vp/Vs versus acoustic impedance (AI) for target zone (Palaeocene) of two North Sea wells. Cluster I is the cap rock shale in both wells, II comprises the brine sandstones in both wells, III and IV are reservoir sandstones in well 1 filled with oil and gas, respectively. V is the upper oil zone in well 2, with Vp/Vs higher than the brine sandstones, and AI lower than the gas sandstones in well 1. This is counter-intuitive, and must be explained by difference in sandstone quality, c.f. Fig. 18.15 (Adapted from Avseth et al. 2009)

18.7 Seismic Reservoir Characterisation Using AVO

More than 20 years ago, William Ostrander published a ground-breaking paper on offset-dependent reflectivity (Ostrander 1984). He showed that gas-saturated sands capped by shales would cause an *amplitude variation with offset* (AVO effect) in pre-stack seismic data (Fig. 18.17). Shortly after, AVO technology became a commercial tool for the oil industry, quickly gaining in popularity as it was now possible to explain seismic amplitudes in terms of rock properties. The technique proved successful for hydrocarbon prediction in many areas of the world, yet in some cases it failed. The technique suffered from ambiguities caused by lithological effects, tuning effects and overburden effects. It turned out that even seismic processing and acquisition effects could cause false AVO anomalies. Application of AVO analysis was therefore reduced. However, in many of these cases it was incorrect use of the technique that was responsible for the failure, not the technique itself. In the last decade we have observed a revival of the AVO technique. This is due to the improvement of 3D seismic technology, better pre-processing routines, more frequent shear-wave logging that can better constrain AVO modelling, improved understanding of rock physics properties, greater data capacity, more focus on cross-disciplinary aspects of AVO and, last but not least, more awareness among users of the potential pitfalls.

The technique provides the seismic interpreter with information about pore fluids and lithologies, which complements the conventional interpretation of seismic facies, stratigraphy and geomorphology. The offset dependent reflectivity, $R(\theta)$, also referred to as the AVO response, is given by the famous, but complex and non-linear Zoeppritz equations (Zoeppritz 1919), which have been approximated and linearised by Shuey (1985), among others, into a two-term expression:

$$R(\theta) \approx R(0) + G \cdot \sin^2 \theta \qquad (18.11)$$

where $R(0)$ is the zero offset reflectivity at a given seismic horizon, controlled by the contrast in acoustic impedance (AI) across this interface:

$$R(0) = \frac{AI_2 - AI_1}{AI_1 + AI_2} \qquad (18.12)$$

Here, AI_1 and AI_2 are the acoustic impedances in the cap rock (Layer 1) and reservoir (Layer 2), respectively.

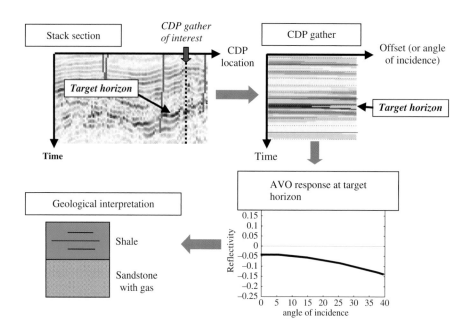

Fig. 18.17 Schematic illustration of the principles in AVO analysis (from Avseth et al. 2005)

The AVO gradient, G, is strongly affected by the contrast in V_p/V_s ratios across the same interface:

$$G = \frac{1}{2}\frac{\Delta V_p}{V_p} - 2\frac{V_s^2}{V_p^2}\left(\frac{\Delta\rho}{\rho} + 2\frac{\Delta V_s}{V_s}\right) \quad (18.13)$$

The most common and practical way to do AVO analysis of seismic data is to make crossplots of the zero-offset reflectivity ($R(0)$) versus the AVO gradient (G). These attributes are estimated from pre-stack seismic gathers using simple least-square regressions, according to Eq. (18.11). Often it is assumed that calibrated near stack seismic data, where the near offset traces have been stacked together, is representative of the zero offset reflectivity. Furthermore, it can be assumed that the difference between the far stack seismic (where the far offset traces have been stacked together) and the near stack seismic is a scaled version of the AVO gradient. Hence, AVO crossplots can be made directly from near and far stack seismic sections. Brine-saturated sands interbedded with shales, situated within a limited depth range and at a particular locality, normally follow a well defined "background trend" in AVO crossplots. A common and recommended approach in qualitative AVO crossplot analysis

is to recognise the "background" trend and then look for data points that deviate from this trend. Deviations from the background trend in an AVO crossplot may be indicative of hydrocarbons.

Figure 18.18 shows a seismic line with two wells indicated as black lines (same wells that are analysed in Fig. 18.16). One well penetrated a thick turbiditic gas sand with a thin oil-leg, whereas the adjacent well penetrated a turbidite sand filled with oil. However, it turned out that the AVO anomaly of the oil sand is stronger than for the gas discovery (Fig. 18.19), and the oil sands show a negative bright near stack response, whereas the gas sands show a dim (i.e. close to zero) near stack response. These observations are counter-intuitive if the reservoir sands are similar. Avseth et al. (2009) demonstrated the importance of rock texture, in particular cement volume, as well as net-to-gross, and how these geological factors affected the seismic signatures of hydrocarbons within these sands.

The deviation from the background trend in an AVO crossplot can be quantified in terms of the fluid factor (Smith and Gidlow 1987, Fatti et al. 1994) as follows:

$$\Delta F = 8/5 R(0) - \kappa \cdot [R(0) - G] \quad (18.14)$$

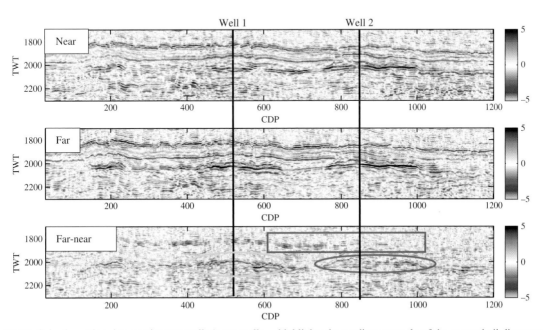

Fig. 18.18 Seismic sections intersecting two wells (same wells that are analysed in Fig. 18.16), including near stack, far stack, and the estimated far-near stack. The *blue square* indicates the window where a background trend is defined. The *yellow ellipse* highlights the gradient anomaly of the gas and oil discovery of Well 1. The *red ellipse* highlights an adjacent oil discovery of Well 2 (Adapted from Avseth et al. 2009)

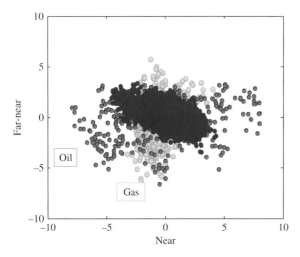

Fig. 18.19 Intercept (Near) versus gradient (i.e. Far-Near) for the seismic stack section in Fig. 18.18. Only data from the selected polygons are included. The *blue points* represent the background trend *right* above the target, the *yellow points* represent data from the gas (and oil) discovery in Well 1 in Fig. 18.18, whereas the *red data points* represent data from the oil discovery in Well 2 in Fig. 18.18 (Adapted from Avseth et al. 2009)

where κ is a constant at a given depth, depending on the background V_p/V_s ratio, χ, and the slope of the V_p versus V_s relationship of the modelled shale trend, m:

$$\kappa = m/\chi \qquad (18.15)$$

Commonly, m is seleced to be the slope of the *Mudrock Line* (Castagna et al. 1985), where $m= 1.16$. Similarly, the background V_p/V_s ratio (χ) is often set equal to 2, giving a κ value of 0.58. However, with modelled shale trends, we can estimate more realistic depth-dependent values of κ and get a better control on the expected background trend to be used in the fluid factor (Avseth et al. 2008). Both gas- and oil-filled sands will cause negative fluid factor anomalies relative to the background brine trend. However, chemical compaction has a significant impact on the absolute value of the fluid factor, which is directly related to the fact that the fluid sensitivity decreases with depth and increasing rock stiffness (Fig. 18.20).

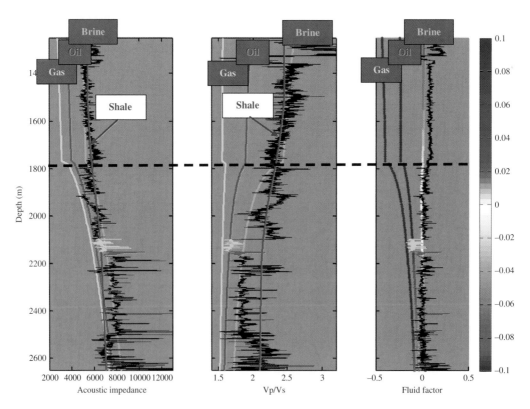

Fig. 18.20 Seismic depth trends and expected fluid factor. Note that both oil and gas sands show large negative fluid factor values down to about 2,300–2,400 m burial depth, whereas brine sands show weak or positive fluid factor values for all depth ranges. Gas sands seem to give negative fluid factor even deeper than the modelled window, beyond 2,600 m. The gas and oil zone in the well log data also show a relatively strong fluid factor anomaly, as expected from the depth trends (Adapted from Avseth et al. 2008)

An improved understanding of rock physics depth trends in sands and shales helps us to better understand how AVO signatures change with depth. By conducting AVO depth trend modelling, we can verify the feasibility of AVO analysis as a function of burial depth and diagenesis, and thereby extrapolate to other burial depths. This can be a useful task in exploration or appraisal, where we want to extend our search to slightly deeper or shallower prospects adjacent to existing discoveries.

We can now estimate the fluid factor attribute for the seismic line shown in Fig. 18.18. The fluid factor can be derived from near- and far-stack amplitudes using the formula:

$$\Delta F = \text{Near} - \eta \cdot (\text{Far} - \text{Near}) \qquad (18.16)$$

Here, η is not the same as κ in the earlier formulation of the fluid factor, since Far–Near is not exactly the same as the AVO gradient. However, the two-term AVO makes a linear relationship between the Far–Near and the AVO gradient, and therefore η and κ can also be easily correlated. Assuming the far stack to be around 30° and the near stack to be at 0° (normally the far stack will be representative for slightly lower angles than 30°, whereas near stack will be representative for angles slightly higher than 0°) we obtain the following approximate relationship between the gradient and the Far–Near:

$$\text{Far} - \text{Near} = R(30) - R(0) = G \cdot \sin^2(30) = G \cdot 0.25 \qquad (18.17)$$

Solving for the background trend, that is when $\Delta F = 0$, we obtain:

$$\text{Near} = \frac{-20\kappa}{8 - 5\kappa}(\text{Far} - \text{Near}) \qquad (18.18)$$

and hence:

$$\eta = \frac{-20\kappa}{8 - 5\kappa} \qquad (18.19)$$

The slope of the background trend in the Near versus Far–Near stack is $1/\eta$, and this shows that we can in fact estimate the expected background trend for uncalibrated (but offset balanced) range-limited stacks using the modelled shale trends as illustrated above

(Fig. 18.13). This can be a useful exercise for verifying the correct amplitude balancing between near- and far-stack data during AVO crossplot analysis.

Based on the shale model trends, we estimate the value of η to be of the order 2.5. This means the background trend of the shale in the AVO crossplot is relatively flat, equalling -0.4, i.e., Far$-$Near$= -0.4$ Near. We observe indeed that the shale background trend is relatively flat (blue cloud in Fig. 18.19). Using this background trend, the resulting fluid factor attribute in Fig. 18.21 shows that both the gas and the oil discoveries stand out as strong fluid factor anomalies, whereas the background data is relatively weak.

We also estimate the difference between intercept and gradient, which has been shown to be close to $\Delta V_s/V_s$ (Fatti et al. 1994, Avseth et al. 2005). This attribute should not be significantly affected by pore fluids, but very sensitive to lithology and cementation. Figure 18.21 shows the fluid factor crossplot and line superimposed with the $\Delta V_s/V_s$ attribute for the seismic line intersecting both well 1 and well 2. As expected, both the gas and the oil discoveries show strong fluid factors, whereas only well 1 (gas discovery) shows a strong change in $\Delta V_s/V_s$ at the top of the reservoir. This tallies with the observations from the well log data, and also with the rock physics estimation of relatively low cement volume at the top of the Heimdal sands in well 2 (see Avseth et al. 2009). It further explains why we have such drastically different AVO signatures in the two wells, while the counter-intuitive AVO anomalies for oil and gas are explained by the local change in rock texture. With two seismic parameters ($R(0)$ and G) it is possible to discriminate the fluid effects from the lithological effects.

A problem when interpreting AVO crossplots is that a given point in the crossplot does not correspond to a unique combination of rock properties. Many combinations of rock properties will yield the same $R(0)$ and G. Moreover, due to natural variability in geological and fluid parameters, one given geological scenario may span a relatively large possible outcome area in the AVO crossplot, not just a discrete point. Hence, a hydrocarbon-like AVO response might occasionally result from a brine-associated reflection, and hydrocarbon-saturated sands might not always produce an anomalous AVO response; this problem was seen in the RPT crossplots in Fig. 18.15. One way to resolve this uncertainty is to create probability crossplots of various categories of lithology and pore fluid scenarios.

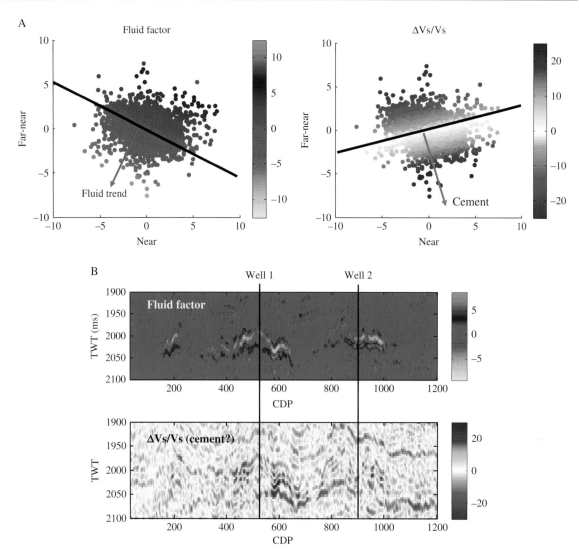

Fig. 18.21 AVO crossplots with fluid factor and ditto with $\Delta Vs/Vs$ as colored attribute, for the seismic section intersecting the two wells in Fig. 18.18. The fluid factor attribute represents the deviation from the local background shale trend. The $\Delta Vs/Vs$ attribute reflects rock stiffness, and it is expected to be an appropriate attribute to detect initial quartz cement in the Heimdal Formation sandstones. The resulting AVO attributes (*lower*) show strong fluid factor values at both well locations, whereas the $\Delta Vs/Vs$ attribute shows a much stronger contrast at well 1 than at well 2. Avseth et al. (2009) documented that the *uppermost part* of Heimdal sandstones in well 2 are less cemented than the Heimdal sandstones in well 1 (c.f. RPT observations in Fig. 18.16), and this can also explain the different AVO responses at the two well locations

These can be based on statistical analysis of well log data and/or rock physics models (Avseth 2000, Avseth et al. 2001, Mukerji et al. 2001). Each category is plotted as a "contour map", almost like topographical maps (Fig. 18.22). Here, the "mountain tops" represent the most likely values of a given class. It is very important to be aware that the contours of different facies and fluids overlap one another. This implies that an observed set of $R(0)$ and G can represent more than one category. This is one reason why AVO analysis, used alone, can give misleading results. In addition, these crossplots are often affected by noise in the seismic data. Nevertheless, as we see in Fig. 18.22, we often observe that we need both intercept ($R(0)$) and gradient (G) in order to separate facies and fluid types from seismic data. Or put another way, we often need to do AVO to discriminate lithology from fluids on seismic data.

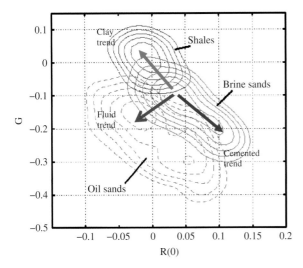

Fig. 18.22 AVO contour plots or probability density functions (pdfs) for main facies and fluids in the Glitne field. Only the iso-probability contours of 50% and larger are included for each group. The R(0) and G pdfs nicely separate the three facies groups, but there are significant overlaps (Adapted from Avseth et al. 2001)

Avseth et al. (2001) applied the contour plots in Fig. 18.22 to conduct an AVO analysis and predict seismic lithofacies and pore fluids from the pre-stack seismic amplitudes in the Glitne Field which are depicted in the amplitude map in Fig. 18.1. As demonstrated in Fig. 18.1, the depositional trends in the Glitne turbidite system do affect the seismic amplitudes. This is also confirmed by deterministic AVO analysis of CDP gathers at the three well locations depicted in Fig. 18.1, see Fig. 18.23. The dramatic variability in the lateral facies distribution, going from a relatively proximal feeder-channel environment to a relatively distal lobe and lobe margin environment, has great impact on the seismic signatures in this turbidite system.

The next step is facies and pore fluid prediction from 3D seismic data. 3D AVO inversion is performed on the turbidite system using Hampson-Russell's AVO software. Again, we focus only on the horizon representing the top of the system (Top

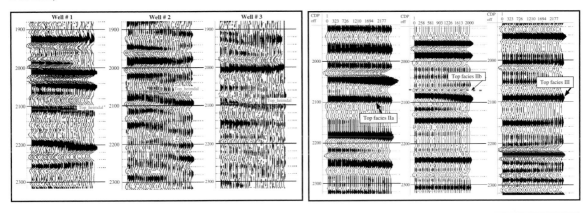

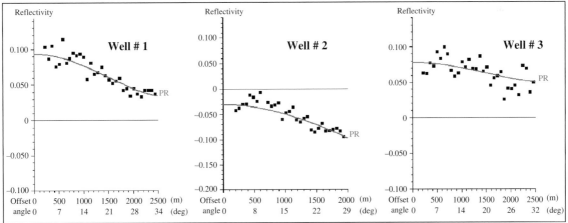

Fig. 18.23 Real CDP gathers (*upper left*), synthetic CDP gathers (*upper right*), and AVO curves (*lower*) for three wells penetrating the Glitne turbidite system (Fig. 18.1) at different locations. Note the drastic change in AVO signature from the feeder-channel (cemented sandstone filled with brine), to the lobe-channel (unconsolidated sand filled with oil), and to the lobe margin (interbedded sand-shale and shaly sandstone) (from Avseth et al. 2001)

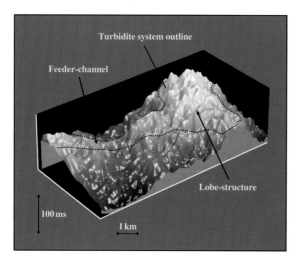

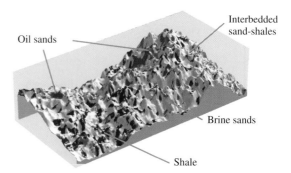

Fig. 18.24 Three-dimensional seismic topography of Top Heimdal horizon (two-way traveltime). The depositional geometry of a feeder-channel and fan lobe is outlined (compare to Fig. 18.1) (Adapted from Avseth et al. 2001)

Fig. 18.26 Lithofacies prediction beneath Top Heimdal seismic horizon, in the Glitne Field. Note the prediction of most likely oil sands in the top lobe structure, as well as in the *upper part* of the feeder channel. The latter is not realistic, knowing that a well is penetrating brine-filled feeder-channel sands *right* outside the amplitude map. However, as we see in the AVO contour plots in Fig. 18.22, we expect some ambiguity and overlap between brine sands and oil sands (Adapted from Avseth et al. 2001)

Heimdal). Figure 18.24 shows the three dimensional topography (in two-way traveltime) of this seismic horizon, where the geometries of the feeder-channel and the lobe structure are outlined. The inversion gives us $R(0)$ and G over the whole area across this horizon slice. Figure 18.25 shows the intercept $R(0)$ and

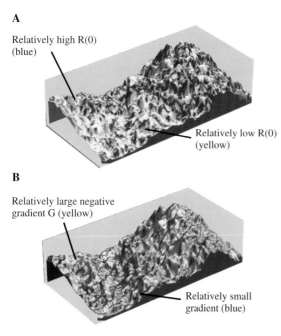

Fig. 18.25 Intercept ($R(0)$) and AVO gradient (G) estimated along Top Heimdal horizon in the Glitne Field (Adapted from Avseth et al. 2001)

Fig. 18.26 shows the gradient G. These plots allow us to predict the most likely seismic lithofacies underlying this horizon. This is done by combining the $R(0)$ and G inverted from the seismic with the $R(0)-G$ contour plots (Fig. 18.22), also referred to as probability density functions (pdfs), which are derived from well log data. Before we can do this, however, the inverted parameters must be calibrated to the well log values (see Avseth et al. 2005 for further details).

We apply the Mahalanobis distance method to calculate the most likely facies group and pore fluid from the seismic data (Avseth et al. 2005). The results are shown in Fig. 18.26. We predict oil-saturated sands in the lobe area where the lobe is structurally highest. The rest of the lobe area is most likely water-saturated according to the prediction. Furthermore, we predict oil-filled sands in the upper feeder-channel. Outside the submarine fan, the most likely facies is predicted to be dominated by shale. The sands are mainly predicted in the channel and lobe areas while oil is predicted in the structurally highest areas of the sand deposits. The depositional patterns predicted from the AVO analysis agrees well with the expected facies associations in the Glitne submarine fan. (For a more detailed facies prediction, see Avseth et al. 2005). This case example demonstrates that AVO and quantitative seismic interpretation can be particularly useful in geological settings where facies prediction is not easily obtained from conventional seismic interpretation.

18.8 Conclusions

In this chapter we have demonstrated how we can apply rock physics tools to relate geological parameters and trends to seismic properties, and thereby predict rock and fluid characteristics from seismic amplitude data. The focus of this chapter has been to present a workflow for quantitative seismic interpretation during petroleum exploration. Rock physics ultimately serves as a toolbox for lithology and fluid substitution, where we can understand local well log observations and extrapolate to certain expected "what if" scenarios, either in terms of depositional or compactional trends, or in terms of changes in pore fluids. The models can also be applied to predict or classify reservoir parameters from sonic and seismic data. We have demonstrated our cross-disciplinary approach on well log and seismic data from the Palaeocene interval of the North Sea where turbidite sands are prone to change drastically, both because of facies variability and diagenetic alterations. We have also demonstrated the importance of understanding the rock physics and seismic properties of shales during seismic reservoir prediction using AVO analysis.

Acknowledgements Thanks to Prof. Gary Mavko and Prof. Tapan Mukerji at Stanford University for collaboration over a long period and contributions to the work presented here. Thanks to Arild Jørstad and Hans Oddvar Augedal at Lundin-Norway for geological input to the Palaeocene sands studied in this chapter; also thanks to Aart-Jan van Wijngaarden, Erik Ødegaard, Torbjørn Fristad and Anders Dræge at Statoil for fruitful discussions and input to the work included in this chapter. I also acknowledge Prof. Ran Bachrach at Tel Aviv University and Prof. Tor Arne Johannesen at University of Bergen for valuable discussions.

Further Reading

Avseth, P. 2000. Combining rock physics and sedimentology for seismic reservoir characterization of North Sea turbidite systems. Ph.D. thesis, Stanford University, 200 pp.

Avseth, P., Dræge, A., van Wijngaarden, A.-J., Johansen, T.A. and Jørstad, A., 2008. Shale rock physics and implications for AVO analysis: A North Sea demonstration. The Leading Edge 27, 788–797.

Avseth, P., Dvorkin, J., Mavko, G. and Rykkje, J. 2000. Rock physics diagnostic of North Sea sands: Link between microstructure and seismic properties. Geophysical Research Letters 27, 2761–2764.

Avseth, P., Flesche, H. and van Wijngaarden, A.-J. 2003. AVO classification of lithology and pore fluids constrained by rock physics depth trends. The Leading Edge 22, 1004–1011.

Avseth, P., Jørstad, A., van Wijngaarden, A.-J. and Mavko, G. 2009. Rock physics estimation of cement volume, sorting and net-to-gross in North Sea sandstones. The Leading Edge 28, 98–108.

Avseth, P., Mukerji, T., Jørstad, T., Mavko, G. and Veggeland, T. 2001. Seismic reservoir mapping from 3-D AVO in a North Sea turbidite system. Geophysics 66, 1157–1176.

Avseth, P., Mukerji, T. and Mavko, G. 2005. Quantitative Seismic Interpretation – Applying Rock Physics Tools to Reduce Interpretation Risk. Cambridge University Press, Cambridge, 376 pp.

Avseth, P., Mukerji, T., Mavko, G. and Tyssekvam, J.A. 2001. Rock physics and AVO analysis for lithofacies and pore fluid prediction in a North Sea oil field. The Leading Edge 20, 429–434.

Batzle, M. and Wang, Z. 1992. Seismic properties of pore fluids. Geophysics 57, 1396–1408.

Bjørlykke, K. and Egeberg, K. 1993. Quartz cementation in sedimentary basins. AAPG Bulletin 77, 1538–1548.

Bjørlykke, K. 1998. Clay mineral diagenesis in sedimentary basins – A key to the prediction of rock properties; Examples from the North Sea Basin. Clay Minerals 33, 15–34.

Box, G.E.P. 1976. Science and statistics. Journal of the American Statistical Association 71, 791–799.

Brevik, I. et al., 2007. Documentation and quantification of velocity anisotropy in shales using wireline log measurements. The Leading Edge 26, 272–277.

Castagna, J.P., Batzle, M.L. and Eastwood, R.O. 1985. Relationships between compressional-wave and shear-wave velocities in clastic silicate rocks. Geophysics 50, 571–581.

Domenico, S.N. 1976. Effect of brine-gas mixture on velocity in an unconsolidated sand reservoir. Geophysics 41, 882–894.

Dræge, A., Jakobsen, M. and Johansen, T.A. 2006. Rock physics modeling of shale diagenesis. Petroleum Geosciences 12, 49–57.

Dræge, A. 2009. Constrained rock physics modeling. The Leading Edge 28, 76–80.

Dvorkin, J., Mavko, G. and Gurevich, B. 2007. Fluid substitution in shaley sediment using effective porosity. Geophysics 72, O1–O8.

Dvorkin, J. and Gutierrez, M. 2002. Grain sorting, porosity and elasticity. Petrophysics 43, 185–196.

Dvorkin, J. and Nur, A. 1996. Elasticity of high-porosity sandstones: Theory for two North Sea datasets. Geophysics 61, 1363–1370.

Fatti, J.L., Smith, G.C., Vail, P.J., Strauss, P.J. and Levitt, P.R. 1994. Detection of gas in sandstone reservoirs using AVO analysis: A 3-D seismic case history using the Geostack technique. Geophysics 59, 1362–1376.

Gassmann, F. 1951. Uber die elastizitat poroser medien. Vierteljahrsschrift der Naturforschende Gesellschaft 96, 1–23.

Giles, M.R., Indrelid, S.L., Beynon, G.V. and Amthor, J. 2000. The origin of large-scale quartz cementation: Evidence from large data sets and coupled heat-fluid mass transport modeling. In: Worden, R.H. and Morad, S. (eds.), Quartz Cementation in Sandstones. Blackwell Science, London, Spec. Publ. Int. Assoc. Sedimentol. 29, 21–38.

Hornby, B.E., Schwartz, L.M. and Hudson, J.A. 1994. Anisotropic effective-medium modeling of the elastic properties of shales. Geophysics 59, 1570–1583.

Johansen, T.A., Jakobsen, M. and Ruud, B.O. 2002. Estimation of the internal structure and anisotropy of shales from borehole data. Journal of Seismic Exploration 11, 363–381.

MacQuaker, J.H.S., Taylor, K.G. and Gawthorpe, R.L. 2007. High-resolution facies analyses of mudstones: Implications for paleoenvironmental and sequence stratigraphic interpretations of offshore ancient mud-dominated successions. Journal of Sedimentary Research 77, 324–339.

Marcussen, Ø., Thyberg, B.I., Peltonen, C., Jahren, J., Bjørlykke, K. and Faleide, J.I. 2009. Physical properties of Cenozoic mudstones from the northern North Sea: Impact of clay mineralogy on compaction trends. AAPG Bulletin 93, 127–150.

Marion, D. 1990. Acoustical, mechanical and transport properties of sediments and granular materials. Ph.D.-thesis, Stanford University, 136 pp.

Mavko, G., Mukerji, T. and Dvorkin, J. 2009. The Rock Physics Handbook (2nd edition). Cambridge University Press, Cambridge, 340 pp.

Middleton, G.V. 1973. Johannes Walther's Law of the correlation of facies. Geological Society of America Bulletin 84, 979–988.

Mindlin, R.D. 1949. Compliance of elastic bodies in contact. Journal of Applied Mechanics 16, 259–268.

Mondol, N.H., Bjørlykke, K., Jahren, J. and Høeg, K. 2007. Experimental mechanical compaction of clay mineral aggregates – Changes in physical properties of mudstones during burial. Marine and Petroleum Geology 24, 289–311.

Mukerji, T., Jørstad, A., Avseth, P., Mavko, G. and Granli, J.R. 2001. Mapping lithofacies and pore-fluid probabilities in a North Sea reservoir: Seismic inversions and statistical rock physics. Geophysics 66, 988–1001.

Ødegaard, E. and Avseth, P. 2004. Well log and seismic data analysis using rock physics templates. First Break 22, 37–43.

Ostrander, W.J. 1984. Plane-wave reflection coefficients for gas sands at non-normal angles of incidence. Geophysics 49, 1637–1648.

Peltonen, C., Marcussen, Ø., Bjørlykke, K. and Jahren, J. 2009. Clay mineral diagenesis and quartz cementation in mudstones: The effects of smectite to illite transformation on rock properties. Marine and Petroleum Geology 26, 887–898.

Ramm, M. and Bjørlykke, K. 1994. Porosity/depth trends in reservoir sandstones: Assessing the quantitative effects of varying pore-pressure, temperature history and mineralogy, Norwegian Shelf data. Clay Minerals 29, 475–490.

Reading, H.G. and Richards, M. 1994. Turbidite systems in deep-water basin margins classified by grain-size and feeder system. AAPG Bulletin 78, 792–822.

Reuss, A. 1929. Berechnung der fliessgrense von mishkristallen. Zeitschrift fur Angewandte Mathematik und Mechanik 9, 49–58.

Ruud, B.O., Jakobsen, M. and Johansen, T.A. 2003. Seismic properties of shales during compaction. SEG Extended Abstract, p. 1294.

Sams, M. and Andrea, M. 2001. The effect of clay distribution on the elastic properties of sandstones. Geophysical Prospecting 49, 128–150.

Shuey, R.T. 1985. A simplification of the Zoeppritz equations. Geophysics 50(4), 609–614.

Smith, G.C. and Gidlow, P.M. 1987. Weighted stacking for rock property estimation and detection of gas. Geophysical Prospecting 35, 993–1014.

Storvoll, V., Bjørlykke, K. and Mondol, N.H. 2005. Velocity-depth trends in Mesozoic and Cenozoic sediments from the Norwegian Shelf. AAPG Bulletin 90, 1145–1148.

Thyberg, B., Jahren, J., Winje, T., Bjørlykke, K. and Faleide, J.I. 2009. From mud to shale; Rock stiffening by micro-quartz cementation. First Break 27, 53–59.

Voigt, W. 1910. Lehrbuch der Kristallphysik. Teubner, Leipzig.

Walderhaug, O., Lander, R.H., Bjørkum, P.A., Oelkers, E.H., Bjørlykke, K. and Nadeau, P.H. 2000. In: Worden, R.H. and Morad, S. (eds.), Quartz Cementation in Sandstones. Special Publication of the International Association of Sedimentologists No 29, Blackwell Science, London, pp. 39–50.

Walker, R. 1978. Deep-water sandstone facies and ancient submarine fans: Models for exploration for stratigraphic traps. AAPG Bulletin 62, 932–966.

Worden, R.H. and Morad, S. 2000. Quartz cementation in oil field sandstones: A review of the key controversies. In: Worden, R.H. and Morad, S. (eds.), Quartz Cementation in Sandstones. Blackwell Science, London, Spec. Publ. Int. Assoc. Sedimentol. 29, 1–20.

Xu, S. and White, R.E. 1995. A new velocity model for clay-sand mixtures. Geophysical Prospecting 43, 91–118.

Yin, H. 1992. Acoustic velocity and attenuation of rocks, isotropy intrinsic anisotropy, and stress induced anisotropy. Ph.D. thesis, Stanford University, 245 pp.

Zoeppritz, K. 1919. Erdbebenwellen VIIIB, Ueber Reflexion and Durchgang seismischer Wellen durch Unstetigkeitsflaechen. Goettinger Nachrichten I, 66–84.

Chapter 19

4D Seismic

Martin Landrø

19.1 Introduction

The term 4D seismic reflects that calendar time represents the fourth dimension. A more precise term is *repeated seismic*, because that is actually what is done: a seismic survey over a given area (oil/gas field) is repeated in order to monitor production changes. *Time-lapse seismic* is another term used for this. For some reason, the term 4D seismic is most common, and we will therefore use it here. It is important to note that if we repeat 2D surveys, it is still denoted 4D seismic according to this definition. Recent examples of such surveys are repeated 2D lines acquired over the Troll gas province.

Currently there are three major areas where 4D seismic is applied. Firstly, to monitor changes in a producing hydrocarbon reservoir. This is now an established procedure being used worldwide. So far, 4D seismic has almost exclusively been used for clastic reservoirs and only rarely for carbonate reservoirs. This is because carbonate reservoirs (apart from those in porous chalk) are stiffer, and the effect on the seismic parameters of substituting oil with water is far less pronounced. Secondly, 4D seismic is being used to monitor underground storage of CO_2. Presently, there is a global initiative to decrease the atmospheric CO_2 content, and one way to achieve this goal is to pump huge amounts of CO_2 into saline aquifers. A third application of 4D seismic (with other geophysical methods) is the monitoring of geohazards (landslides, volcanoes etc.), however this will not be covered here.

The business advantage of using 4D seismic on a given field is closely correlated with the complexity of the field. Figure 19.1 shows a 3D perspective view of the top reservoir (top Brent) interface for the Gullfaks Field. Since the oil is trapped below such a complex 3D surface, it is not surprising that oil pockets can remain untouched even after 10–15 years of production. Since 4D seismic can be used to identify such pockets, it is easy to understand the commercial value of such a tool. However, if the reservoir geometry is simpler, the number of untapped hydrocarbon pockets will be less and the business benefit correspondingly lower.

The first 4D seismic surveys were probably acquired in America in the early 1980s. It was soon realised that heavy oil fields were excellent candidates for 4D seismic. Heavy oil is highly viscous, so thermal methods such as combustion or steam injection were used to increase its mobility. Combustion describes a burning process, which normally is maintained by injection of oxygen or air. As the reservoir is heated, the seismic P-wave velocity decreases, and this results in amplitude changes between baseline and monitor seismic surveys.

The best known example is maybe the one published by Greaves and Fulp in 1987, for which they received the award for the best paper in *Geophysics*. They showed that such thermal recovery methods could be monitored by repeating conventional 3D land seismic surveys (Fig. 19.2). These early examples of seismic monitoring of thermal recovery methods did not immediately lead to a boom in the 4D industry, mainly because they were performed on small and very shallow onshore fields. The major breakthrough for commercial 4D seismic surveys in the North Sea was the Gullfaks 4D study launched by Statoil in 1995. Together with WesternGeco a pilot study was done

M. Landrø (✉)
NTNU Trondheim, Trondheim, Norway
e-mail: martin.landro@ntnu.no

K. Bjørlykke (ed.), *Petroleum Geoscience: From Sedimentary Environments to Rock Physics*,
DOI 10.1007/978-3-642-02332-3_19, © Springer-Verlag Berlin Heidelberg 2010

Fig. 19.1 3D seismic image of the top reservoir interface (top Brent Group) at the Gullfaks Field. Notice the fault pattern and the complexity of the reservoir geometry

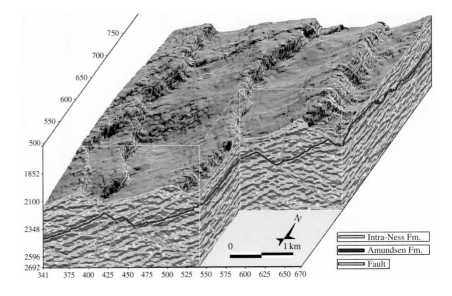

Fig. 19.2 Shows clear 4D amplitude brightening (*top* Holt horizon) as the reservoir is heated. Also note the extended area (marked by *white arrows*) for this amplitude increase (Greaves and Fulp, Geophysics, 52, 1987)

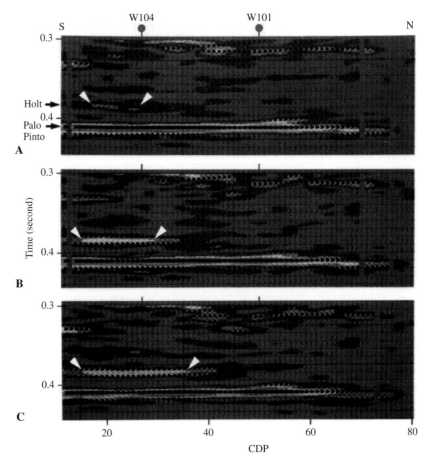

over the major northern part of the field, and the initial interpretation performed shortly after this monitor survey demonstrated a promising potential.

However, the first use of 4D seismic in the North Sea was probably the monitoring of an underground flow (Larsen and Lie 1990). In January 1989, when

Fig. 19.3 Seismic monitoring of the underground flow caused by drilling a deep Jurassic well in 1989 in the North Sea. The new seismic event marked by the *red arrow* on the *middle section* is interpreted as a gas-filled sand body. On the *lower section* (6 months after the drilling event) the areal extent of this event has increased further. Figure provided by Dag O. Larsen, from his SEG-presentation in 1990

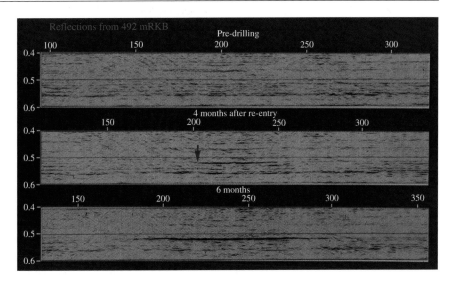

drilling a deep Jurassic well, Saga Petroleum had to shut the well by activating the BOP. The rig was moved off location, and a relief well was spudded. When the rig was reconnected 3 months later, the pressure had dropped, indicating a subsurface fluid transfer. The flow and the temperature measurements indicated a leak in the casing at 1,334 mSS. A strong amplitude increase at the top of a sand layer could be observed on the time-lapse seismic data close to this well (Fig. 19.3) – showing the fluid was gas.

The areal extent of this 4D anomaly increased from one survey to the next. After 4 months this amplitude increase was also observed on shallower interfaces, probably connected to gas migrating upwards to shallower sand layers.

to the seismic parameters. Rock physics provides this link. Both theoretical rock physics models and laboratory experiments are used as important input to time-lapse seismic analysis. A standard way of relating for instance P-wave seismic velocity to changes in fluid saturation is to use the Gassmann model. Figure 19.4 shows one example, where a calibrated Gassmann model has been used to determine how the P-wave velocity changes with water saturation (assuming that the reservoir fluid is a mixture of oil and water).

In addition, various contact models have been proposed to estimate the effective modulus of a rock. Mavko, Mukerji and Dvorkin present some of these models in their rock physics handbook (1998). The Hertz-Mindlin model (Mindlin 1949) can be used to describe the properties of pre-compacted

19.2 Rock Physics and 4D Seismic

When a hydrocarbon field is produced, there are several reservoir parameters that might change, most crucially:

– fluid saturation changes
– pore pressure changes
– temperature changes
– changes in layer thickness (compaction or stretching).

A critical part of all 4D studies is to link key reservoir parameters like pore pressure and fluid saturation

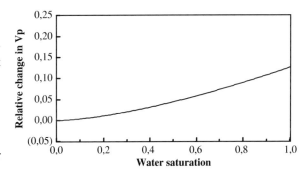

Fig. 19.4 Relative change in P-wave velocity versus water saturation, estimated from a calibrated Gassmann model. Zero water saturation corresponds to 100% oil

granular rocks. The effective bulk modulus of dry random identical sphere packing is given by:

$$K_{\text{eff}} = \left\{ \frac{C_p^2 (1 - \varphi)^2 G^2 P}{18\pi^2 (1 - \sigma)^2} \right\}^{\frac{1}{3}} \qquad (19.1)$$

where C_p is the number of contact points per grain, φ is porosity, G is the shear modulus of the solid grains, σ is the Poisson ratio of the solid grains and P is the effective pressure (that is $P = P_{\text{eff}}$). The shear modulus is given as:

$$G_{\text{eff}} = \left\{ \frac{3C_p^2 (1 - \varphi)^2 G^2 P}{2\pi^2 (1 - \sigma)^2} \right\}^{\frac{1}{3}} \frac{5 - 4\sigma}{5(2 - \sigma)} \qquad (19.2)$$

This leads to:

$$V_{\text{P}} = \sqrt{\frac{K_{\text{eff}} + \frac{4}{3} G_{\text{eff}}}{\rho}} \qquad (19.3)$$

$$V_{\text{S}} = \sqrt{\frac{G_{\text{eff}}}{\rho}} \qquad (19.4)$$

where V_p and V_s are P- and S-wave velocities, respectively, and ρ is the sandstone density. Inserting Eqs. (19.1) and (19.2) into Eqs. (19.3) and (19.4) and computing the V_p/V_s ratio, yields (assuming $\sigma = 0$):

$$\frac{V_{\text{P}}}{V_{\text{S}}} = \sqrt{2}. \qquad (19.5)$$

This means that according to the simplest granular model (Hertz-Mindlin), the V_p/V_s ratio should be constant as a function of confining pressure if we assume the rock is dry. The Hertz-Mindlin model assumes the sand grains are spherical and that there is a certain area of grain-to-grain contacts. A major shortcoming of the model is that at the limit of unconsolidated sands, both the P and S-wave velocities will have the same behaviour with respect to pressure changes, as shown in Eq. (19.5). Combining with the Gassmann (1951) model (i.e. introducing fluids in the pore system of the rock), will ensure that the P-wave velocity approaches the fluid velocity for zero effective stress, and not zero as in the Hertz-Mindlin model (dry rock assumption).

If we assume that the in situ (base survey) effective pressure is P_0 we see from Eq. (19.3) that the relative P-wave velocity versus effective pressure is given as:

$$\frac{V_{\text{P}}}{V_{P_0}} = \left(\frac{P}{P_0} \right)^{\frac{1}{6}} \qquad (19.6)$$

Figure 19.5 shows this relation for an in situ effective pressure of 6 MPa. When such curves are compared to ultrasonic core measurements, the slope of the measured curve is generally smaller than this simple theoretical curve. The cause for this might be multifold: Firstly, the Hertz-Mindlin model assumes the sediment grains are perfect, identical spheres, which is never found in real samples. Secondly, the ultrasonic measurements might suffer from scaling issues, core damage and so on. Thirdly, cementation effects are

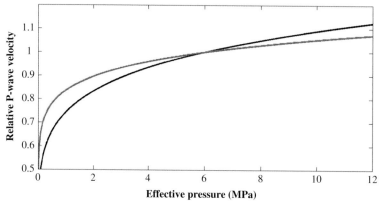

Fig. 19.5 Typical changes in P-wave velocity versus effective pressure using the Hertz-Mindlin model. In this case the in situ effective pressure (prior to production) is 6 MPa, and we see that a decrease in effective pressure leads to a decrease in P-wave velocity. The *black curve* represents the Hertz-Mindlin model (exponent = 1/6, as in Eq. (19.6)), the *red curve* is a modified version of the Hertz-Mindlin model (exponent = 1/10) that better fits the ultrasonic core measurements

not included in the Hertz-Mindlin model. It is therefore important to note that there are major uncertainties regarding the actual dependency between seismic velocity and pore pressure changes.

19.3 Some 4D Analysis Techniques

The analysis of 4D seismic data can be divided into two main categories, one based on the detection of amplitude changes, the other on detecting travel-time changes, see Fig. 19.6. Practical experience has shown that the amplitude method is most robust, and therefore this has been the most frequently employed method. However, as the accuracy of 4D seismic has improved, the use of accurate measurements of small timeshifts is increasingly the method of choice. There are several examples where the timeshift between two seismic traces can be determined with an accuracy of a fraction of a millisecond. A very attractive feature of 4D timeshift measurement is that it is proportional to the change in pay thickness, and this method provides a direct quantitative result. The two techniques are complementary in that amplitude measurement is a local feature (measuring changes close to an interface), while the timeshift method measures average changes over a layer, or even a sequence of layers.

In addition to the direct methods mentioned above, 4D seismic interpretation is aided by seismic modelling of various production scenarios, often combined with reservoir fluid flow simulation and 1D scenario modelling based on well logs.

19.4 The Gullfaks 4D Seismic Study

One of the first commercially successful 4D examples from the North Sea was the Gullfaks study (Landrø et al. 1999). There are two simple explanations for this success: the complexity of the reservoir geometry (Fig. 19.1) means that there will be numerous pockets of undrained oil, and there is a strong correlation between the presence of oil and amplitude brightening as shown in Fig. 19.7.

In addition to these two observations, a rock physics feasibility study was done, and the main results are summarised in Fig. 19.8. It is important to note that the key parameter for 4D seismic is not the relative change in the seismic parameters, but the expected change in reflectivity between the base and monitor surveys. If we assume that the top reservoir interface separates the cap rock (Layer 1) and the reservoir (Layer 2), it is possible to estimate the relative change in the zero offset reflection coefficient (Landrø et al. 1999) caused be production changes in the reservoir layer. Denoting the acoustic impedances (velocity times density) in Layers 1 and 2 by λ_1 and λ_2, respectively, and using B and M to denote base and monitor surveys, we find that the change in reflectivity is given as (Landrø et al. 1999):

$$\frac{\Delta R}{R} = \frac{\lambda_2^M - \lambda_2^B}{\lambda_2^B - \lambda_1^B} = \frac{\Delta \text{AI(production)}}{\Delta \text{AI(original)}}. \qquad (19.7)$$

Here we have assumed that the change in acoustic impedance in Layer 2 is small compared to the actual impedance. Equation (19.7) means that the change in reflectivity is equal to the change in acoustic impedance in Layer 2 divided by the original (pre-production) acoustic impedance contrast between Layers 1 and 2. For Gullfaks, the expected change in acoustic impedance is 8% (assuming 60% saturation change, Fig. 19.8) while the original acoustic impedance contrast for top reservoir was approximately 18%. The estimated relative change in reflectivity is therefore 44% (8/18). This is the basic background for the success of using 4D seismic at Gullfaks,

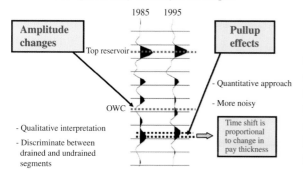

Fig. 19.6 Shows the two 4D analysis techniques. Notice there are no amplitude changes at top reservoir between 1985 and 1995, but huge amplitude changes at the oil-water contact (OWC). Also note the change in travel time (marked by the *yellow arrow*) for the seismic event below the oil-water contact

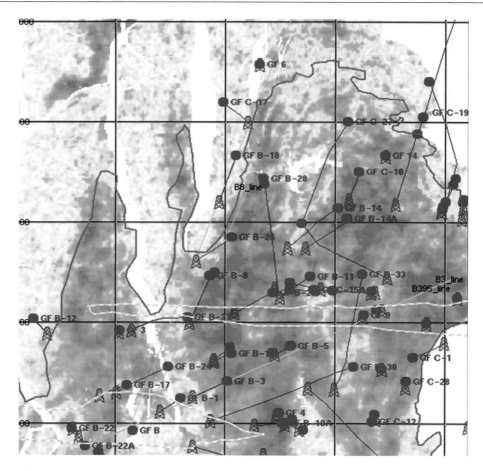

Fig. 19.7 Amplitude map for the top reservoir event at Gullfaks. The *purple solid line* represents the original oil-water contact. Notice the strong correlation between high amplitudes (*red colours*) and presence of oil. The size of one square is 1,250 by 1,250 m

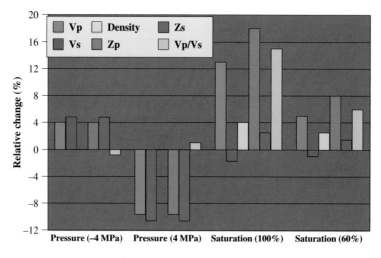

Fig. 19.8 Expected changes (based on rock physics) of key seismic parameters for pore pressure changes (*two left columns*) and for two fluid saturation scenario changes (*two right columns*)

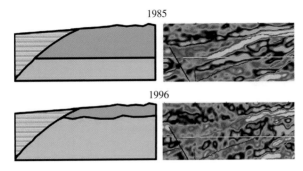

Fig. 19.9 Seismic data from 1985 (*upper right*) and 1996 (*lower right*) and the corresponding 4D interpretation to the *left* (*green* representing oil, and *blue* water)

as shown in Fig. 19.9. The effect of replacing oil with water is evident on this figure, and is further strengthened by the amplitude difference map taken at the original oil-water contact (Fig. 19.10). Several Gullfaks wells have been drilled based on 4D interpretation, and a rough estimate of the extra income generated by using time-lapse seismic data is more than 1 billion USD.

19.5 Repeatability Issues

The quality of a 4D seismic dataset is dependent on several issues, such as reservoir complexity and the complexity of the overburden. However, the most important issue that we are able to influence is the repeatability of the seismic data, i.e. how accurately we can repeat the seismic measurements. This *acquisition repeatability* is dependent on a number of factors such as:

- Varying source and receiver positions (x, y and z co-ordinates)
- Changing weather conditions during acquisition
- Varying seawater temperature
- Tidal effects
- Noise from other vessels or other activity in the area (rig noise)
- Varying source signal
- Changes in the acquisition system (new vessel, other cables, sources etc.)
- Variation in shot-generated noise (from previous shot).

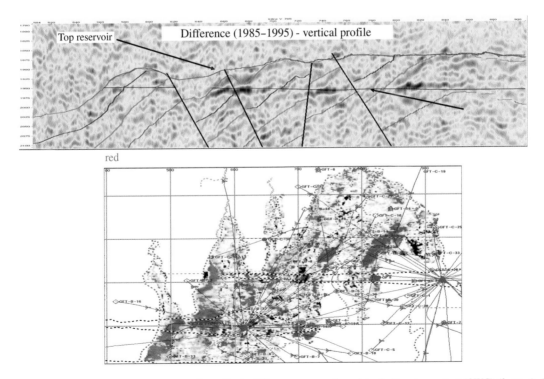

Fig. 19.10 Seismic difference section (*top*) and amplitude difference map at the original oil-water contact (OWC) (*bottom*). *Red colour* indicates areas that have been water-flushed, and *white areas* may represent bypassed oil

Disregarding the weather conditions (the only way to "control" weather is to wait), most of the items listed are influenced by acquisition planning and performance. A common way to quantify repeatability is to use the normalised RMS (root-mean-square)-level, that is:

$$NRMS = 2\frac{RMS(monitor - base)}{RMS(monitor) + RMS(base)}, \quad (19.8)$$

where the RMS-levels of the monitor and base traces are measured within a given time window. Normally, NRMS is measured in a time window where no production changes are expected. Figure 19.11 shows two seismic traces from a VSP (Vertical Seismic Profile) experiment where the receiver is fixed (in the well at approximately 2 km depth), and the source co-ordinates are changed by 5 m in the horizontal direction. We notice that the normalised rms-error (NRMS) in this case is low, only 8%.

In 1995 Norsk Hydro acquired a 3D VSP dataset over the Oseberg Field in the North Sea. This dataset consists of 10,000 shots acquired in a circular shooting pattern and recorded by a 5-level receiver string in the well. By comparing shot-pairs with different source positions (and the same receiver), it is possible to estimate the NRMS-level as a function of the horizontal distance between the shot locations. This is shown in Fig. 19.12, where approximately 70,000

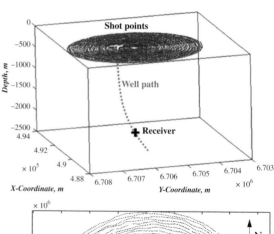

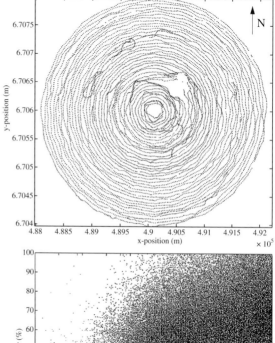

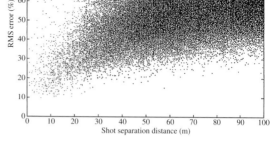

Fig. 19.12 *Top*: The 10,000 shot locations (map view) for the 3D VSP experiment over the Oseberg Field. *Bottom*: NRMS as a function of the source separation distance between approximately 70,000 shot pairs for the Oseberg 3D VSP dataset

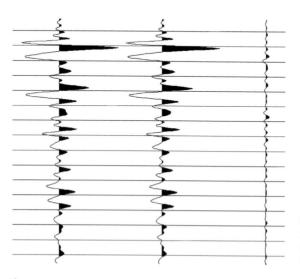

Fig. 19.11 Two VSP-traces measured at exactly the same position in the well, but with a slight difference in the source location (5 m in the *horizontal direction*). The distance between two timelines is 50 ms. The difference trace is shown to the right

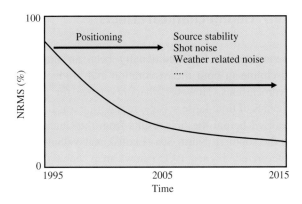

Fig. 19.13 Schematic view showing improvement in seismic repeatability (NRMS) versus calendar time

frequency dependent, so the frequency band used in the data analysis should be given.

During the last two decades the focus on source and receiver positioning accuracy has lead to a significant improvement in the repeatability of 4D seismic data. Some of this is also attributed to better processing of time-lapse seismic data. This trend is sketched in Fig. 19.13. Today the global average NRMS-level is around 20–30%. It is expected that this trend will continue, though not at the same rate because the non-repeatable factors that need to be tackled to get beyond 20–30% are more difficult. In particular, rough weather conditions will represent a major hurdle.

19.6 Fixed Receiver Cables

shot pairs with varying horizontal mis-positioning are shown. The main message from this figure is clear. It is important to repeat the horizontal positions both for sources and receivers as accurately as possible; even a misalignment of 20–30 m might significantly increase the NRMS-level.

Such a plot can serve as a variogram, since it shows the spread for each separation distance. Detailed studies have shown that the NRMS-level increases significantly in areas where the geology along the straight line between the source and receiver is complex. From Fig. 19.12 we see that the NRMS-value for a shot separation distance of 40 m might vary between 20 and 80% and a significant portion of this spread is attributed to variation in geology. This means that it is not straightforward to compare NRMS-levels between various fields, since the geological setting might be very different. Still, NRMS-levels are frequently used, since they provide a simple, quantitative measure. It is also important to note that the NRMS-level is

For marine seismic data, there are several ways to control the source and receiver positions. WesternGeco developed their Q-marine system, where the streamers can also be steered in the horizontal direction (x and y). This means that it is possible to repeat the receiver positions by steering the streamers into predefined positions. The devices are shown attached to the streamers in Fig. 19.14.

Another way to control the receiver positions is to bury the receiver cables on the seabed. One of the first commercial offshore surveys of this kind was launched at the Valhall Field in the southern part of the North Sea in 2003. Since then, more than 11 surveys have been acquired over this field. The buried cables cover 70% of the entire field and the 4D seismic data quality is excellent. The fact that more than 10 surveys have been recorded into exactly the same receiver positions, opens for alternative and new ways (exploiting the multiplicity that more than 2 datasets offers) of

Fig. 19.14 Photographs of the birds (controlling devices) attached to a marine seismic cable at deployment (*left*) and in water operation (*right*). These devices make it possible to steer the cable with reasonable accuracy into a given position (Courtesy of WesternGeco)

analysing time-lapse seismic data. Other advantages offered by permanent receiver systems are:

- cheaper to increase the shot time interval in order to reduce the effect of shot generated noise
- continuous monitoring of background noise
- passive seismic
- possible to design a dedicated monitoring survey close to a problem well at short notice.

Despite all these advantages that permanent systems offer, there has not been a marked increase in such surveys so far, probably because of the relatively high upfront costs and the difficulty in quantifying in advance the extra value it would bring.

19.7 Geomechanical Effects

Geomechanics has traditionally been an important discipline in both the exploration and production of hydrocarbons. However, the importance for geomechanics of time-lapse seismic monitoring was not fully realised before the first results from the Chalk fields in the southern North Sea (Ekofisk and Valhall) were interpreted. Figure 19.15 shows map views of estimated travel-time shifts between base and monitor surveys at Ekofisk (Guilbot and Smith 2002), where the seafloor subsidence had been known for many years. The physical cause for the severe compaction of the Chalk reservoir is two-fold. Firstly, depletion of the field leads to pore pressure decrease, and hence the reservoir rock compacts mechanically due to lack

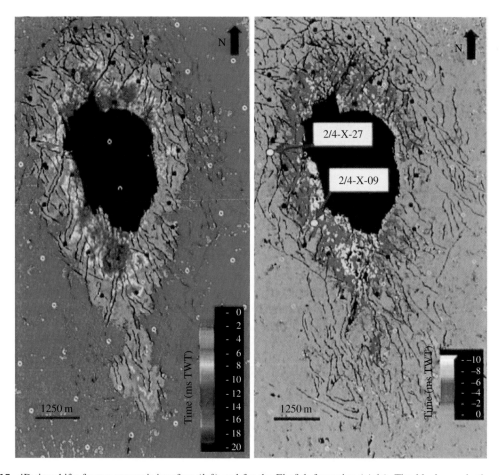

Fig. 19.15 4D timeshifts for top reservoir interface (*left*) and for the Ekofisk formation (*right*). The *black area* in the *middle* is caused by the gas chimney problem at Ekofisk, leading to lack of high quality seismic data in this area (from Guilbot and Smith 2002)

of pressure support. Secondly, the chemical reaction between the chalk matrix and the water replacing the oil leads to a weakening of the rock framework, and a corresponding compaction.

When the reservoir rock compacts, the over- and under burden will be stretched. This stretch is relatively small (of the order of 0.1%). However, this leads to a small velocity decrease that is observable as timeshifts on time-lapse seismic data. Typical observations of seafloor subsidence show that the subsidence is less than the measured compaction for the reservoir. The vertical movement of the seafloor is often approximately 20% less than that of the top reservoir, though is strongly dependent on reservoir geometry and the stiffness of the rocks above and below.

The commonest way to interpret 4D seismic data from a compacting reservoir is to use geomechanical modelling as a complementary tool. One such example (Hatchell and Bourne 2005) is shown in Fig. 19.16. Note that the negative timeshifts (corresponding to a slowdown caused by overburden stretching) continue into the reservoir section. This is due to the fact that the reservoir section is fairly thin in this case and even a compaction of several metres is not enough to shorten the travel time enough to counteract the cumulative effect. Similar effects have also been observed for sandstone reservoirs, such as the Elgin

and Franklin fields. Normally, sandstone reservoirs compact less that chalk reservoirs and the corresponding 4D timeshifts are therefore less, although still detectable. Most of the North Sea sandstone reservoirs show negligible compaction, and therefore these effects are normally neglected in 4D studies. However, as the accuracy of 4D seismic is increasing, it is expected that compaction will be observed for several of these fields as well. For instance, the anticipated compaction of the Troll East field is around 0.5–1 m after 20 years of production, and this will probably be detectable by 4D seismic.

A major challenge when the thickness of subsurface layers is changed during production, is to distinguish between thickness changes and velocity changes. One way to resolve this ambiguity is to use geomechanical modelling as a constraining tool. Another way is to combine near and far offset travel time analysis (Landrø and Stammeijer 2004) to estimate velocity and thickness changes simultaneously. Hatchell et al. and Røste et al. suggested in 2005 using a factor (R) relating the relative velocity change (dv/v) to the relative thickness change (dz/z; displacement)

$$R = -\frac{dv/v}{dz/z}.$$ (19.9)

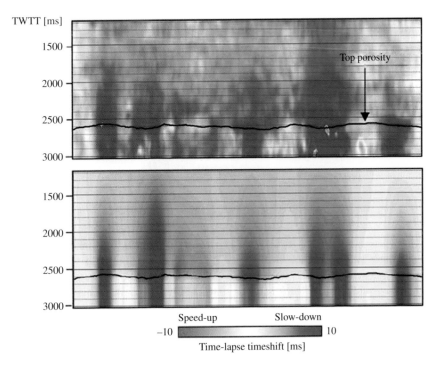

Fig. 19.16 Comparison of measured timeshifts from 4D seismic data (*top*) and geomechanical modelling. The *black solid line* represents the top reservoir. Notice that the slowdown (*blue colours*) continues into the reservoir (from Hatchell and Bourne 2005)

Hatchell et al. found that this factor varies between 1 and 5, and is normally less for the reservoir rock than the overburden rocks. For sandstones and clays it is common to establish empirical relationships between seismic P-wave velocity (v) and porosity (φ). These relations are often of the simple linear form:

$$v = a\varphi + b, \qquad (19.10)$$

where a and b are empirical parameters. If a rock is stretched (or compressed), a corresponding change in porosity will occur. Assuming that the lateral extent of a reservoir is large compared to the thickness, it is reasonable to assume a uniaxial change as sketched in Fig. 19.17. From simple geometrical considerations we see that the relation between the thickness change and the corresponding porosity change is:

$$\frac{dz}{z} = \frac{d\varphi}{1 - \varphi}, \qquad (19.11)$$

In the isotropic case (assuming that the rock is stretched in all three directions), we obtain:

$$\frac{dz}{z} = \frac{d\varphi}{3(1 - \varphi)}. \qquad (19.12)$$

Inserting Eqs. (19.10) and (19.11) into (19.9) we obtain an explicit expression for the dilation factor R:

$$R = 1 - \frac{a + b}{v}, \qquad (19.13)$$

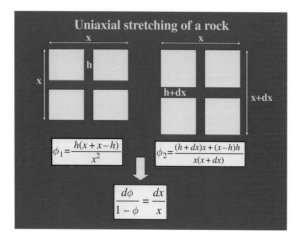

Fig. 19.17 Cartoon showing the effect of increased porosity due to stretching of a rock sample

which is valid for the uniaxial case. Using Eq. (19.12) instead of (19.11) gives a similar equation for the isotropic case.

19.8 Discrimination Between Pore Pressure and Saturation Effects

Although the main focus in most 4D seismic studies is to study fluid flow and detect bypassed oil pockets, the challenge of discriminating between pore pressure changes and fluid saturation changes occurs frequently. From rock physics we know that both effects influence the 4D seismic data. However, as can be seen from Fig. 19.8, the fluid pressure effects are not linked to the seismic parameters in the same way as the fluid saturation effects. This provides an opportunity to discriminate them, since we have different rock physics relations for the two cases. Some of the early attempts to perform this discrimination between pressure and saturation were presented by Tura and Lumley (1999) and Landrø (1999). In Landrø's method (2001) the rock physics relations (based on Gassmann and ultrasonic core measurements) are combined with simple AVO (amplitude versus offset) equations to obtain direct expressions for saturation changes and pressure changes. Necessary input to this algorithm is near and far offset amplitude changes estimated from the base and monitor 3D seismic cubes. This method was tested on a compartment from the Cook Formation at the Gullfaks Field. Figure 19.18 shows a seismic profile (west–east) through this compartment.

A significant amplitude change is observed for the top Cook interface (red solid line in the figure), both below and above the oil-water contact. The fact that this amplitude change extends beyond the oil-water contact is a strong indication that it cannot be solely related to fluid saturation changes, and a reasonable candidate is therefore pore pressure changes. Indeed, it was confirmed that the pore pressure had increased by 50–60 bars in this segment, meaning that the reservoir pressure is approaching fracture pressure. We also observe from this figure that the base reservoir event (blue solid line) is shifted slightly downwards, by 2–3 ms. This slowdown is interpreted as a velocity drop caused by the pore pressure increase in the segment.

Figure 19.19 shows an attempt to discriminate between fluid saturation changes and pressure changes

Fig. 19.18 Seismic section through the Cook Formation at Gullfaks. The *cyan solid line* indicates the original oil-water contact (marked OWC). Notice the significant amplitude change at *top* Cook between 1985 and 1996, and that the amplitude change extends beyond the OWC-level, a strong indication that the downflank amplitude change is caused by pore pressure changes

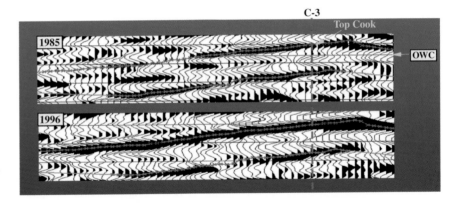

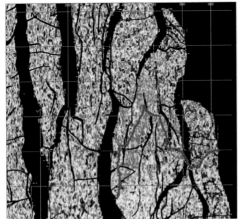
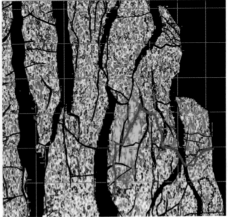

Fig. 19.19 Estimated fluid saturation changes (*left*) and pore pressure changes (*right*) based on 4D AVO analysis for the *top* Cook interface at Gullfaks. The *blue solid line* represents the original oil-water contact. Notice that the estimated pressure changes extend beyond this *blue line* to the west, and terminate at a fault to the east. *Yellow colours* indicate significant changes

for this compartment using the method described in Landrø (2001). In 1996, 27% of the estimated recoverable reserves in this segment had been produced, so we know that some fluid saturation changes should be observable on the 4D seismic data.

From this figure we see that in the western part of the segment most of the estimated fluid saturation changes occur close to the oil-water contact. However, some scattered anomalies can be observed beyond the oil-water contact in the northern part of the segment. These anomalies are probably caused by inaccuracies in the algorithm or by limited repeatability of the time-lapse AVO data. The estimated pressure changes are more consistent with the fault pattern in the region, and it is likely that the pressure increase is confined between faults in almost all directions. Later time-lapse surveys show that the eastern fault in the figure was "opened" some years later.

19.9 Other Geophysical Monitoring Methods

Parts of this section are taken from "Future challenges and unexplored methods for 4D seismics", by Landrø, Recorder (2005).

Up to now, 4D seismic has proved to be the most effective way to monitor a producing reservoir. However, as discussed above, there are several severe limitations associated with time lapse seismic. One of them is that seismic reflection data is sensitive to acoustic impedance (velocity times density). Although 4D time shift can reveal changes in average velocity between two interfaces, an independent measurement of density changes would be a useful complement to conventional 4D seismic. The idea of actually measuring the mass change in a reservoir caused by hydrocarbon production has been around for quite some time. The limiting factor for gravimetric reservoir monitoring has been the repeatability (or accuracy) of the gravimeters. However, Sasagawa et al. (2003) demonstrated that improved accuracy can be achieved by using 3 coupled gravimeters placed on the seabed. This technical success led to a full field programme at the Troll Field, North Sea. Another successful field example was time-lapse gravity monitoring of an aquifer storage recovery project in Leyden, Colorado (Davis et al. 2005), essentially mapping water influx. Obviously, this technique is best suited for reservoirs where significant mass changes are likely to occur, such as water replacing gas. Shallow reservoirs are better suited than deep. The size of the reservoir is a crucial parameter, and a given minimum size is required in order to obtain observable effects. In quite another field, gravimetric measurements might help in monitoring volcanic activity by distinguishing between fluid movements and tectonic activity within an active volcano.

Around 2000 the hydrocarbon industry started to use electromagnetic methods for exploration. Statoil performed a large scale research project that showed that it is possible to discriminate between hydrocarbon-filled and water-filled reservoirs from controlled source electromagnetic (CSEM) measurements. Since this breakthrough of CSEM surveys some years ago (Ellingsrud et al. 2002) this technique has mainly been used as an exploration tool, in order to discriminate between hydrocarbon-filled rocks and water-filled rocks. Field tests have shown that such data are indeed repeatable, so there should definitely be a potential for using repeated EM surveys to monitor a producing reservoir. So far, frequencies as low as 0.25 Hz (and even lower) are being used, and then the spatial resolution will be limited. However, as a complementary tool to conventional 4D seismic, 4D EM might be very useful. In many 4D projects it is hard to quantify the amount of saturation changes taking place within the reservoir, and time-lapse EM studies might be used to constrain such quantitative estimates of the saturation changes. Furthermore, unlike conventional 4D seismic which is both pressure and saturation sensitive, the EM technique is not very sensitive to pressure changes, so it may be a nice tool for separating between saturation and pressure changes.

In a recent paper, Johansen et al. (2005) show that the EM response over the Troll West gas province

is significantly above the background noise level, see Fig. 19.20. If we use the deviation from a smooth response as a measure of the repeatability level, a relative amplitude variation of approximately 0.1– 0.2 (measured in normalised EM amplitude units) is observed. Compared to the maximum signal observed at the crest of the field (2.75) this corresponds to a repeatability level of 4–7%, which is very good compared to conventional time-lapse seismic. This means it is realistic to assume that time-lapse EM data can provide very accurate low-resolution (in x-y plane) constraints on the saturation changes observed on 4D seismic data. Recently, Mittet et al. (2005) showed that depth migration of low frequency EM data may be used to enhance the vertical resolution. In a field data example, the vertical resolution achieved by this type of migration is maybe of the order of a few hundred metres. The EM sensitivity is determined by the reservoir thickness times the resistivity, underlining the fact that both reservoir thickness and reservoir resistivity are crucial parameters for 4D CSEM. This means that the resolution issue with the active EM method is steadily improving, and such improvements will of course increase the value of repeated CSEM data as a complement to conventional 4D seismic.

By using the so-called interferometric synthetic aperture radar principle obtained from orbiting satellites, an impressive accuracy can be attained in measuring the distance to a specific location on the earth's surface. By measuring the phase differences for a signal received from the same location for different calendar times, it is possible to measure the relative changes with even higher accuracy. By exploiting a sequence of satellite images, height changes can be monitored versus time. The satellites used for this purpose orbit at approximately 800 km.

Several examples of monitoring movements of the surface above a producing hydrocarbon reservoir have been reported, though there are of course limitations to the detail of information such images can provide for reservoir management. The obvious link to the reservoir is to use geomechanical modelling to tie the movements at the surface to subsurface movements. For seismic purposes, such a geomechanical approach can be used to obtain improved velocity models in a more sophisticated way. If you need to adjust your geomechanical model to obtain correspondence between observed surface subsidence and reservoir compaction, then this adjusted geomechanical model

Fig. 19.20 Modelled (*open circles*) and measured (*red circles*) relative electrical field strength as a function of offset (*top*). The values are normalised to the response outside the reservoir (*blue circles*). *Bottom* figure shows relative values for a constant offset of 6.5 km along a profile, where the background shows the reservoir model. Notice that the relative signal increases by a factor of 3 above the thickest part of the reservoir (figure taken from Johansen et al. 2005)

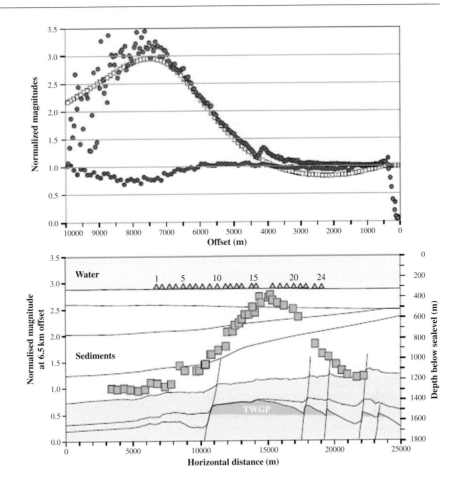

can be used to distinguish between overburden rocks with high and low stiffness for instance, which again can be translated into macro-variations of the overburden velocities. The deeper the reservoir, the lower the frequency (less vertical resolution) of the surface imprint of the reservoir changes, thus this method can not be used directly to identify small pockets of undrained hydrocarbons. However, it can be used as a complementary tool for time-lapse seismic since it can provide valuable information on the low frequency spatial signal (slowly varying in both horizontal and vertical directions) of reservoir compaction.

19.10 CO$_2$ Monitoring

This text is taken from the paper "Quantitative Seismic Monitoring Methods" by Landrø (2008), in ERCIM News 74, pp. 16–17:

Interest in CO$_2$ injection, both for storage and as a tertiary recovery method for increased hydrocarbon production, has grown significantly over the last decade. Statoil has stored approximately 10 million tons of CO$_2$ in the Utsira Formation at the Sleipner Field, and several similar projects are now being launched worldwide. At NTNU our focus has been to develop geophysical methods to monitor the CO$_2$-injection process, and particularly to try to quantify the volume injected directly from geophysical data.

One way to improve our understanding of how the CO$_2$ flows in a porous rock is to perform small-scale flooding experiments on long core samples. Such experiments involve injecting various fluids in the end of a 30–40 cm long core (Fig. 19.21). An example of such a flooding experiment and corresponding X-ray images for various flooding patterns is shown in Fig. 19.21. By measuring acoustic velocities as the flooding experiment is conducted, these experiments

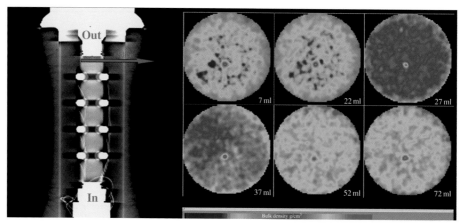

Fig. 19.21 Long core (*left*) showing the location for the X-ray cross-section (*red arrow*). Water injection is 50 g/l. To the *right*: X-ray density maps of a core slice: 6 time steps during the injection process (from Marsala and Landrø, EAGE extended abstracts 2005)

can be used to establish a link between pore scale CO_2-injection and time-lapse seismic on the field scale.

Figure 19.22 shows an example of how CO_2 influenced the seismic data over time. By combining 4D travel time and amplitude changes, we have developed methods to estimate the thickness of CO_2 layers, which makes it possible to estimate volumes. Another key parameter is CO_2 saturation, which can be estimated using rock physics measurements and models. Although the precision in both 4D seismic methods and rock physics is increasing, there is no doubt that precise estimates are hard to achieve, and therefore we need to improve existing methods and learn how to combine several methods in order to decrease the uncertainties associated with these monitoring methods.

19.11 Future Aspects

The most important issue for further improvement of the 4D seismic method, is to improve the repeatability. Advances in both seismic acquisition and 4D processing will contribute to this process. As sketched in Fig. 19.13, it is expected that this will be less pronounced than in the past decade. However, it is still a crucial issue, and even minor improvements might mean a lot for the value of a 4D seismic study. Further improvements in repeatability will probably involve issues like source stability, source positioning, shot time interval, better handling of various noise sources. Maybe in the future we will see vessels towing a superdense grid of sensors, in order to obtain perfect repositioning of the receiver positions.

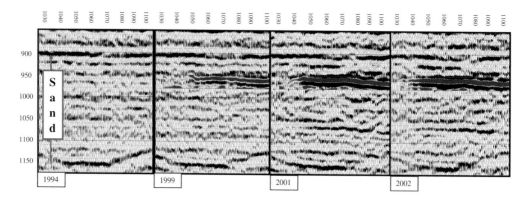

Fig. 19.22 Time-lapse seismic data showing monitoring of the CO_2 injection at Sleipner. The strong amplitude increase (shown in *blue*) is interpreted as a thin CO_2 layer (printed with permission from Statoil)

Another direction to improve 4D studies could be to constrain the time-lapse seismic information by other types of information, such as geomechanical modelling, time-lapse EM, gravimetric data or innovative rock physics measurements.

Our understanding of the relation between changes in the subsurface stress field and the seismic parameters is still limited, and research within this specific area will be crucial to advance the 4D. New analysis methods like long offset 4D, might be used as a complementary technique or as an alternative method where conventional 4D analysis has limited success. However, long offset 4D is limited to reservoirs where the velocity increases from the cap rock to the reservoir rock.

The link between reservoir simulation (fluid flow simulation) and time-lapse seismic will continue to be developed. As computer resources increase, the feasibility of a joint inversion exploiting both reservoir simulation and time-lapse seismic data in the same subsurface model will increase. Despite extra computing power, it is reasonable to expect that the non-uniqueness problem (several scenarios will fit the same datasets) will require that the number of earth models is constrained by models predicting the distribution of rock physical properties based on sedimentology, diagenesis and structural geology.

Acknowledgments Statoil is acknowledged for permission to use and present their data. The Research Council of Norway is acknowledged for financial support to the ROSE (Rock Seismic) project at NTNU. Lasse Amundsen, Lars Pedersen, Lars Kristian Strønen, Per Digranes, Eilert Hilde, Odd Arve Solheim, Ivar Brevik, Paul Hatchell, Peter Wills, Jan Stammeijer, Thomas Røste, Rodney Calvert, Alexey Stovas, Lyubov Skopintseva, Andreas Evensen, Amir Ghaderi, Dag O. Larsen, Kenneth Duffaut, Jens Olav Paulsen, Jerome Guilbot, Olav Barkved, Jan Kommedal, Per Gunnar Folstad, Helene Veire, Ola Eiken, Torkjell Stenvold, Jens Olav Paulsen and Alberto Marsala are all acknowledged for co-operation and discussions.

Further Reading

Amundsen, L. and Landrø, M. December 2007. 4D seismic – Status and future challenges, Part II. GeoExpro, 54–58.

Barkved, O. 2004. Continuous seismic monitoring. 74th Annual International Meeting. The Society of Exploration Geophysicists, 2537–2540.

Barkved, O., Buer, K., Halleland, K.B., Kjelstadli, R., Kleppan, T. and Kristiansen, T. 2003. 4D seismic response of primary production and waste injection at the Valhall field.

65th Meeting. European Association of Geoscientists and Engineers, Extended Abstracts, A22.

Barkved, O.I. and Kristiansen, T. 2005. Seismic time-lapse effects and stress changes: Examples from a compacting reservoir. The Leading Edge 24, 1244.

Calvert, R. 2005. Insights and methods for 4D reservoir monitoring and characterization, EAGE/SEG Distinguished Instructor Short Course 8.

Christensen, N.I. and Wang, H.F. 1985. The influence of pore pressure and confining pressure on dynamic elastic properties of Berea sandstone. Geophysics 50, 207–213.

Davis, K., Li, Y., Batzle, M. and Reynolds, B. 2005. Time-lapse gravity monitoring of an aquifer storage recovery project in Leyden, Colorado. 75th Annual International Meeting. The Society of Exploration Geophysicists, GM1.4.

Eiken, O., Zumberge, M., Stenvold, T., Sasagawa, G. and Nooen, S. 2004. Gravimetric monitoring of gas production from the Troll Field. 74th Annual International Meeting. The Society of Exploration Geophysicists, 2243–2246.

Ellingsrud, S., Eidesmo, T., Johansen, S., Sinha, M.C., MacGregor, L.M. and Constable, S. 2002. Remote sensing of hydrocarbon layers by seabed logging (SBL): Results from a cruise offshore Angola. The Leading Edge 21(10), 972–982.

Foldstad, P.G., Andorsen, K., Kjørsvik, I. and Landrø, M. 2005. Simulation of 4D seismic signal with noise – Illustrated by WAG injection on the Ula Field. 75th Annual International Meeting. The Society of Exploration Geophysicists, TL2.7.

Furre, A-K., Munkvold, F.R. and Nordby, L.H. 2003. Improving reservoir understanding using time-lapse seismic at the Heidrun Field, 65th Meeting. European Association of Geoscientists and Engineers, A20.

Gassmann, F. 1951. Über die elastizität poroser medien. Vierteljahrsschrift der Naturforchenden Gesellschaft in Zürich 96, 1–23.

Gosselin, O., Aanonsen, S.I., Aavatsmark, I., Cominelli, A., Gonard, R., Kolasinski, M., Ferdinandi, F., Kovacic, L. and Neylon, K. 2003. History matching using time-lapse seismic (HUTS), SPE paper 84464.

Guilbot, J. and Smith, B. 2002. 4-D constrained depth conversion for reservoir compaction estimation: Application to Ekofisk Field. The Leading Edge 21(3), 302–308.

Hatchell, P.J. and Bourne, S.J. 2005. Rocks under strain: Strain-induced time-lapse time shifts are observed for depleting reservoirs. The Leading Edge 24, 1222–1225.

Hatchell, P.J., Kawar, R.S. and Savitski, A.A. 2005. Integrating 4D seismic, geomechanics and reservoir simulation in the Valhall oil field. 67th Meeting. European Association of Geoscientists and Engineers, C012.

Johansen, S.E., Amundsen, H.E.F., Røsten, T., Ellingsrud, S., Eidesmo, T. and Bhuiyan, A.H. 2005. Subsurface hydrocarbons detected by electromagnetic sounding. First Break 23, 31–36.

Kommedal, J.H., Barkved, O.I. and Howe, D.J. 2004. Initial experience operating a permanent 4D seabed array for reservoir monitoring at Valhall. 74th Annual International Meeting. The Society of Exploration Geophysicists, 2239–2242.

Koster, K., Gabriels, P., Hartung, M., Verbeek, J., Deinum, G. and Staples, R. 2000. Time-lapse seismic surveys in the

North Sea and their business impact. The Leading Edge 19(03), 286–293.

Kragh, E. and Christie, P. 2002. Seismic repeatability, normalized rms, and predictability. The Leading Edge 21, 640–647.

Landrø, M. 1999. Repeatability issues of 3-D VSP data. Geophysics 64, 1673–1679.

Landrø, M. 2001. Discrimination between pressure and fluid saturation changes from time lapse seismic data. Geophysics 66(3), 836–844.

Landrø, M. 2002. Uncertainties in quantitative time-lapse seismic analysis. Geophysical Prospecting 50, 527–538.

Landrø, M. 2008. The effect of noise generated by previous shots on seismic reflection data. Geophysics 73, Q9–Q17.

Landrø, M., Digranes, P. and Strønen, L.K. 2001. Mapping reservoir pressure and saturation changes using seismic methods – Possibilities and limitations. First Break 19, 671–677.

Landrø, M., Digranes, P. and Strønen, L.K. 2002. Pressure effects on seismic data – Possibilities and limitations: Paper presented at the NPF biannual conference in Kristiansand, Norway, 11–13 March, 2002.

Landrø, M., Digranes, P. and Strønen, L.K. 2005. Pressure depletion measured by time-lapse VSP. The Leading Edge 24, 1226–1232 (December issue).

Landrø, M. and Duffaut, K. 2004. V_p–V_s ratio versus effective pressure and rock consolidation – A comparison between rock models and time-lapse AVO studies. 74th Annual International Meeting. The Society of Exploration Geophysicists, 1519–1522.

Landrø, M., Nguyen, A.K. and Mehdizadeh, H. 2004. Time lapse refraction seismic – A tool for monitoring carbonate fields? 74th Annual International Meeting. The Society of Exploration Geophysicists, 2295–2298.

Landrø, M. and Skopintseva, L. 2008. Potential improvements in reservoir monitoring using permanent seismic receiver arrays. The Leading Edge 27, 1638–1645.

Landrø, M., Solheim, O.A., Hilde, E., Ekren, B.O. and Strønen, L.K. 1999. The Gullfaks 4D seismic study. Petroleum Geoscience 5, 213–226.

Landrø, M. and Stammeijer, J. 2004. Quantitative estimation of compaction and velocity changes using 4D impedance and traveltime changes. Geophysics, The Society of Exploration Geophysicists 69, 949–957.

Landrø, M., Veire, H.H., Duffaut, K. and Najjar, N. 2003. Discrimination between pressure and fluid saturation changes from marine multicomponent time-lapse seismic data. Geophysics, The Society of Exploration Geophysicists 68, 1592–1599.

Larsen, D. and Lie, A. 1990. Monitoring an underground flow by shallow seismic data: A case study. SEG Expanded Abstracts 9, 201 pp.

MacLeod, M., Hanson, R.A., Bell, C.R. and McHugo, S. 1999. The Alba Field ocean bottom cable survey: Impact on development. The Leading Edge 18, 1306–1312.

Mavko, G., Mukerji, T. and Dvorkin, J. 1998. The Rock Physics Handbook. Cambridge University Press, Cambridge, ISBN 0521 62068 6, 147–161.

Mehdizadeh, H., Landrø, M., Mythen, B.A., Vedanti, N. and Srivastava, R. 2005. Time lapse seismic analysis using long offset PS data. 75th Annual International Meeting. The Society of Exploration Geophysicists, TL3.7.

Mindlin, R.D. 1949. Compliance of elastic bodies in contact. Journal of Applied Mechanics 16, 259–268.

Nes, O.M., Holt, R.M. and Fjær, E. 2000. The reliability of core data as input to seismic reservoir monitoring studies. SPE 65180.

Osdal, B., Husby, O., Aronsen, H.A., Chen, N. and Alsos, T. 2006. Mapping the fluid front and pressure buildup using 4D data on Norne Field. The Leading Edge 25, 1134–1141.

Parr, R., Marsh, J. and Griffin, T. 2000. Interpretation and integration of 4-D results into reservoir management, Schiehallion Field, UKCS. 70th Annual International Meeting. The Society of Exploration Geophysicists, 1464–1467.

Rogno, H., Duffaut, K., Furre, A.K., Eide, A.L. and Kvamme, L. 1999. Integration, quantification and dynamic updating – Experiences from the Statfjord 4-D case. 69th Annual International Meeting. The Society of Exploration Geophysicists, 2051–2054.

Røste, T., Stovas, A. and Landrø, M. 2005. Estimation of layer thickness and velocity changes using 4D prestack seismic data. 67th Meeting. European Association of Geoscientists and Engineers, C010.

Sasagawa, G.S., Crawford, W., Eiken, O., Nooner, S., Stenvold, T. and Zumberge, M. 2003. A new sea-floor gravimeter. Geophysics, The Society of Exploration Geophysicists 68, 544–553.

Stovas, A. and Landrø, M. 2004. Optimal use of PP and PS time-lapse stacks for fluid pressure discrimination. Geophysical Prospecting, European Association of Geoscientists and Engineers 52, 301–312.

Tura, A. and Lumley, D.E. 1999. Estimating pressure and saturation changes from time-lapse AVO data. 69th Annual International Meeting, The Society of Exploration Geophysicists, Expanded Abstracts, 1655–1658.

Tøndel, R. and Eiken, O. 2005. Pressure depletion observations from time-lapse seismic data at the Troll Field. 67th EAGE Meeting, Abstract, C037.

Vasco, D.W., Datta-Gupta, A., Behrens, R., Condon, P. and Rickett, J. 2004. Seismic imaging of reservoir flow properties: Time-lapse amplitude changes. Geophysics 69, 1425–1442.

Watts, G.F.T., Jizba, D., Gawith, D.E. and Gutteridge, P. 1996. Reservoir monitoring of the Magnus Field through 4D time-lapse seismic analysis. Petroleum Geoscience 2, 361–372.

Zimmer, M., Prasad, M. and Baggeroer, A.B. 2002. Pressure and porosity influences on V_p–V_s ratio in unconsolidated sands. The Leading Edge 21(2), 178–183.

Chapter 20

Production Geology

Knut Bjørlykke

Production Geology has become an important field and in mature sedimentary basins more geologists and geophysicists are involved in production than exploration. Today more of the global reserves of petroleum are added by increasing the recovery in existing fields than by discovering new fields.

The development of horizontal drilling has made it possible to produce oil much more efficiently than from vertical wells. It is now possible to drill wells up to 8–9 km along a very complex path draining several small reservoir compartments.

During high oil prices much more can be invested in methods to increase oil and gas recovery.

The production of oil and gas requires detailed mapping of the reservoir with respect both to the distribution of petroleum and the flow properties of the reservoir rock.

After the discovery of a new field it may therefore be necessary to drill some delineation wells to acquire more information about the extent of the reservoir and the distribution of reservoir properties. Production wells and injection wells must be planned carefully to secure optimal recovery. Each new production well will provide a large database which needs to be interpreted; this will include information about pressure barriers in the reservoir. In many cases 3D seismic surveys will be repeated during the producing lifetime of the reservoir, providing a fourth dimension (time) to make it a 4D survey. It is then often possible to follow the change in the oil/water contact, gas/oil contact

or gas/water contact by the density and velocity contrast which is a function of petroleum saturation. This may enable poorly drained parts of the reservoir to be identified and the financial feasibility of drilling a new production well to drain this part to be assessed.

Securing optimal production from oil and gas fields is an important challenge both from an economical and environmental perspective. The presence of free gas provides good pressure support for the oil production. The oil must normally be produced before the gas, otherwise production of the gas will reduce the reservoir pressure markedly and distort the oil/gas contact, making it difficult to recover the remaining oil. There is also gas dissolved in the oil and water (Fig. 20.1).

20.1 Capillary Forces

If the wetting angle is more than $90°$ the oil/water contact in the pipe will be below the general OWC, which means the oil is drawn downwards because it is preferentially wetting the inside of the pipe (Fig. 20.2) or with lower wetting angles the OWC is raised.

The pores in a reservoir rock are usually filled with some water in addition to the oil or gas. If we submerge a very thin pipe (e.g. of glass) in oil and water, the water will rise above the oil/water contact. This capillary rise is a function of the radius of the pipe bore (R) and the wetting angle (θ).

The capillary forces ($F1$) must be in equilibrium with the gravitational forces ($F2$) which is the difference in density between oil (ρ_o) and water (ρ_w) over the capillary rise H (Fig. 20.2). We then obtain

K. Bjørlykke (✉)
Department of Geosciences, University of Oslo, Oslo, Norway
e-mail: knut.bjorlykke@geo.uio.no

K. Bjørlykke (ed.), *Petroleum Geoscience: From Sedimentary Environments to Rock Physics*,
DOI 10.1007/978-3-642-02332-3_20, © Springer-Verlag Berlin Heidelberg 2010

Fig. 20.1 Simple petroleum trap with a gas/oil contact (GOC) and an oil/water contact (OWC)

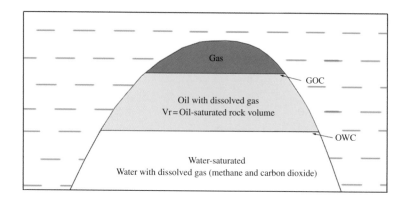

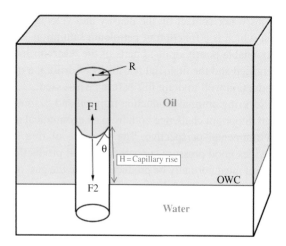

Fig. 20.2 Capillary rise of water in a very thin pipe of water-wet material. The capillary forces acting upwards are $F1 = 2R\pi\gamma\cos\theta$. Here R is the radius of the pore throat, γ the surface tension and θ is the wetting angle. The contact between fluid surface and the grain surface in the pore throat is thus $2R\pi$ and this must be multiplied with the surface tension (γ) and the wetting angle (θ) between the fluid and the grain surface. The gravitational forces acting downwards are $F2 = \pi R^2 gH(\rho_w - \rho_o)$. Here $(\rho_w - \rho_o)$ is the difference between the density of water and the density of oil. At equilibrium the forces acting downwards must be equal to the forces acting upwards so that $F1 = F2$ and we can calculate capillary rise (H) by $H = 2\gamma\cos\theta/Rg(\rho_w - \rho_o)$

$H = 2\gamma\cos\theta/Rg\,(\rho_w - \rho_o)$ and we see that the capillary rise is a function of the interfacial tension (γ) and the wetting angle (θ). This is assuming that the rocks are water-wet ($\theta < 90°$).

γ is the interfacial tension between gas and water, varying within the range 30–70 dynes/cm for gas. For oil, γ is 5–35 dynes/cm (Schowalter 1979). In a sediment, R represents the radius of the intergranular pore throats (Fig. 20.2), θ is the wetting angle between petroleum or water against a mineral surface

(Fig. 20.3). The wetting angle depends on the surface properties of different minerals, as well as on the composition of the petroleum and water, and on the temperature. The capillary resistance is thus also a function of the lithology. In sandstones the mineral surfaces may be dominated by quartz, mica and clay minerals and the system is water-wet. Biodegraded oil may nevertheless wet these mineral surfaces and produce an oil-wet system. Carbonates tend to be more oil-wet than siliceous sandstones and have a higher wetting angle.

Coarse-grained sandstones with little cement, and fractured rocks, have the lowest capillary entry pressures (Fig. 20.4).

We see that the capillary rise H is larger if the pores are small and the wetting angle is low.

In fine-grained sediment the OWC is drawn upwards higher than in coarse-grained sediments. During production the capillary resistance from fine-grained sand will be greater, due to the small pore throats. In fine-grained sandstones or siltstones with small pores, water is drawn (imbibed) upwards due to capillary forces. The difference in OWC which is recorded on the resistivity logs may be as much as a few metres. Since the pores in the reservoir rock usually also contain some water, the percentage of oil or gas in the pores is called the *oil saturation* or the *gas saturation*.

The petroleum saturation depends on the composition of the petroleum phase and its surface properties, on the mineral surfaces and on the pressure difference between the petroleum and the water phase. This pressure difference increases with increasing height above the OWC due to the lower density of the hydrocarbon column (Fig. 20.4b). As a result, petroleum will enter smaller pores higher up in the reservoir, and the

Fig. 20.3 Wetting angle for oil in water against a mineral surface. If the wetting angle is higher than 90° the system is oil-wet. At lower wetting angles it is water-wet

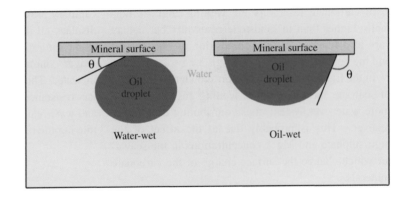

degree of petroleum saturation there will be higher if everything else remains constant.

Most sandstone reservoirs are water-wet and this means that the minerals are lined with a thin layer of water. The smallest pores, i.e. between clay minerals or along the contacts between quartz grains, will then be filled with water. The percentage of water in the pores (water saturation) may be from 80 to 90% in well-sorted sandstones to 50–60% in sandstones with higher clay content. Also clay minerals formed during diagenesis (kaolinite, chlorite and illite) will have rather low water saturation if the system is water-wet because oil can not enter into the smaller pores due to capillary forces. The gas saturation will normally be higher. In reservoirs with biodegraded heavy oil, however, the wetting angle is normally more oil-wet.

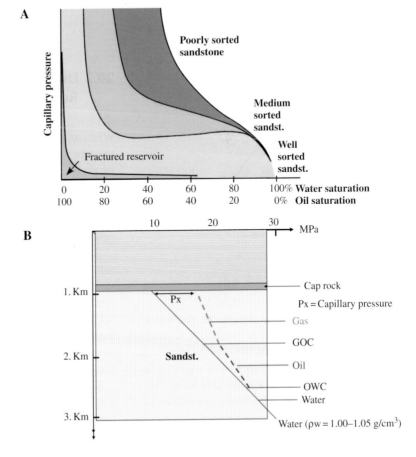

Fig. 20.4 (**a**) Capillary pressure as a function of sediment composition controlling the size of the pore throats. Fractured reservoirs have the largest pore throats and the lowest capillary pressures. (**b**) The pressure in the hydrocarbon phase increases upwards from the oil/water contact (OWC) due to the lower density. The pressure difference between the petroleum phase and the water in the pores is equal to the capillary pressure. As a result petroleum will be able to fill smaller pores (that have higher capillar entry pressure) with increasing height above the OWC

In carbonate reservoirs the wetting angle is normally higher than in sandstone reservoirs but it may vary as a function of the composition of both the oil and the porewater.

In the Ekofisk Field it has been shown that injection of seawater has changed the wetting angle towards a more water-wet system, thus contributing to a higher recovery. This is probably due to the effect of the high sulphate and Mg^{++} concentrations in the seawater which change the surface charge of the carbonate minerals.

When the wetting phase (i.e. water) increases, the contact angle (wetting angle) will increase and reduce the capillary resistance. This is called *imbibition* and a core with high oil saturation may take up (imbibe) water. When the wetting phase decreases, this is called *drainage*. The contact angle forms a hysteresis in water-wet reservoirs as a function of displacing oil and water (Fig. 20.5).

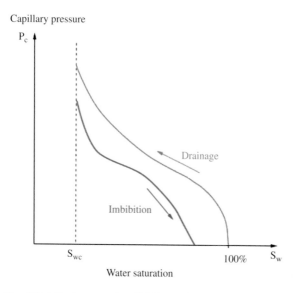

Fig. 20.5 Drainage and imibibition of oil and water as a function of water saturation and capillary pressure (Pc)

Starting with 100% water saturation, the water is displaced by oil so that the water saturation is reduced to a value Swc. This is called the connate water saturation. It is not possible to expel more water (reduce the water saturation) by oil beyond this value due to the high capillary pressures Pc. When water flows back into the rock (imbibition), the capillary pressure curve follows a different path, reaching a point where the

capillary pressure is zero. The water can therefore not displace all the oil and we can not get back to 100% water saturation. This is because the oil phase is no longer continuous and there are only droplets of oil in water. The droplets of oil will meet strong capillary resistance when trying to pass through pore throats and water can then flow through the reservoir without moving the remaining oil (Fig. 20.6).

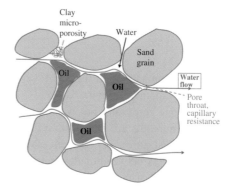

Fig. 20.6 Schematic figure showing how oil may be by-passed as water flows around the oil. Oil must overcome a greater capillary resistance to flow through the pore throats

20.2 Determination of Petroleum Reserves

When an exploration well has detected oil an estimate must be made of the reserves that can be produced. In many cases additional wells (appraisal wells) are needed before the full extent of the field and its financial viability for production can be determined.

The total initial volume of oil in the reservoir is called the volume of oil in place (V_p).

The percentage of that oil which can be produced is called (recovery) or the recovery factor.

The volume of rock above the oil/water contact (*OWC*) and below the cap rock, V_r, is determined by 3D seismic and data from the wells. This requires that the (*OWC*) is correctly defined, but it may not be visible on the seismic records, nor necessarily be at the same depth throughout the reservoir.

Good reservoir intervals are often called *pay* zones, whereas the tight zones are called *non-pay* zones. Intervals (layers) where the porosity is low (<10%) and the permeability below a few milli Darcy (mD) may not be capable of producing significant amounts of oil even if they lie above the oil/water contact. In fractured rocks (e.g. fractured limestones) oil can be produced at very much lower porosities. The ratio between the volume of reservoir rocks (N) that possess sufficient reservoir quality to produce significant amounts of oil (pay), and the total volume (G), is called *net to gross (N/G)*. We then need to estimate the average porosity (φ_a).

The volume of oil or gas in place (V_p) is then:

$$V_p = V_r \cdot N/G \cdot \varphi_a \cdot \text{Sat}$$

Calculation of the volume above the OWC also depends on correct depth conversion of seismic travel times. If the velocity of the overlying rocks changes across the field, this will influence the calculated volume. The oil/water contact may not always be horizontal across a structure and there may be sub-compartments within the field with different OWCs.

The average porosity φ_a is difficult to determine. We may have a very wide range of porosities in the same reservoir, from good reservoir quality to rather poor. Sandstones with less than 10–12% porosity usually have such low permeability that they are not considered to be producing parts (zones) of the reservoir.

We see that the calculation of oil in place is based on factors that are impossible to determine accurately. Each value, such as porosity or oil saturation, may be represented by a probability distribution and the calculation of oil is then based on statistical analysis. If the estimated average porosity in a reservoir is for example 18% we are interested in the probability of the average porosity being significantly lower or higher (e.g. 15 or 21%).

The fraction of the oil in place that can be produced is called the *recovery* and this can vary greatly, from 20–30% to 60–70%. The number of wells and their position are critical for field production, and the recovery can also be stimulated by injection of gas or a chemical. While 20–30 years ago the maximum recovery was considered to be 40–50%, it has since increased due to improved technology. Recovery in highly porous and permeable reservoirs may approach 70%. Towards the end of the production life of a field the wells may produce much more water than oil, but the water is then re-injected.

20.3 Reservoir Energy

A reservoir may be overpressured, giving it the potential to flow to the surface. If the wells are not managed properly a blow-out may occur. Even with reservoirs at hydrostatic pressure in the water phase, the pressures in the oil and gas will be higher due to their buoyancy relative to water and they will flow towards the surface. When oil is rising up a well the pressure is falling and gas may then bubble out of solution in the oil phase, as well as in the water phase. This will reduce the density and further increase the buoyancy effect so that a very high flow rate results. The formation of free gas is an important part of the mechanism that causes blow-outs in wells.

20.4 Relative Permeability

Permeability is a term expressing the resistance to fluid flow. It is a function of the rock properties and is independent of the viscosity of the fluid phase. However, when there are more than two fluid phases present in the pores, the flow and the permeability are different compared with when there is only one fluid phase.

In pores containing two fluid phases, the permeability of one of the fluid phases is expressed as the relative permeability (k_r). This is the permeability with two fluid phases divided by the permeability with only one fluid phase. In the case of oil and water:

$$k_r = k_{o+w}/k_w$$

The relative permeability is a function of the percentage of each fluid phase in the pore space and also of the wetting properties (wetting angle) of the phases.

The permeability with both oil and water present (k_{o+w}) is lower than the permeability with only water (k_w) (Fig. 20.7).

Two phase flow will always reduce the total flow rate (Figs. 20.7 and 20.8) and this is critical for flow near the well and in the pipes.

We see that if the oil saturation is less than about 35% only gas will flow. Even with high oil saturation (80%) there will still be some production of gas. In some cases we may have three phases together (water, oil and gas) and this is very difficult to model.

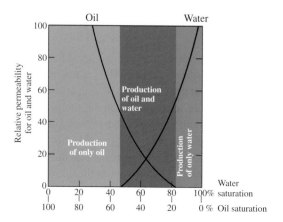

Fig. 20.7 Diagram showing the relative permeability with a mixture of water and oil. Similar relations exist between oil and gas, and between gas and water

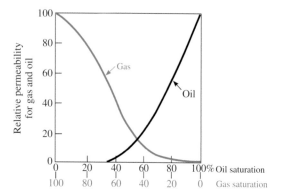

Fig. 20.8 Relative permeability of oil and gas. We see that if the oil saturation is less than about 35% only gas will flow. Even with high oil saturation (80%) there will still be some production of gas

Behaviour of reservoir fluids.

The reservoir fluids obey the phase rules.

For an ideal gas we have:

$$p \cdot V = nRT$$

Here p is the pressure, V is the volume and T is temperature (K). n is the molecular weight of the gas and R is the universal gas constant.

When oil and gas are produced, the pressure within the reservoir is reduced and water then flows into the reservoir rock (Fig. 20.9). The rate of flow from the surrounding rocks controls the rate of pressure drop

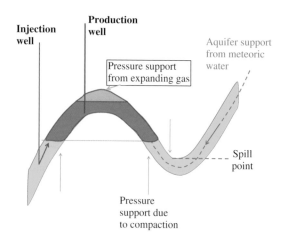

Fig. 20.9 Different types of flow of water into the reservoir providing pressure support. This may be from the volume of sandstones communicating with the reservoir, from adjacent shales and siltstones, from meteoric water and from water injection

during production (Fig. 20.9). If the reservoir is connected to another, very large and extensive, reservoir there will be very significant pressure support from the surrounding rocks, which are often referred to as *the formation*. In a relatively isolated sand body surrounded by low permeability shales the pressure will drop fairly quickly during production. The rate of pressure reduction is thus a function of the volume of sandstones connected to the oil-saturated reservoir. If the sandstone volume is large or if it is communicating with siltstones and sandy shale with significant permeability, there will be more pressure support to the reservoir during production. Changes in the reservoir pressure may also cause changes in the petroleum phase (gas/oil ratio, formation of condensate etc.).

At the critical point gas and liquid have the same properties (Fig. 20.10). This is the temperature and the pressure where liquids can be transformed to gas without any change in volume. For water this is 22.1 MP (221 bar) and 374°C. For hydrocarbons it depends very much on their composition.

A reservoir may be thought of as a large container; during production we can therefore calculate reservoir pressures using material balance equations. If there is little compaction of the reservoir during production, the volume available for the fluids (the porosity) will be almost constant. The main change is due to elastic deformation due to loading or unloading. In some cases as in the Ekofisk Field, where the reservoir rock is mechanically weak, there is significant

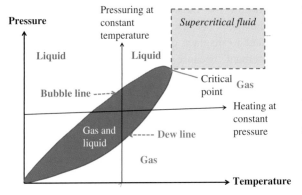

Fig. 20.10 Simple phase diagram showing how gas will form in the reservoir as pressure is reduced during production. When gas is produced, lower temperatures and pressures at the surface may cause liquids to form (condensate). If the pressure is reduced in the reservoir gas, methane in particular will bubble out of solution from both the oil and water phases. This has a feedback effect which helps to maintain reservoir pressure. At the critical point gas and liquid have the same properties. This is a temperature and pressure where liquids can be transformed to gas without any change in volume. For water this is 22.1 MP (221 bar) and 374°C. For hydrocarbons it depends very much on the composition

compaction of the reservoir during production due to increased effective stress or chemical effects of the water flooding.

The equation of state is:

$$V \cdot P/T = C$$

If the temperature is constant, the pressure is a function of the change in volume, thus

$$C = \Delta V / V \Delta p$$

$$\Delta V = C V \Delta p$$

Fluid flow during production can be calculated by applying the conservation of mass. The volume of hydrocarbons produced must be replaced either by fluids flowing into the reservoir from outside or by an expansion of the fluid phases with corresponding reduction in pressure.

When we produce petroleum from a reservoir the volume of produced petroleum can at least partly be replaced by water flowing into the reservoir from surrounding rocks. The drop in volume (ΔV) is then a function of the volume of produced petroleum (V_{pro})

minus the volume of water flowing into the reservoir which is referred to as the *water drive* (V_{wd}).

$$\Delta V = V_{prod} - V_{wd}$$

The drop in pressure due to produced oil, gas and water is a function of:

(1) The compressibility of the fluids present. Oil with a low content of dissolved gas has relatively low compressibility and the pressure will be reduced rather rapidly. If free gas is present as a separate phase above the oil, it will be highly compressible and will expand as the oil is produced and help to maintain the higher pressures in the oil phase. Oil saturated with gas will form a separate gas phase when the pressure is reduced and thus maintain the pressure.

(2) The supply of fluids from outside the reservoir. This may be from meteoric water connecting to the surface where there is a continual supply of water, or by the injection of water or gas.

(3) Supply of water from around the reservoir. Pressure reduction in the reservoir during production creates a potential for water to flow from the surrounding rocks. Reservoir sandstones that are surrounded by low permeability shales will have a very low supply of water (water drive) and the pressure will therefore drop faster during production. This is particularly true when reservoir sandstones are offset laterally by faults. In most cases the permeability of shales is so low as to be almost irrelevant in the time span during which petroleum is produced from a field. However, if the reservoir is part of a thick sandstone sequence, the pressure support provided from a larger sand volume may be very significant. If we only consider the water phase, the pressure drop in a sandstone reservoir is a function of the volume of sandstone in communication with the reservoir.

20.5 Meteoric Water Drive

Reservoirs connected to an aquifer that is part of the meteoric water flow system will experience very little drop in pressure during production because the produced petroleum is quickly replaced by meteoric water from the aquifer.

While the supply of meteoric water may be relatively fast, compaction-driven water flow is many orders of magnitude lower and can be ignored.

In a closed system the response to production and the associated reduction in reservoir pressure is a function of the compressibility of the water (C_w), the compressibility of the oil (C_o) and the compressibility of the gas (C_g). There will also be changes in the reservoir volume (porosity) because reduced fluid pressure will result in higher effective stresses, but this response will normally be rather small unless the reduction is sufficiently large to cause mechanical compaction with grain crushing.

When the pressure is reduced there will be an expansion of both the water and the petroleum in the reservoir, but this is relatively small.

20.6 Gas Expansion Drive

Gas has a high compressibility and if free gas is present in the reservoir this gas phase will expand as the pressure is reduced. The pressure drop relative to the volume of oil produced will therefore be relatively small. In addition, gas that was originally in solution in the oil will bubble out as a separate phase and add to the gas drive.

20.7 Compaction Drive

If the fluid pressure is reduced, the increase in effective stress may cause mechanical compaction, reducing the pore volume. While this will also contribute to the maintenance of high pressure, it may damage the reservoir by grain crushing and closing fractures (if present). In the Ekofisk Field, however, the compaction of the reservoir provides a drive that enhances recovery of the oil.

20.8 Water Injection

Water is injected into the reservoir to replace the produced petroleum and thus maintain high pressure in the reservoir. It will also help to maintain the pressure gradients towards the production wells. Injected water

will follow the most permeable layers at the base of sandstones because it is denser than oil. Injection of water to displace oil or gas is called "immiscible replacement" because oil is not soluble in water. Water will displace oil found in pockets (depressions) against low permeability shales (Fig. 20.13b).

The vertical distribution of permeability and capillary resistance within a bed is important for the flow of oil and water and the total recovery (Figs. 20.11 and 20.12). We see that in coarsening-upwards sequences the highest permeability and lowest capillary resistance are at the top, where oil or gas flow will preferentially occur due to the buoyancy relative to

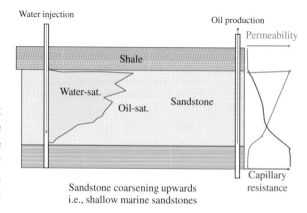

Fig. 20.11 Production from a coarsening-upwards unit, e.g. a marine shoreface sequence. The injected water will sink in to the less permeable layers and displace oil also from there

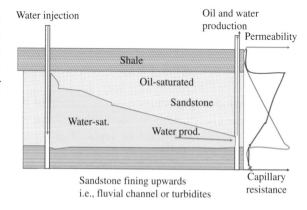

Fig. 20.12 Production from a fining-upwards sequence, e.g. fluvial channel facies. The injected water will sink to the lowermost permeable layers and then may break through to the production well, leaving most of the oil in the less permeable beds behind. Addition of polymers to increase the viscosity of the water may result in a better sweep and produce more oil

water. In fining-upwards sequences the injected water will tend to follow the basal permeable part, causing early water breakthrough to the production well.

20.9 Gas Injection

When gas is injected in to the reservoir it is partly soluble (miscible) in both the oil phase and the water phase. The most common gases used are methane, nitrogen and more recently also carbon dioxide. Gas injection contributes to the maintenance of the pressure in the reservoir. Gas is less dense than oil and is efficient in terms of displacing oil from small micro-traps where pockets of oil are found above water. Gas will therefore sweep the upper parts of sandstone beds in a reservoir and displace the oil downwards, so that it can flow out of the small traps (Fig. 20.13a).

The injected gas may be methane that is produced along with the oil. These gases are also to a large extent dissolved in the oil phase during the injection.

The use of CO_2 is now encouraged for injection in reservoirs because this represents a storage of carbon dioxide, reducing the contributions to global warming.

Carbon dioxide has been applied, for example, in the Sleipner Field in the North Sea, but not in the reservoir.

Tracers may be employed to see if water from the injection well has reached the production well.

20.10 Other Methods for Enhanced Recovery

When there is breakthrough of water from the injection well to the production well, the flow of water may be reduced by adding polymers that increase the viscosity of the water. The pressure gradient will then increase in the oil phase and it may be possible to displace more oil towards the production well.

When the oil saturation is low, particularly towards the end of production, movement of the remaining oil drops is prevented by the capillary forces. By adding surfactants (soap) the oil drops may be broken down into an emulsion which can flow with the water to the production well.

Many enhanced recovery methods are more effective in onshore oil fields where the distance between the wells is much less than offshore. During high oil prices it may be economical to spend more on chemicals to stimulate production.

There are three main enhanced recovery methods:

I. *Thermal processes:* These are processes which increase the temperature of the oil and thus lower its viscosity. The most common are:
   (a) Steam injection. This is the main recovery method for heavy oil and tar sand.
   (b) In situ combustion. Supplying oxygen to partially burn heavy oil or coal to increase the temperature and reduce the viscosity of oil, and to produce gas from coal.

II. *Chemical processes:* These involve injection of chemicals into the reservoir. The most important types are:
   (a) Injection of caustic (alkaline) chemicals which reduce surface tension (surfactants).
   (b) Polymers which are injected into water to increase its viscosity to prevent water breakthrough.

III. Injection of various types of gas to increase the miscibility of hydrocarbon phases. These may include carbon dioxide, neutral gases (e.g. nitrogen) or hydrocarbon gases.

I. *Thermal Processes:* These are only used with reservoirs with highly viscous oil, where an increase in temperature will greatly reduce the viscosity. This will normally be heavy, asphaltic oil which has been

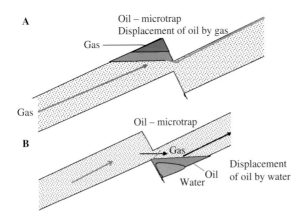

Fig. 20.13 (a) Displacement of oil by gas in small traps during production. (b) Displacement of oil by water in a reservoir with mostly gas

subjected to bacterial breakdown and evaporation. Oil reservoirs of this type are normally fairly shallow (less than 1,500 m). Deeper reservoirs will have a high temperature anyway, so there is less to be gained by artificially heating the oil.

In situ combustion can be induced by pumping air down into the reservoir, and the combustion can be regulated by addition of oxygen. The hydrocarbons which do not burn will then become hotter and less viscous. Burning also produces gases, which may increase production. Special wells can be used for injecting air, and combustion spreads from these to the production wells, forcing a zone of hot water, gas and light oil ahead of it to the production wells. Combustion may also increase the porosity of the reservoir rock, making it possible to produce later on from the wells in which combustion has taken place.

Steam injection is responsible for 90% of all production through secondary recovery methods apart from water injection. One of the major reasons for the low recovery percentage from oil reservoirs is that the oil is too viscous. This is particularly true of oil which has been subjected to bacterial degradation (biodegradation), leaving the viscous asphaltic part of the oil behind as heavy oil (low gravity 16–17 API). The Cretaceous Athabasca tar sand in Alberta, Canada, is the most famous example, and contains huge oil reserves. When oil reservoirs with such heavy biodegraded oil are exposed at the surface, the oil is almost solid at normal temperatures, and is called tar sand. For it to be possible to produce oil, the pressure difference between the oil in the reservoir rock and the well must exceed the resistance to flow exerted by capillary forces (in water-wet reservoirs) and viscosity forces (friction).

Biodegraded oil is normally found at relatively shallow depths, and the temperature of the oil is low and the viscosity consequently high. By injecting steam the temperature can be raised and the viscosity lowered considerably. There are examples of the viscosity of oil being reduced from 300 to 10 cP after steam injection, and biodegraded oil is often too viscous to flow at all without steam injection. Steam injection is particularly important in California, where 300,000 barrels/day are produced using this method. Steam injection must be repeated at regular intervals because the viscosity increases again as the reservoir cools off. Sandstones which are to be subjected to steam injection must also have high oil saturation, because otherwise water and steam can flow past the oil. Steam is recovered from the produced oil, and in some fields one out of every three barrels produced is used for steam production.

II. *Chemical Processes:* When oil flows out of a reservoir rock into a producing well, it must overcome the capillary forces involved in two-phase flow. By reducing the surface tension between oil and the water which is injected into the wells, the capillary forces can be reduced. The addition of surfactants, a sort of soap, will reduce the surface tension so that the *wettability* changes. The wettability can be measured by means of the contact angle, i.e. the angle which the liquid phases make with a mineral surface, for example (Fig. 20.3).

If an oil reservoir is *oil-wet*, it means that oil is more strongly bound to the mineral surface than water. By reducing the surface tension in the water phase, the *wetting angle* is altered and the system may become *water-wet*, and the oil will become more mobile so that the relative permeability to oil is increased. A mixture of different chemicals (*slug*) is injected into oil reservoirs to reduce the surface tension. It includes micro-emulsions, soluble hydrocarbons, micelles and frequently also alcohol and various salts.

One serious drawback of this method is that chemicals can easily be adsorbed onto the reservoir rock. Clay minerals in particular have high ion exchange and adsorption capacities. The specific surface (m^2/g) of rocks is also of significance. Consequently chemicals are often pumped down in advance (*preflush*) to reduce adsorption of the chemicals which are intended to reduce surface tension. This treatment ought also to take into account which clay minerals occur in the rock.

Polymers. In oil reservoirs with an undesirably high relative water permeability, large amounts of oil will remain in the reservoir rock. Water will flow where there is least resistance to its movement, e.g. along cracks and permeable sand beds, and injection of water will then have little effect. This is particularly relevant with oil reservoirs with low oil saturation (high water saturation). If the viscosity is increased so that the pressure gradient in the water phase increases, the gradient in the oil phase will also increase. Polymers reduce the mobility of the water, causing a piston-like replacement mechanism. With increased viscosity, the polymer-bearing water will exert greater shear forces on the oil phase, so that oil drops are more easily carried along. Water containing polymers is no longer a Newtonian liquid, but a pseudoplastic liquid in which the apparent viscosity is reduced by the rate of deformation.

Polymers will also have a tendency to be adsorbed onto the surface of minerals so that their effect decreases with increasing distance from the injection well.

Addition of Alkaline (Caustic) Chemicals. Strong alkaline solutions, for example NaOH, will also reduce the surface tension of aqueous phases. Injection of a mixture of these alkaline solutions may therefore increase both the water-wetness of the reservoir and oil production. The most common chemicals include sodium and potassium hydroxide, sodium orthosilicate, sodium carbonate and sodium phosphate. The method is used largely in sandstone reservoirs. Small amounts of gypsum or anhydrite will cause $Ca(OH)_2$ to precipitation and neutralise caustic soda (NaOH).

The object of *gas injection* is to produce a fluid phase which will dissolve the reservoir oil. The flow of such a fluid through reservoirs with low oil saturation may dissolve the oil and carry it along to the production well. Hydrocarbon gases which are in the liquid phase at reservoir pressure (*Liquid Petroleum Gas – LPG*) will be fully miscible with oil. This requires that the reservoir temperature be below the critical temperature for the gas. At higher temperatures the gas is in gaseous form regardless of pressure, and is not fully miscible with oil. Propane is often used in the oil-expelling mixture, and gas and water are injected subsequently.

CO_2 has a critical temperature of about $31°C$, and is therefore gaseous at all reservoir pressures. Carbon dioxide is very soluble in oil, which increases in volume when it is saturated with respect to CO_2. This creates a sort of gas drive.

Increasing the CO_2 content also reduces the viscosity and density of oil, making it very mobile. The method is used for oil reservoirs with oil saturation of 25–55%. Neutral gases such as nitrogen may also be injected into reservoirs, but far greater pressure is then required to make the gas soluble in oil.

20.11 Changes During Production

With production from sandstone reservoirs, the physical properties of sandstone must be taken into account, e.g. distribution of porosity and permeability. The actual production, however, will lead to changes in the reservoir rock which may have an adverse effect on the reservoir properties, i.e. cause reservoir damage. This may happen in two ways:

1. Chemical reactions between minerals and liquids which are used in drilling or injection of water.
2. Mechanical damage through relatively loose clay minerals or other small grains being carried by the fluid flowing towards the well, and obstructing the pore throats.

As oil flows into a well, the flux (cm^3/cm^2) is inversely proportional to the distance to the well. If we can improve the permeability of the reservoir nearest the well, we may be able to step up the production rate considerably. For this reason chemicals are sometimes injected into the reservoir to improve the permeability. We then need to understand some of the chemical properties of minerals in order to be able to predict the reactions which will result from treatment with acids and other chemicals.

As far as chemical reactions are concerned, clay minerals are important because they have a large specific surface and ion exchange capacity. The specific surface varies with the size and shape of the mineral grains. Kaolinite will typically have a specific surface of 5–30 m^2/g, chlorite 10–50 m^2/g, illite and smectite (montmorillonite) >100 m^2/g. The high specific surface of illite and smectite is due to the fact that they often occur as very thin sheets or fibres.

Kaolinite has very low solubility in water, and strong acid (HF) is required to dissolve it. Kaolinite is therefore chemically stable and its ion exchange capacity is lower than that of other clay minerals. Chlorite, on the other hand, is soluble in acids such as HCL. If acid is used to dissolve minerals like chlorides in order to increase the permeability around the well, it is important to avoid iron being precipitated as iron hydroxide ($Fe(OH)_3$). This could reduce the porosity very severely because it forms pore filling cement. To avoid this, one can add chemicals which form complexes with iron (e.g. citrate).

Illite and smectite are not very soluble in HCL, and hydrofluoric acid (HF) must be added to dissolve these minerals they are sensitive to variations in salinity and in particular to the K^+ content). K^+ and other alkali ions will be held in the interlayer position in smectite so that water is expelled and the mineral contracts in volume. On the other hand, injection of fresh or slightly saline water into a reservoir may lead to uptake of water and expansion of smectite and mixed layer minerals, which may greatly reduce the porosity. Sodium smectites can swell to 5–10 times their original volume. Swelling can be reduced by injecting dilute acid (Hcl) or water with a high salinity (e.g. KCl). Water is also often injected to maintain reservoir pressure. Offshore drilling platforms in particular depend on using seawater. However, seawater may have adverse effects on the reservoir. The high sulphate concentration in seawater may cause precipitation of baryte ($BaSO_4$) which is the least soluble of the common sulphate minerals. Seawater also contains sulphate-reducing bacteria which will start to reduce SO_4^{2-} to H_2S in the reservoir. This is most undesirable since H_2S is a very poisonous gas which also has a highly corrosive effect on steel during production.

To avoid unforeseen chemical damage to the reservoir rock, it is necessary to carry out very thorough mineralogical analyses so as to be able to predict the effect of various types of chemical stimulation.

Physical damage to the reservoir, as previously mentioned, is due to mineral grains being loosened and carried by oil and water, eventually clogging the flow paths. If the reservoir is water-wet, fine-grained minerals will tend to remain suspended in the aqueous phase, and will not come into the oil phase very easily because of surface tension. When considerable quantities of water are produced together with oil, therefore, the aqueous phase is particularly likely to carry clay minerals and other small mineral particles, for example of quartz and felspar, which may block the pore throats. This effect can be reduced by lowering the production rate. In the laboratory this type of mechanical formation damage can be tested in an experimental flow rig. If the measured permeability increases temporarily when the flow direction is reversed, formation damage has been caused by "a moving clay and silt fraction". Sandstones with a high percentage of secondary porosity will have relatively large pores with small pore throats. The ratio between the pore diameter and the pore throat is often referred to as the aspect ratio. Pores with high aspect ratios will be particularly vulnerable to mechanical formation damage. Pore geometry is also important with respect to two-phase flow (oil and water) because of the capillary forces that have to be overcome in the pore throats. This is even more the case with condensate reservoirs, where we may have three phases (oil, gas, water).

The planning of production from a reservoir involves a detailed production strategy and optimal positioning of production and injection wells. The trend production will take with time is simulated on extremely powerful computers which are fed with a very large number of parameters relating to the reservoir. Simulations depend very heavily on the geometry of the reservoir and internal communication, e.g. between sandstone bodies, the exact position of the sealing fault, etc. The internal properties of the reservoir can be measured on cores from the wells, but a three-dimensional representation of the distribution of porosity and permeability and the pore geometry must be constructed, using models for the detailed depositional environment and diagenetic alteration.

20.12 Carbonate Reservoirs and Fractured Reservoirs

Many of the same principles apply to carbonate reservoirs as apply to sandstones. There is a tendency for carbonate reservoirs to be less water-wet than sandstone reservoirs, but this depends on the composition of the oil and on the temperature and pressure. Pores often have a high aspect ratio, particularly in limestones where we often have intragranular porosity inside fossils or vuggy mouldic porosity after selective dissolution of organic fossils. The reservoir properties of bioclastic and reef limestones are largely a result of the palaeo-environment, which determined the distribution of fauna.

Fractures are generally much more important in carbonate than in sandstone reservoirs. Stylolites are sometimes found in sandstone reservoirs that have been buried to at least 3.4–4 km. In limestones stylolites develop at much shallower depths, often forming a thin clay surface that is quite impermeable, and which may divide a reservoir up into compartments or even serve as a cap rock.

In fractured reservoirs there is a high percentage of porosity due to fractures rather than more uniformly distributed porosity. Fractures are most extensively developed in brittle rocks like well-cemented limestones and chert. They may also form in metamorphic and igneous rocks. Fractures are often invaded by drilling mud, causing lost circulation, and this may damage the reservoir near the well. The flow of petroleum through a fractured reservoir during production is very complex, and it is difficult to map out the fracture system of the subsurface in sufficient detail.

As a result of reduced pressure during production, fractures may start to close due to increased net stress. By injecting water at very high pressures (higher than the geostatic pressure) fractures may be widened (hydrofracturing), a method also sometimes used in other reservoirs to create or widen fractures close to wells and improve the permeability of the critical area around the well. The introduction of some coarse-grained material like sand and also other types of granular materials may help to wedge the fractures and prevent them from closing.

Onshore oil fields have much closer well spacing than offshore fields because of the lower drilling costs. This means that many of the enhanced recovery methods can be used much more effectively. At high oil prices mature oil fields can have a very long tail production and be economical even with very high water production (>90% water).

20.13 Monitoring Production

During production, comparison of pressure changes in the production and injection wells will provide information about fluid communication within the reservoir. These data are fed into vary large and complex reservoir simulation models.

4D seismic may be successful in some fields for detecting changes in the gas/oil contact (GOC) and in some instances also the oil/water contact (OWC). Repeated seismic surveys at 1–3 years intervals may detect an upward movement of the OWC as the reservoir is depleted. The difference in impedance may, however, not be as pronounced at the primary OWC because there may be 30–40% oil saturation in the drained intervals, It may then be possible to detect parts of the reservoir which are not drained and which are large enough to justify an additional well.

20.14 Well-to-Well Tracers

Tracers can be added to the water in the injection well and their arrival in the production well can provide important information about the reservoir. If the tracer is recorded in the production well relatively soon after the injection there must be a high permeability connection in the form of an open fault or fracture or a high permeability sand layer. The time of the arrival of the tracers makes it possible to calculate the flow velocity of the water phase. The tracers must remain dissolved in the water and should not be adsorbed on the minerals. Since they become part of the produced water they should be environmentally acceptable. Many of the tracers are radioactive but only with β radiation, such as tritiated water (HTO), $^{36}Cl^-$ and $^{22}Na^+$. Most of them have a relatively short half life, except $^{36}Cl^-$.

It is also possible to use non-radiocative tracers, but extremely low levels must be detectable (in the ppb range) because of the strong dilution. Gas tracers may also be used.

The concentration of the tracer as a function of time during production may provide very important information about the rock properties and the flow in the reservoirs. Different types of tracers may to different degrees interact and be adsorbed onto the reservoir rocks.

20.15 Reservoir Models and Field Analogues

When planning the production of a reservoir, a relatively detailed reservoir model is required.

Seismic data has limited resolution and in offshore reservoirs the spacing between wells may be relatively large, usually several 100 m or a few kilometres. It is difficult to construct a 3D model when the producing units are offset by faults that are below seismic resolution so that they can not be detected. Outcrops of similar reservoir rocks may be used to provide quantitative data on the geometry of sand bodies and the distribution of minor faults. to be used in the reservoir model.

It is however critical that the field analogue was deposited in a similar sedimentary environment and plate tectonic setting. Many factors such as sediment supply, climate, water depth, tidal range, etc. have to be similar.

All reservoir analogues that are exposed have been buried to a certain depth and then uplifted to the surface. It is difficult to reconstruct the burial history in terms of temperature and effective stress so that the diagenetic effects can be compared with reservoirs at their maximum burial depth.

The structures observed in the field analogue must be analysed so that the deformation which occurred during uplift can be distinguished from that produced during subsidence.

20.16 Conclusion

Production geology encompasses many different specialities in geology, geophysics and engineering subjects:

Reservoir sedimentology/diagenesis
Reservoir geophysics.
Reservoir modelling (modelling of fluid flow)
Drilling and well completion.

A team of specialists is therefore required to produce a reservoir efficiently.

Further Reading

Archer, J.S. and Wall, C.G. 1994. Principles of Petroleum Engineering. Principles and Practice. Graham and Trotman, London, 362 pp.

Dandekar, A.Y. 2006. Petroleum Reservoir Rocks and Fluid Flow Properties. CRC Publications, Boca Raton, FL, 460pp.

Dickey, P.A. 1979. Petroleum Development Geology. Petroleum Publishing Co., Tulsa, OK, 398 pp.

Dikkers, A.J. 1985. Geology in petroleum production. Developments in Petroleum Science 20, 239 pp.

Robinson, A., Griffiths, P., Price, S., Hegre, J. and Muggeridge, A. 2009. The Future of Geological Modelling in Hydrocarbon Developments. Geological Society Special Publication, 232 pp.

Tiab, D. and Donaldson, E.C. 2004. Petrophysics. Elsevier, Amsterdam, 889 pp.

Chapter 21

Unconventional Hydrocarbons: Oil Shales, Heavy Oil, Tar Sands, Shale Gas and Gas Hydrates

Knut Bjørlykke

For many decades conventional oil which could be produced at low cost was present in abundance. A low oil price gave no incentive to look for other types of resources. It is now clear, however, that we are gradually running out of new sedimentary basins to explore and that the reserves of conventional oil which can be produced cheaply are limited.

This is the reason why many of the major oil companies now invest in what is often called unconventional hydrocarbons. There are large reserves of these and the main types are oil shales, heavy oil, tar sand and shale gas.

Gas hydrates may also become an important source of hydrocarbons but the development of methods to produce from such accumulations is still at an early stage.

The reserves of unconventional hydrocarbons are very large, probably larger than those of conventional oil. We will therefore not run out of fossil hydrocarbons for a long time but the prices will be higher because these sources are more difficult to produce, particularly if strict environmental standards should be met.

21.1 Heavy Oil and Tar Sands

Heavy oil seeping out on the surface has been known for a long time and this was easy to exploit for use in small quantities. Even after drilling for oil became successful (in 1857 in Pennsylvania), mining for heavy oil continued in many parts of the world including Germany. In southern California (Ventura and Los Angeles basins) oil was mined from the early 1860s to the 1890s because the heavy oil would not flow to the wells (Fig. 21.1).

Tar sands are sandstone reservoirs which have been filled with oil at shallow depth <2 km (<70–80°C) so that the oil has become biodegraded. Reservoir rocks which have been buried more deeply and then uplifted before the oil migration may be sterilised at higher temperatures and are less likely to be biodegraded.

When the sandstone has not been buried to more than about 2 km depth the sand will remain mostly uncemented as loose sand because of a lack of quartz cement.

Tar sand contains asphaltic oil rich in asphaltenes and resins. It has a high content of aromatics and naphthenes compared to paraffins, and a high content of nitrogen, sulphur and oxygen (NSO). Most of the hydrocarbon molecules have more that 60 carbon atoms and the boiling point and viscosity are therefore very high.

The viscosity of the biodegraded oil is very low and the oil must be heated so that the viscosity is reduced before it can be produced by drilling wells. There are transitions between reservoirs with heavy oil and nearly solid bitumen. Heating can be achieved by soaking the reservoir with injected steam so that the heat of condensation to water helps to heat the oil. This is called cyclic steam injection when the steam is injected in the production well and left to soak for a few weeks before production starts when the oil is warmer. The steam may also be injected in a nearby well and driven towards the production well. Much of the energy is used to heat the water in the reservoir and

K. Bjørlykke (✉)
Department of Geosciences, University of Oslo, Oslo, Norway
e-mail: knut.bjorlykke@geo.uio.no

K. Bjørlykke (ed.), *Petroleum Geoscience: From Sedimentary Environments to Rock Physics*,
DOI 10.1007/978-3-642-02332-3_21, © Springer-Verlag Berlin Heidelberg 2010

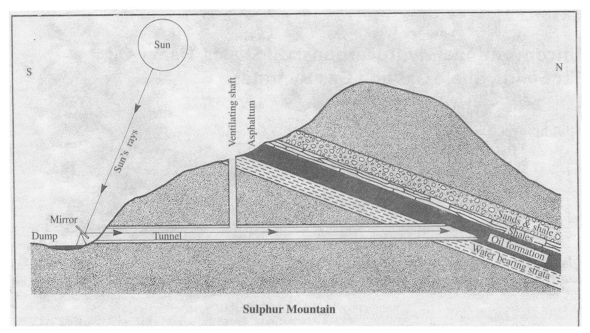

Sulphur Mountain

Fig. 21.1 Mining for heavy oil in the Sulfur Mountain near Santa Paula, South California (from 1860 to 1890). Mine tunnels could be 6–700 m long and a mirror was used to light the tunnel and the sunlight would contribute to the heating of the heavy oil so that it could flow. See the Oil Museum in Santa Paula (http://www.oilmuseum.net/)

it is important to reduce the inflow of water from adjacent rocks. One method includes freezing the ground at a distance from the well to avoid the flow of water towards the well. It is also possible to burn some of the oil in the subsurface to provide heat to heat the oil.

It has also been proposed to heat the oil electrically, possibly powered by a nuclear reactor to reduce the CO_2 emissions from burning oil to produce heat.

The tar sands in Alberta, Canada (Athabasca) are of Middle Cretaceous age (Aptian, 100 million years).

The main reservoir rock is the McMurray Formation, a sandstone representing fluvial to tidal environments, and it is important to find thick sequences of sand with few clay layers. The oil was generated from older source rocks during the Laramide folding of the Rocky Mountains to the west and migrated into the Athabasca sands during the late Cretaceous and early Tertiary. Starting in the Eocene (50 million years) the sand was uplifted and most of the overburden eroded. The oil was then biodegraded by bacteria. These tar sands contain 1.7 trillion bbl (270×10^9 m^3) of bitumen in-place, comparable in magnitude to the world's total proven reserves of conventional *petroleum*. There are currently large mining operations at Athabasca (Fig. 21.2). At surface temperatures, which are low in northern Canada, this tar sand is rather hard (Fig. 21.3)

Fig. 21.2 Athabasca tar sand. The oil-impregnated sand is mined and the heavy oil is separated from the sand with hot water

Fig. 21.3 Cores from the Athabasca tar sand. The oil sand is highly viscous, almost solid, but will flow at high temperatures. Dark intervals are loose sand mostly with more than 30% porosity, filled with biodegraded heavy oil. *Grey layers* are thins shales which will reduce the vertical mobility of the heated oil during production

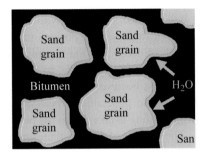

Fig. 21.4 Schematic representation of tar sand. The oil may also stick more closely to the grain surfaces, so the system is more oil-wet

because of the high viscosity of the oil. The oil must be separated from the sand using hot water (Fig. 21.4). The oil (tar) is very viscous and may be denser than water (API <10).

Only about 20% is close enough to the surface to be economically mined and the rest must be heated in place.

A cubic metre of oil, mined from the tar sands, needs 2–4.5 m^3 of water.

Oil may be extracted by steam-assisted gravity drainage (SAGD). Two parallel horizontal wells are drilled so that one is about 5 m below the other. Steam is injected in the shallowest well and the heated, less viscous, oil is drained into the lower well where it is produced. The steam will remain near the top due to its low density but will sink when condensed to water. The shales will reduce the vertical permeability and

the drainage, but the heating will cause some fracturing which may allow more vertical flow. This method requires rather thick and homogeneous sand. Steam can also be injected into one well for a few weeks until the oil is heated and then produced from the same well. This is called cyclic steam stimulation (CSS).

Large amounts of energy and water are required to produce the steam for this method. If petroleum is burned to produce the heat, high emissions of CO_2 will result not only from when the oil is burnt as fuel but also from producing it. The steam may be mixed with solvents (gas) and this may reduce the amount of heat (steam) required. Much of the water used to produce steam can be recovered and used again.

For each 3 bbl of oil produced one barrel is burned to produce the steam required for the production. This heavy oil should also be mixed with gas during the refining and a supply of gas is therefore important.

Heating the oil with electricity has also been proposed and hydroelectric or nuclear power would reduce the CO_2 emissions.

173 billion bbl (27.5×10^9 m^3) of crude bitumen, which is 10% of the total bitumen in place, are economically recoverable using current technology from the three Alberta oil sand areas based on benchmark WTI market prices of $62/bbl in 2006, rising to a projected $69/bbl in 2016.

The environmental problems associated with oil production from tar sand are considerable and very large amounts of water are required both for the production of steam for subsurface operations and for the processing of loose sand excavated from surface pits. As a result the release of CO_2 is higher than with normal petroleum production if the CO_2 is not captured and sequestrated.

In Venezuela there are also very large reserves of heavy oil and tar sand such as the Orinoco tar sand. While the average surface temperature in Northern Alberta is only slightly above 0°C the surface temperature in Venezuela is much higher (>20°C) and the oil therefore needs less heating to reduce its viscosity.

Much of the oil in Venezuela is heavy oil and it occurs in a foreland basin in front of the Cordilleran Mountains in a similar plate tectonic position to the heavy oil in Alberta. Also the oil in the Middle East is located in a similar foreland basin but here the reservoirs have not been uplifted so close to the surface and are therefore in most cases not so biodegraded.

It is possible to make an emulsion between bitumen and water which has low viscosity and which can flow in pipelines and be burned directly to produce electric power. The heavy oil and bitumen is rather rich in sulphur which should be removed to avoid pollution. It can also be mixed with gas so that it can be used to produce normal oil products.

There are other important tar sand deposits in the USA (Utah) and Africa (Congo and Madagascar).

Tar sand has only recently been included in data on world oil reserves and as a result Canada has become one of the nations with the highest oil reserves.

21.2 Oil Shales

Oil shales are source rocks, usually mudstones and shales, with a high organic content (TOC), which have not been buried deeply enough to become sufficiently mature for most of the hydrocarbons to be generated. Although they may contain some hydrocarbons they must be heated in an oven (pyrolysis) to 400–500°C so that most of the petroleum can be generated from the remaining kerogen.

Oil shales must therefore be mined near the surface in quarries and then heated in large ovens so that the petroleum can be distilled off.

Source rocks may be uplifted close to the surface after deeper burial; it is the temperature history that determines how much of the kerogen is altered to oil and gas.

Some source rocks may have been buried to more than 5–6 km (160–170°C) and they have then generated and expelled most of the hydrocarbons, but some oil and particularly gas may remain. The Upper Cambrian alum shale which is found in the Oslo region is a good example of a rich source rock which has been buried to at least 200°C (5–7 km) and lost most of its hydrocarbons during the Caledonian folding in the late Silurian and early Devonian. In Sweden the Upper Cambrian alum shales have not been buried so deeply and therefore contain more oil which can be distilled off by pyrolysis at 400–500°C.

In the Baltic region the lowermost Ordovician shales are even less mature so that more of the kerogen remains and in Estonia this shale is mined for oil on a large scale. Here the reserves are very large (0.6×10^9 sm^3 oil equivalents – o.e).

Oil shales are used in electric power generation and provide 60% of Estonia's stationary energy. It is also used for oil production and refining. The waste is 70–80 Mt. of semi-coke and the mounds exceed 100 m in height. The waste is very fine grained and is alkaline with significant concentrations of sulphides and heavy metals.

On a global basis oil shales represent a reserve almost as large as conventional oil. Conservative estimates: 350 Gt shale oil equivalent to 2.6×10^9 barrels of oil (410×10^9 sm^3 o.e.). This is about equivalent to the estimated world reserves of conventional oil and gas (480 sm^3 o.e.). More than 80% of the well known reserves are located in the US but there are probably other regions with oil shales which have not been recorded and evaluated.

The Green River Shale in Colorado, Utah and Wyoming is a gigantic source of hydrocarbons. This organic-rich mudstone was deposited in very large lakes during the Eocene and the organic matter was mostly freshwater algae (Fig. 21.5a, b).

Oil shales must however be mined and heated in ovens (pyrolysis) to generate the petroleum and the energy for the heating is taken from the burning of the oil shales. The release of CO_2 is therefore high also during production. Oil shales may only contain 5–10% organic matter and the volume of waste will then be 10–20 times the oil produced. The waste consists of coke and also smectite formed in the heating process and is very difficult to store. It is also rich in heavy metals, including vanadium and uranium which are typical of black shales.

Production of oil from oil shales requires very large amounts of water, so in dry areas the water supply may be a limiting factor. There are therefore considerable environmental problems linked to the exploitation of oil shales as a major source of oil.

Some source rocks may have enough permeability to serve as reservoir rocks too. The Miocene Monterey Formation in California is an organic-rich diatomaceous source rock, which is also a reservoir rock. Oil can in this case be produced because of tectonic fracturing which has enhanced the permeability. Source rocks may also be interbedded with thin sandstones or limestones and then only a very short migration is required.

A

B

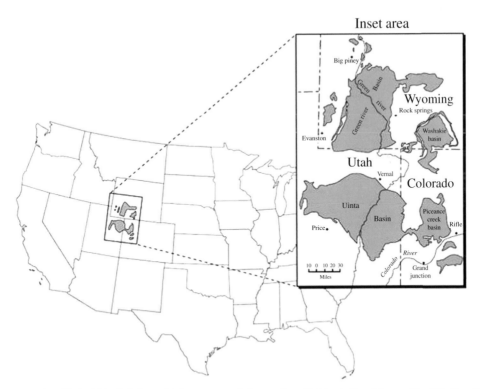

Fig. 21.5 (**a**) Green River Shale. This shale has a high content of organic carbon from algae and contains mostly type 1 kerogen. It is far from mature and the shale must be mined and heated to 400–500°C to form oil by pyrolysis. (**b**) The Green River oil shale was deposited in shallow lakes and is of Tertiary (Eocene) age. This shale is very extensive in Utah, Wyoming and Colorado (adapted from Smith 1980)

21.3 Coal Bed Methane (CBM)

Coal and kerogen rock with a high content of plant material (type 3 kerogen) are the main source rocks for gas which will migrate to a reservoir rock or up to the surface. Relatively large quantities of gas (methane) will, however, be retained in the coal because its microporous structure provides a very large surface area; coal can hold considerable volumes of gas. Much of the gas is adsorbed on the surface and coal has very large surface area. Commercial production of methane from coal is common in the US. Near the surface gas can leak off from coal but where coal is buried a few hundred metres much of the gas is still retained in the coals and can be produced by drilling. Coal has very low permeability and the flow of gas to the wells depends on thin fractures (cleats) developed during uplift. The gas produced from coal is normally very pure methane with low sulphur content and is referred to as sweet gas. Wyoming has very large reserves of coal bed methane in the Powder River, Bighorn, Wind River and Green River basins. These are Cretaceous and Tertiary coals which have been buried more deeply and then uplifted. Gas may be produced from these coals down to depths of 1.5–2 km (5,000 ft). The reserves of CBM are large in the USA (20×10^{12} m^3) and Canada also has major reserves.

Methane is constantly being formed by bacteria at shallow depth and it may be argued that this is at least to some extent a renewable resource but the rate of accumulation is slow compared to our consumption. Artificial growth of algae may produce some oil and gas and consume CO_2.

21.4 Shale Gas

Organic-rich shales which have been buried to depths where most of the oil and gas has been generated and expelled may nevertheless contain considerable amounts of gas. The gas remaining in these shales is present in very small pores and may also be partly adsorbed on remaining organic matter or its residue (coke) and on clay minerals. The shales have been uplifted and may therefore have small extensional fractures, but they must be hydrofractured by water injection to increase the permeability.

Barnett Shale is tight shale of Mississippian age in Texas, containing at least 2.5 trillion cubic feet of gas.

It is referred to as a tight gas reservoir. Much of the gas is in urban areas such as the Dallas-Forth Worth area.

The permeability of the shale matrix is generally very low but there may be thin silty layers and also fractures that increase the effective permeability. Hydraulic fracturing can be carried out to further increase the permeability, and horizontal drilling also helps to produce more gas. The Woodford Shale (Devonian) in Oklahoma can be almost 100 m thick.

Devonian tight gas shales include the Middle Devonian Marcellus Shales in the Appalachians. These are now mostly at 1–2 km depth but have previously been buried much deeper (5–6 km or more).

The Upper Devonian Bakken Shale is another major producer, particularly in North Dakota, and is part of the Williston Basin. It also extends into Canada.

Shale gas is estimated to produce 50% of the gas in North America by 2020.

Higher gas prices will also result in increased interest in shale gas in other parts of the world.

Shale gas represents very large reserves of hydrocarbons which can be used directly as gas, but can also be converted to diesel fuels for cars and trucks. This gas can also be mixed with heavy oil and tar sand to make regular petrol. Production of gas from shales requires much water for fracturing, and produced water may also create environmental problems.

21.5 Gas Hydrates (Clathrates)

Gas hydrates are crystalline solids almost like ice, consisting of gas (mostly methane) surrounded by water. Most common is methane clathrate which is water molecules bonded by hydrogen and with gas trapped

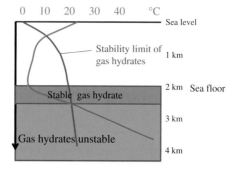

Fig. 21.6 Stability of gas hydrates. Gas hydrates are stable at high pressure and low temperatures and are therefore found beneath the deeper part of the continental slopes. As the temperatures in the sedimentary basins increase with depth, gas hydrates become unstable

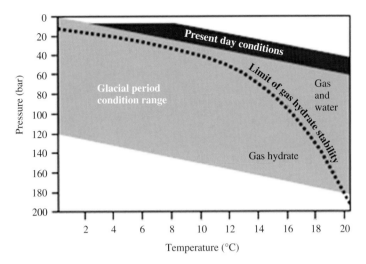

Fig. 21.7 During glacial periods, gas hydrates formed in basins like the North Sea due to the low surface temperatures in front of the ice (from Fichler et al. 2005)

within its structure. It is stable at high pressures and low temperatures (Fig. 21.6). It therefore typically occurs on the slopes and below the seafloor in deep oceans. At about 2.0 km water depth (20 MPa water pressure) the temperature must not exceed about 18°C and at 0.5 km below the seafloor (2.5 km depth – 25 MPa pressure) the temperature must be below 20°C. The temperature near the seafloor is usually less than 2–3°C but the geothermal gradients may be about 30°C/km and it is therefore clear that gas hydrates can only exist a few hundred metres below the seafloor (Fig. 21.7).

When gas hydrates dissolve (melt) one volume of gas hydrate produces 160 volumes of gas. The source of the methane is mostly biogenic, from organic rich sediments, but gas hydrates may also fill the pores in sand beds. During the glaciations gas hydrates were more widespread than now and occurred also beneath the seafloor in basins like the North Sea (Fig. 21.7). Gas hydrates are potentially a very important source of gas.

Production of methane from sediments with gas hydrates may become possible from sand beds cemented with gas hydrates. It is however difficult to predict how economic it would be to produce from gas hydrates.

21.6 Summary

Unconventional oil and gas as discussed above represent very large energy reserves which are much greater than conventional oil. The environmental problems associated with producing these reserves are however great, particularly for oil shale and tar sand.

There is a need for new research to develop methods which can minimise both the local environmental impact and the release of CO_2 into the atmosphere.

Further Reading

Fichler, C., Henriksen, S., Rueslaatten, H. and Hovland, M. 2005. North Sea Quaternary morphology from seismic and magnetic data:indications for gas hydrates during glaciation? Petroleum Geoscience 11, 331–337.

Salvador, A. 2005. ENERGY: An historical perspective and 21st century forecast. AAPG Studies in Geology 54, 33–121.

Williams, A., Larter, S.R., Head, I. et al. 2001. Biodegradation of oil in uplifted basins prevented by deep burial sterilization. Nature 411, 1034–1037.

Chapter 22

Geology of the Norwegian Continental Shelf

Jan Inge Faleide, Knut Bjørlykke, and Roy H. Gabrielsen

22.1 Introduction

In the preceding chapters we have included only a few regional examples and case studies because of space limitations. The present chapter will, however, provide some examples. The North Sea and other parts of the Norwegian continental shelf contain several different petroleum provinces which can illustrate some of the general principles of petroleum geology and geophysics. The geological evolution of these sedimentary basins provides a necessary background to understand the distribution of source rocks and the timing of petroleum migration. The structural history of rifted basins, passive margins and also uplifted basins such as the Barents Sea is critical to the trapping of oil and gas. These basins are very well documented by seismic and well data. The Norwegian Petroleum Directorate also provide information about exploration and production on their web pages (www.npd.no).

22.1.1 Regional Geological Setting

The Norwegian continental shelf comprises three main provinces (Figs. 22.1 and 22.2):

- North Sea
- Mid-Norwegian continental margin
- Western Barents Sea

Prior to continental break-up and the onset of seafloor spreading in the Norwegian-Greenland Sea (Fig. 22.3) these provinces were part of a much larger epicontinental sea lying between the continental masses of Fennoscandia, Svalbard and Greenland. Thus there are many similarities in the stratigraphy (Fig. 22.4) and geological evolution of the various provinces but also important differences – in particular during Cretaceous-Cenozoic times.

The sedimentary basins at the conjugate continental margins off Norway and Greenland and the adjacent shallow seas, the North Sea and the western Barents Sea, developed as a result of a series of post-Caledonian rift episodes until early Cenozoic time, when complete continental separation took place. The Norwegian Margin comprises the mainly rifted volcanic margin offshore mid-Norway (62–70°N) and the mainly sheared margin along the western Barents Sea and Svalbard (70–82°) (Fig. 22.2). Physiographically, the Norwegian margin consists of a continental shelf and slope that vary considerably in width and morphology (Fig. 22.1).

22.1.2 Petroleum Exploration

When gas was found onshore in the Groningen Field in the Netherlands in 1958 it soon generated interest in the North Sea itself. However, it was difficult to say what lay off the Norwegian coast beneath the Quaternary sediments. Moraine material which had been recovered from the seabed during marine geological investigations in the 1950s contained clasts that almost without exception were basement rocks.

J.I. Faleide (✉)
Department of Geosciences, University of Oslo, Oslo, Norway
e-mail: j.i.faleide@geo.uio.no

K. Bjørlykke (ed.), *Petroleum Geoscience: From Sedimentary Environments to Rock Physics*,
DOI 10.1007/978-3-642-02332-3_22, © Springer-Verlag Berlin Heidelberg 2010

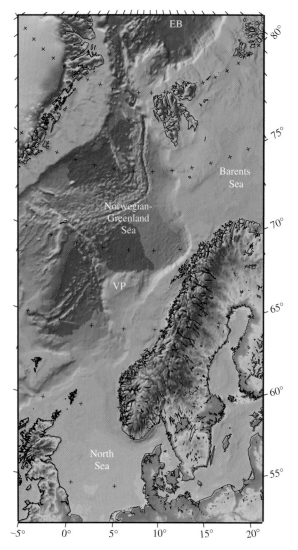

Fig. 22.1 Regional setting (bathymetry/topography) of the Norwegian Continental Shelf and adjacent areas. EB = Eurasia Basin, VP = Vøring Plateau

The pre-Quaternary geology of the Norwegian Continental Shelf was virtually unknown prior to the early 1960s, as there had been no seismic measurements and few other indications of what was beneath the cover of Quaternary, partly glacial, sediments. Before any seismic or borehole data were obtained in the North Sea, one had to extrapolate from the geology of the surrounding land areas. The Mesozoic rocks of northeast England and Denmark were assumed to continue beneath at least parts of the North Sea, and Carboniferous coal seams and Permian sediments were already well known in northeastern England. There

were also a few indications that younger sedimentary rocks were present beyond the Norwegian coast. The moraines on Jæren contained small amounts of probably Cretaceous material, there were erratic blocks of Jurassic age on Froan and pieces of coal from Tun in Verran, off Trøndelag. Much farther north, the fairly thin Jurassic and Cretaceous sequences on Andøya had long been known. It was difficult, though, to know how much reliance could be placed on these scattered observations; not least, one had no idea of how thick such Mesozoic sediments would be.

A confirmation of the presence of thick sediment sequences on the shelf was obtained in seismic refraction studies in the Barents Sea and the central North Sea in the late 1950s, indicating accumulations in the order of several kilometres. Seismic refraction studies supported by potential field data in the Skagerrak, performed in 1963–1964, confirmed the presence of more than 3 km of sediments also in this area. The early seismic data that was shot in the North Sea had very poor quality and it was not until drilling on the Norwegian shelf commenced in 1966 that it was possible to have an informed opinion on the potential for oil and gas.

The Norwegian continental shelf, extending from the baseline to the limit approved by the UN Commission on the Limits of the Continental Shelf, amounts to 2.2 million km^2. About half of this acreage has bedrock in which petroleum may be found, and half of that has been opened for petroleum exploration. The first licences were awarded in 1965 and the first commercial discovery in the Norwegian part of the North Sea (block 2/4) was made in 1969. The motivation for this drilling was to test a possible Permian (Rotliegendes sandstone) reservoir. Instead, however, an unexpected discovery of hydrocarbons was made in a domal structure in Upper Cretaceous Chalk. Continued exploration on this play type resulted in a number of discoveries in the Ekofisk area of the central North Sea.

Large fields in the North Sea, such as Sleipner, Statfjord and Gullfaks, were discovered during the first 10 years after the Ekofisk Field was proven, and all four are still in production. Most of the other large fields on the shelf were found between 1979 and 1984 (see Sect. 22.2.3 for more details on the North Sea exploration history).

Following the exploration success in Norwegian waters south of 62° N, the areas to the north were opened for exploration in 1979. Results and experience gained from the North Sea were utilised when areas

Fig. 22.2 Main structural elements of the Norwegian Continental Shelf and adjacent areas related to different rift phases affecting the NE Atlantic region (modified/updated from Faleide et al. 2008). For more details see close-up maps in Figs. 22.5, 22.10 and 22.14. JMMC = Jan Mayen micro-continent

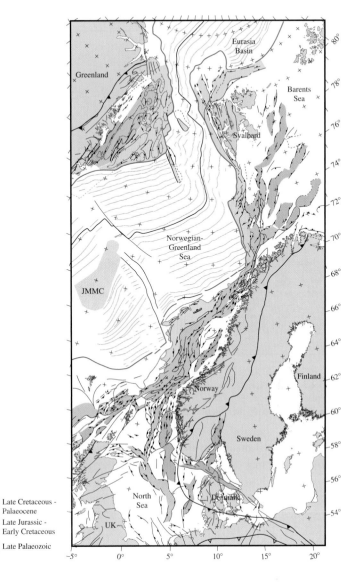

further north were opened, and areas with a similar geological setting to the northern North Sea were tested first (Haltenbanken offshore Mid-Norway and Tromsøflaket in the southwestern Barents Sea). The exploration histories of the areas north of 62° N are more complex, and even today important aspects of the petroleum geology remains enigmatic – partly due to geological complexities.

Large parts of the Norwegian continental shelf are now well explored. Better knowledge of the geology and advances in technology lead to a higher discovery rate, but the finds have mostly been small. Consequently, the growth in resources has been relatively low for the last 20 years. The largest discovery made in this period was Ormen Lange in the

Norwegian Sea in 1997, but the growth in resources since 1997 does not equal the production in the same period.

Technological advances have led to development of discoveries in areas of great water depth and far from shore. The technological advances have also made a higher proportion of the resources profitable to recover. On average, fields on the Norwegian shelf have a recovery factor of 46% for oil. This is high compared with oil provinces in other parts of the world. Continuous research and development of technology are required to raise this further.

Significant volumes remain to be found and produced on the Norwegian continental shelf. Only a third of the total resources have so far been produced,

(a) Present   (b) Chron 13 (c. 33 Ma)   (c) Breakup (c. 55 Ma)

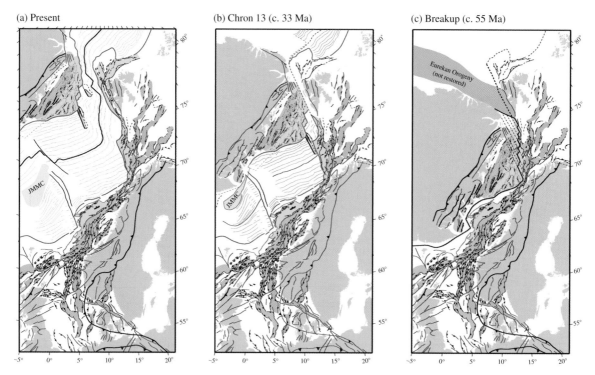

Fig. 22.3 Plate tectonic reconstructions of the NE Atlantic. (**a**) Present (same as Fig. 22.2), (**b**) Reconstruction to chron 13 (c. 33 Ma) using the rotation pole of Gaina et al. 2009,

(**c**) Reconstruction to time of breakup (ca. 55 Ma) based on unpublished work of Faleide et al.

and an estimated quarter of them have still not been discovered. It may still be possible to make large discoveries in less explored areas, such as in deep waters in the Norwegian Sea, in the Barents Sea and in areas that are not yet open. Oil production reached its peak in 2000–2001, but gas production is still increasing. For more facts about the petroleum activity on the Norwegian shelf see http://www.npd.no/en/Publications/Resource-Reports/2009/.

22.2 North Sea

22.2.1 Structure

The North Sea is an example of an intracratonic basin; that is to say a basin which lies on continental crust. The prerequisite for forming major sedimentary basins on continental crust is that the crust (and mantle lithosphere) is thinned, resulting in subsidence to maintain isostatic equilibrium. The North Sea has been subjected to periods of stretching/thinning and

subsidence during late Carboniferous, Permian-Early Triassic and Late Jurassic times. Each rift phase was followed by a thermal cooling stage, characterised by regional subsidence in the basin areas.

The Northern North Sea province is dominated by the Viking Graben, which continues into the Sogn Graben towards the north (Fig. 22.5). These grabens are flanked by the East Shetland Basin and the Tampen Spur to the west, and the Horda Platform to the east (Figs. 22.6 and 22.7). These are Jurassic-Cretaceous features, and the main crustal thinning took place in the late Middle to Late Jurassic, followed by thermal subsidence and sediment loading in the Cretaceous. However, the Viking Graben and its margins are underlain by an older major rift basin of assumed Permian-Early Triassic age. The axis of this rift system is thought to lie beneath the present Horda Platform. It is bounded by the East Shetland Platform in the west and the Øygarden Fault Zone in the east. Structures within this area are characterised by large rotated fault blocks with sedimentary basins in asymmetric half-grabens associated with lithospheric extension and crustal thinning (Fig. 22.6). The area was presumably also

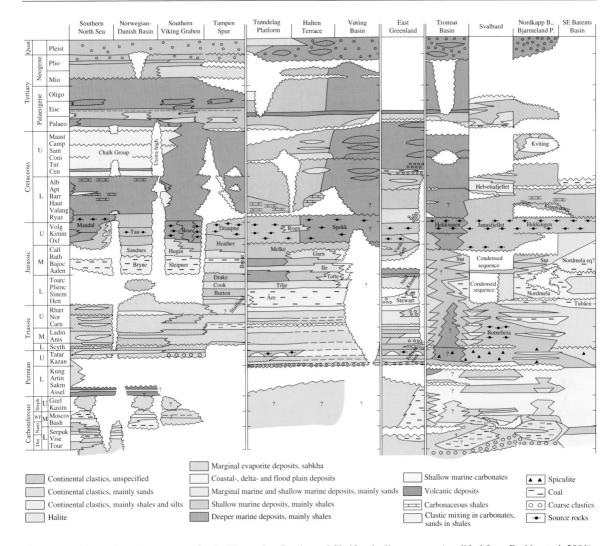

Fig. 22.4 Lithostratigraphic summary for the Norwegian Continental Shelf and adjacent areas (modified from Brekke et al. 2001)

strongly affected by post-orogenic (post-Caledonian) extension in Middle to Late Devonian times.

The Norwegian Central North Sea Province encompasses the northwestern part of the Central Graben (Figs. 22.5, 22.8 and 22.9), which is mainly a Jurassic-Cretaceous structure, and its northeastern margin. The strata of the Central Graben area were affected by halokinesis already in the Triassic, and major structuring accordingly occurred already in pre-Jurassic times. However, salt movements have locally continued into the Tertiary. Jurassic rifting and generation of large, rotated fault blocks resulted in extreme erosion in places. The complex structural pattern established during this stage was further complicated by Cretaceous

inversion. The Norwegian-Danish Basin, east of the Central Graben, also contains many salt structures but has not been affected by significant rifting (Figs. 22.8 and 22.9).

The geometric shape of the sedimentary basin is influenced by the structure in the underlying rocks (basement) and by the thickness of the continental crust. Most of the North Sea is underlain by a Caledonian basement. Long-lived zones of weakness inherited from the Caledonian Orogeny played a role in the later evolution of the North Sea basin. The southeastern North Sea and Skagerrak are underlain by a Precambrian basement covered by a Lower Palaeozoic sedimentary succession. In the south

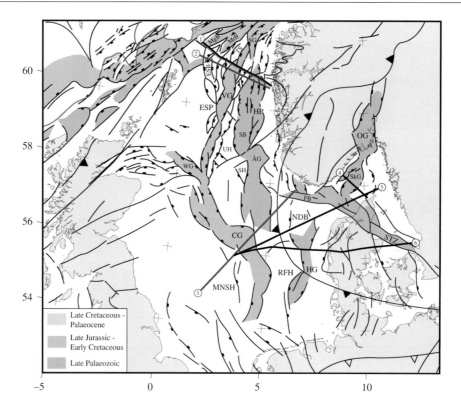

Fig. 22.5 Main structural elements in the North Sea and adjacent areas (close-up of Fig. 22.2 – modified/updated from Faleide et al. 2008). Location of interpreted regional profiles (*black*), crustal transects (*red*) and seismic examples (*blue*) shown in Figs. 22.4–22.9. CG = Central Graben, ESB = East Shetland Basin, ESP = East Shetland Platform, HG = Horn Graben, HP = Horda Platform, MgB = Magnus Basin, MNSH = Mid North Sea High, MrB = Marulk Basin, NDB = Norwegian-Danish Basin, OG = Oslo Graben, RFH = Ringkøbing-Fyn High, SB = Stord Basin, SG = Sogn Graben, SH = Sele High, SkG = Skagerrak Graben, STZ = Sorgenfrei-Tornquist Zone, TS = Tampen Spur, UH = Utsira High, VG = Viking Graben, WG = Witchground Graben, ÅG = Åsta Graben

the North Sea basin is bounded by the Hercynian (Variscan) mountain range which runs E-W through Germany, northern France and southwest England. The contraction occurred during the Carbonifererous-Permian. Uplift in this area resulted in vast amounts of sediment being deposited in the area to the north and this initiated the formation of the North Sea basin.

The offshore part of the Oslo Rift in the Skagerrak is characterised by NE-SW striking half-grabens (Fig. 22.9). Cross-sections across the Skagerrak show tilted half-grabens filled with down-faulted Upper Carboniferous-Lower Permian and Lower Palaeozoic strata that are unconformably overlain by Triassic sedimentary rocks. Although a thick succession of Lower Palaeozoic strata has been preserved in the Skagerrak Graben, a substantial part of this pre-rift unit has been eroded in the central part of the rift. From strata preserved in the onshore Oslo Graben we

know that the Lower Palaeozoic succession consists mainly of Cambrian-Ordovician and Silurian platform series and Upper Silurian-lowermost Devonian clastic sedimentary rocks, with the latter being deposited in the Caledonian foreland basin. In the western part of the Skagerrak, the late Palaeozoic structures of the Oslo Rift link up with the Sorgenfrei-Tornquist Zone (also termed the Fennoscandian Border Zone in Denmark). The Sorgenfrei-Tornquist Zone strikes in a NW-SE direction from the western Skagerrak across northern Jutland and the Kattegat into Scania (Fig. 22.2). If the effects of post-Permian structuring along the Sorgenfrei-Tornquist Zone are restored, late Palaeozoic rift structures similar to those seen in the Oslo Rift are observed. The major NW-SE trending faults are associated with dextral strike-slip movements during Late Carboniferous-Early Permian times.

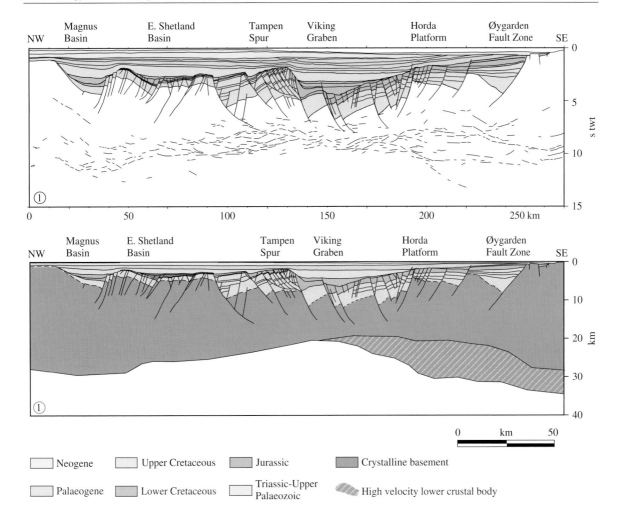

Fig. 22.6 Interpreted regional deep seismic line and crustal transect across the northern North Sea (modified from Christiansson et al. 2000). See Fig. 22.5 for location

22.2.2 Stratigraphy/Evolution

A stratigraphic summary for the North Sea is given in Fig. 22.4.

22.2.2.1 Devonian

The Caledonian Orogeny led to uplift and formation of a major mountain chain along western Scandinavia and Scotland, East Greenland and a southern branch into Poland. Thick red continental sediments, which in Britain are known as Old Red Sandstone, were deposited in Devonian time in response to the extensional collapse of the Caledonides. On the Norwegian

mainland we have Devonian basins in western Norway (Hornelen, Håsteinen, Kvamshesten, Solund) which are filled with thick conglomeratic sediments. There are also Devonian deposits further north, for example on the island of Smøla. The Devonian sandstones of western Norway are metamorphosed and can not be reservoir rocks, but they are somewhat more porous in southern England.

Although Devonian sediments in the northern North Sea have been reached in only a few wells, there are reasons to believe that Devonian sediments are present regionally in the deeper parts of the pre-Triassic half-grabens beneath the Horda Platform, Viking Graben and East Shetland Basin. The presence of Upper Palaeozoic rocks, of both Devonian

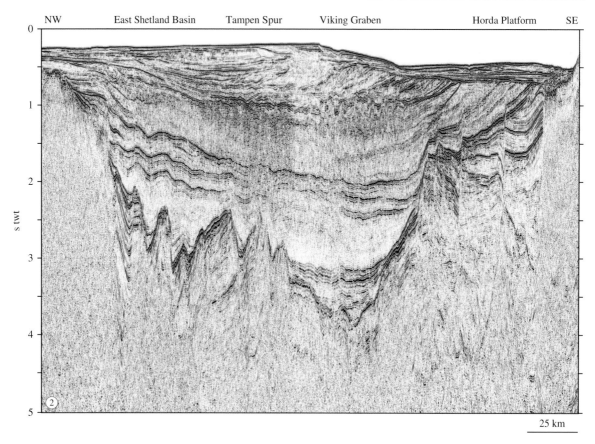

NW   East Shetland Basin   Tampen Spur   Viking Graben   Horda Platform   SE

25 km

Fig. 22.7 Regional seismic line across the northern North Sea (courtesy Fugro and TGS). See Fig. 22.5 for location and Fig. 22.6 for stratigraphic information

and Lower Permian age (Rotliegendes), has also been confirmed by drilling on the East Shetland Platform. Seismic data reveal large sedimentary basins beneath the platform, thought to contain Upper Palaeozoic (Devonian-Carboniferous) rocks. In this context it is important to note that oil is produced from Devonian sandstones in the Embla Field of the Central Graben.

The Caledonian plate movement changed from subduction to Late Devonian lateral (strike-slip) movement between Greenland and Fennoscandia, including along the Great Glen Fault. In Scotland this was marked by active volcanism, especially in the Midland Valley Rift.

22.2.2.2 Carboniferous

Following the markedly dry climate which prevailed through Devonian time in the North Sea region, the Carboniferous period gradually became more humid. Northwest Europe moved northwards from the arid belt of the southern hemisphere into the humid equatorial belt. This provided the basis for all the coal deposits. There was also a marked transgression over large areas. The strike-slip movements along the Greenland/Fennoscandia plate boundary ceased at the transition from Devonian to Carboniferous, and ever since this has been an area of diverging plate movement and rift formation until final continental break-up and onset of seafloor spreading in earliest Eocene time. There was rifting in the Midland Valley of Scotland, continuing along the Forth Approhes into the Witch Ground Graben which was a volcanic centre in the North Sea. Black mud deposited in these graben structures forms source rocks for oil.

There was no rifting in England AT that time and the Carboniferous Limestone was deposited as a shelf carbonate during the Early Carboniferous. North of the carbonate platform and in much of the North Sea, deeper water shales predominated, with some carbonate. Sandstones are encountered higher up in the Lower Carboniferous and these can be important reservoir

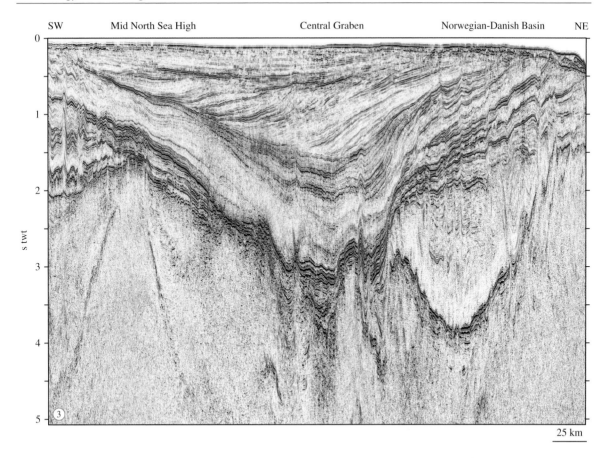

Fig. 22.8 Regional seismic line across the central North Sea (courtesy Fugro and TGS). See Fig. 22.5 for location

rocks in the southern North Sea (Fell Sandstone). At the Lower/Upper Carboniferous boundary is the Yoredale Formation, consisting of cyclic deposits of marine carbonates and shales, and fluvial sandstones. These have been interpreted as the result of eustatic changes in sea level due to glaciations on the southern hemisphere, but local tectonics and even sedimentation processes can also produce cyclicity.

The Hercynian (Variscan) mountain range formed along the subduction zone through Germany and northern France close to south England. It was uplifted as a marked topographic feature, and at its foot (the Variscan foredeep) major sedimentary units were deposited, derived from erosion in the mountains. Thick beds of coal developed from the swamp areas on the deltas. In Britain these sandy delta deposits are known as the Millstone Grit. Contemporary coal deposits are found in the southern North Sea and in the Netherlands, and it is these which provided the source of the gas in this region. Fluvial sandstones between the coal seams can be reservoirs both for

oil and gas. Black marine shales were also deposited in the mid-Carboniferous in the southern part of the North Sea. In the Oslo Region there are thin sandstones and carbonate sediments of the Asker Group yielding Upper Carboniferous fossils, demonstrating a connection with the North Sea basin.

22.2.2.3 Permian

During the early Permian, uplift of the Hercynian (Variscan) mountain range continued and sedimentary basins developed in front of it in the southern North Sea and within subsided areas of the range itself. At the same time, northwest Europe was pushed farther northwards from the equator, into the dry belt of the northern hemisphere. The high mountain range to the south also contributed to severe aridity in the North Sea basin and most of northwest Europe. This region was then located in the middle of a large continent in a similar position to the dry areas to the north of the present Himalaya.

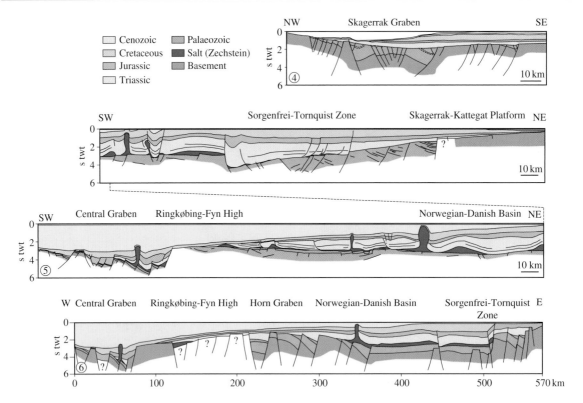

Fig. 22.9 Regional profiles across the SE North Sea, Skagerrak and Kattegat (modified from Heeremans and Faleide 2004). See Fig. 22.5 for location

The latest Carboniferous-Permian extension in the North Sea region (Fig. 22.2) was linked to the Variscan fold belt by strike-slip movements on a series of NW-SE trending zones of weakness (among others the Sorgenfrei-Tornquist Zone). The rifting was associated with widespread and massive igneous activity. This is well known from the onshore part of the Oslo Rift (Oslo Graben) but similar rocks, both extrusives and intrusives, are present in the subsurface offshore (e.g. Skagerrak Graben, Sorgenfrei-Tornquist Zone, Central Graben; Fig. 22.9).

Sedimentation in the North Sea region was dominated by two E-W aligned basins separated by the Mid-North Sea and Ringkøbing–Fyn highs. The basins were infilled with Rotliegend continental sediments, most rapidly in the southern basin which was nearest the mountains and subsided fastest. Alluvial fans and aeolian dune sand accumulated most of the sediment supply from the south. These types of sediment are well-exposed in southwest England near Exmouth but also in northeast England. In the North Sea basin the Upper Rotliegend contains an extensive, mostly aeolian, sandstone called the Auk Formation.

The climate behind the mountain range was dry, and a marine evaporite basin eventually developed, possibly with a narrow passage through the Viking Graben to the open ocean in the north to a seaway between Norway and Greenland. There was, however, an important connection eastwards through Poland. The deposition of the Zechstein salt commenced in the southern Permian basin and spread across most of the North Sea basin during Late Permian time but is absent in the northern North Sea. Sabkha deposits accumulated in areas marginal to the evaporite basins.

22.2.2.4 Triassic

Rifting continued into earliest Triassic time and the Triassic to Middle Jurassic succession reflects a pattern of repeated outbuilding of clastic wedges from the Norwegian and East Shetland hinterlands within a generally evolving post-rift basin. There was still considerable sediment supply from the Variscan mountains to the south and there is also evidence of an uplift of Scandinavia. A broadly similar geometry of the megasequences in both continental Triassic and marine Jurassic successions was related to subsidence-rate variations. Differential subsidence across faults throughout Triassic time has also been reported. The Øygarden Fault Zone, forming the eastern margin of the Permo-Triassic basin, was active throughout most of the time interval. The sedimentation rate was in most cases high enough to keep up with the subsidence, resulting in a rather flat landscape with gently flowing rivers. If the sediment supply had been less, these rift basins would have been marine. Along the rift structure along the middle of the North Sea the thickness of the Triassic sediments may exceed 5 km. The underlying Permian salt started to form diapirs and Triassic sediment was eroded, or not deposited, at the top of the structures.

The climate in northwest Europe during the Triassic was still arid, and continental red beds continued to be predominant. In England these are referred to as the New Red Sandstone because they are similar to the Devonian continental sediments (Old Red Sandstone). The equivalent facies in Germany is the Bunter sandstone of Lower Triassic age (sandstones of the Brent Group and Statfjord Formation). In south England east of Exmouth there are also good exposures of Triassic deposits with sandstones and conglomerates (Sherwood Sandstone Group) and mudstones (Mercia Mudstone Group). The Sherwood Sandstone is a very important groundwater reservoir in large areas of England. In Southern Norway north of Hamar, the Brumundal sandstone is an outlier of a more extensive cover of Permian red bed facies.

In the Upper Triassic we find carbonates (Muschelkalk) and salt deposits (Keuper salt) in the southern part of the North Sea. In the central and northern areas, however, continental clastic sedimentation continued right up to the end of the Triassic (Rhaetian). Sabkha environments fringed the evaporite basins and caliche, that is carbonate precipitation in soil profiles, is typical. Towards the end of the Triassic the climate became less arid with more normal fluvial sedimentation and gradually also marine sedimentation when the Statfjord Formation was deposited.

22.2.2.5 Jurassic

The transition from Triassic to Jurassic approximately coincides with a change from continental to shallow marine depositional environments. The climate also gradually became more humid in the Jurassic as northwest Europe was pushed northward out of the arid belt at about 30° N. A transgression in the Early Jurassic (Lias) led to the accumulation of black shales over large parts of NW Europe and they are exposed in Yorkshire in northeast England and Dorset in southern England. This is because a transgression over an uneven land surface results in a shallow sea with poor vertical circulation in the water. The Upper Lias (Toarcian) in particular contains good source rocks for oil and gas in southern parts of the North Sea. In Britain and much of the North Sea the Lias deposits consist of black shales with thin carbonate and sand beds. Iron-rich layers with siderite ($FeCO_3$) and chamosite (Fe-chlorite) are common. Beds rich in these two minerals were the basis for iron mines which were especially important during the industrial revolution in England, Germany and France.

In the northern part of the North Sea fluvial and partly marine sandstones of Lower Jurassic age (Lunde and Statfjord formations) are important reservoir rocks in the Viking Graben. The Statfjord Formation is succeeded by the Dunlin Group, which is a dark marine shale but normally without enough organic content to become a significant source rock. Then follows the Brent Group sandstone, a prograding delta sequence which forms the main reservoir rock in the northern North Sea. This sandstone was deposited in a delta that drained the central part of the North Sea towards the marine embayment to the north, between the Shetland and Horda platforms. It was sourced from an uplifted area in the south associated with Middle Jurassic (Bajocian-Bathonian) volcanic activity. The volcanic centre was located south of the Viking Graben and east of Scotland (Moray Firth). On the other side of the uplifted area, the sediments were transported southwards, exemplified in good coastal sections in Yorkshire.

The lower part of the Brent Group consists of upward-coarsening beds of mica-rich sandstones that represent prograding delta deposits, especially distributary mouth bars. This is the Etive and Rannoch formations. The middle part is the Ness Formation, comprised mostly of fluvial delta facies with channels, crevasse sand deposits, lagoonal deposits and coal beds. These delta top facies were deposited during a northward progradation of the Brent Delta. At the top of the Brent Group, the Tabert Formation is a generally well-sorted sandstone formed by the reworking of deltaic deposits during a transgression. The sediment supply from the south was no longer able to keep up with the basin subsidence, resulting in a gradual drowning of the Brent Delta. Each of these delta facies has its characteristic reservoir properties, which are a function both of primary sorting and mineralogy and subsequent diagenetic modification.

In the Late Jurassic, volcanism was very much reduced and the areas within and surrounding the rift systems subsided in response to lower geothermal gradients. At the same time, normal faulting along the Viking Graben led to the rotation of basement blocks and their overlying sediments. The shoulders of the tilted fault blocks were exposed to erosion, removing Lower-Middle Jurassic and locally even Upper Triassic strata. The Late Jurassic (Oxfordian) transgression covered the Viking Graben with a thick drape of clayey sediments of the Heather Formation, while the coarser clastics (sand) were deposited as turbidites and in deltas along the basin margins. Some of the deltas appear to have been controlled by the same structures that have determined the location of the fjords in western Norway.

The uppermost Jurassic Kimmeridge Clay Formation is transgressive and often forms a several hundred metres thick rich source rock which on the Norwegian side is called the Draupne Formation. The rift topography produced numerous, locally overdeepened, basins with poor bottom water circulation. Only a relatively small proportion of the organic production was oxidised while the sedimentation rate was fairly high. The organic-rich shales of the Upper Jurassic are thus the prime source rock in the North Sea, and provided the main petroleum source in both the Statfjord and Ekofisk areas. The thickness of the Upper Jurassic sediments along the rift axis may reach 3,000 m. The deposition of organic-rich shales continued into the Early Cretaceous in some of the basins. The edges of the rotated blocks suffered erosion and a few were not buried until the Late Cretaceous. The majority of the faults die out before the Cretaceous but a few continue up into younger beds. The rifting resulted in rotated fault blocks containing sandstone reservoirs of Lower and Middle Jurassic age (sandstones of the Brent Group and Statfjord Formation). Small fan deltas developed along the rift and also deepwater sandstones including debris flows. Some

of these are good reservoir rocks occuring within the Upper Jurassic source rocks (see Chap. 12).

22.2.2.6 Cretaceous

The last phase of rifting in the North Sea in the Late Jurassic was followed by a major transgression, but the uplifted rift stuctures remained dry and were islands for most of the Early Cretaceous. There is a major unconformity between the Cretaceous and the Jurassic except in the deep parts of the rifts where there may have been continuous sedimentation (Figs. 22.6 and 22.7). The Base Cretaceous Unconformity is very well marked on most seismic sections from the North Sea. Fault activity diminished during the Cretaceous, and the Cretaceous subsidence was due primarily to crustal cooling after the Jurassic rifting. The transition from syn- to post-rift configuration was strongly diachronous, suggesting that the thermal state of the system was not homogeneous at the onset of the post-rift stage.

Three stages can be identified in the post-rift Cretaceous development of the northern North Sea: (1) The incipient post-rift stage (Ryazanian–latest Albian) was characterised by different degrees of subsidence. The major structural features inherited from the syn-rift basin (e.g. crests of rotated fault blocks, relay ramps and sub-platforms) had a strong influence on the basin configuration and hence the sediment distribution. (2) In the middle stage (Cenomanian–late Turonian) the internal basin relief became gradually drowned by sediments. This is typical for basins where sediment supply outpaces or balances subsidence, as was the case in the northern North Sea. Thus, the influence of the syn-rift basin topography became subordinate to the subsidence pattern determined by the crustal thinning profile, which in turn relied on thermal contraction and isostatic/elastic response to sediment loading. (3) The mature post-rift stage (early Coniacian–early Palaeocene) was characterised by the evolution into a wide, saucer-shaped basin where the syn-rift features were finally erased. Since thermal equilibrium was reached at this stage, subsidence ceased, and the pattern of basin filling became, to a larger degree, dependent on extra-basinal processes (see Chap. 12).

The Lower Cretaceous shales are black, but only locally do they form good source rocks. Conditions subsequently became more oxidising as bottom circulation improved. In the Early Cretaceous grey to reddish oxidised shales were also deposited, which become increasingly calcareous upwards. The Lower Cretaceous shales (Cromer Knoll Formation) are shallow to deep marine mudstones with little sand.

In the Late Cretaceous the sea attained its transgressive maximum and clastic sedimentation almost ceased across large areas of northwest Europe. Parts of Scandinavia were probably also covered by the Cretaceous sea. Sedimentation was dominated by planktonic carbonate algae (coccolithoporids) which formed a lime mud, the main component of Chalk, though it also includes some foraminifera and bryozoa. The main development of the Chalk was in the Campanian and Maastrichtian, but sedimentation continued up into the Danian of the Palaeocene, for the most part through resedimentation of earlier Chalk deposits which had been uplifted along the central highs. In the Viking Graben the carbonate content diminishes northwards and we do not have pure limestone (Chalk) facies like that in the southern and central part of the North Sea. Instead, shales predominate, though often with a significant carbonate content.

At the close of the Cretaceous and beginning of the Tertiary, compressive movements were felt from the Alpine Orogeny to the south. Part of this movement was accommodated along diagonal fault zones, such as the Sorgenfrei-Tornquist Zone in Scania and the Kattegat. The Polish Basin and parts of the North Sea area were uplifted (inversion) and eroded.

22.2.2.7 Cenozoic

The early Cenozoic rifting, break-up and onset of seafloor spreading in the NE Atlantic gave rise to differential vertical movements also affecting the North Sea area. The sedimentary architecture and breaks are related to tectonic uplift of surrounding clastic source areas, thus the offshore sedimentary record provides the best age constraints on the Cenozoic exhumation of the adjacent onshore areas.

Major depocentres sourced from the uplifted Shetland Platform and areas along the incipient plate boundary in the NE Atlantic formed during Late Palaeocene-Early Eocene times. A local source area also existed in western Norway. Tectonic subsidence accelerated in Palaeocene time throughout the basin,

with uplifted areas to the east and west sourcing prograding wedges, which resulted in large depocentres close to the basin margins. Subsidence rates outpaced sedimentation rates along the basin axis, and water depths in excess of 600 m are indicated. Uplift along the Sorgenfrei-Tornquist Zone caused erosion into top Chalk along the southeastern flank of the North Sea basin.

Prominent ash layers of earliest Eocene age are found throughout the entire North Sea and also further north. Onshore in Denmark the volcanoclastic sediments are known as "moler". The extensive volcanism was related to the opening of the North Atlantic and both the Eocene and Oligocene mudstones are dominated by smectitic clays formed from volcanic ash. These smectitic mudstones are characterised by low seismic velocities also compared to the overlying Neogene sediments. Ash layers rich in volcanic glass (hyaloclastites or tuff) may, however, form hard layers due to diagenesis. They will then be characterised by high seismic velocities, particularly after burial to more than 2 km depth and quartz cementation, giving rise to strong seismic marker horizons (top Balder Formation).

In Eocene times progradation from the East Shetland Platform was dominant and major depocentres developed in the Viking Graben area, with deep water along the basin axis. The Palaeocene and Eocene submarine fans were built up with turbidite sands carried out into the central part of the North Sea with the Utsira High limiting their eastward extent. Parts of Fennoscandia were probably covered by sea during Middle-Late Eocene times.

At the Eocene-Oligocene transition, southern Norway became uplifted. This uplift, in combination with prograding units from both the east and west, gave rise to a shallow threshold in the northern North Sea, separating deeper waters to the south and north. The uplift and shallowing continued into Miocene time when a widespread hiatus formed in the northern North Sea, as revealed by biostratigraphic data. Miocene outbuilding from the north into the southeastern North Sea was massive (Fig. 22.8). Coastal progradation of the Utsira Formation in the Northern North Sea reflects Late Miocene-Early Pliocene uplift and erosion of mainland Norway. This relatively thick sandstone is a good aquifer and is used for injection of CO_2 in the Sleipner Field. It was still relatively warm in the Early Pliocene, compared to the Late Pliocene when mountain glaciation started to develop.

The Late Pliocene basin configuration was dominated by the progradation of thick clastic wedges in response to uplift and glacial erosion of eastern source areas (Figs. 22.7 and 22.8). Considerable Late Pliocene uplift of the eastern basin flank is documented by the strong angular relationship and tilting of the complete Cenozoic succession below the Pleistocene unconformity. The Plio-Pleistocene sediments are partly glacial and partly marine representing reworked glacial sediments and are typically poorly sorted. Such sediments compact readily and periods of glacial advances may contribute to the compaction.

The ice sheets advanced repeatedly out onto the shelf, but often deposited much of their debris load relatively near to land. Only during relatively short periods did the glaciers cover most of the North Sea basin. When major ice sheets developed on the Baltic Shield much of the ice flowed southwards along the valleys to the Oslo region and deepened the Oslofjord. The ice flow continued along the south coast of Norway and eroded a trough in the Tertiary and Mesozoic sediments. This is up to 700 m deep and continues up along western Norway. Thick Pleistocene sedimentary fans were deposited at the slope in front of the bathymetric trough.

At most places on the shelf the Holocene (the last 10,000 years) is represented by only a thin layer of silty sediment, chiefly reworked from Quaternary or older sediments exposed on topographic highs or along the margins of the North Sea. Very little "new" sediment has been supplied from land during the Holocene. This is because the fjords act as very efficient sediment traps, collecting the sediment from the rivers. Because fjords are deep but have shallow thresholds, not much clastic sediment reaches the shelf. On the seafloor there are in many areas frequent depressions which are typically a few tens of metres across and several metres deep. These are called pockmarks and have formed by fluid, generally gas, seeping from deeper layers to emerge at the seafloor.

The Cenozoic sedimentation was relatively rapid and the clayey sediments had little time to compact sufficiently to reduce the water content. Some beds therefore display plastic folding and diapir structures due to the under-compacted clays, especially in the Eocene. Polygonal faults are also common in these mudstones. They form a network which are from several hundred metres to 1 km across.

22.2.3 Exploration History and Petroleum Provinces/Systems

Exploration has been taking place in the North Sea for over 40 years and most of the large Norwegian fields are situated here. This is a mature part of the continental shelf and most of the plays are confirmed. The Norwegian part of the North Sea can be divided into two petroleum provinces characterised by different petroleum systems/plays: (1) the Central Graben area, and (2) the northern North Sea (Fig. 22.5).

The most important source rocks in the North Sea are shales in the Upper Jurassic. Thick beds of shale rich in organic matter were deposited over most of the North Sea area in Late Jurassic time. In the Norwegian sector, these belong to the Draupne Formation in the Viking Group or the Mandal Formation in the Tyne Group. Middle Jurassic coal is another important source rock, mainly for gas. In the southern part of the Norwegian sector, this coal is found in the Bryne Formation in the Vestland Group. In the northern part of the North Sea, these coal seams are found in the Brent Group. Source rocks may possibly be found in Carboniferous or older strata in the southern part of the North Sea, but these have so far not been confirmed.

22.2.3.1 Central Graben

In the Central Graben area rifting took place during Permian-Early Triassic and Middle-Late Jurassic times, and Zechstein salt was deposited north of the Mid North Sea High (Figs. 22.8 and 22.9). When this area was first drilled, the target was the Permian sandstones beneath the salt. It was by accident that oil was discovered in the Upper Cretaceous rocks (Chalk). Ekofisk was the first field to be discovered, but now there is a string of fields with Upper Cretaceous reservoirs, both on the Norwegian side and in the Danish Sector.

There are other sandstone reservoirs too, such as those of Upper Jurassic age in the Ula and Gyda fields. This lead is found south of the Brent delta, which excluded the Brent Group from forming reservoir rocks there. The area was uplifted and Middle Jurassic sediments are mostly absent. Later, oil has been found in Upper Palaeozoic reservoir rocks which had been the original prospecting target in the area.

In the Embla Field the reservoir rock is a sandstone believed to be of Devonian age.

The Ekofisk Field, found in 1969, was the first large oilfield in Europe. The Chalk in the Ekofisk reservoir is of the same type as we have onshore in Denmark and eastern England in stratigraphic levels approaching the Cretaceous/Tertiary boundary (Maastrichtian and Danian). Chalk lithologies had previously been assumed to be far too fine-grained to be reservoir rocks, and Ekofisk was the first large oilfield of this type. The Austin Chalk in Texas is one of the few other occurrences where Chalk forms a reservoir rock, but the fields there are fairly small by comparison.

Chalk is comprised of microscopic (0.001–0.005 mm) skeletons of planktonic algae which floated in the surface waters. After they died they sank to the seafloor to accumulate as a calcareous ooze. First when the scanning electron microscope (SEM) was developed was it possible to study these algae. The coccolithoporid skeleton consists of calcite, which makes the sediment more stable during diagenesis. We do not find as much solution and reprecipitation as would have been the case if the fossils had been aragonitic. There was uplift and some erosion of the Chalk in the Danian (earliest Tertiary) with a degree of resedimentation of Chalk beds along the slopes of the shallower areas, i.e. slumping. These redeposited Chalk sediments have proved to be the ones with the best reservoir characteristics.

The Chalk beds were then buried by Palaeocene and Eocene clays which sealed the Chalk and prevented the freshwater circulation which we otherwise often find in carbonate rocks along continental shelves. These clays have high contents of expanding clay minerals (smectite) and form a layer with very low permeability, so that porewater could not be forced out quickly enough with respect to the subsidence rate. The effective stresses and hence compaction have therefore been strongly reduced, and this is an important factor explaining why the porosity can still be as high as 30–35%. Both mechanical compaction and chemical pressure solution are reduced by the overpressure. The Ekofisk reservoir lies at over 3 km depth and without overpressure there would not have been such good porosity. In the shallower Valhall reservoir the effective stresses are even lower.

The Ekofisk structure is a dome-shaped anticlinal which has formed in response to an underlying salt diapir (Upper Permian). The structure covers c. 50 km^2

and has a closure height of 244 m. The oil column is 306 m in thickness, which means there is oil below the structure's lowest point (spill point). This situation requires an additional diagenetic trap, i.e. low permeability coupled with the capillary forces in the Chalk hinder the oil from seeping out laterally from below the structural trap. The porosity within the reservoir rock varies from 30 to 35% to tighter layers with 0–20%. However, in addition to this primary porosity there is a secondary porosity provided by fissures. Fissures are particularly important in connecting the small pores, so that the permeability increases to circa one to a few millidarcy. Without the fissuring the permeability would be much lower and it would have been impossible to produce the reservoir. The fissures are probably related to the horizontal tension which developed when the beds were bent upwards above the salt structure, so that the cracks are mostly vertical. The high overpressure also aids the fissuring process. The source rock for the hydrocarbons at Ekofisk is the underlying Kimmeridge Clay which attains its optimal maturity at this depth.

Gas injection and water injection have been important for maintaining reservoir pressure and preventing the fissures from closing. The reservoir rock has also been deliberately fissured by hydrofracturing, to increase the permeability. The platform at Ekofisk has sunk several metres during production, and the subsidence has continued despite the pressure being maintained with water injection. According to the laws of soil- and rock-mechanics the subsidence should cease when the effective stresses are not increasing, but here there has probably also been a chemical compaction which is not simply a function of effective stresses. This may be due to water substituting for the oil during production. There are several satellite fields round Ekofisk, including Cod, Albuskjell, Tor, Eldfisk, West Ekofisk and Edda.

22.2.3.2 Northern North Sea

The northern North Sea contains the majority of the giant fields discovered so far in the Norwegian continental shelf. The area is, however, still under intensive exploration, and the existing infrastructure makes even relatively small and complex traps interesting. The province is extensively drilled, particularly along the margins of the basin, and a whole series of trap types have been investigated. The most common trap type is provided by rotated fault blocks along both margins of the Viking Graben (Figs. 22.6 and 22.7), generated during Bathonian-Ryazian extension. The reservoirs are commonly found within sandstones of the Rhaetian-Sinemurian Statfjord Formation, in the Aalenian-Bathonian Brent Group, and locally in the Pliensbachian-Toarcian Cook Formation of the Dunlin Group. Sealing faults are frequently important for this trap type. Also Upper Jurassic (Bathonian-Kimmeridgian) sandstones provide good reservoirs, perhaps particularly in the platform areas, where fault block rotation is moderate (e.g. the Troll Field), and in mixed structural-stratigraphic traps along fault block crests. Post-rift marine sands (Cretaceous and Tertiary) add to the prospectivity of the northern North Sea. Indeed, the Tertiary marine sands of the Frigg type was the major target of early exploration activity.

In the northern North Sea there are no Permian salt deposits, but there are large thicknesses of Permian and Triassic sediments at depth (Fig. 22.6). The large oil fields in the northern part of the North Sea have Brent Group sandstones, and often also Statfjord Formation sandstones, as reservoir rocks. This applies to Statfjord, Gullfaks, Vigdis, Visund, Snorre and Oseberg. The traps consist of rotated fault blocks formed during rifting in the Late Jurassic. They stood as islands in the sea, and much of the Upper Jurassic and Lower Cretaceous is absent through non-deposition or erosion. At the top of the Gullfaks structure we only find a thin shale from the Upper Cretaceous, and the structure probably remained above sea level until then.

The source rock is the Kimmeridge Clay from the Upper Jurassic (Draupne Formation), and the oil has migrated up to the top of the fault blocks, but stratigraphically downwards from Upper to Middle or Lower Jurassic. In the Snorre Field the erosion at the top of the rotated fault blocks has reached right down to sandstones in the Upper Triassic (Lunde Formation) which is the reservoir rock together with the Statfjord Formation. The Statfjord Field was discovered in 1974 and was the first giant oil field on the Norwegian shelf after Ekofisk. The field lies in blocks 33/9 and 13/12, and extends into the British Sector. The Statfjord Field is large, estimated to contain about 550 mill. tons oil and $70 \cdot 10^9$ m^3 gas and NLG. Since the sand has such high porosity and permeability, the production rate was very high (330,000 barrels a day in 1997). It is part of the platform-like

Tampen Spur which is situated north of the Viking Graben and east of the East Shetland Basin, and which incorporates the Statfjord Field (Figs. 22.6 and 22.7). The Tampen Spur is one of several fault blocks which subsided and rotated along low-angle faults in Middle to Upper Jurassic time. The reservoir rocks consist of the Statfjord Formation (Lower Jurassic) and the Brent Group (Middle Jurassic).

The Statfjord Formation is comprised of fluvial sandstones which are replaced towards the top by shallow marine deposits. The Dunlin Shale succeeds the Statfjord Formation, and has enough organic material to be a source rock even if its contribution is very modest compared to the Kimmeridge Clay (Draupne Formation). During the Middle Jurassic there was considerable volcanic activity in the central part of the North Sea (Rattray Formation). This volcanism and the high geothermal gradients caused this area to be uplifted. The Triassic and Permian sandstones were then eroded and the sediments were for the most part transported northwards in a fluvial system which debouched in the Brent delta. The delta built out across almost the entire area between west Norway and Shetland, and sediment was also supplied from these areas to the delta. A several hundred metre thick series called the Brent Group was deposited, consisting of five formations that represent different facies. First the Broom Formation was deposited, mostly on the British side, and the Oseberg Formation on the Norwegian side. These were deposited as local fans related to faults along the sides of the initial rift. The Rannoch Formation consists of a mica-bearing sandstone which represents an upwards-coarsening sequence - from "offshore" to "shoreface" - and represents the actual progradation of the Brent delta. The mica settled out of suspension outside the fluvial channels. The Etive Formation consists of sandstones deposited on the delta front above wave base and is therefore better sorted. The Ness Formation represents the fluvial delta-top facies. Here there are meandering fluvial channels and crevasse splays. The clay and mica contents are higher in the levée and overbank deposits. The Ness Formation represents the maximum northward extent of the delta. The Tarbert Formation is much better sorted, having been deposited as marine delta-front sediments while the delta was being transgressed by the sea. The sediment supply from the Mid-North Sea High was reduced as this area was gradually transgressed and sediment supply to the delta could then no longer keep pace with the subsidence. Part of the

Ness Formation was probably eroded and redeposited as marine sediments as part of the Tarbert Formation.

The reservoir properties are a function of these depositional conditions, both on account of primary sorting and because diagenetic changes are usually determined by the primary mineralogical composition and sorting. Mica, and especially biotite, is mostly altered to kaolinite and then expands, blocking much of the porosity. Carbonate-cemented beds are often associated with marine environments with aragonitic fossils which have dissolved and formed cement.

The Brent delta was transgressed by the mudstones of the Heather Formation towards the end of the Late Jurassic. The main source rock for the area is the Upper Jurassic Kimmeridge Clay. It was deposited over a large part of the North Sea basin and also onshore present Britain. In the northern North Sea it is referred to as the Draupne Formation and the Mandal Formation in the Norwegian Central Graben. It may be more than 500 m thick along the rotated fault blocks which emerged as islands. The irregular bottom topography in the basin gave poor circulation and stagnant conditions with the deposition of a good source rock. The crests of many of the blocks were so heavily eroded during rotation that most of the Jurassic sequence is missing from them.

Shale at less than 3.5 km depth on the shallower parts of the structures is not sufficiently mature to have generated oil or gas in significant quantity. Where the Kimmeridge Clay is downfaulted deeply enough along the listric faults, oil has migrated up into the tectonically overlying but stratigraphically underlying Statfjord and Brent Groups. Most of the oil comes from the area north of the Snorre Field where the source rock (Draupne Formation) is buried to 4–5 km, but there is also some from the areas south of Gullfaks. Such areas that have generated a lot of oil are called kitchen areas.

In the Troll Field the reservoir rock is of Upper Jurassic age and contains mostly gas formed in the deepest parts of the basin (>4.5 km).

The oil began to migrate into the reservoir during early to mid-Tertiary times, when the source rock (Kimmeridge Clay) began to be mature. The oil/water contact in the Statfjord Formation and the Brent Group is at different levels which relate to two different pressure cells. The reservoir lies at 2.5–3 km depth and the sandstones have little quartz cement, such that loose sand is often recovered instead of solid core samples. There is also a danger of some sand coming up during oil production.

During recent years, an increasing focus has been set on the prospectivity of sands in the post-rift sequences, and some discoveries in different settings have been made (e.g. the Agat and Grane fields).

Some of the reservoirs in the northern North Sea leak oil or gas, and the gas in particular is seen in seismic reflection data. The leakages are probably the result of pressure having built up to the point where the cap rock has fractured. This is the case at Gullfaks (block 34/10) where fracture pressure has been reached. This means that some of the oil has migrated up through the Cretaceous shale cap rock and into the Cenozoic succession. Parts of this oil and gas have accumulated in Palaeocene and Eocene sandstones that were deposited as turbidites emanating from the Shetland Platform to the west. After deposition, the eastern side towards the Norwegian coast was uplifted, imparting a gentle westward dip to the top of these sandstone beds. They now form traps, as in the Frigg Field where the reservoir is Eocene turbidite sands. Heimdal, Balder and Grane fields also contain Palaeocene and Eocene sandstones, which are good reservoir rocks. Eocene clays rich in smectite make good cap rocks. These clays are lightly compacted, have low seismic velocities and often form small diapirs. During the Palaeocene a lot of sediment was supplied from the west, but these beds also dip westwards and rarely have closure towards the east.

In the northern North Sea a gas field (Peon) is found in the Quaternary glacial sediments only about 160 m below the seafloor. This shows that compacted glacial sediments may serve as a cap rock and capable of trapping gas.

22.3 Mid-Norwegian Shelf/Margin

22.3.1 Structure

Along strike the mid-Norwegian margin comprises three main segments (Møre, Vøring and Lofoten-Vesterålen), each 400–500 km long, separated by the East Jan Mayen Fracture Zone and Bivrost Lineament/Transfer Zone (Fig. 22.10).

The Møre Margin is characterised by a narrow shelf and a wide/gentle slope, underlain by the wide and deep Møre Basin with its thick Cretaceous fill (Figs. 22.11 and 22.12). The inner flank of the Møre

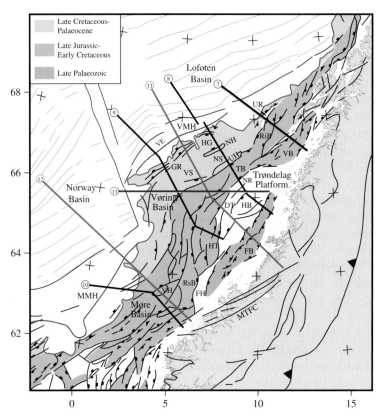

Fig. 22.10 Main structural elements of the mid-Norwegian Margin and adjacent areas (close-up of Fig. 22.2 – modified/updated from Faleide et al. 2008). Location of interpreted regional profiles (*black*), crustal transects (*red*) and seismic examples (*blue*) shown in Figs. 22.11, 22.12 and 22.13. DT = Dønna Terrace, FB = Froan Basin, FH = Frøya High, GR = Gjallar Ridge, HT = Halten Terrace, HB = Helgeland Basin, HG = Hel Graben, MMH = Møre Marginal High, MTFC = Møre-Trøndelag Fault Complex, NH = Nyk High, NR = Nordland Ridge, NS = Någrind Syncline, RiB = Ribban Basin, RsB = Rås Basin, TB = Træna Basin, UH = Utgard High, UR = Utrøst Ridge, VE = Vøring Escarpment, VH = Vigra High, VB = Vestfjorden Basin, VMH = Vøring Marginal High, VS = Vigrid Syncline

Basin is steeply dipping basinward and the crystalline crust thins rapidly from >25 km to <10 km. The sedimentary succession is thickest along the western part of the basin, 15–16 km, decreasing landwards to 12–13 km. The Møre Basin comprises sub-basins separated by intrabasinal highs formed during Late Jurassic-Early Cretaceous rifting. Most of the structural relief was filled in by mid-Cretaceous time. Sill intrusions are widespread within the Cretaceous succession in central and western parts of the Møre Basin, and lava flows cover the western part. Seaward of the Faeroe-Shetland Escarpment, at the Møre Marginal High, thickening of the crystalline crust and shallowing of the pre-Cretaceous sediments and top crystalline basement occur near the continent-ocean transition.

The ~500 km wide Vøring Margin comprises, from southeast to northwest, the Trøndelag Platform, the Halten and Dønna terraces, the Vøring Basin and the Vøring Marginal High (Figs. 22.10, 22.11, 22.12 and 22.13). The Trøndelag Platform has been largely stable since Jurassic time and includes deep basins filled by Triassic and Upper Palaeozoic sediments. The Vøring Basin can be divided into a series of sub-basins and highs, mainly reflecting differential vertical movements during the Late Jurassic–Early Cretaceous basin evolution.

The Vøring Plateau is a distinct bathymetric feature (Fig. 22.1), and includes the Vøring Marginal High and the Vøring Escarpment. The Vøring Marginal High consists of an outer part of anomalously thick oceanic crust, and a landward part of stretched continental crust, covered by thick Early Eocene basalts and underplated by mafic intrusions.

The Bivrost Lineament separates the Vøring and Lofoten-Vesterålen margins, marking the northern

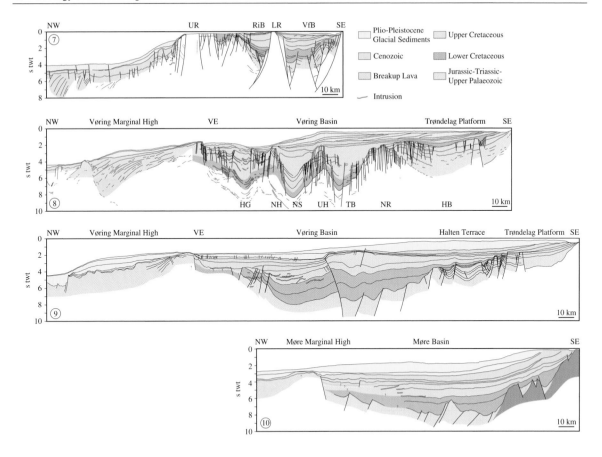

Fig. 22.11 Regional profiles across the Mid-Norwegian Margin (profiles 7–8 modified from Blystad et al. 1995 and profiles 9–10 modified from Tsikalas et al. 2005). See Fig. 22.10 for location and abbreviations

termination of the Vøring Plateau and the Vøring Marginal High, as well as the Vøring Escarpment. The Bivrost Transfer Zone is a major boundary in terms of margin physiography, structure, break-up magmatism and lithosphere stretching; break-up related magmatism is more voluminous south of it while the less magmatic Lofoten-Vesterålen margin was more susceptible to initial post-opening subsidence.

The Lofoten-Vesterålen margin is characterised by a narrow shelf and steep slope (Figs. 22.10, 22.11 and 22.12). The sedimentary basins underneath the shelf are narrower and shallower than on the Vøring and Møre margins. Typically they form asymmetric half-graben structures with changes in polarity bounded by a series of basement highs along the shelf edge. Beneath the slope, break-up-related lavas mask a sedimentary basin whose detailed mapping is hampered by poor seismic imaging. The continental crust on the Lofoten-Vesterålen margin appears to

have experienced only moderate pre-break-up extension, contrasting with the greatly extended crust in the Vøring Basin farther south.

22.3.2 Stratigraphy/Evolution

A stratigraphic summary for the Mid-Norwegian Shelf/Margin is shown in Fig. 22.4. The pre-opening, structural margin framework is dominated by the NE Atlantic-Arctic Late Jurassic–Early Cretaceous rift episode responsible for the development of major Cretaceous basins such as the Møre and Vøring basins off mid-Norway, and the deep basins in the SW Barents Sea (Fig. 22.3). Prior to that, Late Palaeozoic rift basins formed between Norway and Greenland and in the western Barents Sea along the NE-SW Caledonian trend. It has been suggested that the main

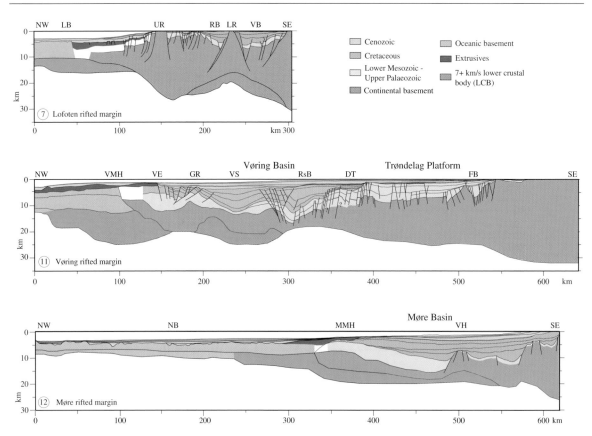

Fig. 22.12 Regional crustal transects across the Mid-Norwegian Margin (modified/updated from Faleide et al. 2008). See Fig. 22.10 for location and abbreviations

Late Palaeozoic–Early Mesozoic rift episodes took place in mid-Carboniferous, Carboniferous–Permian and Permian–Early Triassic times. Sediment packages associated with these movements are poorly resolved, mainly because of overprint by younger tectonism and burial by thick sedimentary strata.

On the mid-Norwegian margin, the Trøndelag Platform (Froan Basin) and Vestfjorden Basin record significant fault activity in Permian–Early Triassic time. Permian–Triassic extension is generally poorly dated, but is best constrained onshore East Greenland where a major phase of normal faulting culminated in the mid-Permian and further block faulting took place in the Early Triassic. The later Triassic basin evolution was characterised by regional subsidence and deposition of large sediment volumes. The Lower-Middle Jurassic strata (mainly sandstones) reflect shallow marine deposition prior to the onset of the next major rift phase.

A shift in the extensional stress field vector to NW-SE is recorded by the prominent NE Atlantic-Arctic late Middle Jurassic–earliest Cretaceous rift episode, an event associated with northward propagation of Atlantic rifting. Considerable crustal extension and thinning led to the development of major Cretaceous basins off mid-Norway (Møre and Vøring basins) and East Greenland, and in the SW Barents Sea (Harstad, Tromsø, Bjørnøya and Sørvestsnaget basins). These basins underwent rapid differential subsidence and segmentation into sub-basins and highs.

By mid-Cretaceous time, most of the structural relief within the Møre and Vøring basins had been filled in and thick Upper Cretaceous strata, mainly fine-grained clastics, were deposited in wide basins. Pulses of coarse clastic input with an East Greenland provenance appeared in the Vøring Basin from early Cenomanian to at least early Campanian times.

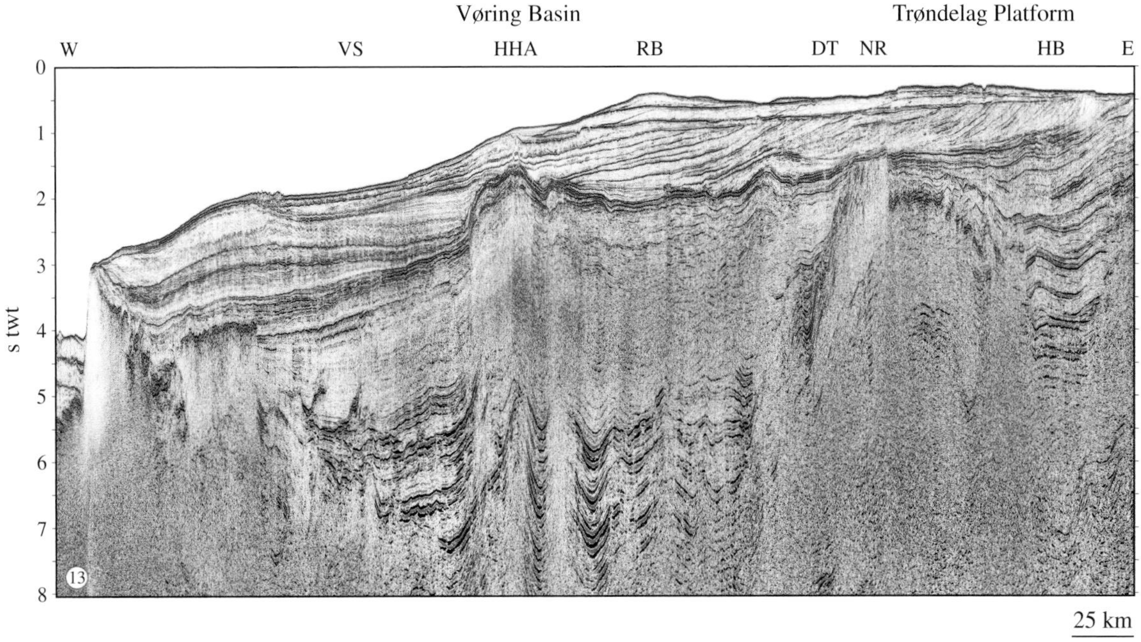

Fig. 22.13 Regional seismic line across the Mid-Norwegian Margin (courtesy Fugro and TGS). See Fig. 22.10 for location and abbreviations and Fig. 22.11 for stratigraphic information

Break-up in the NE Atlantic was preceded by prominent Late Cretaceous–Palaeocene rifting. At the onset of this rifting, the area between NW Europe and Greenland was an epicontinental sea covering a region in which the crust had been extensively weakened by previous rift episodes. The main period of brittle faulting occurred in Campanian time followed by smaller-scale activity towards break-up. The Campanian rifting resulted in low-angle detachment structures that updome thick Cretaceous sequences and sole out at medium-to-deep intracrustal levels on the Vøring and Lofoten-Vesterålen margins.

Late Cretaceous–Palaeocene rifting at the Vøring Margin covers a ∼150 km wide area bounded on the east by the Fles Fault Complex and the Utgard High (Fig. 22.10). Along the outer Møre and Lofoten-Vesterålen margins, most of the Late Cretaceous–Palaeocene deformation is masked by the lavas, but the structures appears to continue seawards underneath the break-up lavas. On the Møre and Vøring margins, the Palaeocene epoch was characterised by relatively deep water conditions. Depocentres in the western Møre and Vøring basins were sourced from the uplifted rift zone in the west. The northwestern corner of southern Norway was also uplifted and eroded, and the

sediments were mainly deposited in the NE North Sea and SE Møre Basin.

Final lithospheric break-up at the Norwegian margin occurred near the Palaeocene–Eocene transition at ∼55 Ma. It culminated in a 3–6 m.y. period of massive magmatic activity during break-up and the onset of early seafloor spreading. At the outer margin (e.g. Møre and Vøring margins), the lavas form characteristic seaward-dipping reflector sequences (Fig. 22.11) that drilling has demonstrated to be subaerially and/or neritically erupted basalts. These have become diagnostic features of volcanic margins. During the main igneous episode at the Palaeocene–Eocene transition, sills were intruded into the thick Cretaceous successions throughout the NE Atlantic margin, including the Vøring and Møre basins. Magma intrusion into organic-rich sedimentary rocks led to formation of large volumes of greenhouse gases that were vented to the atmosphere in explosive gas eruptions forming several thousand hydrothermal vent complexes along the Norwegian margin.

The mid-Norwegian margin experienced regional subsidence and modest sedimentation since Middle Eocene time and developed into a passive rifted margin bordering the oceanic Norwegian-Greenland Sea.

Mid-Cenozoic compressional deformation (including domes/anticlines, reverse faults and broad-scale inversion) is well documented on the Vøring margin, but its timing and significance are debated. The main phase of deformation is clearly Miocene in age but some of the structures were probably initiated earlier in Late Eocene–Oligocene times.

There is increasing evidence on the Norwegian margin for Late Miocene outbuilding on the inner shelf (Molo Formation) indicating a regional, moderate uplift of Fennoscandia. Over the entire shelf there is a distinct unconformity, which changes on the slope to a downlap surface for huge prograding wedges (Fig. 22.13) of sandy/silty muds sourced on the mainland areas around the NE Atlantic and the shelf. This horizon marks the transition to glacial sediment deposition during the Northern Hemisphere glaciation since about 2.6 Ma. Pliocene sedimentation is interspersed with ice-rafted debris signifying regional cooling and formation of glaciers. Large Plio-Pleistocene depocenters formed fans in front of bathymetric troughs scoured by ice streams eroding the shelf.

22.3.3 Exploration History and Petroleum Provinces/Systems

The mid-Norwegian continental shelf was opened for exploration in 1980, when a limited number of blocks on Haltenbanken were announced in the fifth licensing round, and it has had fields in production since 1993. These are situated within the major fault complexes which separate the platform area to the east from the deep Cretaceous basins in the west, and along the margin of the Trøndelag Platform (the Halten and Dønna terraces) (Figs. 22.10 and 22.11). In recent years blocks have also been awarded in greater water depths within the deep Vøring and Møre basins.

22.3.3.1 Haltenbanken

The Haltenbanken area (Halten and Dønna terraces) has good source rocks and reservoir rocks, and also structural traps (Fig. 22.11). On the eastern side of the main fault (the Klakk Fault), water has drained eastwards through the Jurassic sandstones up to the coast so that pressure has not built up much above

hydrostatic pressure; this has been an important factor in hindering leakage. The main problem in the area is that many of the reservoirs are very deep and oil can have escaped due to overpressure. Typical trap types drilled so far on the mid-Norwegian continental shelf are rotated fault blocks of Jurassic age in the horst-and-graben terrane of the Halten Terrace, and complex fault blocks within the major fault zones. Here too the source rock formed during rifting in the Late Jurassic. It is called the Spekk Formation, which approximately equates with the Draupne Formation in the North Sea and the Kimmeridge Formation in England. In the Early Jurassic Åre Formation there are coal beds, which may be the source for the gas and possibly some of the oil.

The most important reservoir rocks are of Early and Middle Jurassic age, with alternating sandstones and shales from the Åre, Tilje, Tofte, Ile and Garn formations (Fig. 22.4). The fields in the east, for example Heidrun Field, are not so very deep, but in the Smørbukk Field the reservoir rocks lie at more than 4 km. Also here it was block rotation during rifting which created the majority of the structures.

West of the Smørbukk Field there is a large fault, and to the west of that there is high overpressure in the Jurassic sandstones. Several wells have been drilled here that show signs of there having once been oil present, and that it has leaked out. This is presumably due to the high pressure having reached fracture pressure. In the Kristin Field, however, some of the oil is still in place. The Lavrans and Kristin fields lie at more than 5 km, where the temperature can reach 170–180°C. Here the porosity and permeability in the reservoir rocks are critical, and it is the degree of quartz cementation which essentially determines whether there is sufficient porosity to produce petroleum. The sand grains are coated with chlorite in much of the Tilje Formation, and in places also in the Garn Formation. This hinders quartz precipitation and thus helps maintain the relatively high porosities (20–25%) despite the great burial depth. Where the chlorite is absent and the sand is pure quartz, as in other parts of the Garn Formation, the porosity is quite low (<10–12%).

The Draugen Field has Upper Jurassic sandstone reservoirs and is fairly shallow, just 1.5 km. The source rock in this area is not mature and the oil found at Draugen has actually had a relatively long migration path from deeper-lying areas to the west.

The subsidence and sedimentation rates were high during the Pliocene and Quaternary, providing a sequence about 1 km thick (Fig. 22.13). This is considerably greater than in most of the North Sea. Thus neither the reservoir nor source rocks have been deeply buried for a geologically long period. This has left the source rocks somewhat immature for their depth, but also means that the reservoir rocks have less quartz cement.

22.3.3.2 Vøring and Møre Basins

This area was first opened for exploration at the end of the 1990s, with great expectation because of the large structures that could be seen on the regional seismic. Inversion structures (mainly huge, gentle anticlines with wavelengths in the order of tens of kilometres) are common in the deep Cretaceous basins, and provide traps of considerable size. These were generated during Tertiary inversion.

The geology proved to be very different from the North Sea and Haltenbanken. There had been considerable magmatic activity associated with the break-up between Norway and Greenland and the initiation of seafloor spreading. Tertiary lavas had flowed out across great tracts of land and intrusive dykes and sills penetrated the Cretaceous sediments. There was some concern that this could have raised the temperature so much that all the oil had become gas. However, it was found that this heating had been quite local and had had little overall effect.

More significant was the extreme thickness of the Cretaceous sequence, up to 6–7 km. This meant that the Upper Jurassic source rocks had already matured by the mid-Cretaceous. This was before the reservoir rocks, of Upper Cretaceous and Lower Tertiary age, had even been deposited, and before the structures we now see (e.g. Ormen Lange Dome, Helland Hansen Arch, Gjallar Ridge) had formed. It thus appears that the structures have only trapped some of the gas formed at depth from the oil. Gas has been found in several prospect in the western Vøring Basin, in Upper Cretaceous sandstones. It has been possible to map the distribution of sands from amplitude analysis of the seismic data.

The large Ormen Lange gas field was proven in 1997. It is located at water depths of 800–1,100 m in the eastern Møre Basin within the Storegga Slide area. The gas is found in Palaeocene and uppermost Cretaceous reservoir rocks (sandstones).

22.4 Barents Sea

22.4.1 Structure

The Barents Sea covers the northwestern corner of the Eurasian continental shelf. It is bounded by young passive margins to the west and north that developed in response to the Cenozoic opening of the Norwegian-Greenland Sea and the Eurasia Basin, respectively (Figs. 22.1 and 22.2).

The western Barents Sea is underlain by large thicknesses of Upper Palaeozoic to Cenozoic rocks constituting three distinct regions (Figs. 22.14, 22.15 and 22.16). (1) The Svalbard Platform is covered by a relatively flat-lying succession of Upper Palaeozoic and Mesozoic, mainly Triassic, sediments. (2) A basin province between the Svalbard Platform and the Norwegian coast is characterised by a number of sub-basins and highs with an increasingly accentuated structural relief westwards. Jurassic-Cretaceous, and in the west Palaeocene-Eocene, sediments are preserved in the basins. (3) The continental margin consists of three main segments: (a) a southern sheared margin along the Senja Fracture Zone; (b) a central rifted complex southwest of Bjørnøya associated with volcanism and (c) a northern, initially sheared and later rifted margin along the Hornsund Fault Zone. The continent-ocean transition occurs over a narrow zone along the line of Early Tertiary break-up and the margin is covered by a thick Upper Cenozoic sedimentary wedge.

The post-Caledonian geological history of the western Barents Sea is dominated by three major rift phases, Late Devonian?-Carboniferous, Middle Jurassic-Early Cretaceous, and Early Tertiary, each comprising several tectonic pulses. During Late Palaeozoic times most of the Barents Sea was affected by crustal extension. The later extension is characterised by a general westward migration of the rifting, the formation of well-defined rifts and pull-apart basins in the southwest, and the development of a belt of strike-slip faults in the north. Apart from

Fig. 22.14 Main structural
elements in the western
Barents Sea and adjacent
areas (close-up of Fig. 22.2 –
modified/updated from
Faleide et al. 2008). Location
of interpreted regional profiles
(*black*), crustal transects (*red*)
and seismic examples (*blue*)
shown in Figs. 22.15, 22.16,
22.17, 22.18 and 22.19.
BB = Bjørnøya Basin
FSB = Fingerdjupet Sub-basin
GH = Gardarbanken High
HB = Harstad Basin
HfB = Hammerfest Basin
HFZ = Hornsund Fault Zone
KFC = Knølegga Fault Complex
KR = Knipovich Ridge
LH = Loppa High
MB = Maud Basin
MH = Mercurius High
MR = Mohns Ridge
NB = Nordkapp Basin
NH = Nordsel High
OB = Ottar Basin
PSP = Polheim Sub-platform
SB = Sørvestsnaget Basin
SFZ = Senja Fracture Zone
SH = Stappen High
SR = Senja Ridge
TB = Tromsø Basin
TFP = Troms-Finnmark Platform
VH = Veslemøy High
VVP = Vestbakken Volcanic Province

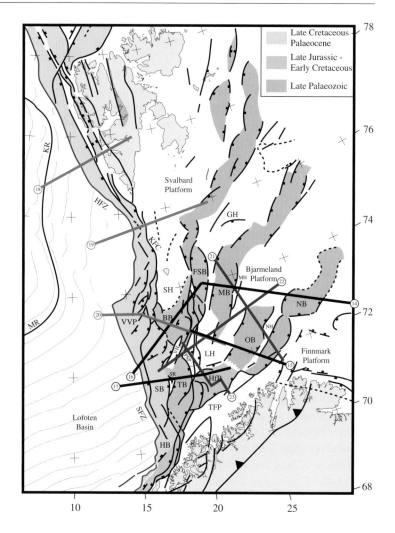

epeirogenic movements which produced the present
day elevation differences, the Svalbard Platform and
the eastern part of the regional basin have been largely
stable since Late Palaeozoic times.

22.4.2 Stratigraphy/Evolution

The Barents Sea is underlain by a thick succession of
Palaeozoic to Cenozoic strata showing both lateral and
vertical variations in thickness and facies – characterised by Upper Palaeozoic mixed carbonates, evaporites and clastics, overlain by Mesozoic-Cenozoic
clastic sedimentary rocks (Fig. 22.4). Direct information on the nature of the crystalline crust beneath
the Barents Sea sedimentary basins is scarce, but

the available, mostly indirect, evidence indicates that
the basement underlying much of its western part
was metamorphosed during the Caledonian Orogeny
and that the structural grain within the Caledonian
basement may have influenced later structural development. The Caledonian crystalline basement has a
NE-SW grain and was folded during Silurian time.

22.4.2.1 Upper Palaeozoic

Within the Caledonian domain Devonian molasse sediments were deposited in intermontane basins undergoing extensional collapse. A Devonian tectonic regime
comprising both extensional and compressional events
is so far only known on Svalbard, located on the
northwestern corner of the Barents Shelf. Here, thick

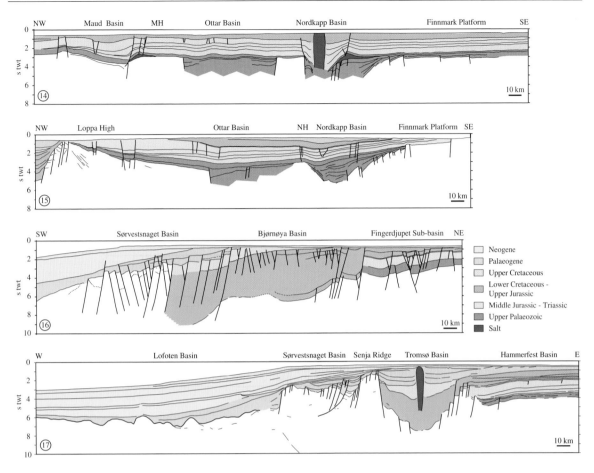

Fig. 22.15 Regional profiles across the western Barents Sea (modified from Faleide et al. 1993, Breivik et al. 1995). See Fig. 22.14 for location and abbreviations

Devonian strata are found in a north-south trending graben structure and the graben fill is discordantly overlain by Carboniferous strata (Fig. 22.17).

Lower to lower Upper Carboniferous strata, mainly clastics, were deposited in extensional basins ranging from wide downwarps to narrow grabens. A 300 km wide rift zone, extending at least 600 km in a north-easterly direction (Fig. 22.14), was formed mainly during mid-Carboniferous times. The rift zone was a direct continuation of the northeast Atlantic rift between Greenland and Norway, but a subordinate tectonic link to the Arctic rift was also established. The overall structure of the rift zone is a fan-shaped array of rift basins and intrabasinal highs with orientations ranging from northeasterly in the main rift zone to northerly at the present western continental margin. The structural style is one of interconnected and segmented basins characterised by half-graben geometries.

The Carboniferous rift phase resulted in the formation of several interconnected extensional basins filled with syn-rift deposits and separated by fault-bounded highs. Structural trends striking northeast to north dominate in most of the southwestern Barents Sea (Fig. 22.14) where the Tromsø, Bjørnøya, Nordkapp, Fingerdjupet, Maud and Ottar basins have been interpreted as rift basins formed at this time (Figs. 22.15, 22.18, 22.19 and 22.20). Fault movements ceased in the eastern areas towards the end of the Carboniferous and the structural relief was gradually infilled and blanketed by a platform succession of Late Carboniferous-Permian age. The lower part of this succession passes upwards from cyclical dolomites and evaporites to massive limestones. It includes a widespread evaporite layer of latest Carboniferous-earliest Permian age mapped regionally in the SW Barents Sea (also on the NE Greenland shelf/margin).

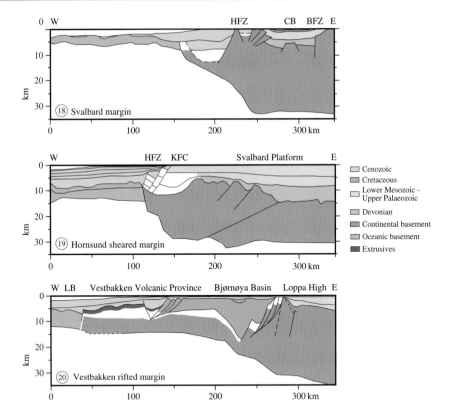

Fig. 22.16 Regional crustal transects across the western Barents Sea-Svalbard margin (modified/updated from Faleide et al. 2008). See Fig. 22.14 for location and abbreviations

The thickness of the salt layer in the Nordkapp Basin (Fig. 22.18), which is inferred to have reached 4–5 km locally, implies that a substantial fault-generated depression was in place or developed during early Late Carboniferous times. Not all of this thickness corresponds to a fault-defined relief, however, because salt was also deposited in the basin during the subsequent phase of differential thermal subsidence. Still, a considerable fault-bounded basin must have existed.

From Bashkirian to Artinskian/Early Kungurian times, carbonate deposition took place on a broad shelf. Carbonate build-ups were common at basin margins and intrabasinal highs controlled by underlying older structures. Following Gzelian-Asselian/Sakmarian deposition of basinal evaporites and growth of marginal carbonate build-ups, a regional shallow-water carbonate platform was established during Sakmarian-Artinskian times. Carbonate sedimentation occurred in the entire region until late Early Permian time when clastic deposition started to

dominate the area (Fig. 22.4). The change to platform type sedimentation marked the initial development of a regional sag basin which continued to subside in the Late Permian during the deposition of cherty limestones and shales.

The western part of the rift system was affected by renewed faulting, uplift and erosion in Permian-Early Triassic times. Normal faulting along the western margin of the Loppa High (Figs. 22.19 and 22.20) as well as the uplift, tilting and erosional truncation of the high itself are of sufficient magnitude to indicate a significant Permian-Early Triassic rift phase affecting the N-S structural trend. Evidence of fault movements is found as far north as the Fingerdjupet Sub-basin. The erosional surface associated with this rift phase extends from the Loppa High to the Stappen High, and Permian tectonic activity is known to have occurred on Bjørnøya and on the Sørkapp High in the southern part of Spitsbergen. Therefore the rift phase probably affected a narrow northerly trending zone along the entire present western margin.

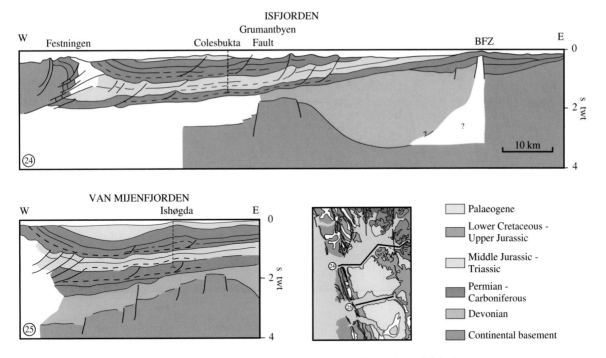

ISFJORDEN

Fig. 22.17 Svalbard geology in seismic lines from fjords (modified from Faleide et al. unpublished)

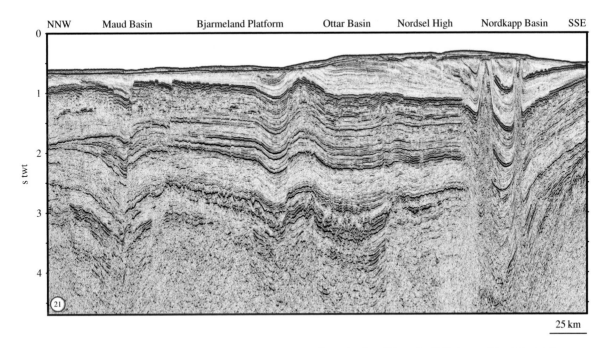

Fig. 22.18 Regional seismic line across the SW Barents Sea (Nordkapp Basin – Bjarmeland Platform – Maud Basin) (courtesy Fugro and TGS). See Fig. 22.14 for location and Fig. 22.15 for stratigraphic information

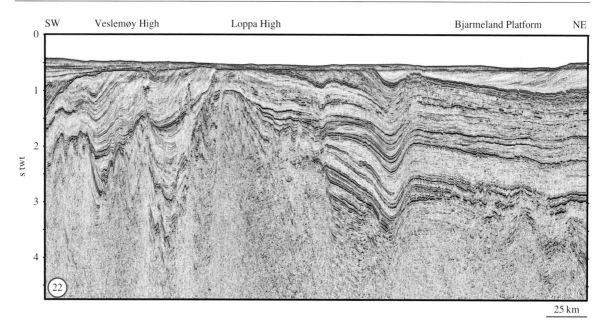

Fig. 22.19 Regional seismic line across the SW Barents Sea (Bjarmeland Platform – Loppa High – Veslemøy High) (courtesy Fugro and TGS). See Fig. 22.14 for location and Fig. 22.15 for stratigraphic information

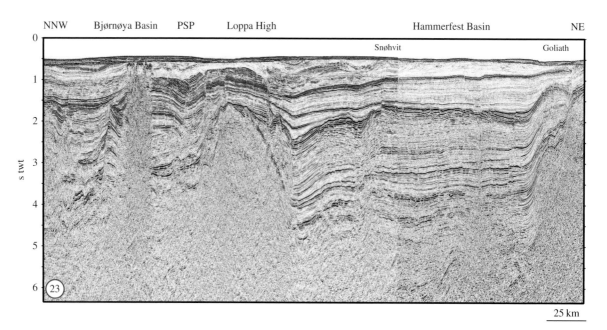

Fig. 22.20 Regional seismic line across the SW Barents Sea (Hammerfest Basin – Loppa High – Bjørnøya Basin) (courtesy Fugro and TGS). See Fig. 22.14 for location and Fig. 22.15 for stratigraphic information

22.4.2.2 Mesozoic

In Early Triassic time a regional deepwater basin covered much of the Barents Sea. During the relatively short Triassic period large amounts of clastic sediments were deposited. The Uralian highland to the east was an important sediment source. Sediments were also shed into the Barents Sea from the Baltic Shield and from other local source areas. Differential loading triggered salt diapirism in the Nordkapp Basin

(Fig. 22.18) and possibly later also in other salt basins in the SW Barents Sea.

The Triassic strata are dominated by shales and sandstones, the vertical and lateral distribution of which is complex. There seems to be an increased content of coarse clastic rocks in the younger intervals. A similar increase is also observed in an easterly and a southerly direction towards the main sediment source area. Marine conditions prevailed in Late Permian and Early Triassic times followed by a shallowing and partial exposure of some areas. The continental regime prevailed in large areas in Middle Triassic time and the northward and westward prograding deltaic system continued to infill the regional basin.

In the central and northern basin areas marine conditions existed throughout the Middle Triassic. A good Middle Triassic source rock is known from Svalbard and this is probably present in large parts of the western Barents Sea. In the Late Triassic the shoreline was moved back to the southern and eastern borders of the SE Barents Sea basin. The Triassic ended with regression and erosion. Late Triassic sediments were also derived from other source areas, mainly in the northwest.

The Lower-Middle Jurassic interval is dominated by sandstones throughout the Barents Sea. These sandstones form the main reservoir in the SW Barents Sea (Hammerfest Basin; Fig. 22.20). They probably also covered the Loppa High and Finnmark Platform but were partly eroded during later tectonic activity. The late Middle Jurassic sequence boundary marks the onset of rifting in the southwestern Barents Sea, whereas unconformities within the Upper Jurassic sequence reflect interplay between continued faulting and sea-level changes. The sequence is so thin that it is generally impossible to resolve these unconformities on the seismic data. The shales and claystones contain thin interbedded marine dolomitic limestone and rare siltstones or sandstones toward the basin flanks, reflecting relatively deep and quiet marine environments.

The Late Jurassic-earliest Cretaceous structuring in the SW Barents Sea was characterised by regional extension accompanied by strike-slip adjustments along old structural lineaments, developing the Bjørnøya, Tromsø and Harstad basins as prominent rift basins (Fig. 22.15). The evolution of these basins was closely linked to important tectonic phases/events in the North Atlantic-Arctic region. Rifting continued in Early Cretaceous time and an important phase of

Aptian faulting is documented in the SW Barents Sea. Several phases of Late Mesozoic and Early Cenozoic rifting gave rise to very deep basins in the SW Barents Sea. A tentative Middle Jurassic horizon corresponding to the top of the shallow marine sandstones, also marking the onset of late Mesozoic rifting, can be followed from the seafloor down to 9 s twt (16–17 km) over a relatively short distance (Fig. 22.15).

The Lower Cretaceous comprises three sedimentary units from Valanginian to Cenomanian. Shales and claystones dominate, with thin interbeds of silt, limestone and dolomite. These strata make up the main basin fill in the deep SW Barents Sea basins and they dominate the regional subcrop pattern. The marine environments throughout the deposition of these units are dominated by distal conditions with periodic restricted bottom circulation.

In Early Cretaceous time the northern Barents Sea was characterised by widespread magmatism without any signs of faulting. Extrusives and intrusives (sills and dykes) belonging to a regional Large Igneous Province (LIP) in the Arctic are documented from both onshore (Svalbard and Franz Josef Land) and offshore areas in the northern Barents Sea. The magmatism was most extensive during Barremian?-Aptian times but both older and younger dates have been reported. The present distribution in the northern Barents Sea is modified by later uplift and erosion. The Early Cretaceous magmatism within the Arctic LIP had important palaeogeographic implications. It caused regional uplift and the Neocomian in the north and east was characterised by southward sediment progradation. This depositional regime is difficult to document in the western Barents Sea due to later uplift and erosion. The magmatism and regional uplift was related to rifting and break-up in the Amerasia (Canada) Basin and the formation of the Alpha Ridge.

Little or no Upper Cretaceous sediments were deposited in the Barents Sea except in the SW Barents Sea which continued to subside in response to faulting in a pull-apart setting. The Wandel Sea Basin is a NE Greenland equivalent where Late Cretaceous strike-slip faulting and pull-apart basin formation is well documented (Figs. 22.2 and 22.3). The Upper Cretaceous succession varies in thickness and completeness. In the Tromsø Basin a 1,200 m shale succession has been drilled while seismic data indicate that the sequence reaches 2,000–3,000 m in rim synclines in the central basin. Whereas the Tromsø and

Sørvestsnaget basins were depocentres through most of this period, areas further east were either transgressed only during maximum sea-level and/or display only condensed sections.

22.4.2.3 Cenozoic

The Cenozoic structuring was related to the two-stage opening of the Norwegian-Greenland Sea and the formation of the predominantly sheared western Barents Sea continental margin (Fig. 22.3). Continental break-up and onset of sea-floor spreading were preceded by a phase of rapid late Palaeocene subsidence. The Palaeogene succession rests unconformably on the Cretaceous and this depositional break at the Cretaceous-Tertiary transition (Maastrichtian-Danian) occurs throughout the southwestern Barents Sea.

The western Barents Sea-Svalbard margin developed from a megashear zone which linked the Norwegian-Greenland Sea and the Eurasia Basin during the Eocene opening. The first-order crustal structure along the margin and its tectonic development is mainly the result of three controlling parameters: (1) the pre-break-up structure, (2) the geometry of the plate boundary at opening and (3) the direction of relative plate motion. The interplay between these parameters gave rise to striking differences in the structural development of the different margin segments of sheared and/or rifted nature. The continent-ocean boundary is clear along the sheared Senja margin. The central rifted margin segment southwest of Bjørnøya was associated with magmatism in the Vestbakken Volcanic Province both during break-up at the Palaeocene-Eocene transition and later in the Oligocene. In the north, strike-slip movements between Svalbard and Greenland gave rise to compressional (transpressional) deformation within the Spitsbergen Fold and Thrust Belt (Fig. 22.17). Compressional deformation is also observed in the Barents Sea east of Svalbard, showing that stress related to transpression at the plate boundary west of Svalbard was transferred over large distances. Domal structures observed in the eastern Barents Sea may be related to this compressional regime.

The opening of the Greenland Sea was complex, involving jumps in the location of the spreading axis and the splitting off of microcontinents. Since earliest Oligocene time (magnetic chron 13) Greenland moved

with North America in a more westerly direction relative to Eurasia (Fig. 22.3). This gave rise to extension, break-up and onset of seafloor spreading also in the northern Greenland Sea west of Svalbard, and a deep-water gateway between the North Atlantic and Arctic was finally established sometime in the Miocene. This had large implications for the palaeo-oceanography and climate in the region.

The late Cenozoic evolution was characterised by subsidence and burial of the margins by thick sediments in a clastic wedge derived from the uplifted Barents Sea area. Late Cenozoic uplift and erosion of the Barents Sea has removed most of the Cenozoic sediments, and even older strata. The subcrop pattern is dominated by Mesozoic units. The erosion seems to have been most extensive in the western Barents Sea, especially in the northwestern part including Svalbard where more than 3,000 m of sedimentary strata have been removed. In the SW Barents Sea the erosion estimates are in the range 1,000–1,500 m. In the eastern Barents Sea, little work has so far been done on Cenozoic uplift and erosion, however, erosion has been calculated to be between 250 and 1,000 m. High seismic velocities at the top of the bedrock in the NW Barents Sea, decreasing south- and eastwards, reflect the pattern of differential uplift and erosion. The Neogene and Quaternary, resting unconformably on Palaeogene and Mesozoic rocks, thickens dramatically in the huge sedimentary wedge at the margin. The glacial sediments are dated as Late Pliocene to Pleistocene/Holocene in age.

22.4.3 Exploration History and Petroleum Provinces/Systems

More than three decades of research and petroleum exploration in the Barents Sea have revealed a deep and complex sedimentary basin system affected by a variety of geological processes. A large petroleum potential has been proven including multiple source and reservoir intervals. However, there are still many uncertainties and problems linked to understanding the prospectivity of the Barents Sea. Areas in the southern Barents Sea were opened for exploration in 1980 and the first discovery was made in 1981.

A great variety of traps (fault and salt structures, stratigraphic pinchout) and sealing mechanisms exist in the Barents Sea area, and several different play models have proven hydrocarbon accumulations. The discoveries are dominated by gas and gas condensate accumulations (e.g. Snøhvit Field). Potential reservoirs are distributed across several stratigraphic levels from Devonian to Tertiary and comprise both sandstone and carbonate lithologies. Jurassic sandstones are the main proven reservoir but hydrocarbons have also been found in sandstones of Triassic and Cretaceous age. Several proven and potential source rock units are present in the Barents Sea. The most important are the Upper Jurassic and Triassic organic-rich (anoxic) shales. Lower Cretaceous, Permian and Carboniferous shales, as well as Lower Permian evaporites and Lower Jurassic coals also have source potential.

About 25 discoveries have been made in the Barents Sea, most of them in the Hammerfest Basin (Fig. 22.20) where their reservoirs are in sandstones, mainly Jurassic, as in the Snøhvit Field. Deeper discoveries have also been made, such as in Triassic sandstones in 7122/7-1 (Goliath Field) and 7125/4-1 (Nucula Field). Oil and gas have also been found in Triassic sandstones in the Nordkapp Basin, which is dominated by salt tectonics (Fig. 22.18). Gas and oil have been found in carbonates of Carboniferous to Permian age on the Finnmark Platform.

The most significant exploration problem in the Western Barents Sea relates to the severe uplift and erosion of the area that took place during the Cenozoic. The quantity of sediments removed, and the timing of this removal, are still a matter of debate, but it is agreed that the uplift and erosion have had important implications for oil and gas exploration in the Barents Sea. Residual oil columns found beneath gas fields in the Hammerfest Basin indicate that the structures were once filled, or partially filled, with oil. The removal of up to 2 km of sedimentary overburden from the area has had critical consequences for these accumulations: exsolution of gas from the oil, and expansion of the gas due to the decrease in pressure, resulted in expulsion of most of the oil from the traps. Seal breaching and spillage probably also occurred as a result of the uplift and tilting. A further consequence of these late movements was the cooling of the source rocks in the area, which effectively caused most hydrocarbon generation to cease. Thus, little new oil was available to fill available trapping space. These mechanisms may explain the predominance of gas over oil in the Barents Sea.

Further Reading

Bergh, S.G., Eig, K., Kløvjan, O.S., Henningsen, T., Olesen, O. and Hansen, J.A. The Lofoten Vesterålen continental margin: A multiphase Mesozoic-Palaeogene rifted shelf as shown by offshore-onshore brittle fault-fracture analysis. Norwegian Journal of Geology 87, 29–58.

Blystad, P., Brekke, H., Færseth, R.B., Larsen, B.T., Skogseid, J. and Tørudbakken, B. 1995. Structural elements of the Norwegian continental shelf, Part II: The Norwegian Sea Region. Norwegian Petroleum Directorate Bulletin 8.

Breivik, A., Gudlaugsson, S.T. and Faleide, J.I. 1995. Ottar Basin, SW Barents Sea: A major Upper Paleozoic rift basin containing large volumes of deeply buried salt. Basin Research 7, 299–312.

Breivik, A.J., Faleide, J.I. and Gudlaugsson, S.T. 1998. Southwestern Barents Sea margin: Late Mesozoic sedimentary basins and crustal extension. Tectonophysics 293, 21–44.

Brekke, H. 2000. The tectonic evolution of the Norwegian Sea continental margin with emphasis on the Vøring and Møre basins. In: Nøttvedt, A. et al. (eds.), Dynamics of the Norwegian Margin. Geological Society Special Publication 167, pp. 327–378.

Brekke, H., Sjulstad, H.I., Magnus, C. and Williams, R.W. 2001. Sedimentary environments offshore Norway – An overview. In: Martinsen, O. and Dreyer, T. (eds.), Sedimentary Environments Offshore Norway – Paleozoic to Recent, NPF Special Publication 10, pp. 7–37 (Norwegian Petroleum Society and Elsevier Science B.V.).

Christiansson, P., Faleide, J.I. and Berge, A.M. 2000. Crustal structure in the northern North Sea – An integrated geophysical study. In: Nøttvedt, A. (ed.), Dynamics of the Norwegian Margin. Geological Society Special Publication 167, pp. 15–40.

Doré, A.G., Lundin, E.R., Jensen, L.N., Birkeland, O., Eliassen, P.E. and Fichler, C. 1999. Principal tectonic events in the evolution of the northwest European Atlantic margin. In: Fleet, A.J. and Boldy, S.A.R. (eds.), Petroleum Geology of Northwest Europe: Proceedings of the 5th Conference. Geological Society, pp. 41–61.

Doré, A.G., Cartwright, J.A., Stoker, M.S., Turner, J.P. and White, N.J. 2002. Exhumation of the North Atlantic margin: Introduction and background. In: Doré, A.G., Cartwright, J.A., Stoker, M.S., Turner, J.P. and White, N.J. (eds.), Exhumation of the North Atlantic Margin: Timing, Mechanisms and Implications for Petroleum Exploration. Geological Society Special Publication 196, pp. 1–12.

Ebbing, J., Lundin, E., Olesen, O. and Hansen, E.K. 2006. The mid-Norwegian margin: A discussion of crustal lineaments, mafic intrusions, and remnants of the Caledonian root by 3D density modelling and structural interpretation. Journal of the Geological Society 163, pp. 47–60.

Eldholm, O., Tsikalas, F. and Faleide, J.I. 2002. Continental margin off Norway 62–75 N: Paleogene tectono-magnetic segmentation and sedimentation. In: Jolley, D.W. and Bell, B. (eds.), North Atlantic Igneous Province: Stratigraphy, Tectonic, Volcanic and Magmatic Processes. Geological Society Special Publication 197, pp. 38–68.

Evans, D., Graham, C., Armour, A. and Bathurst, P. (eds.). 2003. The Millenium Atlas: Petroleum geology of the central and northern North Sea. Published by the Geological Society.

Faleide, J.I., Gudlaugsson, S.T. and Jacquart, G. 1984. Evolution of the western Barents Sea. Marine and Petroleum Geology 1, 123–150.

Faleide, J.I., Vågnes, E. and Gudlaugsson, S.T. 1993. Late Mesozoic-Cenozoic evolution of the southwestern Barents Sea in a regional rift-shear tectonic setting. Marine and Petroleum Geology 10, 186–214.

Faleide, J.I., Solheim, A., Fiedler, A., Hjelstuen, B.O., Andersen, E.S. and Vanneste, K. 1996. Late Cenozoic evolution of the western Barents Sea-Svalbard continental margin. Global and Planetary Change 12, 53–74.

Faleide, J.I., Kyrkjebø, R., Kjennerud, T., Gabrielsen, R.H., Jordt, H., Fanavoll, S. and Bjerke, M.D. 2002. Tectonic impact on sedimentary processes during the Cenozoic evolution of the northern North Sea and surrounding areas. In: Dore et al. (eds.), Exhumation of the Circum-Atlantic Continental Margins: Times, Mechanisms and Implications for Petroleum Exploration. Geological Society Special Publication 196, pp. 235–269.

Faleide, J.I., Tsikalas, F., Breivik, A.J, Mjelde, R., Ritzmann, O., Engen, Ø., Wilson, J. and Eldholm, O. 2008. Structure and evolution of the continental margin off Norway and the Barents Sea. Episodes 31, 82–91.

Færseth, R.B. 1996. Interaction of Permo-Triassic and Jurassic extensional fault-blocks during the development on the Northern North Sea. Journal of the Geological Society 153, 931–944.

Færseth, R., Gabrielsen, R.H. and Hurich, C.A. 1995. Influence of basement in structuring of the North Sea Basin, offshore southwest Norway. Norsk Geologisk Tidsskrift 75, 105–119.

Færseth, R.B. and Lien, T. 2002. Cretaceous evolution in the Norwegian Sea – A period characterized by tectonic quiescence. Marine and Petroleum Geology 19, 1005–1027.

Gabrielsen, R.H. and Doré, A.G. 1995. The history of tectonic models on the Norwegian continental shelf. In: Hanslien, S. (ed.), Petroleum Exploration and Exploitation in Norway - Past Experiences and Future Challenges. A Celebration of 25 Years. Norwegian Petroleum Society Special Publication 4, pp. 341–375.

Gabrielsen, R.H., Færseth, R.B., Jensen, L.N., Kalheim, J.E. and Riis, F. 1990. Structural elements of the Norwegian continental shelf, Part II: The Norwegian Sea Region. Norwegian Petroleum Directorate Bulletin 6.

Gabrielsen, R.H., Færseth, R.B., Steel, R.J., Idil, S. and Kløvjan, O.S. 1990. Architectural styles of basin fill in the northern Viking Graben. In: Blundell, D.J. and Gibbs, A.D. (eds.), Tectonic Evolution of the North Sea Rifts. Clarendon, Oxford, pp. 158–179.

Gabrielsen, R.H., Odinsen, T. and Grunnaleite, I. 1999. Structuring of the Northern Viking Graben and the Møre Basin; the influence of basement structural grain and the particular role of the Møre-Trøndelag fault Complex. Marine and Petroleum Geology 16, 443–465.

Gabrielsen, R.H., Kyrkjebø, R., Faleide, J.I., Fjeldskaar, W. and Kjennerud, T. 2001. The Cretaceous post-rift basin configuration of the northern North Sea. Petroleum Geoscience 7, 137–154.

Gaina, C., Gernigon, L. and Ball, P. 2009. Palaeocene–Recent plate boundaries in the NE Atlantic and the formation of the Jan Mayen microcontinent. Journal of the Geological Society 166, 601–616.

Gernigon, L., Ringenbach, J.-C., Planke, S., LeGall, B. and Jonquet-Kolstø, H. 2003. Extension, crustal structure and magmatism at the Outer Vøring Basin, North Atlantic Margin (Norway). Journal of the Geological Society 160, 197–208.

Gernigon, L., Ringbach, J.C., Planke, S., Planke, S. and Le Gall, B. 2004. Deep structures and breakup along rifted margins: Insights from integrated studies along the outer Vøring Basin (Norway). Marine and Petroleum Geology 21, 363–372.

Gudlaugsson, S.T., Faleide, J.I., Johansen, S.E. and Breivik, A. 1998. Late Palaeozoic structural development of the southwestern Barents Sea. Marine and Petroleum Geology 15, 73–102.

Heeremans, M. and Faleide, J.I. 2004. Permo-Carboniferous rifting in the Skagerrak, Kattegat and the North Sea: Evidence from seismic and borehole data. In: Wilson, M., Neumann, E.-R., Davies, G., Timmerman, M.J., Heeremans, M. and Larsen, B.T. (eds.), Permo-Carboniferous Rifting in Europe. Geological Society Special Publication 223, pp. 159–177.

Hjelstuen, B.O., Sejrup, H.P., Haflidason, H., Nygård, A., Ceramicola, S. and Bryn, P. 2005. Late Cenozoic glacial history and evolution of the Storegga Slide area and adjacent slide flank regions, Norwegian continental margin. Marine and Petroleum Geology 22, 57–69.

Hjelstuen, B.O., Eldholm, O. and Faleide, J.I. 2007. Recurrent Pleistocene mega-failures on the SW Barents Sea margin. Earth and Planetary Science Letters 258, 605–618.

Johansen, S.E., Ostisty, B.K., Birkeland, Ø., Fedorovsky, Y.F., Martirosjan, V.N., Bruun Christensen, O., Cheredeev, S.I., Ignatenko, A.A. and Margulis, M. 1993. Hydrocarbon potential in the Barents Sea region: Play distribution and potential. In: Vorren, T.O. et al. (eds.), Arctic Geology and Petroleum Potential, NPF Special Publication 2. Elsevier, New York, NY, pp. 273–320.

Jordt, H., Faleide, J.I., Bjørlykke, K. and Ibrahim, M.T. 1995. Cenozoic stratigraphy of the central and northern North Sea Basin: Tectonic development, sediment distribution and provenance areas. Marine and Petroleum Geology 12, 845–879.

Lundin, E.R. and Doré, A.G. 1997. A tectonic model for the Norwegian passive margin with implications for the NE Atlantic: Early Cretaceous to break-up. Journal of the Geological Society 154, 545–550.

Lundin, E.R. and Doré, A.G. 2002. Mid-Cenozoic post-breakup deformation in the "passive" margins bordering the Norwegian-Greenland Sea. Marine and Petroleum Geology 19, 79–93.

Mjelde, R., Raum, T., Breivik, A., Shimamura, H., Murai, Y., Takanami, T., Brekke, H. and Faleide, J.I. 2005. Crustal structure of the Vøring Margin, NE Atlantic derived from OBS-data: Geological implications. In: Doré, A.G. and Vining, B. (eds.), Petroleum Geology: North West Europe and Global Perspectives – Proceedings of the 6th Petroleum Geology Conference. Geological Society, pp. 803–813.

Mosar, J., Eide, E.A., Osmundsen, P.T., Sommaruga, A. and Torsvik, T. 2002. Greenland-Norway separation: A geodynamic model for the North Atlantic, Norwegian Journal of Geology 82, 281–298.

Neumann, E.-R., Wilson, M., Heeremans, M., Spencer, E.A., Obst, K., Timmerman, M.J. and Kirstein, L.A. 2004. Carboniferous-Permian rifting and magmatism in Southern Scandinavia, the North Sea and northern Germany: A review. In: Wilson, M., Neumann, E.-R., Davies, G.R., Timmerman, M.J., Heeremans, M. and Larsen, B.T. (eds.), Permo-Carboniferous Magmatism and Rifting in Europe. Geological Society Special Publication 223, pp. 11–40.

Nøttvedt, A., Gabrielsen, R.H. and Steel, R.J. 1995. Tectonostratigraphy and sedimentary architecture of rift basins; with reference to the northern North Sea. Marine and Petroleum Geology 12, 881–901.

Odinsen, T., Christiansson, P., Gabrielsen, R.H., Faleide, J.I. and Berge, A.M. 2000a. The geometries and deep structure of the northern North Sea rift system. In: Nøttvedt, A. (ed.), Dynamics of the Norwegian Margin. Geological Society Special Publication 167, pp. 41–57.

Olaussen, S., Larsen, B.T. and Steel, R.J. 1994. The Upper Carboniferous-Permian Oslo Rift; basin fill in relation to tectonic development. In: Embry, A.F., Beauchamp, B. and Glass, D.J. (eds.), Pangea; Global Environments and Resources. Canadian Society of Petroleum Geologists, Calgary, Memoir 17, pp. 175–197.

Osmundsen, P.T., Sommaruga, A., Skilbrei, J.R. and Olesen, O. 2002. Deep structure of the mid-Norway rifted margin, Norwegian Journal of Geology 82, 205–224.

Osmundsen, P.T. and Ebbing, J. 2008. Styles of extension offshore mid-Norway and implications for mechanisms of crustal thinning at passive margins. Tectonics 27, TC6016.

Ritzmann, O. and Faleide, J.I. 2007. Caledonian basement of the western Barents Sea. Tectonics 26, TC5014.

Ren, S., Faleide, J.I., Eldholm, O., Skogseid, J. and Gradstein, F. 2003. Late Cretaceous-Paleocene development of the NW Vøring Basin. Marine and Petroleum Geology 20, 177–206.

Ro, H.E., Stuevold, L.M., Faleide, J.I. and Myhre, A.M. 1990. Skagerrak Graben; the offshore continuation of the Oslo Graben. Tectonophysics 178, 1–10.

Roberts, A.M., Lundin, E.R. and Kusznir, N.J. 1997. Subsidence of the Vøring Basin and the influence of the Atlantic continental margin. Journal of the Geological Society 154, 551–557.

Ryseth, A., Augustson, J.H., Charnock, M., Haugerud, O., Knutsen, S.-M., Midbøe, P.S., Opsal, J.G. and Sundsbø, G. Cenozoic stratigraphy and evolution of the Sørvestsnaget Basin, southwestern Barents Sea. Norwegian Journal of Geology 83, 107–130.

Skogseid, J., Pedersen, T. and Larsen, V.B. 1992. Vøring Basin: Subsidence and tectonic evolution. In: Larsen, R.M. et al. (eds.), Structural and Tectonic Modelling and Its Application to Petroleum Geology; Proceedings, Norwegian Petroleum Society Special Publication 1, pp. 55–82.

Skogseid, J., Planke, S., Faleide, J.I., Pedersen, T., Eldholm, O. and Neverdal, F. 2000. NE Atlantic continental rifting and volcanic margin formation. In: Nøttvedt, A. (ed.), Dynamics of the Norwegian Margin. Geological Society Special Publication 167, pp. 295–326.

Steel, R.J. 1993. Triassic-Jurassic megasequence stratigraphy in the northern North Sea: Rift to post-rift evolution. In: Parker, J.R. (ed.), Petroleum Geology of Northwest Europe. Proceedings of the 4th Conference. Geological Society, pp. 299–315.

Tsikalas, F., Faleide, J.I., Eldholm, O. and Wilson, J. 2005. Late Mesozoic-Cenozoic structural and stratigraphic correlations between the conjugate mid-Norway and NE Greenland continental margins. In: Doré, A.G. and Vining, B. (eds.), Petroleum Geology: North West Europe and Global Perspectives – Proceedings of the 6th Petroleum Geology Conference. Geological Society, pp. 785–801.

Tsikalas, F., Eldholm, O. and Faleide, J.I. 2005. Crustal structure of the Lofoten-Vesterålen continental margin, off Norway. Tectonophysics 404, 151–174.

Tsikalas, F., Faleide, J.I. and Eldholm, O. 2001. Lateral variations in tectono-magmatic style along the Lofoten-Vesterålen margin off Norway. Marine and Petroleum Geology 18, 807–832.

Ziegler, P.A. 1982. Geological Atlas of Western and Central Europe. Shell International, The Hague.

Ziegler, P.A. 1988. Evolution of the Arctic-North Atlantic and the western Tethys. American Association of Petroleum Geologists Memoir 43, 198 pp.

Ziegler, P.A. 1992. North sea rift system. Tectonophysics 208, 55–75.

Subject Index

K. Bjørlykke (ed.), *Petroleum Geoscience: From Sedimentary Environments to Rock Physics*,
DOI 10.1007/978-3-642-12968-1, © Springer-Verlag Berlin Heidelberg 2010

Printing and Binding: Stürtz GmbH, Würzburg